中国植物染色体研究 40 年

陈瑞阳 等 编著

南开大学出版社
天津

图书在版编目(CIP)数据

中国植物染色体研究 40 年 / 陈瑞阳等编著. —天津：南开大学出版社，2019.10
ISBN 978-7-310-05903-4

Ⅰ.①中… Ⅱ.①陈… Ⅲ.①植物－染色体－研究－中国 Ⅳ.①Q943

中国版本图书馆 CIP 数据核字(2019)第 245420 号

版权所有　侵权必究

中国植物染色体研究 40 年
ZHONGGUO ZHIWU RANSETI YANJIU SISHINIAN

南开大学出版社出版发行
出版人：陈　敬
地址：天津市南开区卫津路 94 号　邮政编码：300071
营销部电话：(022)23508339　营销部传真：(022)23508542
http://www.nkup.com.cn

三河市同力彩印有限公司印刷　全国各地新华书店经销
2019 年 10 月第 1 版　2019 年 12 月第 1 次印刷
260×185 毫米　16 开本　43 印张　4 插页　1082 千字
定价：298.00 元

如遇图书印装质量问题，请与本社营销部联系调换，电话：(022)23508339

内容简介

本书是作者对改革开放40年来我国140科656属1563种植物染色体基础研究的总结。从作者发表的论文和著作中选取了80篇，按方法篇、研究结果篇、探索与发现篇、应用篇、人才培养篇和附录进行重新编写。本书对从事生物学、农学的教师、大学生、研究生和欲从事植物染色体基础研究与产业化的人士具有重要参考价值。

Plant Chromosome Studies of China in the Past 40 years: A Centennial Celebration of the 100th Anniversary of Nankai University

Edited by Chen Ruiyang

Nankai University Press
Tianjin

作者简介

陈瑞阳，男，1934年生，河北省迁西县新集镇人。1960年南开大学生物学系植物学专业遗传专门组毕业，同年留校，在生物学系任教。1983年任副教授，1988年获日本广岛大学理学博士学位。1992年任南开大学教授，同年被国务院学位办批准为博士生导师。1993—1996年任生物学系系主任。研究方向为植物细胞遗传学与细胞生物学。先后发表学术论著138篇（部），代表著作为《中国主要经济植物基因组染色体图谱》（Ⅰ～Ⅴ册），包括：

Ⅰ.《中国果树及其野生近缘植物染色体图谱》（1994年，万国学术出版社）

Ⅱ.《中国农作物及其野生近缘植物染色体图谱》（2003年，科学出版社）

Ⅲ.《中国园林花卉植物染色体图谱》（2003年，科学出版社）

Ⅳ.《中国竹类染色体图谱》（2003年，科学出版社）

Ⅴ.《中国药用植物染色体图谱》（2009年，科学出版社）

曾获国家自然科学奖二等奖1项，教育部科技进步奖二等奖2项，农业部科技进步奖二等奖1项，天津市科技进步奖一等奖1项、二等奖2项，天津市自然科学奖二等奖1项、三等奖1项。曾获天津市"植物遗传学专家"授衔专家称号，2012年获南开大学特殊贡献奖，批准为南开大学荣誉教授，获2019年度南开大学科学研究杰出贡献奖，享受国家政府特殊津贴。

作者名单

陈瑞阳　宋文芹　李秀兰　陈成彬

梁国鲁　陈　力　王春国　张力鹏

Contributors

Chen Ruiyang   Song Wenqin   Li Xiulan   Chen Chenbin
Liang Guolu   Chen Li   Wang Chunguo   Zhang Lipeng

序
——中国植物染色体研究 40 年

自 1888 年 W. Waldeyer 命名染色体（chromosome）以来，有关染色体与染色体组或基因组（genome）结构与功能的研究，一直是生命科学最活跃的研究领域之一。每种生物的体细胞都具有一定数目的染色体，并且染色体的大小和形态都各具特点，具有种的特异性，它们是不同物种基因组简单明了的形象表现。染色体研究是与遗传学同步发展起来的，自 1926 年摩尔根（Morgan）确立基因连锁的遗传理论，提出基因在染色体上呈直线排列，染色体是基因的载体之后，染色体研究达到了巅峰时期。现在人们已清楚染色体的功能有三个方面：第一是贮存、复制和传递遗传物质；第二是对基因活动的调解，控制基因活动；第三个功能是调节有性后代的基因重组频率。染色体是基因的载体，是基因组的存储单元，它将基因组分为若干个基本单位，是研究任何一个物种基因组的前提，基因组所获得的浩如烟海的测序结果，只有回到染色体上来与染色体研究相结合，才能确切知道基因在染色体上的排列，如何受到上下游基因的调控，只有在染色体水平上认识基因的分布和表达调控规律，才能真正了解基因组遗传信息的生物学含义。因此，染色体和染色体功能的研究，不仅发展了细胞生物学、细胞遗传学、细胞分类学、细胞地理学和物种生物学，同时它也是功能基因组学研究中的基础。目前日本、美国等许多国家已将染色体研究列为 21 世纪生命科学研究的重点之一。

一、研究背景

20 世纪五六十年代，随着各种染色体新技术的发明与应用，在国际上，染色体研究达到了鼎盛时期，T. C. Hsu（徐道觉）历时 10 年（1967—1977）主编了世界第一部哺乳动物染色体图谱（Atlas of Mammalian Chromosomes），使涉及以染色体为基础的学科得到了极大的发展，也为后来的分子生物学和基因的染色体定位奠定了良好的基础。然而，植物染色体虽然在 20 世纪三四十年代有过先驱作用，但自 1950 年以后，与动物染色体研究相比发展缓慢，还没有一本可与哺乳动物染色体图谱相比的植物染色体图谱出版。

中国是世界栽培植物起源最早、最大的中心之一，据不完全统计，起源我国的栽培植物在 200 种以上，现在不仅有极其丰富的栽培品种，还蕴藏着许多珍贵的野生近缘植物。然而，我国植物染色体研究，1976 年以前除台湾省外，基本处于空白，这与我国的地位和占有的丰富植物资源极不相称。以前，外国人采走中国植物种子，由外国人研究报道，例如英国人 Wilson 和 Forest 在 20 世纪初从我国西南地区采去大量杜鹃花，引种于英国爱丁堡皇家植物园，多达 400 种，由 Janaki Ammal 等（1950）研究报道了其中近 300 种的染色体数目。可喜的是，改革开放后，南开大学在染色体研究技术上取得突破，掌握了植物染色体的核心技术，建立了一套具有我国特色的植物染色体研究方法，并先后在南开大学和全国各地举办了各种学习班，全国有 300 多个单位派人来学习，大大促进了我国植物染色体研究的蓬勃发展。改革开放 40 年来，我们课题组自 1983 年开始共获得国家自然基金资助项目 26 项，对我国植物染色体研究的发展做出了开创性和奠基性贡献，使我国植物染色体研究达到国际先进水平，2003 年获得国家自然科学奖二等奖。

二、研究成果

1. 中国植物染色体研究。40 年来我们共对我国 194 科（占全国 85%），693 属（占全国 23.1%），1579 种（占全国 5.2%）的植物染色体数目和倍性进行了报道，其中 90% 为首次报道。1979 年和 1980 年在国内首次报道了普通小麦和水稻的染色体数目和核型，从此填补了中国植物染色体在国际上的空白。

2. 出版了世界上第一部《中国主要经济植物基因组染色体图谱》（Ⅰ～Ⅴ 册），共 3631 页，收载了 1389 种植物染色体图像。包括：

①第Ⅰ册：中国果树及其野生近缘植物染色体图谱，29 科 52 属 261 种 1140 品种；

②第Ⅱ册：中国农作物及其野生近缘植物染色体图谱，21 科 59 属 243 种；

③第Ⅲ册：中国园林花卉植物染色体图谱，64 科 205 属 356 种；

④第Ⅳ册：中国竹类植物染色体图谱，1 亚科 33 属 185 种；

⑤第Ⅴ册：中国药用植物染色体图谱，123 科 403 属 563 种。

3. 新发现 478 种多倍体，其中四倍体 222 种，六倍体 182 种，三倍体 37 种，其他倍性 37 种，其中半夏、三脉紫菀、芦苇、木兰科、竹亚科为多倍体复合体。

4. 首次鉴定报道了我国银杏、芦笋、菠菜的性别机制为 ZW 型和 XY 型。

5. 对三脉紫菀、芦苇、半夏多倍体复合体细胞地理学进行了研究。

6. 创建了植物染色体标本制备的酶解、去壁、低渗法，简称 WDH 法，它与 1950 年建立的人类染色体的低渗法具有同等重要意义。

7. 首次获得多种植物染色体 G-带，特别是川百合，蚕豆，黑麦，一粒小麦染色体 G 带，至今尚无人突破。

8. 从 1997 年就开展了植物 B-染色体分子生物学及人工 B-染色体载体的研究，曾 4 次获得国家基金资助，获得 B-染色体特异序列，为进一步人工合成 B-染色体打下了基础。

9. 建立和改进了 45S rDNA、5S rDNA 荧光原位杂交的方法，并成功地在多种植物上进行了应用。

10. 建立了具有南开大学特色的中国植物染色体研究技术体系，即：材料采集→去壁低渗法制备染色体标本→①半自动化核型分析，②Giemsa C-带，③N-带，④G-带，⑤FISH，⑥rDNA 原位杂交→核型标准化数据分析。

11. 创建了在普通光学显微镜 100×15=1500 倍油镜下、分离、切割染色体的新方法，可准确识别、分离任何一种植物的任一条染色体，切割精度可达到 0.2 μm，建立了单条染色体基因组测序方法，为单条染色体基因组测序产业化奠定了基础。

12. 创建了以丹参为主的"基因组三倍化与杂交相结合"的遗传改良与新品种培育的育种平台。

13. 创建了多种兰花遗传改良、种质创新方法与种苗工厂化生产技术。

14. 基于对多倍体特别是异源多倍体染色体结构大数据分析，我们发现自然界除多倍体优势和杂交优势以外，可能还存在第三种优势。第三种优势是由"基因组三倍化与杂交相结合"产生的，它是将多倍体优势、杂交优势、远缘优势固定在一起的三位一体的强大优势，我们称之为第三种优势。在自然界，在第三种优势作用下形成了一些超级物种，如北美红杉（世界爷）、西双版纳歪脚龙竹、巴布亚新几尼亚大香蕉、普通小麦等。凡是经过基因组三倍化形成的异源六倍体，都是在第三种优势作用下形成的。因此，我们认为"基因组三倍化与杂交相结合，在高等植物进化中起了重要作用"。第三种优势将给人类创造巨大的财富。

15. 植物染色体基础研究与产业化。40 年的植物染色体基础研究结果表明，它与产业化密切相关。由于 1978 年以前我国植物染色体研究基本空白，所以与染色体相关的产业化也就无从谈起。经过这 40 年的中国植物染色体基础研究，我们不仅搞清了中国 30000 种高等植物染色体的家底，而且提出了与染色体相关的产业化方向：①建立单条染色体基因组测序与文库构建，现在，正在启动地球生物基因组测序计划，建立单条染色体 DNA 测序，特别是在异源多倍体基因组测序方面有着广泛的用途；②建立"基因组三倍化与杂交相结合"育种平台，利用第三种优势培育出一批超级物种，如异源六倍体超级稻、异源六倍体超级大豆、异源六倍体、十二倍超级甜高粱、异源六倍体水仙花、三倍体、六倍体郁金香等。20~30 年后，将会改变现在的农业品种种植结构。这就是基础研究的魅力所在，只有基础研究才能产生原始创新、产生核心技术。

16. "水稻杂种优势固定与异源多倍体水稻的培育及种子安全生产"。随着研究杂种优势的产生，便产生了杂种优势"固定"的研究，但 100 多年来，还没有一个公认的好"固定"方法产生。我们依据异源四倍体、异源六倍体减数分裂时，当基因型为 $A_1A_1A_2A_2$、$A_1A_1A_2A_2A_3A_3$ 时只产生一种配子——A_1A_2、$A_1A_2A_3$，后代不分离，成为永久杂种，从而固定了杂种优势的原理，提出通过异源双二倍体法，进行杂种优势固定的研究，并应用中国特有的籼、粳稻双二倍法进行水稻杂种优势固定。结果表明，籼粳稻双二倍体法完全可以"固定"水稻杂种优势。另外，异源六倍体水稻也应可以固定水稻杂种优势，还可能不存在结实率问题，尚待进一步研究。

17. 总结染色体实验室对中国植物染色体研究做出的贡献，可以归纳为①中国植物染色体研究的开拓者和奠基基地；在 20 世纪 80 年代初，为中国 300 多个单位培训了植物染色体研究技术人材；在国内最先发表了水稻、小麦、大麦等染色体分析资料；创建了以酶解、去壁、低渗法（WDH 法）；建立了中国植物染色体标准化分析方法体系，使植物染色体研究达到了人类和哺乳动物染色体研究水平。②对中国3000多种植物染色体进行了研究，出版了世界首部《中国主要经济植物基因组染色体图谱》（Ⅰ～Ⅴ册），积累了大量染色体科基础数据。③对中国植物染色体及其相关领域进行了研究，使中国植物染色体研究达到了世界先进水平，并在植物染色体G-带研究方面，居领先地位。另外，提出自然界可能存在第三种优势，为人类培育超级物种开辟了一条新途径。④在水稻杂种优势固定与异源四倍体、异源六倍体水稻培育方面取得新进展。⑤使南开大学在20世纪八九十年代成为中国植物染色体研究中心和对外交流窗口，陈瑞阳成为世界报道植物染色体数目（IPCN）最多的人。

陈瑞阳

2019年3月8日

前　言
——南开大学染色体实验室发展史记

一

本书的资料选取自40年来本实验室发表的300篇论文、专著，按现在的顺序：方法篇→研究结果篇→探索与发现篇→应用篇→人才培养篇→附录，进行了整理。为简化，删除了原文的中文摘要、关键词、大部分图和表。另有部分内容取自本实验室尚未发表的论文。在文后均注明作者和原载期刊信息，便于读者查看原文。各时期期刊格式差别很大，这次书中尽可能按统一规范修正。因为过去发表的论文都是随机的，没有连惯性，此次整理，为了把原来零散的论文组合起来，形成有机的体系，我们在有关的章、节前，增写了"提要"，在"空间"上把它们联接起来。为减少篇幅，对《中国主要经济植物基因组染色体图谱》（Ⅰ～Ⅴ册）的1605种植物染色体研究结果，我们只取了核型分析数据和部分染色体图像资料，供参考。本书共选取了80篇论文和7本专著的部分内容。

二

1. 染色体实验室的前身是陈瑞阳的"显微及亚显微技术课"实验室，再前身是吴小航教授的"切片课"实验室，1978年以前，已具有一定的细胞学基础。染色体实验室是1978年以后自然形成的，不知从何时起就这么叫了。

2. 改革开放后，我们不清楚应该如何做科研。原来受的教育，综合性大学应该搞基础理论研究。但文革十年，我们对国外发展情况一无所知。经过几个月去北京科学院情报所查阅资料，查到一项号称70年代Giemsa分带技术的文章，像磁铁一样吸引着我们。可什么是Giemsa，不知道，什么叫分带技术也不清楚，是基础理论吗？两眼一摸黑。后去请教吴旻院士，他给我看了一张人染色体的Giemsa分带片子，我才豁然开朗。1978年我们以这项70年代Giemsa分带技术为切入点，开展了我们的植物染色体研究。

3. 经过一年多的奋斗，我们就突破了这项新技术。1979年在昆明召开的中国植物学会45周年大会，在细胞学分组会上，我介绍了我们的植物染色体Giemsa分带研究结果，得到了老一代细胞学家吴素萱、郑国锠、郝水、朱澂教授的好评，他们并倡议中国植物学会应委托南开大学举办全国Giemsa分带技术培训班，于是1981年，由中国植物学会主持，朱澂先生负责，在南开大学举办了改革开放后，全国第一个染色体分带技术培训班，对这项新技术进行了推广。来自全国300多个单位的技术人员受到了免费培训。

从此，中国植物染色体研究在全国开展起来。截至1984年6月我国发表论文227篇，分别隶属于163科625属1259种（徐炳声，1985）。因此，南开大学染色体实验室在我国植物染色体研究中起了重要的先驱带头作用。

4. 植物染色体Giemsa分带成功后，我们并没有满足于这项70年代水平的研究成果。因为Giemsa分带方法完全是从文献上抄来的，没有自己的方法，而且它本身有很多不足，很难推广。因此，我们在研究Giemsa分带的同时，抓住了植物染色体研究的关键问题——染色体标本制作。植物染色体一直沿用着1921年的压片法，与人类及哺乳动物染色体研究相比落后了几十年。人类染色体的低渗法虽然是偶然发现的，但它是迄今为止最完善的染色体研究方法。如何将人类染色体的低渗法移植到植物上来，我们虽不是第一人，但我们是最先成功者。

5. 植物染色体的酶解、去壁、低渗法（WDH法）的创建。经过一年多的努力，我们终于将人类染色体的低渗法移植到植物上来，获得成功，并在37科477种（含种下）植物上进行了试验。于1979年和1982年正式发表了这个方法。这对植物染色体研究来说，是自1921年的醋酸洋红压片法以来，具有里程碑意义的方法。

6. 有了我们自己的去壁低渗法后，我们立刻转入研究中国特有植物上来。Н.И.瓦维洛夫提出：中国是世界栽培植物起源最早和最大的中心，起源我国的栽培植物有200多种，至今在全国各地还分布着野生大豆、野生茶和各种野生果树。于是，就这样，我们与中国植物染色体研究结下了不解之缘，一干就是40年，也正好赶上改革开放40年的好时代，成就了我们的中国梦。40年来我们不仅填补了中国植物染色体研究的空白，而且使中国植物染色体研究达到了世界先进水平，并在某些方面走在前面。

7. 染色体实验室自1983年获得国家自然科学基金项目开始，40年来共获得62项各类科研项目（见附录2）。

8. 实验室历届共获得9项科学技术奖励（详见附录3）。所有奖项都是与染色体相关的，是对我们染色体研究的肯定。2003年获得国家自然科学奖二等奖，是通过：杨弘远院士、翟中和院士、郑国锠院士、董玉琛院士、陈俊愉院士推荐上去的项目。评审委员会主任是北京大学校长许智宏院士，评审委员对我们的工作给予了高度肯定，陈家宽委员说：这是你的终生成就奖！

9. 1983～2018年染色体实验室共培养研究生122名，包括硕士生77名、博士生45名，博士后3名（详见附录7）。他们当中已有30多名成为教授级高级人才。

10. 1979～2018年染色体实验室共发表论文300多篇（见附录6）。

11. 1993～2003年美国密苏里植物园（*Index to Plant Chromosome Numbers*（IPCN））共收录了Ruiyang Chen（陈瑞阳）等报导的植物染色体总数为780种（2004年以后待收录）。

12. 2002年陈瑞阳退休后，转向理论联系实际的研究，并获得了一批发明专利（详见附录4）。

13. 染色体实验室的发展得到诸多单位和专家的支持与帮助，染色体实验室曾先后得到了华南植物园，昆明植物园，西双版纳植物园，安吉竹种园，辽宁兴城果树研究所，河北昌黎果树研究所，中国医学科学院药用植物研究所（广西分所、云南景洪分所、新疆分所），中国农业科学院品种资源研究所和特产研究所，云南农业科学院粮食作物研究所和蔬菜研究所，吉林农业科学院大豆研究所，广东农业科学院水稻研究所、果树研究所和粮食作物研究所等几十个单位的大力支持和提供实验材料。另外，染色体实验室的发展得到了我国老一代专家的大力支持和帮助，他们是：武汉大学杨弘远院士和周嫦教授，东北师范大学郝水院士，兰州大学郑国锠院士，中科院植物所洪德元院士，北京大学翟中和院士和朱澂教授，中科院副院长北大校长许智宏院士，上海植物生理研究所洪孟民院士，中国农业科学院董玉琛院士，北京林业大学陈俊愉院士，上海细胞生物学研究所王亚辉教授，中山大学刘良式教授，南开大学吴小航教授、周之杭教授、张自立教授、钟贻诚教授、胡志衡教授、蒋如章教授、白玉华教授、唐廷贵教授、张丰德教授，河北农业科学研究院昌黎果树研究所陈景新、陈永利，中国科学院生物物理研究所王又明研究员，日本广岛大学田中隆庄校长、谷口研至教授，奥地利Greilhuber教授等，他们先后为本实验室的发展提出宝贵建议、推荐意见和鉴定材料，对实验室的发展起了重要指导作用，值得永久感谢！

14. 染色体实验室主持召开了两次国内学术会议，两次中日学术会议。①1984年8月由洪德元、李懋学和陈瑞阳主持，在兴城召开了第一届中国植物染色体学术讨论会，代表80人，收到论文83篇。②2002年8月实验室主持，在南开大学召开了"基因在染色体上的FISH定位及其在基因组研究中的应用学术讨论会"，郝水院士、洪德元院士、李懋学、程祝宽、宋运醇等100多人参加了会议。③1987年10月在北京香山卧佛寺召开了第一届中日植物染色体学术讨论会，中方：主席郝水、秘书长洪德元、组织委员长陈瑞阳，日方：主席田中隆庄、秘书长中田正司、组织委员长谷口研至。参加会议有100多人，会议开得非常成功，中日科学家之间非常友好，还正赶上八月十五中秋节，中日科学家之间的团结友好前所未有，特别是中方一些老一代科学家也出席了会议，如汤佩松、王伏雄、郑国锠、郝水、陈心启、罗鹏等。④1992年11月在南开大学召开了第二届中日植物染色体学术讨论会。两届会议都出版了论文集。

15. 1988年11月16日至12月21日陈瑞阳赴日本广岛大学参加博士论文答辩，以"数种高等植物特别是百合染色体G带的研究"为题，经过答辩和考试，以29票全票通过的成绩，于1988年12月19日（昭和63年12月19日）获得日本广岛大学理学博士学位（一七七八号）。

16. 实验室参加过两届6年的"中日共有植物"国际合作研究，由日本文部省资助，田中隆庄主持。中方参加单位是南开大学和中科院植物所系统与进化开放实验室。我们负责的题目是：中国紫菀属多倍体复合体的细胞地理学研究。第二次中日合作，由近腾胜彦主持，改为中国野生稻调查研究，1993年陈瑞阳和岛田多喜子对海南岛野生稻进行调查、采集，野生稻种在南开大学进行研究，岛田教授回国时要将野生稻带回去，我们不同意，因为这违反了中国的规定，为此，我们退出了中日合作研究。

17. 为朝鲜培养7名进修生，其中金春相为硕士研究生。1992年9月我们访问朝鲜金日成综合大学，并进行了野外采集。

18. 1996年7~10月陈瑞阳参加了香港大学兰花合作研究，得到孙梅教授多方帮助。并对香港植物进行了野外采集。1998年孙梅教授对南开大学进行访问。

19. 1986年陈瑞阳第一次访问日本时，向有关单位介绍"中国植物染色体研究"，他们就很吃惊，原来他们认为中国没人研究植物染色体。到1987年在北京召开第一次中日植物染色体学术讨论会后，日本朋友就认为南开大学是中国植物染色体研究的中心、对外交流的窗口。经过这40年的努力，南开大学在植物染色体研究上取得了显著成绩。世界上唯一收录植物染色体的（IPCN）机构，已收录了我们700多种植物染色体资料。"染色体图书馆网站"创始人谷口研至博士认为陈瑞阳博士在植物染色体数目报道上已排世界第一。南开大学在植物染色体研究方面的成绩是南开人的百年积淀，我们仅以此祝福：南开百年、青春永驻、人才辈出、成果累累！

三

参加此项研究的还有：陈永利、谷口研至、张培新、纵微星、马小军、汪卫星、林汝顺、冯学林，还有染色体实验室历届毕业生：苏建英、郑坚瑜、谷春艳、崔香芹、纵微星、郭歌、韩健、翟晓玲、梁思源、彭永康、孙德岭、张荣信、王晓梅、孙易、龙鸿、阎国荣、于建春、张峰、毛英伟、江赐忠、祁仲夏、曾辉、刘松、李凌、史岸冰、刘博、兰添颖、杨宇、张勇、董风平、赵换、张力鹏、刘少华、徐悦凡、林军、王凤波、李金岚、陈琳、祁燕君等染色体实验室全体学生。宋霖、李天宇、成璐路、滕彦娇、王宏鹏在染色体图的修改、文献的编辑和文献的检索等方面协助做了大量的工作，在此一并表示感谢！

李懋学、林盛华、安祝平生前参加了部分工作，在此谨表示哀悼！

本书出版得到了染色体实验室实验室全体师生的赞助，在此表示感谢，预祝大家在不同的岗位上取得更多的成绩！

陈瑞阳

2019年5月5日

国内外学生捐款名单

李凯军	陈瑞阳	宋文芹	阎国荣	高英堂	张世光	孙德岭	李华兵	史岸冰	
陈成彬	陈 力	任晨春	张荣信	纵微星	龙 鸿	王 娟	古 瑜	王 冰	
汪 泉	程 华	王 彬	苗旭红	胡宝全	王晓梅	江赐忠	江汉民	王顺启	
李丽红	韩春乐	耿美娟	李尚乘	杨美玲	任鲁风	杨 宇	马钰婕	张 勇	
王春国	李 慧	董天皓	王 菁	牛 凯	韦 韬	刘 博（男）		于建春	
陈小强	郝 擎	赵 磊	刘少华	李 爱	饶冠华	王 楠	赵风治	秦二军	
周亚楠	李 宽	张青丽	郑 翔	李德超	安 冬	金 川	张 勐	薛振毅	
赵 换	刘志伟	王明明	张银兴	张庆霞	王 玉	高永红	徐嘉艺	张云秀	
刘 超	张欢欢	贺丽霞	王 勃	王 霞	郑书行	张文惠	韩祥艳	王洪彬	
贾英澜	武 美	张力鹏	赵瑞红	于得水	苑吉叶	李 聪	滕艳娇	苏建英	
郑坚瑜	王 峰	刘 松	张 峰	兰添颖	祁仲夏	曾 辉	董凤平	宋春娟	
刘 博（女）									

陈瑞阳
2019 年 8 月 2 日

目 录

第一章 植物染色体研究方法 1

第一节 植物染色体制备方法 1
    1.1.1 植物染色体的压片技术 2
    1.1.2 植物有丝分裂染色体标本制作的新方法 5
    1.1.3 植物染色体标本制备的去壁、低渗法及其在细胞遗传学中的意义 7
    1.1.4 染色体标本制备的蒸气干燥法 14

第二节 植物染色体分带技术 14
    1.2.1 植物染色体 Giemsa 分带技术的研究 14
    1.2.2 植物染色体 C-带法 19
    1.2.3 植物染色体 N-带法 21
    1.2.4 植物染色体 G-显带方法 22
    1.2.5 植物染色体带型分析法 25

第三节 染色体的荧光原位杂交与基因定位 27
    1.3.1 5S rDNA 在植物染色体上的荧光原位杂交方法 28
    1.3.2 45S rDNA 在植物染色体上的荧光原位杂交方法 30
    1.3.3 植物基因组原位杂交（GISH） 32
    1.3.4 BAC 克隆在植物染色体上的定位 33

第四节 植物染色体电子显微镜标本制备技术 36
    1.4.1 植物染色体扫描电镜标本制备方法 36
    1.4.2 透射电子显微镜的染色体表面展开法 37

第五节 植物有丝分裂染色体同步化方法 40
    1.5.1 利用 HU 和 APM 双阻断法诱导高频率植物根尖细胞有丝分裂同步化研究 41
    1.5.2 小麦根尖细胞有丝分裂同步化诱导与中期染色体分离 43
    1.5.3 APM 对小麦根尖分生组织细胞异常有丝分裂的诱导 46
    1.5.4 大蒜根尖细胞有丝分裂同步化诱导与中期染色体分离 48

第六节 植物核型分析的标准化 50
    1.6.1 关于植物核型分析的标准化问题 51
    1.6.2 植物染色体的半自动化核型分析法 56
    1.6.3 应用同一细胞进行核型与带型的分析方法 65
    1.6.4 植物染色体分子核型技术研究 67

第七节 植物染色体分离与 PCR 扩增方法 74
    1.7.1 植物染色体的分离方法 74

1.7.2 植物染色体 G-带特异区聚合酶链反应技术的初步研究 76

第二章　中国植物染色体研究结果

第一节　中国果树及其近缘植物基因组染色体研究结果 83
2.1.1 部分果树染色体 84
2.1.2 中国果树及其近缘植物染色体数据分析 85

第二节　中国农作物及其野生近缘植物基因组染色体研究结果 99
2.2.1 部分农作物染色体 100
2.2.2 中国农作物及其近缘植物基因组染色体数据分析 140

第三节　中国园林花卉植物基因组染色体研究结果 153
2.3.1 部分园林植物染色体 154
2.3.2 中国花卉植物基因组染色体数据分析 155

第四节　中国竹类植物基因组染色体研究结果 176
2.4.1 部分竹类植物染色体 177
2.4.2 中国竹类植物基因组染色体数据分析 181

第五节　中国药用植物基因组染色体研究结果 193
2.5.1 部分药用植物染色体 194
2.5.2 中国药用植物基因组染色体数据分析 195

第三章　探索与发现

第一节　植物 B 染色体 228
3.1.1 B 染色体分子生物学 228
3.1.2 黑麦 B 染色体显微切割和微克隆 232
3.1.3 玉米 B 染色体特异 RAPDs 标记的分离与染色体定位 235
3.1.4 黑麦 B 染色体端粒相关序列的克隆 237
3.1.5 黑麦 A，B 染色体着丝粒区同源性的荧光原位杂交分析 241
3.1.6 黑麦 B 染色体 DNA 组成特征 243
3.1.7 植物 B 染色体序列分离及关键功能组件的研究 251

第二节　植物染色体 G-带研究 292
3.2.1 利用胰蛋白酶和尿素诱导植物染色体 G-带 292
3.2.2 百合（*Lilium davidii* L.）G-带核型分析 298
3.2.3 植物染色体 G-带的深入研究 307
3.2.4 植物染色体高分辨 G-带的研究 310
3.2.5 黑麦 G-带和 C-带核型分析 314

第三节　染色体分离及应用 317
3.3.1 黑麦染色体的显微分离与 PCR 扩增 317
3.3.2 黑麦 1R 染色体的微切微克隆研究 320
3.3.3 水稻染色体的显微分离与克隆 324

3.3.4 水稻第 9 号染色体 DNA 文库的构建 328
3.3.5 水稻 4 号染色体的分离与 DNA 文库的构建 330
3.3.6 蚕豆大 M 染色体长臂端部的显微切割与 PCR 扩增 335
3.3.7 黑杨 1 号染色体的显微分离、克隆和定位 337
3.3.8 银杏 W 染色体显微分离及荧光原位杂交分析显示性染色体不同
杂交带型 346
3.3.9 通过微切、扩增建立植物（Vicia faba）染色体区段特异性基因文库
研究初探 352
3.3.10 黑杨 1 号染色体抗病基因同源序列的克隆与鉴定 357

第四节 荧光原位杂交与基因定位 369
3.4.1 荧光原位杂交技术的研究进展 369
3.4.2 菠菜 rDNA 及端粒多色荧光原位杂交分析 376
3.4.3 豆科三属八种植物的核型及 rDNA 定位研究 379
3.4.4 30 种植物 45S rDNA 在中期染色体上的物理定位 384
3.4.5 rDNA 序列在多种蔬果类植物染色体上的定位 389
3.4.6 rDNA 和端粒重复序列在七种杨属植物中的多色荧光原位杂交
定位 398
3.4.7 25S rDNA 在杨属植物染色体上的定位 403
3.4.8 利用组合标记探针对华北落叶松进行多色荧光原位杂交和详细的
核型分析 409
3.4.9 应用二色荧光原位杂交和 DAPI 带对四种落叶松属植物进行分子
细胞遗传学分析 414
3.4.10 32 种植物端粒组构信息的初步研究 422
3.4.11 细胞质雄性不育辣椒育性恢复基因在染色体上的定位 423
3.4.12 孤雌生殖卤虫基因组的特异 DNA 片段在染色体上的定位 424
3.4.13 长春花端粒相关序列荧光原位杂交定位 425
3.4.14 应用 BAC-FISH 系统鉴定芝麻（Sesamum indicum L.）染色体 442

第五节 植物性别机制的研究 453
3.5.1 银杏性别染色体的研究 453
3.5.2 芦笋性别染色体的研究 459
3.5.3 应用 45S rDNA FISH 鉴定雌雄异株菠菜的性染色体 462

第六节 细胞地理学 465
3.6.1 三脉紫菀多倍体复合体的细胞地理学和物种形成 465
3.6.2 半夏多倍体复合体及其细胞地理学研究 473
3.6.3 芦苇及其近缘种种内多倍体的染色体研究 479

第四章 应用篇 505
第一节 药用植物的遗传改良与新品种培育 505

　　　　4.1.1 丹参的遗传改良——白花丹参同源四倍体的诱导与鉴定 506
　　　　4.1.2 三倍体丹参的培育及其可持续利用研究 509
　　　　4.1.3 三倍体丹参新品种介绍 515
　　第二节 兰花种质资源的遗传改良与种质创新 522
　　　　4.2.1 兰花种质资源的收集与细胞学鉴定 522
　　　　4.2.2 兰花有性杂交育种平台的建立与种苗工厂化生产技术 525
　　　　4.2.3 兰花种子非共生萌发、原球茎诱导、分化与成苗移栽方法 525
　　　　4.2.4 兰花原球茎快速增殖方法 527
　　　　4.2.5 兰花多倍体的诱导方法 528
　　　　4.2.6 十三种兰花有性杂交育种技术平台 535
　　　　4.2.7 铁皮石斛种苗简易工厂化生产 539

第五章　人才培养 542

附　录 586
　　附录1　三位中国科学院院士为《中国经济植物基因组染色体图谱》写的序言......586
　　郝水　序 586
　　杨弘远　周嫦　序 590
　　洪德元　序 593
　　附录2　实验室历年科研项目汇总 595
　　附录3　实验室获奖汇总 597
　　附录4　实验室获得的授权专利汇总 598
　　附录5　密苏里植物园 *Index to Plant Chromosome Numbers*（IPCN）收录的
　　　　　　Ruiyang Chen 报道的中国植物染色体部分索引 599
　　附录6　1979～2019年实验室发表的论文 636
　　附录7　南开大学生科院染色体实验室招收培养研究生情况简表 651
　　附录8　实验室主持和参加国内外学术会议部分照片 657

"厚"望 666

第一章 植物染色体研究方法

提要：染色体研究方法在染色体研究中一直占有重要地位，是从事染色体研究者首先必须解决的问题，方法决定结果。在19世纪染色体研究的初始阶段，使用的是切片法，1921年Belling建立了压片法，1950年人类及哺乳动物染色体研究开始使用低渗法。压片法虽然在染色体计数方面起了很大作用，但仍然有很多错误，典型的例子是："人类染色体是48条"。该错误结论统治了近百年人类的认识，因此，在使用1950年以前的染色体结果时要慎重，特别是植物染色体，直到1979年以前，一直在使用压片法，有的国家发表的文章连染色体照片都没有，只提供了人工画的染色体图像，这些结果只能作为参考。自1950年人类染色体研究发明了低渗法，使染色体研究步入正规。十年间召开了四次国际会议，制定了人类染色体研究的国际标准，使人类染色体的基础研究和应用研究都取得了巨大的进步。我们1979～1981年建立的植物染色体酶解去壁低渗法，实际上是将人类染色体的低渗法移植到植物染色体标本制备上来，获得成功。而人类染色体低渗法是染色体自发现以来最完善、最完美的研究方法，是染色体研究的里程碑。我们将它移植到植物染色体研究上来，是正确的，实际也证明用酶解去壁低渗法制备的植物染色体标本，完全可以使植物染色体研究达到人类染色体研究的水平。只是它正赶上分子生物学时代，使用的人还不多。相信随着分子生物学的发展还会回归到整体上来，正如没有化学、物理学的进步，分子生物学是揭不开生命本质的，就像没有生命科学的进步，医学只能打外围战一样，生命科学的发展必须和其他学科相结合。染色体是基因的载体，是遗传的本质，是生命科学的永恒主题，基因组测序的结果必须回到染色体上定位才算真正完成，因此，染色体研究还有很多工作需要去完成。

第一节 植物染色体制备方法

提要：染色体标本制备方法分压片法与酶解去壁低渗法两种。一般染色体计数时可使用压片法，核型分析和各种分带处理要使用酶解去壁低渗法。植物染色体的酶解去壁低渗法，实际上是将人类染色体标本制备的低渗方法移植到植物上来，所以，主要困难在于去除植物细胞壁，不同的植物，不同的组织，不同材料的大小都会影响去壁的时间，初学者必须控制好每种条件。我们虽然不是应用酶解去壁的第一人，但我们是建立该方法的成功者和幸运者，也是使用该方法最多的人。去壁低渗法制备的染色体标本，适用于染色体的各种分带处理、荧光原位杂交和染色体分析，使植物染色体研究达到了国际先进水平，使植物染色体与人类和哺乳动物染色体研究达到同一水平。因此，酶解去壁低渗法的建立具有划时代的意义。

1.1.1 植物染色体的压片技术

植物染色体制片技术是植物细胞遗传学、染色体工程、植物细胞生物学、植物细胞分类学和物种生物学等众多学科的基本实验技术。植物染色体的研究由于和植物遗传育种工作的结合而获得良好的发展。因此，植物染色体的技术在这些研究工作中具有特别重要的意义。从 Belling（1921）配制醋酸洋红用于染色体压片开始，至今已有 90 多年的历史了，经过人们长期实践和不断的改进，这一技术至今仍是细胞遗传和细胞分类学研究中应用最为普遍的基本技术。

1.1.1.1 取材

凡是能进行细胞分裂的植物组织或单个细胞都可以作为观察染色体的材料，如植物的顶端分生组织根尖和茎尖、居间分生组织，禾本科的幼茎和叶鞘、愈伤组织和胚乳、大小孢子母细胞的减数分裂时期以及从小孢子发育成雄配子过程中的两次细胞分裂等，都是常作为观察染色体的适宜材料。但是，并不是随时取来材料都是合适的，植物细胞的分裂活动既有自身发育的阶段性，又受到环境条件的影响，因此只有充分掌握植物组织的一般结构或生长发育的特性规律，才能够取到比较合适的材料。准确的取材是我们获得优良制片的基础。

制备植物细胞的染色体标本，在取材上必须选择细胞分裂较旺盛，而且取材较方便的组织作为实验材料。高等植物有丝分裂主要发生在根尖、茎尖生长点及幼叶等器官的分生组织（分生区），其中根尖是最常用的材料 ①根尖取材容易，操作和鉴定也比其他器官与组织方便；②实验室内采用种子萌发后所长出的新鲜幼嫩根尖，不受植物生长季节的影响和限制，并且可以大量获得；③对于某些珍稀的试验材料，取用自然条件下生长植株的根尖，比取用茎尖、花器等对材料的伤害要小得多；④采用实验室内种子发根，切取根尖后的种苗通常还可以进行正常种植，利于后续研究进行。

1.1.1.2 预处理

预处理的作用在于阻止纺锤体的形成，但并不妨碍分裂前期的正常运行，由于没有纺锤体的牵引，因而使细胞分裂被阻断于中期阶段，这样可以获得较多的中期分裂相。预处理的另一个目的是使染色体收缩变短。

（1）预处理药物

可用于预处理的化学药物有生物碱、醌类、酚类及其他物质，最为常用的、比较有效的预处理药物如下。

①秋水仙素：秋水仙素是从百合科秋水仙属的秋水仙的种子和鳞茎中提取出来的一

种生物碱，是一种微管特异性药物，可以与微管蛋白亚基结合，导致细胞内微管网络的解体。秋水仙素水溶液的浓度与其作用的效力是呈正相关的，各种不同的植物和不同的组织对药物的反应是不同的，一般所用的浓度范围从 0.01%～0.2%，温度控制在 8～16℃。

②对二氯苯：对二氯苯为一种苯的衍生物，难溶于水，药物配制方便，一般配成饱和水溶液，也可在其中加入几滴 α-溴代萘效果更佳。对二氯苯的使用范围极广，对较大的染色体、中等大小染色体和小染色体的作用都很有效。而且价格低廉，便于广泛使用。

③8-羟基喹啉：所用浓度范围在 0.002～0.004 mol/L 之间，它特别适用于具有较小染色体的植物的处理，经 8-羟基喹啉处理后的染色体，缢痕区比较清晰。

④α-溴代萘：α-溴代萘为萘酚的一种衍生物，微溶于水。常用方法有：a）在 100 mL 蒸馏水中加入 1 滴 α-溴代萘，充分振摇，配成饱和水溶液使用；b）取 1 mL α-溴代萘加入 100 mL 无水乙醇中，配成 1%的酒精贮存液。使用时取 1 mL 酒精贮存液加入 100 mL 蒸馏水中，充分振摇后使用；c）与对二氯苯混合使用，即在 100 mL 对二氯苯饱和水溶液中加入 1 滴 α-溴代萘，充分振摇后使用。

（2）处理方法

①离体处理：将处理的器官或组织从植株上切取下来，直接投入预处理液中进行处理。

②非离体处理：将种子连同萌发的根尖整个投入预处理液中进行处理，常用于种子和根都很小的材料。或者是将种子或鳞茎上长出的根或幼小植株的根浸入预处理液中进行培养。

（3）处理时间

处理时间包括两方面内容，一方面指在一天中什么时候取样处理合适，另一方面指材料在处理液中的持续时间。

关于什么时候取材处理的问题，我们认为植物细胞分裂是不同步的，一天中任何时候取样都是可行的。

关于预处理持续时间的长短，根据实际情况而定。

①染色体的大小：染色体大的材料处理时间长，染色体小的材料处理时间宜短。

②材料的大小：大者应长，小者宜短。

③处理方法的不同：离体处理则短，非离体处理则长。

④处理液浓度：高浓度宜短，低浓度可长。

⑤温度：高温宜短，低温可长。

根据我们的经验，一般情况下，大染色体材料处理 4～5 h；小染色体材料，以 1～2 h 为宜。

（4）处理液温度

总的原则是以低温长时间处理为宜，温度范围在 10～20℃之间。

1.1.1.3 固定

固定的目的是利用化学药物把细胞迅速杀死，使蛋白质变性和沉淀，并尽量保持各种结构的原有状态。染色体压片材料的固定，一般要求固定作用迅速，固定液尽可能不含金属离子，以排除对后续染色的可能干扰。

（1）固定液

最主要的用卡诺（Carnoy）1886年创用的两种固定液配方。

Carnoy Ⅰ：冰醋酸：无水乙醇=1：3

Carnoy Ⅱ：冰醋酸：氯仿：无水乙醇=1：3：6

以第Ⅰ配方应用最为广泛，1978年，对此固定液有些改动，以甲醇代替乙醇，效果更佳。第Ⅱ配方常用于某些含油脂类物质较多的材料以及某些需要更加硬化的组织的固定。

（2）固定时间：常用的时间范围是2～24 h。

（3）固定温度：室温固定即可，以低温（冰箱）固定的效果较好。

（4）保存：经固定后的材料，通常换至70%乙醇中于冰箱中可长期保存。一般长期保存过的材料在压片前用固定液重新固定一次，效果较好。

1.1.1.4 解离

根尖和茎尖等需经过某些处理，除去细胞之间的果胶层并使细胞壁软化，这种细胞分离合软化后的组织才便于压片。最常用的是酸水解和酶处理两种。

（1）酸水解

固定后的材料在蒸馏水中洗涤后，转入1 mol/L 盐酸中于60℃下处理5～20 min，或者于45℃下处理45 min。

（2）酶处理

固定后的材料在蒸馏水中洗涤后，用1%～2%的果胶酶和纤维素酶处理。

1.1.1.5 染色

（1）染料及其配制

压片常用的染料有：醋酸洋红、地衣红、树脂蓝、甲苯胺蓝、碱性品红和苏木精。其中碱性品红染色快速而且染色效果好，是近年最常用的优良染料。

碱性品红用作染色体染色有两个最重要的配方：其一是石碳酸—品红，其二是锡夫（Schiff）试剂。其中石碳酸—品红又称卡宝品红（Carbol fuchsin），是最常用的压片染料，其配方如下。

原液A：称取3 g碱性品红溶于100 mL 70%乙醇中（此液可以长期保存）。

原液B：取10 mL原液A加入90 mL 5%的石碳酸（苯酚）水溶液中（2周内使用）。

染色液：55 mL原液B加6 mL冰醋酸和6 mL 37%的甲醛。

此染色液适合于植物细胞原生质培养中的细胞核和核分裂的染色，因为它是水溶液而且含有较多的甲醛，不能使组织软化，所以不太适合于一般植物组织的染色体压片染色。后来，在此基础上加以改良，改良后的染色液可以普遍地用于一般植物组织的染色体压片的染色。改良后的配方如下：

取染色液2～10 mL加90～98 mL 45%醋酸和1.8 g山梨醇。

此染色液配制后为淡红色，如果立即使用，染色较淡，放置2周后，染色能力显著增强，而且放置时间越久，染色效果越好。此液在室温下可存放 2 年，2 年内染色液保持稳定，无沉淀也不褪色。

（3）染色操作

盐酸解离对于卡宝品红染色是关键的操作步骤，盐酸处理时间太短，染色不清晰，细胞质也染上深浅不同的颜色；处理时间过长，染色体染色极淡或不染色。一般 1 mol/L 盐酸 45℃下处理 45 min 对大多数植物材料都是比较适合的。

经解离后的材料用蒸馏水洗几次，在蒸馏水中浸泡 10 min，取 100 μL 染料于 0.2 mL 离心管中，把材料置于染料中染色 2 h 以上，对于一些不易着色的材料可以染色过夜。

1.1.1.6 压片操作

植物染色体压片的操作方法和所用的工具，并无一定的规格，各人的手法不尽相同。现只根据我们常用的操作法介绍如下。

用具包括一把尖细的不锈钢镊子（钟表游丝镊子），一把一次性竹筷子（一端削尖），双面刀片，滤纸。

操作时，用镊子小心从染料中取出根尖置于载玻片上，用刀片截除根冠和伸长区部分，只留 1 mm 长的分生区，用镊子取少量染料到载玻片的材料上，盖 22 mm×22 mm 盖片后马上用右手食指指甲在材料处轻敲一下，于显微镜下检查材料中是否有分裂相，如果有则可进行以下操作。左手中指和食指戴上指套（也可戴乳胶手套），用左手中指和食指轻轻压住盖片左边两角，以免盖片错动，将双面刀片轻轻插入盖片的右上角，这样染料左右就形成一个斜面，用竹签的尖端轻轻敲击盖片有材料的地方，使材料逐渐散开，直到看不到大的组织块为止，轻轻把刀片取出，左手按紧盖片左边两角，再次用竹签尖端敲击盖片所有的部分，盖片上加一片滤纸用右手食指先轻轻用力然后逐渐加力垂直按压吸出多余染料，最后用大拇指将盖片压实。

1.1.1.7 镜检及显微摄影

在显微镜下仔细进行镜检，选择分散良好的完整细胞，用 CCD 拍照。

（作者：陈成彬。原载：陈成彬博士、硕士论文）

1.1.2 植物有丝分裂染色体标本制作的新方法

1.1.2.1 材料和方法

1. 材料

AR1 黑麦品种（*Secele cereale* 2*n*=14）（由中国农科院提供），"天津 1 号"裸大麦（*Hordeum vulgare* 2*n*=14），"津丰 1 号"普通小麦（*Triticum aestivum* 2*n*=42）。

2. 方法

（1）材料培养：种子均在 25℃温箱内发芽培养。

（2）前处理：在有丝分裂高峰前 2~3 h，大麦用 0.05% 秋水仙碱或 0.002 mol/L 8-羟基喹啉，黑麦、小麦使用 0.2% 秋水仙碱进行处理（陈瑞阳等，1979）。

（3）前低渗：切取分裂旺盛部分根尖，放在 0.075 mol/L KCl 低渗液中，在 20~25℃ 条件下处理 30 min。

（4）去壁：倒去 KCl 溶液，加入 2.5% 混合酶液（纤维素酶与果胶酶各占 2.5%），25℃ 温箱内处理 5 h 左右，在处理过程中最好将材料瓶轻轻摇动数次，促使反应充分。酶液的量与材料要有一定的比例，酶液过少，会造成消化不足，100 条小麦根尖（只切取分裂区）加入 1 mL 酶液为宜。

（5）后低渗：倒去酶液，用 20~25℃ 蒸馏水慢慢冲洗 2~3 次，然后在蒸馏水中停留 5~10 min，进行后低渗处理。

（6）制备细胞悬液：倒去蒸馏水，用镊子立即将细胞挟碎，制成悬液。

（7）固定：向细胞悬液中加入新配制的甲醇：冰醋酸（3:1）固定液 4~5 mL。

（8）去沉淀：静止片刻，使大块组织沉淀，然后倒取上层细胞悬液，去掉沉淀物。

（9）去上清液：将上层细胞悬液，静置 20~30 min，可见细胞沉淀，用吸管吸去上层清液，留约 1 mL 细胞悬液，制备标本。

（10）标本制备：取一张预先在蒸馏水中冷冻的清洁载片，用滴管滴 2~3 滴细胞悬液于其上，立即将载片一端台起，并轻轻吹气，使细胞迅速分散，然后在酒精灯火焰上微微加热烤干。

（11）染色：干燥的片子以 40:1 Giemsa 染液（用 pH7.2 1/15 mol/L 磷酸缓冲溶液稀释）染色 4~5 min，蒸馏水冲洗，空气干燥后 Damer 树胶封片，制成永久标本。

1.1.2.2 结果和讨论

（1）应用酶解去壁、低渗、火焰干燥制备植物染色体有丝分裂标本，在黑麦、大麦、小麦上均获成功，表明这个方法具有普遍意义。

（2）这个方法与传统的压片法相比具有若干优点：首先，用这个方法获得的染色体形态比较完整，接近正常状态。染色体各组成部分——长臂、短臂、着丝点、随体、染色单体显示的都比较清楚，有利于染色体的测量和分析。这个方法不但可使中期染色体散开，而且也可使前中期，甚至前期的染色体散开，在黑麦材料中我们还观察到出现频率很低的 B-染色体（松田忠男，1975）。其次，方法简便，效率高，不需要特殊设备，一般实验室都能进行此项工作。

（3）几个问题：①低渗处理：在人类染色体标本制备过程中，多数以 0.075 mol/L KCl 或蒸馏水作为低渗液，这对植物材料也是适用的。Kurata（1978）在去壁前使用 0.075 mol/L KCl 进行低渗处理，据我们试验，这个步骤是不可缺少的。去壁以后以蒸馏水冲洗材料，这同时起到低渗作用，为区别于去壁前 KCl 低渗处理，我们称之为后低渗。后低渗是否适宜，常常是试验成败的关键，材料在蒸馏水中停留时间过长，会引起原生质体的解体。②纤维素酶的浓度与时间：我们使用了各占 1%、2.5%、5% 纤维素酶和果胶酶的混合液，进行了不同时间的消化处理，发现三种不同浓度的酶液均可部分地除掉细胞壁。我们初步认为，使用较低浓度（2.5%）的酶液，较长时间的消化更好。③分裂相：材料在前处理后即行固定，可以很准确的获得较高频率的中期分裂相，但是此方法在前处理后，还需一段时间的酶解和低渗处理才能固定，这样有丝分裂高峰就较难掌握，有待于进一步研究解决。

（作者：陈瑞阳、宋文芹、李秀兰。原载：植物学报，1979，21（3）297-298）

1.1.3 植物染色体标本制备的去壁、低渗法及其在细胞遗传学中的意义

1.1.3.1 材料和方法

我们对 37 科 105 种植物（包括 327 个品种）进行了试验，其中蕨类植物 2 种，裸子植物 4 种，被子植物 99 种。大部分材料为广东、云南、吉林省农科院提供，部分材料是在本系实验基地种植和野外采集得到。去壁、低渗、火焰干燥制备植物染色体标本的具体方法，如前报道，其简要流程流程如下：

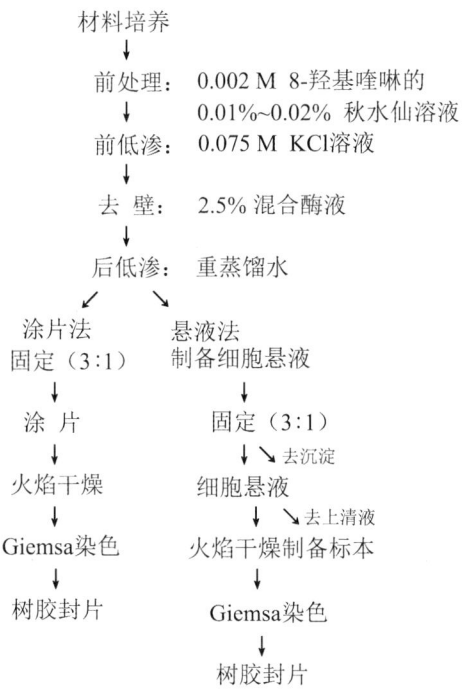

植物染色体 Giemsa 分带方法：按 BSG 法、HSG 法和 N-带法处理（陈瑞阳等，1979，1981）。

扫描电镜染色体标本的制备：用上述方法制备的染色体标本，Giemsa 染色后，首先在光学显微镜下挑选染色体分散好的细胞，做上标记，再按照扫描电镜样品台大小，将载片切割开，喷金（100～200Å）后，用 JSM-25S 型扫描电镜观察拍照，电压 2.5～12 kV，标本倾斜角为 15～25 度。

1.1.3.2 结果与讨论

使用低渗、火焰干燥法制备染色体标本的优越性，20世纪50年代早已在人类及哺乳动物上得到证明。我们将这种方法移植到植物材料上来，经过对37科105种植物的试验，结果表明，除蕨类和裸子植物未能取得很好的结果外，其他大部分被试材料都取得了良好的结果（详见表1.1）。因此我们认为，用纤维素酶和果胶酶将植物细胞壁去除后，再按当前哺乳动物上通常使用的低渗、火焰干燥法制备植物染色体标本，即去壁、低渗、火焰干燥法，在植物染色体研究中，同样具有重要意义。它比压片法具有若干优点，是当前植物染色体研究中的一个重要方法。在植物染色体计数、组型分析、Giemsa分带以及扫描电镜观察方面都取得了较好的结果。

1. 去壁、低渗法在细胞遗传学中的意义

（1）在染色体计数和组型分析方面的应用：去壁、低渗法可使染色体分散的比较好，即使是染色体数目多的材料，如甘薯、甘蔗、棉花、菠萝等，染色体也能分散开，并且在染色体铺展平整，便于计数和测量。我们利用此方法，曾对我国野生大豆、栽培大豆、野生稻（陈瑞阳等，1981）、栽培稻（陈瑞阳等，1980）、陆地棉（陈瑞阳，1981）、甘蔗等20多种植物的染色体组型进行了研究，都取得了比较好的结果。另外应用去壁、低渗法，在一部分材料中，如栽培稻、野生稻、栽培大豆、野生大豆、甘蔗、谷子、柿子、桑、无花果等材料，经Giemsa染色后，在前期—晚前期染色体的一定部位上显示有一种染色深的区域。这种着色深的区域与Giemsa C-带不同，为与其相区别，我们称此区域为染色体Giemsa染色区（chromosome Giemsa region）。Giemsa染色区具有以下特点：①着色区域比较大，一般不呈带状或点状，有的涉及染色体的一个臂，或整个染色体，容易识别。②染色区比较稳定，处于分裂晚前期或前中期的色体，大部分都能显示出这种染色区。③同源染色体所显示的Giemsa染色区基本上是相近的。根据这些特点，我们认为，Giemsa染色区，在核型分析时可以作为识别同源染色体的一种指标，从而可提高核型分析的准确性（陈瑞阳等，1980）。

（2）在Giemsa分带方面的应用：在我们所研究的材料中，应用去壁、低法制备的染色体标本，不经其他药物处理，直接Giemsa染色，有四种结果：①显示Giemsa带，如玉米、蓖麻、黄瓜、甜橙、黑麦等。玉米所显示的带纹与我们过去报道用BSG法所显示的Giemsa C-带基本一致（陈瑞阳等，1979）。黑麦则只显示端带，着丝点带较浅，中间带则不明显。蓖麻所有的染色体均显示有明显的着丝点带。②在间期核里显示间期带，在研究的大多数材料间期核内，都清楚的显示有染色中心（chromocenter），如马铃薯、向日葵、香蕉、酸桔、黑麦、玉米、萝卜、明开夜合、中国槐、桑等，经过低处理后，细胞核被膨胀大了，染色中心更清楚可数。染色中心是染色体上的大型异染色质区，在细胞周期中，保持持久的浓缩状态，它与染色体倍数正相关，因此它在染色体倍数的鉴定上和间期核结构的研究上有利用价值。③前期染色体显示有Giemsa染色区，Giemsa染色区具有生物种的特异性，因此利用前期染色体的形态和Giemsa染色区也可进行核型分析。④染色体全着色，类似一般染色。如小麦、大麦、棉花等。对于这些材料的染色体标本，我们曾按BSG法和HSG法进行了分带处理，但都未能取得满意的结果。只有大麦、小麦按 N 带法进行分带处理，得到了良好的结果，带纹不但清晰明显而且提高了分带的可重复性（陈瑞阳等，1981）。

（3）在扫描电镜研究植物染色体方面的应用：染色体的螺旋结构在光学显微镜下已发现了近百年，从而建立了染色体螺旋结构学说（汤浅明，1980）。但在电子显微镜下，长期以来都不能清楚地看到这种螺旋结构。有人指出（饭野晃启等，1980）：残留在染色体周围的部分细胞质妨碍了对染色体细微结构的观察。我们用去壁、低渗法制备的植物染色体标本进行扫描电镜观察，清楚地看到了植物染色体的螺旋结构，从而将光镜与电镜结果统一起来，实现了光镜和电镜对同一细胞的观察。

2. 去壁、低渗法中几个主要因素的分析

去壁、低渗法虽然方法简单，不需要特殊设备，但由于材料在整个处理过程中是处在活体状态，因此有些环节较难掌握。

（1）前低渗处理。此步的时间范围变化较大，一般掌握在 30～60 min 即可，未经前低渗处理的材料，细胞核膨胀得较小，染色体不易散开。

（2）酶解去壁问题。包括几个因素：① 酶液浓度：提高酶液浓度可以缩短去壁时间，但耗费大。据我们的经验，国产纤维素酶（上海酒精二厂出品）的浓度控制在 2.5%～5%，根尖材料为 2.5%，树木幼叶、幼芽可用 5%。② 酶解时间：酶解时间是个变化的因素，要根据材料的老嫩、大小和种类作适当的调整。材料过大，会使酶解时间成倍增加，材料过多会导致消化不足，酶液量过多会使酶解时间不易稳定。③ 酶解温度必须恒定（25～26℃）。④ 酶液的配制与 pH：A. Mouras（1978）使用 0.2 mol/L 盐溶液（0.15 mol/L 山梨醇＋0.05 mol/L 柠檬酸钠）配制酶溶液。但我们按照这种配法未能取得满意的结果。我们使用蒸馏水（pH5.5）配制国产酶液，效果很好。在酶液浓度、酶解温度一定的条件下，可用酶解时间的长短来调节酶解的合适程度。

（3）后低渗处理。这与动物细胞的低渗处理相同，低渗的目的在于使细胞吸收水分而膨胀，染色体分散到细胞质中，在制片时便于染色体分散开，这是一个关键步骤。后低渗处理过度，使细胞在低渗液中胀破，造成材料解体；后低渗处理不足，染色体分散不好。酶解不足的材料，可适当延长后低渗的时间；反之，酶解过度的材料要缩短后低渗的时间。涂片法可比悬液法的低渗时间加长。

（4）细胞悬液的制备与制片方法的选择 A. Mouras（1978）等用机械力将酶解的材料捣碎后，采用 70～200 μm 网过筛，100 g 离心的方法获得悬液。此法不但手续麻烦，效果也并非很好。用静置片刻去沉淀和静置 20～30 min 去上清液的方法，既简单且细胞收率也高，在上清液中有大量的被酶解离的细胞碎片，采用离心的方法，这部分碎片很容易与细胞混在一起，妨碍染色体铺展开。对于制片方法的选择，我们认为，材料少，染色体数目少而小或酶解过度的材料可用涂片法；取材容易，染色体比较大，数目多而难散开的材料可用悬液法。悬液法要求条件比较严，酶解和后低渗的时间必须合适。

（5）Giemsa 染色体与 pH 值的调整 在我们所研究过的材料中，还没有发现对 Giemsa 染色体不适合的材料，但着色的速度差别很大，从几分钟至数小时不等（详见表 1.1）。进化上处于低等的类型有易着色的趋势。染液的 pH 值多为 6.8，少数材料，如山楂、柿子等需调整至 pH 7.2～7.4。

表 1.1　不同植物去壁，低渗法处理时间

Table 1.1　The treatment periods of wall degradation, hypotonic method different plant

植物种名 Species of Plants	材料 Material	染色体数目 Chromosome number	前处理时间 Pretreatment Period (min)	前低渗时间 0.075 M KCl Pre-hypotonic period (min)	酶解时间 Hydrolyze with enzyme period (min)	后低渗时间 Post-hypotonic Period (min)	Giemsa 染色时间 Giemsa staining Period (min)	结果 Result ++[1]	+++[2]
水杉 Metasequoia glyptostroboids Hu et Cheng.	A	$2n=22$	240	30	240	6	30	++	
银杏 Ginko biloba L.	A	$2n=24$	240	30	300	10	30	++	
普通小麦 Triticum aestivum L.	B	$2n=42$	180	30	240	10	15		+++
大麦 Hordeum vulgare L.	B	$2n=14$	180	30	240	10	15		+++
黑麦 Secale cereale L.	B	$2n=14$	180	30	240	10	15		+++
谷子 Setaria italita L.Beauv.	B	$2n=18$	180	30	180	156	240		+++
高粱 Sorghum vulgare Pers.	B	$2n=20$	180	30	180	20	30		+++
玉米 Zea mays L.	B	$2n=20$	680	30	240	30	60		+++
甘薯 Ipomoea batatas Lam.	C	$2n=90$	180	30	240	7	240		+++
野生甘薯 K₁₂3 I.trifida (H.B.K.) G.Don.	C	$2n=90$	300	30	210	20	360		+++
马铃薯 Solanum tuberosum L.	C	$2n=48$	180	30	210	10	360		+++
黄麻圆果种 Corchorus capsularis L.	B	$2n=14, 28$	240	30	180	10	240		+++
黄麻长果种 Corchorus olitorius L.	B	$2n=14, 28$	240	30	180	15	240		+++
栽培大豆 Glycine max (L.) Merr.	B	$2n=40$	480	30	210	15	240		+++
野生大豆 Glycine soja (L.) Sieb et Zucc.	B	$2n=40$	360	30	120	15	600		+++
半野生大豆 Glycine Gracilis sp.Nov.	B	$2n=40$	360	30	120	15	600		+++
绿皮蔓豆 Glycine viridula	B	$2n=40$	360	30	120	15	600		+++
白花野生豆 Glycine soja albiflora	B	$2n=40$	360	30	120	10	600		+++
赤豆 Phaseolus angularis Wight. et al.	B	$2n=22$	240	30	180	10	600		+++
绿豆 Phaseplus angularis Wight. et al.	B	$2n=22$	280	30	240	10	600		+++
花生 Arachis hypogaea L.	B	$2n=40$	180	30	240	10	240		+++
芝麻 Sesamum indicum L.	B	$2n=26$	360	30	240	5	600		+++

续表

植物种名 Species of Plants	材料 Material	染色体数目 Chromosome number	前处理时间 Pretreatment Period (min)	前低渗时间 0.075 M KCl Pre-hypotonic period (min)	酶解时间 Hydrolyze with enzyme period (min)	后低渗时间 Post-hypotonic Period (min)	Giemsa 染色时间 Giemsa staining Period (min)	结果 Result ++[1]	++[2]
蓖麻 Ricinus communis L.	B	2n=20	600	30	150	5	300		+++
向日葵 Lelianthus annus L.	B	2n=34	240	30	300	5	120		+++
萝卜 Raphanus sativus L.	B	2n=18	210	30	120	8	240		+++
大茴香 Foeniculum vulgare Mill.	B	2n=22	210	30	120	8	240		+++
甘蓝 Brassica caulorapa Pasq.	B	2n=18	210	30	240	7	600		+++
番茄 Lycopersicum esculenium mill.	B	2n=24	180	30	120	8	600		+++
黄瓜 Cucumis sativus L.	B	2n=14	180	30	300	10	240		+++
丝瓜 Luffa cylindrical (L.) Roem.	B	2n=24	360	30	150	8	240		+++
冬瓜 Benincasa hispida Cogn.	B	2n=24	210	30	120	7	600		+++
含羞草 Mimosa pudica L.	B	2n=52	180	30	150	8	240		+++
栝楼 Trichosanthes kirilowii Maxim.	A, B	2n=?	360	30, 60	180	8	180		+++
籼稻 Oryza sativa subsp.hsien Ting	B	2n=24	360	30	120	10	300		+++
粳稻 Oryza sativa subsp.Reng Ting.	B	2n=24	360	30	120	10	300		+++
普通野生稻 Oryza perennis Moench.	C	2n=24	180	30	180	10	300		+++
疣粒野生稻 Oryza meyeriana Bail.	C	2n=24	180	30	180	10	300		+++
药用野生稻 Oryza officinalis Wall.	C	2n=24	180	30	180	10	300		+++
中国竹蔗 Saccharum sinense Roxb Jeswet.	C	2n=80	360	30	150	20	30		+++
热带蔗 Saccharum officinarum L.	C	2n=80, 81	360	30	90	25	30		+++
割手密 Saccharum spontaneum L.	C	2n=40, 80	360	30	120	15	30	++	
陆地棉 Gossypium hirsutum L.	B	2n=52	150	30	300	5	120		+++
甜橙 Citrus sinensis Osbeck.	A	2n=18	360	30	120	10	660		+++
酸桔 Citrus sunki Hort et Tanaka.	A	2n=18	360	30	120	10	660		+++
渡萝 Ananas comsus (L.) Merr.	C	2n=50	360	30	180	10	240		+++
野生香蕉 Musa balisiana Colla.	C	2n=22	180	30	180	6	240		+++

续表

植物种名 Species of Plants	材料 Material	染色体数目 Chromosome number	前处理时间 Pretreatment Period (min)	前低渗时间 0.075 M KCl Pre-hypotonic period (min)	酶解时间 Hydrolyze with enzyme period (min)	后低渗时间 Post-hypotonic Period (min)	Giemsa 染色时间 Giemsa staining Period (min)	结果 Result ++[1]	结果 Result +++[2]
野生香蕉 ABB 型 *Musa paradisiaca* L.	C	2n=33	180	30	180	6	240		+++
阿加蕉 *Musa acuminate* Coll	C	2n=33	180	30	180	10	240		+++
梨 *Pyrus ussuriensis* Maxim.	A	2n=34	240	60	240	10	120		+++
丁香 *Syringe oblate* Lindl.	A	2n=46	240	60	240	10	120		+++
柿 *Diospyros kaki* L.	A	2n=90	240	60	240	10	120		+++
海棠花 *Malus spectabilis* (Ait) Borkh.	A	2n=34[3]	240	60	240	10	120		+++
无花果 *Ficus carica* L.	A	2n=26	240	60	240	10	120		+++
法国梧桐 *Platanus orientalis* L.	A	2n=40[3]	240	60	240	10	120		+++
毛白杨♀ *Populus tomentosa* Carr.	A	2n=38	240	60	240	10	120		+++
臭椿 *Ailanthus altissima* Swingle.	A	2n=64[3]	240	60	240	10	120		+++
槐树 *Sophora japonica* L.	A	2n=28	240	60	240	10	120		+++
榆叶梅 *Prunus triloba* Lindl.	A	2n=64	240	60	240	10	120		+++
珍珠梅 *Sorbaria kirilowii* Maxim.	A	2n=36[3]	240	60	270	10	120		+++
黄杨 *Buxus microphylla* Sieb.et zucc.	A	2n=28[3]	240	60	270	10	120	++	
明开夜合 *Euonymus bungeana* maxim.	A	2n=32[3]	240	60	270	10	120	++	
迎春 *Jasminum nudiflorum* Lindl.	A	2n=52	240	60	270	10	120		+++
榆树 *Ulmus pumila* L.	A	2n=28	240	60	270	10	120	++	
日本樱花 *Prunus yedoensis* Matsum. Yu	A	2n=16	240	60	270	10	120	++	
月季 *Rosa chinensia* Jacq.	A	2n=21	240	60	270	10	120	++	
旱柳 *Salix matsudana* Koidz.	A	2n=?	240	60	270	10	120	++	
桑树 *Morus alba* L.	A	2n=28	240	60	270	10	120		+++
吊兰 *Chlorophytum capense* Kuntze	D	2n=28	420	30	130	8	30		+++
水仙 *Narcissus tazetta* L. var.*chinensis* Roem.	C	2n=30	360	30	180	10	20	++	

A: 叶芽 leaf bud; B: 幼根 seed root; C: 不定根 aeventitious root; D: 气生根 respiratory root.
1) 良 better; 2) 好 good; 3) 染色体数目由本文作者初步确定。Chromosome number was determinded preliminarily by the writer of this paper. Wall Degradation Hypotonic Method of Preparing Chromosome Samples in Plant and Its Significance in the Cytogenetics

3. 存在的问题

（1）染色体末端解螺旋。使用这个方法，在一部分材料中，如水稻、大豆、玉米、大麦等，染色体末端（尤其是晚前期染色体末端）易发生解螺旋，因而使染色体末端发虚，形态轮廓不清，影响染色体的准确测量。

（2）Giemsa 分带此方法制备的植物染色体标本其 Giemsa 显带的条件，不同于以往的压片法，有待进一步研究。另外，此方法使一些植物染色体显带的机制还不清楚。

Wall Degradation Hypotonic Method of Preparing Chromosome Samples in Plant and Its Significance in the Cytogenetic

Abstract: In this paper, the wall degradation hypotonic method of preparing chromosome samples in plant has been studied. From the chromosome count, karyotype analysis, Giemsa banding and SEM observation of chromosome obtained by testing 105 materials of 37 branches, it is shown that this method is of great significance. Some of the main factors and processes of this method are narrated as follows:

1. Meterial culture: Roots from germinating seeds growing at 25℃.

2. Pretreatment: Treating the attached root with 0.01%-0.2% colchicine or 0.002M 8-hydroxyquinoline for 3-4 hours prior to the peak time of mitotic activity.

3. Pre-hypotonic treatment: Excising the good root tips and immersing them in 0.075mol/L KCl solution for 30 minutes at 20-25℃.

4. Wall degradation: Treating with 2.5% enzyme mixture for 2-4 hours at 25-26℃.

5. Post-hypotonic treatment: Rinsing the material 2-3 time in 25-30℃ redistilled water and then immersing it in the distilled water for 5-10 minutes.

Preparation of samples:

I. Suspension of samples:

I 6. Preparation of cell suspension: Pour out the water. Tear the material with a pair of forceps to make well a cell suspension.

I 7. Fixation: Add 2-3ml of fresh prepared methanol: acetic acid（3:1）to the suspension.

I 8. Removing of precipitate: Keep the solution for a moment to precipitate the residues. Use the upper suspension.

I 9. Removing of supernatant: Keep the suspension for 20-30 minutes. Remove the supernatant with a pipette and take 1ml of concentrated suspension for chromosome preparation.

I 10. Preparation of chromosomes: Drop 2-3 drops of the suspension on a chilled slide kept in distilled water. Lift one end of the slide and blow gently to spread the cells. Flame the slide gently above an alcohol burner to have it dry.

I 11. Staining: Stain the dried slides in a 1:20 or 1:40 Giemsa solution for 4-5 minutes, then rinse them in distilled water, and have them air-dried and mounted in damar balsam.

II. Smear method:

II 6. Fixation: Pour out the distilled water. Add fresh prepared methanol: acetic acid（3:1） to the material.

II 7. Smear: Put on 3-4 root tip on the slide which was kept in the freezed distilled water. Add drop fixation solution to them. Tear the material with a pair of forceps and Remove the supernatant.

II 8. Flame-drying: Add 2 drops fixation solution on the slide, then bake it on the alcohol burner.

II 9. Staining: see I.11

（作者：陈瑞阳、宋文芹、李秀兰。原载：遗传学报，1982，9（2）151-159）

1.1.4 染色体标本制备的蒸气干燥法

蒸气干燥法系作者提出的一种改进的染色体标本制备方法，它以蒸气干燥代替原来的酒精灯火焰干燥，其方法是：取 500 mL 三角烧瓶一个，加一合适橡皮塞，在皮塞中打一洞，倒插一玻璃滴管（小口端在外），在烧瓶内加约 20 mL 自来水，放在电磁炉上加热，至从玻璃管中喷出蒸气流，温度可达 60～100℃，气流的大小，由玻璃管出气口直径控制，可自行调节，选一合适口径，一般以听见嘘嘘的蒸气流声即可。滴好的片子，材料面向上，立即在蒸气流上通过 2～3 次，材料即可干燥。实验证明，上述蒸气干燥法不但方法简单易行，且染色体分散良好，保持了火焰干燥法中染色体的分散程度。此法，因载片受热均匀，温度恒定，片间差异小，分带可重复性高，是制备染色体分带标本的一种好方法。

（作者：陈瑞阳、李秀兰）

第二节　植物染色体分带技术

提要：植物染色体 Giemsa 分带技术是 20 世纪 60 年代和 70 年代初发展起来的一项研究新技术，在核型分析和染色体个性识别中发挥了重要作用。然而，我国在这个历史时期对国外的发展一无所知。改革开放焕发了科学工作者的智慧和青春，本实验室在 1978～1979 年短短一年的时间里，就攻克了植物染色体 Giemsa 分带技术难关。为此，1979 年中国植物学会委托南开大学在 1979 年底举办了全国第一个染色体分带技术培训班，为全国 300 多个单位培训 Giemsa 分带技术人员。

1.2.1 植物染色体 Giemsa 分带技术的研究

1.2.1.1 材料和方法

将洋葱（*Allium cepa* 2n=16）、蚕豆（*Vicia faba* 2n=12）、黑麦（*Secale cereale* 2n=14）、

玉米（*Zea mays* 2*n*=20）、大麦（*Hordeum vulgare* 2*n*=14）、普通小麦（*Triticum aestivum* 2*n*=42）和八倍体小黑麦（*Triticale* 2*n*=56）放在 24℃温箱内发芽培养，用 0.01%～0.2% 秋水仙碱进行前处理 2～4 h，然后用卡诺氏液（1 份冰醋酸和 3 份无水酒精）固定 5～24 h。材料用 45%醋酸压片，待材料散开后在酒精灯上微热，以去掉部分细胞质，载片用液态 CO_2 或半导体致冷器迅速冰冻，将底片与盖片分开，移入 95%～100%酒精中脱水，室温下空气干燥。Giemsa 分带处理，我们使用两种方法。

（1）BSG 法（Barium hydroxide/Saline/Giemsa）我们基本上使用了 Vosa 等（Gill, 1974；Hadlaczky, 1975；Linde-Laursen, 1975；Sachan, 1976；Vosa, 1971, 1974）的方法，步骤如下：将空气干燥的载片用 5% Ba（OH）$_2$ 水溶液进行变性处理，然后用 2×SSC（0.3 mol/L NaCl + 0.03 mol/L $Na_3N_6H_5O_7$）溶液进行复性处理（表 1.2），蒸馏水冲洗，风干，Giemsa 染色（用 0.5 g Giemsa 粉 + 33 mL 甘油 + 33 mL 甲醇配制成贮存母液，用 1/15 mol/L pH6.8 磷酸缓冲液稀释），用蒸馏水冲洗，空气干燥，直接用 Damer 胶或中性树胶封片。

表 1.2 不同材料 2×SSC 处理时间
Table 1.2 The period of 2×SSC treatment in different materials

材料 Material	BSG 法 BSG method		HSG 法 HSG method	
	温度（℃）Temperature	时间（分）Period（min）	温度（℃）Temperature	时间（分）Period（min）
黑麦 *Secale cereale*	60	45～60	60	30～35
小麦 *Triticum aestivum*	60	60	60	30～35
小黑麦 *Triticale*	60	45～60	60	30～35
大麦 *Hordeum vulgare*	60	30～40	60	30～40
玉米 *Zea mays*	60	120	60	60
洋葱 *Allium cepa*			60～66	15～30
蚕豆 *Vicia faba*			60～66	30～50

（2）HSG 法（Hydrochloric acid Saline Giemsa）用 Fiskesjö 在洋葱上所使用的方法，将空气干燥的载片用 0.2M HCl 在室温下（18～25℃）变性处理，然后用 2×SSC 复性处理，蒸馏水冲洗，Giemsa 染色（用 1/15 mol/L pH7.2 的磷酸缓冲液稀释）。

1.2.1.2 结果和讨论

1. 关于 HSG 法的显带效果

1974 年 Fiskesjö 使用 0.2 M HCl 变性预处理，再经 2×SSC 复性处理，在洋葱上获得了 Giemsa C-带图型，但在其他植物上没见到相同的报道，为此，我们在原方法的基础上，根据不同植物，对显带条件进行了修改，在蚕豆、黑麦、玉米、大麦、小麦和小黑麦等植物上也获得了 Giemsa C-带图型。用 HSG 法所显示的蚕豆 M 染色体上的带与 Vosa 等（1972）用 BSG 法和 Döbel 等（1973）用尿素法所显示的带基本是一样的。

因此，我们认为 0.2M HCl 变性处理显带方法，在植物染色体 Giemsa 分带技术中具有普遍意义，是植物染色体 Giemsa 分带的一种较为优越的新方法，我们将这种方法简称为 HSG 法。

2. 植物染色体 Giemsa 分带条件

由于植物染色体分带的可重复性较低，使当前植物染色体分带的研究进展很慢（谷口研至，1977），根据我们的试验认为，分带的可重复性取决于对分带条件的控制，使每次重复试验的条件都保持基本相同，则失败的机率是比较低的。

（1）前处理和染色体有丝分裂的时期　所有的材料均使用秋水仙碱进行前处理，方法是在有丝分裂高峰 3～4 h 前，在培养皿中滴加一定浓度的秋水仙碱，在原培养的温度下再继续培养一段时间。秋水仙碱浓度过高或处理时间过长会使染色体收缩的太短，对分带有一定影响。我们在不同材料上所使用的秋水仙碱浓度和处理时间见表 1.3。

表 1.3　不同材料秋水仙碱前处理的浓度和时间
Table 1.3　Different materials pretreated with colchicine at different cocentrations for different periods

材料 Material	秋水仙碱浓度（%） Colchicine Concentration（%）	前处理时间 Pretreatment hours	从发芽到固定所需时间 The period from germination tofixation（h）
黑麦 *Secale cereale*	0.2	8∶00～11∶00	39
小麦 *Triticum aestivum*	0.2	9∶00～11∶00	39
小黑麦 *Triticale*	0.2	8∶00～11∶00	39
大麦 *Hordeum vulgare*	0.05	8∶00～11∶00	39
玉米 *Zea mays*	0.001	15∶00～2∶00	46
洋葱 *Allium cepa*	0.02	12∶00～16∶00	48～72
蚕豆 *Vicia faba*	0.05	8∶00～23∶00	39

在有些实验中，染色体不显带并不是由于分带处理条件所造成的。如在同一张标本中，不是所有处于分裂中期的染色体都能分出完好一致的带，而是有很大差异。造成这种差异的原因可能与染色体在有丝分裂过程中所处的时期和状态有关。据观察，处于有丝分裂早中期的染色体最适于分带，处于晚中期染色体的带常常较小。其原因尚有待研究。

（2）解离试验表明，压片前的酸解不仅对材料的解离起作用，而且对以后的分带也产生很大影响。在洋葱中，0.1M HCl 60℃水解 5 min，显示末端带，当水解延长到 8 min 时，末端带着丝点带可以同时显示出来。当 HCl 的浓度增加到 1M 时，带则完全消失了。所以，并不是任何一种解离液对分带标本制备都是适用的。在 BSG 方法中材料水解多用 lN HCl 60℃水解数 10 s。HSG 方法中，洋葱使用了 0.1M HCl 60℃水解 5 min，但这种水解时间在禾谷类根尖染色体标本制备上是比较困难的。为了解决材料压散问题，我们在酸解之前增加了酶解（表 1.4）。

表 1.4　不同材料的解离液和解离时间
Table 1.4　The hydrolyzing agents and hydrolyzing period for different materials

材料 Material	处理 Treatment					
	BSG 法　BSG method			HSG 法　HSG method		
	水解剂 Hydrolysing agent	温度（℃）Temperature（℃）	时间（分）Period（min）	水解剂 Hydrolysing agent	温度（℃）Temperature（℃）	时间（分）Period（min）
黑麦 Secale cereale	1%酶液[1]	T[2]	120	1%酶液	T[2]	120
	1% Enzyme +1M HCl	60	0.4	1% Enzyme + 0.1M HCl	60	5
小麦 Triticum aestivum	1%酶液	T[2]	120	1%酶液	T[2]	120
	1% Enzyme +1M HCl	60	0.4~0.5	1% Enzyme + 0.1M HCl	60	5
小黑麦 Triticale	1%酶液	T[2]	120	1%酶液	T[2]	120
	1% Enzyme +1M HCl	60	0.4	1% Enzyme + 0.1M HCl	60	5
大麦 Hordeum vulgare	1M HCl	60	0.5	0.1M HCl	60	5
玉米 Zea mays	1% 酶液	T[2]	150	0.1M HCl	60	5
洋葱 Allium cepa				0.1M HCl	60	5~8
蚕豆 Vicia faba				0.1M HCl	60	8

1）1% 酶液为：果胶酶+纤维素酶 1∶1。
2）T 为室温（Room Temperature）。

（3）风干脱水后的标本，一般都需要经过一定时间的空气干燥（成熟）。在我们使用的材料中，洋葱是最典型的。标本不经过 5 d 以上的贮存，则不显带或呈现浅而模糊的带，这与 Fiskesjö（1974）的结果是一样的。关于贮存条件 Fiskesjö 曾作过详细的描述。我们的实验表明，空气干燥时间长短是主要的。不同的材料，不同的显带方法所需空气干燥的时间是不一样的（表 1.5）。

表 1.5　不同材料空气干燥时间
Table 1.5　The periods of air drying for different materials

材料 Material	空气干燥时间（d）The air drying in days	
	BSG 法　BSG method	HSG 法　HSG method
黑麦 Secale cereale	1~15	1~7
小麦 Triticum aestivum	1~30	1~14
小黑麦 Triticale		1~7
大麦 Hordeum vulgare	4~7	2~7
玉米 Zea mays	1~10	1~10
洋葱 Allium cepa		5~60
蚕豆 Vicia faba		1~3

（4）变性处理　变性处理是显带的关键。分带常因变性处理的温度、时间掌握不严而失败。变性时间过长，细胞膨胀，模糊不清，染色体不着色或染成蓝绿色，有时甚至没有细胞结构。如果变性不足，整个染色体全着色，带显示不出来。不同的材料，不同方法，具体变性时间是不一样的（表 1.6）0.02 M HCl 处理时间比 $Ba(OH)_2$ 处理时间范围

大，较易掌握。BSG 法中 Ba(OH)$_2$ 处理的时间、温度是严格的。但大麦 Ba(OH)$_2$ 处理时间有一个很大的范围，从 60-80 min 都可以获得好的结果。谷口研至等（1977）报道了玉米 Ba(OH)$_2$ 处理为 50℃ 5 min，按这个条件在我们实验室则显不出带。在我们的实验室条件下，经过反复试验 Ba(OH)$_2$ 处理则需要 50℃，55 min。变性剂处理后，必须用蒸馏水彻底冲洗，特别是 BSG 法，即使是有微量的 Ba(OH)$_2$ 都会对以后的分带造成不良影响。

表 1.6　不同材料变性剂处理温度和时间
Table 1.6　The temperature and periods denaturant treatment for different materials

材料 Material	变性剂 Denaturant			
	5% Ba(OH)$_2$		0.2 M HCl	
	温度（℃）Temperature	时间(min)Period(min)	温度（℃）Temperature	时间(min)Period(min)
黑麦 Secale cereale	18～22	5	T*	10
小麦 Triticum aestivum	18～22	6	T*	25～30
小黑麦 Triticale	18～22	5	T*	10
大麦 Hordeum vulgare	25～30	60～80	T*	60
玉米 Zea mays	50	55	24～30	180
洋葱 Allium cepa			T*	30
蚕豆 Vicia faba			22～23	70

*室温（Room Temperature）。

（5）Giemsa 染色 根据我们试验的七种不同厂家批号的 Giemsa 染色结果表明，Giemsa 的质量有很大差别。在同一条件下有的根本不着色，有的着色很浅或颜色不正。就是着色正常的所需时间（表 1.7）与当前文献中报道的也不同，需要重新摸索。

表 1.7　不同材料的 Giemsa 染色时间
Table 1.7　The periods of Giemsa staining for different materials

材料 Material	BSG 法 BSG method		HSG 法 HSG method	
	Giemsa[1]浓度 Concentration	染色时间（min）Staining period(min)	Giemsa*浓度 Concentration	染色时间（min）Staining period(min)
黑麦 Secale cereale	10∶1	120	10∶1	15
小麦 Triticum aestivum	10∶1	120	10∶1	15
小黑麦 Triticale	10∶1	120	10∶1	15
大麦 Hordeum vulgare	10∶1	60	10∶1	15～20
玉米 Zea mays	50∶1	12	50∶1	18
洋葱 Allium cepa			10∶1	15
蚕豆 Vicia faba			10∶1	13～15

*我们使用的是上海试剂总厂第三分厂 710222 批号的 Giemsa 粉。

3. 植物染色体的 Giemsa C-带

使用以上两种分带方法，我们在黑麦、小麦、小黑麦、大麦、玉米、洋葱和蚕豆七种植物上成功地获得了 Giemsa C-带图像。

（1）黑麦染色体的 Giemsa C-带。黑麦染色体 C-带的特点是明显的：①每条染色体的短臂末端都有染色很深大而明显的端粒带；②着丝点带较浅；③第七条染色体上有副缢痕带；④大多数染色体长臂上有中间带。

（2）小麦染色体 Giemsa C-带。小麦 21 对染色体均显带，大部分属于中间带，着丝点带有的较浅，只有少数染色体有末端带，颗粒较小，与黑麦染色体的末端带从大小上很容易区别。

（3）小黑麦用 HSG 法成功地获得了八倍体小黑麦的 Giemsa C-带图像。图中清楚地显示出了黑麦和小麦两种染色体，依据它们各自所特有的带型特点，很容易将两种染色体区分开。

（4）大麦染色体 Giemsa C-带。大麦所有染色体均显示有明显的着丝点带，中间带较少，大部分集中在着丝点附近。

（5）玉米。据我们观察玉米自交系（墩子黄）Giemsa C-带，只有一半染色体显带，其中有四对染色体显示末端带，一对染色体显示副缢痕带，另外五对染色体未发现有带。

（6）洋葱染色体 Giemsa C-带。洋葱的每条染色体上都有明显的末端带，着丝点带较浅，副缢痕和随体被染成很深的带，间期颗粒明显。

（7）蚕豆染色体 Giemsa C-带。蚕豆六对染色体上都有明显的带。M 染色体上副缢痕和着丝点附近的三条带是明显的。S 染色体群中，除 Sa 短臂有一条带外，其余的带均分布在长臂上。

（作者：陈瑞阳、宋文芹、陈晓、李德成、徐悦凡、范海平、周显昌。

原载：植物学报，1979，21（1）11-18）

1.2.2　植物染色体 C-带法

1.2.2.1　设备、药品和材料

1. 设备和药品

（1）显微镜及显微照相冲洗放大设备。

（2）液态 CO_2 或半导体致冷器。

（3）万分之一天秤，小台秤。

（4）温箱。

（5）恒温水浴锅。

（6）其他玻璃器皿与用具　试剂瓶、容量瓶、量筒、烧杯、滴瓶、染色缸、载玻片、盖玻片、剪刀、镊子、刀片、铅笔、滤纸、牙签、普通玻璃板，切片盒、载玻片架。

（7）Giemsa 母液。

（8）Sörensen 磷酸缓冲液。

（9）2×SSC 盐溶液（0.3 mol/L NaCl+0.03 mol/L 柠檬酸钠）：称取 NaCl 17.53 g，柠檬酸钠 8.823 g，置于容量瓶中，加蒸馏水至 1000 mL。

（10）当量盐酸用滴定管量取浓盐酸配成所需当量浓度的盐酸：

欲配当量浓度	相对密度 1.19 HCl（mL）	相对密度 1.16 HCl（mL）	加水（mL）
0.1 M HCl	8.25	9.83	1000
0.2 M HCl	16.5	19.66	1000
1 M HCl	82.5	98.3	1000

（11）其他药品：氢氧化钡、秋水仙素、纤维素酶、果胶酶、甲醇、冰醋酸。

2. 材料

黑麦种子由中国农科院作物所提供。

1.2.2.2　操作规程

（1）材料培养

种子浸种后排列在铺有滤纸的培养皿内，25℃温箱培养，根长 2～3 mm 时，进行预处理。

（2）预处理

上午 8∶30～11∶30 进行预处理，用滴管向培养皿中加入 0.2%秋水仙碱，使根尖都能均匀的接触到药液。

（3）固定

用 3∶1 甲醇∶冰醋酸固定液固定 4～24 h，后换入 70% 乙醇中冰箱保存。

（4）解离

用酸解离材料，不仅使根尖组织软化，而且对以后染色体显带也有直接关系，因此解离时，酸的浓度、温度、处理时间都要严格控制。固定后的材料先经 1% 纤维素酶、果胶酶处理 2～3 h，然后在 1M HCl 60℃解离 25～30 s。

（5）压片

将解离的根尖放在载玻片中央，切取分生组织部分，滴上几滴 45% 冰醋酸，用镊子轻轻敲打，使材料成一片均匀薄层，在酒精灯上微热，然后将载玻片放在平整的桌子上，最好是钢化玻璃板上，以左手中指和母指固定盖玻片的位置，勿使其移动，再用铅笔一类东西在盖玻片上敲打至细胞破裂染色体散开，在显微镜下挑选染色体分散而完整的片子，用干冰或半导体致冷器冰冻片子后揭开盖玻片，于室温下空气干燥。

（6）空气干燥

空气干燥对未经染色的染色体标本起着一种缓慢的但显著的影响，一般需干燥 1 天至 1 周，干燥 1 个月只显着丝点带。我们还发现贮存六个月至一年的白片子，不经别的处理也能显带。

（7）显带处理

将干燥后的片子用氢氧化钡饱和水溶液（5%～8%）在 18～21℃条件下处理 5 min→40℃左右热蒸馏水冲洗→流水冲洗至片子透明→2×SSC 溶液 65℃保温 2 h→蒸馏水冲洗→空气干燥。

（8）Giemsa 染色

将片子放在摆有牙签的玻璃板上，有材料面向下，滴加 5% pH 6.8 Giemsa 染液，染色 30 min 到 2 h→流水冲洗→空气干燥。

(9) 树胶封片

压片标本必须经用 Damer 树胶封片后镜检。

1.2.2.3 讨论

(1) 植物染色体 Giemsa C-带技术的主要问题是分带的可重复性较低，其原因我们认为不在于显带处理，而在于压片标本，每张片子很难重复一致，压片时材料受到的破坏程度不一。因此，后来的显带条件就难于统一，所以显带结果因染色体标本而异，甚至同一张染色体标本不同位置显带结果也不同。我们认为这是当前 Giemsa C-带可重复性低的主要原因之一。

(2) 当前用于植物染色体分带的方法很多，但就其结果而言，大同小异，不同方法之间虽有带纹数量上的差别，但就带的性质都属于异染色质区的 C-带范围。难以理解的是，完全使用人类染色体胰酶 G-带法，在植物上仍然只显示 C 带。为使植物染色体研究向前推进一步，期待早日突破植物染色体 Giemsa G-带技术。

(作者：陈瑞阳。原载：植物细胞学研究方法，科学出版社，1985，306-314)

1.2.3 植物染色体 N-带法

1.2.3.1 实验药品和材料

1. 设备和试剂

二孔水浴锅一台。

1M NaH_2PO_4：称取 NaH_2PO_4 156.01 g 溶于容量瓶中，加水至 1000 mL。

2. 材料

用去壁、低渗法于一周内制备的小麦属、大麦属、山羊草属植物染色体标本若干张。

1.2.3.2 操作规程

(1) 标本制备：参阅 1.2.2 方法制备染色体标本。

(2) 染色：用 20∶1～40∶1 Giemsa 染色。

(3) 冲洗：蒸馏水冲洗。

(4) 贮存：空气干燥 3～5 d。

(5) 镜检：选取染色体分散良好的细胞，最好为早中期染色体，做好标记。

(6) 拍照：分带前可先进行核型拍照。

(7) 显带处理：1 mol/L NaH_2PO_4 94～96℃处理。小麦：10 min，大麦：12 min，山羊草：9 min。

(8) 冲洗：40℃左右热蒸馏水冲洗，然后自来水冲洗 30 min。

(9) 干燥：室温下空气干燥。

（10）染色：用 20∶1 pH 6.8 Giemsa 染液染色 30～60 min。染色时间因处理时间不同而异，NaH_2PO_4 处理过头，要加长染色时间，反之则短。

（11）冲洗：自来水冲洗，并使其迅速干燥。

（12）镜检：空气干燥后镜检拍照。

1.2.3.3 讨论

（1）关于 N-带的专一性。有的学者证明 N-带技术与 NOR 并不存在有相关联系，这要具体情况具体分析，当前用 N-带技术研究的植物种类尚少，在玉米上，则表现出典型的 NOR 专一性。在很多禾本科植物上，NOR 区与其他带纹则不能同时显现。

（2）N-带与 C-带的比较。如果将小麦、大麦的 N 带与 C-带进行比较，会发现有许多带纹是相同的。如普通小麦都是 B 组染色体和 A 组的 4A 和 7A 显带，大麦 7 对染色体都是着丝点区域显带，只是在个别染色体带纹的数量上不完全一致，这是由于显带条件所致还是两种方法的差别，尚待研究。

（3）N-带的可重复性。据我们统计 N-带的可重复性比 C-带高得多，在我们应用去壁、低渗法制备的染色体标本，尤其是大麦材料，显带是稳定的，基本都是可重复的，这是 BSG C-带法所不具备的。因此，我们认为，N-带技术至少在小麦属、大麦属和山羊草属植物染色体分带研究中，是一个实用的方法。

（作者：陈瑞阳。原载：植物细胞学研究方法，科学出版社，1985，268-318）

1.2.4 植物染色体 G-显带方法

十几年来，围绕植物染色体为什么不显 G-带提出了种种推论和假设。Greilhuber（1977）认为如果不是染色体标本制备方法不恰当，便是植物界中根本就不存在 G-带。Nagl（1976）则认为植物染色体之所以不显 G-带，仅仅是因为植物染色体浓缩度太大，中等重复序列 DNA 比率过高。Mckay（1973）认为各种带纹的显示是由于染色体中染色质浓缩度的差别，带纹是染色体上原来就有的结构。各种分带处理只是加强了这种原有带纹的显示而不是人为地产生带纹（Comings，1973），减数分裂粗线期的染色粒是染色体上天然存在的结构，染色粒的分布和有丝分裂染色体的 G-带带纹有很好的一致性，这已在中国仓鼠和人的染色体上得到证实（Okada，1974；Luciani，1975）。既然植物染色体也和动物一样有染色粒的结构，那么，植物染色体上也应该存在 G-带。

1982 年，Drewry（1982）以红松（*Pinus resinosa*）染色体 G-带为标题，报道了他的试验结果，然而他只指出 C-带的位置，并未明确标明 G-带的位置和数目，图片模糊，带纹数目少且不清晰。虽然以往少数作者也有关于植物染色体 G-带的报道，但从结果看，仍属 C-带，而非分布于染色体全长上的带纹。因此，当前植物 G-带仍然是一个悬而未决的难题。

作者对植物染色体 G-带进行了研究，旨想证实植物染色体 G -带的真实性与其普遍程度，并旨探索一个既适合动物亦适合植物染色体 G-带的方法。

1.2.4.1 材料和方法

1. 实验材料

供试材料计有：华山松（*Pinus armardii*），川百合（*Lilium davidii*）和七叶一枝花（*Paris polyphylla*）。

2. 试验方法——染色体标本制备流程

（1）取材：切取上述材料的生长旺盛根尖，约 5 mm。

（2）前处理：将根尖置 0.1%～0.01%秋水仙素溶液中于室温下处理 2～4 h。

（3）前低渗：用 0.075 mol/L KCl 溶液洗去前处理液，并在该溶液中停留 30 min。

（4）酶解：材料在 2.5%混合酶液（2.5%纤维素酶+2.5%果胶酶）中，于 20～25℃条件下酶解 0.5～1.5 h。

（5）后低渗：用双蒸水低渗 15～30 min。

（6）固定：以新配制的甲醇∶冰醋酸（3∶1）固定 2～24 h。

（7）制片：将材料涂或滴在预先于蒸馏水中冷冻的载玻片上，用蒸气干燥法制备染色体标本。

3. G-带的流程

贮存 2～7 d 的片子，在 0.05%～0.2%的胰酶（Difocs 250∶1）以 D'hanks 配制，0.3%缓血酸胺调 pH 到 7～8 之间，百合一般处理 10～60 s，华山松处理 1～3 min，七叶一枝花处理 1～2 min。胰酶处理后的片子要立即放入 0.85%的 NaCl 溶液中，充分洗去胰酶，从 NaCl 中取出玻片，以 3% Giemsa（以 Sörenson 磷酸缓冲溶液配制，pH = 7.4）染色 15 min 左右，蒸馏水冲洗，风干，镜检。

4. C-带流程

贮存一周左右的片子，在 0.1M HCl 60℃解离 8 min，流水冲洗 10～15 min，5% Ba(OH)$_2$ 20～24℃ 60 min，45～50℃蒸馏水冲洗，2×SSC 60℃ 60 min，蒸馏水冲洗，5% Giemsa 染色，自来水冲洗，空气干燥，镜检。

5. 带纹的自动光谱分析

使用 Reicherf 产 Univar 扫描显微分光光度计，用 Commodore 8032-sk 微型计算机控制（带有染色体扫描程序），光度计波长范围 40～700 nm，扫描步距为 0.05 μm，经处理可得到染色体上每条带纹的宽度、面积和积分光密度，可绘出染色体带纹模拟三维结构图和柱形图。

1.2.4.2 结果与讨论

1. 几种植物染色体 G-带观察

从川百合（*Lilium davidii*）的核型、C-带和 G-带可以看出。C-带带纹很少，除第 3 号染色体长臂远端有一条中间带外，其他染色体均不显中间带，只在着丝点处显带，这与 Preben（1976）在麝香百合（*L. longiflorum*）中观察的 C-带结果接近。经胰酶处理后，百合全部染色体都显示有清楚的带纹，带的数目很多，分布在整个染色体的长、短臂上，带区与非带区界线清楚，前期染色体带呈颗粒状，早中期、中期染色体带呈明显的带状，与哺乳动物染色体 G-带很相似。

华山松（*Pinus armardii*）染色体 G-带带纹数目亦很多。带纹清晰，5 对中着丝点大染色体长、短臂均显带，这与 Drewry（1982）报道的红松（*P. resinosa*）染色体 G-带完全不同。

七叶一枝花（*Paris polyphylla*）$2n=10$，C-带只在近着丝点区和末端显带，每条染色体只有 2~3 条带，G-带带纹则比较丰富。

2. G-带与染色体时期的关系

植物染色体之所以难显 G-带，是由于植物染色体比动物染色体结构致密，如果把人类有丝分裂染色体浓缩度视为 1，则植物染色体浓缩度为 3.5~16.5。在有丝分裂早中期人类染色体染色粒间的距离约为 0.6 μm，而植物染色体中染色粒的距离只有 0.1μm，甚至更小。因此，由于植物染色体浓缩度过高，使相邻带纹互相融合连成一片，在光学显微镜下无法分辨。

作者观察到，随细胞分裂时期向前推移，由中期→早中期→晚前期→前期，带纹数目明显增多。为精确测得上述各时期带纹数目，作者使用 Uinvar 型显微扫描分光光度计，对百合各时期染色体 G-带的数目、峰值和面积进行了测量，峰值与染色体上 G-带带纹数目完全一致，表明此方法的可靠性。前期染色体带纹数目为中期染色体的三倍左右，这与 Rohme（1982）在赤鹿中、Yunis（1978）在人类中观察到的结果是很接近的。

目前，染色体的 G-带分带水平已进入早中期、前期，正力求进入间期 Yunis（1978）。为了进一步提高带纹的分辨率，增加带纹的清晰度、对比度，我们必须尽量使染色体拉长，保持在前期水平。

G_2 期放线菌素 D（AMD）处理（Rybak，1982），S 期 BUdR 处理（Hsu，1961）都可阻止染色体收缩变短，但 AMD 在高浓度时，会诱发染色体粘连，断裂，严重抑制有丝分裂活性，并影响染色体分散。BUdR 虽可引起染色体伸长，但这种伸长是不均匀的。我们在实验中，采用缩短前处理时间和降低秋水仙碱浓度，同样得到了满意的结果。

哺乳动物细胞停留在前期的时间很短，一般只有 2~3 min。多年来，动物学家们试图寻找一种前期抑制剂，以获得更多的前期分裂相，但迄今为止还没有一种理想的前期抑制剂。在这方面，植物细胞倒是具有得天独厚的优点，它的前期比动物长，如洋葱根尖有丝分裂停留在前期的时间达 71 min，豌豆根 78 min，紫露草雄蕊毛 103 min，为前期染色体的获得提供了便利条件。

3. 胰酶法在植物染色体 G-带中的应用

1971 年 Dutrillux 等人用蛋白质分解酶（胰酶）处理染色体标本，能够得到与 ASG 法相同的带型，这个方法后来被 Seabright（1971）Wang 和 Federoff（1972）等所发展和充实，成为现在被广泛使用的胰酶 G-带法。作者使用胰酶法（Diffocs 250∶1 Trypsin D'hanks 配制 Tris 调 pH =7.8）。在上述几种植物上获得 G-带，说明胰酶法是一种既适合动物亦适用于植物 G-带的分带技术。以往此法失败的原因，除染色体时期不适合外，可能不在胰酶法本身，而在于染色体处理技术。染色体标本制备的酸解法，对 Q-带、G-带的产生都是有害的，这已在哺乳动物中得到证明。因此，在当前的植物染色体 G-

带研究中,应尽量避免使用 C-带标本制备方法。当然不同的植物产生带纹的条件会有很大差异,不同的实验室也会得到不同的结果。然而,我们可以坚信,如果找到正确适当的条件和方法,在其他植物染色体上也一定能出现 G-带,因为,所有真核生物染色体结构和行为的主要特征都相同。

4. 植物染色体 G-带与 Q-带的关系

Holm(1976)及 Kongsuwan 和 Smyth(1977)分别报道了 *Lilium longiforcum* 和 *L.pardalinum* 的 Q-带和 C-带带型,指出 Q-带与 C-带不同,有其自身的特性,通过 Q-带带型与染色体结构的比较,揭示出 Q-带与染色粒之间的相关性,它们都是有秩序地分布在染色单体上,并且在姊妹染色单体的同源区常是相同的。然而,Q-带与 G-带在植物中是不完全相同,尚待研究。

1.2.4.3 问题和展望

与动物染色体研究相比,植物有其自身的很多特点,如:细胞周期中,停留在前期的时间长,植物类群中有很多大染色体种类,如:裸子植物染色体均较大;被子植物中芍药属(*Paeonia*)、百合属(*Lilium*)、延龄草属(*Trillium*)、紫露草属(*Tradescantia*)、桑寄生科(*Krameriacaee*)染色体都很大,长达几十微米,有的超过裸子植物,这些大染色体结构可能相对疏松,适合 G-带显示。

我们测得川百合第一对染色体中,一条中期染色体上有 14 条带,前期则有 41 条,相当于前者的三倍,在人类染色体中,当早中期的带纹数目相当于中期四倍的时候就已达到了高分辨带的水平。因此,随着染色体新技术的不断改进,植物染色体也一定会跨入高分辨带的时期。

植物染色体不仅在裸子植物而且在被子植物中也可以显示 G-带,但它的普遍规律和不同植物的具体条件,目前还没有完全掌握,有待继续深入研究。然而,可以相信,不久的将来,植物染色体 G-带一定会取得长足进展。

(作者:陈瑞阳、安祝平、宋文芹、李秀兰、苏健英。原载:武汉植物学研究,1986,4(2)111-118)

1.2.5 植物染色体带型分析法

所谓带型分析,是借助于常染色质与异染色质区,经特殊处理后在染色体不同区段显现出的带纹,来鉴别染色体或染色体组的一种更精确的核型分析方法。

本文仅就植物染色体 C-带所显示的四种带纹分析方法,简述如下。

1.2.5.1 设备、药品和材料

1. 设备和药品

(1)设备:照相显微镜、放大机、测微尺、2~4 号放大纸。其他用具:眼科剪刀、量角规、游标卡尺、台纸、胶水。

(2)药品:显影液、定影液。

D-19 显影液：水 50℃ 750 mL，米妥尔 2 g，无水亚硫酸钠 96 g，对苯二酚 9 g，无水碳酸钠 48 g，溴化钾 5 g，加水至 1000 mL。

D-72 显影液：水 50℃ 750 mL，米妥尔 3 g，无水亚硫酸钠 45 g，对苯二酚 12 g，无水碳酸钠 67.5～80 g，溴化钾 2 g，加水至 1000 mL。

F-5 定影液：水 52℃ 600 mL，海波 240 g，无水亚硫酸钠 15 g，醋酸 28% 45 mL，硼酸 7.5 g，硫酸铝钾 15 g，加水至 1000 mL。

2. 材料

选具备染色体分散良好且带纹清晰的标准细胞。

1.2.5.2 操作规程

（1）拍照

选取 3～5 个染色体分散良好、带纹清晰的细胞，进行显微摄影，一般分带的材料着色较浅，反差小，拍摄时底片需要厚一些。

（2）冲卷与放大

为增加反差，可用 D-19 冲卷。放大前应先用镜台测微尺底片校正放大机的放大倍数，例如：镜台测微尺每一小格为 1/100 mm，欲放大 2000 倍，1/100×2000 =20 mL，测微尺每一小格放大 20 mm 即为 2000 倍，余此类推，可以得出照片的准确放大倍数。根据底片的厚薄选用不同型号的放大纸，放大纸号越大，反差越强，一般较薄底片应选用反差强的放大纸，一张好的显微照片不仅反差好，同时层次也要好。分带的细胞洗相时，常常需要增大反差，以 3 号、4 号放大纸为宜。

（3）剪贴

用眼科剪刀，沿染色体边缘将每一条染色体剪下来，存放在小培养皿内。

（4）配对

根据带纹的位置、宽窄、大小、数量进行同源染色体配对。与一般形态核型分析不同，带型分析不必进行测量计算，单依据带纹就可达到同源染色体的精确配对，如黑麦染色体 C-带。普通小麦染色体根据 N-带带纹可精确鉴别 B 组染色体），但最好核型与带型分析使用同一细胞，更准确无误。

（5）绘图

根据多个显带细胞带纹的位置、宽窄、数量与深浅按比例绘制带型模式图。

（6）描述

当前植物染色体分带研究中，不管使用 BSG 法、N-带法还是胰酶法，显示的均为 C-带。所谓植物染色体 C-带，主要包括以下四种带。

①着丝粒带（centromeric band）指着丝粒及其附近的带，大部分植物染色体都显示有这种带。

②中间带（intercary band）分布在染色体两臂上的带叫中间带。这部分带表现比较复杂，不是所有染色体都具有中间带，往往一部分染色体才有中间带。它们有的分布在短臂上，有的分布在长臂上有的在基部，有的在远端，并且在同源染色体之间可能表现

有多态带。显带条件不佳或处于不同分裂时期的染色体，中间带常常显现不一，分析时必须注意。

③末端带（telornere band）位于染色体两臂末端的带，黑麦尤为明显。

④核仁缢痕带（nucleolor constriction band）位于次缢痕区的带纹，是核仁染色体专一带。玉米染色体经 N-带处理只显示这专一带。

当前关于植物染色体 Giemsa C-带型尚没有统一的分类标准，我们根据 C-带所显的四种带，并以它们的英文字头为代表，将植物染色体 C-带分为两大类。

完全带类型：同时具有以上四种带的类型叫完全带，以 CITN 表示；如黑麦染色体 C-带就属于这种类型。

不完全带类型：只显示三种以下的带，又可细分为 CIN 型：不具有末端带类型，如大麦染色体属于此种类型。CTN 型：不具有中间带类型，如洋葱。TN 型：只有末端带和缢痕带，如 BSG 法处理的玉米染色体。N 型：只有缢痕带，如 N-带法的玉米染色体。

为表示中间带和末端带在染色体上的分布可用"+"表示，如果只分布在短臂上，则在字母的右上角划"+"表示（I⁺T⁺），如果只分布在短臂上，则在字母的右下角划"+"表示（I₊T₊）。不标明"+"表示长短臂上都有带。同类染色体数目，以符号前的数字表示，不显带的染色体以数字表示。按上述规定，一种植物染色体带型可以用公式书写，如黑麦染色体 C-带可以写成下式：

$$2n=14=CITN=2CT+4CI^+T_++6CI_+T+2CI_+TN$$

带型分析实际是更精确的一种核型分析，此处为方便起见单立一节叙述，在应用时应与核型分析合二为一。

（作者：陈瑞阳。植物细胞学研究方法，科学出版社，1985，298-305）

第三节　染色体的荧光原位杂交与基因定位

提要：荧光原位杂交（Fluorescence In Situ Hybridization，FISH）技术是一种遗传学实验术，其基本原理是：荧光素直接与寡聚核苷酸结合或者采用间接法以生物素、地高辛等标记，获得寡聚核苷酸探针，探针与变性后的染色体、细胞或者组织中的核酸按照碱基互补配对原则进行杂交，经变性—退火—复性—洗涤形成靶 DNA 与核酸探针的杂交复合体，经直接检测或通过免疫荧光系统检测，最后在荧光显微镜下观察或显影，对待测 DNA 进行定位、定性或相对定量分析。简单地说，染色体原位杂交技术就是在核酸分子碱基互补配对原则的基础上，利用放射性或非放射性物质标记外源核苷酸制备探针，探针与变性处理后的染色体单链 DNA 杂交，利用添加不同滤光片的荧光显微镜分别观察染色体和杂交信号并进行高质量拍照，再通过图像处理软件进行图片叠加，将探针核酸序列在染色体上的位置显示出来。

随着应用的推广和技术方法的更新，FISH 技术也在逐步完善。从杂交的探针数目来看，从单色 FISH 发展到多色 FISH，从一次杂交一个探针到同时杂交显示不同杂交信号颜色的多个探针；从杂交对象来看，从早期的中期染色体玻片标本到减数分裂时期的粗线期染色体玻片标本，甚至是伸展 DNA 纤丝玻片标本。根据使用探针类型的不同，FISH 还从早期的重复序列探针杂交，衍生出了基因组原位杂交（Genome In Situ Hybridization, GISH）、细菌人工染色荧光原位杂交（BAC-FISH）和酵母人工染色体荧光原位杂交（YAC-FISH）等技术（Shi et al., 1998; Zhang et al., 2004; Hao et al., 2006）。近年的研究结果表明，FISH 研究中已经很少使用同位素探针，同时，随着冷 CCD 拍照系统的改进及电脑软件的发展与运用，FISH 技术的灵敏性和分辨率都得到了显著提高。在经典的 45S rDNA、5S rDNA 原位杂交方法中，质粒探针要进行探针标记、探针变性、杂交液成分复杂，杂交时间长，杂交后洗脱费事，杂交信号需要抗体检测等过程，操作起来比较烦琐、费时费力。现有的杂交方法远远满足不了染色体定位的需求。随着荧光原位杂交的发展，多色荧光标记已成为可能，利用多色荧光原位杂交技术可以同时将多种不同标记的探针定位到染色体上，就可以分析一些形态上难以区分的染色体，但是由于步骤烦琐，费时、费力而且每次的实验结果不稳定等诸多因素使该技术发展远滞后于其他技术的发展。因此，改进现有的探针标记方法和杂交的方法，提高杂交的效率是急需解决的关键问题。

本实验室在原荧光原位杂交的基础上发明了一种 45S rDNA、5S rDNA 在植物染色体上荧光原位杂交的方法，它主要包括染色体标本制备、45S rDNA、5S rDNA 探针制备、染色体荧光原位杂交、杂交信号检测等步骤，重点解决了现有荧光原位杂交探针标记复杂烦琐的方法。发明的探针为寡核苷酸探针在合成时可加入荧光集团修饰，不用在进行标记，与现有技术相比方法简单、成本低廉。利用这种技术能够在三小时内完成 45S rDNA、5Sr DNA 在染色体上的定位。

1.3.1　5S rDNA 在植物染色体上的荧光原位杂交方法

5S rDNA 探针一般采用质粒或 PCR 扩增获得，再进行直接或间接标记用于 FISH，目前还没有利用荧光基团标记的的寡核苷酸探针在植物染色体上进行 5S rDNA FISH 定位的报道。

1.3.1.1　染色体标本制备

（1）按常规植物材料培养，待种子或植株根尖长至 0.5～1cm；
（2）切下生长旺盛的根尖，用饱和对二氯苯溶液室温下预处理 3～5 h；

（3）吸去预处理液，根尖用蒸馏水或 0.075 mol/L 氯化钾低渗 30min，然后用 3∶1（甲醇∶冰醋酸）固定 20~60 min；

（4）用蒸馏水洗三次，每次 5 min，充分洗去固定液；

（5）用果胶酶和纤维素酶按重量分数比 1∶1 混合的酶 2.5%（W/W），在 37℃解离根尖 60 min；

（6）蒸馏水洗去酶液，用 0.075 mol/L 氯化钾后低渗 15~30min，用 3∶1 甲醇∶冰醋酸固定液固定 20 min；

（7）取根尖于预先在 4℃蒸馏水中冷冻的洁净载玻片上，滴一滴固定液，用镊子将根尖充分捣碎，再加 1 滴固定液，将细胞吹散，空气干燥；

（8）用 1∶30 稀释的 Giemsa 染色液染色 10 min，自来水冲洗凉干；

（9）镜检，选择分散良好的染色体标本放在 -20℃冰箱中保存待用。

1.3.1.2　5S rDNA 探针制备

从拟南芥 5S rDNA 序列中选取一段 20 个左右碱基的高度保守序列，合成时在探针的 5' 或 3' 端连接一个荧光基团：

5'- CTGATGGGATCCGGTGCTTT -3'，5' 端带红色荧光标记 TAMRA

1.3.1.3　染色体荧光原位杂交

1. 染色体标本预处理

（1）在荧光显微镜上记下已标记分裂相的坐标，并用玻璃刀在载玻片背面标记分裂相的位置。

（2）将载玻片在 45%醋酸中浸泡 5 min 褪色，空气干燥。

（3）染色体标本变性：

① 在标记处加 30 μL 70%去离子甲酰胺/2×SSC，盖上盖玻片，于 PCR 仪或烘箱中 70℃处理 2 min；

② 去掉盖片，-20℃的 70%、85%、100%冷乙醇脱水，每级 3 min，空气干燥。

2. 杂交

（1）用 2×SSC 稀释探针至 5 ng/μL；

（2）每张标本加 10 μL 探针，并盖 18 mm×18 mm 的盖玻片；

（3）于湿盒中 37℃杂交 1~2 h。

1.3.1.4　杂交后洗脱

（1）在 4×SSC，0.2% Tween 20 中室温下避光洗涤 10 min；

（2）蒸馏水冲洗片刻，避光空气干燥。

1.3.1.5　杂交信号检测

（1）滴加 3 μL 含有 DAPI 的防荧光淬灭剂，盖上盖玻片；

（2）Nikon 80i 荧光显微镜观察，按照所记录的标本坐标在紫外光激发下可观察到

蓝色的染色体,在绿色激发光激发下可观察到红色的 5S rDNA 杂交信号;

(3) SPOT RT KE 冷 CCD 进行图像采集,SPOT 4.1 软件进行图像合成。

(作者:陈成彬、王春国、宋文芹。原载:授权专利:ZL 2013 1 0346485.5)

1.3.2 45S rDNA 在植物染色体上的荧光原位杂交方法

在经典的 45S rDNA 原位杂交方法中,质粒探针要进行探针标记、探针变性、杂交液成分复杂,杂交时间长,杂交后洗脱费事,杂交信号需要抗体检测等过程,操作起来比较烦琐、费时费力。现有的杂交方法远远满足不了染色体定位的需求。随着荧光原位杂交的发展,多色荧光标记已成为可能,利用多色荧光原位杂交技术可以同时将多种不同标记的探针定位到染色体上,就可以分析一些形态上难以区分的染色体,但是由于步骤烦琐,费时、费力而且每次的实验结果不稳定等诸多因素使该技术发展远滞后于其他技术的发展。因此,改进现有的探针标记方法和杂交的方法,提高杂交的效率是急需解决的关键问题。

1.3.2.1 染色体标本制备

(1) 按常规植物材料培养,待种子或植株根尖长至 0.5~1 cm;
(2) 切下生长旺盛的根尖,用饱和对二氯苯溶液室温下预处理 3~5 h;
(3) 吸去预处理液,根尖用蒸馏水或 0.075 mol/L 氯化钾低渗 30 min,然后用 3∶1(甲醇∶冰醋酸)固定液固定 20~60 min;
(4) 用蒸馏水洗三次,每次 5 min,充分洗去固定液;
(5) 用果胶酶和纤维素酶按重量分数比 1∶1 混合的酶 2.5%(W/W),在 37℃解离根尖 60 min;
(6) 蒸馏水洗去酶液,用 0.075 mol/L 氯化钾后低渗 15~30 min,用 3∶1 甲醇∶冰醋酸固定液固定 20 min;
(7) 取根尖于预先在 4℃蒸馏水中冷冻的洁净载玻片上,滴一滴固定液,用镊子将根尖充分捣碎,再加 1 滴固定液,将细胞吹散,空气干燥;
(8) 用 1∶30 稀释的 Giemsa 染色液染色 10 min,自来水冲洗凉干;
(9) 镜检,选择分散良好的染色体标本放在-20℃冰箱中保存待用。

1.3.2.2 45S rDNA 探针制备

从拟南芥 45S rDNA 序列中选取一段 20 个左右碱基的高度保守序列,合成时在探针的 5'或 3'端连接一个荧光基团:

5'-TCGTAACAAGGTTTCCGTAG -3'(SEQ NO:1),5'端带绿色荧光标记 FAM(羧基荧光素)

1.3.2.3 染色体荧光原位杂交

1. 染色体标本预处理

（1）在荧光显微镜上记下已标记分裂相的坐标，并用玻璃刀在载玻片背面标记分裂相的位置；

（2）将载玻片在 45%醋酸中浸泡 5 min 褪色，空气干燥。

2. 染色体标本变性

（1）在标记处加 30 μL 70%去离子甲酰胺/2×SSC，盖上盖玻片，于 PCR 仪或烘箱中 70℃处理 2 min；

（2）去掉盖片，−20℃的 70%、85%、100%冷乙醇系列脱水，每级 3 min，空气干燥。

3. 杂交

（1）用 2×SSC 稀释探针至 5 ng/μL；

（2）每张标本加 10 μL 探针，并盖 18 mm×18 mm 的盖玻片；

（3）于湿盒中 37℃杂交 1～2 h。

4. 杂交后洗脱

（1）在 4×SSC，0.2% Tween 20 中室温下避光洗涤 10 min；

（2）蒸馏水冲洗片刻，避光空气干燥。

5. 杂交信号检测

（1）滴加 3 μL 含有 DAPI（4′,6-二脒基-2-苯基吲哚）的防荧光淬灭剂，盖上盖玻片；

（2）Nikon 80i 荧光显微镜观察，按照所记录的标本坐标在紫外光激发下可观察到蓝色的分裂相，在蓝色激发光激发下可观察到绿色的 45S rDNA 杂交信号；

（3）SPOT RT KE 冷 CCD 进行图像采集，SPOT 4.1 软件进行图像合成。

本发明公开的 45S rDNA 在植物染色体上的荧光原位杂交方法与现有技术相比所具有的积极效果在于：

（1）本发明的探针为寡核苷酸探针在合成时可加入荧光基团修饰，不用再进行标记，与现有技术相比方法简单、成本低廉。

（2）寡核苷酸探针很容易渗透进入染色体，所以染色体标本预处理步骤可简化。

（3）寡核苷酸探针不需要变性，杂交液的成分仅为 2×SSC，比经典方法成分简单很多。

（4）杂交时间仅需 1 h，不需要过夜。

（5）杂交后洗脱步骤简便，不需要甲酰胺。

（6）杂交信号检测可直接观察，不需要抗体孵育等信号检测步骤。

（作者：陈成彬、王春国、宋文芹。原载：授权专利：ZL2013 1 0346495.9）

1.3.3 植物基因组原位杂交（GISH）

基因组原位杂交（Genomic In Situ Hybridization，GISH）技术是 20 世纪 80 年代末 90 年代初发展起来的一种原位杂交技术，主要是利用一个物种的基因组 DNA 作探针对其他物种染色体进行荧光原位杂交，其杂交信号检出设计专门用于检出重复 DNA 序列，因此可以直观地显示不同物种间共同的重复序列在染色体上的分布情况。最初应用于动物方面的研究，但很快在植物中得到推广应用。用于检测植物杂交后代是否存在或渗入的外源 DNA 情况，后来逐渐成为一种基因组进化分析的新手段，用于研究不同种间的物种进化和亲缘关系研究，是研究不同基因组间重复序列的变异性和保守性的一种简便方法。它用来自一个物种的总基因组 DNA 作为探针，以适当浓度的另一物种总基因组 DNA 进行封阻，在靶染色体上进行原位杂交。具体操作方法如下。

1.3.3.1 探针标记

探针标记采用随机引物法标记。

（1）纯化后的 DNA 1～3 μg 在沸水中变性 10 min。

（2）将变性 DNA 立即放至冰水中至少 5 min。

（3）加入 2 μL 六碱基随机引物，2 μL dNTP（dATP，1 mmol/L；dCTP，1 mmol/L；dGTP，1 mmol/L；dTTP，0.65 mmol/L；DIG-11-dUTP，0.35 mmol/L，pH 7.5）（Takara）。

（4）加入 Klenow 酶 2U（Takara），加水补足 20 μL。

（5）37℃过夜标记 20 h。

（6）加 2 μL 0.2mol/L EDTA pH 8.0，2.5 μL 4 mol/L LiCl，75 μL 冰冷无水乙醇混匀。

（7）-20℃沉淀 2 h 后，12000 rpm 离心 15 min。

（8）70%冷乙醇洗涤沉淀 2 次。

（9）加入 20 μL TE 溶解沉淀。

1.3.3.2 染色体标本预处理

（1）每张载玻片上滴加 100 μL RNA 酶（Rnase）（100 μg/mL），盖 parafilm 膜，37℃温育 1 h。

（2）2×SSC 洗载玻片 3×5 min。

（3）-20℃下 70%→85%→100%乙醇系列脱水各 5 min 后，室温下干燥。

（4）每张载玻片上滴加 200 μL 胃蛋白酶（Pepsin）（Sangon）（含 0.01%胃蛋白酶的 10 mmol/L HCl），盖 parafilm 膜，37℃温育 10 min。

（5）1×PBS 洗涤 2×5 min，10×PBS 中包含：8% NaCl、0.2% KCl、1.44% Na_2HPO_4 和 0.24% KH_2PO_4。

（6）用含 50 mmol/L $MgCl_2$ 的 1×PBS 洗涤玻片 5 min。

（7）用含 50 mmol/L $MgCl_2$ 和 1%甲醛的 1×PBS 固定标本 5 min。

（8）2×SSC 洗玻片 2 min。

（9）在含 70%去离子甲酰胺的 2×SSC 中 65℃处理 2 min。

（10）−20℃下 70% →85% →100%乙醇系列脱水各 5 min 后，室温下干燥。

1.3.3.3 杂交

（1）配制杂交液，其组分为：50%去离子甲酰胺，2×SSC，50 mmol/L 磷酸钠，10%硫酸聚糖（dextran sulfate, DS），0.1% SDS 和 2 ng/μL 探针 DNA。

（2）杂交液在沸水中变性 10 min，然后迅速冰浴至少 5 min。

（3）将杂交液滴加到染色体标本上（10 μL/片），盖上盖玻片，置于 37℃湿盒中杂交过夜。

1.3.3.4 杂交后洗脱

（1）在含 30%去离子甲酰胺的 2×SSC 中去掉盖玻片，37℃洗片 5 min。

（2）2×SSC，37℃洗片 2×5 min。

（3）0.2×SSC，0.1%SDS，37℃洗片 3×5 min。

1.3.3.5 杂交信号检测

（1）在冲洗缓冲液（Washing Buffer）（含 0.2% Tween-20 的 1×PBS）中洗片 3 min。

（2）在载玻片滴加封堵缓冲液（Blocking Buffer）（含 1×Blocking Reagent 的 1×PBS）100 μL，盖 parafilm 膜。

（3）以 1∶25 稀释抗体（Anti-DIG-Fluorescein, Fab fragments）于封堵缓冲液中（1 μg/mL）。每张载玻片上加 50 μL 抗体缓冲液，盖 parafilm 膜，37℃温育 1 h。

（4）冲洗缓冲液（Washing Buffer）洗片 3×5 min。

（5）1×PBS 短暂冲洗。

（6）PI（Propidium Iodle）（10 ng/mL）染色 5 min。

（7）用去离子水短暂冲洗后，黑暗中晾干。

（8）在载玻片上加 1 滴 VectaShield（Vecta），Nikon 80i 荧光显微镜下观察并照相。

（作者：陈成彬。原载：陈成彬博士论文 p40-41）

1.3.4 BAC 克隆在植物染色体上的定位

在早期的 FISH 研究当中因为检测信号分布较少，低拷贝或单拷贝的 DNA 序列直接定位是很难的。随着基因组测序的大规模开展，很多物种构建了用于基因组测序的细菌人工染色文库和酵母人工染色体文库。因而可以通过间接法利用该基因所在的 BAC 克隆的荧光原位杂交实现。一般来说，在理想的杂交条件和检出条件下，靶 DNA 序列越长，探针与靶 DNA 的接触几率越高，FISH 的灵敏度随之提高。由于 BAC 和 YAC 含有大的插入片段，扩大了探针与杂交位点相遇的机会，使功能基因的染色体定位效率显著提高。对于 BAC 来说，BAC 克隆在宿主菌中拷贝数只有 1 至 2 个，可稳定遗传，很少出现缺失、重组及嵌合现象，转化和提取 DNA 方法比较成熟，采用一般碱变性法即可获取质粒 DNA 制备探针，应用起来非常方便。同时，由于片段较大，BAC 克隆数量相对较少，有利于构建真核生物基因组的物理图谱。因此，BAC-FISH 和 YAC-FISH 技术在重要功能基因定位研究中具有重要意义。

1.3.4.1 植物材料与试剂配置

（1）质粒 DNA 小量纯化试剂盒购自 Qiagen 公司（德国）。

（2）DNA Marker、pd（N）6、Klenow DNA 聚合酶、鲑鱼精 DNA 等购自于宝生物工程有限公司（TaKaRa Biotechnology，大连）。T4 连接酶购自 NEB 公司。

（3）Tetramethyl-Rhodamine-5-dUTP、Fluorescein-12-dUTP、dNTPs、和 4',6-二脒基-2-苯基吲哚（DAPI）等探针标记相关试剂购自于罗氏公司（Roche，德国）。

（4）50%硫酸葡聚糖：称取 2.5 g 硫酸葡聚糖，加 4 mL 灭菌蒸馏水，85℃加热溶解，最后体系为 5 mL，分装-20℃保存备用。

（5）杂交缓冲液（HB）：含 200 μL 50%硫酸葡聚糖；500 μL 去离子甲酰胺；100 μL 20×SSC；100 μL 打断的鲑鱼精 DNA（5mg/mL），震荡混匀，共 900 μL，使用时比例为 9 μL 杂交缓冲液中加 1 μL 探针。

（6）杂交工作液：5 μLBAC 探针标记体积加入杂交缓冲液 HB 45 μL。

（7）10 mg/mL DAPI：10 mg DAPI 加入蒸馏水 1 mL，充分溶解，-20℃保存。

（8）DAPI 工作液：用抗褪色剂 Vectashield 将 DAPI 稀释为 4 μg/mL，作为染色体复染染色液。

1.3.4.2 BAC 探针标记体系配制（三步法）

（1）配制 10×Rho/ FITC-dUTP/dNTP mix，30 μL 总体积中含有：10mM dATP 3 μL；10 mM dCTP 3 μL；10 mM dGTP 3 μL；10 mM dTTP 1.95 μL；1 mM Rho-dUTP/FITC-dUTP 10.5 μL；ddH$_2$O 8.55 μL。

（2）配制 RPmix，90 μL 总体积中含有 pd（N）6 1 μg/μL 30 μL；10×Rho/FITC-dUTP/dNTP mix 30 μL；10× Klenow Buffer 30 μL。

（3）BAC 探针标记体系组成：BAC DNA 模板 300 ng 到 1 μg；RP mix 3 μL；Klenow Fragment 酶 1 μL；ddH$_2$O 补足 10 μL。

1.3.4.3 BAC 探针制备

质粒 DNA 经检测合格后，取 20 μL，于压力 0.1 MPa、温度 120℃条件下处理 DNA 5 min，对质粒 DNA 进行物理法打断，普通琼脂糖凝胶电泳检测碎片长度范围。随机引物法进行 BAC 探针标记，标记体系如下：质粒 DNA 300~1000 ng；dATP/ dGTP/ dCTP 各 0.1 mM，dTTP 为 0.065 mM，Tetramethyl-Rhodamine-5-dUTP(红色荧光)或 Fluorescein-12-dUTP（绿色荧光）为 0.035 mM；Random Primer pd（N）6 100 ng；Klenow Fragment 2.5 U，1×Klenow Buffer，总体积 10 μL。37℃避光反应 20 h。用杂交缓冲液 HB 对标记产物进行 10×稀释，即得杂交工作液。

1.3.4.4 染色体标本制备

（1）根尖培养：选取健康芝麻种子，播于铺有湿滤纸的培养皿中，21℃暗培养，直至根长为 1.2~1.8 cm。

（2）根尖处理：切取 4~5 mm 含分生区的根尖置于玻瓶中，浸入 0.002 M 8-羟基喹啉于 21℃暗处理 1.5 h，倒出液体，直接加入固定液中，4℃固定 1 h；取出根尖于蒸馏水中清洗 2~3 次，切取根尖分生区于蒸馏水中浸泡 10~20 min，然后按每 10 个根尖分生区加入 20 μL 含 2.5%纤维素酶和 2.5%果胶酶的混合酶液，于 37℃水浴中酶解 2 h 40 min；取出根尖分生区，蒸馏水洗去残留酶液，然后蒸馏水中浸泡 10~20 min；然后将根尖分生区再次移至固定液中 4℃固定 1 h。

（3）涂片用胶头滴管吸取根尖分生区，置于已于 4℃预冷的干净玻片上，用尖头镊子尖端将根尖分生区快速敲碎，制成局部细胞悬液；然后滴加固定液使悬液分散于整个玻片，自然晾干。

（4）染色：玻片倒置于玻璃板上，两端用毛细管支起，使玻片与玻璃板之间留有空隙，将吉姆萨染液注入此空隙，染色 15 min，然后自来水冲洗，自然晾干。

（6）镜检：在 Nikon 80i 荧光显微镜下，挑选处于有丝分裂中期的染色体制片，在背面标记目的染色体所在范围，即标记区，记录其在玻片上对应的显微镜坐标位置。及时使用或置于-20℃备用。

1.3.4.5 BAC 荧光原位杂交

（1）染色体标本处理

为使染色体能顺利与 BAC 探针杂交，并增加染色体与玻片的附着力，使在重复杂交过程中染色体不从玻片上脱落，试验设置了 A、B、C 三种处理。

试验设置三组不同条件进行比较，分别为 4%多聚甲醛处理、RNase A 处理+胃蛋白酶处理+1%多聚甲醛处理和不做任何处理，具体处理步骤如下：

首先将染色体玻片标本浸泡于 45%乙酸中 1~2 min，取出后自然晾干，褪去 Giemsa 染液颜色，然后分别实施 A、B、C 三组处理。

A：37℃下用含 100 μg/mL RNase A 的 2×SSC 处理玻片标记区 1 h；然后用 2×SSC 洗片 3 次，每次 5 min；使用 70%、85%及无水乙醇对制片逐级脱水，每级 3~5 min；自然晾干后，37℃下用 1%胃蛋白酶处理制片标记区 30min，1×PBS 洗 2 次，每次 5 min；然后用 1%多聚甲醛处理制片标记区 10 min，2×SSC 洗 2 次，每次 5 min，自然晾干。

B：然后用 1%多聚甲醛处理制片标记区 10min，2×SSC 洗 2 次，每次 5min，自然晾干。

C：不进行任何处理。

经 A、B、C 步骤分别处理的标本，在制片标记区加 70%去离子甲酰胺，加盖片，70℃变性 2 min；将变性后的制片依次置于-20℃预冷的 70%、85%及无水乙醇中逐级脱水，每级 3~5 min，自然晾干备用。

（2）探针变性

取 20 µL 工作液，95℃变性 10 min，迅速置冰水中 5 min 以上，备用。

（3）杂交

制片标记区加探针工作液 8 µL（双色荧光探针同时杂交同一玻片标本时，二者各加 4 µL），加盖片，胶水封边，37℃杂交过夜。

（4）信号检测

室温下 2×SSC 液浸泡制片，洗去盖片，含 0.1% SDS 的 2×SSC 浸洗 3 次，每次 5 min；然后用蒸馏水冲洗一遍，避光晾干；在制片标记区滴入 4 µL 含 4 µg/mL DAPI 的 Vectashield H1000，加盖片置于 Nikon 80i 荧光显微镜下，根据制片坐标确定目标染色体，观察杂交信号，冷光源 CCD 下进行图像采集，采用 Spot Rtke 4.1 软件进行图像合成，并利用 Adobe Photoshop 7.0 软件对图像进行调整。

（作者：赵瑞红。原载：赵瑞红博士论文，2018 年 pp50-57）

第四节　植物染色体电子显微镜标本制备技术

提要：随着染色体标本制备方法的改进，新的实验技术很快应用于染色体结构的研究，其中之一就是扫描电镜（SEM）。当扫描电镜研制成后不久很快将它应用于染色体研究方面。

扫描电镜用于染色体研究，有三个突出特点：①焦点深度大，可以得到光学显微镜与透视显微镜不能得到的染色体三维立体图像。一般光镜放大 500 倍时，其焦点深度为 1 µm，而扫描电镜可达 200 µm，比透射式电镜（TEM）高 10 倍。在 x、y、z 轴可同时获得清晰的图像，所以染色体的立体感非常强。②具有较高的分辨率，现一般商品扫描电镜分辨率为 50～100 Å，一旦扫描电镜的分辨率提高到投射式电镜水平，人类的微观知识将飞跃般的扩大。③因电子射线不是透过标本而是在样品表面扫描，因此通常用与光学显微镜的染色体载玻片标本，可直接进行扫描电镜观察，并且可以实现光学显微镜与扫描电镜对同一细胞染色体的观察。

1.4.1　植物染色体扫描电镜标本制备方法

1.4.1.1　设备、药品和材料

扫描电镜，具照相设备的显微镜，钻石刀或钻石笔。

染色体标本制备及药品同第一节 1.1.2。

材料：用去壁、低渗法制备的植物染色体玻片标本若干张。

1.4.1.2 操作规程

（1）染色　制备的染色体标本，先经 20∶1 pH 6.8 Giemsa 染色 30 min，后流水冲洗，再空气干燥。

（2）镜检　光学显微镜检查，找到染色体分散良好、且染色体周围无细胞质的细胞，用绘图墨水点上标记，并进行显微摄影。

（3）切割　将有标记的载玻片标本，用玻璃刀在有材料的反面按扫描电镜样品台规定切割一定大小。

（4）粘贴　用导电胶将切割好的样品粘贴在样品台上，记好材料位置。

（5）喷镀　在镀膜机上喷镀 100～200 Å 厚的白金。

（6）观察　在低倍镜下先找好标记，然后转换高倍观察。因染色体厚度很小，常常需要调整样品台角度，图像才能显示清楚。

（7）拍照　用于拍摄的细胞不要过长时间的观察，避免样品损坏。

1.4.1.3 讨论

染色体载玻片标本，用于扫描电镜观察有两个不利点：①在载玻片上展开的染色体没有相当的厚度，不能充分发挥扫描电镜焦点深的特点。②残留在染色体周围的部分细胞质，妨碍了对染色体标本细微结构的观察。我们利用去壁、低渗法制备染色体标本，可使染色体完全裸露出来，在多种植物染色体观察中都取得了较好的结果。如甘蔗（*Saccharum officinarum*，$2n$=80）染色体，清楚的显示出染色体螺旋结构，染色体图立体感很强，展示了扫描电镜观察染色体的优越性，但也不是每次观察都能成功，喷金过厚，显示不出结构，常常是失败的主要原因。

1.4.2　透射电子显微镜的染色体表面展开法

早在透射式电镜发明后不久，人们就用超声波破碎、涂抹等方法，在电镜下研究细胞核与染色体的构造，但所得结果与原有光学显微镜结果面貌皆非，于是人们对用电镜研究染色体产生了怀疑。到 20 世纪 50 年代，超薄切片机发明后，用超薄切片法研究染色体又风行一时，但仍进展不大。在电镜下染色体显示出均一致的结构，而看不到光学显微镜下所见的螺旋结构。后来 Gall（1963）利用容器内水表面张力将细胞各组成成分在水表面展开，然后再将其捞在喷有碳膜的载网上，进行电镜观察的方法，即空气、水面表面展开发明之后，电镜技术才广泛用于高等生物的染色质与染色体的研究。其中 Dupraw（1965、1966）所提出的染色体折叠纤维模型（folde fiber made）学说，引人注目。

染色体表面展开法，实际可分为：水表面展开法与空气表面展开法。为方便起见，以下按照这两种方法介绍。

1.4.2.1　设备、药品和材料

1. 设备及用具

（1）投射式电镜及喷镀仪。

(2) 烘箱、温箱、离心机、显微镜。

(3) 蒸馏水重蒸器。

(4) 用具: 载网、游丝指钳、镊子、刀片、剪刀、载玻片、尼龙网、染色架、量筒、称量瓶、布氏漏斗、容量瓶、试剂瓶、表面皿、培养皿（9 cm）、微量注射器（5~50 μm）、刻度离心管、滴管。

2. 药品

醋酸氨、乙二胺四钠（EDTA）、三羟基甲基甲烷、细胞色素 C、盐酸、氢氧化钠、三氯甲烷、醋酸戊脂、乙醇、醋酸双氧铀、铗酸、Formvar、Parlodion、双蒸水、滑石粉、活性炭、石蜡、分子筛。染色体制备药品同第一节。

3. 实验材料

大麦种子，试验基地种植，大鼠，动物饲养中心购买。

1.4.2.2 操作规程

因为实验精细复杂，植物染色体又难以提取，在做植物染色体展开之前，可先用大鼠骨髓细胞染色体练习此方法，待表面展开技术熟练后再转入植物染色体表面展开法。以下按水表面展开法与空气表面展开法二种方法分别介绍:

1. 水表面展开法

水表面展开法有两种溶液，一种是上相液或展层液，一种是下相液。上相液是研究材料（大分子、染色质、染色体）的提取液和碱性蛋白等组成的高浓度盐溶液，下相液是由低浓度盐溶液或双蒸水组成。

(1) 溶液配制

①贮存液

溶液	配制量（mL）	称取药量（g）
2 mol/L NH$_4$Ac	200	30.836
500 mmol/L EDTA	100	18.6125
10 mmol/L Tris-HCl	100	0.12114

用 HCl 调至 pH8.0

EDTA 很难溶解，需加入一定量的 NaOH 才能促使其溶解，也可按 0.4 mol/L Na$_2$-EDTA 与 0.4 mol/L NaOH 1:1 等量混合，完全溶解，pH 值亦为 8.0。

②上相液: 用微量注射器吸取下列物质，存于离心管的底部待用。

2 mol/L NH$_4$Ac　　　　　25 μL;

500 mmol/L EDTA　　　　0.4 μL;

细胞色素 C 5mg/mL　　　4 μL;

10 mmol/L Tris-HCl　　　0.5 μL;

提取材料+双蒸水　　　　70 μL /100 μL。

③下相液: 取 2 mol/L NH$_4$Ac 25 mL，加重蒸水至 200 mL，配置成 0.25 mol/L NH$_4$Ac。

④染色液: 醋酸双氧铀的饱和 70%乙醇溶液或 0.2%醋酸双氧铀水溶液。

⑤Formvar 溶液: 取 Formvar 0.2 g，三氯甲烷 100 mL（事先用分子筛吸水）。

（2）载网的清洗与膜的制备

①载网的清洗

新买来的铜网可用乙醇清洗，干燥后备用。用过的铜网先用醋酸戊酯浸泡1~3 d，溶去支持膜，再用乙醇清洗。也可将用过的铜网放在盛有浓硫酸的具盖的称量瓶中浸泡。不断搅动至铜网发亮，倒掉硫酸。用清水洗净，最后用乙醇洗过，干燥后即可再用。

②Formvar膜的制备

Formvar膜的化学名称为聚乙烯醇缩甲醛。用三氯甲烷配制的聚乙烯醇缩甲醛溶液贮于冰箱备用。将清洁的载玻片垂直浸入上述溶液中（约浸入载玻片的1/2），静置片刻，取出后垂直立于-滤纸上，使多余的溶液流下。干燥后，用锋利的刀片沿载玻片边沿将膜划一刻痕，然后将载玻片具膜的一端水平放入直径10 cm盛有蒸馏水的蒸发皿中，薄膜即和载玻片脱离，飘浮于水面。如将载玻片两面的膜均划出刻痕，垂直插入水中，这时载玻片两边的薄膜可同时脱离载玻片，得到两张Formvar膜。根据水面上薄膜呈现的干涉颜色，选取暗灰色的膜（黄色的浅，红色的太后），将铜网排放在其上，用一比膜面积稍大的滤纸盖于膜上，小心将滤纸提起，晾干备用。

（3）具体操作

①染色体的提取：提前1 h向大鼠腹腔内注射秋水仙素，乙醚麻醉，取出骨髓与离心管内，加入0.075 M KCl溶液低渗30min（37℃），离心（1000转/min）10 min，加入少许双蒸水，制备适当浓度提取液。

②用微量注射器吸取70 μL材料提取液，加入到上相液备用。

③在直径9 cm清洁的培养皿内（培养皿内壁涂有黑色石蜡，蜡中加入活性炭）注入下相液。

④将一张清洁脱脂的载玻片一端浸入下相液中，另一端搭在培养皿边上。用滤纸条将水面清扫干净，并撒上少许滑石粉，以识别膜在水面上展开的程度。

⑤用微量注射器吸取展层液，滴在接近下相液表面的倾斜载玻片上。当展层液接触下相液表面时，包括染色体在内的蛋白质形成一很薄的膜，在水面展开。在材料与水面接触的一瞬间，细胞由于受到浸透压与表面张力的作用而破裂。核内的染色体、染色质纤维等从细胞内溶出，与蛋白质同时形成一很薄的界面膜，从滑石粉被推开的程度可以判断膜扩展的大小。

⑥用事先制备好带有Formvar膜的载网，膜面朝下，慢慢接近展开的膜，水平蘸取，然后用滤纸小心地沿载网边缘吸取水分。

⑦固定、脱水：材料经95%乙醇固定，脱水2~3次。

⑧染色：在表面皿内加入数滴醋酸双氧铀染色液，将带有材料的载网以材料朝下扣在染液上。染色15~30 min。

⑨脱水：染色后用无水乙醇冲洗2-3次。电镜下可显示出不同等级的染色质纤维。

⑩电镜观察

2. 空气表面展开法

（1）染色体提取：培养的大麦根尖按第一节1.1.2去壁、低渗悬液法制备。甲醇：冰醋酸3∶1固定30 min以上，去沉淀后，上清液离心（1000转/min）3 min，去上清液，在细胞提取液中加入新鲜固定液。

（2）Parlodion 膜的制备：将脱脂光洁的载玻片浸入 1% Parlodion 醋酸戊脂溶液中片刻，静置干燥后喷上一层碳膜，然后迅速用刀片刮去载玻片四周的薄膜，用力向膜呼口气，以 45°角将载玻片慢慢插入水中，让 Parlodion 膜剥离载玻片而赋予水面上备用。

（3）展开：将（1）染色体提取液换入乙醇：冰醋酸为 1：1 固定液中，用滴管立即将样品滴在浮于水面（水温 50℃左右）的 parlodion 膜上。

（4）镜检：用光洁的载玻片将有材料的膜捞起，在光学显微镜（最好是相差显微镜）下检查，选取染色体铺展好的膜。

（5）捞取：选好的膜再脱回布氏漏斗水中，在布氏漏斗底部铺上尼龙网，其上排好电镜铜网，然后用排水下降法将有材料的 Parlodion 膜贴附于铜网上。

（6）干燥：将铜网在白炽灯光下烤干，并使 Parlodion 膜与铜网贴牢。

（7）染色：经醋酸双氧铀染色 30 min。

（8）脱水：用无水乙醇冲洗 2～3 次。

（9）电镜观察：经空气表面展开法，在 Parlodion 膜上的染色体与光学显微镜下形态相似，因此它可用于染色体显带的电镜观察。

1.4.2.3 讨论

植物染色体表面展开法的关键在于染色体的提取，应用去壁、低渗悬液法提取染色体虽有一定的效果，但此项技术尚有很多问题值得进一步研究。

（作者：陈瑞阳。原载：植物细胞学研究方法，科学出版社，1985，pp108-119）

第五节　植物有丝分裂染色体同步化方法

提要：有关高等植物细胞周期同步化诱导的方法已有很多报道，但都存在有丝分裂中期指数（MI）不高，效果不稳定，重复性不好等问题。最近发展起来的利用羟基脲（hydroxyurea，简称 HU）和甲基氨草磷（amiprophos-methyl，简称 APM）双阻断法诱导植物细胞有丝分裂同步化，获得了高频率有丝分裂中期指数（Dolezel，1992；Pan，1993）。HU 是一种 DNA 合成抑制剂，可暂时阻止细胞周期进程，使细胞积累在合成期（S 期），一旦除去 HU 作用，细胞就同步进入下一个时期。APM 是一种干扰细胞微管蛋白合成的特异性药物，能阻止中期细胞纺锤丝的产生而使细胞停留在中期。目前国内在这方面的研究报道很少，仅邵启全（1981）、周钟信（1985）等人做过有丝分裂同步化的诱导。作者利用 HU 和 APM 对多种高等植物根尖细胞进行活体诱导处理，获得了高频率同步化的有丝分裂中期染色体分裂相。这对研究染色体的结构、染色体原位杂交和染色体显微切割等具有重要的意义。

1.5.1 利用 HU 和 APM 双阻断法诱导高频率植物根尖细胞有丝分裂同步化研究

1.5.1.1 材料和方法

1. 材料

本实验所用材料见表 1.8。

表 1.8　材料名称及来源
Table 1.8　The name and origin of materials

种名 Specific name	来源 Origin	染色体数 Chromosome number
小麦 *Triticum aestivum*	南开大学生物系遗传研究室	$2n=42$
大麦 *Hordeum vulgare*	邯郸市农科所	$2n=14$
黑麦 *Secale cereale*	由 Dr.R.Nel Jones 惠赠	$2n=14$
蚕豆 *Vicia faba*	购自市场	$2n=12$
玉米 *Zea mays*	购自市场	$2n=20$
水稻 *Oryza sativa*	天津市农科院	$2n=24$

2. 方法

本实验主要参照 Pan（1993）的方法，并略加改变。种子萌发后置于浸过 1.25 mmol/L HU 的滤纸上处理 18 h，之后用蒸馏水冲洗三次移到浸过蒸馏水的滤纸上培养 1～6 h，然后再转移到浸过 4 μmol/L APM 的滤纸上继续生长 2～4 h（上述各处理均在 24℃温箱中避光进行）。经上述处理的材料用蒸馏水漂洗三次后放入冰水中，在 4℃保存 24 h。

切下根尖转入 2.5% 纤维素酶和果胶酶混合水溶液中，在室温下酶解 2～3 h。倾去酶液，于双蒸水中低渗 20 min 后，再将根尖转移至固定液中固定 1 h。将根尖置于载玻片上，除去根冠只留下一小块分生组织，用卡宝品红染色后压片。镜检并统计有丝分裂中期指数，有丝分裂中期指数的计算参照 Pan（1993）介绍的计算方法。

1.5.1.2 结果与讨论

采用 HU 结合 APM 对种子根尖进行诱导可获得高效的同步化细胞，各种材料的最佳同步化条件可以通过调整 HU、水培、APM 的处理时间而获得，表 1.9 列出了几种供试材料的最佳同步化处理条件及分裂指数（MI），可见 MI 均在 50% 以上，其中小麦经 HU 处理 18 h，水培 4 h，APM 处理 4 h，冰水处理 24 h 后 MI 高达 65%（见图 1.1）。各种植物的根尖压片照片见图 1.2，可见有丝分裂正常，中期染色体占 80%～90%，染色体无畸变现象。

表 1.9　各种材料的处理条件及分裂指数
Table 1.9　Exposure conditions and MI of several materials

种名 Specific name	处理条件 Deal with condition				分裂指数% Metaphase index
	HU/h	水 water/h	APM /h	冰水 ice water/h	
小麦 *Triticum aestivum*	18	4	4	24	65
大麦 *Hordeum vulgare*	18	4	4	24	51
黑麦 *Secale cereale*	18	2	4	24	50
蚕豆 *Vicia faba*	18	1.5	5	24	60
玉米 *Zea mays*	18	5	3	24	53
水稻 *Oryza sativa*	18	3	2	24	50

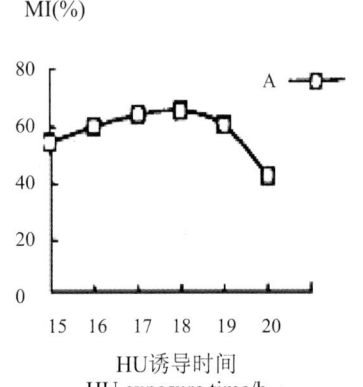 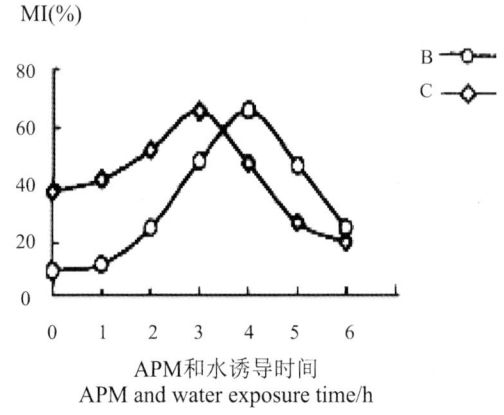

图 1.1　HU、水培、APM 时间对小麦根尖细胞有丝分裂指数的影响
Fig. 5.1　MI in *Triticum aestivum* root-tip after treatment with HU，water，and APM
A：HU 处理 15～20 h，水培 3 h，APM 处理 4 h；B：HU 处理 18 h，水培 3h，APM 处理 0～6h；
C：HU 处理 18 h，水培 0～6 h，APM 处理 4 h

羟基脲（HU）是一种 DNA 合成阻断剂，它通过抑制核糖核酸还原酶的活性来阻断 DNA 的合成，对处于合成期（S 期）的细胞起作用，而对其他各期细胞不起作用，因此当进行 HU 处理时，首先 S 期细胞的 DNA 合成受阻，而其他各期细胞继续运转，当经间期 1（G1 期）进入 S 期开始合成 DNA 时即被阻断，再经适当时间后大部分细胞则集中于前 S 期（Clain，1980）。我们采用 1.25 mmol/L 的 HU 处理根尖，既可有效阻断 DNA 的合成又可在去除 HU 后迅速恢复 DNA 的合成，以达到同步化的目的。理论上 HU 处理时间越长同步化效率越高，但实际上对细胞的恢复有影响。从图 1.1A 中可看出不同 HU 处理时间对小麦 MI 的影响，18 h 以前 MI 呈上升趋势，超过 18 h MI 降低。值得注意的是在动、植物细胞中已经发现 HU 处理超过 24 h 后可诱发染色体断裂和姊妹染色单体的不可逆互换（Anderson，1983）。APM 是一种直接干扰植物微管合成的特异性药物（Morejohn，1984），其作用类似于秋水仙素，起到收集中期染色体的作用，不同 APM 处理时间对有丝分裂中期指数有一定的影响，从图 1.1B 中可见小麦 4 h 为最佳处理时间。

水培时间对同步化的影响，由表 1.9 可见不同植物水培时间相差很大，这可能与两个方面有关系：其一各种植物由 S 期到中期的时间不同；其二为不同植物对 HU 的敏感性不同，在去除 HU 后恢复分裂的快慢不同（见图 1.1C）。

在植物有丝分裂同步化过程中，染色体粘连在一起不容易散开是一个关键问题。尽管获得了大量中期染色体，但由于染色体粘连在一起，给以后的工作带来很大的困难，为此我们做了一系列缓解染色体粘连现象的尝试，发现冰水处理效果显著。与 Pan（1993）的方法相比，采用蒸馏水代替了成分复杂的培养基，并且对多种植物的诱导都获得了很好的结果（图 1.2），使得此方法更简便，更易于推广使用。同步化不仅是获得大量中期染色体的有效方法，也为细胞周期的研究提供了手段，这样把细胞同步到不同的时期，从而可以对不同时期的细胞进行生理生化的分析。在我们的实验中发现：经 HU 处理后，如水培时间缩短，则会得到相当部分的前期细胞；如不经 APM 处理，则可得到大量后期细胞，对它们的结构和功能的研究将另文发表。

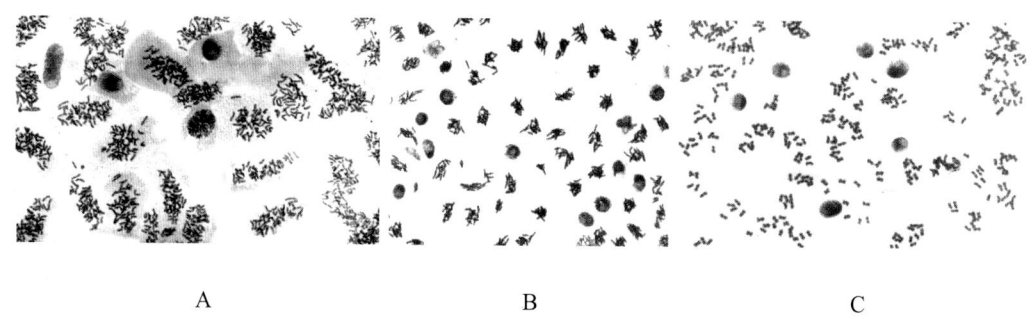

图 1.2　经同步化诱导的小麦、蚕豆、大麦根尖中期分裂相
Fig. 1.2　Metaphase cells of several plant species after synchronization
A. *Triticum aestivum*；B. *Vicia faba*；C. *Hordeum vulgare*

（作者：陈成彬、宋文芹。原载：南开大学学报（自然科学），1999，32（4）28-31）

1.5.2　小麦根尖细胞有丝分裂同步化诱导与中期染色体分离

1.5.2.1　材料与方法

1. 有丝分裂中期同步化诱导

小麦（*Triticum aestivum*）种子在室温下浸泡过夜，然后于 23℃下 1.25 mmol/L HU 中处理 18 h，蒸馏水洗净后转入水培 3 h，然后在 4 μmol/L APM 中连续培养 1～6 h。将同步化根尖分成两份，一份固定于 70 %乙醇中，供检测细胞中期有丝分裂指数（Met. I）用另一份迅速置染色体分离缓冲液中，-20℃下保存，供分离染色体用。

2. 细胞 Met. I 检测和中期染色体分离

将固定根尖置 1M HCl 中（60 ± 1）℃下酸解 6 min 后，用卡宝品红染色，压片，光镜下检查并统计细胞 Met. I。中期染色体的分离参照 Schubert（1993）和 Dolezel（1992）

等介绍的方法，先将经同步化的根尖分生组织（约 0.2 cm）置染色体分离缓冲液（15 mmol/L Tris-HCl、80 mmol/L KCl、20 mmol/L NaCl、2 mmol/L EDTA、0.5 mmol/L 精胺、0.1 Triton×100、15 mmol/L 巯基乙醇（pH7.5）中。用组织匀浆器匀浆破壁，匀浆液分别用 50 μm 两层尼龙布过滤，以除去大部分未裂解细胞和间期核，用注射器来回抽动匀浆液数十次，进一步破膜使其释放染色体。小心将滤液置 40 % 蔗糖分离液，200 g 离心 30 min，进一步去核，取悬浮液，1000 g 离心 20 min，镜检，如染色体纯度不够，可反复数次以达一定纯度为止。取分离液 1 μL，置干净载玻片上，风干后滴一滴卡宝品红染色数分钟，盖片并轻压，光镜检查并照相。

1.5.2.2 结果与讨论

1. 高频率中期有丝分裂同步化诱导

以往对于植物细胞有丝分裂中期同步化的诱导多采用 HU 和秋水仙素的双阻断法进行，并且多以悬浮细胞系为材料。已知 HU 能够抑制核苷酸还原酶（ribonucleotide reductase）的活性，因此，能够将细胞分裂阻止在 G1-S 期（Griesbach，1987）。秋水仙素能够扰乱纺锤丝的正常功能，使细胞分裂阻止在中期。但已有一些研究表明，秋水仙素诱导中期同步化，由于使用浓度过高，因此染色体的凝集作用十分明显，这对进行染色体形态结构研究造成困难，Met. I 也偏低。因此，近 10 多年来，很多研究者常采用 APM 来替代秋水仙素。AMP 是一种抑制微管蛋白形成的药物（Morejohn，1984），它可以扰乱分裂细胞中纺锤丝行使正常的功能，因此，使细胞分裂在中期累积。我们用 HU 和 APM 双阻断法诱导得到的小麦中期有丝分裂同步化，实验结果表明，在 1.25 mmol/L HU 溶液中，23℃下培养 18 h，再经水培 3.5 h 以解除 HU 作用，使其恢复正常生长后再用 4 μmol/L APM 分别处理 1~6 h 后发现，在 4~5 h APM 处理其诱导效果最为理想。Met. I 可接近 50%，但超过 5 h 时，Met. I 开始下降。从经同步化诱导后的根尖分生组织细胞压片检查，此时的分裂细胞几乎都处于中期，并且染色体缩短、变粗，形态较好。

2. 前期、后期和末期的部分同步化诱导

迄今为止，我们还没有看到在植物根尖中直接进行前期、后期和末期细胞有丝分裂同步化诱导的报道。本研究中，我们在进行中期同步化诱导时，发现通过调节 HU 诱导的时间，可以较成功地将前期、后期和末期细胞进行部分同步化诱导。一些研究者曾用 0.75 μmol/L HU 处理洋葱根尖，研究其对细胞有丝分裂的影响，结果表明，当用 HU 处理 8 h 后，分裂细胞中约有 80 % 在 S 期累积，表明 HU 不影响细胞从 G1→S 期运行，而对处于 G2 期的细胞则有延缓使其进入 M 期的特性（Navarrete，1979）。这一实验结果表明，如果掌握合适的 Hu 处理时间和浓度，可望将植物细胞有丝分裂同步在前期。本实验结果表明，当用 1.25 mmol/L HU 处理已萌动的小麦种子 18 h 后，去除 HU 作用，使其恢复生长 4~12h 后，小麦中陆续出现了前期分裂细胞，尤其在恢复生长 8 h 时，前期细胞已高达 11.5 %。因此，利用 HU 处理可以对小麦根尖分生组织细胞进行前期的部分同步化诱导。

小麦根尖分生组织的后期和末期有丝分裂同步化诱导比较困难，我们试图了解具有较高频率的中期同步化细胞在解除 APM 的药物抑制作用后，这些中期细胞是否会恢复生长，然后进入后期和末期，而实际得到的结果表明，由 HU 和 APM 双阻断后同步在中期的细胞去除 APM 作用转入水培恢复正常生长后，根尖中很难检测到正常的后期和末期分裂细胞，而是出现了大量分裂异常的中期细胞。这表明，与 HU 作用机理不同，APM 扰乱纺锤丝行使正常功能的药物效应似乎是不能恢复的。经 APM 处理过的细胞中，纺锤丝并不能进一步将中期细胞中的染色体均等地拉向两极，正常进入后期、末期，而是出现多极、辐射状分布的染色体，表明纺锤丝已失去了正常的功能。

但我们同时发现，在仅经 HU 处理的细胞，去除其药物作用，水培使根尖恢复生长后 1～24 h 连续取材，在小麦根尖中检测到有较高频率的后期（7.3%）、末期（6.5%）细胞。这些后期和末期细胞形态正常，推测它们中至少有一部分是由经 HU 同步在 G1-S 期的细胞在去除 Hu 的抑制作用恢复正常生长后发育而来的。HU 对细胞分裂作了第一次阻断，使分裂细胞累积在 G1-S 期（Griesbach，1987）。当去除 HU 作用后，细胞恢复正常周期运行，因而出现较多的后期、末期细胞，而未经 HU 处理的根尖中，其后期和末期细胞的出现频率均不足 1%，可能是在未经处理的根尖中，由于没有经 HU 的阻断，故没有大量的细胞在 G1→S 期累积，对细胞的分裂没有起到同步化诱导作用。这样看来，HU 对后期和末期细胞的同步化分裂方面也起到一定作用。

3. 中期染色体分离

中期染色体的分离在动物中较易成功，但植物中，由于具有坚硬的细胞壁，去壁进而裂解细胞难度大，而常用的酶解去壁法留下大量的细胞碎片，给进一步纯化染色体增加了困难，所以，以往植物中染色体的分离主要以悬浮细胞系为材料。但悬浮细胞培养过程复杂，培养过程中核型不易稳定（Karp，1985；Lee，1988）而从不稳定的核型细胞中分离得到的同源染色体在 DNA 相对长度和含量上都有明显差异（Sgorbati，1991）。因此，以悬浮细胞为材料分离得到的染色体用于形态结构、生化特性分析和构建单条染色体基因文库都会给实验结果带来误差。相反，根尖细胞可以从很多植物中得到，同步化诱导方便且核型稳定，因此是分离染色体的理想材料。

染色体分离能否取得成功，关键是中期有丝分裂同步化诱导频率。我们发现，当 Met. I 低于 30% 时，所得到的染色体悬浮液中细胞碎片和间期核污染较严重，在进一步地分离、纯化染色体过程中染色体丢失严重。本实验中，我们改用 APM 作为中期有丝分裂阻断剂替代秋水仙素，使中期同步化频率达到 50%，中期分裂细胞的增多，减少了染色体分离悬浮液中细胞碎片和间期核的污染。另外还避免了秋水仙素诱导中常出现的由于染色体凝集而产生的染色体"簇"，而在以往的用秋水仙素诱导中期同步化的工作中，由于所使用秋水仙素的浓度较高，染色体成簇现象是很普遍的，这些成簇的染色体很难与较大的细胞碎片分开。用 APM 作为中期阻断剂，由于使用浓度低，在合适的诱导时间内（不超过 5h），染色体很少凝集，这为进一步分离出较纯和较高得率的染色体提供了方便。机械破壁和用注射器来回抽打，以破核膜释放中期染色体已在蚕豆根尖中期染色体分离中获得成功（Schubert，1993；Dolezel，1992）。这一方法的优点是避免了酶解去壁法中所产生的大量与染色体大小相接近的细胞碎片，而这些碎片较难与染色体分开，缺点是机械去壁法不能象酶法那样将根尖中的细胞都游离开，悬浮液中会有很多较大的多细胞片段存在，因此造成染色体损失。但由于本研究中，中期同步化频率较高，加之大片段易与染色体分开，为纯化染色体提供了方便，所以从根尖中分离染色体，目前这种方法还是较适合的。我们分离所得到的小麦中期染色体，染色体形态正常，分散性较好，细胞碎片和未裂解间期细胞核已基本去净。

（作者：彭永康，赵建，陈瑞阳。原载：天津师范大学学报，1998，18（4）52-57）

1.5.3 APM 对小麦根尖分生组织细胞异常有丝分裂的诱导

1.5.3.1 供试材料及培养

将在室温萌动过夜的小麦（*Triticum aestivum* L.）种子分成 2 组，第 1 组内分别加入，$c=1\sim4$ μmol/L 的 APM 溶液，（23±1）℃温箱内萌发 6 h，第 2 组内先将萌动种子在 HU 中处理 16 h，然后停止 HU 作用，再分别加入 4 μmol/L、5 μmol/L APM 溶液，（23±1）℃温箱内处理 6~7 h。迅速切取萌发的小麦主胚根（1~0.2 cm），70 %乙醇中固定，将固定根尖用蒸馏水冲洗干净，然后置 c（HCl）= 1 mol/L，60℃ 酸解 6 min，后将根尖用蒸馏水冲洗干净，置 2.5%纤维素酶：2.5 %果胶酶（1∶1）混合液中酶解 30 min。取根尖，卡宝品红染色，压片，然后在光镜下观察，并记录根尖分生组织细胞有丝分裂指数和间期、中期、后期、末期细胞异常有丝分裂结果。

1.5.3.2 结果与讨论

1. 不同 c（APM）对小麦根尖分生组织细胞有丝分裂的影响

我们发现，用不同 c（APM）单独处理正在萌发的小麦种子，经 5 h 后明显提高根尖分生组织细胞的有丝分裂指数，尤其可提高前期和中期细胞的分裂频率（表 1.10），与 CK 相比，1~4 μmol/L 各处理细胞有丝分裂指数增加，但增加频率与 c 无明显线型关系；从所检测到的细胞有丝分裂指数看，1~4 μmol/L 下各测定值接近。

APM 对植物根尖分生组织有丝分裂有明显的中期同步化效应，一般经 c（APM）= 4 μmol/L 处理细胞 5 h，细胞分裂在中期累积的频率高达 50%，有些文献报道可高达 80%（Pan，1993）左右，并且几乎找不到前期和后-末期细胞。但这些结果是在 HU 和 APM 双阻断基础上获得的。我们是用 APM 单独处理小麦萌发种子，根尖分生组织细胞中没有大量的中期分裂细胞的累积，但与对照相比，在用 1~4 μmol/L APM 处理后根尖分生组织细胞中有丝分裂指数明显提高，说明 APM 有刺激小麦根尖作用。

2. 间期细胞的微核效应和中期细胞染色体聚合现象

APM 处理小麦萌发的种子后，根尖间期细胞中可明显检测到数目不等的微核，间期细胞中微核的出现与 APM 处理的浓度和时间有直接关系，一般在 $c>4$ μmol/L，$t<5$ h 时，间期细胞中很难找到微核，但当 $t>5$ h 或将 c（APM）>5 μmol/L、$t=5$ h 时，间期细胞中微核现象非常普遍；如处理时间延长至 7 h，很多间期细胞核甚至被诱导成多个微核。

表 1.10　1~4μmol/L APM 处理 5 小时后小麦根尖分生组织细胞有丝分裂频率*
Table 1.10　Frequencies of cell mitotis of T.aestivum root tip meristems treated with 1~4μmol/L APM for 5 h

c（APM）/μmol·L^{-1}	前期（%） Prophase（%）	中期（%） Metaphases（%）	后-末期（%） Ana-telophases（%）	有丝分裂指数 （M.I）（%）
CK	1.45	1.16	0.17+0.38	3.61
1	3.05	4.52	0.33	7.90
2	3.60	4.13	0.27	8.0
3	3.15	4.75	0	7.90
4	2.90	4.90	0	7.80

*根据 10 条根尖至少 5000 个细胞统计所得。(Based on a scoring of at least 5000 cells from about 10 root tip meristems.)

APM 对小麦根尖分生组织细胞中期有丝分裂影响的一个最明显的特征是染色体的聚合现象。小麦种子萌发后先经 c(HU) = 1.25 mmol/L 预处理 16 h，再分别在 c（APM）为 4 μmol/L、6 μmol/L 中处理 5 h 后检测到的根尖分生组织细胞中期有丝分裂图像。HU 预处理可以得到高频率分裂一致的前期细胞，因为 HU 的作用是将细胞分裂阻止在 G1-S 期（Navarrete，1979）。由于这里所使用的 HU 浓度低和处理时间短，因而不会引起细胞生理功能的损伤和染色体变异（Anderson，1983），当停止 HU 的作用后细胞会恢复正常的周期运行。由于 HU 预处理后有大量分裂细胞累积在前期，这样的根尖移入 APM 中处理后可得到大量分裂一致的中期细胞，便于实验分析。预处理的 HU 浓度和处理时间不足以引起染色体的变异，因此在中期细胞中检测到的染色体变异可归因为是 APM 作用所致。用 4 μmol/L APM 经 5 h 处理，根尖染色体没有产生明显变异，但将 c（APM）提高至 6 μmol/L 时，染色体相互缠绕呈"螺旋状"结构，很显然，这种现象是由 APM 处理引起的，表明在小麦根尖分生组织中，用 4 μmol/L APM 处理 5 h 以上或者 6 μmol/L APM 处理 5 h 足以引起这种现象。

3. 纺锤丝功能的抑制效应

按照细胞周期运行的规律，这些累积在中期的细胞要进入后期、末期，但我们观察到，APM 可明显阻止细胞分裂正常进入后期、末期，如将处理时间延长至 6 h，此时在 4 μmol/L APM 和 5 μmol/L AMP 下处理的根尖中均未能到正常的后期、末期细胞，但可观察到较高频率分裂异常的细胞，这种异常分裂的细胞可以分为不均等分裂和多极化分裂两种类型。

APM 被认为是一种可以阻止和抑制微管蛋白合成的药物（Kiermayer，1977），它可以干扰植物细胞中纺锤丝的形成或抑制纺锤丝行使正常的功能，而高等植物细胞分裂从中期进入后期和末期是靠纺锤丝的牵引作用，纺锤丝将排列在赤道板上的中期染色体均等的拉向两极，形成两个后期细胞，但在本实验中观察到的"不均等"分裂显然是纺锤丝不能均等的将染色体拉向两极，而出现一极多一极少的现象。本实验中观察到的另一种细胞异常分裂现象"多极化分裂"，则可以明显看到纺锤丝无规则的将染色体拉向多极。在这两种类型的异常分裂细胞中，纺锤丝的正常功能被扰乱了，因此，最终都不能形成正常的后期和末期细胞。我们的实验结果初步表明，APM 处理在超过临界浓度和临界处理时间（在小麦中我们认为是 c= 4 μmol/L，t=5 h）对作物是有害的。因此，APM

作为一种除草剂在使用时掌握合适的浓度和处理时间是必须的，否则会造成农业环境的污染。

（作者：彭永康、于建春、赵建、宋文芹、陈瑞阳。原载：应用与环境生物学报，1997，3（3）：204-207）

1.5.4 大蒜根尖细胞有丝分裂同步化诱导与中期染色体分离

1.5.4.1 材料和方法

1. 细胞有丝分裂中期同步化诱导

大蒜（*Allium sativum*）蒜瓣洗净后 23℃下 1.25 mmol/L HU 中处理 16～24 h，蒸馏水洗净后转入水培 5 h，然后在 4 μmol/L APM 中继续培养 3～6 h。将同步化根尖分成 2 份，1 份固定于 70 % 乙醇中，另 1 份迅速置染色体分离缓冲液中。

2. 细胞有丝分裂样品制备和中期染色体分离

将固定根尖用碳酸品红染色，压片后光镜下检查并统计细胞有丝分裂中期指数（Met. I）。中期染色体的分离参照 Schubert（1993）介绍的方法，先将根尖分生组织（约 0.2 cm）置染色体分离缓冲液中（15 mmol/L Tris-HCl，80 mmol/L KCl，20 mmol/L NaCl，2 mmol/L EDTA，0.5 mmol/L 精胺，0.1 % Triton × 100，15 mmol/L 巯基乙醇，pH 7.5），用组织捣碎器匀浆破壁，匀浆液分别用 300 目、400 目三层不锈钢网抽滤，除去未裂解细胞和间期核。用注射器来回抽动匀浆液数十次，破膜使释放部分染色体，取滤液加至 40% 蔗糖分离液中，350 rpm 离心 20 min，进一步去核。取悬浮液，1000 g 离心 15 min，取沉淀，再次用染色体悬浮液悬起，如染色体纯度不够，可反复数次以达一定纯度为止。取分离悬浮液 1～2 μL，置干净载玻片上风干后滴一滴碳酸品红液染数分钟，盖片并轻压，显微镜观察并照相。

1.5.4.2 结果与讨论

1. 用羟基脲（HU）和甲酰氨草磷（APM）双阻断法诱导大蒜根尖细胞有丝分裂中期同步化。以往对于植物细胞有丝分裂中期同步化诱导多采用 HU 和秋水仙素双阻断法，并且多以悬浮细胞系为材料，但用这种方法诱导，其细胞有丝分裂中期指数（Met. I）偏低（周钟信等，1985），并且悬浮细胞培养所需时间长，因此进展缓慢。一些研究者为探讨一种简便且 Met. I 较高的诱导方法，曾直接以根尖为材料，用 HU 和秋水仙素双阻断法诱导。采用这种方法尽管 Met. I 有所提高，但部分分裂细胞并未完全被阻断在中期，在经中期同步化的根尖细胞中尚可见到不少前期和后、末期细胞，因此，同步化效果并不理想。本研究参照 Pan（1993）等在大麦、春小麦中采用的方法，用 HU 和 APM 双阻断法诱导大蒜根尖细胞有丝分裂中期同步化，得到较为理想的结果。我们发现，当在 23℃下用 1.25 mmol/L HU 处理 18 h，室温下 4 μmol/L APM 处理 5.5 h，其 Met. I 高达 35 %（图 1.3）。APM 是一种干扰细胞微管蛋白合成的特异性药物（Louis，1984）。因此，它能阻止中期细胞纺锤丝的产生而使细胞分裂停留在中期。近来，国外一些研究者将 APM 用于大麦、春小麦等根尖细胞有丝分裂中期同步化诱导中，但大蒜中还没有正式报道。利用 HU 和 APM 双阻断后大蒜根尖细胞有丝分裂中期可以清楚的看到，细胞分裂基本被阻断在中期，除可见到部分间期细胞外，未发现前期和后、末期细胞，表明利用 HU-APM 双阻断法可有效的将大蒜根尖分生组织细胞阻断在中期。

2. 细胞前、后、末期有丝分裂部分同步化诱导

一些研究者曾用 0.75 mmol/L HU 处理洋葱根尖，研究其对细胞有丝分裂的影响，结果表明，当用 HU 处理 8 h 后，分裂细胞中约有 80%（Navarrete, 1979）在 S 期累积，但 HU 不影响 G1 期细胞的分裂，而对处于 G2 期的细胞则有使其延缓进入 M 期的特性。这个实验结果表明，如果掌握合适的 HU 浓度与时间，可望将植物细胞有丝分裂同步在前期。我们在 23℃下 1.25 mmol/L HU 处理 18 h 后检测到的前期分裂细胞。统计结果表明，其频率已达到 27%。很显然，HU 在诱导大蒜根尖有丝分裂前期部分同步化方面起着重要的作用。尽管在动物中已有后、末期有丝分裂同步化的报道，但在植物中还没有见到。我们曾推测，具有较高频率的中期同步化细胞在解除了 APM 药物抑制后可能会有部分中期细胞恢复沿周期运行进入后、末期，但实际结果与推测不同，由 HU、APM 双阻断后同步在中期的细胞去除 APM 转水培使根尖恢复正常生长后再检测根尖中的细胞。其后，末期细胞出现的频率很低（1.5%）。但我们同时注意到，经 HU 部分同步在前期的细胞，去除 HU 作用，水培使根尖恢复正常生长后却可检测到高达 17% 左右的后、末期细胞。这些后、末期细胞形态正常，它们中间有一部分可能是用 HU 同步在前期的分裂细胞在去除 HU 抑制作用后恢复正常生长的结果，因为 HU 对细胞分裂作了第一次阻断，使许多细胞分裂暂时累积在 S 期，当去除 HU 作用后，细胞恢复生长，因此有较多的后、末期细胞，而对照中，未能检测到如此高的后、末期细胞，因为没有经 HU 的第一次阻断。这样看来，似乎 HU 对后、末期细胞的同步化方面也起着一定的作用。

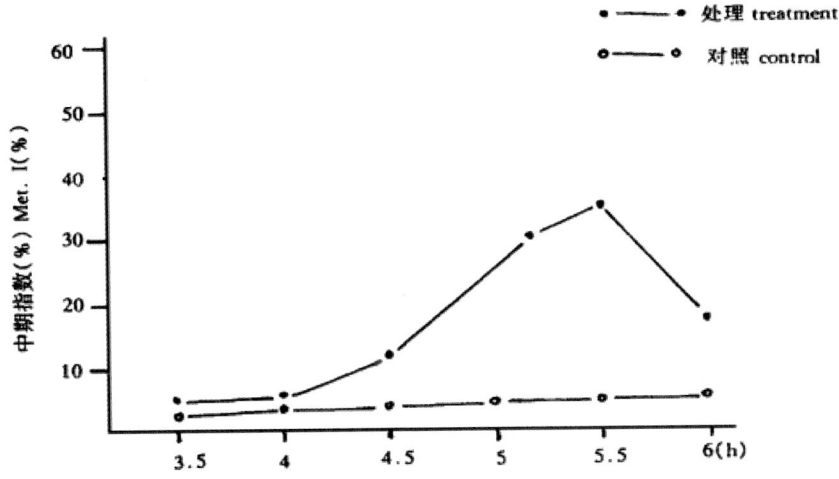

图 1.3　在 1.25 mol/L HU 下培养 18 h 后的大蒜经 4 μmol/L APM 处理 1～6 h 后根尖细胞中的百分率。

Fig.1.3　Percentage of metaphase root tip in *Alium sativum* after treatment with 1.25 mmol/L HU for 18 hr and 4 μmol/L APM for 1～6 hr

APM 的作用是破坏纺锤丝的形成，但被 APM 阻断在中期的细胞在去除 APM 的抑制作用后，这些细胞是否还能恢复生长？这方面至今尚未见到正式报道。本文中，当我们将 APM 处理后的具较高 Met.I 的根尖去除 APM 的抑制作用，恢复生长后 1~24 h 内连续取材观察，这些中期细胞似乎并没有能进入后、末期，因为我们未能检测到较高频率的后、末期细胞，但这一问题尚需作更多的工作才能定论。

3. 中期染色体分离

中期染色体的分离与纯化在动物中较易成功，但在植物中由于细胞具有坚硬的细胞壁，去壁进而裂解细胞难度大，加之去壁后留下大量的细胞碎片，为进一步纯化染色体增加了难度，因而进展缓慢。当今，尽管分离高等植物染色体已有一些报道，但多以悬浮细胞系为材料。本研究采用的分离中期染色体的方法是参照 Schubert（1993）介绍的植物根尖分生组织细胞中的分离方法，实验中省去了酶解步骤，改用将根尖置染色体分离悬浮液中，然后用机械方法离散根尖细胞，去壁进一步裂解细胞释放染色体，未破碎细胞、间期核和较大细胞碎片通过双层不锈钢网过滤弃去，滤液中的染色体和细胞碎片再经分步离心进一步分开。我们从 HU-APM 双阻断法同步化的大蒜根尖中期细胞中分离的部分染色体。光镜观察表明，分离的染色体样品中，没有看到未裂解细胞和间期核，细胞碎片也不多。统计结果表明，每微升染色体分离液中约含有 170 条染色体，这一得率稍低于 Schubert（1993）在蚕豆根尖中所得到的结果。Pan（1993）等利用 HU、APM 的双阻断法在大麦等根尖分生组织细胞中期同步化诱导中得到高达 50% Met.I，并将 APM 对中期细胞的诱导效果与秋水仙素作了比较，得出较有意义的实验结果，但他们没有探讨双阻断过程中细胞的前期和后、末期部分同步化现象，并且也未对中期同步化细胞中的染色体进行分离。本工作利用 Pan（1993）介绍的方法在大蒜根尖分生组织中得到 35% 的中期同步化细胞，并且还对前期和后、末期细胞的部分同步化进行了观察和统计分析，这对于进一步探讨 HU、APM 对细胞有丝分裂的影响、细胞有丝分裂不同时期同步化诱导条件探讨都有参考意义。本工作还分离了根尖分生组织细胞中期染色体，这为进一步研究染色体的结构、功能、生化特性提供了实验材料。因此，也是有价值的。

（作者：彭永康、赵建、陈瑞阳。原载：植物研究，1999，19（3）302-307）

第六节　植物核型分析的标准化

提要： 核型分析是细胞遗传学研究的基本方法，是研究物种演化、分类以及染色体结构、形态与功能之间的关系所不可缺少的重要手段。不同的物种染色体都有各自特定的形态结构（包括染色体的长度、着丝点位置、臂比、随体大小等）特征，而且这些特征是相对稳定的。但是这些信息量远远不能满足染色体分析的需要。所提供的信息量极其有限，特别是对于基因组内染色体形态相近的物种以及基因组间分化程度较小的物种之间，这种核型分析无法提供更多有效而可靠的信息。近年来，荧光原位杂交及荧光分带等细胞遗传学技术的发展提供了更多的染色体信息，使科研工作者能从更精细水平上对植物进行核型分析，更好地分辨基因组内各条非同源染色体，以及进行分化程度小的基因组间的比较，这种核型分析技术为研究染色体行为和变异、单个基因及转基因在基因组内的定位等方面开辟了新的途径。本章主要提供一个从核型、带型分析的标准，使从事这方面研究的科研工作者有一个共同的约束。另外，本章还总结了实验室改进的半自动化核型分析方面的一些方法，最后介绍了最近几年发展起来的分子核型技术。

1.6.1 关于植物核型分析的标准化问题

我国的植物染色体研究工作进展较快，并取得了显著的成绩。但在研究工作中仍存在一个迫切需要解决的问题，即核型分析的标准化问题。由于国际上尚无植物核型分析的共同标准，因此，有关染色体的统计、测量、命名、图表格式等等，各人所采用的方法和标准也不尽相同。这种状况，对核型资料的比较分析及对研究结果的评价都带来不便。有鉴于此，1984年8月在辽宁兴城召开的第一届全国植物染色体学术讨论会上，李懋学和陈瑞阳联名做了"关于植物核型分析的标准化问题"的报告，经过与会代表的充分讨论，现根据大多数代表的意见，修改整理如下（需要说明的是，这是大家约定的标准，并非具有约束力的法规，仅供国内有关工作者参考）

1.6.1.1 核型分析——染色体数目

一般以体细胞染色体数目为准。减数分裂细胞，由于价体分析难以保证准确，所以，除苔藓和蕨类等因材料所限而用减数分裂细胞计数染色体外，其他则一般只宜作为辅助计数材料。统计的细胞数目应在30个以上。其中85%以上的细胞具有恒定一致的染色体数，即可认为是该植物的染色体数目。如果观察材料是混倍体，则应如实记录其染色体数的变异范围和各类细胞的数量或百分比。

1.6.1.2 染色体形态

作为核型分析的染色体，一般以体细胞分裂中期的染色体作为基本形态。此外，如果减数分裂粗线期的染色体分散良好，着丝粒清晰者，也可用做核型分析。

1. 染色体长度

（1）绝对长度（或实际长度）：均以微米表示。一般宜在放大的照片或图像上进行测量，然后按下式换算：

$$绝对长度 = \frac{放大的染色体长度（毫米）}{（或实际长度）放大倍数} \times 1000$$

绝对长度值只在某些情况下有相对的比较价值，在许多情况下，它不是一个可靠的比较数值。由于预处理条件和染色体缩短的程度不同，所以，即使同一种植物，不同实验者所测得的绝对长度值也往往有明显差异，这是无法避免的。

（2）相对长度：均以百分比表示。计算相对长度值的方法，在过去的文献中也有多种公式，现以 Levan（1964）的公式为准。即：

$$相对长度 = \frac{染色体长度}{染色体组总长度} \times 100$$

相对长度值则是稳定的可比较的数值。在近年国内外多数核型研究的文献中，往往只用相对长度值，这种简化是可取的。

（3）染色体相对长度系数（I、R、L）：这是郭幸荣等（1972）提出的对染色体长度进行分类的方法。即：

$$染色体相对长度系数 = \frac{染色体长度}{全组染色体平均长度}$$

I、R、L < 0、76，为短染色体（S）

0、76 ≤ I、R、L ≤ 100，为中短染色体（M1）

1、0.1 ≤ I、R、L ≤ 125，为中长染色体（M2）

I、R、L ≥ 126，为长染色体（L）

（4）染色体长度比：这是指核型中最长染色体与最短染色体的比值。即：

$$染色体长度比 = \frac{最长染色体长度}{最短染色体长度}$$

在 Stebbins（1971）的核型分类系统中，它是衡量核型对称或不对称的两个主要指标之一。

2. 臂比

臂比计算公式如下：

$$臂比 = \frac{长臂}{短臂}$$

3. 着丝粒位置

以上述臂比值确定。参照 Levan 等（1964）的命名，经过讨论，略加改动，即均取用小数点后两位数值，以便严格区分，如表 1.11 所示。

表 1.11 着丝粒位置

臂比值	着丝粒位置	简写
1.00	正中着丝粒（medianpoint）	M
1.0～1.70	中部着丝粒区（median region）	m
1.71～3.00	近中部着丝粒区（submedian region）	sm
3.01～7.00	近端部着丝粒区（subterminal region）	st
7.01 以上	端部着丝粒区（terminal region）	t
∞	端部着丝粒区（terminal region）	T

此命名规则的特点是将染色体的一半长度分为两粒四区,这四区是等分的。Levan 等在分析了其他各种关于着丝粒的命名法之后指出,这种命名法是比较合理的。现已被全世界广泛采用。此外,为了便于阅读文献时参考,下面把其他常用的着丝粒命名法,如着丝粒指数和长短臂差值 Levan 等(1964)的命名规则对照列于表 1.12 中。

表 1.12 着丝粒命名规则对照表

着丝粒位置	臂比值(r)	差值(d)	着丝粒指数(i)
M	1.00	0.0	50.0
m	1.05	0.5	47.5
m	1.22	1.0	45.0
m	1.35	1.5	45.2
m	1.50	2.0	40.0
	1.67	2.5	37.5
sm	1.86	3.0	35.0
sm	2.08	3.5	32.5
sm	2.33	4.0	30.0
sm	2.64	4.5	27.5
	3.00	5.0	25.0
st	3.44	5.5	22.5
st	4.00	6.0	20.0
st	4.71	6.5	17.5
st	5.67	7.0	15.0
	7.00	7.5	12.5
t	9.00	8.0	10.0
t	12.33	8.5	7.5
t	19.00	9.0	5.0
t	39.00	9.5	2.5
T	∞	10.0	0.0

臂比 = $\dfrac{长臂}{短臂}$ ($r=\dfrac{L}{S}$),差值 = 长臂 − 短臂 ($d=L-S$,染色体全长为 10)。

着丝粒指数 = $\dfrac{短臂}{染色体全长}\times 100$ ($i=\dfrac{100S}{C}$)。

三者的换算公式分别为:$d=\dfrac{10(r-L)}{r+L}$;$r=\dfrac{10+d}{10-d}$;$i=\dfrac{100}{r+L}5(10-d)$。

4. 臂指数(或称 N.F.值)

把具中部和近中部着丝粒的"V"形染色体计算为两个臂;把距近端和端部着丝粒的"J"或"I"形染色体计算为一个臂。以此来统计核型中的总臂数。

最后,关于模式核型应分析的细胞数目,从考虑实际的需要和工作量情况出发,经过讨论,一致同意以 5 个以上的细胞为准。这里要强调的是,核型分析的准确性,不仅要求分析一定数量的细胞,更要求有高质量的染色体图象,这是保证核型分析准确的基础。

1.6.1.3 核型的表述格式

1. 表格:核型分析中各项测定的平均数值,应列表报道,列表内容要力求简明和实用,其格式和项目如表 1.13 所示。

表1.13 ×××染色体相对长度、臂比和类型列表示例

序号	相对长度%(短臂+长臂=全长)	臂比(长/短)	类型

表 1.13 中的染色体序号一律用阿拉伯字母,相对长度和臂比值均取小数点后两位数。染色体绝对长度值的变异范围、染色体度比、核型类别等,在表下单列说明。

随体的长度是否计算,不作统一规定,不过都应在表下加以文字注明。具随体(或次缢痕)的染色体,在表中应以星号"*"标记。

2. 染色体序号:一律按染色体全长顺序编号。如两对染色体长度完全相等,则按短臂长度顺序排列,长者在前短者排后。性染色体和 B 染色体一律排在最后。如为二型核型(bimodal karyotype),如中国水仙、芦荟等植物,则长染色体群按 $L_{1,2}...$顺序排列,短染色体群按 $S_{1,2}......$顺序排列。对于向普通小麦等异源多倍体植物,其系统发生的亲本来源已清楚,则应根据其亲本的染色体组分别排列,如普通小麦是按 A、B、D 三组分别编号排列,而不是全部 21 对染色体统一顺序排列。如果核型中有差异明显而恒定的杂合染色体对时,则应分别测量每一成员的长度值和臂比值,分别列于表中,编号可在任选其中一成员为准,并附加说明。

3. 模式照片:一般每种材料应附一张有代表性的中期染色体的完整照片。并标出一个以微米为长度单位的标尺,便于目测染色体大小。尽量少用放大倍数。

4. 核型图:将与模式照片同一细胞的染色体剪下,参照染色体长度和臂比值,进行同源染色体"配对",然后按表格中的染色体序号顺序排列,此即为该细胞的核型图(karyogram)。

5. 核型模式图(idiogram):以表中所列各染色体的相对长度平均值绘制。构图有各种形式,现推荐一种较好的形式,如图 1.4 所示。

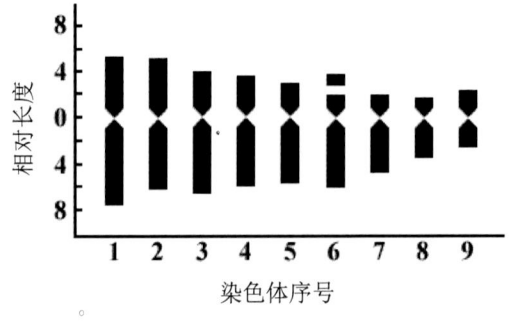

图1.4 核型模式图

6. 核型公式：即综合核型分析的结果，将核型的主要特征以公式表示。它简明扼要，便于记忆和进行比较。例如：芍药（*Paeonia lactiflora*）：2n=2x=10=6m+2sm+2st（SAT）。

7. 核型分类：Stebbins（1971）参照生物界现有的核型资料，根据核型中染色体的长度比和臂比两项主要特征，用以区分核型的对称和不对称程度，并将其分为 12 种类型，如表 1.14 所示。

表 1.14 核型分类

最长/最短	臂比大于 2:1 的染色体的百分比			
	0.0	0.01～0.5	0.51～0.99	1.0
<2:1	1A	2A	3A	4A
（2:1）～（4:1）	1B	2B	3B	4B
>4:1	1C	2C	3C	4C

该分类法在分析和讨论核型进化的一个方面是有参考价值的，可作为核型表述的一项内容。

1.6.1.4 关于具小染色体的植物核型分析

所谓小染色体，是指其长度在 2 μm 以下而又不易分辨着丝粒的染色体。植物界中具此类染色体的种类居多，有的整个属或甚至整个科均有。以往，这类植物所提供的唯一细胞学信息就是染色体数目。为了扩大核型研究的范围，对这类植物提供比单一的数目更多一些有用的核型信息，初步拟定以下几个方面进行核型的分析和比较。

（1）染色体数目。

（2）具随体染色体的数目（如可见的话）。

（3）每对染色体的相对长度值。

（4）染色体长度比。

（5）如含有大小差别明显的染色体，可分大、小群分别统计其数量和长度，以及各自所占染色体组全长的百分比。

1.6.1.5 凭证标本

除一般栽培植物外，其他植物则无论是只报道染色体数目，还是进行核型或分带研究，均需有凭证标本，并注明学名、采集地、采集者、标本鉴定人、凭证标本号和标本存放地点等。

最后，需要加以说明的是，此次会议所讨论和约定的上述标准与当前国际上大多数学者所采用的标准是基本上一致。但是，由于国际上尚无一个共同标准，所以，以往和今后国外的有关文献，都有或可能有与本标准不一致之处，对此，应取辩证的灵活态度，如与本标准基本相符，为便于与原作者的研究进行比较，可按原作者的格式，如差异较大而又不合理者则摒弃而采用本标准。如为首次报道则希望用本标准。那种过分强调应与前人工作保持标准一致而不问其是否合理的态度，是不可取的。因为照此处理，统一标准便无法推行，各行其是的混乱局面将延续下去。事实上，核型研究的历史也表明，随着科学的发展，核型的概念、内容、计算的要求和着丝粒命名等，也都曾有过变动，使之更合理和更具科学性。我们也是本着这种精神商定本标准的，但是，不足或不妥当之处在所难免，希望全国从事植物染色体研究的工作者，通过自己的实践提出宝贵意见。

（作者：李懋学、陈瑞阳。原载：武汉植物学研究，1985，3（4）297-302）

1.6.2 植物染色体的半自动化核型分析法

本实验室建立的以数码照相为基础的半自动化植物核型分析技术，是当前植物核型分析的最佳方法，具有广泛的使用价值。

20世纪80年代中期，由实验室本科生叶晨同学设计的第一个植物核型分析软件，曾起过很大作用，被很多单位使用。但该软件当时配套的286机型，后虽经数学系骆家舜教授改编，提高了机型，但该软件依据的核型分析数据是由人工用圆规测量的染色体长短臂数据，再由人工输入计算机计算出染色体数据表，该软件比过去完全由人工计算进行核型分析提高了数十倍效率，但仍然非常消耗人力和时间，完成一种植物核型分析，从头到尾至少要1~2天时间，而且，最后给出的结果、核型图和染色体数据表是分开的，染色体图版需要人工制作，出书时还需要高分辨率扫描仪进行扫描制版。现在建立的半自动化核型分析，是以数码照相为基础，染色体数字采集借助人类染色体数据测量软件自动测量获得，在此基础上采用人工同源染色体配对，然后转换成由本实验室建立的植物核型分析软件计算，获得染色体数据表和核型分析的十二项核型指标。该方法除同源染色体的人工配对需要占用一定时间（占用时间长短由染色体数目和染色体标本质量决定）外，其他各项都是自动生成的，该方法给出的核型分析结果图和染色体数据表是一体的，可以直接用于出版社出书制版，省去了很多中间环节。应用本方法，一个人一天可完成数十个细胞的核型分析。又比原来的核型分析软件提高了数倍效率，具有广泛应用前景。具体实验步骤如下：

1. 染色体图像用CCD采集，直接输入计算机，可同时加上标尺。
2. 采用Karyotype核型分析软件对染色体进行分割、配对、数据测量并直接输出核型图版。

（8）测量染色体长度。

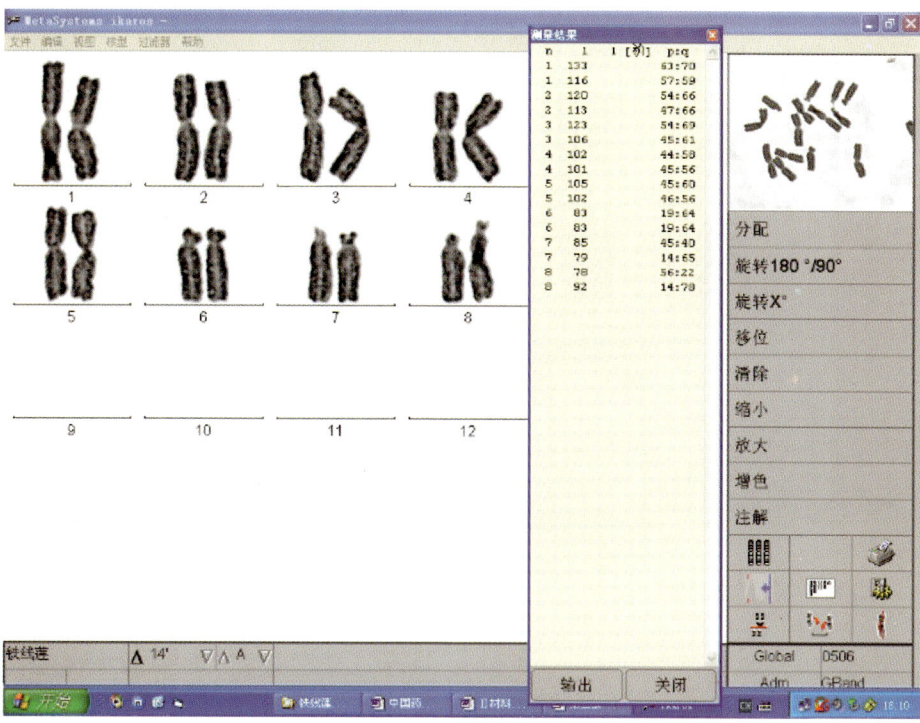

（9）根据染色体总长度和长短臂比例调整染色体配对和排列位置。

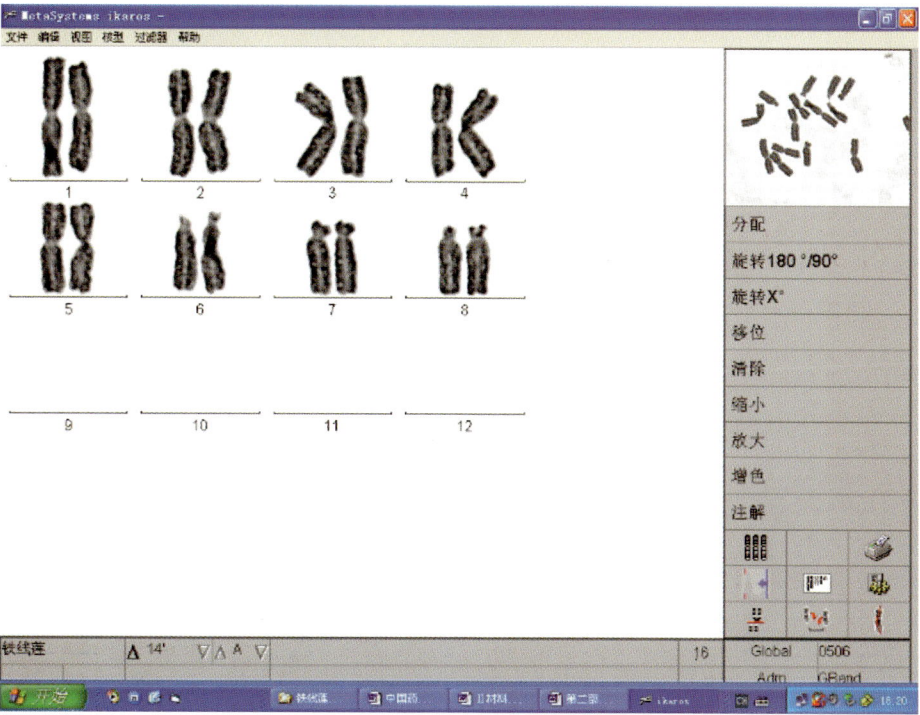

（10）测量染色体长度并输出。

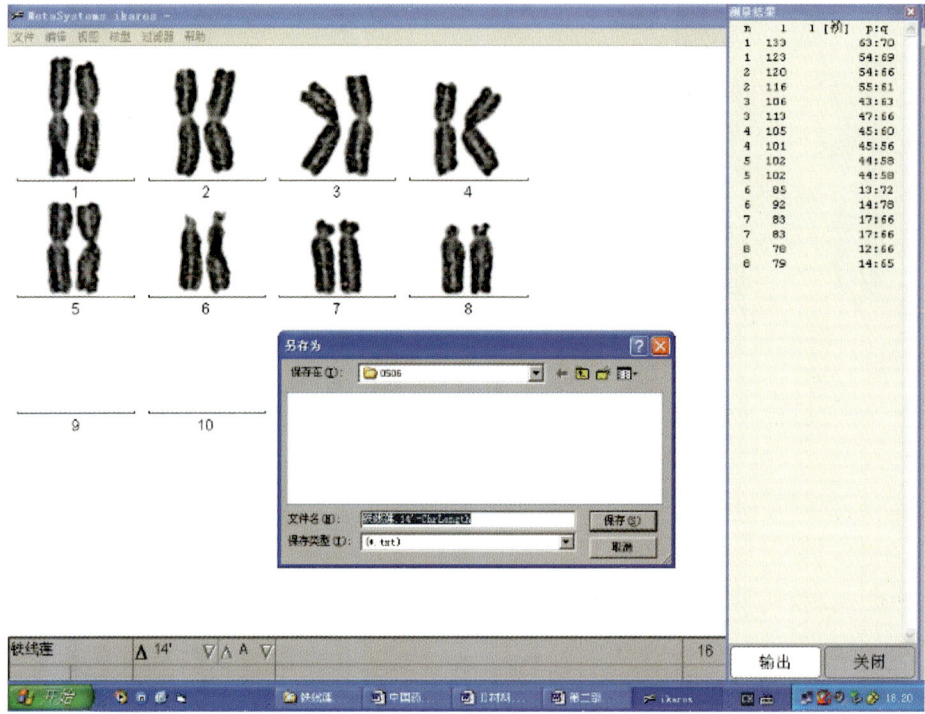

（11）输出核型分析图版。

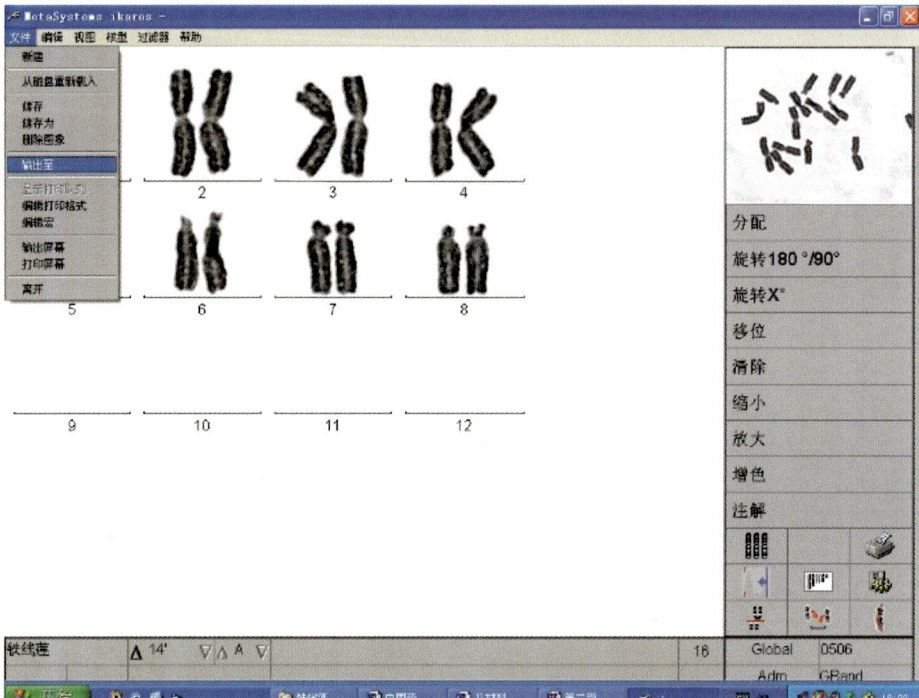

3. 采用自己设计的 NK-Karyotype 3.0 核型分析计算软件,计算并直接输出核型分析数据表。

（1）将输出的染色体长度数据文件在 Word 中编辑。

（2）运行 NK-Karyotype 3.0 核型分析计算软件,点击输入数据并调用编辑好的染色体长度数据文件。

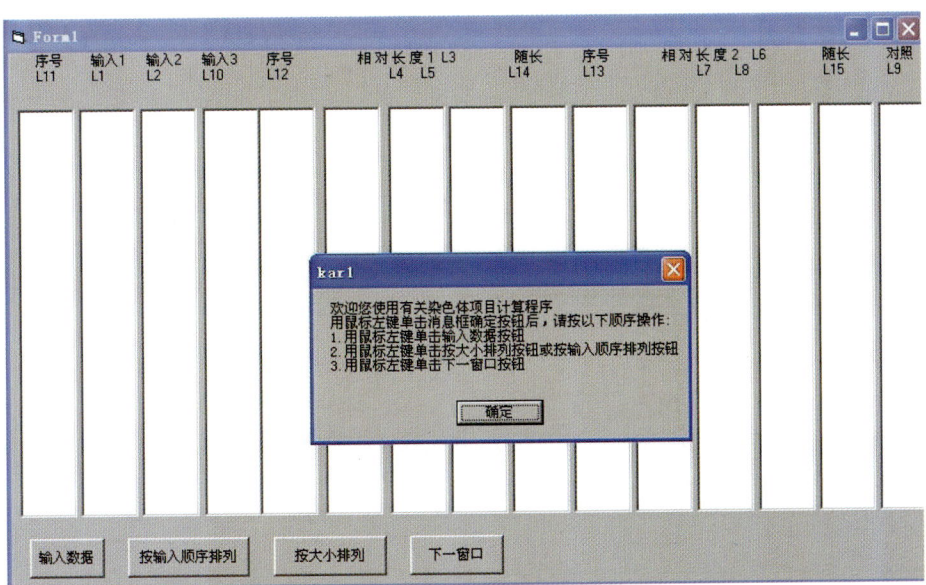

（3）点击"按输入顺序排列"或"按大小排列"按钮。

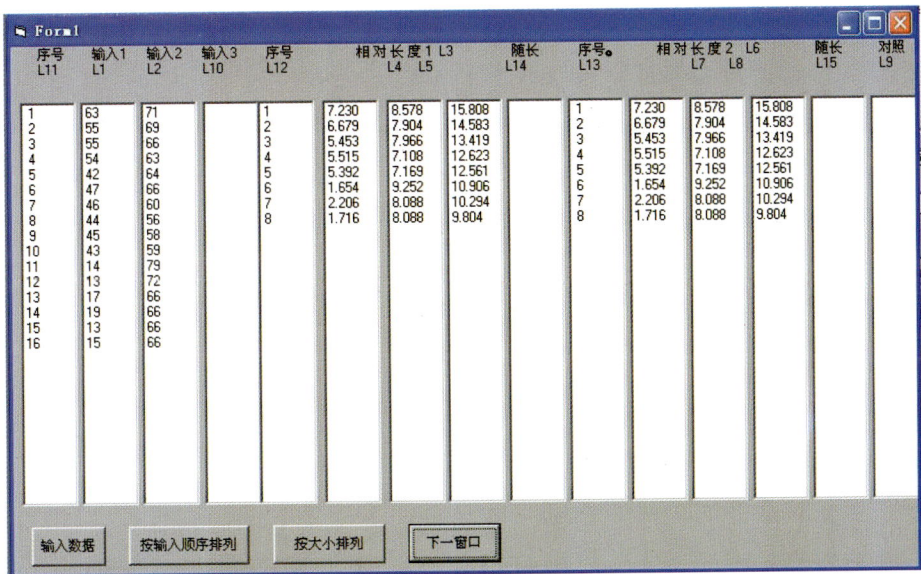

（4）点击"下一窗口"按钮，再点击"计算一"按钮。

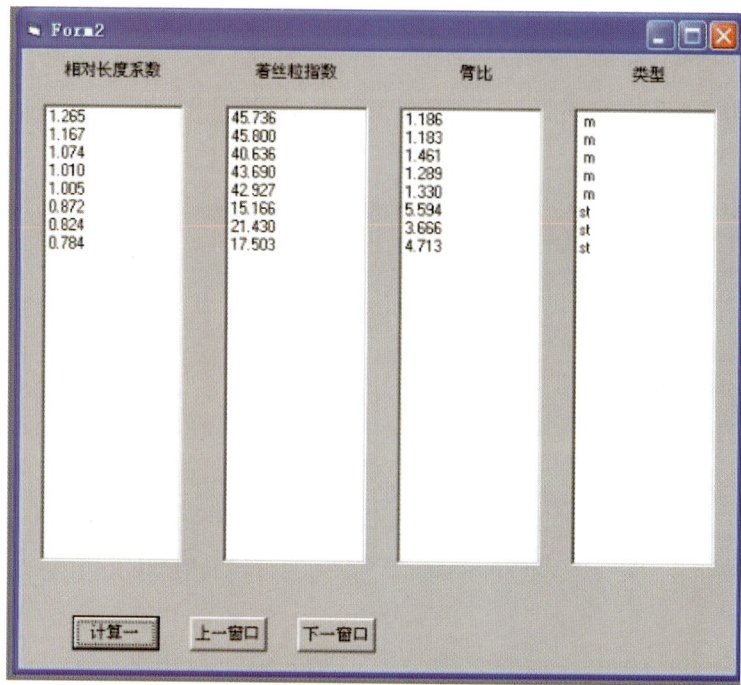

（5）点击"下一窗口"按钮，再点击"计算二"按钮。

（6）输入表编号、植物名及学名后，点击"输出结果到文本框并输入存放结果的文件名"按钮，将核型分析数据表输出到文件并保存。

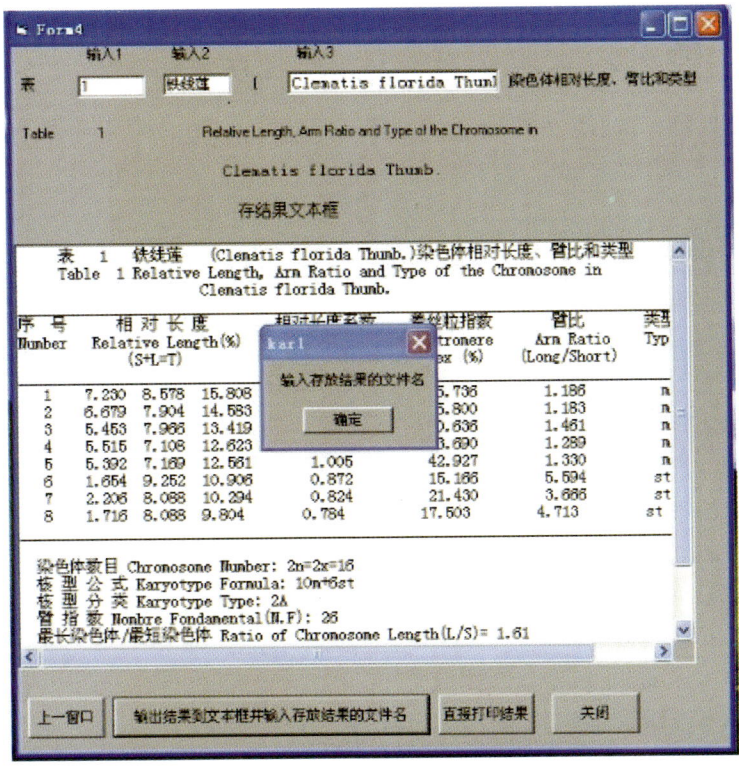

另外，以前的染色体资料大都是以黑白底片为基础的，存放在胶片袋中，查阅起来非常不便。现在可以通过底片透射扫描仪，把这些黑白底片上的图像转化为数字图像，存储于计算机中便于查阅和编辑，扫描得到的图像还可直接输入核型分析软件进行核型分析。在我们的工作中发现在扫描时采用彩色模式和较高的分辨率，扫描完成后再转化为灰度图像，这样得到的图像的效果较好。

（作者：陈成彬。原载：陈成彬博士论文，pp34-40）

1.6.3 应用同一细胞进行核型与带型的分析方法

1.6.3.1 方法步骤

1. 首先应用植物有丝分裂染色体标本制备新方法制备染色体标本。
2. 染色：以无钙、镁离子蒸馏水稀释 Giemsa 染液，20∶1 染色 30 min 左右。试验证明，金属离子，特别是 Ca^{2+}、Mg^{2+} 等离子对 Giemsa 显带有明显影响，所以，在第一次核型分析染色时，应尽量消除这些金属离子的影响。我们通常采取用蒸馏水或者在自来水中加少量 EDTA 以螯合金属离子。

3. 摄影：选取染色体分散良好处于早中期的细胞，做好标记，用油浸系物镜拍照。染色体长时间浸在香柏油中，对分带也有影响，可用水临时封片再拍照。如使用了香柏油，显微拍照后要立即用二甲苯将香柏油彻底冲洗干净，然后空气干燥。

4. 分带处理：空气干燥 48 h 至一周以内的标本，进行分带处理，下列条件可供参考：

A．C-带流程（以玉米为例）：

（1）贮存一周以内的染色体标本，经 0.1 M HCl 60℃处理 6 h 左右；
（2）自来水冲洗 3～5 min；
（3）蒸馏水冲洗 5 min；
（4）5% $Ba(OH)_2$ 45℃处理 8～10 min；
（5）45～50℃水冲洗至 $Ba(OH)_2$ 完全除掉；
（6）2×SSC 60～66℃ 30 min；
（7）蒸馏水冲洗 3～5 min；
（8）pH 5.8 磷酸缓冲溶液 20∶1 Giemsa 染色 40 min 左右；
（9）自来水冲洗。空气干燥。

B．N-带流程（以小麦为例）

（1）贮存一周以内的染色体标本；经 1 M NaH_2PO_4 ±1℃处理 8～10 min；
（2）50℃左右蒸馏水冲洗；
（3）自来水冲洗 30 min；
（4）pH 6.8 磷酸缓冲溶液 20∶1 Giemsa 染色 30～60 min；
（5）冲洗：自来水冲洗，并使其迅速干燥。

5. 带型拍照：镜检选取核型已经拍照的细胞，再进行带型拍照。将每条对应的染色体做好标记，分带处理染色体。形态轮廓变得不清，有些不显带的染色体难识别，采用此法，每对同源染色体都能一一识别（见图1.5）。

1.6.3.2 讨论

此方法成功的关键，作者认为有两方面：

1. 金属离子对显带的影响。试验表明，溶液中的钙、镁等金属离子对显带起着不良影响。因此，在显带前应排除这些离子，另外，香柏油中一些未知因素对显带亦有显著作用，应设法减少香柏油与染色体接触时间。

2. 分带的可重复性。当前植物染色体分带的主要问题是分带的可重复率较低。而解决这一问题的关键，作者认为在于染色体标本制备方法的改进，去壁低渗法比通常的压片法可以提高分带的可重复性已被证明。但去壁、低渗法中的火焰干燥处理"火候"常常不容易掌握，降低了分带的可重复性。为解决此问题，最近作者应用蒸气干燥法代替通常所使用的酒精灯火焰干燥法，取得了满意结果，解决了制片时"火候"的控制问题。方法是，在普通家庭做饭用的压力锅中加约 1/2 水，在电炉上煮沸，不加陀，让水蒸气从喷嘴中喷出，用这种高温蒸气流代替酒精灯火焰干燥制片，染色体分散良好。与酒精灯火焰干燥法无异样。此外，因每次从压力锅嘴中喷出的气流温度（根据染色体分散程度可以选择适当气流高度，近嘴处可达 90℃，一般在 60~80℃温度处即可）都是一定的。所以制片时每片受热程度基本是一致的。因此，分带可重复率高。此方法简单易行，是值得推广的一点技术改进。

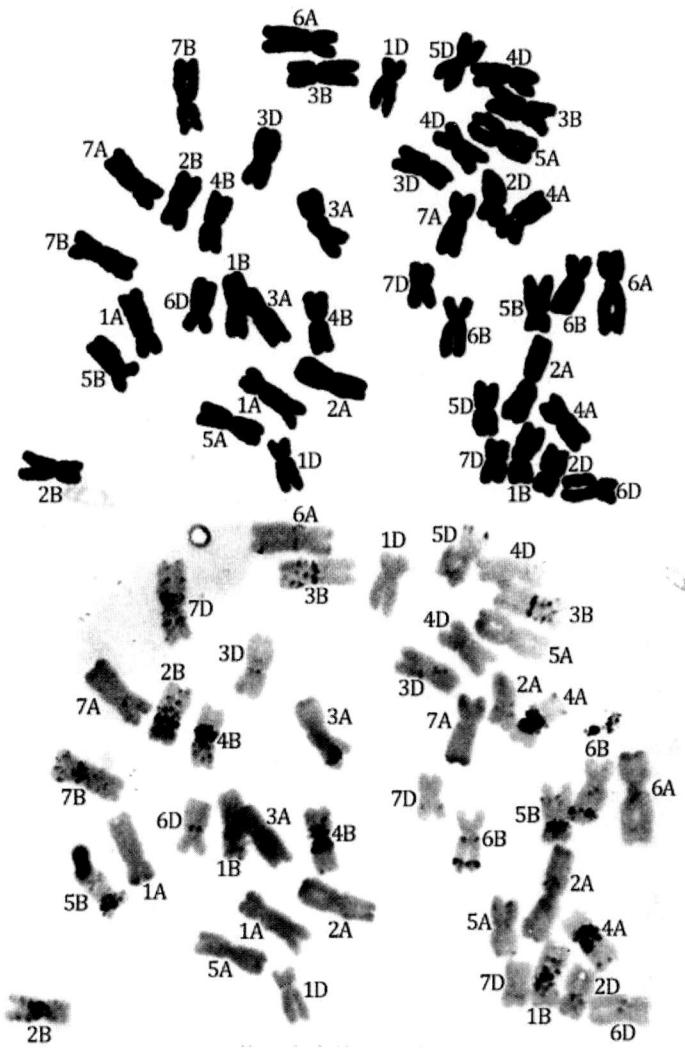

图 1.5 应用同一细胞进行分带处理的小麦
Fig. 1.5 The *Triticum aestivum* cv. Chinese spring of Banding Analysis by Using the Same Cell

（作者：陈瑞阳、宋文芹、李秀兰。原载：武汉植物学研究，1985, 3（4）457-459.）

1.6.4 植物染色体分子核型技术研究

核型分析是细胞遗传学、染色体工程学、细胞分类学等学科的基本研究方法。细胞遗传学和分子细胞遗传学的建立和发展，都是以染色体研究技术的创新和应用为其先导的。常规核型分析所用的数据均来源于体细胞染色体在普通光学显微镜下可测定的表型特征，信息量极其有限，特别是对于基因组内染色体形态相近的物种以及基因组间分化程度较小的物种之间，这种核型分析无法提供更多有效而可靠的信息。荧光原位杂交（Fluorescence In Situ Hybridization，FISH）技术和荧光分带（fluorochrome banding）技术可以使染色体呈现特定的杂交信号和带型，为识别和分析染色体提供标记。由于这些标记都是基于染色体特定位点区域客观的 DNA 序列或结构组成特点，使其比来源于染色体形态的测量数据更为准确而可靠。无疑，这种加入了特定位点信息的染色体研究，已经广泛应用于细胞遗传学领域，使核型分析技术进入了新的时代。1997 年，Brown 等在利用 FISH 技术对 2 种云杉（*Picea*）的基因组结构的研究中，使用了"分子核型（molecular karyotype）"这一概念。刁英（2004）也在其综述中把这种将分子生物学技术与传统的细胞生物学技术相结合的新兴技术概括为"分子核型"由于同样是应用于染色体研究但是主要基于脉冲场凝胶电泳（Pulsed-Field Gel Electrophoresis，PFGE）技术的凝胶核型（gel karyotype）已经使用了"分子核型"这一概念，为与其相区别，将以荧光原位杂交技术及荧光分带技术为基础的核型分析方法全称为"染色体分子核型"现对近年来各种染色体分子核型的研究进展及其在植物细胞遗传学领域的应用进行了综述。

1.6.4.1 以荧光原位杂交为基础的染色体分子核型

20 世纪 80 年代出现的以非放射性荧光物标记探针的原位杂交技术（Langer，1981），以其简便快捷、灵敏度高、相比放射性标记探针具有良好的安全性，并可通过多色标记探针进行多种序列同时定位的特点，逾 30 年来取得了长足的进展和广泛的应用。在植物分子细胞遗传学研究领域，以 FISH 技术为基础发展起来的多色荧光原位杂交（multicolor Fluorescence In Situ Hybridization，mFISH）、基因组原位杂交（Genomic In Situ Hybridization，GISH）和染色体绘染（Chromosome Painting，CP）技术都已成功应用于植物染色体分子核型研究。第一种方法主要用于分辨同一基因组内不同染色体，后两种方法则多用于辨别不同基因组来源的染色体，以下将结合应用实例对这三种方法分别加以介绍。

1. 多色荧光原位杂交（mFISH）

许多植物种类的染色体或者很大，或者很小，且彼此间常具有非常相似的形态特征。这就必须通过若干染色体标记的辅助才能对染色体进行确定无误的辨认，包括同源染色体的正确配对和区别同一基因组内各条非同源染色体。对基因组内各条染色体的正确识别是研究染色体进化和变异、染色体行为、基因组序列定位以及相近植物种属分类和系统进化的重要前提。利用多色荧光原位杂交技术在染色体上呈现的杂交信号，不仅可以辅助识别染色体，并且那些染色体或染色体区域特异性的位点还能够作为以上所提到的后续研究的重要参考标记。

应用多色荧光原位杂交技术构建的染色体分子核型也称为 FISH 核型（FISH karyotype），一般用多种不同的串联重复序列作为探针。探针的选择主要依据其在不同物种基因组内信号的数目和在染色体上的分布模式。在植物中应用最普遍的是在各植物种类间高度保守的基因家族：核糖体 RNA 基因（rDNA）包括 18S-5.8S-25S rDNA（简称 45S rDNA），或其转录单元内的一部分（如 25S rDNA），以及 5S rDNA，此外还有端粒序，端粒相关序列（telomere-associated sequences，TAS 和着丝粒区域特异性重复序列。通过以上 2 种或多种探针的组合，使很多植物染色体的准确辨认成为可能。如棉花（*Gossypium mustlinum*）（Wu，2013）、牛尾菜（*Smilax rufescens*（Pizzaia，2013）、柳树（*Salix viminalis*）（Nemeth，2013）、百合（*Lilium*）（Sultana，2013）等的部分植物。尽管在这些植物中有些并未能将所有染色体逐一区分，但已经大大提高了核型分析的精细程度，杂交信号作为染色体标记已经为研究这些植物的多倍体组成、系统发育等提供了重要信息。为了更好的区分各条染色体许多研究者针对特定植物种类筛选出位于其他染色体区域的卫星 DNA 序列（stellite DNA），为分子核型提供更多的标记。Dou 等（2009）利用一个微卫星（micro-satellite）AAG 和 Afa-family 重复序列，结合基因组原位杂交技术（GISH）分析了披碱草属（Elymus nutans）3 个基因组 St、H、Y 的亲缘和进化关系。课题组曾利用 25S rDNA、5S rDNA、LPD 序列，端粒序列克隆 LTA 14 和弥散型重复序列克隆 LRA427 等 5 种探针分组混合后（Liu，2006，2007），分 2 次与相同的细胞进行多色荧光原位杂交。各种探针在不同染色体上产生了不同排列组合的标记模式。使几乎全部的染色体可以被区分。首次将组合标记探针方法应用于落叶松属（*Larix*）的分子核型的构建（图 1.6，图 1.7）。

构建多色 FISH 核型，除选择串联重复序列作为探针，另一个策略则是以多个低拷贝序列为探针，低拷贝序列在基因组中一般只有 1 个或几个位点，更可能作为染色体特异性标记；此外，低拷贝序列也为探针的选择提供了丰富的资源（Kim，2002）。为了产生可辨认的 FISH 信号这些低拷贝序列探针一般为带有较大插入片段的克隆（Dong，2000）。Shibata 等（2013）从向日葵（*Helianthus annuus* L.）的栽培种 cv.HA89 等 2 个基因组文库中筛选出 44 个可以产生特定杂交位点的克隆（24 BACs，20BIBACs），并利用这些杂交信号位点将这些克隆划分为 18 个连锁群，构建了传统 RFLP 和 SSR 分子标记遗传图谱所无法区分的 4 个连锁群成功整合了遗传连锁图谱和物理图谱。

对于多色 FISH 核型，除了探针的选择，另一个重要问题是如何在同一染色体标本上同时呈现多种不同颜色的标记。为了扩充探针标记容量，出现了组合标记（combinatorial labeling）探针和比例标记（ratio-labeling）探针。前者是将 1 种探针用 2 种或以上可产生不同颜色的标记物进行标记，原则上可标记最多达 $2n-1$（n 为不同标记物的数目）种探针（Ried，1992）后者是将不同标记物以一定比例标记同一种探针，标记容量进一步增加（Nederlof，1992）。但是，由于荧光染料本身和显微镜技术的限制，后一种标记技术尚存在分辨率低等缺点，因而，未得到广泛的应用（Levsky，2003）。在植物分子核型研究中，Hasterok 等（2002）将组合标记探针用于黑麦、小麦等多种植物染色体的识别，可一次性同时清楚地检测 3 种标记。课题组也将此方法用于构建落叶松属（*Larix*）、杨属（*Populus*）的分子核型（Liu，Dong，2007）。不过，在多数研究中，对 2 种以上序列位点的检测仍多是通过重复杂交（Lengerova，2004）、FISH 信号的直接检测和间接检测相结合（Vischi，2003）等策略实现的。

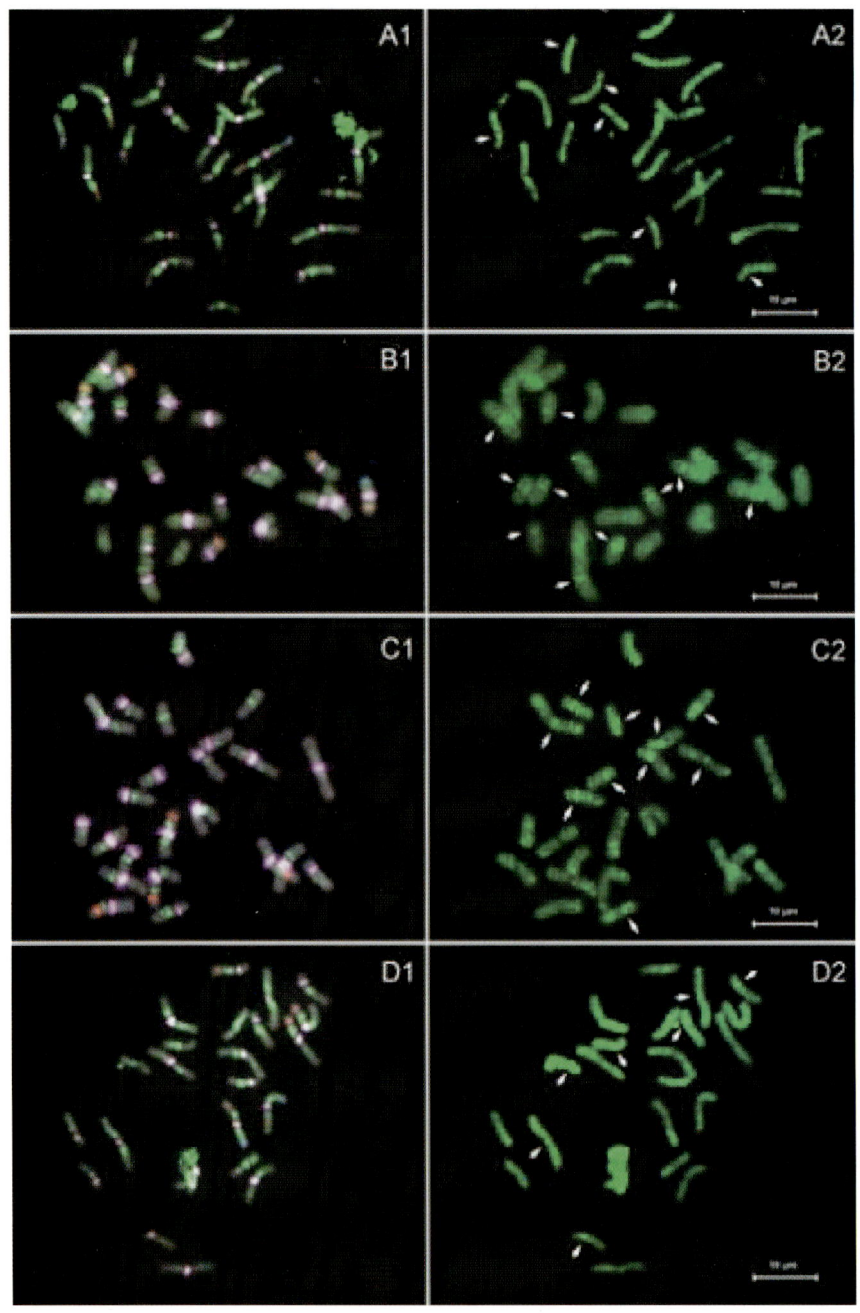

图 1.6 四种落叶松属植物染色体多色 FISH 结果

Fig. 1.6 Multi-color FISH on metaphase chromosomes of four Larix spp.

A：兴安落叶松；B：华北落叶松；C：日本落叶松；D：长白落叶松；1：染色体与 25S rDNA
（红色）、5S rDNA（蓝色）、LPD 序列（紫色）和 LTA14（绿色）探针信号的合成图像；2：探针 LRA427 的信号图像，
箭头示特异信号位点

A：L. gmelinii；B：L. principis-rupprechtii；C：L. kaempferi；D：L. olgensis；1：chromosome spreads probed with 25S rDNA
（red signals），5S rDNA（blue），LPD sequences（magenta）and LTA14（green）；2：signal image produced by LRA426 probe，
and the arrows show specific hybridization locations

(1) 将染色体图像输入 Karyotype 核型分析软件。

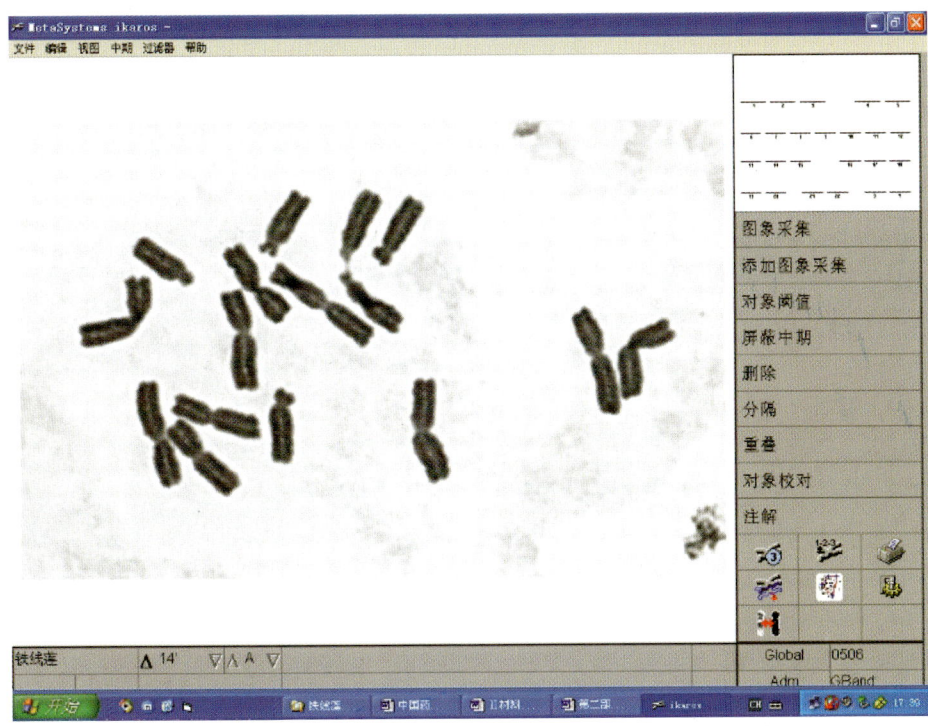

(2) 调整对象阈值，去除图像背景。

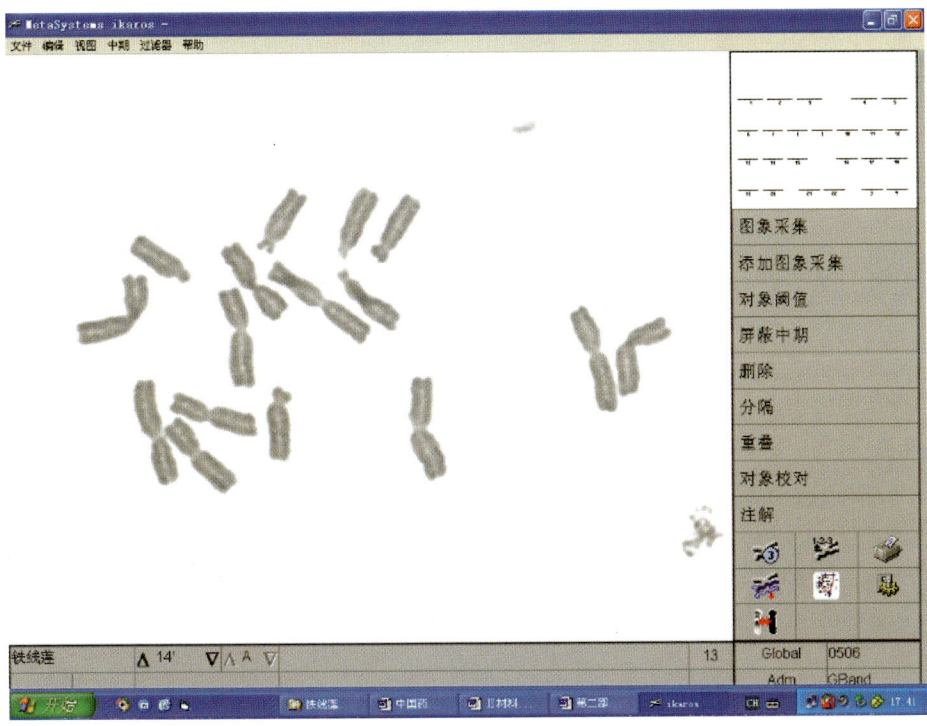

(3) 对象校对，分割染色体。

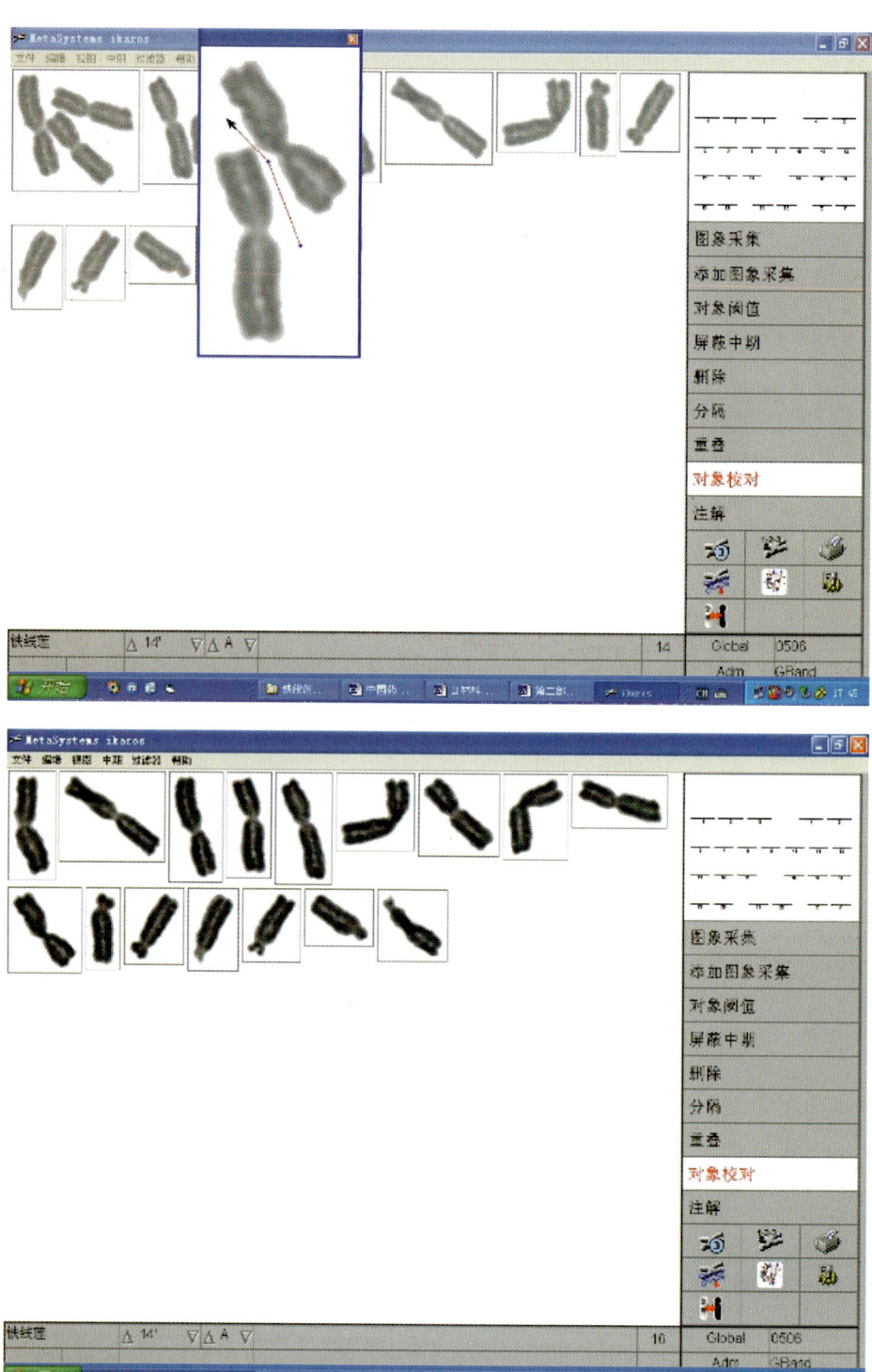

(4) 分配染色体。

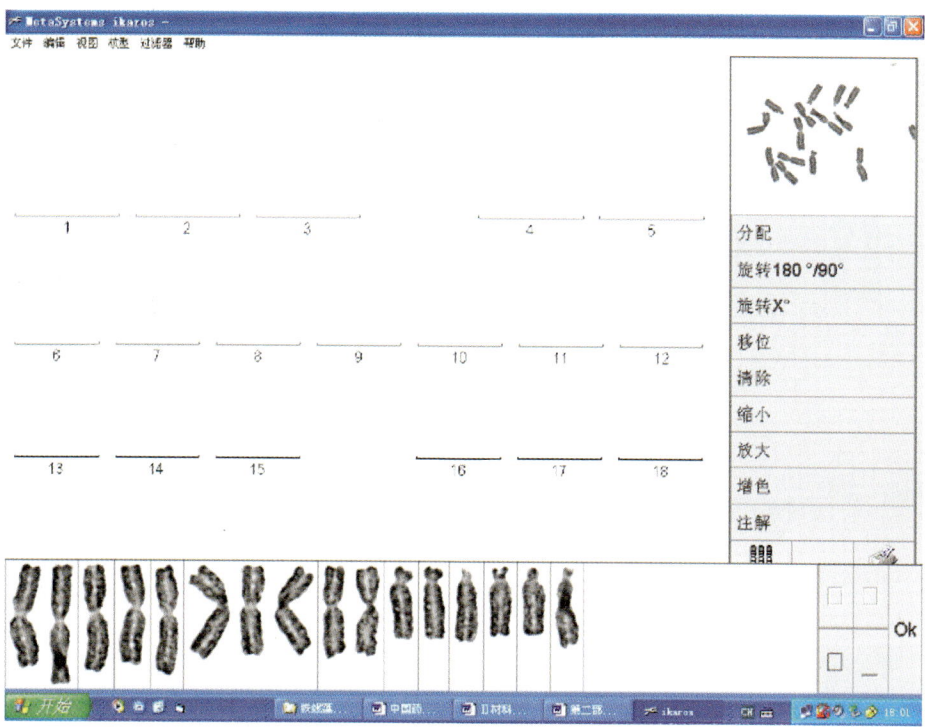

(5) 选择核型表，根据染色体数选择合适的核型表。

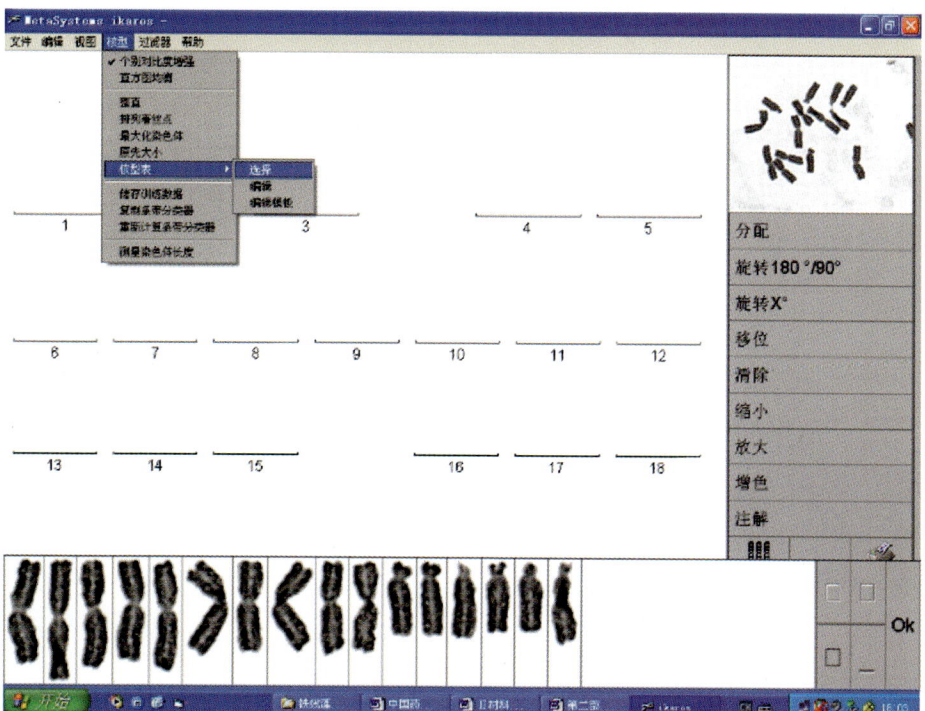

（6）在数字上方双击鼠标，染色体自动排列到数字上方。

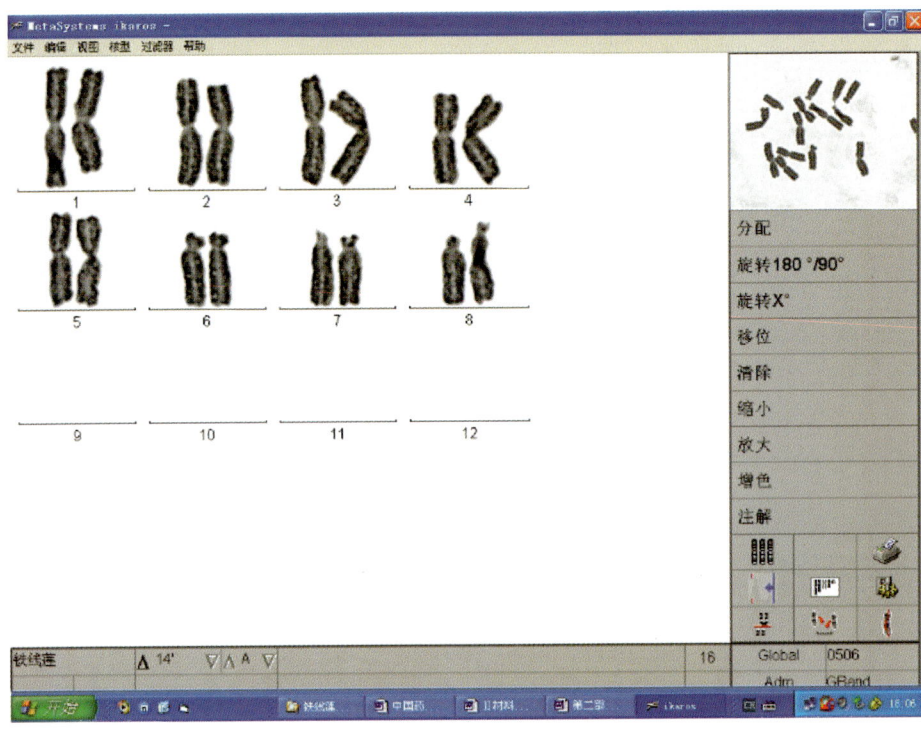

（7）排列着丝点位置，用鼠标点击红色线条调整到着丝点位置。

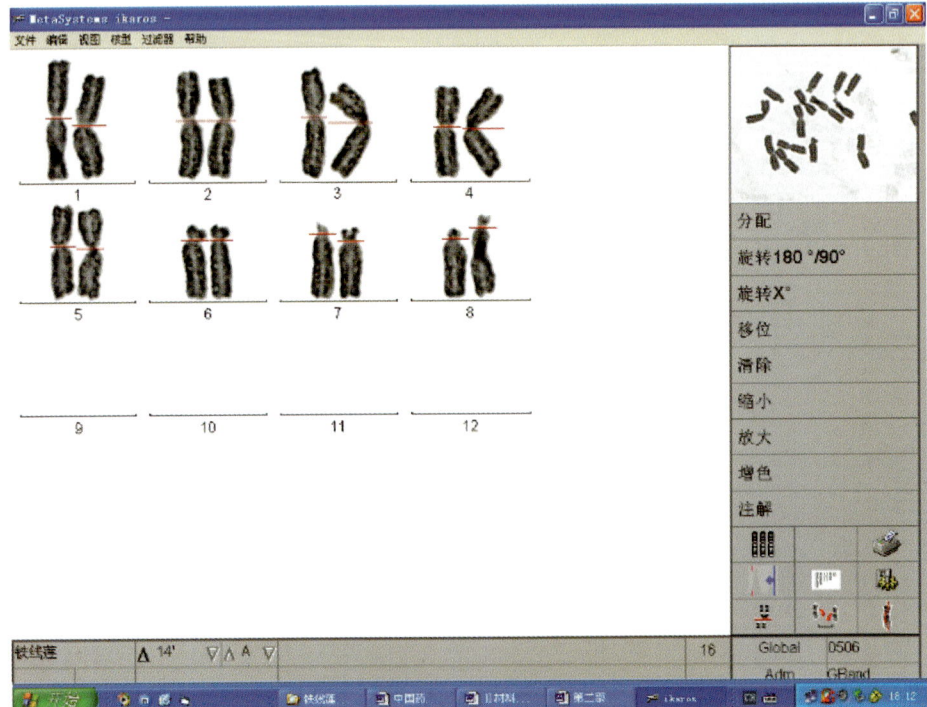

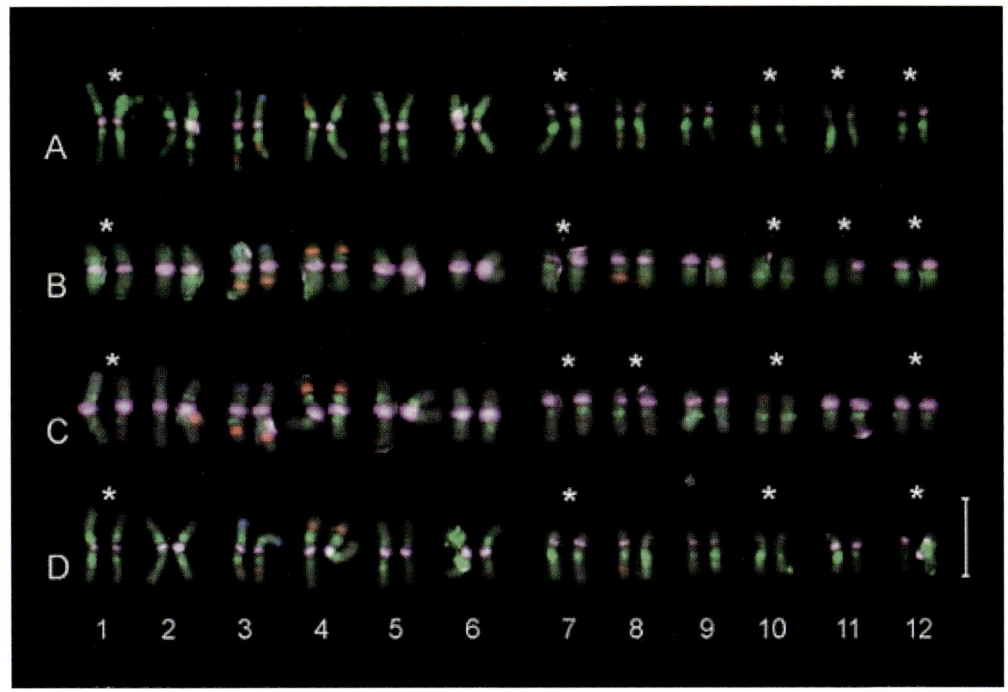

图 1.7 4 种落叶松属植物的分子核型
Fig. 1.7 Four kinds of Larix spp. Molecular karyotype
A：兴安落叶松；B：华北落叶松；C：日本落叶松；D：长白落叶松；
星号示具有 LRA427 特异性位点的染色体对；比例尺：10 μm
A：L. gmelinii；B：L. principis-rupprechtii；C：L. kaempferi；D：L. olgensis；asterisks show chromosome pairs bearing specific LRA426 locus；bar = 10 μm.

2. 基因组原位杂交

尽管多色 FISH 核型已广泛应用于多种植物，而最早应用荧光标记辨别植物染色体的方法却是基因组原位杂交（Schwarzacher，1989）。这种方法可以直接分辨种间杂种（interspecific hybrids）、异源多倍体和种间渐渗系（interspecific introgression lines）中来自不同亲本基因组的染色体，并对染色体的组成和变异进行分析（Neelam，2011）。

婆罗门参（Tragopogon miscellu）是一个自然形成的异源四倍体，距今已传 40 余代，但是细胞生物学研究证明，在其进化过程中子代存在持久的染色体变异事件。Chester 等（2012）以二倍体亲本 T.pratensis 和 T.dubius 基因组为探针的 GISH 发现子代 76% 个体具有不同情况的染色体易位，69% 个体为非整倍体，并阐述了多倍体进化过程中出现的染色体重排现象。此外，芸薹属（Brassica）的 3 个异源四倍体：B.carinata、B.juncea、B.napus 和它们的 3 个二倍体祖先种（B.nigra、B.oleracea、B.campestris）组成了一个非常适用于研究多倍体的模式系统 Maluszynska 等（2005）通过以 B.nigra 和 B.campestris 的基因组为探针的双色 GISH 结合 25S 和 5S rDNA 标记，成功区分了 B.juncea 中的 A、B、基因组，并分别标记出了 36 条染色体中的 28 条。小麦作为异源多倍体，也有许多研究者应用 GISH 技术对其和其他小麦族成员的基因组进行研究。Han 等（2003，2004）应用改进的多色 GISH 技术，清楚地将组成六倍体小麦基因组的 A、B、和 D 基因组区分开。在该技术中，Han 等用荧光素标记 *Thinopyrum intermedium* 和 *Triticum urartu* 的基因组总 DNA，用生物素标记 *Aegilops tauschii* 的基因组总 DNA，*Aegilop speltoides* 的基因组 DNA 作为封阻。这种多色 GISH 技术为研究异源多倍体植物基因组结构与变异提供了重要的技术基础。

但是，尽管 Han 等在区分小麦不同基因组来源的染色体研究中取得了成功，更多像小麦一样，其亲本基因组的亲缘关系非常相近的多倍体，则很难通过 GISH 区分不同来源的染色体。因此，GISH 技术有其无法避免的局限性。

另一方面，Zoller 等（2001）在用亲缘关系较远的植物基因组 DNA 作为探针，进行基因组原位杂交试验中获得了另一种效果。针对小麦、黑麦、大麦、洋葱等单子叶植物和蚕豆、甜菜等双子叶植物，Zoller 等（2001）均采用拟南芥的基因组 DNA 作为 GISH 探针。杂交信号位于核仁组织区（Nucleolus Organizing Regions，NORs）、含卫星 DNA 的 N 带区域，及其他含有与拟南芥同源的保守序列区域，如端粒。Zoller 等（2001）的试验为构建多种不同植物的分子核型提供了新的策略。其优点在于拟南芥基因组探针的通用性，使研究者无需再针对某一植物单独筛选探针，而且杂交信号的强度和可重复性良好，无需进行信号的放大，也无需进行封阻。

3. 染色体绘染

"染色体绘染"（Chromosome Paintin，CP）这一术语最早由 Pinkel 等（1988）提出，用以概括利用 FISH 将一个染色体组中特定的染色体或染色体片段进行原位呈现的技术。虽然此后 CP 在动物和人类细胞遗传学研究中得到广泛应用，在植物中却很难获得成功。这很可能是由于植物基因组中含有大量弥散分布于各条染色体的重复序列的原因。尽管如此，一些研究者还是通过改进研究策略取得了一定的突破。与哺乳动物中以单、低拷贝序列作为 CP 探针的策略相反，针对植物中富含重复序列这一特点，Houben 等（2001）富集了染色体特异的重复序列作为探针，较为成功的标记了黑麦（*Secale cereale*）、*Allium schoenoprasum*、*Brachycome dichromosomatic* 的 B 染色体。不过在 A 染色体上仍有较弱的弥散信号无法消除。Yu 等（2013）利用以来自于 Cot-1 文库的 11 个串联重复序列为探针产生的多色荧光杂交信号，准确清晰的区分出四倍体苜蓿（*Medicago sativa* L.）的 2 个染色体组，并结合常规染色体测量数据，成功构建了分子核型。Lysak 等（2003）利用含染色体特异序列的 BAC 克隆，标记出了拟南芥的 1、2 和 4 号染色体，这种标记使这些染色体在减数分裂和有丝分裂各个时期都可以被分辨，从而可以进行不同细胞周期染色体行为的研究。Lysak 等的成功的一个重要原因是拟南芥的基因组较小所含重复序相对很少。此外，在菜豆（Almeida，2013）、番茄（Budiman，2004）、高粱（Islam，2002）、和马铃薯（Dong，2000）中，也利用染色体特异的 BAC 克隆，使部分染色体得以分辨，但是仍需要加入少量的 Cot I DNA 以抑制重复序列的杂交背景，在含有大型基因组的植物中，因所含重复序列显著增加，则需要改变探针选择的策略。Zhang 等（2004）为区分六倍体小麦中 3 个不同来源的染色体组，分别从 A 组和 D 组供体种的 BAC 克隆文库中筛选出若干含有基因组特异重复序列的克隆，这些克隆所产生的 CP 信号弥散于其所特异的染色体组的所有染色体上，因此可用于检测形成异源多倍体后染色体的易位情况。但是也因其供体种的亲缘关系较近的原因使得 Zhang

（2004）等无法通过 GISH 进行异源多倍体和渐渗系的染色体分辨的研究。

1.6.4.2　荧光分带技术（fluorochrome banding）在分子核型中的应用

与开创荧光分带技术的 Q 带技术不同，现今的植物染色体分析更多的是采用色霉素 A3，CMA）、4',6-二脒-2 苯基吲哚(4',6-diamidino-2-phenylindole，DAPI)、Hoechst33258 等荧光染料。尤其是 DAPI 由于作为 FISH 试验的复染剂被广泛使用，且能在很多植物染色体的特定区域呈现清晰的阳性或阴性带型（Nakamura，2001），DAPI 带型常常被作为分子核型的重要标记。尽管 Brown、刁英等在对分子核型的概括中并未提到荧光分带技术，但因其作用机理与 DNA 分子的碱基组成相关（Liu 2004），更重要的是其在许多植物染色体分析中，都能提供一定的染色体标记信息，所以，将该技术也列入到染色体分子核型技术中，以作为 FISH 技术的重要补充。

在一些基因组分化程度高、染色体变异较大的植物分类群中，甚至仅靠荧光分带就可以完成分子核型分析。Yurkevich 等（2013）利用 DAPI 能够区分所有 *Adenolnum* 的全部染色体，证实了该物种染色体模式的相似性，并在此基础上结合 Ag-NOR、FISH 及 RAPD 分析发现 *Adenolnum* 与 *Linum* 其他种存在较远的进化关系；Urdampilleta 等（2013）运用 CMA 和 DAPI 分带检测到地铃属（*Cardiospermum*）种间染色体的变异，并结合 FISH 定位技术进行了系统发育和进化分析；Karlov 等（2003）根据 DAPI 阳性带的有无识别了蛇麻草（*Humulus lupulus*）中的性染色体，结合 rDNA 定位分析，构建了该种的分子核型；Liu（2004）等将 DAPI 染色与三维图像软件分析手段相结合，在大麦、水稻、小麦、*Lilium brownii* 和蚕豆中构建了稳定的 DAPI 带型，可用于识别特定的染色体和染色体区域。如果这一技术能够进一步应用于更多的植物种类，将以荧光分带技术为基础的染色体分析迈上一个新的台阶，在染色体的结构分析和基因的物理定位研究中将具有广泛的应用价值。

1.6.4.3　结论与展望

综上所述，根据作用机理、研究的广泛性与实用性，将 mFISH、GISH、CP 和荧光分带这些技术手段都包括进构建染色体分子核型的技术范畴。这些细胞遗传学技术的发展为植物基因组研究提供了新的工具，可以获得更精确的信息研究染色体结构和行为。越来越多植物基因组测序的完成及 BAC 克隆文库的构建，将为选择基因组特异探针、染色体或染色体片段特异的探针提供更为丰富的资源，从而使染色体分子核型技术在植物基因组功能的研究领域中发挥更大的作用。

（作者：张力鹏、刘博、陈成彬。原载：北方园艺，2014（8）179-183）

第七节　植物染色体分离与 PCR 扩增方法

提要： 染色体显微切割（chromosome microdissection）技术是细胞遗传学与分子遗传学相结合的一项桥梁技术。染色体显微切割与微克隆技术是指在显微操作条件下对特定染色体进行显微切割、分离，并通过构建特定染色体或区域特异性 DNA 或 cDNA 文库以进行目的基因的分离克隆方法。本课题组从 1997 年开始对植物微切和微克隆的技术进行探究，由于植物染色体比较小，通常都需要在油镜镜下进行观察，但是，由于显微镜的限制，用于染色体显微切割的方法是通过玻璃针或激光束对染色体进行切割，这两种方法都是借助于安装在倒置显微镜上的显微操作仪或激光发生器进行的，倒置显微镜一般只能使用低倍（20×，40×）干燥物镜进行切割，本实验室发明了一种简易的微分离染色体的装置，主要是通过升高物镜前焦点平面的高度，利用自制的玻璃台架，实现了从 100× 油浸物镜下进行显微切割，大大提高了染色体显微切割的精度和准确度，可以分离任一种植物基因组中的任一条染色体，解决了植物染色体的微切微分离技术难点，利用该技术分离了多种植物的染色体。

1.7.1　植物染色体的分离方法

提要： 本试验室利用改进的普通光学显微镜进行染色体显微切割的方法。先使显微镜物镜焦点平面升高约 7 mm，将制在盖玻片上的染色体标本反扣在自制的玻璃台架上，从物镜下方进行染色体切割，这样就解决了在 100× 油浸物镜下不能切割染色体的困难，实现了在高放大倍数（放大率 1000×~1500×）条件下进行染色体显微切割的问题，大大提高了显微切割的精度和准确度。具体改进方法如下。

1.7.1.1　染色体标本的制备

利用改进的去壁低渗法制备植物染色体标本，但材料不是制在载玻片上，而是制在 50 mm×24 mm 的盖玻片上。

1.7.1.2　显微切割玻璃针的制备

取外经 0.6 mm、长 10 cm 的实芯毛细玻璃管，用竖直式拉制器（Narishi gepp83）拉成 0.5~1 μm 左右的微细玻璃针，并将玻璃针先端弯曲成 120 度，保存备用。显微切割台架的制作如图 1.8 所示，取 7 mm 厚的玻璃裁成 25 mm×75 mm 两条，在两条玻璃的左端上方用一张载玻片加白乳胶粘合起来，形成一个中空的正方形或演台架（575 m）（图 1.8（B））

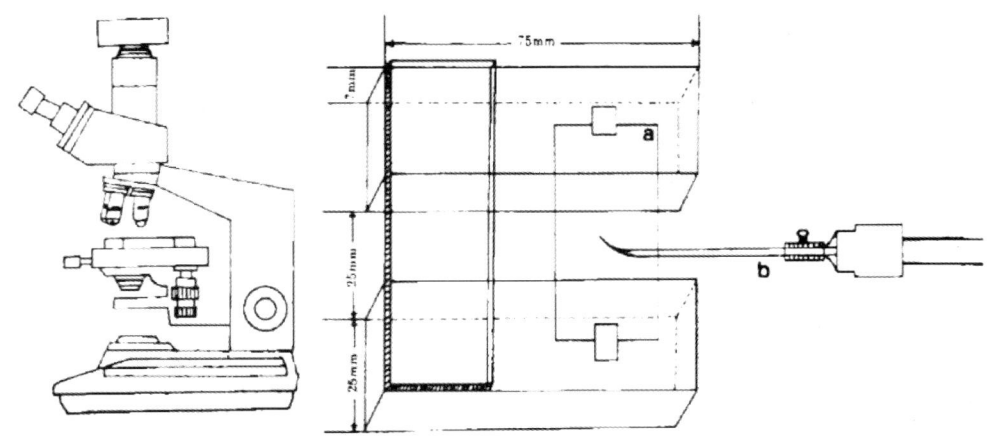

图 1.8　显微切割台架的制作
A. 玻璃台架在显微镜镜台上的位置；B. 自制的玻璃台架

Fig. 1.8　A. The lecation of self-made glass frame on the microscopestage；B. self-made glass frame
a：盖玻片标本　specimen of coverslip；b. 切割时的玻璃针　glass needle of microdissection

1.7.1.3　显微操作器底座的改装

显微操作器的进给量一般都是以 μm 级计算，对每次切割后重新调整微切针的位置很不方便，作者将日本产 NARISHIGE 被压转动式显微操作仪的臂手按装在一台旧显微镜的镜台上，借助显微镜的粗螺旋进行升降调节，以镜台推进器作前、后左、右移动，这样在毫米级进给量条件下，很容易移动和调整微切针的位置，从更换一个新微切针到调整到合适的微切位点只需要数十秒，而且准确、操作方便。

1.7.1.4　显微镜焦点的重新调整

当使用图 1.8 显微切割玻璃台架时，需将显微镜物镜前焦点平面与显微镜镜台之间的距离加大约 7 mm，即用显微镜粗螺旋降低显微镜镜台或升高镜架，重新调准显微镜焦点。

1.7.1.5　染色体显微切割操作

将制好的染色体盖玻片标本，材料面向下反扣在自制的玻璃台架上并用胶布粘牢两端，先在低倍物镜下找到待切割的细胞，然后分三步调节安装在显微操作器上的玻璃针到位。第一步：用粗螺旋升降调节显微操作器至适当高度；第二步：用改装的显微镜镜台推进器 X、Y 轴调节显微操作器微切针到被切割染色体的大约位置；第三步：操纵显微操作器的微调螺旋杆，使玻璃针对准待切的染色体部位，然后再换成 100×油镜，在目镜视野的监视下进行显微切割。

1.7.1.6　染色体显微切割中应注意的事项

（1）显微切割使用的容器、试剂及整个操作过程必须在无菌条件下进行，以免切取的染色体被其他 DNA 污染。

（2）显微切割过程中应避免震动，在放大 1000× 以上切割时主要问题是防止震动，显微切割的仪器最好按装在一楼或防震台上。

（3）为防止收集过程中染色体的脱落，玻璃针可不经硅化，未经硅化的玻璃针与被切取的染色体粘附的比较牢，不易脱落。在完成切割后，将染色体与玻璃针一同折断在微量离心管的底部，密封好，放入低温保存备用，

本研究利用普通光学显微镜代替普通倒置显微镜，通过自制的玻璃台架和改装的显微操作仪，在 100× 油浸物镜下进行显微切割，可使染色体的识别与切割同步进行，不仅可以分离任一种植物基因组中的任一条染色体，而且可使切割片段缩小到 0.3 μm 以下。Fukui（1992）报道应用美国 Meridian 公司生产的激光共聚焦扫描显微镜（ACAS470），切割大麦、水稻染色体只获得 0.5~1 μm 大小的片段。本研究建立的方法为染色体显微切割技术的普及和应用提供了一套简便、快速、实用的方法，在基因组研究中具有重要的应用前景。

（作者：李秀兰、宋文芹、许文胜、陈瑞阳。原载：南开大学学报，1998，31（2）102-105）

1.7.2 植物染色体 G-带特异区聚合酶链反应技术的初步研究

The typical procedure of isolating and studying a gene begine with the study on the gene product. However, many genes have no products have not been found. Therefore, because of the lack of markers, it is difficult to isolate them by current strategies such as chromosome walking（Cross and Lettle, 1986）even though the linkage location has been determined by genetic linkage studies. RFLP is a kind of useful marker in chromosome walking, but they are too sparse to saturate chromosomes. Microdissction and microcloning can provide a large a mount of markers coming from a specific chromosome region. The DNA fragments obtained by this method can also be used in other studies, e.g. in the study of the mechanism of banding and fragile sites.

The pioneer of this method is Scalenghe（1981）. He microdissected the polytene chromosomes of Drosophila and cloned the dissected segments into a bacteria phage vector in1981.Later, this method was applied to mammalian metaphase chromosomes(Rohme et al., 1984) and to haman chromosomes（Bates, 1986）. Ludecke et al., （1989）used PCR amplification to make this method more practical. Recently, several other PCR methods have been developed, such as nonspecific primered PCR（Cedric et al., 1989）and PCR through attached adaptors(Johnson, 1990). On the other hand, the use of laser beam made this method more precise and quick（monajembashi et al., 1986）.

However, the microdissection and microcloning technique so far has rarely been applied to plant chromosomes. Not until 1992 did Fukui microdissect Barley chromosomes by argonion laser beam. In the same year, Jung et al.（1992）dissected *Beta patellaris* chromosomes.

We applied this method to *Lilium lancifolium* and used G-banded chromosomes to focus on a special region. Our purpose is to establish a specific G- band DNA libraries of plant chromosome.

Materials and Methods Metaphase Chromosome Preparation Bulbs of *Lilium lancifolium* were cultured in mosist wood scraps till roots were 1-2cm long. The root tips were excised and treated with 0.02% colchicine solution for 2 hours. After being immersed in 0.075M KCl at 20-25℃ for 30 minutes, they were fixed in 3∶1（methanol∶acetic acid）fixative for 2-5 minutes and then washed with distilled ater. The enzymic digestion was accomplished in 1% enzyme mixture [1g cellulose（"Onozuka" R-10, from Kinki Yakult MFG. Co, Ltd, Japan）and 1g pectinase（from Aspergillus niger, fro, serva, New York）in 100ml distilled water] at 25℃ for 50 minutes. The material was washed with 0.075M KCl twice and incubated at 25℃ for 10minutes. One or two root tips were placed on each slide, which was kept in ice. With a drop of fixative, they were torn apart with a pair of forcips. Two drops of the fixative were added to each slide, which was then dried with steam from the other side of the slide, and then placed in 60-80℃ oven for two hours. The slide was stored at room temperature till it was used to show bands 1-2 day later.

Banding Treatment The chromosome were put in to 1% Typesin（Difco, in 0.02% EDTA-CMF solution）at 4-10℃ for 5 seconds. Then they were washed twice with 0.85% NaCl and treated with 5.3Murea（2 part of 8Murea in 0.02% EDTA and one part of 1/15mol phosphate buffer adjusted to pH7.4）at 4-10℃ for 3-5 seconds. After they were washed twice with 0.85% NaCl again, the samples were stained with 1∶40 phosphate buffer（pH7.4）for 10 minutes.Finally, the samples were washed with flowing water and dried in air.

Microdissection The microdissection was performed with aglass microneedle under an inverted microscope with the aid of a rifitment micromanipulator（Narishige）. The microneedle was siliconied and drawn on a microforge（MF-9 Narishige）. A special G-band was cut off and collected in 100 nl collection solution（10mM Tris-HCl; pH 7.5; 10mM NaCl; 0.5mg/ml proteinase K; 0.1% SDS; 1% glycerol）which was covered with paraffin to prevent evaporation. After several segments had been collected, another 100 nl collection solution was added. The collection solution and material together with some paraffin was then transferred into a Eppendorf tube with a micropipette. The DNA was digested in 37℃ for q hour and extracted with phenol and chloroform, and then twice extracted with saturated diethyl ether. Two units of restriction enzyme Kpn 1 was added to digest the DNA at 37℃ for 6-8 hours. Kpn 1 was also extracted with phenol and chloroform. The digested DNA was dissolved in 5 μL TE（pH 7.6）. The volume was increased to 20 μL by the addition of 4 μL 10×ligase buffer （0.5M Tris-HCl, pH 7.4）. 1M MgCL; 0.1M DTT; 1mg/ml BSA; 10mM ATP）, 1μL T4 ligase（6μg/μL）, 0.5 OD 20mer oligonucleotides（5'CGGGAATTCTGCTCGAGATC 3'）and 12μL TE（pH 7.6）. The mixture was stored at 23℃ for 10 hours.

PCR Amplification After ligation, 5μL 10×buffer, 4μL dNTP, 19μL ddwater were added into the mixture, which was then incubated at 72℃ for 5minutes followed by another 10 minutes incubation at 72℃ to fill the 3'ends. PCR procedure was performed as table 1.

Results and Discussion The electrophoresis analysis showed that the DNA increased from pg to μg.The length of the DNA fragments was 300-500bp. Although the DNA segments were very small, they were long enough to be markers in chromosome walking and other gene studying. Larger DNA segments were seldom amplified because of the degradation of DNA the preference to small DNA segments in PCR procedure.

Table 1.15   The conditions of PCR

Condition Ciroles	Denaturing		Anealing		DNA Synthesis	
	Temperature	Time	Temperature	Time	Temperature	Time
1	93℃	5′	50℃	3′	72℃	3′
2-5	93℃	1.5′	50℃	3′	72℃	3′
6-30	93℃	1.5′	50℃	3′	72℃	3′

In our experiment, we used G-banded chromosomes in microdissection to improve the precision. There were at least 30 bands on one chromosome. Therefore, we could fix on a region of 1/30 of the whole chromosome lengh. In other words, 30 microlibraries could be obtained by this method.

Compard with the conventional acetic squashing method the enzymatic digestion of shortly (several seconds) fixed roots was shown to be a more efficient procedure.This method decreased depurination of the DNA by acetic acid to reduce thedegradation of DNA.

In the microdissection, we used 20mer oligonucleotide to be ligased to the k pn 1 digested ends and to be as PCR primer as well.It was not necessary to microdissect a great deal of DNA from the chromosomes to ensure the ligasion reaction. In our experiment 4-5 DNA segments were enough to be PCR amplified to 10-20μg DNA.

(Authors: Song Wenqin, Cui Xiangqin, Li Xiulan, Xu Wensheng, An Zhuping, Wang Shurong, Chen Ruiyang. Published in: Proc.Sec.Sino-Jpn.Symposium PI.Chromos. 1992. Plant Chromosome Research p111-114.)

参 考 文 献

陈瑞阳, 宋文芹, 李秀兰. 植物有丝分裂染色体标本制作的新方法, 植物学报, 1979, 21（3）: 297-298.
陈瑞阳, 宋文芹, 李秀兰, 等. 植物染色体 Giemsa 分带技术的研究, 植物学报, 1979, 21（1）: 11-18.
刁英. 染色体核型研究的方法及应用. 渝西学院学报, 2004, 3 :55-58.
邵启全, Carlson P. 小麦、大麦、黑麦染色体分离技术. 中国科学院遗传研究所年报, 1981, 211-216.
谷口研至. 植物染色体的分染法. 细胞, 1977, 9（4）: 126-131.
董凤平, 韩素英, 张守功等. 25SrDNA 杨属植物染色体上的定位. 云南植物研究, 2007, 29(4): 423-428.
宋文芹, 李秀兰, 许文胜, 等. 黑麦染色体的分离与 PCR 扩增. 植物学报, 1998, 40（2）: 158-162.
松田忠男. Heterchromatin of B-chromosomes in plants. 细胞 1979, 7（10）: 454-466.

施立明. 染色体分带技术的回顾与展望。动物学研究，1984，5（1）：1-12.

周钟信，米景九，玉米细胞系中期染色体的分离及其生化分析. 作物学报, 1985, 11(2): 89- 94

Anderson HC. Hydroxyurea induces sister chromatid exchanges in G2: implications for the formation of chromosomal aberrations. Hereditas, 1983, 98: 61-64.

Anna Viktória Németh, Dénes Dudits, Márta Molnár-Láng, et al. Molecular cytogenetic characterisati on of *Salix viminalis* L. using repetitive DNA sequences. J Appl Genetics, 2013, 54:265-269.

Auhltana S, Lee SH, Hang JW, et al. Physical mapping of rRNA gene loci and interspecific relationships in wild *Lilium distributed* in Korea. Plant boil, 2013, 53 : 433-443.

Brown GR, Carlson JE. Molecular cytogenetics of the genes encoding 18S-5.8S-26S rRNA and 5S rRNA in two species of spruce (*Picea*). Theor Appl Genet, 1997, 95 :1-9.

Budiman MA, Chang S, Lee S, et al. Localization of *jointless-2* gene in the centromeric region of tamato chromosome 12 based on high resolution genetic and physical mapping. Theor Appl Genet, 2004, 108 :190-196.

Chester M, Gallagher JP, Symonds VV, et al. Extensive chromosomal variation in a recently formed natural allopolyploid species, Trugopogoi miscellus(Asteraceae). PNAS, 2012, 6:1176-1181.

Cı'cero Almeida, Andrea Pedrosa-Harand. High macrocollinearity between lima bean (*Phaseolus lunatus* L.) and the common bean (*P. vulgaris* L.) as revealed by comparative cytogeneticmapping. Theor Appl Genet, 2013, 126:1909-1916.

Clain E, Brulfert A. Hydroxyure-induced mitotic synchronization in *Allium sativum* root meristems. Planta,1980, 150: 26-23.

Comings DE. The mechanism of C-and G-banding of chromosomes, Exptl Cell Res, 1973,77: 489-493.

Dolezel J, Cihalikova J, Lucretti S. A high yield procedure for isolation of metaphase chromosome f rom root tips of *Vicia faba*. Planta, 1992, 188: 93-98.

Dong F, Song J, Naess SK, et al. Development and applications of a set of chromsome-specific cytogenetic DNA markers in patato. Theor Appl Genet, 2000, 101 :1001-1007.

Dolezd J, Cihalikova J, Lucretti S. A High-Yield Procedure for Isolatio n of Metaphase Chromosomes from root tips of *Vicia faba* L.Planta,1993,(188):93-98.

Dobel P. The Giemsa banding patterns of standard and four reconstructed karyotypes of *Vicia faba* chromosoma, 1973, 43: 409-422.

Dou QW, Chen ZG, Liu YA, et al. High frequency of karyotype variation revealed by sequential FISH and GISH in plateau perennial grass forage *Elymus nutans*. Breeding Science, 2009, 59: 651-656.

Drewry A. G-banded chromosomes in *Pinus resinoso*, The Journal of Heredity. 1982, 75, 305-306.

Dutrillaux B, Lejeune J. Sur une nouvelle technique d'analyse du caryotype humain. C R Acad Sci Paris, 1971, 272: 2638-2640.

Feng Yu, Yunting Lei, Yuan Li, et al. Cloning and characterization of chromosomal markers in alfalfa(*Medicago sativa* L.). Theor Appl Genet, 2013, 126:1885-1896.

Fiskesjö G. Two types of constitntive heterochromatin made visible in *Allium* by a rapid C-banding method. Hereditas, 1974,78:153-156.

Fukui K，Minezawa M, Kamisugi Y, et al. Microdissection of plant chromosomes by argonion laser beam. Theor Appl Genrt,1992, 84:778-791.

Fukashi Shibata, Kiyotaka Nagak I, Etsuko Yokota et al. Tobacco karyotyping by accurate centromere identification and novel repetitive DNA localization. Chromosome Res, 2013, 21:375-381.

Gill BS,.Kimer G. Giemsa C-banding and the evolution of wheat. Proc Nat Acad Sci, 1974,71: 4086-4090.

Griesbach RJ. Chromosomes mediated Transformation Via Microinjection. Plant Sci, 1987, (50): 69 -77.

Greilhuber J. Why plant chromosomes do pot show G-baad. Theer Appl Genet,1977,50:121-124.

Hadlaczky GY, Kalmaz L. Diserimination of homologous chromosomes of maize with Giemsa staining. Heredity,1975,35: 371-374.

Han F, Fedak G, Benabdelmouna A, et al. Characterization of six wheat × *Thinopyrum intermedium* derivatives by GISH, RFLP and multicolor GISH. Genome, 2003, 46 :490-495.

Han F, Liu B, Fedak G, et al. Genomic constitution and variation in five partial amphiploids of wheat-*Thinopyrum intermedium* as revealed by GISH, multicolor GISH and seed storage protein analysis. Theor Appl Genet, 2004, 109 :1070-1076.

Hao L, Song GL,Li BL, et al. Karyotype analysis of Gossypium arboretum × G.bickii by genome in situ hybridization. Acta Genetica sinica, 2006, 33(6):565-572.

Hasterok R, Langdon T, Taylor S, et al. Combinatorial labeling of DNA probes enables multicolour fluorescence in situ hybridization in plants. Folia Histochemica Et Cytobiologica, 2002,40(3) :319-323.

Holm PB. The C and Q banding patterns of the chromosomes of *Lilium longiflorum* (Thunb.), Carlsbery Res Commun, 1976,41: 217-224.

Hsu TC, Somers CE. Proc Natl Acad Sci USA,1961,47:396-401.

Islam-Faridi MN, Childs KL, Klein PE, et al. A molecular cytogenetic map of sorghum chromosome 1: fluorescence *in situ* hybridization analysis with mapped bacterial artificial chromosomes. Genetics, 2002, 161 : 345-353.

Houben A, Field BL, Saunders VA. Microdissection and chromosome painting of plant B chromosomes Methods in Cell Science, 2001, 23 :115-124.

Karp A , Bright AWJ. On the Cause and Origins of Somaclonal Variation. Oxford Surv Plant Mil Cell Biol, 1985, (2): 99-234.

karlov GI, Danilova TV, Horlemann C, et al. Molecular cytogenetics in hop (*Humulus lupulus* L.) and identification of sex chromosomes by DAPI-banding. Euphytica, 2003, 132 :185-190.

Kier O, Festke C. Strong anti-microtubule action of amiprophos-methyl (APM) in micasterias. Protoplasma.1977, 92:163-166.

Kim J, Childs KL, Islam-Faridi MN, et al. Integrated karyotyping of sorghum by in situ hybridization of landed BACs. Genome, 2002, 45 :402-412.

Kongsuwan K,Smyth DR. Q-bands in *Lilium* and their relationship to C-banded heterochromatin. Chromosoma (Berl.), 1977,60:169-178.

Kumari Neelam, Nidhi Rawat, Vijay K, et al. Introgression of group 4 and 7 chromosomes of *Ae. peregrina* in wheat enhances grain iron and zinc density. Mol Breeding , 2011, 28:623-634.

Kuo SR, Wang TT, Huang TC. Karyotype analysis of some *Farmosan kyntzlasperms*. Taiwanis, 1972, 17, 1: 66-80.

Kurata N, Omura T. Karyotype analysis in rice 1. A new method for identifying all chromosome pairs. Japan J Genetics, 1978, 53(4): 251-255.

Langer PR, Waldrop AA, Ward DC. Enzymatic synthesis of biotin-labeled polynucleotides: novel nucleic acid affinity probes. Proc Natl Acad Sci USA, 1981, 78 :6633-6637.

Levsky JM, Singer RH. Fluorescence in situ hybridization: past, present and future. Journal of Cell Science, 2003, 116 :2833-2838.

Lee M , Phillips RL. The Chromosomal Basis of Somaclonal Varition. Annu Rew Plant Physiol and Plant Mol Bio, 1988, (39): 413-417.

Lengerova M, Kejnovsky E, Hobza R et al. Multicolor FISH mapping of the dioecious model plant, Silene latifolia. Theor Appl Genet, 2004, 108 :1193-1199.

Levan A, Fredga K, Sandberg AA. Nomenclactur for centromeric position on chromosomes. Hereditas, 1964, 52: 201-220.

Linde-Laursen I. Giemsa C-banding of the chromosomes of"Emir"Berley. Hereditas,1975,81,285-289.

Liu B, Qi L, Chen R, et al. Multicolor fluorescence in situ hybridization with combinatorial labeling probes enables a detailed karyotype analysis of *Larix* principis-rupprechtii. Biological Research, 2007, 40: 23-28.

Liu B, Zhang S, Zhang Y, et al. Molecular cytogenetic analysis of four *Larix* species by bicolor fluorescence in situ hybridization and DAPI banding. International Journal of Plant Sciences, 2006, 167(2): 367-372.

Liu J Y, She CW, Hu ZL, et al. A new chromosome fluorescence banding technique combining DAPI staining with image analysis in plants. Chromosoma, 2004, 113 :16-21.

Louis C, Morejohn L, Fosket E. Inhibition of plant microtuble polymerization in vitro by the phosphoricamid her-bicide amiprophos-methyl. Science (Washington.D.C.), 1984, 224:874-876.

Luciani JM, Morazzani MR, Stahl A. Identification of pachytene bivalents in human male meiosis using G-banding technique. Chromosoma, 1975,52: 275-282.

Lysak MA, Pecinka A, Schubert I. Recent progress in chromosome painting of *Arabidopsis* and related species. Chromosome Res, 2003, 11 :195-204.

Maluszynska J, Hasterok R. Identification of individual chromosomes and parental genomes in *Brassica juncea* using GISH and FISH. Cytogenetic and Genome Research, 2005, 109 :310-314.

Mekay RDG. The mechanism of G and C banding in mammalian metapbase chrorosomes.Chromosoma, 1973,44:1-7.

Michael Chester, Joseph P, Gallagher V, Vaughan Symonds, et al. Extensive chromosomal variation in a recently formed natural allopolyploid species, *Tragopogon miscellus* (Asteraceae). PNAS, 2012, 1176-1181.

Morejohn LC, Fosket DE. Inhibition of plant microtubule polymeri zation in vitro by the phosphoric amid herbicid amiprophos-methyl. Science, 1984, 224: 847-876.

Navarrete MH, Perez-Villami, Lopez JF. Analysis of the interphase accumlation induced by hydroxyurea on prolifer-ating plant cells. Exp Cell Res, 1979, 124: 151-157.

Nakamura R, Kitamura S, Inoue M, et al. Karyotype analysis of *Nicotiana kawakamii* Y. Ohashi using DAPI banding and rDNA FISH. Theor Appl Genet, 2001, 102 :810-814.

Nederlof PM, van der Flier S, Vrolijk J, et al. Fluorescence ratio measurements of double-labeled probes for multiple *in situ* hybridization by digital imaging microscopy. Cytometry, 1992,839-845.

Neelam K, Rawat N, Tiwari VK, et al. Introgression of group 4 and 7 chromosomes of *Ae.perekrina* in wheat enhances grain iron and zinc density. Mol Breeding,2011, 28: 623-634.

Nemeth AV, Dudits D, Molnar-Iand M, et al. Molecular cytogenetic characterization of *Salix vimirkilis* I. using repetitive DNA sequences. J Appl Genetics, 2013, 54:365-269.

Olga Yu. Yurkevich, Alena A, Naumenko-Svetlova, Nadezhda L. Bolsheva, et al. Investigation of genome polymorphism and seed coat anatomy of species of section *Adenolinum* from the genus *Linum*. Genet Resour Crop Evol , 2013, 60: 661-676.

Okada TA, Comings DE. Mechanisms of chromosome banding III. Similarity between G-bands of mitotic chromosomes and chromomeres of meiotie chromosomes. Chromosoma, 1974,48: 65-71.

Pan WH, Hocben A, Schlegel R. Highly effective cell synchronization in plant roots by hydroxyurea and amiprohos methyl or colchicine. Genome, 1993, 36:383-390.

Pizzaia D, Oliveira VM, Martins AR, et al. Karyotype characterization reveals active 45S rDNA sites located on chromosome termini in *Smilax rufescens* (Smilacaceae). Genet Mol Res, 2013, 12 (2): 1303-1310.

Pinkel D, Langegent J, Collins C, et al. Fluorescence *in situ* hybfidization with human chromosome-specific libraries: Detection of trisomy 21 and translocations of chromosome 4. Proc Natl Acad Sci USA, 1988, 85 : 9138-9142.

Preben Bach Holm, WK. The C and Q banding pattern of the chromosome of *Lilium longiflorum* (Thunb.). Carlsberg Res Commum, 1976, 41(5): 217-224.

Qiong Wu, Fang Liu, Shaohui Li, et al. Uniqueness of the *Gossypium mustelinum* genome revealed by GISH and 45S rDNA FISH. Journal of Integrative Plant Biology, 2013, 55(7): 654-662.

Quan-Wen Dou, Zhi-Guo Chen, Yong-An Liu, et al. High frequency of karyotype variation revealed by sequential FISH and GISH in plateau perennial grass forage *Elymus nutans*. Breeding Science, 2009, 59: 651-656.

Ried T, Baldini A, Rand TC et al. Simultaneous visualization of seven different DNA probes by *in stiu* hybridization using combinatorial fluorescence and digital imaging microscopy. Proc Natl Acad Sci USA, 1992, 89 :1388-1392.

Rohme,D, Heneen WK. Premature chromosome cordensation,application in basic, Clinical, and Mutation Research (Ed.P.N.)Pao.R.T.Johnson,K.Sperling pp.131-153.

Sachan JK, Tanmaka S. A banding method method for Zea chromosomes. Japan J Genetics, 1976, 51: 139-141.

Sayeda Sultana, Sang-Hoon Lee, Jae-Wook Bang et al. Physical mapping of rRNA gene loci and inter-specific relationships in wild *Lilium* distributed in Korea. J Plant Biol, 2013, 53:433-443.

Schubert Iy, Oolezel J, Houben A, Scherthan H, Wanner G. Refined examination of plant metaphase chromosome structure at different levels made feasible by new isolation methoeds.Chromosoma, 1993, 102:96-101.

Schwarzacher T, Leitch AR, Bennett MK et al. *In situ* localization of parental genomes in a wide hybrid. Ann Bot, 1989, 315-324.

Schubert I, Dolezel J, Houben A, et al. Refined examination of plant metaphase chromosome structure at different levels made feasible by new Isolation methods. Chromosoma.1993,102:96-101.

Seabright MA. A rapid banding technique for human chromowome. Lancet, 1971,30: 971-972.

Sgorbati S , Sparvoli E, Levi M, et al. Cell Cycle Kinetic Analysis with Flow Cytometry in Pea Root Meristem Synchrowized with *Aphidicolin*. Physiol Plant , 1991, (81):507-512.

Shi Q, Zhang J, Pan S, et al., The study of preparing probes for fluorescence in situ hybridization(FISH) from YAC clones by universal primer PCR. Acta Genetia Sinica, 1998, 25(5): 403-408.

Stebbins GL. Chromosomal evolution in higher plants London: Edward Arnold Ltd. 1971.

Shibate F, Nagaki K, Yokota E, et al. Tobecco karyotyping by sccurate centromere identification and novel repetitive DNA localization. Chromosome Res, 2013, 21: 375-378.

Urdampilleta J D, Coulleri J P, Ferrucci MS, et al. Karyotype evolutio n and phylogenetic analyses in the genus *Cardiospermum* L. (Paullinieae, Sapindaceae). Plant Biology, 2013, 15: 868-881.

Van Dilla M, Deaven L. Constraction of Gene Libratries for Each Human Chromosome Cytometry, 1990 (11): 208-218.

Vischi M, Jurman I, Bianchi G, et al. Karyotype of Norway spruce by multicolor FISH. Theor Appl Genet, 2003, 107 :591-597.

Vosa CG, Marchai P. Quinacrine fluorescence, Giemsa staining and heterochromatin in plant. Nature,1971, 237: 191-192.

Vosa C, Marchisp G. On the quinaerine fluorescence and Giemsa staining patterns of the chromosomes of *Vicia faba*. Giorn Bot Ital, 1972,106: 15-159.

Vosa CG. The basic karyotype of rye (*Seacle cereale*) analysed with Giemsa and fluorescence method. Heredity, 1974, 33: 403-408.

Wangn HC, Federoff S. Banding in human chromosomes,treated with trypsin. Nature New Biol, 235: 52-54.

Wu Q, Liu F, Li SH, et al. Uniqueness of the *Gossypinum mustelinum* genome revealed by GISH and 45S rDNA FISH. Journal of Integrative Plant Biology, 2013, 55, 7: 654-662.

Yu F, Lei YT, Li Y, t al. Cloning and characterization of chromosomal markers in alfalfa (*Medicago sativa* L.). Theor Appl Genet, 2013, 126: 1885-1896.

Yurkevich Y, Naumenko Svetlova AA, Bolsheva NI, et al. Investigation of genome polymorphism and seed coat anatomy of species of section *Adenolinum* from the genus *Linum* Genet Resour Crop Evol, 2013, 60: 661-676.

Yunis J J. The characterization of High-Resolution G-banded chromosomes of man. Chromosoma(berl.),1978, 67: 293-307.

Zhang P, Li W, Friebe B, et al. Simultaneous painting of three genomes in hexaploid wheat by BAC-FISH. Genome, 2004, 47 : 979-987.

Zoller JF, Yang Y, Reinhold G, et al. Comparative genomic in situ hybridization (cGISH) analysis on plant chromosomes revealed by labeled *Arabidopsis* DNA. Chromosome Res, 2001, 9 :357-375.

第二章 中国植物染色体研究结果

提要： 这一部分内容主要是 40 年来对我国 194（85%）科 693 属（23.1%）1579（5.26%）种植物染色体研究的有关论文和《中国主要经济植物基因组染色体图谱》Ⅰ~Ⅴ册的研究结果。但因为篇幅所限，我们在这本书里只提供了染色体数据，以供查阅。为了分析和评价我国植物染色体研究水平，我们在第二章"中国农作物及其野生近缘植物染色体图谱"里提供了部分农作物的染色体图像，以供参考。从这些图像中不难看出：①酶解去壁低渗法制备的植物染色体标本，完全达到了人类染色体标本的水平；②除了核型分析还进行了 Giemsa 分带，使染色体研究达到了 γ 水平；③同一细胞可以进行核型和带型分析；④部分染色体使用了 FISH 技术。这充分显示了中国植物染色体研究达到了世界先进水平。

第一节 中国果树及其近缘植物基因组染色体研究结果

提要： 本书收录了我们历经 12 年对我国 29 科 52 属 261 种果树及其近缘植物染色体研究结果和 1000 多个栽培品种的染色体倍性鉴定资料。瓦维洛夫早就指出，起源我国的果树有 50 多种，因此，我国也被称为"园林之母"。果树多为异花授粉植物，其种子不适合做染色体材料，要取其营养生长的材料（幼叶），所以，果树类染色体研究难度很大，国外报道甚少，早期的报道错误较多。我们完全采用去壁、低渗法，以幼叶为材料制备染色体标本，使我国在果树及其近缘植物染色体研究方面居国际领先地位。在我国果树中发现了一批珍贵的多倍体，有四倍体 44 种、三倍体 44 种、六倍体 3 种、八倍体 2 种、十二倍体 1 种，共 96 种多倍体资源。但在重要的栽培品种苹果、梨、桃、杏、樱桃中，却没有发现多倍体。

据报道在美国栽培苹果中约三分之一为三倍体。20 世纪 70 年代以前，由于我国植物染色体研究的空白，果树多倍体育种工作基本没有开展，20 世纪七八十年代又赶上分子生物学时代，所以果树多倍体育种是我国果树育种中的一块短板，而我国果树资源中有非常丰富的多倍体，如分布在华北地区的三倍体安梨，就是直接由野生酸梨选育的，抗性好、产量高。果树多倍体育种是一项长期效应项目，关系到子孙后代，关系到我国果树的持续发展。因为篇幅所限，本书只选了几种果树染色体图像，以便查找。

2.1.1 部分果树染色体

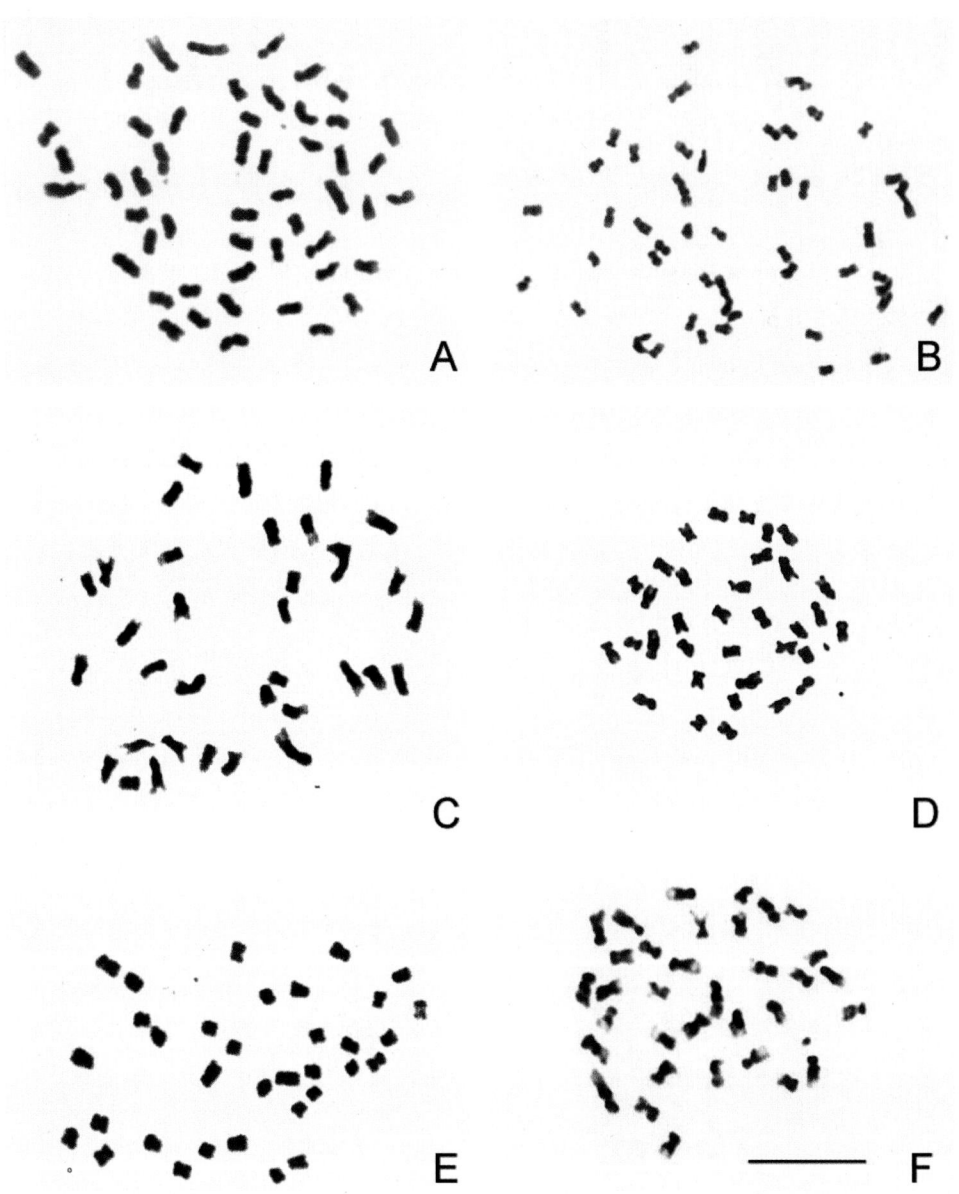

A.八棱海棠（*Malus robusta*），B.安梨（*Pyrus ussuriensis*），C.金冠苹果（*Malus pumila* cv.*jingguan*），
D.早香一号（*Pyrus ussuriensis* cv.*zaoxiangyihao*），E.新疆野苹果（*Malus sieversii*），
F.白梨（*Pyrus bretschneideri*）

2.1.2 中国果树及其近缘植物染色体数据分析

植物核型分析（果树及其野生近缘植物）

科	属	种	染色体数目	核型公式	核型分类	臂指数	最长染色体/最短染色体	臂比>2的染色体比例	核型不对称系数	染色体相对长度组成
I 苹果亚科 Maloideae	1.苹果属 Malus Mill.	1-1 山荆子 Malus baccata (L.) Borkh.	$2n=2x=34$	4M+16m+12m（2SAT）+2st	2A	66	2	0.29	61.34	4L+14M2+14M1+2S
		1-1-1 山荆子 Malus baccata (L.) Borkh.R*	$2n=2x=34$	22m+12sm（4SAT）	2A	68	1.8	0.35	62.15	2L+14M2+18M1
		1-2 毛山荆子 Malus mandshurica (Maxim.) Kom.	$2n=2x=34$	26m+6sm（2SAT）+2st	2A	66	2	0.24	60.12	2L+14M2+16M1+2S
		1-3 丽江山荆子 Malus rockii Rehd.	$2n=2x=34$	28m+6sm（2SAT）	2A	68	1.6	0.12	57.6	2L+10M2+22M1
		1-3-1 丽江山荆子 Malus rockii Rehd.	$2n=2x=34$	28m+6sm（2SAT）	2A	68	1.6	0.12	57.6	2L+10M2+22M1
		1-3-2 丽江山荆子 Malus rockii Rehd. R.3X	$2n=3x=51$	21m+30sm	2A	102	1.8	0.29	63.43	3L+15M2+33M1
		1-4 锡金海棠 Malus sikkimensis (Hook.f.)Koehne. Koehne.B.4X	$2n=4x=68$	28m（2SAT）+4sm+2st（2SAT）	2B	66	2	0.18	59.38	6L+8M2+20M1
		1-5 湖北海棠（泰山）Malus hupehensis (Pamp.) Rehder.	$2n=4x=68$	2M+60m+6sm	2B	136	2.1	0.03	57.75	4L+30M2+28M1+6S
		1-6 垂丝海棠 Malus halliana Koehne.B.3X	$2n=3x=51$	45m+6sm	2A	102	1.9	0.06	59.16	3L+21M2+27M1
		1-6-1 垂丝海棠 Malus halliana Koehne.R.	$2n=3x=51$	33m+18sm	2A	102	2	0.06	60.42	3L+15M2+30M1+3S
		1-7 苹果（金冠）Malus pumila Mill.cv.jinguan B.2X	$2n=2x=34$	20m（SAT）+14sm	2A	68	1.6	0.18	59.54	2L+14M2+18M1
		1-7-1 道生苹果 Malus pumila var.praecox Pall.	$2n=2x=34$	18m+6sm	2B	68	2	0.18	63.37	2L+12M2+18M1+2S
		1-7-2 乐园苹果	$2n=2x=34$	26m+8sm（2SAT）	2A	68	1.6	0.12	60.07	16M2+18M1
		1-7-3 乔纳金 Malus pumila Mill.cv.Jonagold.	$2n=2x=34$	26m+8sm（2SAT）	2B	68	1.6	0.06	58.27	2L+16M2+16M1
		1-7-4 香果 Malus pumila Mill.cv.xiangguo.B.	$2n=3x=51$	18m+33sm（6SAT）	2B	102	2.2	0.47	65.49	3L+18M2+24M1+6S
		1-8 森林苹果 M. sylvestris Mill.	$2n=3x=51$	3M+42m+6sm（3SAT）	2A	102	2	0.06	56.59	3L+24M2+21M1+3S
		1-9 新疆野苹果 Malus sieversii (Ledeb.) Roem.	$2n=2x=34$	26m+8sm	2A	68	1.7	0.06	58.04	2L+14M2+18M1
		1-9-1 新疆野苹果 Malus sieversii (Ledeb.) Roem.R.	$2n=2x=34$	24m+10sm	2B	68	2.3	0.18	61.04	6L+6M2+20M1+2S
		1-10 花红 Malus asiatica Nakai	$2n=2x=34$	12M+20m+2sm	2A	68	1.9	0.06	55.57	4L+12M2+18M1
		1-10-1 花红 Malus asiatica Nakai	$2n=4x=68$	4M+52m+12sm（2SAT）	1A	68	1.8	0	57.91	4L+12M2+18M1
		1-10-2 槟子 Malus asiatica var.rinki Asami	$2n=2x=34$	30m+4sm	2A	68	1.9	0.09	57.75	4L+22M2+40M1+2S
			$2n=2x=34$	28m（2SAT）+6sm	2B	68	2.1	0.06	59.1	6L+8M2+18M1+2S

续表

植物核型分析（果树及其野生近缘植物）

科	属	种	染色体数目	核型公式	核型分类	臂指数	最长染色体/最短染色体	臂比>2的染色体比例	核型不对称系数	染色体相对长度组成
		1-10-3 马蹄奈子 Malus asiatica Nakai cv.matinaizi	2n=3x=51	36m（3SAT）+15sm	2A	102	1.7	0.06	61.23	3L+18M2+27M1+3S
		1-10-4 蜜果 Malus asiatica Nakai cv.miguo.	2n=4x=68	42m+24sm（2SAT）+2st（2SAT）	2B	134	2.1	0.06	60.31	10L+16M2+36M1+6S
		1-10-5 黄甜果 Malus asiatica Nakai cv.huangtianguo	2n=4x=68	60m+8sm	2B	136	2.1	0.06	58.18	10L+20M2+32M1+6S
		1-10-6 魁果 Malus asiatica Nakai cv.kuiguo.	2n=2x=34	24m+10sm	2A	68	2	0.06	60.54	2L+16M2+10M1+6S
		1-10-7 小红奈子 Malus asiatica Nakai cv.xiaohongnaizi.B.	2n=2x=34	22m+8sm+4st（2SAT）	2A	64	1.9	0.18	62.21	4L+10M2+18M1+2S
		1-11 楸子 Malus prunifolia（Willd.）Borkh.	2n=2x=34	30m+4sm	2A	68	1.9	0.12	60.17	2L+14M2+18M1
		1-11-1 楸子 Malus prunifolia（Willd.）Borkh.B.3x	2n=3x=51	39m（3SAT）+12s（3SAT）	2A	102	2	0.12	59.97	3L+12M2+33M1+3S
		1-12 海棠花 Malus.spectabilis（Ait.）Borkh.	2n=2x=34	6M+20m（2SAT）+8sm	2A	68	1.5	0.18	58.39	2L+12M2+20M1
		1-13 西府海棠 Malus × micromalus Makino B.2x	2n=2x=34	2M+20m+12sm	2A	68	1.9	0.12	60.83	4L+8M2+20M1+2S
I 苹果亚科 Maloideae	1.苹果属 Malus Mill.	1-13-1 西府海棠 Malus × micromalus Makino	2n=2x=34	6M+26m+2sm	2B	68	2	0.06	56.07	4L+12M2+16M1+2S
		1-13-2 铁海棠 Malus × micromalus Makino cv.tiehaitang.B.2x	2n=2x=34	2M（SAT）+22m+10sm	2B	68	2.1	0.06	59.46	6L+12M2+12M1+4S
		1-14 三叶海棠 Malus sieboldii（Regal）Rehd.	2n=3x=51	39m+12sm	2A	102	1.8	0,12	60.4	6L+15M2+30M1
		1-14-1 三叶海棠 Malus sieboldii（Regal）Rehd. R.3x	2n=3x=51	45m+6sm	2A	102	1.7	0.12	56.9	3L+30M2+18M1
		1-15 陇东海棠 Malus kansuensis（Batalin）C.K. Schneid.	2n=2x=34	18m+14sm+2st	2B	66	2.2	0.35	63.57	2L+14M2+16M1+2S
		1-16 山楂海棠 Malus komarovii（Sarg.）Rehder B.2x	2n=2x=34	32m+2sm	2B	68	2.1	0.06	57.39	4L+16M2+12M1+2S
		1-17 变叶海棠 Malus toringoides(Rehder)Hughes	2n=2x=34	4M+26m+4sm	2A	68	2	0.06	57.65	4L+12M2+16M1+2S
		1-17-1 变叶海棠 Malus toringoides（Rehder）Hughes R.2x	2n=2x=34	2M+26m+6sm	1A	68	1.5	0	58.09	18M2+16M1
		1-17-2 变叶海棠 Malus toringoides（Rehder）Hughes.	2n=4x=68	2M+54m+10sm+2st	2B	134	2.2	0.12	56.85	4L+24M2+36M1+4S
		1-17-3 变叶海棠 Malus toringoides（Rehder）Hughes.	2n=3x=51	42m+9sm	1A	102	1.9	0	57.2	9L+15M2+24M1+3S

续表

植物核型分析（果树及其野生近缘植物）

科	属	种	染色体数目	核型公式	核型分类	臂指数	最长染色体/最短染色体	臂比>2的染色体比例	核型不对称系数	染色体相对长度组成
I 苹果亚科 Maloideae	1.苹果属 Malus Mill.	1-18 花叶海棠 Malus transitoria (Batalin) C.K. Schneid.	2n=2x=34	12m+22sm (2SAT)	2A	68	1.9	0.18	63.27	4L+14M2+14M1+2S
		1-18-1 花叶海棠 Malus transitoria (Batalin) C.K. Schneid.	2n=2x=34	4M+m+4sm	1A	68	1.8	0	55.74	2L+12M2+20M1
		1-8-2 花叶海棠（青海）Malus transitoria (Batalin) C.K. Schneid.	2n=3x=51	39m+12sm	2A	102	2	0.12	60.78	6L+21M2+21M1+3S
		1-19 川滇海棠 Malus prattii (Hemsl.) Schneid.	2n=2x=34	32m+2sm	1A	68	1.7	0	58.06	4L+10M2+20M1
		1-20 沧江海棠 Malus ombrophila Hand.-Mazz. B.2x	2n=2x=34	20m+12sm+2st (2SAT)	2A	66	1.9	0.24	62.31	2L+10M2+20M1+2S
		1-20-1 沧江海棠 Malus ombrophila Hand.-Mazz.	2n=2x=34	2M+22m+10sm (2SAT)	2A	68	1.8	0.12	59.46	2L+14M2+18M1
		1-21 河南海棠 Malus honanensis Rehder	2n=2x=34	18m+12sm (2SAT) +4st	2B	64	2.1	0.35	65.94	2L+10M2+18M1+4S
		1-22 滇池海棠 Malus yunnanensis (Franch.) C.K. Schneid.	2n=2x=34	20m (2SAT) +8sm+6st (2SAT)	2B	62	2.3	0.18	62.34	4L+14M2+12M1+4S
		1-23 台湾林檎 Malus formosana Kawak.et Koidz.B.2x	2n=2x=34	2M+24m+8sm	2A	68	1.7	0.06	58.54	4L+12M2+18M1
		1-24 尖嘴林檎 Malus melliana (Hand.-Mazz.) Rehd.	2n=2x=34	6M+24m+4sm	2A	68	2	0.06	55.74	4L+10M2+14M1+6S
		1-25 小金海棠 M.xiaojinensis Chen.R.4	2n=4x=68	6M+58m (2SAT) +4sm	2B	136	2	0.03	55.68	8L+18M2+42M1
		1-26 八棱海棠 Malus robusta Carr.	2n=3x=51	27m (3SAT) +15sm (3SAT) +9st	2A	93	1.8	0.47	64.98	3L+21M2+27M1
	2.梨属 Pyrus Lindl.	2-1 秋子梨 Pyrus ussuriensis Maxim.	2n=2x=34	14m+16sm+4st	2B	64	2	0.41	65.5	6L+14M2+8M1+6S
		2-1-1 安梨 Pyrus ussuriensis Max.cv.anli	2n=3x=51	27m+24sm	2A	102	1.7	0.18	62.22	3L+21M2+27M1
		2-1-2 早香一号 Pyrus ussuriensis Max.cv.zaoxiangyihao	2n=2x=34	16m+16sm (2SAT) +2st (2SAT)	2A	66	1.8	0.35	63.95	4L+12M2+18M1
		2-2 河北梨 Pyrus hopeiensis Yü	2n=2x=34	16m+18sm (2SAT)	2A	68	2	0.35	64.35	4L+10M2+18M1+2S
		2-3 新疆梨 Pyrus sinkiangensis Yu	2n=2x=34	18m+16sm (2SAT)	2A	68	1.9	0.18	62.71	2L+12M2+20M1
		2-4 麻梨 Pyrus serrulata Rehd.	2n=2x=34	14m+20sm	2A	68	1.8	0.29	64.09	4L+12M2+18M1
		2-5 洋梨 Pyrus communis L.	2n=2x=34	18m+16sm (2SAT)	2A	68	1.9	0.24	62.33	2L+16M2+16M1

续表

植物核型分析（果树及其野生近缘植物）

科	属	种	染色体数目	核型公式	核型分类	臂指数	最长染色体/最短染色体	臂比>2的染色体比例	核型不对称系数	染色体相对长度组成
I 苹果亚科 Maloideae	2.梨属 Pyrus Lindl.	2-6 杏叶梨 Pyrus armeniacaefolia Yu	2n=3x=51	24m+27sm	2A	102	1.8	0.41	64.73	3L+21M2+27M1
		2-7 木梨 Pyrus xerophila Yu	2n=2x=34	22m+10sm+2st (SAT)	2A	66	1.8	0.24	62.83	2L+14M2+18M1
		2-8 白梨 Pyrus bretschneideri Rehd	2n=2x=34	16m+18sm (2SAT)	2A	68	1.6	0.41	64.39	16M2+18M1
		2-8-1 鸭梨 Pyrus bretschneideri Rehd.cv.yali B.2x	2n=2x=34	2M+24m+6sm+2st	2B	66	2.2	0.06	59.73	4L+10M2+16M1+4S
		2-8-2 鸭梨 Pyrus bretschneideri Rehd.cv.yali	2n=4x=68	4M+50m (2SAT)+10sm+4st	2A	132	1.9	0.09	59.76	6L+26M2+34M1+2S
		2-8-3 鸭梨 Pyrus bretschneideri Rehd.cv.yali	2n=3x=51	48m+3st (3SAT)	2B	99	2.1	0.06	55.85	3L+18M2+30M1
		2-8-4 茭老梨 Pyrus bretschneideri Rehd.cv.yinglaoli	2n=2x=34	8m+22sm (4SAT)+4st	2A	64	1.9	0.41	67.13	2L+12M2+18M1+2S
		2-8-5 蜜梨 Pyrus bretschneideri Rech.cv.mili	2n=2x=34	8M+18m+6sm+2st (2SAT)	2A	66	1.7	0.24	59.08	2L+10M2+22M1
		2-9 砂梨 Pyrus pyrifolia (Burm.) Nakai	2n=2x=34	20m+12sm (2SAT)+2st (2SAT)	2A	66	1.8	0.18	61.88	2L+14M2+18M1
		2-10 杜梨 Pyrus betulifolia Bunge.	2n=2x=34	22m+12sm	2A	68	1.8	0.06	61.62	4L+10M2+20M1
		2-11 褐梨 Pyrus phaeocarpa Rehd.	2n=2x=34	22m+12sm	2A	68	1.9	0.12	61.29	2L+10M2+22M1
		2-12 豆梨 Pyrus calleryana Dcne.	2n=2x=34	26m+8sm	2B	68	2	0.12	60.83	4L+8M2+20M1+2S
		2-13 川梨 Pyrus pashia Buch.-Ham.	2n=2x=34	18m (2SAT)+16sm	2A	68	1.6	0.29	63.46	2L+12M2+20M1
	3.山楂属 Crataegus Lindl.	3-1 山楂 Crataegus pinnatifida Bge.	2n=2x=34	28m+6sm	1A	68	1.8	0	58.03	2L+12M2+20M1
		3-1-1 伏山楂 Crataegus pinnatifida Bge.cv.fushanzha.	2n=3x=51	51m	1A	102	1.5	0	55.27	30M2+21M1
		3-1-2 昌黎紫肉 Crataegus pinnatifida Bge.cv.changlizirou.	2n=2x=34	32m+2sm	1A	68	1.9	0	55.8	2L+14M2+18M1
		3-2 湖北山楂 Crataegus hupehensis Sarg.	2n=2x=34	28m+6sm (4SAT)	2A	68	1.4	0.06	60.76	22M2+12M1
		3-3 华中山楂 Crataegus wilsonii Sarg.	2n=2x=34	32m (2SAT)+2sm	1A	68	1.9	0	56.92	4L+10M2+18M2+2S
		3-4 毛山楂 Crataegus maximowiczii Schneid.	2n=3x=51	51m	1A	102	1.6	0	55.41	27M2+24M1
		3-5 辽宁山楂 Crataegus sanguinea Pall.	2n=4x=68	2M+62m+4sm	2A	136	2	0.06	57.02	2L+28M2+34M1+4S
		3-6 甘肃山楂 Crataegus kansuensis Wils.	2n=2x=34	34m	1B	68	2	0	53.38	2L+16M2+14M1+2S
		3-7 阿尔泰山楂 Crataegus altaica (Loud.) Lange.	2n=4x=68	62m+6sm	2B	136	2.5	0.06	56.62	4L+32M2+28M1+4S

续表

植物核型分析（果树及其野生近缘植物）

科	属	种	染色体数目	核型公式	核型分类	臂指数	最长染色体/最短染色体	臂比>2的染色体比例	核型不对称系数	染色体相对长度组成
	3.山楂属 Crataegus Lindl.	3-8 准格尔山楂 Crataegus songarica C.Koch.	$2n=4x=68$	66m+2sm	2A	136	1.8	0.03	57.61	2L+26M2+38M1+2S
		3-9 云南山楂 Crataegus scabrifolia (Franch.) Rehd.	$2n=2x=34$	26m+8sm	2A	68	1.6	0.12	59.4	4L+10M2+20M1
		3-10 野山楂 Crataegus cuneata Sieb. et Zucc.	$2n=2x=34$	26m+4sm（2SAT）+4st（2SAT）	2A	64	1.8	0.12	60.22	4L+10M2+20M1
	4.木瓜属 Chaenomeles Lindl	4-1 皱皮木瓜 Chaenomeles speciosa Nakai	$2n=2x=34$	24m+10sm	2A	68	1.6	0.18	59.92	2L+10M2+22M1
	5.榅桲属 Cydonia Lindl	5-1 榅桲 Cydonia oblonga Mill.	$2n=2x=34$	24m+10sm（2SAT）	2A	68	1.9	0.12	62.12	4L+12M2+16M1+2S
		6-1 桃 Prunus persica L.	$2n=2x=16$	6m+10sm（2SAT）	2A	32	1.9	0.38	62.89	2L+4M2+10M1
		6-1-1 油桃 Prunus persica var. nucipersica L.	$2n=2x=16$	14m+2sm	2B	32	2.1	0.13	59	2L+2M2+10M1+2M1
		6-1-2 蟠桃 Prunus persica L. var. compressa Bean.	$2n=2x=16$	16m（2SAT）	1A	32	2	0	58	2L+4M2+10M1
I 苹果亚科 Maloideae		6-1-3 寿星桃 Amygdalus persica L. var.densa Makino	$2n=2x=16$	14m+2sm	1B	32	2.3	0	58.07	2L+4M2+6M1+4S
		6-1-4 碧桃 Amygdalus persica L. var. duplex Rehd.	$2n=2x=16$	16m	1A	32	1.7	0	56.5	2L+2M2+12M1
		6-1-5 碧桃 Amygdalus persica L. var. duplex Rehd.	$2n=3x=24$	21m+3sm	1B	48	2.1	0	59.21	3L+6M2+12M1+3S
		6-2 甘肃桃 Amygdalus kansuensis Skeels	$2n=2x=16$	10m+6sm（2SAT）	2B	32	2.3	0.13	60.68	2L+2M2+12M1
	6.桃属 Amygdalus L.	6-3 山桃 Amygdalus davidiana（Carr）Yu	$2n=2x=16$	16m（2SAT）	1B	32	2.1	0	57.21	2L+6M2+6m1+2s
		6-4 光核桃 Amygdalus mira（Koehne.）Kov.et Kost.	$2n=2x=16$	10m+6sm	2A	32	1.9	0.38	64.17	2L+4M2+10M1
		6-5 新疆桃 Amygdalus ferganensis Kov.et Kost	$2n=2x=16$	12m+4sm	2B	32	2.3	0.25	61.15	4L+2M2+6M1+4S
		6-6 扁桃 Amygdalus communis L.	$2n=2x=16$	12m+4sm	1A	32	1.7	0	59.73	2L+4M2+10M1
		6-7 矮扁桃 Amygdalus nana L.	$2n=2x=16$	14m+2sm（2SAT）	1B	32	2.5	0	58.56	2L+4M2+8M1+2S
		6-8 榆叶梅 Amygdalus triloba（lindl）Ricker.	$2n=8X=64$	2M+42m+20m（2SAT）	2B	128	2.4	0.13	60.22	8L+16M2+36M1+4S
		6-9 西康扁桃 Amygdalus tangutica Korsh.	$2n=2x=16$	10m（2SAT）+6sm	2B	32	2.3	0.25	61.97	2L+4M2+8M1+2S
		6-10 蒙古扁桃 Amygdalus mongolica（Maxim.）Yu	$2n=1x=16$	9m+7sm	2C	32	4.2	0.19	60.86	3L+32M2+7M1+4S

续表

植物核型分析（果树及其野生近缘植物）

科	属	种	染色体数目	核型公式	核型分类	臂指数	最长染色体/最短染色体	臂比>2的染色体比例	核型不对称系数	染色体相对长度组成
	6.桃属 Amygdalus L.	6-11 长硬扁桃 Amygdalus pedunculata Pall. B.12x	2n=12x=96							
		7-1 杏 Armeniaca vulgaris Lam.	2n=2x=16	14m (2SAT) +2sm	1A	32	1.8	0	58.26	2L+4M2+10M1
	7.杏属 Armeniaca Mill.	7-1-1 山杏 Armeniaca vulgaris Lam.var.ansu (Maxim.) Yu	2n=2x=16	12m+4sm	2B	32	2.3	0.13	61.64	4L+2M2+6M1+4S
		7-2 西伯利亚杏 Armeniaca sibirica (L.) Lam.	2n=2x=16	10m+6sm	2A	32	1.9	0.13	61.65	2L+2M2+12M1
		7-3 东北杏 A. mandshurica (Maxim.) Skvortz.B.2x	2n=2x=16	10m (2SAT) +6sm	2B	32	2	0.25	60.26	2L+14M1
		7-4 梅 Armeniaca mume Sieb.	2n=2x=16	12m+4sm (2SAT)	2B	32	2.4	0.13	60.59	2L+4M2+8M1+2S
		7-4-1 杏梅 Armeniaca mume var.bungo Makino. B.2x	2n=2x=16	10m+6sm	2B	32	2.5	0.13	62.5	2L+4M2+8M1+2S
		7-4-2 杏梅 Armeniaca mume var. bungo Makino	2n=3x=24	12m+12sm	2B	48	2.3	0.38	64.68	3L+6M2+12M1+3S
		8-1 李 Prunus salicina Lindl.	2n=2x=16	10m+6sm	2B	32	2.4	0.13	61.67	2L+2M2+10M1+2S
		8-1-1 李 Prunus salicina Lindl.	2n=3x=24	15m+9sm (3SAT)	2B	48	2.2	0.25	62.03	3L+9M2+9M1+3S
	8.李属 Prunus Lindl.	8-2 杏李 Prunus simonii Carr.	2n=2x=16	10m+6sm	2B	32	2	0.13	61.4	2L+8M2+4M1+2S
		8-3 欧洲李 Prunus domestica L.	2n=6x=48	36m+12sm	2B	96	2.7	0.13	59.98	6L+10M2+26M1+6S
		8-4 樱桃李 Prunus cerasifera Ehrh.	2n=2x=16	12m+4sm	2B	32	2.2	0.13	59.45	2L+6M2+6M1+2S
		8-5 美洲李 Prunus americans Marsh.	2n=2x=16	10m+6sm (2SAT)	2B	32	2.5	0.13	62.32	2L+2M2+10M1+2S
		8-6 乌苏里李 Prunus ussuriensis Kov.et Kost	2n=2x=16	10m+6sm (2SAT)	2B	32	2.1	0.13	61.69	2L+6M2+6m1+2s
I 苹果亚科 Maloideae		9-1 樱桃 Cerasus pseudocerasus (Lindl.) G.Don	2n=2x=16	14m+2st (2SAT)	2B	30	2.2	0.13	60.93	2L+2M2+10M1+2S
		9-1-1 樱桃 Cerasus pseudocerasus (Lindl.) G.Don B.4x	2n=4x=32	24m+8sm	2B	64	2.4	0.13	59.26	4L+8M2+16M1+4S
	9.樱桃属 Cerasus Mill.	9-2 山樱桃 Cerasus serrulata (Lindl.) G.Don	2n=2x=16	12m (2SAT) +4sm	2B	32	2.2	0.13	60.13	2L+4M2+8M1+2S
		9-3 东京樱花 Cerasus yedoensis (Mats.) Yu	2n=2x=16	6m+10m (2SAT)	2A	32	1.7	0.25	62.78	2L+2M2+12M1
		9-4 欧洲甜樱桃 Cerasus avium (L.) Moench.	2n=2x=16	6m+10sm	2A	32	1.9	0.25	61.98	2L+2M2+12M1
		9-5 欧洲酸樱桃 Cerasus vulgaris Mill.	2n=2x=16	12m+4sm (4SAT)	2B	32	2	0.25	60.94	2L+2M2+12M1
		9-5-1 欧洲酸樱桃 Cerasus vulgaris Mill.	2n=4x=32	30m+2sm	2B	64	2.3	0.06	57.53	4L+6M2+20M1+2S
		9-6 马哈利樱桃 Cerasus mahaleb (L.) Mill.	2n=2x=16	6m+8sm (2SAT) +2st	2A	30	1.9	0.38	66.05	2L+2M2+12M1

续表

植物核型分析（果树及其野生近缘植物）

科	属	种	染色体数目	核型公式	核型分类	臂指数	最长染色体/最短染色体	臂比>2的染色体比例	核型不对称系数	染色体相对长度组成
I 苹果亚科 Maloideae	9.樱桃属 Cerasus Mill.	9-6-1 马哈利樱桃 Cerasus mahaleb (L.) Mill.	2n=3x=24	15m+6sm+3st (3SAT)	2A	45	1.8	0.13	62.85	3L+3M2+18M1
		9-7 毛樱桃 Cerasus tomentosa (Thunb.) Wall.	2n=2x=16	12m (2SAT) +4sm	1B	32	2.5	0	58.38	2L+6M2+4M1+4S
		9-8 欧李 Cerasus humilis (Beg.) Sok.	2n=2x=16	8m+8sm (2SAT)	2B	32	2.5	0.25	61.6	2L+2M2+10M1+2S
II 鼠李科 Rhamnaceae	10.枣属 Ziziphus Mill.	10-1 枣 Ziziphus jujuba Mill.	2n=2x=24	16m (2SAT) +8sm	2A	48	1.6	0.25	60.67	2L+8M2+14M1
		10-1-1 酸枣 Ziziphus jujuba Mill. var. spinosa Hu.	2n=2x=24	16m (2SAT) +2st	2A	46	1.8	0.08	62.2	2L+8M2+14M1
		10-2 团枣 Ziziphus jujuba Mill.cv.tuanzao.	2n=4x=48	28m+20sm	2B	96	2.1	0.17	60.26	4L+14M2+28M1
III 葡萄科 Vitacvae	11.葡萄属 Vitis Lindl.	11-1 葡萄 Vitis vinifera L.	2n=2x=38	6M+24m+8sm (4SAT)	2A	76	1.5	0.11	58.52	20M2+18M1
		11-2 山葡萄 Vitis amurensis Rupr.	2n=2x=38	6M+28m+4sm (2SAT)	1A	76	1.7	0	54.07	2L+16M2+20M1
		11-3 葛藟葡萄 Vitis flexuosa Thunb.	2n=2x=38	8M+24m+6sm (2SAT)	2B	76	2.2	0.16	57.36	4L+16M2+14M1+4S
		11-4 美洲葡萄 Vitis Labrusca L.	2n=2x=38	4M+32m+2sm	2A	76	2	0.05	56.33	6L+8M2+22M1+2S
		11-5 河岸葡萄 Vitis riparia Michx.	2n=2x=38	6M+32 (2SAT)	1A	76	1.6	0	54.01	2L+16M2+20M1
IV 猕猴桃科 Actinidiaceae	12.猕猴桃属 Actinidia Lindl.	12-1 中华猕猴桃 Actinidia chinensis Planch.	2n=2x=58	4M+48m+6sm	1A	116	1.8	0	57.14	4L+20M2+32M1+2S
		12-2 美味猕猴桃 Actinidia chinensis Planch.								
		12-3 狗枣猕猴桃 Actinidia kolomikta (Rupr.et Maxim.) Maxim.	2n=2x=58	52m+6sm	2A	116	1.7	0.07	58.74	2L+26M2+30M1
		12-4 葛枣猕猴桃 Actinidia polygama Miq	2n=2x=58	2M+30m (2SAT) +24sm (2SAT) +2st (2SAT)	2A	114	1.9	0.31	61.87	4L+24M2+26M1+4S
		12-5 毛花猕猴桃 Actinidia eriantha Benth. B.2x	2n=2x=58	2M+50m+6sm (2SAT)	2A	116	1.6	0.03	57.06	2L+24M2+32M1
		12-6 软枣猕猴桃 Actinidia arguta (Sieb. & Zucc) Planch.	2n=4x=116	8M+84m+24sm	2B	232	2.3	0.07	57.82	6L+46M2+56M1+8S
		12-7 对萼猕猴桃 Actinidia valvata Dunn								
		12-8 大籽猕猴桃 Actinidia macrosperma C. F. Liang	2n=4x=116	6M+70m (2SAT) +38sm (2SAT) +2st	2B	230	2.2	0.24	60.75	6L+48M2+58M1+4S
		12-9 美丽猕猴桃 Actinidia melliana Hand.-Mazz. B.2x	2n=2x=58	48m+10sm (4SAT)	2A	116	1.9	0.07	57.74	4L+28M2+26M1
		12-10 长叶猕猴桃 Actinidia hemsleyana Dunn	2n=2x=58	2M+52m+4sm	1A	116	1.7	0	55.58	6L+20M2+32M1

续表

植物核型分析（果树及其野生近缘植物）

科	属	种	染色体数目	核型公式	核型分类	臂指数	最长染色体/最短染色体	臂比>2的染色体比例	核型不对称系数	染色体相对长度组成
IV 猕猴桃科 Actinidiaceae	12. 猕猴桃属 Actinidia Lindl.	12-11 阔叶猕猴桃 Actinidia latifolia (Gardn. & Champ.) Merr. B.2x	2n=2x=58	56m+2sm	1A	116	1.8	0	55.49	28M2+28M1+2S
		12-12 异色猕猴桃 Actinidia callosa Lindl. var. discaor	2n=2x=58	52m (2SAT) +6sm	2A	116	1.8	0.07	59.02	8L+12M2+36M1+2S
	13. 悬钩子属 Rubus Lindl	13-1 茅莓 Rubus parvifolius L.	2n=4x=28	24m+4sm	1A	56	1.2	0	60.25	18M2+10M1
		13-2 牛迭肚 Rubus crataegifolius Bge.	2n=2x=14	10m+4sm (4SAT)	1A	28	1.4	0	59.41	8M2+6M1
		13-2-1 牛迭肚 Rubus crataegifolius Bge.	2n=4x=28	20m+8sm (4SAT)	2A	56	1.4	0.07	59.83	6M2+22M1
		13-2-2 牛迭肚 Rubus crataegifolius Bge.	2n=3x=21	9m+9sm+3st (3SAT)	2A	39	1.3	0.29	63.4	12M2+9M1
		13-3 黑树莓 Rubus occidentalis L.	2n=2x=14	12m+2sm	1A	28	1.4	0	59.96	8M2+6M1
		13-4 欧洲红树莓 Rubus idaeus L.	2n=2x=14	14m	1A	28	1.3	0	59.17	8M2+6M1
		13-4-1 欧洲红树莓 Rubus idaeus L.	2n=3x=21	18m+3sm	1A	42	1.7	0	59.34	3L+6M2+12M1
		13-5 绿叶悬钩子 Rubus komarovi Nakai	2n=2x=14	12m+2sm	1A	28	1.4	0	60.71	6M2+8M1
		13-5-1 绿叶悬钩子 Rubus komarovi Nakai	2n=4x=28	26m+2sm	1A	56	1.3	0	58.36	14M2+14M1
		13-6 库叶悬钩子 Rubus sachalinensis Levl.	2n=2x=14	10m+4sm	1A	28	1.5	0	61.33	6M2+8M1
V 蔷薇科 Rosaceae		13-6-1 库叶悬钩子 Rubus sachalinensis Levl.	2n=4x=28	24m (2SAT) +4sm	2A	56	1.5	0.07	55.05	2L+4M2+22M1
		13-7 美洲树莓 Rubus alleghensiensis Porter.	2n=2x=14	12m+2sm	1A	28	1.4	0	58.3	8M2+6M1
		13-8 黄树莓 Rubus xanthocarpus Bur.et Franch.	2n=2x=14	10m+4sm (2SAT)	2A	28	1.6	0.29	62.86	2L+4M2+8M1
	14. 草莓属 Fragaria Lindl.	14 凤梨草莓 Fragaria ananassa Duch.	2n=8x=56	6M+48m+2sm	1A	112	1.7	0	55.13	4L+26M2+24M1+2S
	15. 醋栗属 Ribes Lindl.	15-1 欧洲黑穗醋栗 Ribes nigrum L.	2n=2x=16	14m+2sm (2SAT)	2A	32	1.3	0.13	55.61	6M2+10M1
		15-2 东北茶藨 Ribes mandshuricum Kom.	2n=2x=16	12m (2SAT) +4sm	1A	32	1.4	0	58.77	8M2+8M1
		15-3 红穗醋栗 Ribes rubrum L.	2n=2x=16	16m (2SAT)	1A	32	1.9	0	56.99	2L+4M2+8M1+2S
		15-4 芳香茶藨 Ribes odoratum Wendl.	2n=2x=16	12m+4sm	1A	32	1.4	0	58.67	4M2+12M1
		15-5 美洲醋栗 Ribes hirtellum Michx.	2n=2x=16	16m	1A	32	1.5	0	55.97	10M2+6M1
		15-6 醋栗 Ribes burejense Fr.Schmidt	2n=2x=16	16m (2SAT)	1B	32	2.1	0	54.87	2L+4M2+8M1+2S

续表

植物核型分析（果树及其野生近缘植物）

科	属	种	染色体数目	核型公式	核型分类	臂指数	最长染色体/最短染色体	臂比>2的染色体比例	核型不对称系数	染色体相对长度组成
V 蔷薇科 Rosaceae	16.越桔属 Vaccinium Lindl.	16-1 越桔 Vaccinium vitis-idaea L.	$2n=2x=24$	22m+2sm（2SAT）	1A	48	1.2	0	56.91	10M2+14M1
		16-2 高丛越桔 Vaccinium corymbasum L.	$2n=4x=48$	40m（2SAT）+8sm（2SAT）	1A	96	1.5	0	58.16	24M2+24M1
VI 桑科 Moraceae	17.桑属 Morus Lindl	17-1 桑 Morus alba L.	$2n=2x=28$	22m+6sm	1B	56	2.6	0	57.33	2L+12M2+10M1+4S
		17-2 蒙桑 Morus mongolica Schneid.	$2n=2x=28$	2M+24m（2SAT）+2sm	2A	56	2	0.07	57.3	4L+8M2+16M1
		17-3 华桑 Morus cathayana Hemsl.	$2n=2x=28$	2M+24m+2sm	2B	56	3.7	0.07	54.89	2L+8M2+16M1+2S
		17-4 山桑 Morus australis Poiret B.2x	$2n=2x=28$	4M+20m+4sm	2B	56	2.8	0.07	55.47	2L+8M2+16M1+2S
		17-5 鲁桑 Morus multicaulis Perr.	$2n=2x=28$	4M+20m+4sm	1B	56	2.3	0	56.71	2L+6M2+20M1
	18.无花果属 Ficus Lindl.	18 无花果 Ficus carica L.	$2n=2x=26$	14m+10sm+2st	2B	50	2.3	0.15	62.98	6L+2M2+14M1+4S
VII 柿树科 Ebenaceae	19.柿属 Diospyros Lindl	19-1 柿 Diospyros kaki L.f.	$2n=6x=90$	74m+16sm	2B	180	2.4	0.02	59	8L+26M2+54M1+2S
		19-2 油柿 Diospyros oleifera Cheng	$2n=2x=30$	22m+8sm	1A	60	1.6	0	59.31	2L+10M2+18M1
		19-3 君迁子 Diospyros lotus L.	$2n=2x=30+1$	18m+14sm（2SAT）	2B	64	2.7	0.13	61.34	4L+10M2+14M1+4S
		19-4 毛柿 Diospyros discolor Willd.	$2n=2x=30$	18m+12sm	2B	60	2	0.13	61.63	2L+12M2+14M1+2S
		19-5 老鸦柿 Diospyros rhombifolia Hemsl.	$2n=2x=30$	26m+4sm	2A	60	1.9	0.07	58.16	4L+10M2+14M1+2S
		19-6 浙江柿 Diospyros glaucifolia Metcalf.	$2n=2x=30$	30m（2SAT）	1A	60	1.8	0	57.58	4L+6M2+20M1
VIII 石榴科 Punicaceae	20.石榴属 Punica Lindl.	20 石榴 Punica granatum L.	$2n=2x=16$	2M+14m	1B	32	2.1	0	55.06	2L+2M2+10M1+2S
	21.沙棘属 Hippophae L.	21 沙棘 Hippophae rhamnoides Linn.	$2n=2x=24$	20m+4sm	2A	48	2	0.08	57.47	6L+2M2+12M1+4S
	22.五味子属 Schisandra Michx.	22-1 五味子 Schisandra chinensis（Turcz.）Baill.	$2n=2x=28$	24m+4sm	2A	56	1.5	0.14	57.11	2L+14M2+12M1
IX 胡颓子科 Elaeagnaceae		22-2 南五味子 Kadsura japonica Dunal.	$2n=2x=28$	2M+24m+2sm	1A	56	1.5	0	55.85	14M2+14M1
	23.接骨木属 SambucusLindl.	23 接骨木 Sambucus racemosa DC.	$2n=2x=36$	2M+10m+12sm+12st（2SAT）	2A	60	2	0.44	68.06	2L+18M2+12M1+4S
X 葫芦科 Cucurbitaceae	24.西瓜属 Citrullus Neck	24-1 西瓜 Citrullus vulgaris Schrad.	$2n=2x=22$	20m+2sm	1A	44	1.9	0	55.49	2L+8M2+10M1+2S
		24-2 西瓜 Citrullus vulgaris Schrad.	$2n=4x=44$	4M+38m+2sm	1A	88	1.6	0	54.08	22M2+22M1

续表

植物核型分析（果树及其野生近缘植物）

科	属	种	染色体数目	核型公式	核型分类	臂指数	最长染色体/最短染色体	臂比>2的染色体比例	核型不对称系数	染色体相对长度组成
X 葫芦科 Cucurbitaceae	25.甜瓜属 Cucumis Lindl.	25 甜瓜 Cucumis melo L.	$2n=2x=24$	2M+18m+4sm	1A	48	1.6	0	57.42	2L+6M2+16M1
	26.板栗属 Castanea Mill.	26-1 板栗 Castanea mollissima Bl.	$2n=2x=24$	18m+6sm（2SAT）	2B	48	2.2	0.08	59.33	4L+2M2+16M1+2S
		26-1-1 野板栗 Castanea mollissima Bl.sp.	$2n=2x=24$	14m+10sm（2SAT）	2B	48	2	0.08	60.12	2L+8M2+12M1+2S
		26-2 茅栗 Castanea seguinii Dode B.2x	$2n=2x=24$	14m+10sm（2SAT）	2B	48	2.1	0.17	62.25	4L+6M2+12M1+2S
		26-3 美洲栗 Castanea dentata Borkh.	$2n=2x=24$	18m+6sm（2SAT）	2B	48	2	0.08	61.46	2L+8M2+12M1+2S
		26-4 日本栗 Castanea crenata S. et Z.	$2n=2x=24$	16m+8sm	2B	48	2.2	0.08	61.64	4L+6M2+12M1+2S
XI 壳斗科 Fagaceae		26-4-1 朝鲜栗 Castanea crenata var.dulcis	$2n=2x=24$	16m（2SAT）+8sm	2B	48	2.4	0.08	61.9	2L+8M2+12M1+2S
	27.榛属 Corylus Lindl.	27 榛 Corylus heterophylla Fisch.	$2n=2x=22$	2M+18m+2sm（2SAT）	2A	44	1.9	0.09	55	4L+6M2+10M1+2S
	28.核桃属 Juglans Lindl.	28-1 核桃 Juglans regia L.	$2n=2x=32$	2M+24m（4SAT）+6sm	2B	64	2	0.06	58.19	4L+10M2+12M1+6S
		28-2 核桃楸 Juglans mandshurica Maxim.	$2n=2x=32$	26m+6sm	2B	64	2.6	0.06	61	4L+14M2+8M1+6S
		28-3 姬核桃 Juglans sieboldiana Maxim.var.cordiformis Mak.	$2n=2x=32$	26m+6sm	2B	64	2.6	0.06	61.36	8L+4M2+14M1+6S
		28-4 野核桃 Juglans cathayensis Dode	$2n=2x=32$	26m+6sm	2B	64	3	0.06	62.02	4L+8M2+14M1+6S
	29.山核桃属 Carya Nutt.	29-1 山核桃 Carya cathayensis Sarg.	$2n=2x=32$	26m+6sm	2B	64	2	0.13	59.67	4L+10M2+16M1+2S
		29-2 薄壳山核桃 Carya illinoinensis Koch.	$2n=2x=32$	2M+20m（2SAT）+10sm（2SAT）	2A	64	1.9	0.06	58.42	4L+8M2+18M1+2S
XIII 银杏科 Ginkgoaceae	30.银杏属 Ginkgo Lindl.	30-1 银杏（♀）Ginkgo biloba L.	$2n=2x=24$	4m+10sm（2SAT）+10st	3A	38	1.9	0.77	78.62	4L+4M2+18M1
		30-2 银杏（♂）Ginkgo biloba L.	$2n=2x=24$	4m+10sm（2SAT）+10st	3A	38	1.9	0.77	78.69	4L+4M2+18M1
XIV 禾本科 Gramineae	31.甘蔗属 Saccharum Lindl.	31 甘蔗 Saccharum officinarum L.R	$2n=80$	4M+62m（2SAT）+12sm+2st（2SAT）	2B	158	3.6	0.13	58.33	14L+12M2+44M1+10S
XV 芸香科 Rutaceae	32.枸桔属 Poncirus Raf.	32-1 普通枳 Poncirus trifoliata (L.) Raf.	$2n=2x=18$	12m+6sm（2SAT）	2A	36	1.8	0.11	59.9	8M2+9M1+1S
		32-2 枳 Poncirus trifoliata (L.) Raf	$2n=2x=18$	13m（2SAT）+5sm	2A	36	1.6	0.11	59.82	1L+11M2+6M1
		32-3 南京枳 Poncirus trifoliata (L.) Raf.cv.nanjingzhi	$2n=2x=18$	9m+8sm（1SAT）+1st（1SAT）	2A	35	1.6	0.33	63.32	2L+6M2+10M1

第二章 中国植物染色体研究结果 | 95

续表

植物核型分析（果树及其野生近缘植物）

科	属	种	染色体数目	核型公式	核型分类	臂指数	最长染色体/最短染色体	臂比>2的染色体比例	核型不对称系数	染色体相对长度组成
	32.枸桔属 Poncirus Raf.	32-4 枣阳积 Poncirus trifoliata Raf.cv.zaoyangzhi	$2n=2x=18$	12m（2SAT）+6sm	1A	36	1.8	0	61.25	4L+5M2+7M1+2S
		32-5 枣阳积 Poncirus trifoliata Raf.cv.zaoyangzhi	$2n=4x=36$	27m+9sm（4SAT）	2A	72	1.8	0.17	61.34	2L+18M2+16M1
		32-6 飞龙积 Poncirus trifoliata cv.feilongzhi	$2n=2x=18$	14m+4sm（2SAT）	2A	36	1.9	0.11	58.92	2L+6M2+9M1+1S
	33.金桔属 Fortunella Swingle	33-1 罗浮 Fortunella margarita (Lour.) Swingle	$2n=2x=18$	13m+5sm（2SAT）	2A	36	1.9	0.22	59.76	2L+6M2+9M1+1S
		33-2 罗纹 Fortunella japonica Thunb.Swingle	$2n=2x=18$	11m+7sm（3SAT）	2A	36	2	0.33	60.83	2L+9M2+5M1+2S
		33-3 长寿金柑 Fortunella obovata Tannka	$2n=2x=18$	11m（1SAT）+7sm（2SAT）	2A	36	1.8	0.28	60.83	1L+10M2+7M1
		33-4 金豆 Fortunella hindsii var.chintou Swingle	$2n=2x=18$	7m+11sm（3SAT）	2A	36	1.9	0.39	64.17	2L+7M2+6M1+3S
		33-5 金豆 Fortunella hindsii var.chintou Swingle	$2n=2x=18$	12m（2SAT）+6sm（1SAT）	2A	36	2	0.06	59.1	2L+8M2+6M1+2S
XV芸香科 Rutaceae		34-1 红河大翼橙 Citrus hongheensis Y.L.D.L.	$2n=2x=18$	7m（2SAT）+10sm（2SAT）+1st	3B	35	3.1	0.56	63.41	3L+3M2+7M1+5S
		34-2 红河大翼橙 Citrus hongheensis Y.L.D.L.	$2n=2x=18$	7m（3SAT）+11sm（1SAT）	2B	36	3.1	0.44	64.9	4L+6M2+2M1+6S
		34-3 来檬 Citrus aurantifolia (Chrit.) Swingle	$2n=2x=18$	11m+7sm（2SAT）	2A	36	1.6	0.22	62.04	2L+4M2+12M1
		34-4 枸橼 Citrus medica L.	$2n=2x=18$	8m（2SAT）+10sm（2SAT）	2B	36	2.3	0.22	62.22	2L+8M2+4M1+4S
		34-5 枸橼 Citrus medica L.	$2n=2x=18$	11m（3SAT）+6sm（1SAT）+1st	2B	35	2.1	0.28	63.49	2L+8M2+4M1+4S
	34.柑橘属 Citrus Lindl.	34-6 尤力克柠檬 Citrus limon (L.) Burm.f.cv.youlikeningmeng	$2n=2x=18$	5m+13sm（4SAT）	3A	36	1.8	0.67	68.07	2L+4M2+12M1
		34-7 柠檬 Citrus limon (L.) Burm. f.	$2n=2x=18$	8m（2SAT）+10sm（2SAT）	2A	36	1.8	0.11	61.82	2L+6M2+10M1
		34-8 八柑檬 Citrus limon (L.) Burm. f.cv.bagangmeng	$2n=2x=18$	5m+13sm（2SAT）	2A	36	2	0.33	65.2	4L+2M2+11M1+1S
		34-9 八柑檬 Citrus limon (L.) Burm. f.cv.bagangmeng	$2n=2x=18$	4m+11sm（3SAT）+3st	3A	33	2	0.67	67.65	2L+5M2+11M1
		34-10 红黎檬 Citrus limonia Osbeck.	$2n=2x=18$	12m（4SAT）+6sm	2A	36	1.7	0.06	60.95	4L+43M2+11M1
		34-11 红黎檬 Citrus limonia Osbeck.	$2n=2x=18$	10m（2SAT）+8sm（2SAT）	2A	36	1.7	0.22	62.19	1L+9M2+8M1
		34-12 垫江柚 Citrus grandis Osbeck.cv.dianjiangyou	$2n=2x=18$	13m+5sm（2SAT）	2A	36	1.9	0.06	59.7	2L+4M2+12M1
		34-13 梁平柚 Citrus giandis Osbeck.cv.liangpingyou	$2n=2x=18$	9m+9sm（4SAT）	2A	36	2	0.39	62.22	2L+6M2+8M1+2S

续表

植物核型分析（果树及其野生近缘植物）

科	属	种	染色体数目	核型公式	核型分类	臂指数	最长染色体/最短染色体	臂比>2的染色体比例	核型不对称系数	染色体相对长度组成
		34-14 晚白柚 Citrus giandis Osbeck.cv.wanbaiyou	$2n=2x=18$	6m+12m (4SAT)	2A	36	2	0.44	62.99	2L+4M2+12M1
		34-15 麻豆文旦 Citrus giandis Osbeck.cv.madouwendan	$2n=2x=18$	8m+10sm (4SAT)	2A	36	2	0.28	62.51	3L+5M2+9M1+1S
		34-16 沙田柚 Citrus giandis Osbeck.cv.shatianyou	$2n=2x=18$	14m+4sm (2SAT)	2A	36	1.9	0.17	58.74	2L+4M2+8M1+4S
		34-17 葡萄柚 Citrus paradisii Macf.	$2n=2x=18$	5m+13m (4SAT)	2A	36	1.9	0.44	64.54	1L+7M2+7M1+3S
		34-18 邓肯葡萄柚 Citrus paradisii Macf.cv.dengken	$2n=2x=18$	7m+11sm (3SAT)	2A	36	2	0.44	63.48	3L+5M2+9M1+1S
		34-19 酸橙 Citrus aurantium L.	$2n=2x=18$	6m+12m (2SAT)	2A	36	1.9	0.44	63.68	4L+4M2+9M1+1S
		34-20 摩洛哥酸橙 Citrus aurantium L.cv.moluogesuancheng	$2n=2x=18$	12m+6sm (2SAT)	2A	36	1.8	0.17	59.87	3L+5M2+9M1+1S
		34-21 三宝柑 Citrus sulcata Tanaka	$2n=2x=18$	14m (2SAT) +4sm	2A	36	2	0.11	59.83	2L+7M2+7M1+2S
		34-22 甜橙 Citrus sinensis Osbeck.	$2n=2x=18$	10m (2SAT) +8sm	2A	36	1.7	0.17	62.11	2L+4M2+12M1
		34-23 锦橙 Citrus sinensis Osbeck.cv.jincheng	$2n=2x=18$	12m (1SAT)+6sm (1SAT)	2A	36	2	0.17	60.66	2L+6M2+8M1+2S
		34-24 罗伯逊脐橙 Citrus sinensis Osbeck.cv.robertsonnavel	$2n=2x=18$	13m (1SAT)+5sm (1SAT)	2B	36	2.1	0.11	58.27	3L+6M2+7M1+2S
		34-25 血橙 Citrus sinensis Osbeck.cv.bloodorange	$2n=2x=18$	8m+10sm (2SAT)	2A	36	1.9	0.33	62.11	2L+7M2+7M1+2S
XV 芸香科 Rutaceae	34.柑橘属 Citrus Lindl.	34-36 实美橙 Citrus sinensis Osbeck.cv.shimeicheng	$2n=2x=18$	12m+6sm (2SAT)	2A	36	1.8	0.06	58.75	2L+7M2+7M1+2S
		34-27 雪柑 Citrus sinensis Osbeck.cv.xuegan	$2n=2x=18$	1M+10m (2SAT) +7sm (1SAT)	2A	36	2	0.22	59.42	2L+5M2+9M1+2S
		34-28 夏橙 Citrus matsudaidai Hayata.cv.valancia	$2n=2x=18$	8m+10sm (2SAT)	2A	36	1.9	0.33	62.46	2L+8M2+7M1+1S
		34-29 宜昌橙 Citrus ichangensis Swingle.	$2n=2x=18$	6m+12m (2SAT)	2A	36	1.7	0.33	64.09	2L+7M2+7M1+2S
		34-30 宜昌橙 Citrus ichangensis Swingle.	$2n=2x=18$	8m+8sm (3SAT) +2st (1SAT)	2B	36	2.3	0.22	63.8	4L+4M2+7M1+3S
		34-31 香园 Citrus wilsonii Tanaka	$2n=2x=18$	6m+10sm+2st (1SAT)	3A	34	1.8	0.67	65.74	2L+6M2+8M1+2S
		34-32 皱皮柑 Citrus vessucosa Hort.	$2n=2x=18$	10m(1SAT)+8sm(1SAT)	2A	34	1.7	0.39	60.79	2L+5M2+11M1
		34-33 瓯柑 Citrus suavissima Hort.et Tanaka	$2n=1x=18$	12m+6sm	2A	36	1.8	0.28	59.91	2L+7M2+9M1
		34-34 温州蜜柑 Citrus unshu Marc.	$2n=2x=18$	12m (1SAT) +6sm	2A	36	1.6	0.28	61.37	12M2+6M1

续表

植物核型分析（果树及其野生近缘植物）

科	属	种	染色体数目	核型公式	核型分类	臂指数	最长染色体/最短染色体	臂比>2的染色体比例	核型不对称系数	染色体相对长度组成
XV芸香科 Rutaceae	34.柑橘属 Citrus Lindl.	34-35 椪柑 Citrus poonensis Hort.et Tanaka	$2n=2x=18$	12m+6sm（2SAT）	2A	36	1.7	0.28	61.04	2L+7M2+9M1
		34-36 土柑 Citrus chuana Hort.	$2n=2x=18$	12m+6sm（2SAT）	2B	36	2	0.17	59.01	2L+4M2+11M1+1s
		34-37 红桔 Citrus tangerita Hort.	$2n=2x=18$	10m+8sm（2SAT）	2A	36	1.8	0.17	61.16	2L+4M2+12M1
		34-38 克里曼丁桔 Citrus clementina Hort.	$2n=2x=18$	12m（2SAT）+6sm	2A	36	1.8	0.33	58.92	2L+6M2+10M1
		34-39 威尔金桔 Citrus reticulata Blanco	$2n=2x=18$	9m+9sm（2SAT）	2A	36	2	0.22	61.44	2L+7M2+8M1+1S
		34-40 蕉柑 Citrus reticulata Blanco	$2n=2x=18$	13m+5sm（1SAT）	2A	36	1.8	0.17	59	2L+7M2+8M1+1S
		34-41 朱桔 Citrus erythrosa Tanaka cv.zhuju	$2n=2x=18$	14m（1SAT）+4sm（1SAT）	2A	36	1.7	0.22	60.26	2L+6M2+10M1
		34-42 八月桔 Citrus erythrosa Tanaka cv.bayueju	$2n=2x=18$	14m（2SAT）+4sm	2A	36	1.7	0.11	60.97	2L+5M2+9M1+2S
		34-43 蔓桔 Citrus tardiferox Hort.	$2n=2x=18$	12m+6sm	2A	36	1.8	0.22	60.5	2L+4M2+12M1
		34-44 年桔 Citrus chachiensis var.oleocarpa Hort.	$2n=2x=18$	12m+6sm	2A	36	1.8	0.06	60.8	2L+6M2+9M1+1S
		34-45 道县野桔 Citrus daoxianensis	$2n=2x=18$	10m+8sm（2SAT）	2A	36	1.5	0.11	62.51	10M2+8M1
		34-46 金桔 Citrus madurensis Lour.	$2n=2x=18$	10m（1SAT）+8sm（2SAT）	2A	36	1.9	0.11	60.84	1L+10M2+5M1+2S
		34-47 四季桔 Citrus microcarpa Bunge	$2n=2x=18$	11m（1SAT）+7sm（1SAT）	2A	36	1.9	0.22	61.94	2L+8M2+6M1+2S
	35.属间杂种 Hybrid between genus	35-1 枳橙 Citrange（P. trifoliata × C. sinensis）	$2n=2x=18$	12m+6sm（4SAT）	2A	36	1.7	0.11	60.83	2L+6M2+10M1
		35-2 枳柚 Citrange（P. trifoliata × C. grandis）	$2n=2x=18$	12m（SAT）+6sm（1SAT）	2B	36	2.1	0.28	59.93	2L+5M2+11M1
		35-3 蜜柚 Citrus Reticulata × Citrus grangis	$2n=2x=18$	12m+6sm（3SAT）	2B	36	2	0.22	59.89	2L+9M2+5M1+2S
XVI蔷薇科 Rosaceae	36.枇杷属 Eriobotrya Lindl.	36 枇杷 Eriobotrya japonica Lindl.	$2n=2x=34$	34m	1A	68	1.6	0	54.5	18M2+16M1
XVII无患子科 Sapindaceae	37.荔枝属 Litchi Sonn.	37 荔枝 Litchi chinensis Sonn.	$2n=2x=30$	24m+6sm	2A	60	1.7	0.07	58.03	2L+12M2+16M1
	38.龙眼属 Dimocarpus Lour.	38 龙眼 Dimocarpus longana Lour.	$2n=2x=30$	16m+12sm+2st	2B	58	2	0.2	61.7	4L+6M2+18M1+2S
XVIII橄榄科 Burseraceae	39.橄榄属 Canarium Lindl.	39 橄榄 Canarium album Raeusch.	$2n=2x=48$	2M+38m（2SAT）+8sm	2B	96	2.4	0.08	58.49	14L+6M2+18M1+10S
		39-1 乌榄 Canarium pimela Koen.	$2n=2x=48$	2M+44m+2sm（2SAT）	2B	96	2.1	0.04	55.62	4L+14M2+28M1+2S
		39-2 越南榄 Canarium tonkinense Engl.	$2n=2x=48$	36m+10sm+2st（2SAT）	2A	94	1.8	0.04	59.79	2L+20M2+26M1

续表

植物核型分析（果树及其野生近缘植物）

科	属	种	染色体数目	核型公式	核型分类	臂指数	最长染色体/最短染色体	臂比>2的染色体比例	核型不对称系数	染色体相对长度组成
XIX 漆树科 Anacardiaceae	40.芒果属 Mangifera Lindl.	40 芒果 Mangifera indica L.	$2n=2x=40$	2M+26m+10sm+2st (2SAT)	2A	78	1.8	0.25	59.85	4L+14M2+20M1+2S
XX 樟科 Lauraceae	41.鳄梨属 Persea Mill.	40-1 野芒果 Mangifer sylvatica Roxb.	$2n=2x=40$	32m+8sm (2SAT)	2A	80	1.6	0.2	58.84	4L+12M2+24M1
		41 鳄梨 Persea americana Mill.	$2n=2x=24$	4M+18m+2sm	2A	48	1.8	0.08	55.92	2L+8M2+12M1+2S
XXI 桑科 Moraceae	42.木波罗属 Artocarpus Forst.	42 木波罗 Artocarpus heteroyllus Lam.	$2n=2x=56$	2M+36m+16sm (2SAT) +2st	2A	110	1.8	0.21	60.86	4L+20M2+30M1+2S
XXII 芸香科 Rutaceae	43.黄皮属 Clausena Burm.f.	43 黄皮 Clausena lansium (Lour.) Skeels	$2n=2x=18$	16m+2sm	1A	36	1.4	0	55.8	2L+4M2+12M1
XXIII 桃金娘科 Myrtaceae	44.番石榴属 Psidium Lindl.	44 番石榴 Psidium guajava L.	$2n=2x=22$	22m	1B	44	2	0	57.19	6L+2M2+12M1+2S
XXIV 番荔枝科 Annonaceae	45.番荔枝属 Annona Lindl.	45 番荔枝 Annona squamosa L.	$2n=2x=18$	4M+10m+4sm	2A	36	1.5	0.22	57.21	2L+4M2+12M1
		45-1 刺果番荔枝 Annona muricata L.	$2n=2x=14$	4m+10sm	3A	28	1.8	0.71	63.25	2L+6M2+6M1
XXV 番薇科 Rosaceae	46.番木瓜属 Carica Lindl.	46 番木瓜 Carica Papaya L.	$2n=2x=18$	12m+6sm	2A	36	1.7	0.11	60.36	2L+6M2+10M1
XXVI 葫芦科 Cucurbitaceae	47.油瓜属 Hodgsonia Hook.f.et.Thoms.	47 油瓜 Hodgsonia macrocarpa (Bl.) Cogn. var. capnicorpa	$2n=2x=18$	10m+8sm	2B	44	1.8	0.22	60.97	4L+4M2+8M1+2S
XXVII 芭蕉科 Musaceae	48.芭蕉属 Musa Lindl.	48（AA）大蕉 Musa paradisiaca L.	$2n=2x=22$	15m+7sm	2A	44	2.2	0.09	61.24	1L+4M2+17M1
		48-1（BB）大蕉 Musa paradisiaca L.	$2n=2x=22$	14m+8sm (2SAT)	2A	44	1.5	0.18	62.03	12M2+10M1
		48-2 大蕉 Musa paradisiaca L.	$2n=3x=33$	12m+9sm+12st	2A	54	1.5	0.36	67.2	3L+9M2+21M1
XXVIII 凤梨科 Bromeliaceae	49.凤梨属 Ananas Mill.	49 凤梨 Ananas comosus (L.) Merr.	$2n=2x=50$	2M+48m	1A	100	1.4	0	55.65	26M2+24M1
XXVIV 棕桐科 Palmae	50.椰子属 Cocos Lindl.	50 椰子 Cocos nucifera L.	$2n=2x=32$	18m (2SAT) +12m+2st	2B	62	2.4	0.25	61.41	4L+14M2+10M1+4S
XXX 桃金娘科 Myrtaceae	51.蒲桃属 Syzygium Gaertn.	51.蒲桃 Syzygium jambos Alston	$2n=4x=44$	28m+14sm (2SAT) +2st	2B	86	2.1	0.23	61.67	8L+10M2+24M1+2S

第二节　中国农作物及其野生近缘植物基因组染色体研究结果

提要： 本图谱收录了作者对我国 21 科 59 属 243 种农作物及其野生近缘植物核型分析结果，总计有 450 版染色体图。为显示我国植物染色体研究水平，本书选取了部分染色体图版供参考。

中国是世界栽培植物起源最早和最大的中心，不仅有极丰富的农作物栽培品种，而且蕴藏着许多珍贵的农作物野生近缘植物，如野生大豆、野生稻、野生大麦、野生茶等，这在世界上都是罕见的，这些野生资源将是人类改造现有栽培作物的重要基因库。我国在过去不同历史阶段，对这些农作物资源从不同层面进行过诸多的研究，对农业发展起到重要作用，但是，在细胞学水平对栽培作物进行深入的研究，不论在我国和世界各国都只见零散报道，没有系统的研究。作者历经 20 多年的不懈努力，终于完成了我国农作物及其野生近缘植物核型分析工作。其中，在小麦属（*Tirtcicum*）、山羊草属（*Aegilops*）、大麦属（*Hordeum*）、黑麦属（*Secale*）、玉黍属（*Zea*）、狗尾草属（*Staria*）、葱属（*Allium*）等染色体研究中，应用 20 世纪 70 年代后发展起来的染色体 C-带，N-带技术，使核型分析达到了 γ 水平。在大麦属、黑麦属、野豌豆属（*Vicia*）和百合属（*Lilium*）染色体研究中，在国际上首次获得了植物染色体高分辨 G-带图像。在小黑麦属（*Triticale*）、黑麦属、棉属（*Gossypium*）、甘薯属（*Ipomoea*）染色体研究中还应用了荧光原位杂交（FISH）技术，使图谱的染色体研究达到了当今国际发展的先进水平。本图谱在取材方面以我国农作物资源为主，同时也兼顾世界分布的野生资源，如野生稻、野生棉。在核型分析方法上，采用了全国统一的核型分析标准，并由计算机编程统一计算，结果准确。每种作物都经过反复试验，在大量分裂细胞中力求选取标准细胞，即染色体分散良好，各组分显示清楚的提供给读者。本书所提供的所有染色体资料均为作者亲自创作，没有编辑或引用他人资料。因此，这些原始染色体信息资料对研究我国栽培植物分类鉴定、起源进化、杂交育种具有重要理论意义和应用价值。

2.2.1 部分农作物染色体

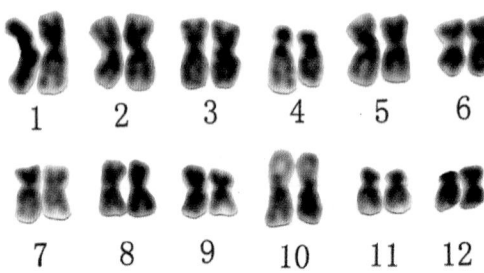

陆稻核型
Karyotype of *Oryza sativa* L.

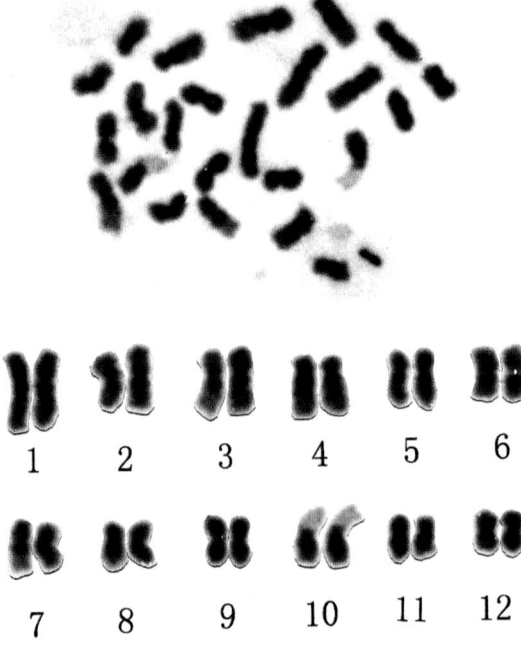

普通野生稻核型（A350）
Karyotype of *Oryza rufipogon* cv.A350

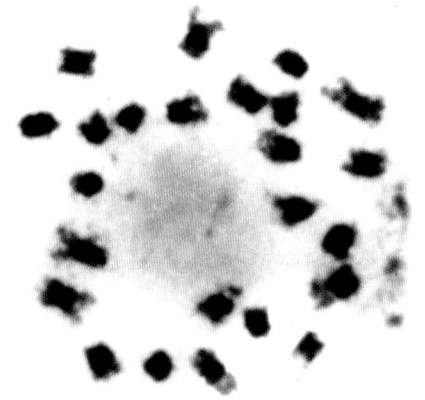

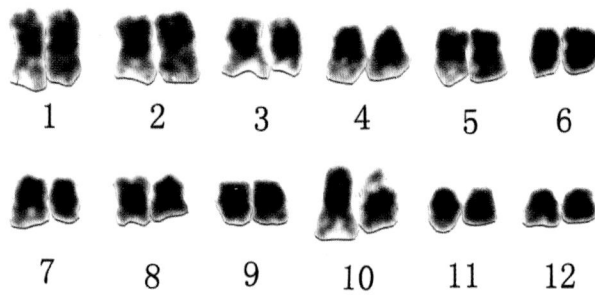

疣粒野生稻核型
Karyotype of *Oryza meyeriana* (Zoll. Et Mor. Ex steud.) Baill

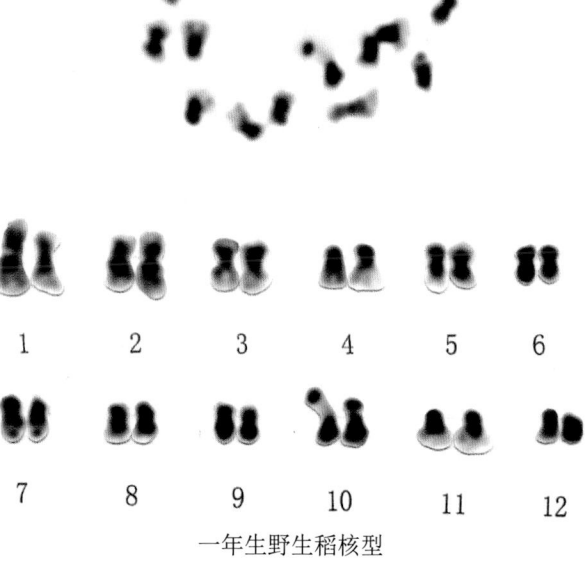

一年生野生稻核型
Karyotype of *Oryza nivara* Shama et Shastry

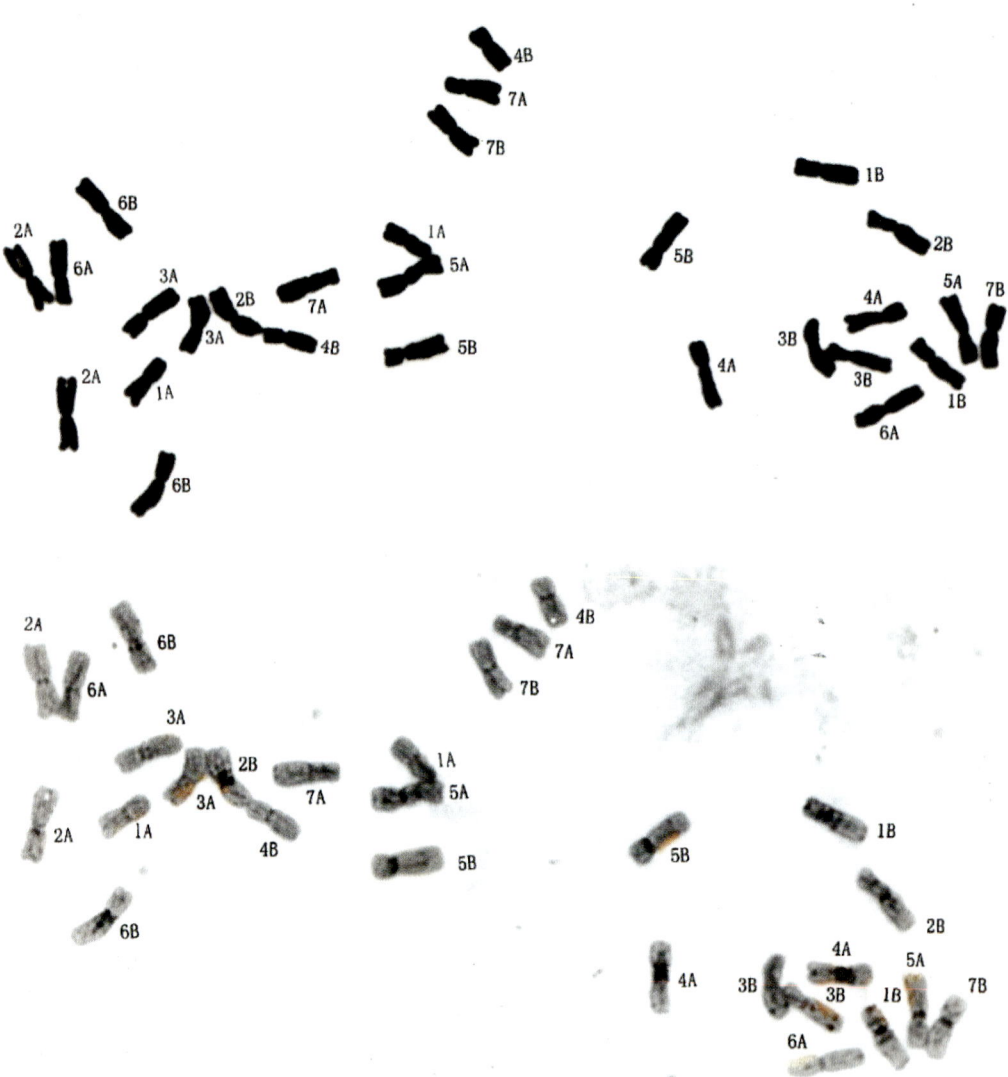

野生二粒小麦核型与分带
Karyotype and banding of *Triticum dicoccoid* Korn.

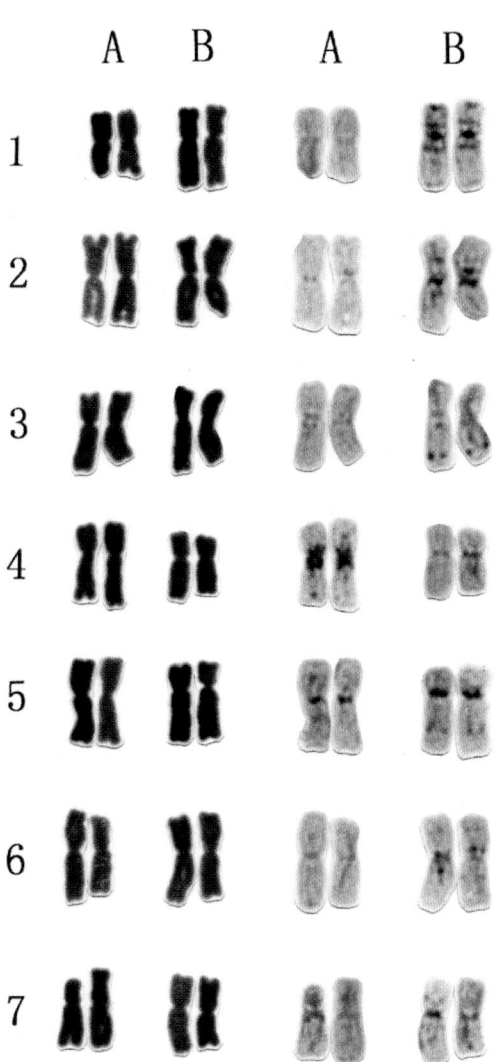

野生二粒小麦核型与带型
Karyotype and banding pattern of *Triticum dicoccum* Shubl.

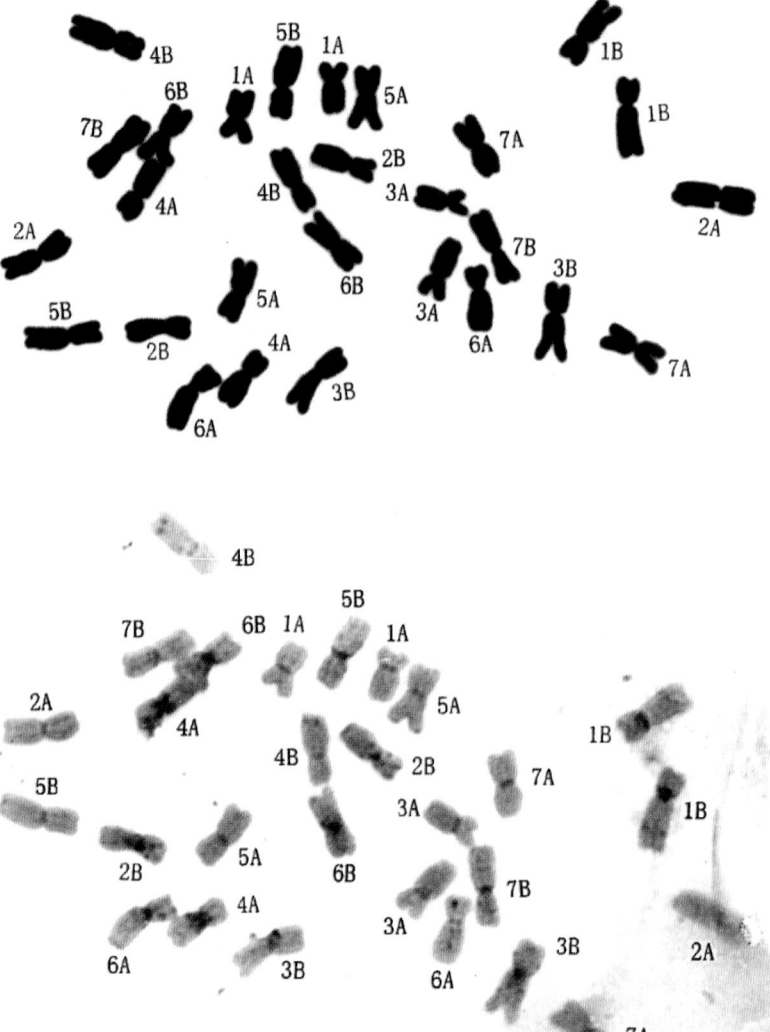

栽培二粒小麦核型与分带
Karyotype and banding of *Triticum dicoccum* Shubl.

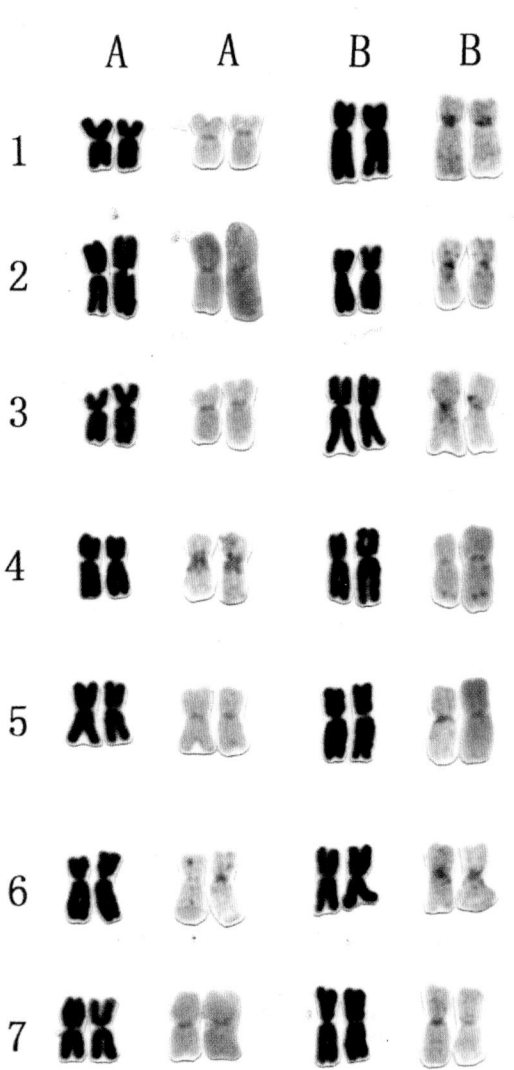

栽培二粒小麦核型与带型
Karyotype and banding pattern of *Triticum dicoccum* Shubl.

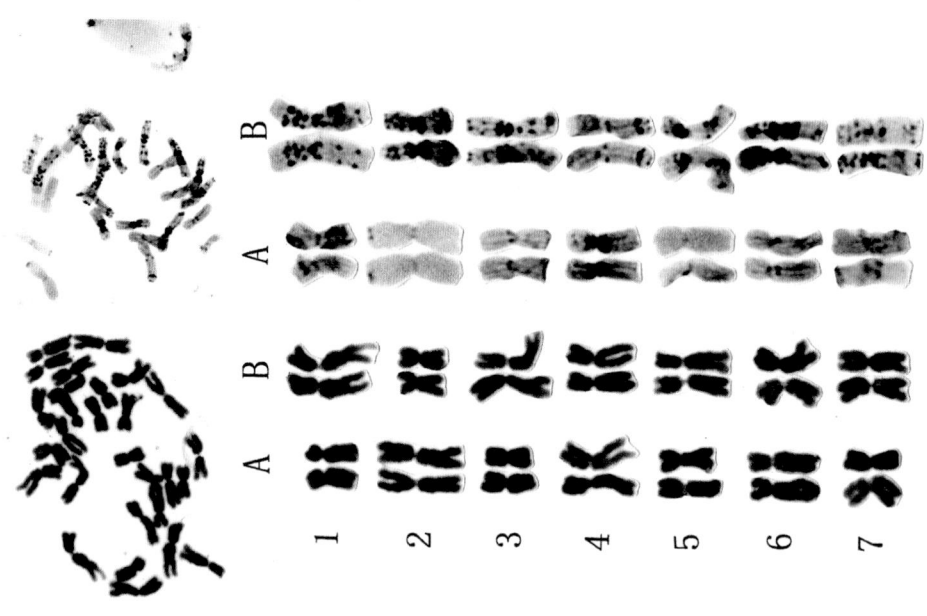

硬粒小麦核型与带型
Karyotype and banding pattern of *Triticum dicoccum* Dest.

东方小麦核型与带型
Karyotype and banding pattern of *Triticum dicoccum* Jakubz.

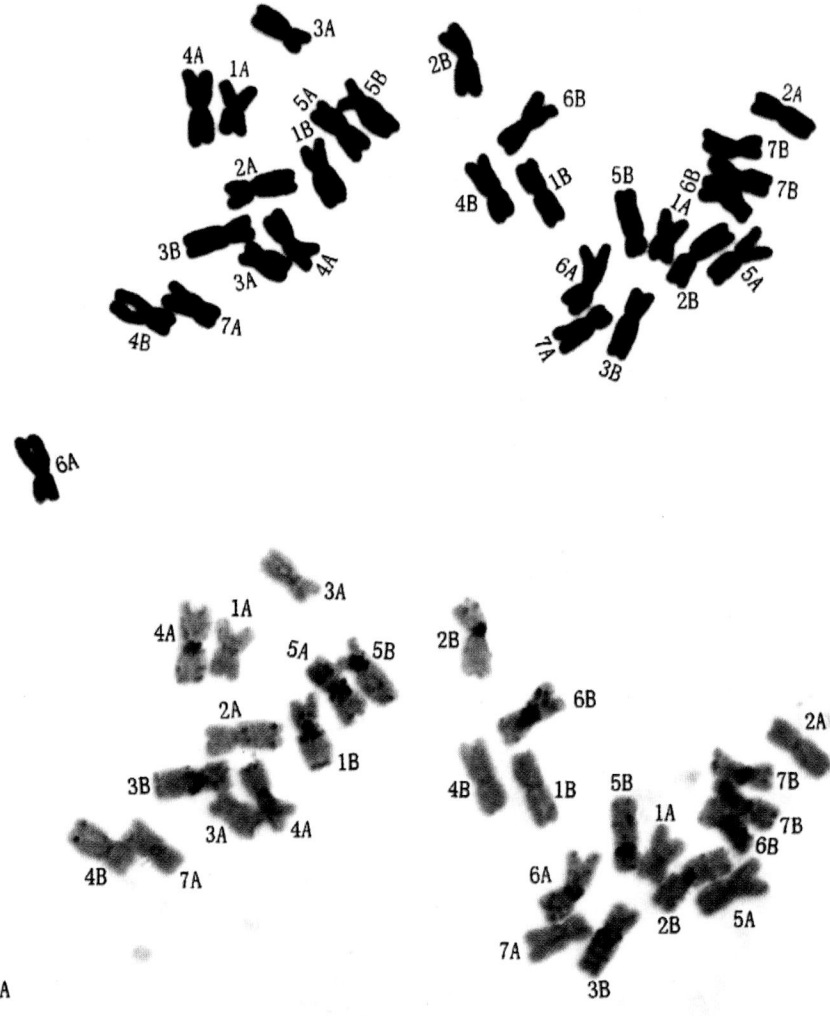

波斯小麦核型与分带
Karyotype and banding of *Triticum Carthlicum* Nevski.

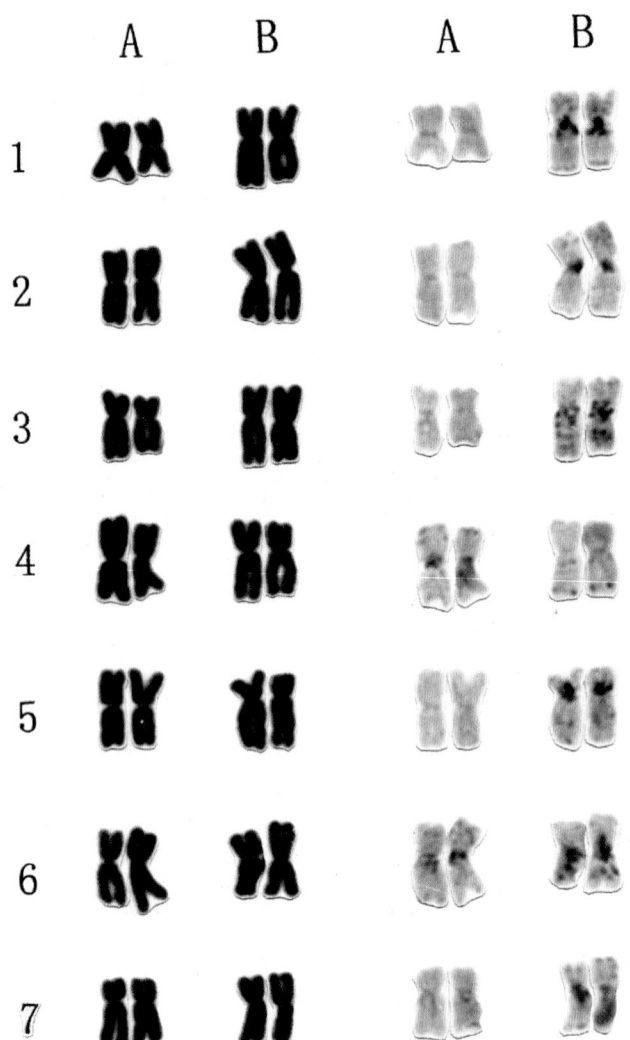

波斯小麦核型与带型
Karyotype and banding pattern of *Triticum Carthlicum* Nevski.

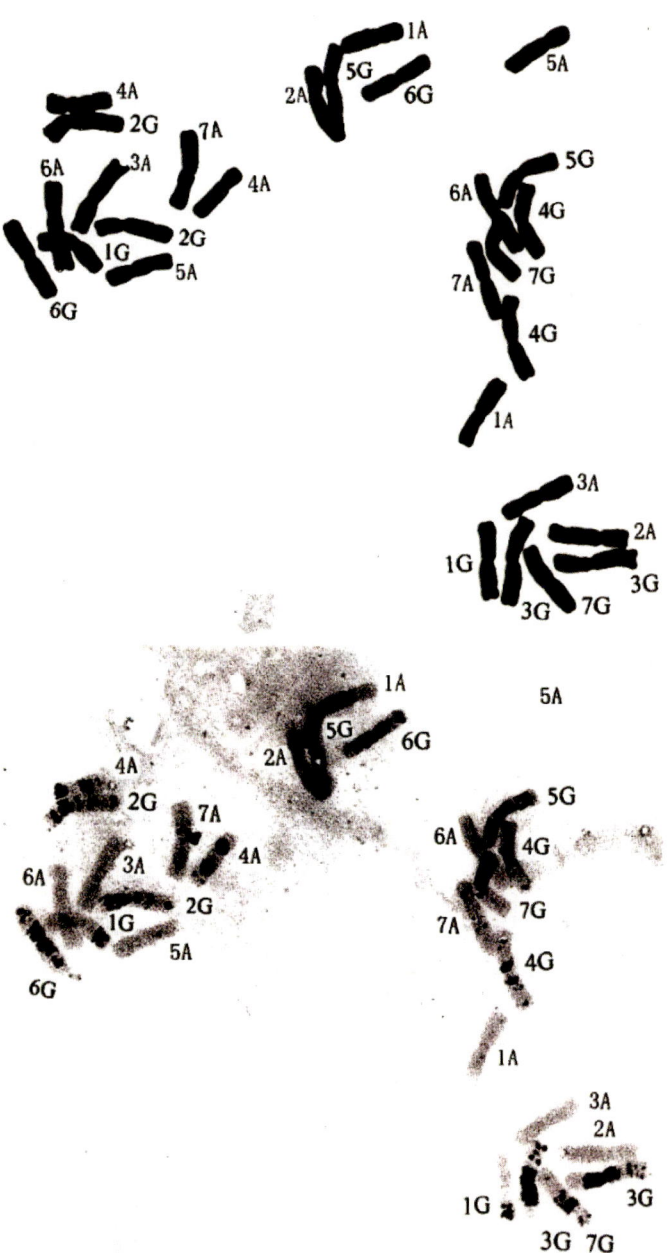

提莫非维小麦核型与分带
Karyotype and banding of *Triticum timopheevii* Zhuk.

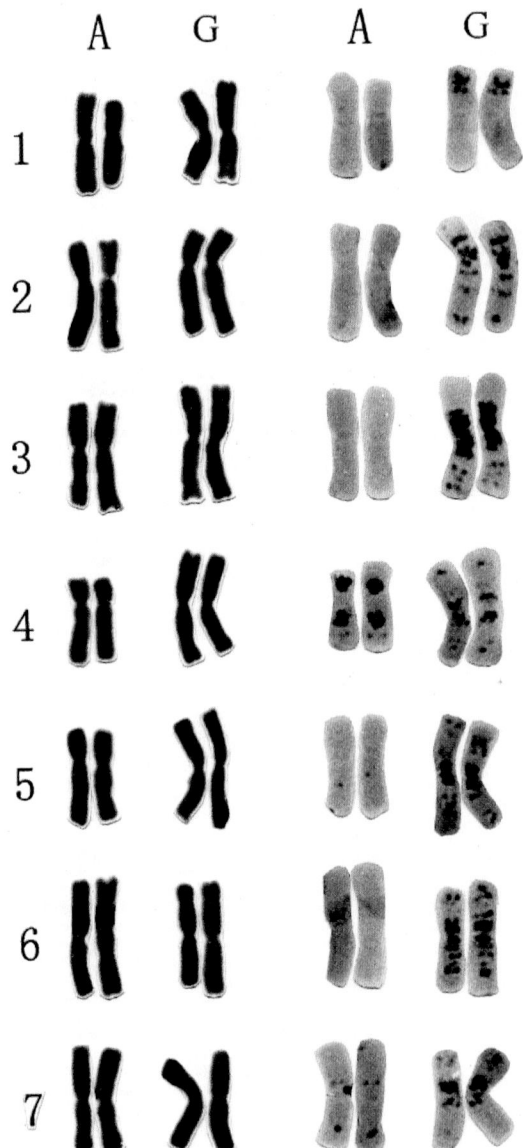

提莫非维小麦核型与带型
Karyotype and banding pattern of *Triticum timopheevii* Zhuk.

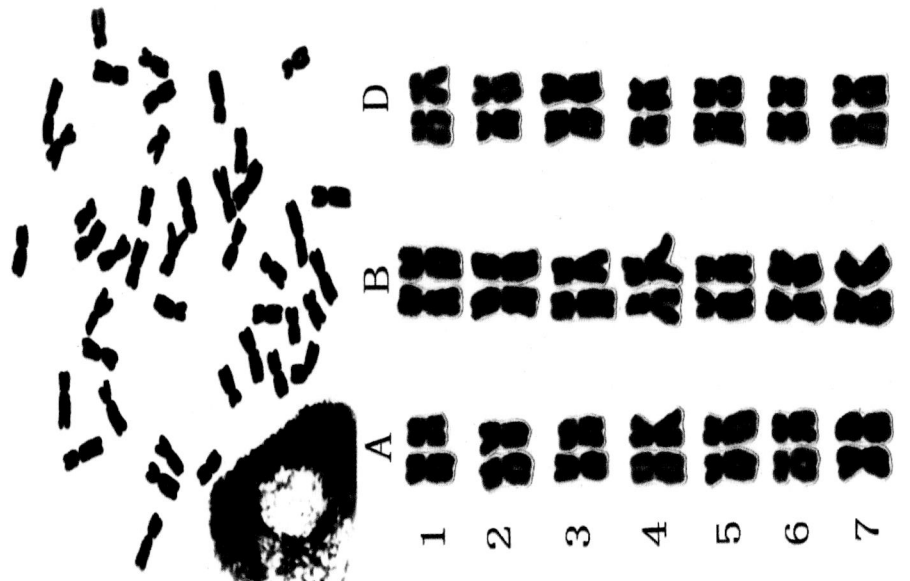

印度圆粒小麦核型

Karyotype of *Triticum sphaerococcum* Perc.

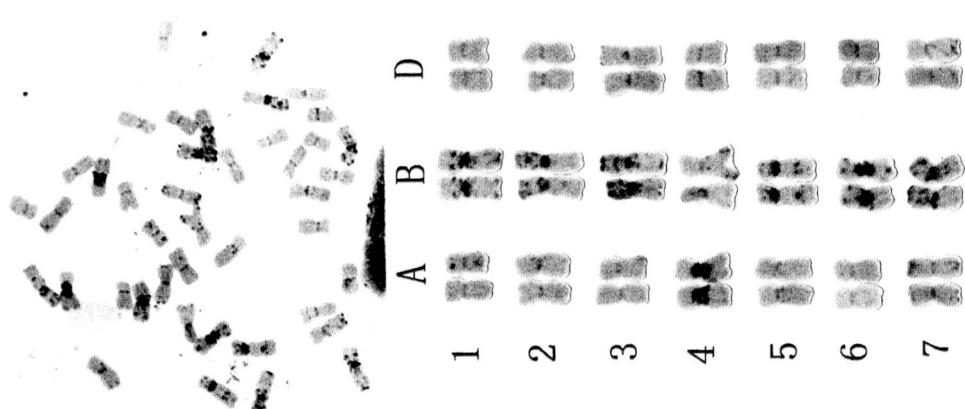

印度圆粒小麦带型

Banding pattern of *Triticum sphaerococcum* Perc.

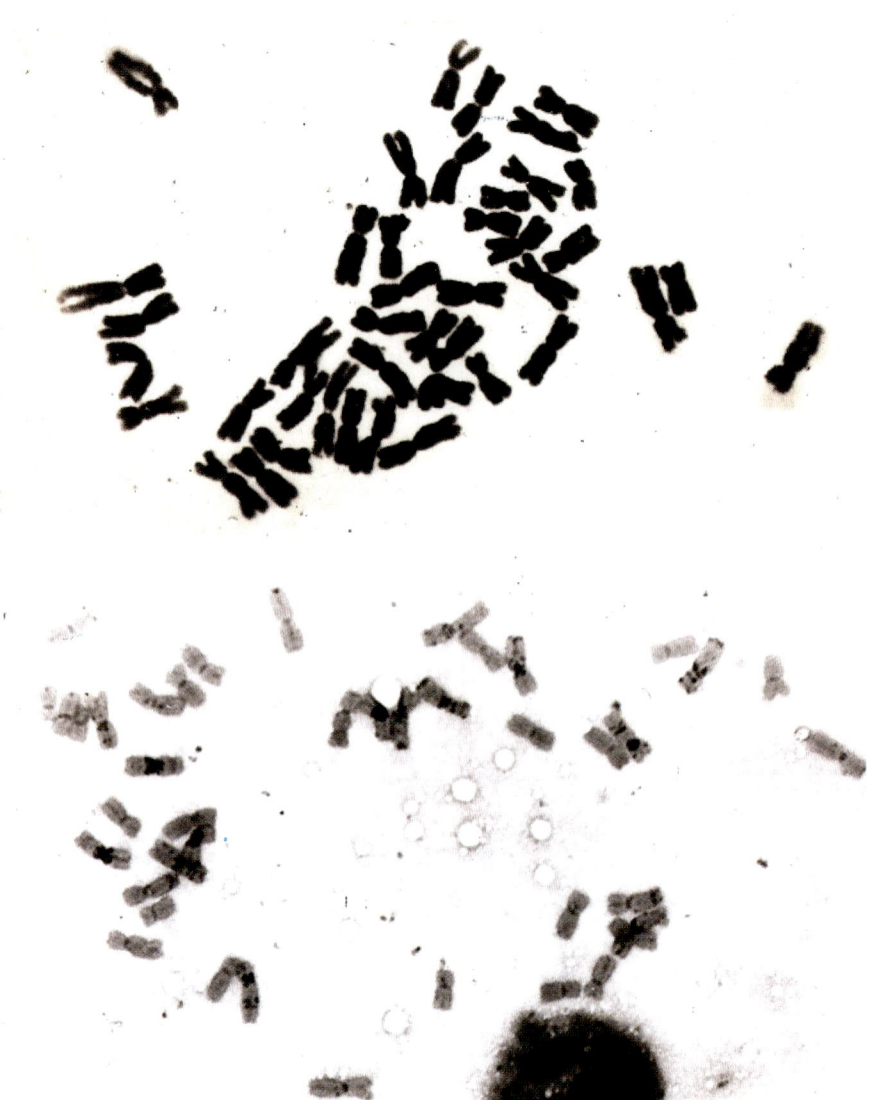

瓦维洛夫小麦核型与分带
Karyotype and banding of *Triticum Vavilovii* Jakubz.

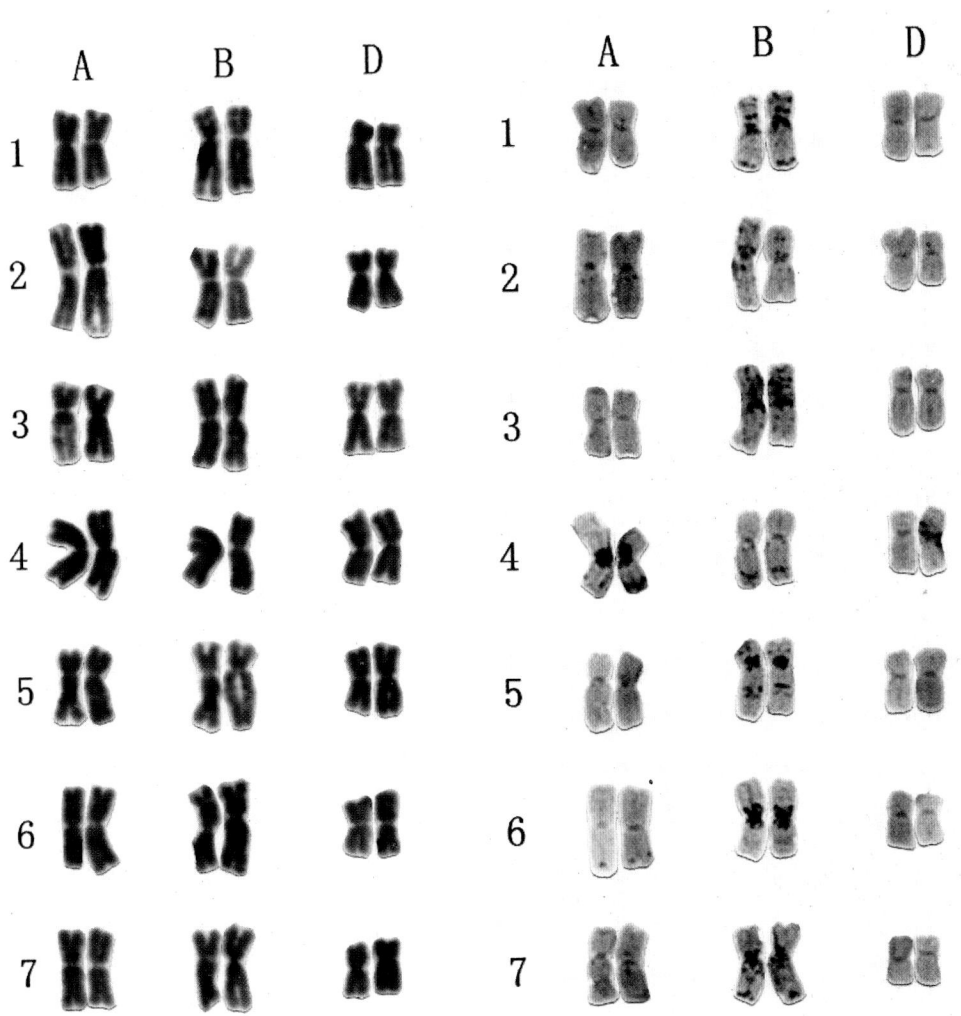

瓦维洛夫小麦核型　　　　　　　　　　瓦维洛夫小麦带型
Karyotype of *Triticum Vavilovii* Jakubz.　　Banding pattern of *Triticum Vavilovii* Jakubz.

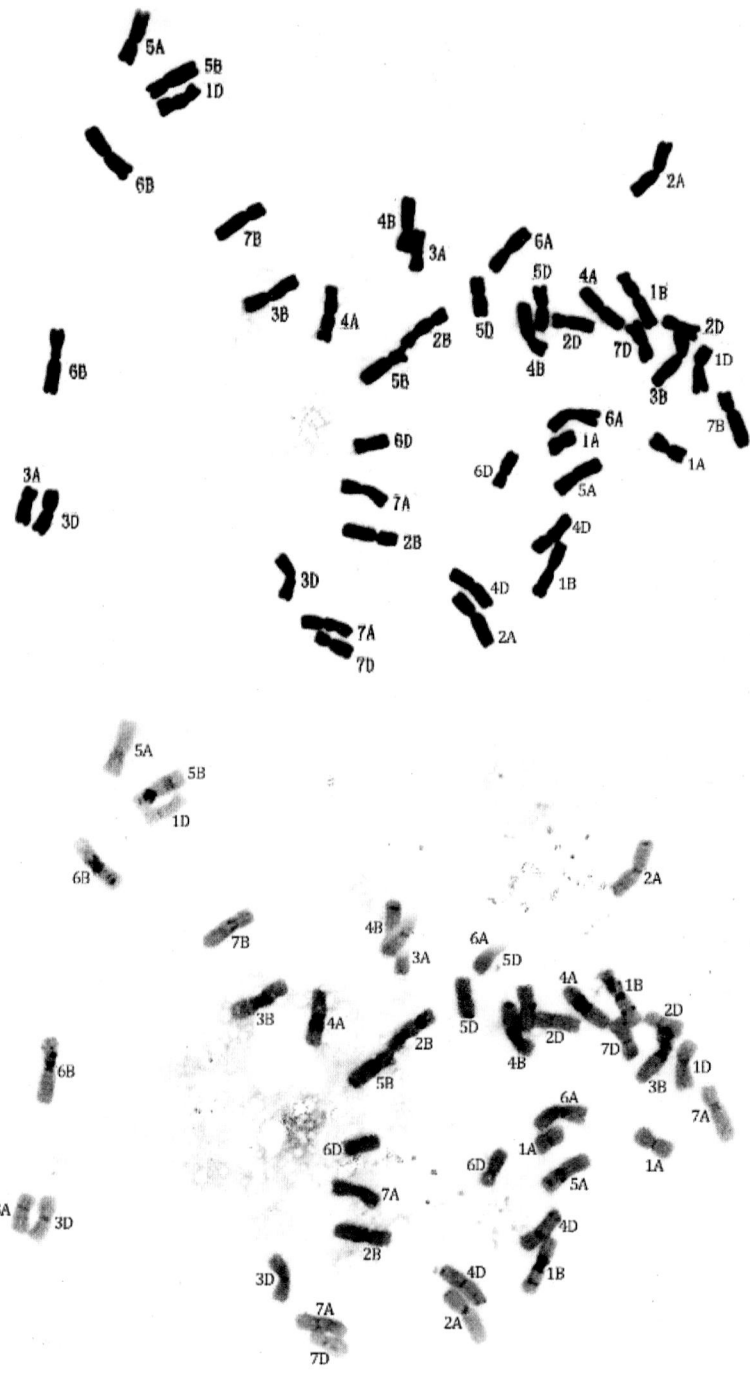

新疆小麦核型与分带

Karyotype and banding of *Triticum Petropavlovskyi* Udacz et Migusch

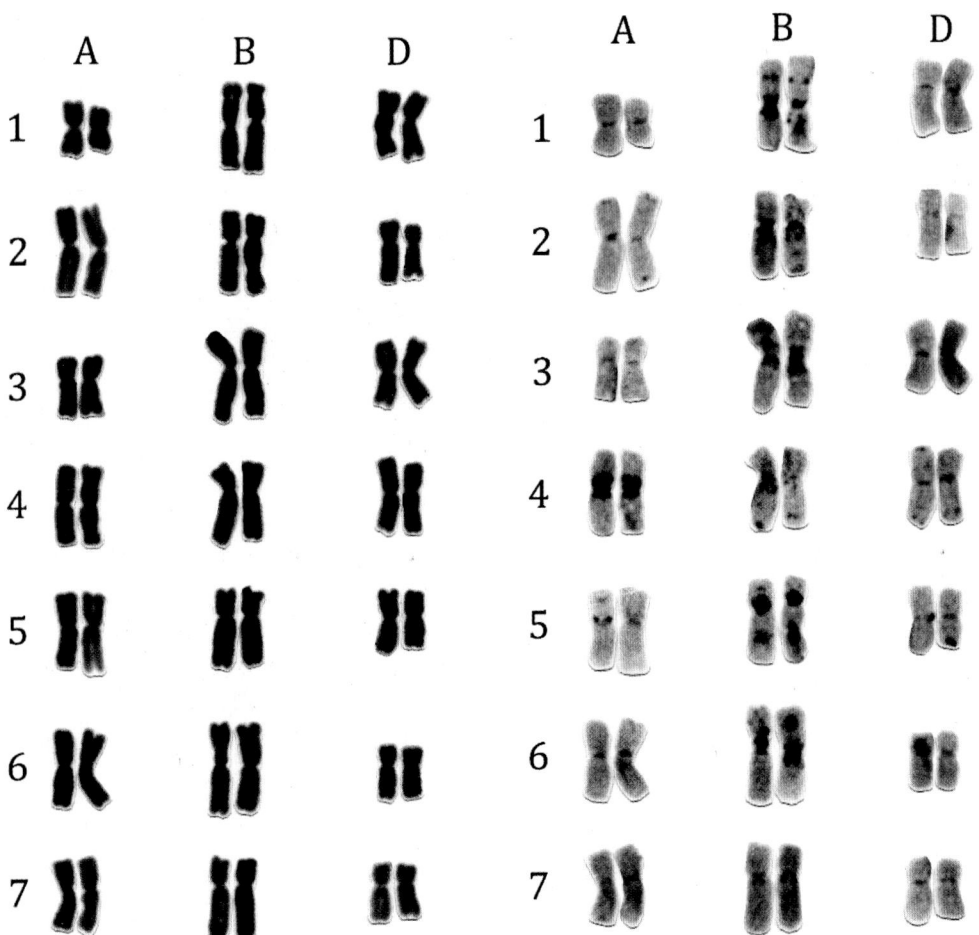

新疆小麦核型
Karyotype of *Triticum Petropavlovskyi* Udacz et Migusch

新疆小麦带型
Banding pattern of *Triticum Petropavlovskyi* Udacz et Migusch

西藏半野生小麦染色体
Shromosome of *Triticum aestivum ssp. tibetanum* Shao.

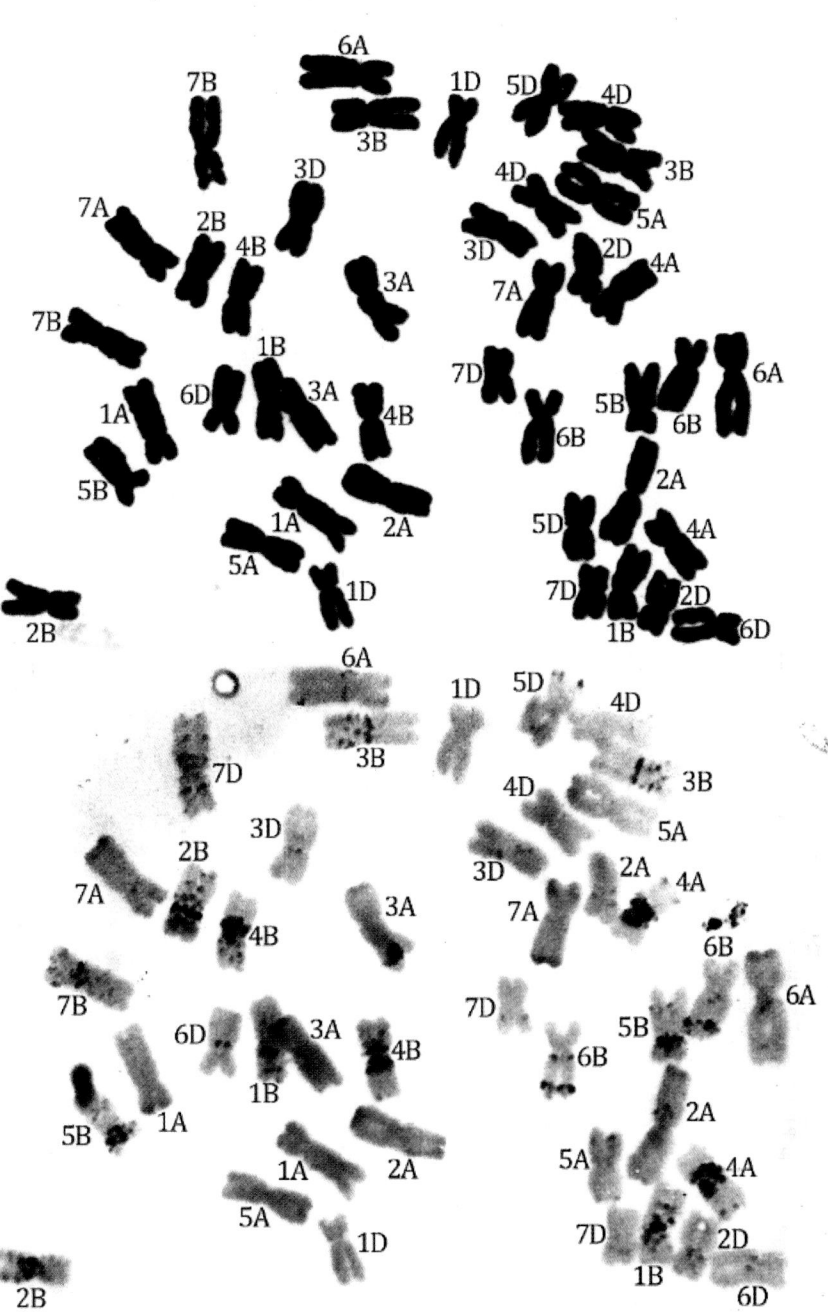

普通小麦核型与分带
Karyotype and banding of *Triticum aestivum* ev. Chinese spring

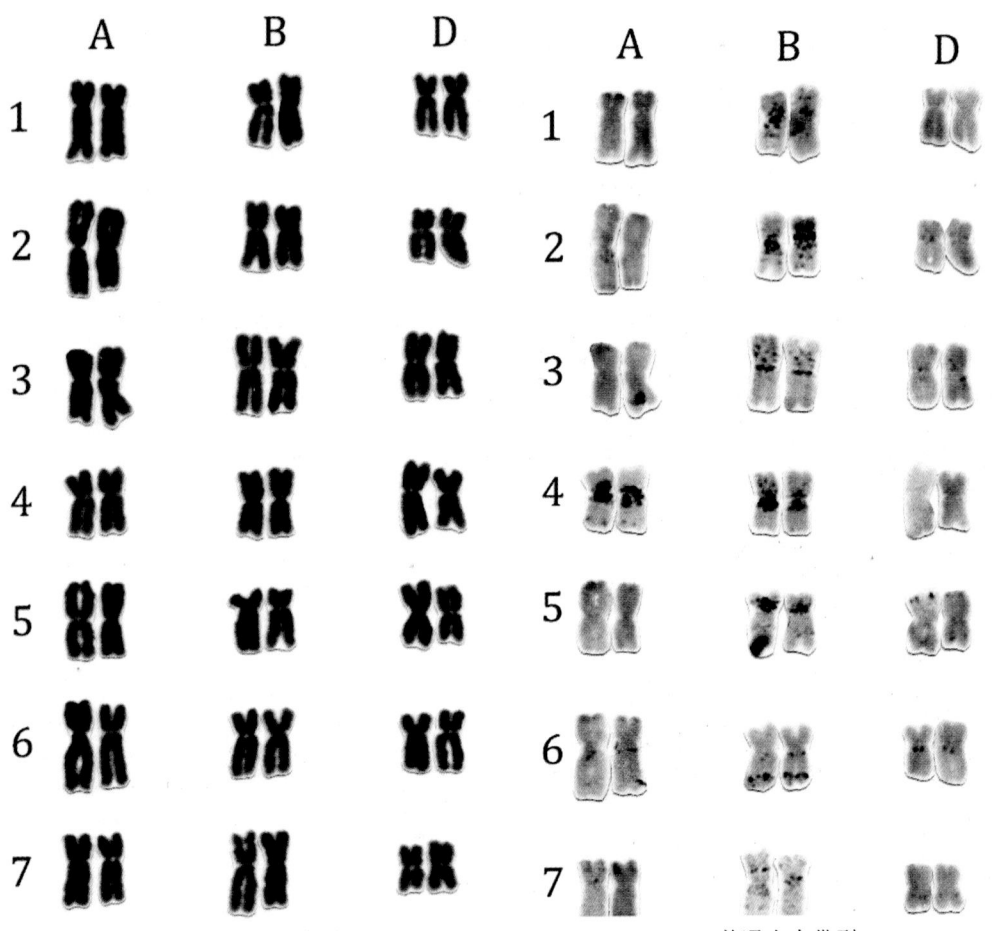

普通小麦核型
Karyotype of *Triticum aestivum* ev. Chinese spring

普通小麦带型
Banding pattern of *Triticum aestivum* L. ev. Chinese spring

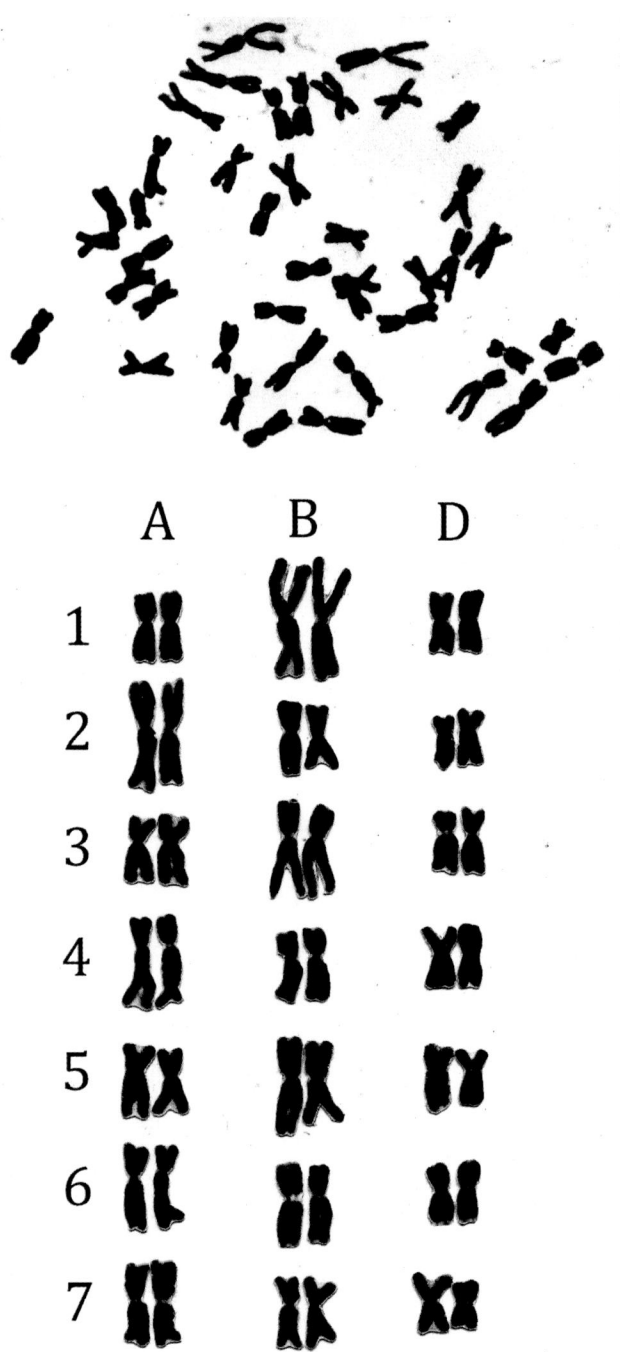

普通小麦津丰 1 号品种核型
Karyotype of *Triticum aestivum* cv. Jinfeng 1

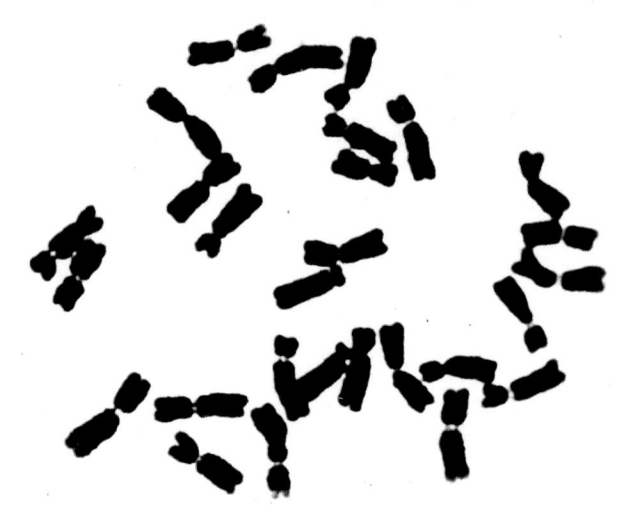

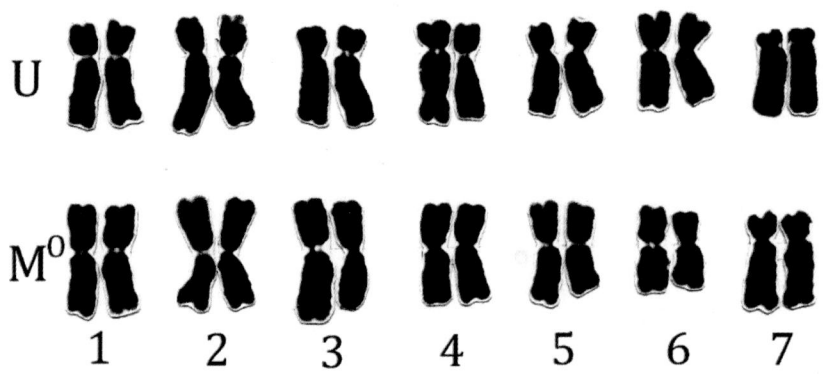

卵园山羊草核型

Karyotype of *Aegilops Ovata* L.

三芒山羊草带型
Banding pattern of *Aegilops triaristata* Wild.

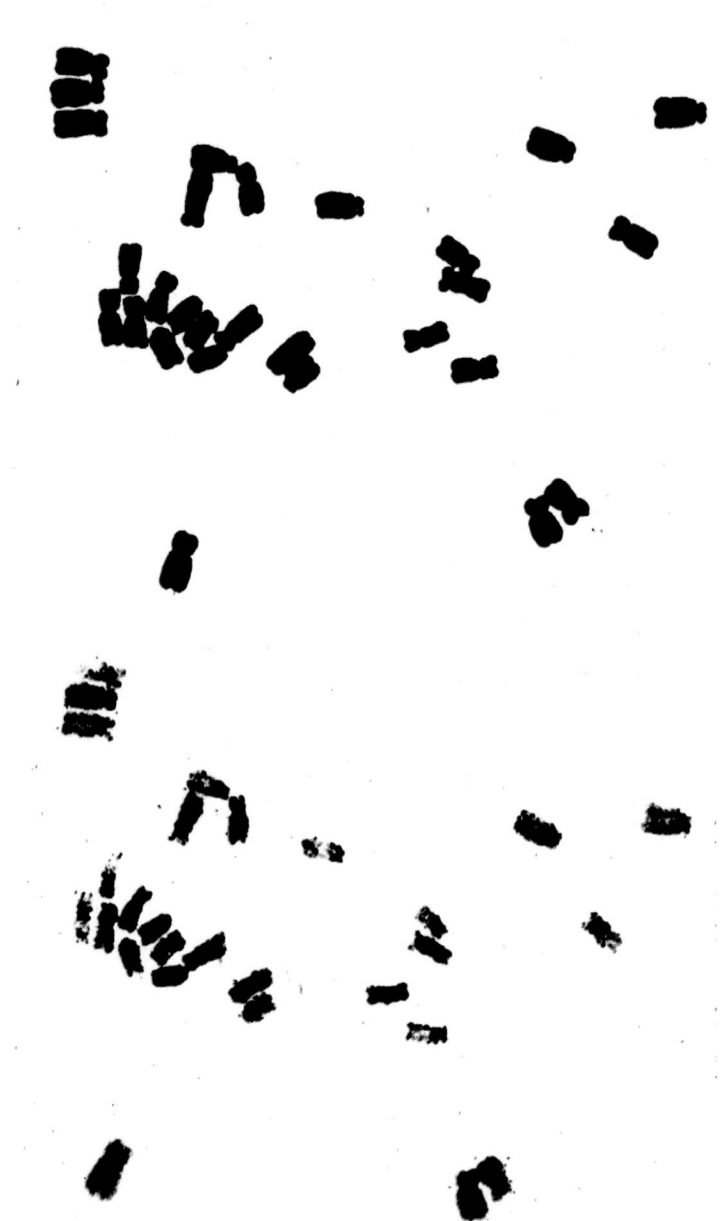

欧山羊草核型与分带
Karyotype and banding of *Aegilops biuncialis* (Vill) Vis.

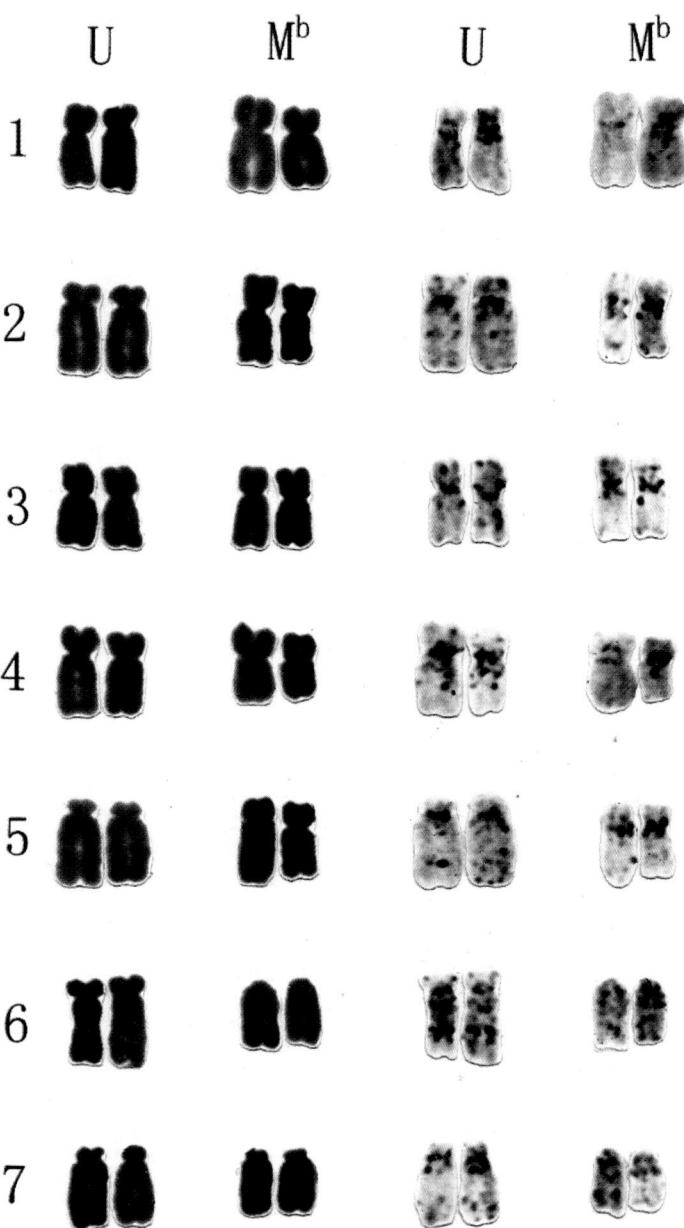

欧山羊草核型与带型
Karyotype and banding pattern of *Aegilops biuncialis* (Vill) Vis.

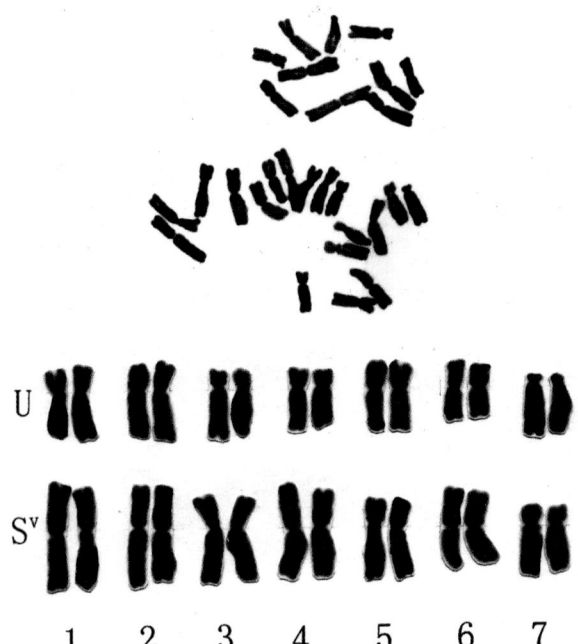

可兹山羊草核型
Karyotype of *Aegilops kotschyi* Boiss

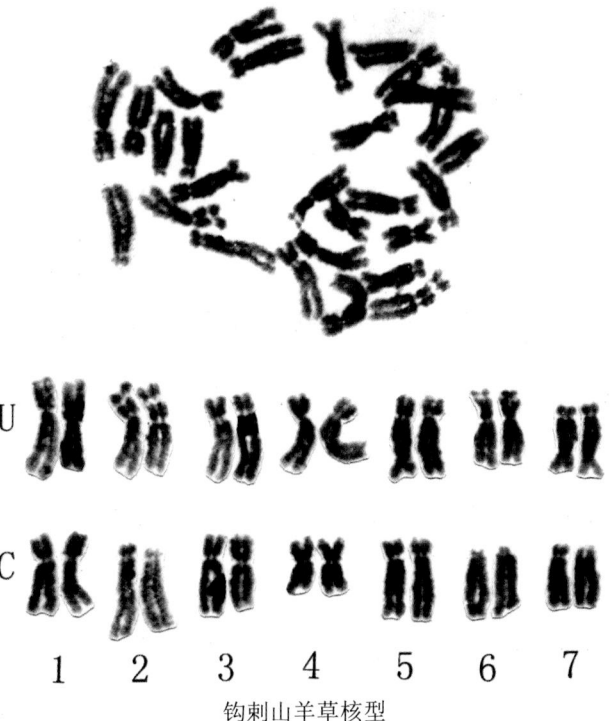

钩刺山羊草核型
Karyotype of *Aegilops triuncialis* L.

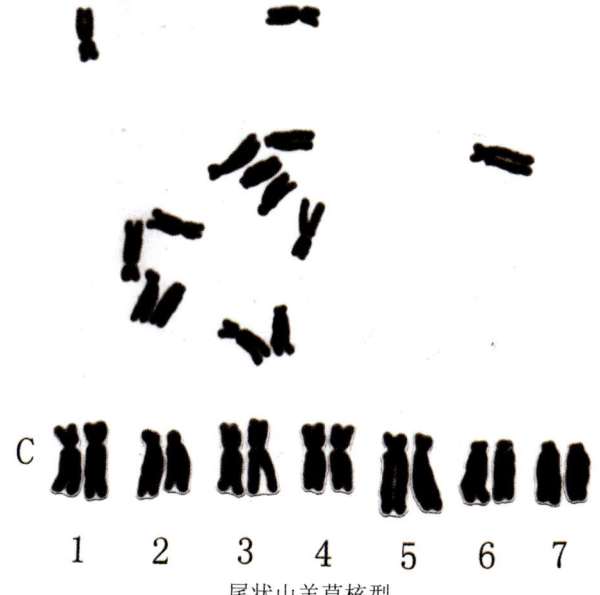

尾状山羊草核型
Karyotype of *Aegilops candata* L.

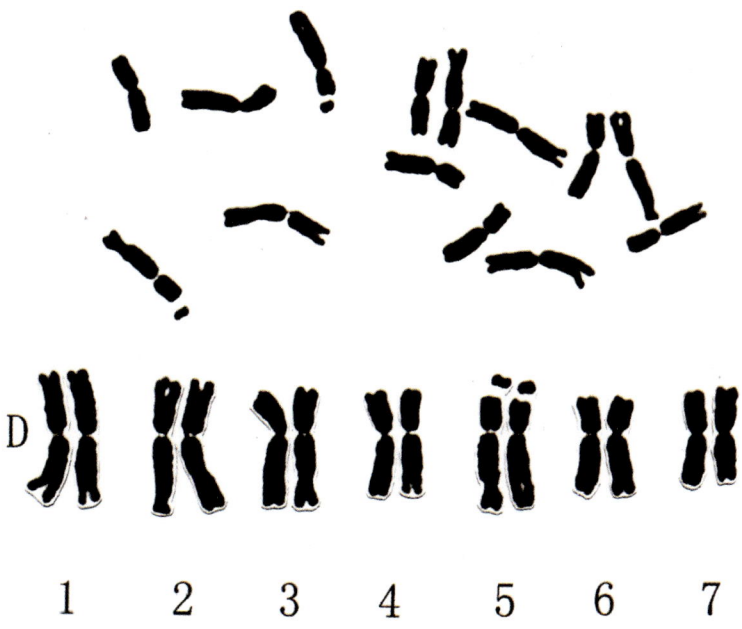

粗山羊草核型与带型
Karyotype and banding pattern of *Aegilops squarrosa* L.

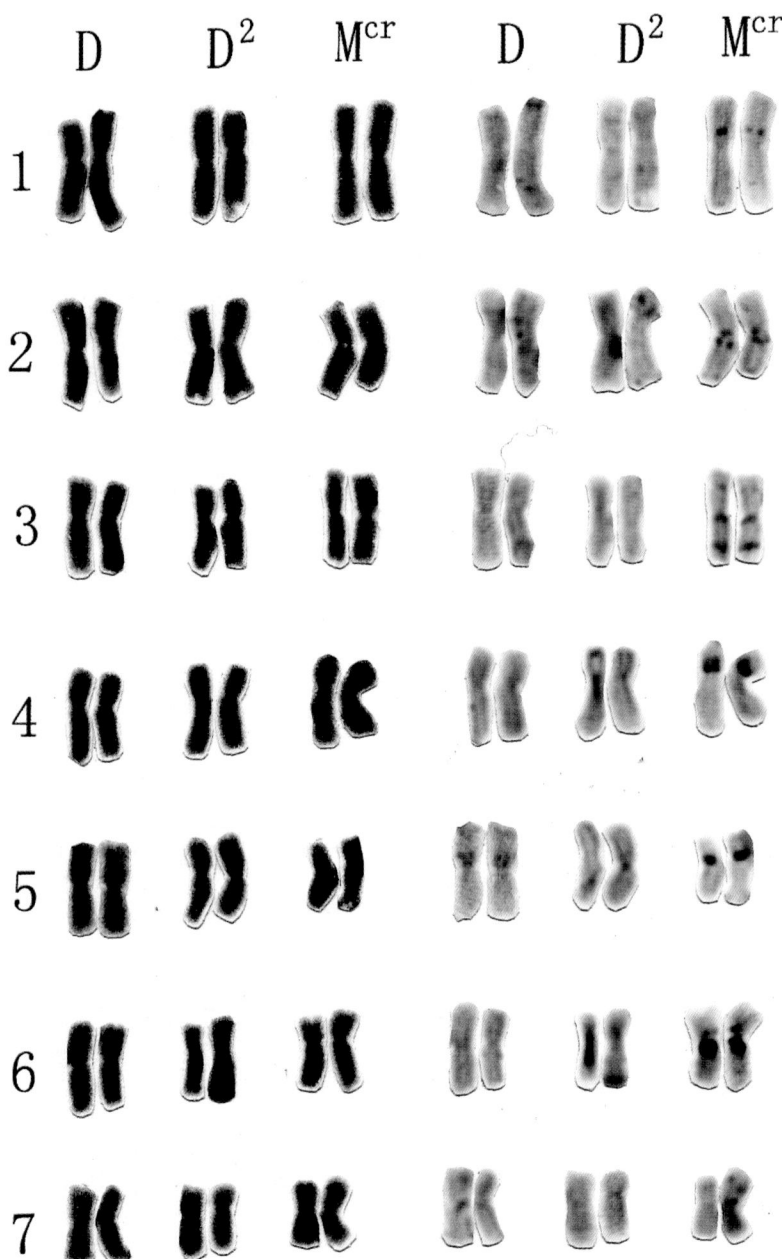

粗厚山羊草核型与带型
Karyotype and banding pattern of *Aegilops crassa* Boiss.

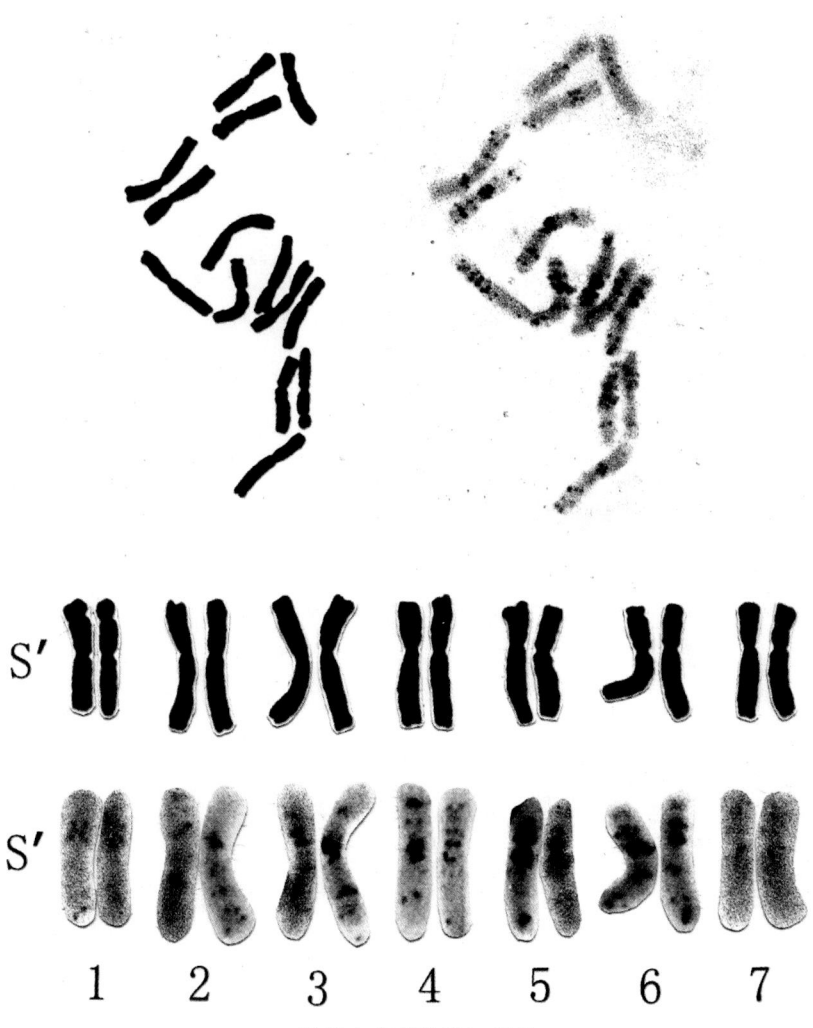

沙融山羊草核型与带型
Karyotype and banding pattern of *Aegilops sharonensis* Eig.

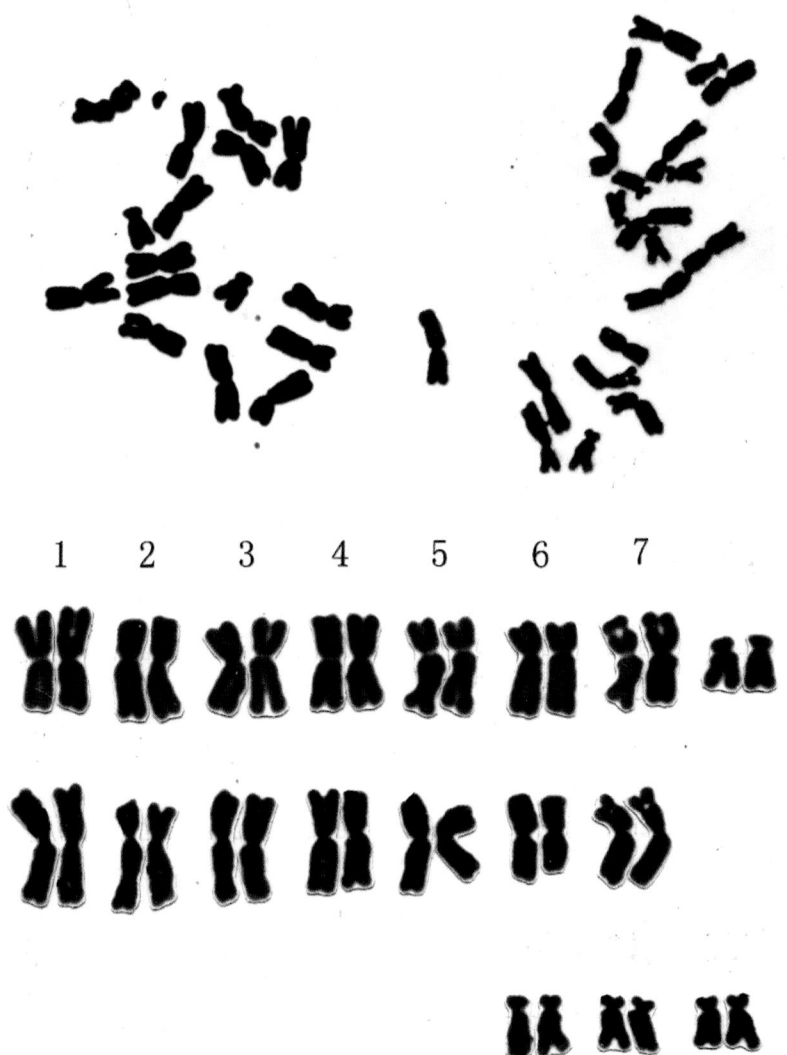

黑麦核型（+B-染色体）
Karyotype (+B-chromosome) of *Secale cereale* L.

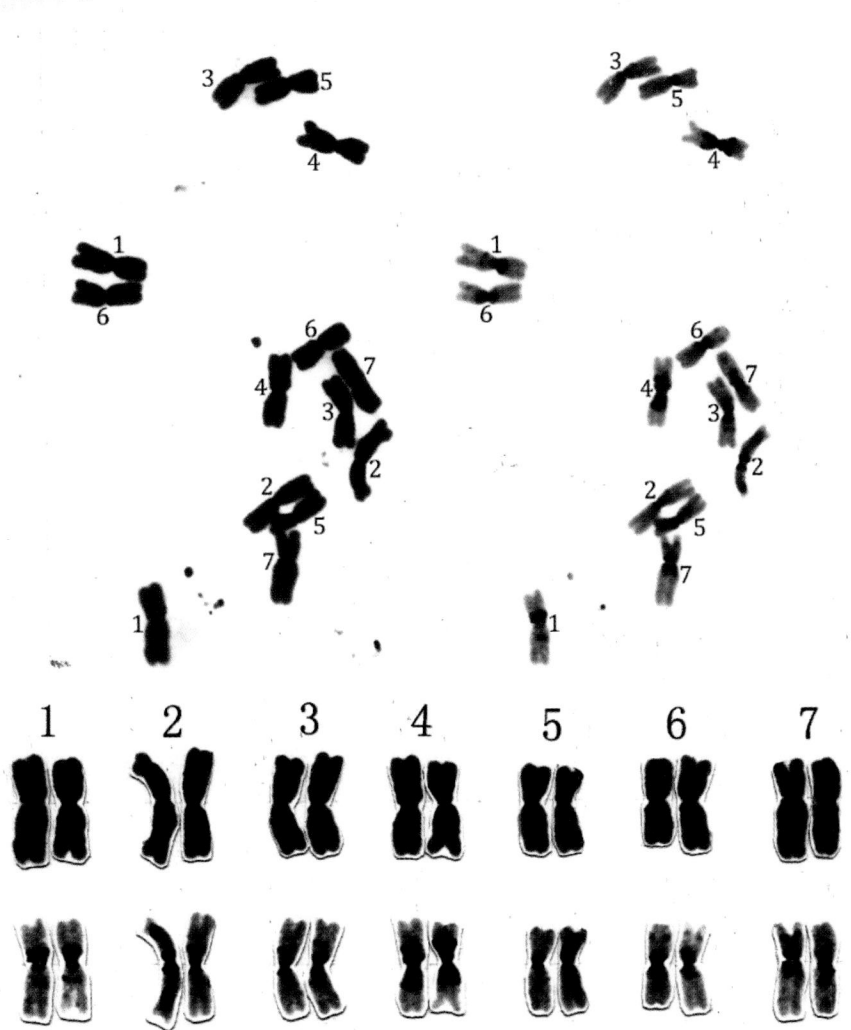

野生二棱皮大麦核型与带型
Karyotype and banding pattern of *Hordeum spontaneum* var. ilthabaurense

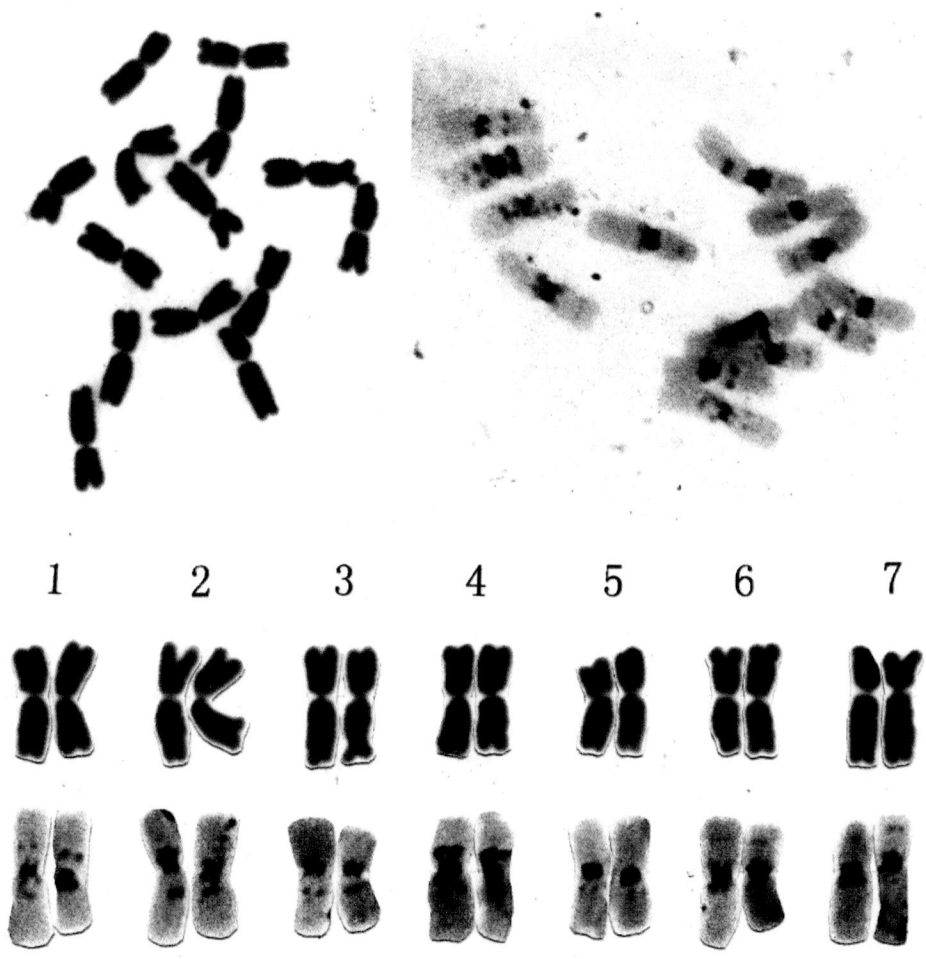

野生二棱裸粒大麦核型与带型
Karyotype and banding pattern of *Hordeum spontaneum* var. ischnatherum (cosson) Thell.

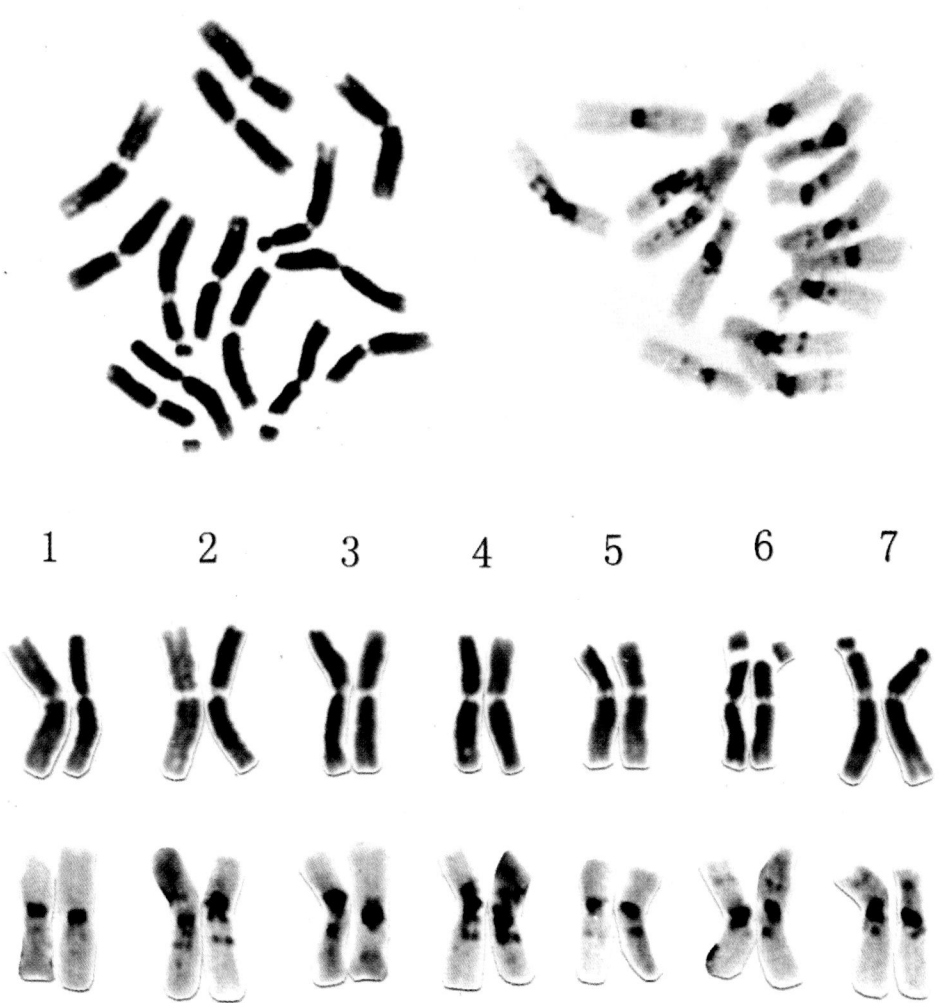

野生二棱白皮大麦核型与带型
Karyotype and banding pattern of *Hordeum spontaneum* var. proskowetzii C. Koch

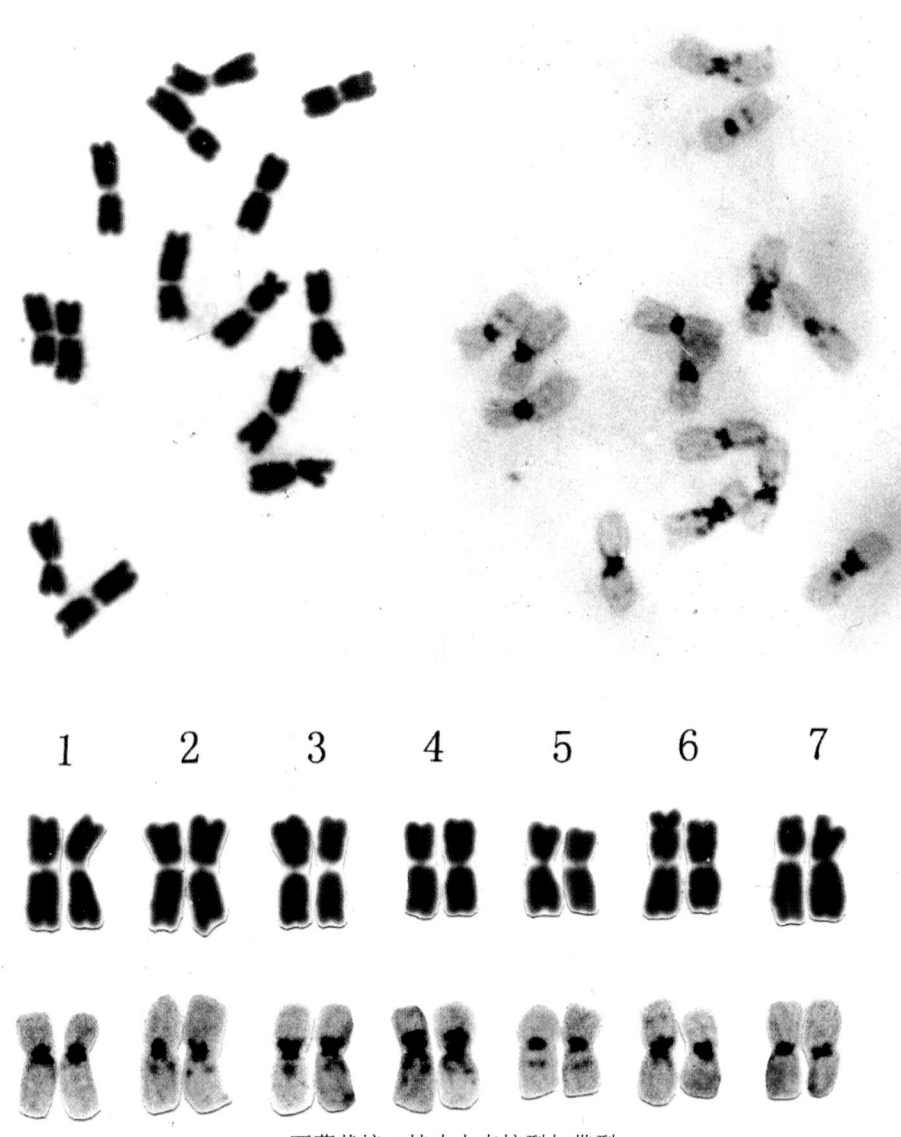

西藏栽培二棱皮大麦核型与带型
Karyotype and banding pattern of *Hordeum distichon* L.

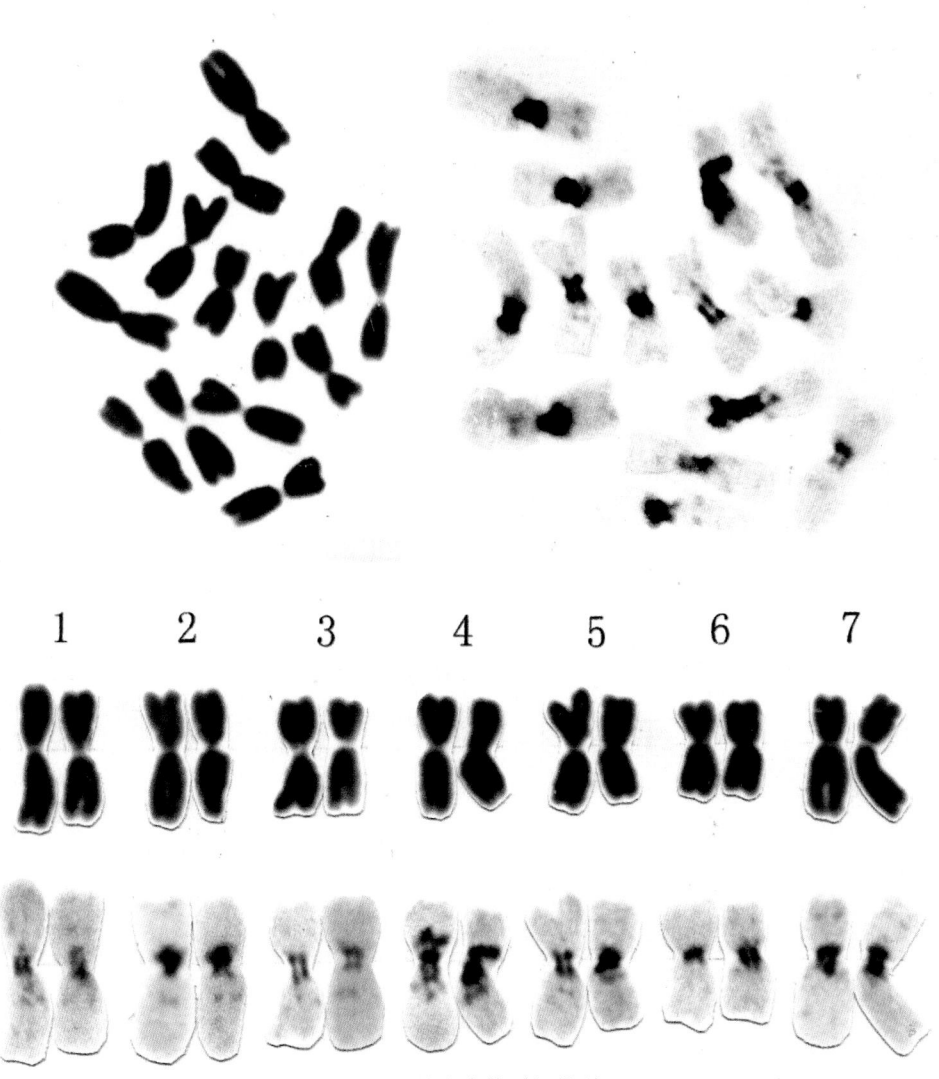

野生四棱裸粒大麦核型与带型
Karyotype and banding pattern of *Hordeum vulgare* L. ssp. *tetrastichon*

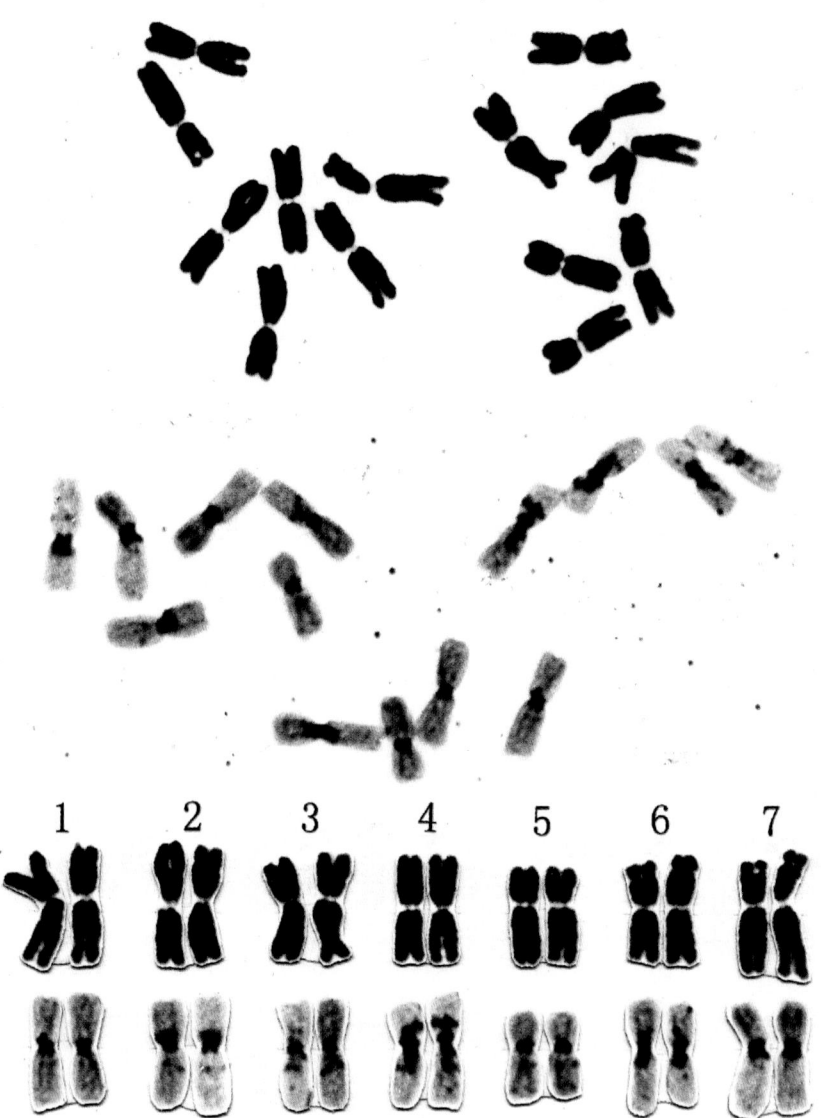

野生四棱皮大麦核型与带型
Karyotype and banding pattern of *Hordeum vulgare* L. ssp. *tetrastichon*

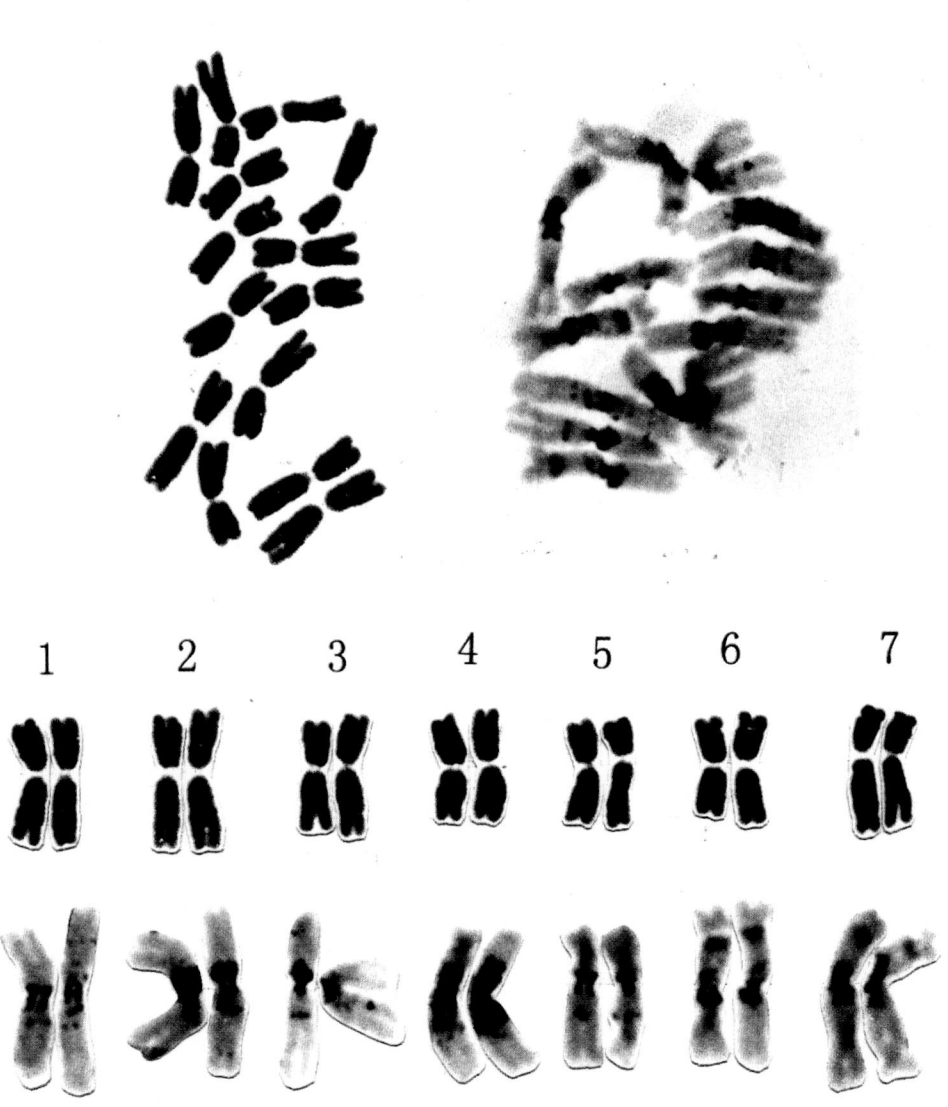

野生六棱皮大麦核型与带型
Karyotype and banding pattern of *Hordeum agriocrithon* E. Aberg

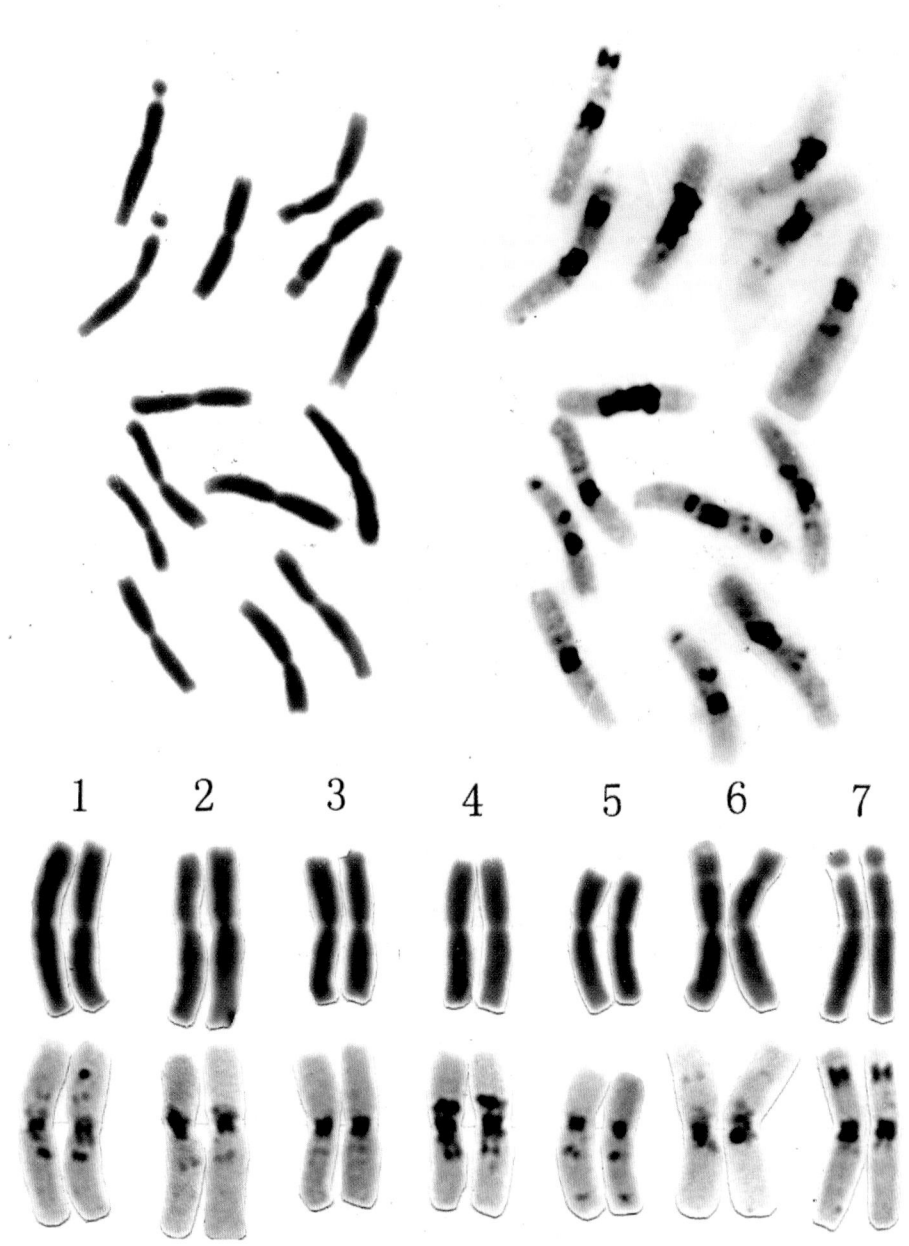

野生六棱瓶形大麦核型与带型
Karyotype and banding pattern of *Hordeum lagunculiforme* Bakht.

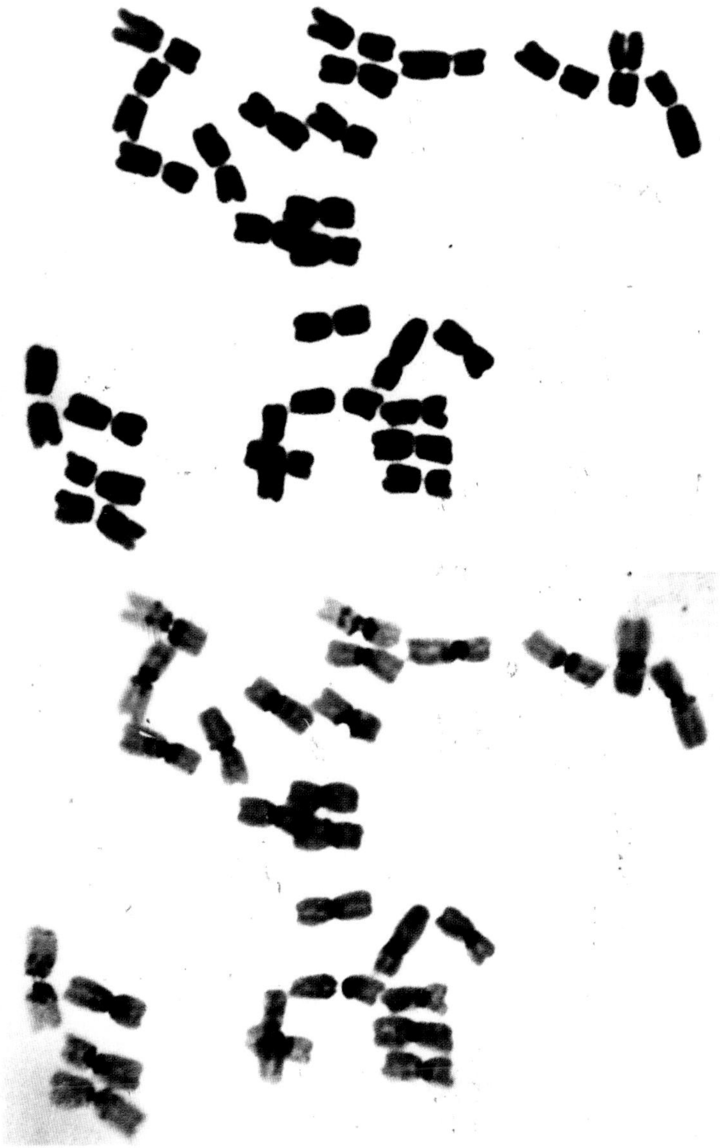

四倍体大麦核型与带型
Karyotype and banding pattern of *Hordeum vulgare* L.

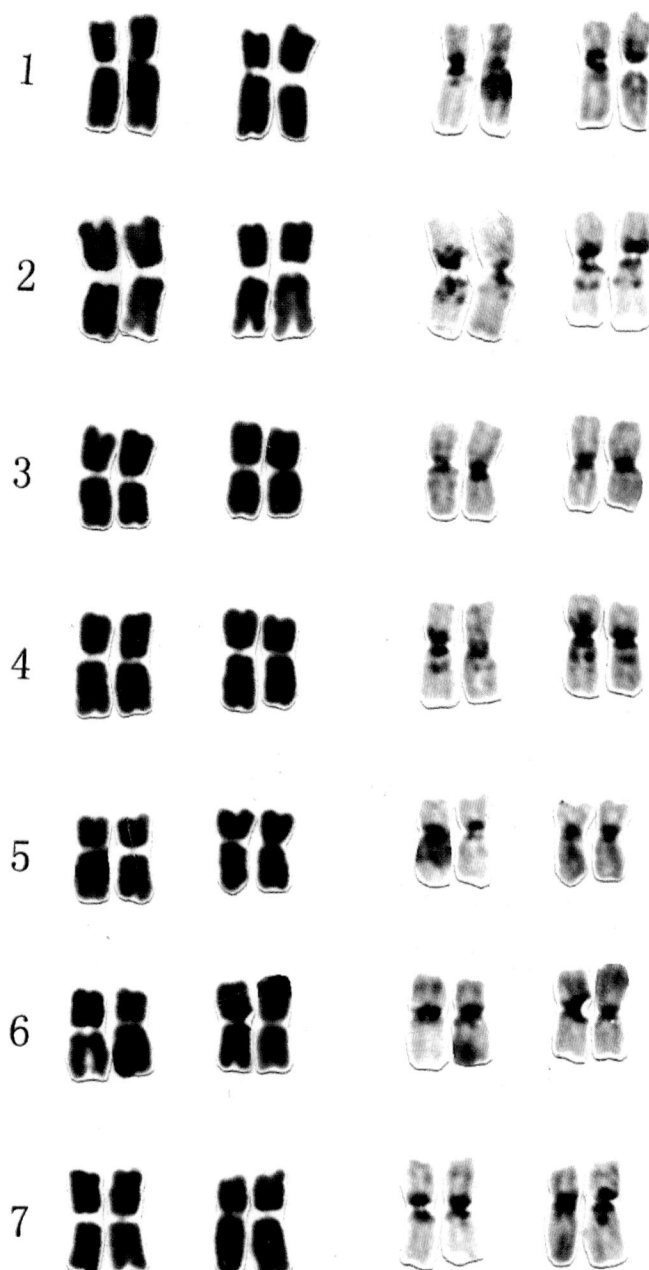

四倍体大麦核型与带型
Karyotype and banding of *Hordeum vulgare* L.

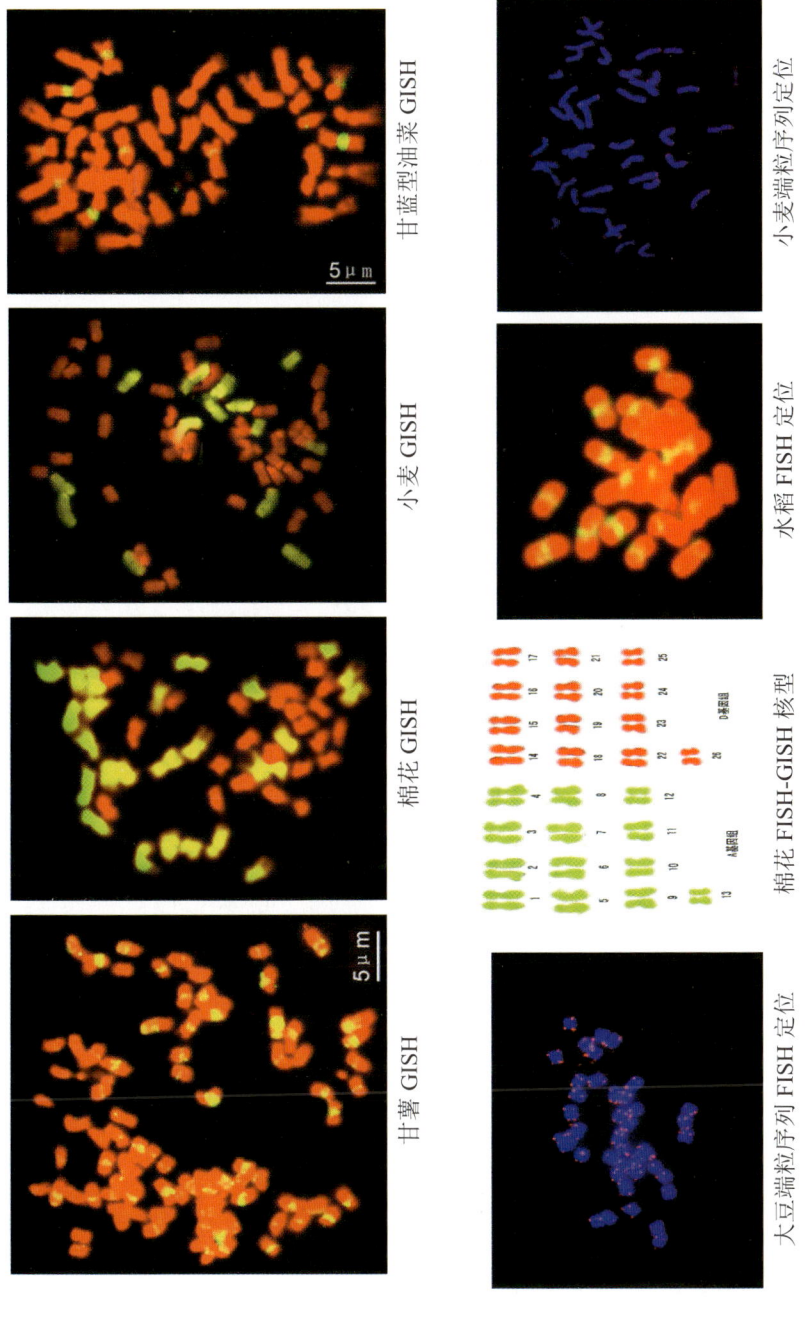

2.2.2 中国农作物及其近缘植物基因组染色体数据分析

植物核型分析（农作物及其近缘植物）

科	属	种	染色体数目	核型公式	核型分类	臂指数	最长染色体/最短染色体	臂比>2的染色体比例	核型不对称系数	染色体相对长度组成
I 禾本科 Gramineae	1.稻属 Orzya L.	1-1 籼稻 Oryza sativa L. subsp. Sativa	$2n=2x=24$	16m+4sm+4st（SAT）	2B	44	2.3	0.17	61.17	4L+6M2+6M1+8S
		1-2 粳稻 Oryza sativa L. subsp. japonica Kato	$2n=2x=24$	20m+2sm+2st（2SAT）	2B	46	2.5	0.17	60.52	2L+10M1+10M2+2S
		1-3 陆稻 Oryza sativa L.	$2n=2x=24$	18m+4sm+2st（2SAT）	2B	46	2.2	0.17	60.76	4L+6M2+12M2+2S
		1-4 普通野生稻 Oryza rufipogon Griff	$2n=2x=24$	18m+4sm+2st（2SAT）	2B	46	2.1	0.25	60.81	2L+8M2+12M1+2S
		1-5 疣粒野生稻 Oryza meyeriana（Zoll.et Mor.ex Steud.）Baill	$2n=2x=24$	12m+12sm（2SAT）	2B	48	2.1	0.25	62.97	2L+10M2+10M1+2S
		1-6 药用野生稻 Oryza officinalis Wall ex wstt	$2n=2x=24$	10m+14sm（2SAT）	2B	48	2.2	0.08	61.78	6L+4M2+8M1+6S
		1-7 一年生野生稻 Oryza nivara Sharma et Shastry	$2n=2x=24$	16m（2SAT）+4sm+4st（2SAT）	2A	44	1.6	0.25	64.35	4L+8M2+12M1
		1-8 非洲栽培稻 oryza glaberrima Steud.	$2n=2x=24$	16m+6sm+2st（2SAT）	2B	46	2	0.25	60.56	4L+8M2+10M1+2S
		1-9 南方野生稻 Oryza meridionalis Ng	$2n=2x=24$	18m+sm	2B	48	1.9	0.25	61.25	4L+10M2+6M1+4S
		1-10 展颖野生稻 Oryza glumaepatula Steud.	$2n=2x=24$	20m+4sm	2A	48	1.7	0.08	57.64	6L+4M2+14M1
		1-11 澳洲野生稻 Oryza australiensis Domin.	$2n=2x=24$	18m+4sm+2st	2A	46	1.5	0.25	60.07	12M2+12M1
	2.小麦属 Triticum L.	2-1 野生一粒小麦 Triticum boeoticum Boiss.	$2n=2x=14$	14m	1A	28	1.6	0	56.07	2L+4M2+8M1
		2-2 乌拉尔图小麦 Triticum urartu Thum ex Gandil	$2n=2x=14$	14m	1A	28	1.4	0	57.89	6M2+8M1
		2-3 栽培一粒小麦 Triticum monococcum L.	$2n=2x=14$	14m	1A	28	1.5	0	58.1	6M2+8M1
		2-4 辛斯卡亚小麦 Triticum sinskajae A.Filat et Kork.	$2n=2x=14$	12m+2sm	1A	28	1.5	0	58.59	2L+4M2+8M1
		2-5 野生二粒小麦 Triticum dicoccoid korn.	$2n=4x=28$	22m+6sm（4SAT）	2A	56	1.4	0.14	60.05	14M2+14M1
		2-6 栽培二粒小麦 Triticum dicoccum Schubl.	$2n=4x=28$	20m+8sm（4SAT）	2A	56	1.4	0.07	60.82	12M2+16M1
		2-7 圆锥小麦 Triticum turgidum L.	$2n=4x=28$	20m+6sm（2SAT）+2sa	2A	54	1.4	0.07	60.67	14M2+14M1
		2-8 硬粒小麦 Triticum durum Desf.	$2n=4x=28$	20m+8sm（4SAT）	2A	56	1.9	0.21	61.35	2L+14M2+10M1+2S
		2-9 东方小麦 Triticum turanicum Jakubz.	$2n=4x=28$	18m+10sm（4SAT）	1A	56	1.5	0	59.65	20M2+8M1
		2-10 波兰小麦 Triticum polonicum L.	$2n=4x=28$	20m（2SAT）+8sm（2SAT）	2A	56	1.5	0.14	61.17	16M2+12M1
		2-11 波斯小麦 Triticum carthlicum Nevski.	$2n=4x=28$	20m+8sm（4SAT）	2A	56	1.5	0.14	60.16	16M2+12M1

续表

植物核型分析（农作物及其近缘植物）

科	属	种	染色体数目	核型公式	核型分类	臂指数	最长染色体/最短染色体	臂比>2的染色体比例	核型不对称系数	染色体相对长度组成
I 禾本科 Gramineae	2.小麦属 Triticum L.	2-12 提莫非维尔小麦 Triticum timopheevii Zhuk.	$2n=4x=28$	20m（4SAT）+8sm	2A	56	1.4	0.07	58.79	18M2+10M1
		2-13 斯卑尔脱小麦 Triticum spelta L.	$2n=6x=42$	36m+6sm（4SAT）	1A	84	1.9	0	58.45	2L+22M2+14M1+4S
		2-14 马卡小麦 Triticum macha Dek. et Men.	$2n=6x=42$	32m+10sm	1A	84	1.8	0	58.17	4L+16M2+22M1
		2-15 印度圆粒小麦 Triticum sphaerococcum Perc.	$2n=2x=42$	38m（2SAT）+4sm（2SAT）	2A	84	1.7	0.05	58.57	2L+22M2+18M1
		2-16 瓦维洛夫小麦 Triticum vavilovii Jakubz.	$2n=6x=42$	30m+12sm（4SAT）	2A	84	2	0.05	58.93	2L+22M2+14M1+4S
		2-17 新疆半野生小麦 Triticum petropavlovskyi Udacz et Migusch.	$2n=6x=42$	32m（2SAT）+10sm（2SAT）	2A	84	1.7	0.05	59.54	4L+20M2+16M1+2S
		2-18 西藏半野生小麦 Triticum aestivum ssp.tibetanum Shao.	$2n=6x=42$	38m（2SAT）+4sm（2SAT）	2A	84	1.8	0.1	58.76	26M2+14M1+2S
		2-19 云南小麦 Triticum yunnanense king.	$2n=6x=42$	38m（2SAT）+4sm（2SAT）	2A	84	2	0.05	58.44	2L+18M2+20M1+2S
		2-20 普通小麦 Triticum aestivum L.cv.Chinese spring	$2n=6x=42$	32m+10sm	2A	84	1.8	0.05	59.22	2L+16M2+22M1+2S
		2-20-1 普通小麦 Triticum aestivum L.cv.Afu	$2n=6x=42$	34m+8sm（4SAT）	2A	84	1.8	0.05	58.63	2L+20M2+20M1
		2-20-2 普通小麦 Triticum aestivum L.cv.jinfeng1	$2n=6x=42$	34m（2SAT）+8sm（2SAT）	1A	84	2	0	59.19	6L+8M2+26M1+2S
		2-20-3 普通小麦 Triticum aestivum L.cv.Xue 81	$2n=6x=42$	30m+12sm（4SAT）	2B	84	2	0.1	59.34	4L+14M2+22M1+2S
		2-20-4 普通小麦 Triticum aestivum L.cv.Nongda 139	$2n=6x=42$	2M+32m+8sm（4SAT）	2A	84	1.9	0.1	58.66	2L+18M2+20M1+2S
		2-20-5 普通小麦 Triticum aestivum L.cv.Xiaoyan 4	$2n=6x=42$	32m+10sm	2B	84	2.1	0.05	58.51	2L+22M2+14M1+4S
		2-20-6 普通小麦 Triticum aestivum L.cv.Xiaoyan 5	$2n=6x=42$	32m+10sm（4SAT）	2A	84	1.7	0.05	58.47	24M2+16M1+2S
		2-20-7 普通小麦 Triticum aestivum L.cv.Xiaoyan 6	$2n=6x=42$		2A	84	1.8	0.05		
		2-20-8 普通小麦 Triticum aestivum L.cv.Xiaoyan 697	$2n=6x=42$	34m（4SAT）+8sm	1A	84	1.8	0	58.28	2L+20M2+18M1+2S
		2-21 密穗小麦 Triticum compacutum Host.	$2n=6x=42$	30m+12sm（2SAT）	2A	84	1.7	0.1	59.15	22M2+18M1+2S
	3.山羊草属 Aegilops L.	3-1 卵穗山羊草 Aegilops ovata Linn	$2n=4x=28$	2M+12m+12sm（2SAT）+2st	2A	54	1.4	0.36	62.61	16M2+12M1

续表

植物核型分析（农作物及其近缘植物）

科	属	种	染色体数目	核型公式	核型分类	臂指数	最长染色体/最短染色体	臂比>2的染色体比例	核型不对称系数	染色体相对长度组成
1 禾本科 Gramineae	3.山羊草属 Aegilops L.	3-2 三芒山羊草 Aegilops triaristata Wild.	$2n=6x=42$	2M+20m+10sm+8st+2t	2A	74	1.5	0.43	65.17	26M2+16M1
		3-3 欧山羊草 Aegilops biuncialis（Vill）Vis.	$2n=4x=28$	8m+8sm+8st+4t	3A	44	1.4	0.64	71.91	18M2+10M1
		3-4 可兹山羊草 Aegilops kotschyi Boiss.	$2n=4x=28$	12m+14sm（2SAT）+2st	2B	54	2.2	0.29	61.81	4L+8M2+14M1+2S
		3-5 钩刺山羊草 Aegilops triuncialis L.	$2n=4x=28$	4m+16sm（4SAT）+6st+2t	3A	48	1.7	0.57	71.83	2L+16M2+8M1+2S
		3-6 尾状山羊草 Aegilops caudata L.	$2n=2x=14$	4m+2sm+6st+2t	3A	20	1.4	0.71	72.99	8M2+6M1
		3-7 柱穗山羊草 Aegilops cylindrica Host	$2n=4x=28$	2M+18m+2sm+2st+4t	2A	50	1.5	0.29	63.92	12M2+16M1
		3-8 粗山羊草 Aegilops squarrosa L.	$2n=2x=14$	2M+10m+2sm（2SAT）	1A	28	1.4	0	58.07	6M2+8M1
		3-9 粗厚山羊草 Aegilops crassa Boiss.	$2n=6x=42$	40m+2sm	1A	84	1.7	0	57.18	2L+20M2+18M1+2S
		3-10 偏凸山羊草 Aegilops ventricosa Tausch	$2n=4x=28$	2M+12m（2SAT）+8sm+6st	2A	50	1.9	0.36	63.9	18M2+8M1+2S
		3-11 拟斯卑尔脱山羊草 Aegilops speltoides Tausch	$2n=2x=14$	10m+4sm（2SAT）	2A	28	1.4	0.14	57.92	6M2+8M1
		3-12 沙融山羊草 Aegilops sharonensis Eig.	$2n=2x=14$	12m（2SAT）+2sm（2SAT）	1A	28	1.4	0	59.62	6M2+8M1
	4.黑麦属 Secale L.	4-1 黑麦 Secale cereale L.	$2n=2x=14$	8m+6sm	2A	28	1.4	0.14	59.75	8M2+6M1
	5.小黑麦属 Triticale	5-1 六倍体小黑麦 Hexaploid Triticale	$2n=6x=42$	36m+6sm	1A	84	1.8	0	58.28	2L+18M2+22M1
		5-2 八倍体小黑麦 Octoploid Triticale	$2n=8x=56$	56m（6SAT）	1A	112	1.8	0	55.54	4L+22M2+28M1+2S
	6.大麦属 Hordeum L.	6-1 野生二棱皮大麦 Hordeum spontaneum var.ilthabaurense	$2n=2x=14$	10m+4sm（4SAT）	1A	28	1.4	0	58.62	10M2+4M1
		6-2 野生二棱裸粒大麦 Hordeum spontaneum（cosson）Thell.	$2n=2x=14$	10m+4sm（4SAT）	1A	28	1.4	0	58.62	10M2+4M1
		6-3 野生二棱白皮大麦 Hordeum spontaneum var.proskowetzii C.Koch	$2n=2x=14$	12m（2SAT）+2sm（2SAT）	1A	28	1.4	0	57.75	6M2+8M1
		6-4 栽培二棱大麦 Hordeum distichon L.	$2n=2x=14$	12m（2SAT）+2sm（2SAT）	2A	28	1.4	0.14	57.5	8M2+6M1
		6-5 西藏栽培二棱皮大麦 Hordeum distichon L.	$2n=2x=14$	12m（2SAT）+2sm（2SAT）	1A	28	1.3	0	55.75	6M2+8M1

续表

植物核型分析（农作物及其近缘植物）

科	属	种	染色体数目	核型公式	核型分类	臂指数	最长染色体/最短染色体	臂比>2的染色体比例	核型不对称系数	染色体相对长度组成
I 禾本科 Gramineae	6.大麦属 Hordeum L.	6-6 野生四棱裸粒大麦 Hordeum vulgare L.ssp.tetrastichon	$2n=2x=14$	12m（2SAT）+2sm（2SAT）	2A	28	1.5	0.14	57.02	6M2+6M1+2S
		6-7 野生四棱皮大麦 Hordeum vulgare L.ssp.tetrastichon	$2n=2x=14$	12m（2SAT）+2sm（2SAT）	1A	28	1.3	0	58.23	8M2+6M1
		6-8 栽培四棱裸粒大麦 Hordeum vulgare L.ssp.tetrastichon	$2n=2x=14$	12m（2SAT）+2sm（2SAT）	2A	28	1.4	0.14	56.86	8M2+6M1
		6-9 野生六棱皮大麦 Hordeum agriocrithon E.Aberg	$2n=2x=14$	10m+4sm（4SAT）	2A	28	1.6	0.14	58.79	6M2+8M1
		6-10 野生六棱瓶型大麦 Hordeum lagunculiforme Bakht.	$2n=2x=14$	14m（4SAT）	1A	28	1.3	0	56.2	6M2+8M1
		6-11 栽培六棱裸粒大麦 Hordeum vulgare L.	$2n=2x=14$	10m+4sm（4SAT）	1A	28	1.4	0	58.62	10M2+4M1
		6-12 四倍体大麦 Hordeum vulgare L.	$2n=4x=28$	10m+4sm（4SAT）	1A	28	1.33	0	58.25	6M2+8M1
	7.燕麦属 Avena L.	7-1 燕麦 Avena sativa L.	$2n=6x=42$	24m+16sm（2SAT）+2st	2A	82	1.8	0.29	61.56	6L+16M2+18M1+2S
		7-2 裸燕麦 Avena nuda L.	$2n=6x=42$	22m+20sm（8SAT）	2A	84	1.8	0.38	62.7	4L+18M2+20M1
	8.玉蜀黍属 Zea L.	8-1 玉米 Zea mays L.	$2n=2x=20$	10m+10sm（2SAT）	2A	40	1.7	0.2	61.5	2L+6M2+10M1+2S
		8-1-2 硬粒型玉米 Zea mays L.indurata Sturt	$2n=2x=20$	14m（2SAT）+6sm	2A	40	1.9	0.3	59.94	4L+2M2+12M1+2S
		8-1-3 马齿型玉米 Zea mays L.indentata Sturt	$2n=2x=20$	14m（2SAT）+6sm	2B	40	2.2	0.2	60.17	2L+10M2+2M1+6S
		8-1-3 半马齿型玉米 Zea mays L.semidentata Kulesh	$2n=2x=20$	10M（2SAT）+10sm	2B	40	2.1	0.4	62.07	2L+8M2+8M1+2S
		8-1-4 粉质型玉米 Zea mays L.amylacea Sturt	$2n=2x=20$	14m（2SAT）+6sm	2B	40	1.8	0.2	58.45	2L+6M2+10M1+2S
		8-1-5 糯质玉米 Zea mays L.ceratina Kulesh	$2n=2x=20$	14m（2SAT）+6sm	2A	40	2	0.2	58.92	2L+8M2+8M1+2S
		8-1-6 甜质型玉米 Zea mays L.saccharata Sturt	$2n=2x=20$	12m（2SAT）+8sm	2B	40	2.1	0.3	60.4	4L+6M2+8M1+2S
		8-1-7 爆裂型玉米 Zea mays L.everta Sturt	$2n=2x=20$	14m+6sm（2SAT）	2A	40	1.8	0.3	61.17	4L+6M2+10M1+2S
		8-1-8 有稃型玉米 Zea mays L.tunicata Sturt	$2n=2x=20$	14m（2SAT）+6sm	2B	40	2.2	0.2	58.59	2L+6M2+10M1+2S
	9.薏苡属 Coix L.	9-1 薏苡 Coix lacryma-jobi L.	$2n=2x=20$	16m（4SAT）+4sm	2A	40	1.8	0.1	57.33	2L+6M2+10M1+2S
	10.高粱属 Sorghum Moench	10-1 高粱 Sorghum vulgare Pers.	$2n=2x=20$	18m+2sm（2SAT）	2A	40	1.8	0.1	58.74	2L+8M2+8M1+2S

续表

植物核型分析（农作物及其近缘植物）

科	属	种	染色体数目	核型公式	核型分类	臂指数	最长染色体/最短染色体	臂比>2的染色体比例	核型不对称系数	染色体相对长度组成
Ⅰ 禾本科 Gramineae	11.甘蔗属 Saccharum L.	11-1 甘蔗 Saccharum officinarum L.	2n=81	55m（2SAT）+18sm+8st（2SAT）	2C	156	4.2	0.15	60.7	16L+18M2+25M1+22S
	12.狗尾草属 Setaria Beauv.	12-1 谷子 Setaria italica (L.) Beauv.	2n=2x=18	16m+2sm（2SAT）	2A	36	1.8	0.11	56.31	2L+8M2+8M1
		12-2 狗尾草 Setaria viridis (L.) Beauv.	2n=2x=18	14m+2sm+2st	2A	34	1.9	0.22	59.85	2L+4M2+12M1
		12-3 金色狗尾草 Setaria lutescens (Weigel) F.T.Hubb.	2n=4x=36	26m+10sm（4SAT）	2B	72	2.8	0.28	59.15	10L+4M2+14M1+8S
	13.狼尾草属 Pennisetum Rich.	13-1 御谷 Pennisetum glaucum (L.) R.Br.	2n=2x=18	14m+2sm+2st	2A	34	1.4	0.22	60.52	8M2+10M1
	14.黍属 Panicum L.	14-1 稗 Panicum crusgalli L.var.frumentaceum Hook.f.	2n=6x=54	42m+4sm+8st	2A	100	1.6	0.19	58.96	2L+26M2+26M1
Ⅱ 蓼科 Polygonaceae	15.荞麦属 Fagopyrum Mill.	15-1 普通荞麦 Fagopyrum esculentum Moench.	2n=4x=32	32m	1A	64	1.5	0	53.29	14M2+18M1
		15-2 鞑靼荞麦 Fagopyrum tataricum Gaertn.	2n=2x=16	10m（2SAT）+6sm（4SAT）	2A	32	1.4	0.13	60.87	6M2+10M1
		15-3 荞麦种间杂种 Fagopyrum esculentum×Fagopyrum tataricum	2n=4x=32	26m（4SAT）+6sm	2B	64	2.4	0.06	56.92	2L+12M2+16M1+2S
Ⅲ 豆科 Leguminosae	16.大豆属 Glycine L.	16-1 栽培大豆 Glycine max (L.) Merrill.	2n=2x=40	20m+18sm+2st	2A	78	1.6	0.3	63.41	2L+20M2+18M1
		16-2 野生大豆 Glycine soja Sieb. et Zucc.	2n=2x=40	36m（2SAT）+4sm	2A	80	1.5	0.05	56.8	22M2+18M1
	17.菜豆属 Phaseolus L.	17-1 红小豆 Phaseolus angularis Wight.	2n=2x=22	20m+2sm	2A	44	1.8	0.09	57.42	2L+6M2+12M1+2S
		17-2 菜豆 Phaseolus vulgaris L.	2n=2x=22	20m+2sm	1B	44	2.1	0	56.99	2L+8M2+10M1+2S
		17-3 多花菜豆 Phaseolus coccineus L.	2n=2x=22	8m（2SAT）+12sm（2SAT）+2st	2A	42	1.8	0.45	65.58	12M2+8M1+2S
	18.豌豆属 Pisum L.	18-1 豌豆 Pisum sativum L.	2n=2x=14	4m+8m（4SAT）+2t	2A	38	1.4	0.36	63.48	10M2+12M1
	19.豇豆属 Vigna Savi.	19-1 眉豆 Vigna cylindrica (L.) Skeels	2n=2x=22	22m	1A	26	1.6	0.43	66.85	2L+4M2+8M1
		19-2 豇豆 Vigna unguiculata Walp.ssp.sesquipedalis (L.) Verdc.	2n=2x=22	20m+2sm	2A	44	1.8	0	55.17	4L+2M2+16M1
	20.刀豆属 Canavalia Dc.	20-1 刀豆 Canavalia gladiata (Jacq.) DC.	2n=2x=22	14m+8m（4SAT）	2A	44	1.8	0.09	54.75	2L+6M2+14M1
			2n=2x=22	14m+8sm（4SAT）	2A	44	1.5	0.27	61.57	2L+8M2+12M1

续表

植物核型分析（农作物及其近缘植物）

科	属	种	染色体数目	核型公式	核型分类	臂指数	最长染色体/最短染色体	臂比>2的染色体比例	核型不对称系数	染色体相对长度组成
III豆科 Leguminosae	21.扁豆属 Dolichos L.	21-1 扁豆 Dolichos lablab L.	2n=2x=22	18m+2sm+2st（2SAT）	2A	42	1.8	0.18	59.59	2L+6M2+14M1
	22.野豌豆属 Vicia L.	22-1 蚕豆 Vicia faba L.	2n=2x=12	2m（2SAT）+2sm+2st+6t	3B	16	2.1	0.83	81.31	2L+2M2+8M1
	23.落花生属 Arachis L.	23-1 花生 Arachis hypogaea L.	2n=4x=40	30m（2SAT）+10sm	2B	80	2.3	0.2	58.45	22M2+14M1+4S
IV胡麻科 Pedaliaceae	24.胡麻属 Sesamum L.	24-1 芝麻 Sesamum indicum L.	2n=2x=56	16m+6sm+4st	2A	48	1.6	0.31	63.08	4L+8M2+14M1
		25-1 大白菜 Brassica campestris L.ssp. Pekinensis Olsson	2n=2x=20	12m+6sm+2st（2SAT）	2B	38	2.3	0.3	62.34	4L+4M2+8M1+4S
		25-2 白菜 Brassica campestris L.ssp. Chinensis Makino	2n=2x=20	18m+2st（2SAT）	2B	38	2	0.1	60.12	2L+8M2+10M1
	25.芸薹属 Brassica (L.) Boiss.	25-3 芜菁 Brassica campestris L.ssp.rapifera Matzg	2n=2x=20	10m+8sm+2st（2SAT）	2B	38	2.5	0.3	63.65	6L+4M2+4M1+6S
V十字花科 Cruciferae		25-4 结球甘蓝 Brassica oleracea var. capitata L.	2n=2x=18	10m+6sm+2st（2SAT）	2A	34	1.6	0.44	62.93	8M2+8M1+2S
		25-5 甘蓝型油菜 Brassica napus L.ssp.oleifera DC.	2n=4x=38	22m+10sm+4st（4SAT）+2t	2B	70	2.7	0.42	63.17	6L+8M2+18M1+6S
		25-6 花椰菜 Brassica oleracea L. var. botrytis L.	2n=2x=18	12m+6sm（2SAT）	2A	36	1.6	0.22	59.91	10M2+6M1+2S
	26.萝卜属 Raphanus L.	26-1 萝卜 Raphanus sativus L	2n=2x=18	16m+2st	2A	34	1.9	0.11	56.69	2L+8M2+4M1+4S
VI大麻科 Cannabinaceae	27.大麻属 Cannabis L.	27-1 大麻 Cannabis sativa L.	2n=2x=20	18m（2SAT）+2sm（2SAT）	2B	40	2.1	0.1	56.05	2L+8M2+8M1+2S
	28.冬瓜属 Benincasa Savi.	28-1 冬瓜 Benincasa hispida（Thunb.）Cogn.	2n=2x=24	16m+6sm+2st（2SAT）	2A	46	1.7	0.25	60.29	16M2+6M1+2S
	29.南瓜属 Cucurbita L.	29-1 黑子南瓜 Cucurbita ficifolia Bouche.	2n=2x=40	24m+12sm（2SAT）+4st	2B	76	2.1	0.35	61.33	4L+12M2+20M1+4S
		29-2 西葫芦 Cucurbita pepo L.	2n=2x=40	26m+6sm+8st	2B	72	2.2	0.3	64.19	4L+12M2+22M1+2S
VII葫芦科 Cucurbitaceae	30.甜瓜属 Cucumis L.	30-1 黄瓜 Cucumis sativus L.	2n=2x=14	12m（2SAT）+2sm	2A	28	1.5	0.14	59.43	8M2+6M1
		30-2 西双版纳黄瓜 Cucumis sativus L.var.xishuangbannanesis Qi et Yuan	2n=2x=14	10m+4sm	2A	28	2	0.14	60.68	2L+6M2+2M1+4S
	31.葫芦属 Lagenaria Ser.	31-1 葫芦 Lagenaria siceraria（Molina）Standl.	2n=2x=22	18m+2sm+2st（2SAT）	2A	42	1.5	0.18	57.68	12M2+10M1
	32.丝瓜属 Luffa L.	32-1 丝瓜 Luffa cylindrica（L.）Roem.	2n=2x=26	26m（2SAT）	1A	52	1.4	0	54.2	10M2+16M13

续表

植物核型分析（农作物及其近缘植物）

科	属	种	染色体数目	核型公式	核型分类	臂指数	最长染色体/最短染色体	臂比>2的染色体比例	核型不对称系数	染色体相对长度组成
Ⅶ葫芦科 Cucurbitaceae	33.苦瓜属 Momordica L.	33-1 苦瓜 Momordica charantia L.	2n=2x=22	18m+4sm	2A	44	1.5	0.18	56.68	8M2+14M1
	34.栝楼属 Trichosanthes L.	34-1 蛇瓜 Trichosanthes anguina L.	2n=2x=22	20m（2SAT）+2sm	2A	44	1.9	0.09	57.32	4L+4M2+12M1+2S
	35.辣椒属 Capsicum L.	35-1 辣椒 Capsicum annuum L.	2n=2x=24	20m+2sm（2SAT）+2st（2SAT）	2B	46	2.1	0.17	55.58	4L+6M2+12M1+2S
	36.番茄属 Lycopersicon Mill.	36-1 番茄 Lycopersicon esculentum（L.）Mill.	2n=2x=24	20m（2SAT）+2sm+2t	2A	46	1.8	0.17	61.07	2L+10M2+12M1
Ⅷ茄科 Solanaceae	37.茄属 Solanum L.	37-1 茄 Solanum melongena L.	2n=2x=24	24m（2SAT）	1A	48	1.7	0	56.38	2L+8M2+14M1
		37-2 马铃薯 Solanum tuberosum L.	2n=4x=48	30m+16sm（2SAT）+2st（2SAT）	2B	94	2.2	0.17	62.41	6L+12M2+28M1+2S
	38.烟草属 Nicotiana L.	38-1 烟草 Nicotiana tabacum L.	2n=4x=48	32m+16sm	2A	96	2	0.33	60.36	4L+16M2+26M1+2S
Ⅸ旋花科 Convolvulaceae	39.甘薯属 Ipomoea L.	39-1 甘薯 Ipomoea batatas（L.）Lam.	2n=6x=89	66m（2SAT）+10sm+12st+2t	2B	166	2.3	0.24	59.57	8L+30M2+48M1+4S
		39-2 百花野牵牛 Ipomoea leucantha Jacq.	2n=2x=30	26m（4SAT）+4sm	2A	60	1.9	0.07	59.47	6L+8M2+14M1+2S
		39-3 三浅裂野牵牛 Ipmoea trifida（H.B.K）DON	2n=6x=90	72m（2SAT）+18sm	2B	180	2.5	0.11	58.68	12L+26M2+40M1+12S
Ⅹ薯蓣科 Dioscoreaceae	40.薯蓣属 Dioscorea L.	40-1 山药 Dioscorea batatas Decne.	2n=140	78m+24sm+18st+2T	2B	121	3.3	0.37	67.23	9L+20M2+38M1+3S
Ⅺ天南星科 Araceae	41.芋属 Colocasia Schott	41-1（a）芋 Colocasia esculenta（L.）Schott var.cormosus Chang	2n=2x=28	18m+8sm+2st	2A	54	1.8	0.21	60.18	2L+12M2+14M1
		41-2（b）芋 Colocasia esculenta（L.）Schott var. cormosus Chang	2n=2x=42	26m+12sm（2SAT）+4st	2B	80	2.1	0.33	61.29	4L+18M2+18M1+2S
Ⅻ睡莲科 Nymphaeaceae	42.莲属 Nelumbo Adans	42-1 莲 Nelumbo nucifera Gaertn.	2n=2x=16	12m+4sm	2B	32	2.8	0.25	62.57	2L+4M2+6M1+4S
	43.莼菜属 Brasenia Schreb.	43-1 莼菜 Brasenia schreberi Gmel.	2n=72	48m+18sm+6st（2SAT）	2B	138	2，7	0.31	60.31	14L+18M2+28M1+12S
XIII藜科 Chenopodiaceae	44.甜菜属 Beta L.	44-1（a）甜菜 Beta vulgaris L.var.saccharifera Alef.	2n=2x=18	18m（2SAT）	1A	36	1.6	0	54.89	10M2+6M1+2S
		44-1（b）甜菜 Beta vulgaris L.var.saccharifera Alef.	2n=3x=27	24m（3SAT）+3sm	2A	54	1.5	0.11	57.31	12m2+15M1

续表

植物核型分析（农作物及其近缘植物）

科	属	种	染色体数目	核型公式	核型分类	臂指数	最长染色体/最短染色体	臂比>2的染色体比例	核型不对称系数	染色体相对长度组成
XIII 藜科 Chenopodiaceae	44.甜菜属 Beta L.	44-1（c）甜菜 Beta vulgaris L.var.saccharifera Alef.	2n=4x=36	28m（4SAT）+8sm	2B	72	2.1	0.11	58.02	4L+12M2+16M1+4S
	45.菠菜属 Spinacia L.	45-1（a）菠菜 Spinacia oleracea L.var.inermis Peterm	2n=2x=72	2m+3sm+7st（4SAT）	3A	22	1.47	0.86	71.06	8M2+4M1
		45-1（b）菠菜 Spinacia L. oleracea L.var.inermis Peterm	2n=2x=12	2m+4sm+6st（2SAT）	3A	18	1.6	0.83	73.41	6M2+6M1
XIV 苋科 Amaranthaceae	46.苋属 Amaranthus L.	46-1 苋 Amaranthus tricolor L.	2n=2x=32	22m+2sm+4st+4t	2A	56	2	0.31	61.43	2L+12M2+16M1+2S
		47-1 茶 Camellia sinensis O. Ktze	2n=2x=30	20m+6sm（2SAT）+2st+2t	2A	56	1.7	0.33	61.65	2L+10M2+16M1+2S
		47-1-1 从江大树茶 Camellia sinensis Congjiang	2n=2x=30	20m（2SAT）+8sm+2st	2A	58	1.6	0.33	59.41	14M2+14M1+2S
		47-1-2 贞丰大坡柳茶 Camellia sinensis Zhenfeng	2n=2x=30	22m+6sm（2SAT）+2st	2A	58	1.7	0.2	59.12	2L+12M2+16M1
		47-1-3 望谟八步茶 Camellia sinensis Wangmo	2n=2x=30	20m+8sm（2SAT）+2st	2A	58	1.6	0.13	60.28	2L+8M2+20M1
		47-1-4 金沙大树茶 Camellia sinensis Jinsha	2n=2x=30	24m+4sm（2SAT）+2st	2A	58	1.7	0.13	59.77	2L+12M2+16M1
		47-1-5 屏山大树茶 Camellia sinensis Pingshan	2n=2x=30	22m（2SAT）+8sm（2SAT）	2A	60	1.7	0.2	58.94	14M2+14M1+2S
		47-1-6 珙县大树茶 Camellia sinensis Gongxian	2n=2x=30	20m+8sm+2st	2A	58	1.8	0.33	60.09	2L+10M2+16M1+2S
		47-1-7 南川德龙大树茶 Camellia sinensis Nanchuan	2n=2x=30	20m+10sm（2SAT）	2A	60	1.7	0.2	59.89	12M2+18M1
XV 山茶科 Theaceae	47.山茶属 Camellia L.	47-1-8 云南腾冲大树茶 Camellia sinensis cv.Yunnantengchong	2n=2x=30	24m+4sm+2st	2A	58	1.8	0.07	58.96	4L+10M2+14M1+2S
		47-1-9 贵州大叶茶 Camellia sinensis cv.Guizhou	2n=2x=30	20m+10sm（2SAT）	2A	60	1.6	0.13	58.45	16M2+14M1
		47-1-10 云南大叶茶 Camellia sinensis cv.Yunnan	2n=2x=30	24m+4sm（2SAT）+2st（2SAT）	2B	58	2	0.2	61.64	2L+14M2+12M1+2S
		47-1-11 黔湄 419 Camellia sinensis Qianmei 419	2n=2x=30	20m+8sm+2st（2SAT）	2A	58	1.7	0.2	61.14	16M2+12M1+2S
		47-1-12 黔湄 502 Camellia sinensis cv.Qianmei 502	2n=2x=30	22m+6sm+2st（2SAT）	2A	58	1.7	0.07	60.19	16M2+12M1+2S
		47-1-13 雅安大叶茶 Camellia sinensis cv.Yaan	2n=2x=30	24m+4sm+2t（2SAT）	2A	58	1.5	0.2	59.85	18M2+12M1
		47-1-14 川茶 Camellia sinensis cv.Sichuan	2n=2x=30	20m+8sm（2SAT）+2st	2A	58	1.4	0.27	60.32	12M2+18M1
		47-1-15 黔湄 101 Camellia sinensis cv.Qianmei 101	2n=2x=30	22m+8sm（2SAT）	2B	60	2.1	0.2	60.58	2L+14M2+12M1+2S

续表

植物核型分析（农作物及其近缘植物）

科	属	种	染色体数目	核型公式	核型分类	臂指数	最长染色体/最短染色体	臂比>2的染色体比例	核型不对称系数	染色体相对长度组成
XV 山茶科 Theaceae	47. 山茶属 Camellia L.	47-1-16 福鼎大白茶 Camellia sinensis cv.Fuding	2n=2x=30	22m+8m（2SAT）	2A	60	1.7	0.13	59.31	2L+12M2+16M1
		47-2 望漠八步红花茶 Camellia sinensis var.rouella	2n=2x=30	22m+8sm	2A	60	1.7	0.2	59.35	14M2+14M1+2S
		47-3 榕江茶 Camellia yungkiangensis Chang	2n=2x=30	20m+8sm+2st（2SAT）	2A	58	1.8	0.27	61.31	2L+12M2+14M1+2S
		47-3-1 盘县大苦茶 Camellia yungkiangensis Panxian	2n=2x=30	20m+10sm	2A	60	1.6	0.27	59.72	18M2+12M1
		47-4-1 习水秃房茶 Camellia gymnogynoides Xishui	2n=2x=30	24m（4SAT）+6sm	2A	60	1.8	0.2	59.25	2L+14M2+12M1+2S
		47-4-2 怀仁秃房茶 Camellia gymnogynoides Renhuai	2n=2x=30	18m+12sm（2SAT）	2A	60	1.8	0.4	61.44	2L+12M2+14M1+2S
		47-5 秃房茶 Camellia gymnogyna Chang	2n=2x=30	20m+8sm+2st	2A	58	1.5	0.27	59.93	14M2+16M1
		47-5-1 崇庆枇杷茶 Camellia gymnogynoides Chongqing	2n=2x=30	20m+10sm	2A	60	1.6	0.2	59.24	18M2+10M1+2S
		47-5-2 怀仁大茶树 Camellia gymnogynoides Renhuai	2n=2x=30	20m+8sm（2SAT）+2st	2A	58	1.6	0.2	61.59	16M2+12M1+2S
		47-5-3 雷山大树茶 Camellia gymnogynoides Leishan	2n=2x=30	24m（2SAT）+4sm+2st	2A	58	1.7	0.2	58.59	2L+10M2+16M1+2S
		47-5-4 叙永大树茶 Camellia gymnogynoides Xuyong	2n=2x=30	20m+8sm（2SAT）+2st	2A	58	1.7	0.33	61.16	2L+12M2+16M1
		47-5-5 江津大树茶 Camellia gymnogynoides Jiangjin	2n=2x=30	18m+10sm（2SAT）+2st	2A	58	1.6	0.4	61.31	2L+10M2+18M1
		47-5-6 南川大树茶 Camellia gymnogynoides Nanchuan	2n=2x=30	20m+10sm	2A	60	1.8	0.27	59.87	2L+14M2+12M1+2S
		47-5-7 黄山苦茶 Camellia gymnogynoides Huangshan	2n=2x=30	20m+8sm（2SAT）+2st	2A	58	1.6	0.27	60.24	14M2+16M1
		47-5-8 赤水大树茶 Camellia gymnogynoides Chishui	2n=2x=30	22m（2SAT）+6sm（2SAT）	2A	58	1.8	0.2	59.7	18M2+10M1+2S
		47-5-9 三都高树茶 Camellia gymnogynoides Sandu	2n=2x=30	20m+10sm（4SAT）	2A	60	1.6	0.27	58.76	12M2+16M1+2S
		47-5-10 独山高树茶 Camellia gymnogynoides Dushan	2n=2x=30	22m+8sm（4SAT）	2A	60	1.5	0.27	57.38	14M2+16M1
		47-5-11 务川大树茶 Camellia gymnogynoides Wuchuan	2n=2x=30	24m（2SAT）+4sm+2st（2SAT）	2A	58	1.5	0.13	58.17	14M2+16M1

续表

植物核型分析（农作物及其近缘植物）

科	属	种	染色体数目	核型公式	核型分类	臂指数	最长染色体/最短染色体	臂比>2的染色体比例	核型不对称系数	染色体相对长度组成
		47-5-12 桐梓大树茶 Camellia gymnogynoides Tongzi	2n=2x=30	18m（2SAT）+12sm（4SAT）	2A	60	1.8	0.27	60.51	4L+10M2+14M1+2S
		47-5-13 道真大树茶 Camellia gymnogynoides Daozhen	2n=2x=30	22m+8sm	2A	60	1.8	0.27	58.96	2L+12M2+14M1+2S
		47-5-14 习水大树茶 Camellia gymnogynoides Xishui	2n=2x=30	18m+6sm+6st（4SAT）	2A	54	1.7	0.4	63.15	2L+10M2+18M1
		47-6-1 安龙大树茶 Camellia quinquelocularis Anlong	2n=2x=30	28m+2sm	2A	60	1.7	0.07	57.32	18M2+8M1+4S
		47-6-2 兴义大树茶 Camellia quinquelocularis Xingyi	2n=2x=30	24m+6sm	2A	60	1.9	0.13	58.42	2L+14M2+12M1+2S
		47-6-3 兴仁大树茶 Camellia quinquelocularis Xingren	2n=2x=30	20m+10sm（4SAT）	2A	60	1.8	0.33	62.05	2L+16M2+10M1+2S
		47-7-1 晴隆大苦茶 Camellia taliensis Qinglong	2n=2x=30	22m+4sm+4st	2A	56	1.5	0.27	59.49	14M2+16M1
		47-7-2 荥经大黑茶 Camellia taliensis Xingjing	2n=2x=30	22m+6sm+2st	2A	58	1.5	0.07	59.3	14M2+16M1
		47-8 普安四球茶 Camellia tetracocca Puan	2n=2x=30	20, +10sm（2SAT）	2A	58	1.5	0.27	57.61	12M2+18M1
XV 山茶科 Theaceae	47.山茶属 Camellia L.	47-9-1 南糯山野生茶 Camellia assamica Nannuoshan	2n=2x=30	22m+8sm（2SAT）	2A	60	1.8	0.33	59.34	2L+14M2+12M1+2S
		47-9-2 大黑山茶 Camellia assamica Daheishan	2n=2x=30	20m（2SAT）+2st	2A	58	1.8	0.27	58.35	2L+14M2+12M1+2S
		47-9-3 高县大树茶 Camellia assamica Gaoxian	2n=2x=30	20m（2SAT）+10sm	2A	58	1.6	0.27	61.24	16M2+12M1+2S
		47-9-4 南川新田湾大树茶 Camellia assamica Nanchuan	2n=2x=30	20m+8sm（2SAT）+2st（2SAT）	2A	60	1.7	0.33	59.95	18M2+10M1+2S
		47-9-5 綦江大树茶 Camellia assamica Qijiang	2n=2x=30	16m+14sm（4SAT）	2A	58	1.9	0.2	60.48	2L+12M2+14M1+2S
		47-9-6 古蔺大树茶 Camellia assamica Gulin	2n=2x=30	18m+8sm（2SAT）+4st（2SAT）	2A	60	1.9	0.4	61.34	4L+10M2+14M1+2S
		47-9-7 筠连大木茶 Camellia assamica Junlian	2n=2x=30	20m+10sm（2SAT）	2A	56	1.7	0.33	62.14	18M2+10M1+2S
		47-10 南川茶 Camellia nanchuanica cv.Nanchuanhualin	2n=2x=30	24m（2SAT）+2sm（2SAT）	2A	60	1.8	0.33	60.86	2L+10M2+16M1+2S
XVI 锦葵科 Malvaceae	48 棉属 Gossypium L.	48-1 草棉 Gossypium herbaceum L.	2n=2x=26	22m+4sm（4SAT）	2A	52	1.8	0.08	58.03	14M2+10M1+2S
		48-2 中棉 Gossypium arboreum L.	2n=2x=26	22m+4sm（4SAT）	2A	52	1.5	0.15	57.61	10M2+16M1

续表

植物核型分析（农作物及其近缘植物）

科	属	种	染色体数目	核型公式	核型分类	臂指数	最长染色体/最短染色体	臂比>2的染色体比例	核型不对称系数	染色体相对长度组成
XVI 锦葵科 Malvaceae	48 棉属 Gossypium L.	48-3 异常棉 Gossypium anomalum Wawra ex Wawra, Peyritch	2n=2x=26	22m（2SAT）+2sm+2st（2SAT）	2A	50	1.5	0.08	59.48	14M2+12M1
		48-4 长须棉 Gossypium barbosanum Phill.et Clem.	2n=2x=26	2m（2SAT）+2sm+2st（2SAT）	2A	50	1.5	0.08	58.63	14M2+12M1
		48-5 斯特提棉 Gossypium sturtianum Willi	2n=2x=26	22m（2SAT）+4sm	2A	52	1.8	0.15	57.49	14M2+10M1+2S
		48-6 南岱华棉 Gossypium Sturtianum var.nandewarense（Der）Fryx.	2n=2x=26	18m+8sm（4SAT）	2A	52	2	0.15	58.22	2L+14M2+6M1+4S
		48-7 鲁滨逊棉 Gossypium robinsonii F.von Muell.	2n=2x=26	22m（2SAT）+4sm（2SAT）	2A	52	1.9	0.15	58.43	2L+12M2+10M1+2S
		48-8 澳洲棉 Gossypium australe F.von Muell.	2n=2x=26	18m+8sm（4SAT）	2A	52	1.6	0.31	60.64	2L+12M2+12M1
		48-9 杨叶棉 Gossypium populifolium（Bent.）F.von Muell.ex Todaro	2n=2x=26	14m（2SAT）+12sm（2SAT）	2A	52	2.3	0.08	60.6	2L+14M2+6M1+4S
		48-10 奈尔逊棉 Gossypium nelsonii Fryxell	2n=2x=26	14m+12sm（4SAT）	2A	52	1.5	0.08	60.57	10M2+16M1
		48-11 瑟伯棉 Gossypium thurberi Todaro	2n=2x=26	24m（2SAT）+2sm（2SAT）	2A	52	1.5	0.08	56.13	10M2+16M1
		48-12 戴维逊棉 Gossypium davidsonii Kellog	2n=2x=26	20m+6sm（4SAT）	1A	52	1.7	0	58.7	2L+12M2+10M1+2S
		48-13 克劳茨基棉 Gossypium klotzschianum	2n=2x=26	22m+4sm（4SAT）	2A	52	1.5	0.15	59.34	12M2+14M1
		48-14 旱地棉 Gossypium aridum（Rose, Standley）Skovsted	2n=2x=26	24m（2SAT）+2sm	2A	52	1.3	0.08	56.17	12M2+14M1
		48-15 雷蒙德棉 Gossypium raimondii Ulbrich	2n=2x=26	22m（2SAT）+4sm（2SAT）	2A	52	1.5	0.15	56.86	14M2+12M1
		48-16 野生拟棉 Gossypium gossypioides（Ulbrich）Standley	2n=2x=26	22m（2SAT）+4sm（2SAT）	1A	52	1.8	0	58.35	2L+10M2+12M1+2S
		48-17 三裂棉 Gossypium trilobum（Mocino, Sesse ex DC）Skovsted	2n=2x=26	22m+4st	2A	48	1.8	0.15	57.63	2L+8M2+14M1+2S
		48-18 松散棉 Gossypium laxum Philips	2n=2x=26	22m（2SAT）+4sm	2A	52	1.8	0	58.96	2L+12M2+10M1+2S
		48-19 斯托克棉 Gossypium stocksii Masters	2n=2x=26	22m+2sm+2st（2SAT）	2A	52	1.3	0.08	56.11	14M2+12M1
		48-20 索马里棉 Gossypium somalense（Gurke）Hutchinson	2n=2x=26	22m（2SAT）+4sm	2A	52	1.4	0.15	58.95	16M2+10M1
		48-21 长萼棉 Gossypium longicalyx Hutchinson, Lee	2n=2x=26	18m+8sm（4SAT）	2A	52	1.7	0.15	61.5	14M2+10M1+2S

续表

植物核型分析（农作物及其近缘植物）

科	属	种	染色体数目	核型公式	核型分类	臂指数	最长染色体/最短染色体	臂比>2的染色体比例	核型不对称系数	染色体相对长度组成
XVI锦葵科 Malvaceae	48 棉属 Gossypium L.	48-22 比克棉 Gossypium bickii Prokhanov	$2n=2x=26$	14m（2SAT）+12sm（2SAT）	2A	52	1.9	0.08	61.71	2L+12M2+10M1+2S
		48-23（a）陆地棉 Gossypium hirsutum L.	$2n=4x=52$	34m+16sm（2SAT）+2st	2B	102	2.9	0.23	59.44	12L+14M2+18M1+8S
		48-23（b）陆地棉 Gossypium hirsutum L.	$2n=4x=52$	40m（2SAT）+10sm+2st	2A	102	2.3	0.19	58.03	6L+20M2+16M1+10S
		48-24 阔叶棉 Gossypium hirsutum "race Latifolium"	$2n=4x=52$	42m+10sm（2SAT）	2B	104	2.2	0.04	60.24	6L+18M2+22M1+6S
		48-25 帕默尔棉 Gossypium hirsutum "race parmeri"	$2n=4x=52$	42m（2SAT）+10sm（4SAT）	1B	104	2.3	0	61.21	8L+16M2+22M1+6S
		48-26 尖斑棉 Gossypium hirsutum "race punctatum"	$2n=4x=52$	38m（2SAT）+14sm（2SAT）	1B	104	2.2	0	60.23	6L+20M2+18M1+8S
		48-27 尤卡坦棉 Gossypium hirsutum "race yucatanense"	$2n=4x=52$	34m（2SAT）+18sm（4SAT）	2B	104	2.1	0.04	61.48	8L+16M2+18M1+8S
		48-28 玛利加朗特棉 Gossypium hirsutum "race marie-galante"	$2n=4x=52$	34m（2SAT）+18sm	2B	104	2.3	0.04	62.05	10L+16M2+18M1+8S
		48-29 莫利尔棉 Gossypium hirsutum "race morilli"	$2n=4x=52$	32m（2SAT）+20sm（2SAT）	2B	104	2.2	0.12	62.22	10L+16M2+20M1+6S
		48-30 墨西哥棉 Gossypium hirsutum "race mexicanum"	$2n=4x=52$	44m（6SAT）+8sm	1B	104	2.3	0	0.39	8L+16M2+22M1+6S
		48-31 雷奇蒙德棉 Gossypium hirsutum "race richmondii"	$2n=4x=52$	32m（2SAT）+20sm	2B	104	2.1	0.12	62.82	4L+22M2+18M1+8S
		48-32 海岛棉 Gossypium barbadense L.	$2n=4x=52$	44m（2SAT）+8sm	2B	104	2.3	0.08	57.54	6L+20M2+22M1+4S
		48-33 达尔文棉 Gossypium darwinii Watt	$2n=4x=52$	42m（2SAT）+10sm（4SAT）	2B	104	2	0.08	59.84	6L+18M2+20M1+8S
		48-34 辣根棉 Gossypium armourianum Kearney	$2n=2x=26$	26m（2SAT）	1A	52	1.6	0	55.53	2L+10M2+14M1
		48-35 亚蕾西棉 Gossypium areysianum Deflers	$2n=2x=26$	24m+2sm（2SAT）	2A	52	1.4	0.08	55.8	16M2+10M1
		48-36 黄褐棉 Gossypium mustelinum Miers ex Watt	$2n=4x=52$	38m（2SAT）+14sm	2B	104	2.5	0.19	59.08	10L+12M2+20M1+10S
		48-37 绿顶棉 Gossypium capitis-viridis Maller	$2n=2x=26$	20m+6sm（2SAT）	2A	52	1.4	0.15	57.69	12M2+14M1
	49.苘麻属 Abutilon Miller	49-1 苘麻 Abutilon theophrasti Medicus	$2n=42$	28m+4sm+6st（2SAT）+4t（2SAT）	2B	74	2.6	0.29	62.52	10L+2M2+22M1+8S

续表

植物核型分析（农作物及其近缘植物）

科	属	种	染色体数目	核型公式	核型分类	臂指数	最长染色体/最短染色体	臂比>2的染色体比例	核型不对称系数	染色体相对长度组成
XVI锦葵科 Malvaceae	50.木槿属 Hibiscus L.	50-1 洋麻 Hibiscus cannabinus L.	2n=2x=20	6m+10sm+2st（2SAT）+2t	2A	36	1.7	0.5	68	10M2+8M1+2S
	51.百合属 Lilium L.	51-1 川百合 Lilium davidii Duchartre	2n=2x=24	2m+2sm+8st+12t	3B	28	2.1	0.83	81.48	4L+4M2+14M1+2S
XVII百合科 Liliaceae	52.葱属 Allium L.	52-1 洋葱 Allium cepa L.	2n=2x=16	14m+2st	2A	30	1.4	0.13	59.3	10M2+6M1
		52-2 葱 Allium fistulosum L.var.giganteum Makino.	2n=2x=16	10m+4sm+2st（2SAT）	2A	30	1.4	0.38	60.93	6M2+10M1
		52-3 大蒜 Allium sativum L.	2n=2x=16	14m+2sm	2A	32	1.3	0.13	55.87	8M2+8M1
		52-4 野韭菜 Allium ramosum L.	2n=2x=16	14m+2st	2A	30	1.8	0.13	56.21	10M2+4M1+2S
		52-5 韭菜 Allium tuberosum Rottl. ex Spr.	2n=4x=32	28m+4sm	2A	64	1.8	0.13	55，51	2L+14M2+12M1+4S
	53.天门冬属 Asparagus L.	53-1 石刁柏 Asparagus officinalis L.	2n=2x=20	10m（2SAT）+10sm	2A	40	2	0.45	70.26	4L+6M2+10M1+2S
		53-2 石刁柏 Asparagus officinalis L.	2n=2x=20	11m（2SAT）+9sm	2B	40	2.3	0.27	69.06	8L+4M2+4M1+6S
XVIII伞形科 umbelliferae	54.胡萝卜属 Daucus L.	54-1 胡萝卜 Daucus carota L. var. sativa Hoffm.	2n=2x=18	8m+8sm+2st（2SAT）	2A	34	1.6	0.44	65.75	10M2+6M1+2S
	55.旱芹属 Apium L.	55-1 芹菜 Apium graveolens L.	2n=2x=22	2m+18st+2t	3A	24	1.7	0.91	81.65	12M2+6M1+4S
	56.向日葵属 Helianthus L.	56-1 向日葵 Helianthus annus L.	2n=2x=34	30m+4st	2A	64	1.6	0.12	58.92	16M2+18M1
XIX菊科 Compositae	57.莴苣属 Lactuca L.	57-1 长叶莴苣 Lactuca sativa var. longifolia Lam.	2n=2x=18	10m+8sm	2A	36	1.8	0.44	63.32	2L+6M2+8M1+2S
		57-2 结球莴苣 Lactuca sativa L. var. capitata L	2n=2x=18	8m+10sm（2SAT）	2B	36	2	0.44	63.43	4L+4M2+8M1+2S
XX大戟科 Euphorbiaceae	58.木薯属 Manihot Mill.	58-1 木薯 Manihot esculenta Crantz	2n=2x=30	22m+8sm	2A	60	2	0.13	60.32	4L+10M2+14M1+2S
XXI椴树科 Tiliaceae	59 黄麻属 Corchorus L.	59-1 黄麻 Corchorus capsularis L.	2n=2x=14	10m+4sm（2SAT）	2A	28	1.7	0.29	60.26	2L+4M2+8M1
		59-2 长蒴黄麻 Corchorus olitorius L.	2n=2x=14	12m+2st（2SAT）	2B	26	2.1	0.14	56.71	2L+6M2+4M1+2S

第三节　中国园林花卉植物基因组染色体研究结果

提要：本书收录了中国园林花卉植物64科205属356种的核型分析结果，总计有402版染色体图。新发现四倍体23种、六倍体25种、三倍体7种、多倍体复合体3种、八倍体1种、非整倍体4种，共计63种。这些多倍体类型，基本是自然形成的。染色体大数据分析显示，我国很多名贵花卉资源还都是原始状态的二倍体种群，如：兰花、梅花、香水月季、郁金香等，说明有很大的开发空间。如：我国是郁金香的原产地，在本书中收集到8种，经染色体研究，都是珍贵的原始二倍体种群（$2n=2x=24$），如进一步诱变成 $4x$、$3x$、$6x$、$8x$ 倍体，则可能形成很好的郁金香产业。还有中国水仙花，是世界名花，核型分析结果表明，是十分珍贵的异源三倍体，至今尚未发现其亲本来源，如诱变成 $6x$，则可能形成一个非常稳定的异源六倍水仙花产业。1978年前我国植物染色体研究，基本处于空白状态，对30000多种高等植物染色体家底完全不清楚，因此，也就谈不上遗传改良与开发，这就是基础研究与开发的关系，我们经过40年的植物染色体基础研究，才明白这个道理，所以必须重视基础研究，只有丰富了基础研究，才会掌握核心技术，才会走在别人前面，立于不败之地。以往对花卉的分类多从实用角度进行分类，本书首次引入染色体基础资料，从亲缘关系角度进行分类，对花卉资源的遗传改良、新品种培育与产业化会有促进作用。为减少篇幅，本书选取了部分染色体原始照片仅供参考。

2.3.1 部分园林植物染色体

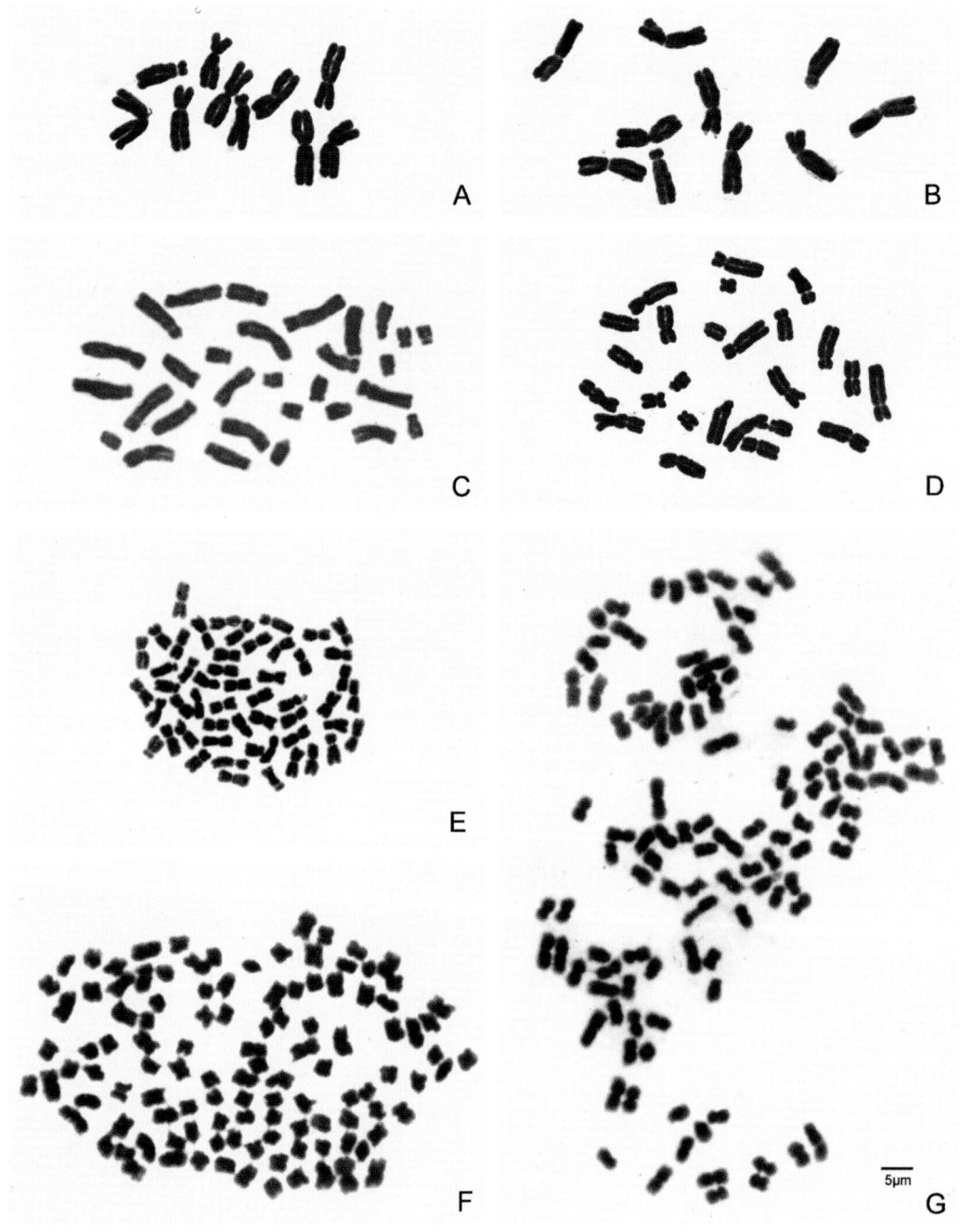

A. 芍药（*Paeonia lactiflora*），B 牡丹（*Paeonia suffruticosa*），C 中国水仙（*Narcissus tazetata*），D 丁香水仙（*Narcissus jonquilla*），E 嫦娥奔月（*Dendranthema morifolium*），F 荷花玉兰（*Magnolia grandiflora*），G 狭叶荷花玉兰（*Magnolia grandiflora var.lanceolata*）

2.3.2 中国花卉植物基因组染色体数据分析

植物核型分析（中国园林花卉植物）

科	属	种	染色体数目	核型公式	核型分类	臂指数	最长染色体/最短染色体	臂比>2的染色体比例	核型不对称系数	染色体相对长度组成
I 龙舌兰科 Agavaceae	1.晚香玉属 Polianthes L.	1-1 晚香玉 Polianthes tuberosa L.	2n=2x=16	14m (2SAT) +2sm	2A	32	1.53	0.13	56.14	8M2+8M1
	2.虎尾兰属 Sanseevieria Thunb.	2-1 虎皮掌 Sansevieria trifasciata Prain	2n=40	34m+4sm+2st	2B	78	2.47	0.15	59.1	6L+6M2+26M1+2S
	3.苋属 Amaranthus L.	3-1 雁来红 Amaranthus tricolor	2n=34	14m+10sm+6st+4t	2B	58	2	0.41	66.48	8L+10M2+10M+6S
	4.莲子草属 Alternanthera Forsk.	4-1 红穗苋 Alternanthera ficoides (L.) R.Br.var.versicolor Lem.	2n=32	16m (2SAT) +8sm+8st	2B	56	2.09	0.44	65.88	2L+12M2+16M1+2S
II 苋科 Amaranthaceae	5.青葙属 Celosia L.	5-1 红宝石 Celosia argentea L.var cristata (L.) O.Ktze	2n=2x=34	30m+4sm (2SAT)	2B	68	2.57	0.12	58.16	2L+16M2+12M1+4S
		5-2 鸡冠花 Celosia cristata L.	2n=36	36m	1A	72	1.56	0	54.93	4L+16M2+16M1
		5-3 凤尾鸡冠花 Celosia argentea var.pyramidalis	2n=36	34m+2sm	1A	72	1.66	0	57.34	2L+20M2+12M1+2S
	6.千日红属 Gomphrena L.	6-1 千日红 Gomphrena globosa L.	2n=38	30m+6sm+2st (2SAT)	2B	74	2.91	0.05	57.99	8L+8M2+10M1+12S
	7.六出花属 Alstroemeria L.	7-1 六出花 Alstroemeria aurantiaca D.Don	2n=25	1M+7m+3sm +4st+10t (3SAT)	3B	36	3.76	0.64	71.49	5L+3M2+10M1+7S
	8.朱顶红属 Amaryllis L.	8-1 朱顶红 Amaryllis vittata Ait.	2n=44	18m+12sm+14st	3B	74	2.09	0.55	68.43	6L+20M2+10M1+8S
	9.君子兰属 Clivia Lindl.	9-1 大花君子兰 Clivia miniata Reg.	2n=2x=22	4m+18sm	2B	44	2.52	0.45	65.91	6L+4M2+6M1+6S
		9-2 垂笑君子兰 Clivia nobilis Lindl.	2n=2x=22	12m+10sm	2B	44	2.07	0.27	62.64	4L+8M2+4M1+6S
III 石蒜科	10.石蒜属 Lycoris Herb.	10-1 忽地笑 Lycoris aurea Herb.	2n=2x=16	6m+10t	3B	22	2.96	0.63	69.75	6L+10S
		10-2 石蒜 Lycoris radiata Herb.	2n=2x=22	2sm+10st+10t	4A	24	1.64	1	86.6	2L+6M2+14M1
	11.水仙属 Narcissus L.	11-1 丁香水仙 Narcissus jonquilla L.	2n=25	7M+14sm+4st	3B	46	3.25	0.64	68.29	7L+5M2+7M1+6S
		11-2 喇叭水仙 Narcissus pseudonarcissus L.	2n=28	10m+18sm	2B	56	2.51	0.5	65.62	4L+12M2+8M1+4S
		11-3 中国水仙 Narcissus tazetta var. chinensis Roem.	2n=3x=30	2m+16sm (1SAT) +12st	3B	48	3.3	0.83	74.41	10L+6M2+2M1+12S
IV 夹竹桃科 Apocynaceae	12.长春花属 Catharanthus G.Don.	12-1 长春花 Catharanthus roseus G. Don.	2n=2x=16	2m+14sm (2SAT)	2A	32	1.28	0.38	66.52	8M2+8M1

续表

植物核型分析（中国园林花卉植物）

科	属	种	染色体数目	核型公式	核型分类	臂指数	最长染色体/最短染色体	臂比>2的染色体比例	核型不对称系数	染色体相对长度组成
IV夹竹桃科 Apocynaceae	13.夹竹桃属 Nerium L.	13-1 夹竹桃 Nerium indicum Mill.	$2n=2x=22$	16m（2SAT）+6sm	2A	44	1.55	0.09	59.57	10M2+10M1+2S
	14.广东万年青属 Aglaonema Schott	14-1 广东万年青 Aglaonema modestum Schott	$2n=60$	28m+30sm（4SAT）+2st	2B	118	2.99	0.3	62.14	8L+24M2+20M1+8S
	15.火鹤芋属	15-1 火鹤芋 Anthurium andraeanum Schott	$2n=30$	8m+18sm+4st（2SAT）	2B	56	2.29	0.33	66.04	4L+8M2+12M1+6S
		15-2 白条花烛 Anthurium scherzerianum var. albistriatum	$2n=30$	6m+24sm	2B	60	2.35	0.4	65.43	4L+6M2+16M1+4S
	15.花烛属 Anthurium Schott	15-3 暗红花烛 Anthurium scherzerianum var. atrosanguineum	$2n=30$	8m（2SAT）+16sm+6st（2SAT）	2B	54	2.34	0.33	66.25	4L+6M2+16M1+4S
		15-4 花烛 Anthurium scherzerianum var.wardianum	$2n=30$	6m（2SAT）+20sm+4st（2SAT）	2B	56	2.53	0.47	65.31	4L+8M2+14M1+4S
V天南星科 Araceae	16.花叶芋属 Caladium Vent.	16-1 花叶芋 Caladium bicolor（Ait.）Vent.	$2n=52$	44m（2SAT）+6sm+2st	2A	102	1.49	0.12	56.99	24M1+28M1
	17.龟背竹属 Monstera Adans	17-1 龟背竹 Monstera deliciosa Liebm.	$2n=60$	34m（2SAT）+22sm+4st	2B	116	2.79	0.27	61.13	10L+14M2+26M1+10S
	19.喜林芋属 Philodendron Schott	18-1 心形喜树芋 Philodendron oxycandium	$2n=30$	24m+6sm	2B	60	2.04	0.2	59.82	4L+14M2+10M1+2S
	19.藤芋属 Scindapsus Schott	19-1 绿萝 Epipremnum aureus（linden et Andre'）Engler	$2n=56$	32m+22sm+2st	2B	110	2.67	0.21	61.73	8L+14M2+28M1+6S
	20.马蹄莲属 Zantedeschia Spreng	20-1 马蹄莲 Zantedeschia aethiopica（L.）Spreng	$2n=2x=32$	6m+16sm+10st	3A	54	1.69	0.81	69.42	2L+14M2+14M1+2S
		20-2 黄花马蹄莲 Zantedeschia elliottiana Engl.	$2n=2x=32$	18m+12sm+2st	2A	62	1.52	0.13	61.05	16M2+16M1
VI凤仙花科 Balsaminaceae	21.凤仙花属 Impatiens L.	21-1 凤仙 Impatiens balsamina L.	$2n=2x=14$	8m+6sm	1A	28	1.59	0	62.25	2L+4M2+8M1
		21-2 紫花凤仙 Impatiens sp	$2n=2x=14$	6m+8sm	2A	28	1.87	0.29	63.87	2L+4M2+6M1+2S
VII小檗科 Berberidaceae	22.小檗属 Berberis L.	22-1 日本小檗 Berberis thunbergii DC.	$2n=2x=28$	26m+2sm（2SAT）	1A	56	1.54	0	54.2	16M2+12M1
		22-2 紫叶小檗 Berberis thunbergii DC. var. atropurpurea Chenault	$2n=2x=28$	26m+2sm（2SAT）	1A	56	1.73	0	55.3	16M2+10M1+2S
VIII紫葳科 Bignoniaceae	23.凌霄花属 Campsis Lour	23-1 凌霄花 Campsis grandiflora（Thunb.）Loisel	$2n=40$	32m+6sm+2st	2A	78	1.57	0.15	58.83	18M2+20M1+2S

续表

植物核型分析（中国园林花卉植物）

科	属	种	染色体数目	核型公式	核型分类	臂指数	最长染色体最短染色体	臂比>2的染色体比例	核型不对称系数	染色体相对长度组成
IX 木棉科 Bombacaceae	24.瓜栗属 Pachira Aubl.	24-1 发财树 Pachira macrocarpa Walp.	$2n=2x=26$	14m+8sm+4st（2SAT）	2B	48	2.06	0.38	63.92	4L+8M2+10M1+4S
X 凤梨科 Bromeliaceae	25.果子蔓属 Guzmania Ruiz & Pav.	25-1 星凤梨 Guzmania lingulata (L.) Mez.	$2n=50$	38m+12sm	2A	100	1.69	0.12	60.01	2L+18M2+30M1
XI 黄杨科 Buxaceae	26.黄杨属 Buxus L.	26-1 黄杨 Buxus microphylla Sieb.et Zucc.var.sinica Rehd.et Wils	$2n=4x=56$	22m+24sm+10st（2SAT）	2B	102	2.49	0.43	65.67	12L+12M2+24M1+8S
	27.白檀属 Chamaecereus Britt.&Rose	27-1 钟掌 Chamaecereus silvestrii (Speg.) Br.et R.	$2n=2x=20$	20m	1A	40	1.32	0	54.12	8M2+12M1
XII 仙人掌科 Cactaceae	28.令箭荷花属 Nopalxochia Britt.et Rose	28-1 令箭荷花 Nopalxochia ackermannii (Haw.) F.M.Kunth	$2n=2x=22$	22m	1A	44	1.57	0	54.78	8M2+14M1
	29.仙人掌属 Opuntia Mill.	29-1 仙人掌 Opuntia dillenii (Ker-Gawl.) Haw.	$2n=40$	38m+2sm	1A	80	1.8	0	55.54	2L+14M2+22M1+2S
	30.蟹爪兰属 Zygocactus K.Schum	30-1 蟹爪兰 Zygocactus truncatus K.Schum	$2n=2x=22$	14m+6sm（2SAT）+2st	2A	42	1.65	0.36	62.47	2L+10M2+8M1+2S
XIII 腊梅科 Calycanthaceae	31.腊梅属 Chimonanthus Lindl.	31-1 山腊梅 Chimonanthus nitens Oliv.	$2n=2x=22$	18m+2sm+2st	2A	42	1.75	0.09	56.91	2L+8M2+8M1+4S
		31-2 腊梅 Chimonanthus praecox (L.) Link.	$2n=2x=22$	12m+10sm	2A	44	1.36	0.27	60.57	12M2+10M1
		31-3 大果腊梅 Chimonanthus sp.	$2n=2x=22$	20m+2sm	2A	44	1.32	0.09	56	12M2+10M1
XIV 桔梗科 Campanulaceae	32.风铃草属 Campanula L.	32-1 风铃草 Campanula medium L.	$2n=2x=32$	32m	1A	64	1.43	0	56.09	16M2+16M1
	33.桔梗属 Platycodon A.DC.	33-1 桔梗 Platycodon grandiflorus A. DC.	$2n=2x=18$	14m（2SAT）+4sm	1A	36	1.75	0	60.34	2L+4M2+12M1
XV 美人蕉科 Cannaceae	34.美人蕉属 Canna L.	34-1 大花美人蕉 Canna generalis Bailey	$2n=2x=18$	14m+2sm+2st	2A	34	1.75	0.22	57.89	2L+6M2+10M1
XVI 忍冬科 Caprifoliaceae	35.忍冬属 Lonicera L.	35-1 金银花 Lonicera japonica Thunb	$2n=2x=18$	8m+8sm+2st	2A	34	1.8	0.44	66.69	4L+4M2+8M1+2S
	36.蝟实属 Kolkwitzia Graebn	36-1 蝟实 Kolkwitzia amabilis Graebn.	$2n=2x=32$	18m（2SAT）+12sm+2st	2A	62	1.97	0.38	61.52	4L+10M2+14M1+4S

续表

植物核型分析（中国园林花卉植物）

科	属	种	染色体数目	核型公式	核型分类	臂指数	最长染色体/最短染色体	臂比>2的染色体比例	核型不对称系数	染色体相对长度组成
XVI 忍冬科 Caprifoliaceae	37.荚迷属 Viburnum L.	37-1 天目琼花 Viburnum sargentii Koehne	$2n=2x=18$	16m（2SAT）+2sm	2A	36	1.63	0.11	55.85	10M2+6M1+2S
XVII 石竹科 Caryophyllaceae	38.石竹属 Dianthus L.	38-1 须苞石竹 Dianthus barbatus L.	$2n=2x=30$	30m	1A	60	1.31	0	57.21	18M2+12M1
		38-2 香石竹 Dianthus caryophyllus L.	$2n=2x=30$	28m+2sm	2A	60	1.45	0.07	54.7	16M2+14M1
	39.丝石竹属 Gypsophila L.	39-1 满天星 Gypsophila elegans var.alba	$2n=2x=26$	26m（2SAT）	1A	52	1.36	0	54.75	16M2+10M1
	40.雪轮属 Silene L.	40-1 高雪轮 Silene armeria L.	$2n=2x=24$	24m	1A	48	1.54	0	57.81	2L+12M2+10M1
XVIII 卫矛科 Celastraceae	41.卫矛属 Euonymus L.	41-1 冬青卫矛 Euonymus japonicas L.	$2n=31$	26m（1SAT）+5sm	2B	62	2.16	0.03	59.82	3L+8M2+19M1+1S
	42.鸭跖草属 Commelina L.	42-1 耳苞鸭跖草 Commelina auriculata B1	$2n=58$	22m+16sm+18st+2t	3B	96	2.93	0.59	70.99	12L+2M2+32M1+12S
		42-2 鸭跖草 Commelina communis L.	$2n=62$	20m+36sm+4st+2t	2B	118	3.64	0.39	69.23	12L+6M2+28M1+16S
XIX 鸭跖草科 Commelinaceae	43.紫竹梅属 Stcreasea K.Schum.et Sydow	43-1 紫竹梅 Setcreasea purpurea B.K.Boom	$2n=2x=24$	22m+2sm	1A	48	1.43	0	55.21	10M2+14M1
	44.紫露草属 Tradescantia L.	44-1 白花紫露草 Tradescantia fluminensis Vell.	$2n=70$	22m+28sm+14st+6t	3B	120	3.49	0.57	71.84	14L+8M2+36M1+12S
	45.吊竹梅属 Zebrina Schnizl.	45-1 吊竹梅 Zebrina pendula Schnizl.	$2n=2x=24$	6m+2sm+6st（2SAT）+10t	3B	32	2.69	0.75	75.06	4L+2M2+14M1+4S
	46.打碗花属 Calystegia R.Br.	46-1 打碗花 Calystegia hederacea Wall.	$2n=2x=22$	8m+12sm（2SAT）+2st	3B	42	2.04	0.55	63.38	2L+8M2+10M1+2S
XX 旋花科 Convolvulaceae	47.番薯属 Ipomoea L.	47-1 牵牛花 Ipomoea hederacea Jacq	$2n=2x=30$	26m（4SAT）+4sm	2A	60	1.9	0.07	59.71	4L+8M2+14M1+4S
	48.牵牛属 Pharbitis Choisy	48-1 大花牵牛 Pharbitis nil Choisy	$2n=2x=30$	24m+6sm（2SAT）	2A	60	1.92	0.07	58.24	2L+16M2+10M1+2S
	49.茑萝属 Quamoclit Mill.	49-1 茑萝 Quamoclit pennata（Desr.）Bojer.	$2n=2x=30$	18m+8sm+4st	2A	56	1.78	0.13	63.47	2L+10M2+18M1
XXI 菊科 Compositae	50.胜红蓟属 Ageratum L.	50-1 大花藿香蓟 Ageratum conyzoides L.	$2n=2x=38$	24m+14sm	2B	76	2.19	0.16	59.94	4L+12M2+20M1+2S
	51.紫苑属 Aster L.	51-1 荷兰菊 Aster novibelgii L.	$2n=48$	30m+18sm（2SAT）	2B	96	2.3	0.17	62.44	6L+16M2+24M1+2S

续表

植物核型分析（中国园林花卉植物）

科	属	种	染色体数目	核型公式	核型分类	臂指数	最长染色体最短染色体	臂比>2的染色体比例	核型不对称系数	染色体相对长度组成
	51.紫菀属 Aster L.	51-2 三脉紫菀 Aster ageratoides Turcz	$2n=2x=18$	18m（2SAT）	1A	36	1.46	0	56.86	10M2+8M1
			$2n=4x=36$	36m（4SAT）	1A	72	1.61	0	56.37	16M2+20M1
			$2n=6x=54$	46m+8sm（4SAT）	2A	108	1.7	0.07	58.09	6L+20M2+28M1
	52.雏菊属 Bellis L.	52-1 雏菊 Bellis perennis L.	$2n=2x=18$	8m+10sm（1SAT）	1B	36	2.05	0	62.86	2L+6M2+8M1+2S
			$2n=2x=18$	12m（2SAT）+5sm（1SAT）+1st（1SAT）	2B	35	2.8	0.17	61.32	3L+6M2+5M1+4S
	53.金盏花属 Calendula L.	53-1 金盏菊 Calendula officinalis L.	$2n=2x=32$	12m+18sm（2SAT）+2st	2B	62	2.29	0.44	64.94	4L+16M2+4M1+8S
	54.翠菊属 Callistephus Cass.	54-1 翠菊 Callistephus chinensis Nees.	$2n=2x=18$	18m（2SAT）	1A	36	1.65	0	56.16	2L+6M2+10M1
	55.矢车菊属 Centaurea L.	55-1 矢车菊 Centaurea cyanus L.	$2n=2x=24$	5m+6sm+10st+3t	3A	35	1.9	0.71	73.96	2L+8M2+12M1+2S
	56.金鸡菊属 Coreopsis L.	56-1 蛇目菊 Sanvitalia tinctoria Nutt.	$2n=2x=24$	12m+10sm+2st	2A	46	1.51	0.25	61.93	2L+8M2+14M1
XXI 菊科 Compositae	57.秋英属 Cosmos Cav.	57-1 硫华菊 Cosmos sulphureus Cav.	$2n=2x=24$	4sm+8st+12t	3A	28	1.82	0.92	84.69	2L+10M2+10M1+2S
	58.大丽花属 Dahlia Cav.	58-1 大丽菊 Dahlia pinnata Cav.	$2n=65$	49m+14sm+2st	2A	128	2.12	0.18	58.75	6L+24M2+34M1+1S
		58-2 小丽菊 Dahlia pinnate cv.	$2n=64$	48m+14sm+2st	2A	126	1.77	0.16	60.41	2L+32M2+28M1+2S
	59.菊属 Dendranthema (DC.) Des Moul.	59-1 野菊 Dendranthema indicum（L.）Des Moul.	$2n=2x=18$	16m+2sm	1A	36	1.41	0	57.78	12M2+6M1
			$2n=4x=36$	32m+4sm	2A	72	1.75	0.11	56.42	20M2+14M1+2S
		59-2 甘菊 Dendranthema lavandulifolium（Fisch.ex Trautv.）Ling et Shih var.lavandulifolium	$2n=2x=18$	14m+4sm	2A	36	1.61	0.22	58.84	8M2+8M1+2S
		59-3 栽培种 Dendranthema morifolium（Ramat.）Tzvel.	$2n=54$	38m+12sm+4st	2A	104	1.72	0.15	61.5	2L+28M2+22M1+2S
		59-3-1 独立寒秋 Dendranthema morifolium Tzvel.cv.Dulihanqiu	$2n=58$	40m+12sm+6st	2A	110	1.81	0.17	61.85	4L+24M2+26M1+4S

续表

植物核型分析（中国园林花卉植物）

科	属	种	染色体数目	核型公式	核型分类	臂指数	最长染色体/最短染色体	臂比>2的染色体比例	核型不对称系数	染色体相对长度组成
XXI 菊科 Compositae	59.菊属 Dendranthema (DC.) Des Moul.	59-3-2 橙黄托桂小菊 Dendranthema morifolium Tzvel.cv.Chenghuangtuoguixiaoju	$2n=6x=54$	40m+10sm（2SAT）+4st	2B	106	2.06	0.15	59.89	30M2+20M1+4S
		59-3-3 黄白托桂小菊 Dendranthema morifolium Tzvel.cv.Huangbaituoguixiaoju	$2n=6x=54$	40m+10sm（2SAT）+4st	2B	104	2.81	0.19	59.38	32M2+16M1+6S
		59-3-4 粉红托桂小菊 Dendranthema morifolium Tzvel.cv.Fenhongtuoguixiaoju	$2n=6x=53$	41m+12sm	2A	106	1.72	0.11	58.11	2L+29M2+18M1+4S
		59-3-5 早白菊 Dendranthema morifolium Tzvel.cv.Zaobaiju	$2n=6x=55$	47m+6sm+2st	2A	108	1.74	0.14	58.26	2L+20M2+33M1
		59-3-6 苍龙爪 Dendranthema morifolium Tzvel.cv.Canglongzhua	$2n=6x=56$	44m+12sm	2A	112	1.92	0.18	58.98	4L+22M2+28M1+2S
		59-3-7 十丈珠帘 Dendranthema morifolium Tzvel.cv.Shizhangzhulian	$2n=6x=55$	37m+14sm（2SAT）+4st	2A	106	1.66	0.21	59.94	4L+24M2+25M1+2S
		59-3-8 长风万里 Dendranthema morifolium Tzvel.cv.Changfengwanli	$2n=6x=50$	32m+16sm（2SAT）+2st	2B	98	2.24	0.24	60.1	2L+20M2+26M1+2S
		59-3-9 绿牡丹 Dendranthema morifolium Tzvel.cv.Lvmudan	$2n=6x=90$	44m+16sm	2A	120	1.48	0.17	58.68	28M2+32M1
		59-3-10 红托桂 Dendranthema morifolium Tzvel.cv.Hongtuogui	$2n=6x=51$	40m+11sm	2A	102	1.76	0.15	59.53	4L+24M2+23M1
		59-3-11 西厢待月 Dendranthema morifolium Tzvel.cv.Xixiangdaiyue	$2n=6x=55$	39m（2SAT）+12sm（2SAT）+4st	2B	106	3.32	0.21	59.87	2L+24M2+26M1+3S
		59-3-12 嫦娥奔月 Dendranthema morifolium Tzvel.cv.Changebenyue	$2n=6x=52$	40m+10sm（2SAT）+2st	2A	102	1.75	0.12	58.76	28M2+22M1+2S
		59-3-13 五九菊 Dendranthema morifolium Tzvel.cv.Wujiuju	$2n=6x=50$	36m+8sm（2SAT）+6st	2A	94	1.67	0.2	60.3	26M2+22M1+2S
		59-3-14 帅旗 Dendranthema morifolium Tzvel.cv.Shuaiqi	$2n=6x=58$	42m+12sm+4st	2B	112	2.41	0.21	59.31	2L+34M2+20M1+2S
		59-3-15 光辉 Dendranthema morifolium Tzvel.cv.Guanghui	$2n=6x=52$	38m+10sm+4st	2A	100	1.27	0.19	60.67	2L+24M2+26M1
		59-3-16 醒狮图 Dendranthema morifolium Tzvel.cv.Xingshitu	$2n=6x=55$	40m+15sm	2A	110	1.46	0.18	59.02	32M2+23M1
		59-3-17 凤凰振羽 Dendranthema morifolium Tzvel.cv.Fenghuangzhenyu	$2n=6x=55$	46m+7sm+2st	2A	108	1.55	0.07	57.87	2L+26M2+27M1

续表

植物核型分析（中国园林花卉植物）

科	属	种	染色体数目	核型公式	核型分类	臂比指数	最长染色体/最短染色体	臂比>2的染色体比例	核型不对称系数	染色体相对长度组成
	60.勋章菊属 Gazania	60-1 勋章菊 Gazania rigens Moench	$2n=2x=30$	18m+2sm	2A	40	1.43	0.1	56.87	10M2+10M1
	61.向日葵属 Helianthus L.	61-1 多色向日葵 Helianthus annuus L.var.citrinus	$2n=2x=34$	20m（2SAT）+8sm+6st	2A	62	1.55	0.24	62.39	2L+10M2+22M1
	62.腊菊属 Helichrysum L.	62-1 麦秆菊 Helichrysum bracteatum Andr	$2n=2x=24$	22m+2sm	1B	48	2.43	0	58.27	6L+2M2+14M1+2S
	63.天人菊属 Gaillardia Follger	63-1 天人菊 Gaillardia pulchella Follg.	$2n=2x=24$	12m（2SAT）+10sm（2SAT）+2st	2A	46	1.56	0.33	62.85	2L+10M2+12M1
	64.单冠菊属 Haplopapus	64-1 单冠菊 Haplopapus gracllis	$2n=2x=4$	2m+2st（2SAT）	2A	6	1.55	0.5	67.83	2M2+2M1
	65.蛇鞭菊属 Liatris Schreb.	65-1 麒麟菊 Liatris ligulistylis	$2n=2x=18$	14m（2SAT）+4sm	1A	36	1.25	0	58.98	6M2+12M1
XXI菊科 Compositae	66.皇帝菊属 Melampodium	66-1 皇帝菊 Melampodium paludosum	$2n=2x=24$	20m+4sm（2SAT）	2A	48	1.49	0.17	59.33	2L+8M2+14M1
	67.金光菊属 Rudbeckia L.	67-1 黑心菊 Rudbeckia hirta L.	$2n=2x=38$	36m（2SAT）+2sm	1A	76	1.57	0	57.8	2L+18M2+18M1
		67-2 金光菊 Rudbeckia laciniata L.	$2n=2x=38$	38m（2SAT）	1A	76	1.67	0	55.12	2L+18M2+18M1
	68.万寿菊属 Tagetes L.	68-1 万寿菊 Tagetes erecta L.	$2n=2x=24$	10m+14sm（2SAT）	2B	48	2.14	0.5	65.48	4L+8M2+M1+4S
		68-2 孔雀草 Tagetes patula L.	$2n=2x=20$	4m+14sm+2st	2B	38	2.2	0.4	67.06	2L+10M2+6M1+2S
	69.蒲公英属 Taraxacum Weber	69-1 蒲公英 Taraxacum mongolicum Hand.-Mazz.	$2n=3x=24$	18m（3SAT）+6sm	2A	48	1.41	0.13	60.07	9M2+15M1
	70.肿柄菊属 Tithonia Desf.ex Juss.	70-1 肿柄菊 Tithonia diversifolia A. Gray	$2n=2x=34$	24m+14sm（2SAT）	2A	68	1.33	0.06	58.61	20M2+14M1
	71.旱花属 Xeranthemum L.	71-1 千花菊 Xeranthemum annum L.	$2n=2x=12$	2m+4sm+6st（2SAT）	3A	18	1.96	0.83	71.94	2L+4M2+4M1+2S
	72.百日菊属 Zinnia L.	72-1 百日草 Zinnia elegans Jacq.	$2n=2x=24$	24m（2SAT）	1A	48	1.15	0	54.32	12M2+12M1
XXII山茱萸科 Comaceae	73.梾木属 Cornus L.	73-1 红瑞木 Cornus alba L.	$2n=2x=22$	8sm+14st（2SAT）	3B	30	2.55	0.91	76.68	4L+6M2+6M1+6S
XXIII十字花科 Cruciferae	74.紫罗兰属 Matthiola R.Br.	74-1 紫罗兰 Matthiola incana R. Br.	$2n=2x=14$	6m+8sm	2A	28	1.43	0.43	62.35	2L+12M1

续表

植物核型分析（中国园林花卉植物）

科	属	种	染色体数目	核型公式	核型分类	臂指数	最长染色体/最短染色体	臂比>2的染色体比例	核型不对称系数	染色体相对长度组成
XXIII 十字花科 Cruciferae	75.诸葛菜属 Orychophragmus L.	75-1 二月兰 Orychophragmus violaceus (L.) O. E. Schulz	2n=2x=24	18m (2SAT) +6sm	2A	48	1.44	0.08	58.35	10M2+14M1
XXIV 葫芦科 Cucurbitaceae	76.葫芦属 Lagenaria Ser.	76-1 小葫芦 Lagenaria siceraria var. microcarpa Hara	2n=2x=22	16m+6sm	2A	44	1.99	0.18	60.76	2L+10M2+8M1+2S
XXV 杜鹃花科 Ericaceae	77.杜鹃花属 Rhododendron L.	77-1 杜鹃花 Rhododendron simsii Planch	2n=2x=26	20m+2sm+4st (2SAT)	2A	48	1.39	0.23	63.58	14M2+12M1
XXVI 大戟科 Euphorbiaceae	78.大戟属 Euphorbia L.	78-1 一品红 Euphorbia pulcherrima Willd.	2n=2x=28	14m+14sm	2B	56	2.08	0.21	62.46	2L+14M2+10M1+2S
	79.草原龙胆属 Eustoma	79-1 洋桔梗 Eustoma grandiflorum (Raf.) Shinners	2n=2x=18	18m	1A	36	1.67	0	57.68	2L+6M2+10M1
XXVII 龙胆科	80.龙胆属 Gentiana L.	80-1 龙胆 Gentiana scabra Bunge	2n=2x=14	2sm+12st	4A	16	1.61	1	78.87	2L+4M2+8M1
XXVIII 牻牛儿苗科 Geraniaceae	81.天竺葵属 Pelargonium L.	81-1 香叶天竺葵 Pelargonium graveolens L'Herit.	2n=2x=30	20m+10sm	2A	60	1.5	0.13	58.88	18M2+12M1
		81-2 天竺葵 Pelargonium hortorum Bailey	2n=2x=18	8m+10sm	2A	36	1.57	0.22	62.82	8M2+8M1+2S
XXIX 苦苣苔科 Gesneriaceae	82.大岩桐属 Sinningia Ness	82-1 大岩桐 Sinningia speciosa Benth.et Hook.	2n=2x=16	14m+2sm (2SAT)	1A	32	1.38	0	56.73	8M2+8M1
XXX 鸢尾科 Iridaceae	83.香雪兰属 Freesia Klatt	83-1 香雪兰 Freesia refracta Klatt.	2n=2x=24	8m+4sm+8st+4t	3B	36	3.8	0.58	70.36	4L+5M2+8M1+7S
	84.鸢尾属 Iris L.	84-1 马兰花 Iris ensata Thunb.	2n=40	34m+4sm+2st	2A	78	1.92	0.1	60.27	2L+16M2+18M1+4S
		84-2 日本鸢尾 Iris japonica Thunb.	2n=2x=18	12m (2SAT) +6sm	2A	36	1.65	0.22	59.69	10M2+6M1+2S
		84-3 鸢尾 Iris tectorum Maxim.	2n=2x=38	12m+16sm (2SAT)	2A	56	1.82	0.36	64.08	16M2+8M1+4S
	85.鞘蕊属 Coleus Lour.	85-1 彩叶草 Coleus blumei Benth.	2n=48	24m+12sm+10st+2t (2SAT)	2B	84	3.8	0.5	66.11	8L+8M2+16M1+16S
	86.贝壳花属 Moluccella	86-1 贝壳花 Moluccella laevis L.	2n=2x=34	28m+6sm (2SAT)	2A	68	1.7	0.12	58.01	16M2+16M1+2S
XXXI 唇形科 Labiatae	87.罗勒属 Ocimum L.	87-1 罗勒 Ocimum basilicum L.	2n=52	28m+10sm+14st	2B	90	2.68	0.46	63.8	12L+10M2+18M1+12S
	88.鼠尾草属 Salvia L.	88-1 一串兰 Salvia farinacea Benth	2n=2x=18	8m+6sm+4st	3B	32	2.38	0.56	67	4L+4M2+4M1+6S
		88-2 一串红 Salvia splendens Ker-Gawler	2n=2x=44	34m+2sm+8st	2A	80	1.85	0.23	62.41	4L+20M2+12M1+8S

续表

植物核型分析（中国园林花卉植物）

科	属	种	染色体数目	核型公式	核型分类	臂指数	最长染色体最短染色体	臂比>2的染色体比例	核型不对称系数	染色体相对长度组成
	89.合欢属 Albizia Durazz	89-1 合欢 Albizia julibrissin Durazz	$2n=2x=26$	16m+10sm（2SAT）	2A	52	1.63	0.15	61.85	2L+12M2+12M1
	90.紫穗槐属 Amorpha L.	90-1 紫穗槐 Amorpha fruticosa L.	$2n=40$	32m（1SAT）+8sm	1A	80	1.7	0	58.27	4L+18M2+16M1+2S
	91.云实属 Caesalpinia L.	91-1 金凤花 Caesalpinia pulcherrima Sw.	$2n=2x=24$	8m+12sm+4st	2A	44	1.77	0.5	67.21	2L+6M2+16M1
	92.锦鸡儿属 Caragana Fabr.	92-1 红花锦鸡儿 Caragana rosea Turcz.	$2n=2x=32$	24m+8sm	2A	64	1.48	0.06	58.29	18M2+14M1
	93.紫荆属 Cercis L.	93-1 紫荆 Cercis chinensis Bge.	$2n=2x=14$	10st（2SAT）+4t	4A	14	1.46	1	84.37	2L+4M2+8M1
	94.香豌豆属 Lathyrus L.	94-1 香豌豆 Lathyrus odoratus L.	$2n=2x=14$	6m+8sm	2A	28	1.44	0.29	63.6	6M2+8M1
	95.银合欢属 Leucaena Benth.	95-1 银合欢 Leucaena glauca（L.）Benth.	$2n=2x=36$	32m（4SAT）+2sm+2st	2B	70	2.85	0.11	57.6	2L+16M2+16M1+2S
	96.含羞草属 Mimosa L.	96-1 含羞草 Mimosa pudica L.	$2n=52$	40m（1SAT）+6sm+6st	2A	98	1.56	0.23	60.84	24M2+28M1
XXXII 豆科 Leguminosae	97.刺槐属 Robinia L.	97-1 毛刺槐 Robinia hispida L.	$2n=30$	10m+12sm+6st（2SAT）+2t（2SAT）	3B	52	2.06	0.6	66.41	6L+6M2+12M1+6S
		97-2 洋槐 Robinia pseudoacacia L.	$2n=2x=22$	4m+8sm+10st（4SAT）	3B	34	2.38	0.82	71.67	4L+2M2+14M1+2S
		97-3 无刺洋槐 Robinia pseudoacacia L.f.inermis（Mirb.）Rehd.	$2n=2x=22$	2m+8sm+12st	3A	32	1.89	0.73	71	4L+4M2+14M1
	98.槐属 Sophora L.	98-1 槐树 Sophora japonica L.	$2n=28$	18m+10sm（2SAT）	2B	56	2.04	0.29	61.7	4L+6M2+18M1
		98-2 五叶槐 Sophora japonica L.f.oligophylla Franch.	$2n=28$	14m+10sm+4st	2B	52	2.38	0.29	63.48	2L+10M2+14M1+2S
		98-3 龙爪槐 Sophora japonica L. f. pendula Loud.	$2n=28$	18m+8sm+2st	2B	54	2.34	0.36	61.38	2L+8M2+16M1+2S
		98-4 红花槐 Sophora Rubriflora Tsoong.	$2n=3x=21$	6m+6sm+6st（1SAT）+3t	3A	22	1.72	0.57	69.97	2L+4M2+8M1
		98-5 黄金槐 Sophora xanthoantha C.M.Ma.	$2n=28$	14m+12sm+2st0	2A	54	1.97	0.21	62.38	2L+4M2+22M1
	99.车轴草属 Trifolium Limn.	99-1 白车轴草 Trifolium repens L.	$2n=32$	28m+4sm	2A	64	1.67	0.06	56.72	2L+12M2+18M1
	100.紫藤属 Wisteria Nutt.	100-1 紫藤 Wisteria sinensis（Soms.）Sweet	$2n=4x=32$	28m+4sm	2A	64	1.37	0.13	58.17	18M2+14M1

续表

植物核型分析（中国园林花卉植物）

科	属	种	染色体数目	核型公式	核型分类	臂指数	最长染色体/最短染色体	臂比>2的染色体比例	核型不对称系数	染色体相对长度组成
	101.芦荟属 Aloe L.	101-1 日本芦荟 Aloe arborescens Mill.	$2n=2x=14$	8sm+6st（2SAT）	3B	22	3.16	0.86	75.86	6L+2M2+6S
		101-2 皂素芦荟 Aloe saponaria (Ait.) Haw.	$2n=2x=14$	8sm+6st（2SAT）	3B	22	3.82	0.86	76.28	6L+2M2+6S
		101-3 刺芦荟 Aloe thomcroftii	$2n=2x=14$	1m+6sm（3SAT）+7st（3SAT）	3C	21	4.31	0.86	75.71	8L+6S
		101-4 花叶芦荟 Aloe variegate L.	$2n=2x=14$	6sm+8st（4SAT）	4B	20	3.64	1	81.63	8L+6S
		101-5 美国芦荟 Aloe vera Linn.	$2n=2x=14$	8sm+6st	3B	22	3.16	0.86	75.62	6L+2M2+6S
		101-6 芦荟 Aloe vera L. var.chinensis (Haw.) Berg.	$2n=2x=14$	8sm+6st	3B	22	3.23	0.57	74.69	8L+6S
	102.天门冬属 Asparagus L.	102-1 文竹 Asparagus plumosus Baker	$2n=2x=20$	12m+8sm	2B	40	2.14	0.4	63.39	2L+10M2+8S
	103.蜘蛛抱蛋属 Aspidistra Ker-Gawl.	103-1 蜘蛛抱蛋 Aspidistra elatior Blume	$2n=2x=38$	24m+2sm+12st	2C	64	5.37	0.37	67.88	14L+2M2+2M1+20S
XXXIII百合科 Liliaceae	104.吊兰属 Chlorophytum Ker-Gawl	104-1 金边吊兰 Chlorophytum capense（L.）Druce var.variegatum Hort.	$2n=2x=28$	6m+14sm+8st	2A	48	1.85	0.43	67.45	4L+10M2+12M1+2S
		104-2 吊兰 Chlorophytum comosum（Thunb.）Baker.	$2n=2x=28$	10m+8sm+10st	2B	46	2.09	0.43	66.78	2L+10M2+12M1+4S
	105.铁线莲属 Clematis L.	105-1 大叶铁线莲 Clematis heracleifolia DC.	$2n=2x=16$	10m+4st（2SAT）+2t	2A	26	1.64	0.38	63.52	2L+6M2+8M1
		105-2 长瓣铁线莲 Clematis macropetala Ledeb.	$2n=2x=16$	10m+4st(2SAT)+2t(2SAT)	2B	26	2.13	0.38	62.64	6L+2M2+4M1+4S
	106.龙血树属 Dracaena Vand.ex L.	106-1 银叶龙血树 Dracaena sanderiana Hort.	$2n=40$	32m+8sm	2B	80	2.62	0.15	59.27	6L+10M2+18M1+6S
	107.萱草属 Hemerocallis L.	107-1 萱草 Hemerocallis fulva (L.) L.	$2n=2x=22$	12m+10sm	2A	44	1.82	0.45	63.44	4L+4M2+14M1
	108.玉簪属 Hosta Tratt.	108-1 玉簪 Hosta plantaginea Aschers.	$2n=60$	20m+10sm+16st+14t	3C	90	5.54	0.67	76.54	14L+2M2+10M1+34S
	109.风信子属 Hyacinthus L.	109-1 风信子 Hyacinthus orientalis L.	$2n=2x=30$	12m（1SAT）+14sm（4SAT）+4st	2C	56	4.31	0.37	60.13	12L+3M2+2M1+13S
	110.百合属 Lilium L.	110-1 麝香百合 Lilium longiflorum Thunb.	$2n=2x=24$	6m+2sm+4st+12t（4SAT）	3B	32	2.25	0.67	78.91	4L+4M2+14M1+2S
		110-2 王百合 Lilium regale Wilson	$2n=2x=24$	2m+2sm+10st+10t	3A	28	1.98	0.83	80.94	4L+4M2+14M1+2S

续表

植物核型分析（中国园林花卉植物）

科	属	种	染色体数目	核型公式	核型分类	臂指数	最长染色体/最短染色体	臂比>2的染色体比例	核型不对称系数	染色体相对长度组成
	111.葡萄风信子属 Muscari Mill.	111-1 蓝壶花 Muscari botryoides Mill.	$2n=36$	24m+6sm+2st+4t	2B	66	2.59	0.17	64.78	6L+8M2+10M1+12S
	112.重楼属 Paris L.	112-1 七叶一枝花 Paris polyphylla Sm.	$2n=2x=10$	6m+2st+2t	2A	16	1.46	0.4	64.33	4M2+6M1
	113.万年青属 Rohdea Roth	113-1 万年青 Rohdea japonica Roth	$2n=2x=28$	6m+14sm+8st	3B	48	2.16	0.57	67.6	2L+10M2+12M1+4S
	114.绵枣属 Scilla L.	114-1 绵枣儿 Scilla scilloides (Lindl.) Druce	$2n=2x=16+4B$s	6m+6sm+4st (2SAT) +4t	3C	32	8.42	0.7	68.41	4L+6M2+6M1+4S
		115-1 阿尔泰郁金香 Tulipa altaica Pall.ex Spreng	$2n=2x=24$	2m+4sm+18st	3A	30	1.82	0.92	76.01	4L+6M2+12M1+2S
		115-2 毛蕊郁金香 Tulipa dasystemon (Regel.) Regel.	$2n=2x=24$	6sm+18st	3A	30	1.66	0.92	77.61	2L+12M2+6M1+4S
XXXIII 百合科 Liliaceae	115.郁金香属 Tulipa L.	115-3 郁金香 Tulipa gesneriana L.	$2n=2x=24$	2m+8sm+14st	3A	34	1.78	0.83	72.14	4L+6M2+12M1+2S
		115-4 异叶郁金香 Tulipa heterophylla Backer	$2n=2x=24$	2m+8sm+14st	3B	34	2.12	0.92	75.42	2L+10M2+8M1+4S
		115-5 伊犁郁金香 Tulipa iliensis Regel	$2n=2x=24$	4sm+20st	4A	28	1.67	1	77.94	2L+8M2+14M1
		115-6 垂蕾郁金香 Tulipa patens Agardh.ex Schult	$2n=2x=24$	14sm+10st	3A	38	1.8	0.92	75.73	4L+6M2+14M1
		115-7 新疆郁金香 Tulipa sinkiangensi Z. M. Mao	$2n=2x=24$	8sm+16st	4B	32	2.03	1	78.34	4L+8M2+8M1+4S
		115-8 天山郁金香 Tulipa tianschanica Regel	$2n=2x=24$	2m+4sm+18st	3A	30	1.83	0.92	77.77	4L+8M2+10M1+2S
	116.丝兰属 Yucca L.	116-1 丝兰 Yucca smalliana Fern								
XXXIV 亚麻科 Linaceae	117.亚麻属 Linum L.	117-1 亚麻 Linum sp.	$2n=2x=18$	16m+2st	2A	34	1.61	0.11	58.37	12M2+4M1+2S
XXXV 千屈菜科 Lythraceae	118.紫薇属 Lagerstroemia L.	118-1 紫薇 Lagerstroemia indica L.	$2n=2x=48$	28m+18sm+2st	2B	94	2.13	0.29	61.74	6L+16M2+20M1+6S
	119.长蕊木兰属 Alcimandra Dandy.	119-1 长蕊木兰 Alcimandra cathcartii Dandy	$2n=2x=38$	26m (2SAT) +4sm+8st (2SAT)	2B	68	2.45	0.32	62.58	6L+6M2+24M1+2S
XXXVI 木兰科 Magnoliaceae	120.鹅掌楸属 Liriodendron L.	120-1 鹅掌楸 Liriodendron chinense (Hemsl.) Sarg	$2n=2x=38$	30m (2SAT) +6sm+2st (2SAT)	2B	74	2.22	0.16	60.8	6L+14M2+14M1+4S
		120-2 北美鹅掌楸 Liriodendron tulipifera L.	$2n=2x=38$	32m+4sm+2st	2A	74	1.97	0.11	59.85	4L+10M2+24M1

植物核型分析（中国园林花卉植物）续表

科	属	种	染色体数目	核型公式	核型分类	臂指数	最长染色体/最短染色体	臂比>2的染色体比例	核型不对称系数	染色体相对长度组成
	121.木莲属 Manglietia Bl.	121-1 香木莲 Manglietia aromatica Dandy	$2n=2x=38$	32m（2SAT）+4sm+2st	2A	74	1.98	0.11	59.54	4L+16M2+14M1+4S
		121-2 灰木莲 Manglietia glauca Blume	$2n=2x=38$	32m+4sm+2st（2SAT）	2B	74	2.04	0.11	58.06	4L+12M2+20M1+2S
		121-3 海南木莲 Manglietia hainanensis Dandy	$2n=2x=38$	32m+4sm+2st	2B	74	2.11	0.11	57.77	6L+14M2+12M1+6S
		121-4 中缅木莲 Manglietia hookeri Cubitt et Smith	$2n=2x=38$	32m+4sm+2st	2B	74	2.18	0.16	59.01	6L+6M2+24M1+2S
		121-5 毛桃木莲 Manglietia moto Dandy	$2n=2x=38$	32m+4sm+2st（2SAT）	2A	74	1.94	0.11	57.43	4L+10M2+22M1+2S
		121-6 乳源木莲 Manglietia yuyanensis Law.	$2n=2x=38$	32m+4sm+2st（2SAT）	2B	74	2.24	0.16	57.73	4L+10M2+22M1+2S
	122.华盖木属 Manglietiastrum Law.	122-1 华盖木 Manglietiastrum sinicum Law.	$2n=2x=38$	28m+4sm+6st（6SAT）	2B	70	2.86	0.21	64.13	8L+6M2+12M1+12S
XXXVI 木兰科 Magnoliaceae	123.木兰属 Magnolia L.	123-1 天目木兰 Magnolia amoena Cheng	$2n=2x=38$	32m+4sm+2st（2SAT）	2B	74	2.57	0.11	58.46	6L+10M2+16M1+6S
		123-2 夜香木兰 Magnolia coco（Lour.）DC.	$2n=2x=38$	32m+4sm+2st（2SAT）	2B	74	2.18	0.16	60.5	6L+10M2+18M1+4S
		123-3 山玉兰 Magnolia delavayi Franch	$2n=2x=38$	32m+4sm（2SAT）+2st（2SAT）	2A	74	1.73	0.16	58.68	4L+12M2+18M1+4S
		123-4 玉兰 Magnolia denudata Desr.	$2n=4x=76$	60m+10sm+6st（2SAT）	2B	146	2.11	0.18	59.71	14L+16M2+42M1+4S
		123-5 荷花木兰 Magnolia grandiflora L.	$2n=6x=114$	96m（2SAT）+12sm+6st	2B	222	2.08	0.14	57.27	16L+30M2+54M1+14S
		123-6 柴叶荷花木兰 Magnolia grandiflora var. lanceolata Ait.	$2n=6x=114$	96m+12sm+6st（2SAT）	2B	222	2.65	0.14	56.62	22L+22M2+60M1+10S
		123-7 紫玉兰 Magnolia liliflora Desr.	$2n=4x=76$	58m+8sm+10st（2SAT）	2B	142	2.85	0.16	60.26	16L+16M2+34M1+10S
		123-8 厚朴 Magnolia officinalis Rehd. et Wils.	$2n=2x=38$	32m+4sm+2st（2SAT）	2B	74	2.13	0.11	57.53	4L+10M2+20M1+4S
		123-9 凹叶厚朴 Magnolia officinalis subsp. biloba（Rehd.et Wils.）Law.	$2n=2x=38$	32m+4sm+2st（2SAT）	2B	74	2.29	0.11	59.68	6L+10M2+16M1+6S
		123-10 长叶木兰 Magnolia paenetalauma Dandy	$2n=2x=38$	30m+4sm+4st（2SAT）	2B	72	2.03	0.16	57.13	10L+4M2+22M1+2S

续表

植物核型分析（中国园林花卉植物）

科	属	种	染色体数目	核型公式	核型分类	臂指数	最长染色体/最短染色体	臂比>2的染色体比例	核型不对称系数	染色体相对长度组成
	123. 木兰属 Magnolia L.	123-11 二乔木兰 Magnolia soulangeana Soul.-Bod	2n=4x=76	56m（2SAT）+12sm+8st	2B	144	2.49	0.24	60.58	12L+14M2+42M1+8S
		123-12 红花山木兰 Magnolia sp	2n=2x=38	30m+6sm+2st（2SAT）	2B	74	2.48	0.21	57.86	8L+6M2+18M1+6S
		124-1 白兰 Michelia alba DC.	2n=2x=38	34m+4sm（2SAT）	2B	76	2.05	0.11	56.98	8L+6M2+22M1+2S
		124-2 黄兰 Michelia champaca L.	2n=2x=38	32m+6sm	2B	76	2.54	0.11	57.94	8L+10M2+14M1+6S
		124-3 含笑花 Michelia figo（Lour.）Spreng.	2n=2x=38	32m+6sm	2A	76	1.81	0.16	56.51	6L+8M2+20M1+4S
		124-4 多花含笑 Michelia floribunda Finet et Gagn.	2n=2x=38	30m+8sm	2B	76	2.38	0.16	57.23	4L+14M2+18M1+2S
		124-5 金叶含笑 Michelia foveolata var. foveolata Merr. ex Dandy	2n=2x=38	34m（2SAT）+4sm	2A	76	1.92	0.11	57.17	6L+10M2+18M1+4S
		124-6 长蕊含笑 Michelia longistamina Law	2n=2x=38	34m+6sm	2B	76	1.79	0.16	56.87	8L+8M2+18M1+4S
	124. 含笑属 Michelia L.	124-7 薄叶含笑 Michelia macclurei Dandy	2n=2x=38	34m（2SAT）+4sm	2B	76	2.07	0.11	56.71	6L+8M2+22M1+2S
		124-8 展毛含笑 Michelia macclurei var. macclurei Dandy.	2n=2x=38	32m（2SAT）+4sm+2t	2B	74	2.22	0.11	58.37	6L+8M2+20M1+4S
XXXVI 木兰科 Magnoliaceae		124-9 深山含笑 Michelia maudiae Dunn	2n=2x=38	32m+6sm	2B	76	2.06	0.16	57.86	6L+8M2+22M1+2S
		124-10 阔瓣含笑 Michelia platypetala Hand.-Mazz	2n=2x=38	32m+6sm（2SAT）	2A	76	1.96	0.16	58.79	6L+10M2+16M1+6S
		124-11 石碌含笑 Michelia shiluensis Chun. et Y. F. Wu	2n=2x=38	32m+6sm	2B	76	2.3	0.16	58.51	6L+10M2+20M1+2S
		124-12 野含笑 Michelia skinneriana Dunn	2n=2x=38	30m+8sm	2B	76	2.41	0.16	57.32	4L+12M2+18M1+4S
		124-13 峨眉含笑 Michelia wilsonii Finet et Gagn.	2n=2x=38	30m+8sm（2SAT）	2B	76	2.67	0.21	58.21	6L+10M2+16M1+6S
	125. 拟单性木兰属 Parakmeria Hu et Cheng	125-1 乐东拟单性木兰 Parakmeria lotungensis（Chun et C.Tsoong）Law.	2n=6x=114	96m+12sm（2SAT）+6st（2SAT）	2B	222	2.02	0.12	57.57	20L+16M2+70M1+8S
		125-2 峨眉拟单性木兰 Parakmeria omeiensis Cheng	2n=4x=76	56m+16sm+4st（2SAT）	2B	148	2.07	0.21	59.46	12L+20M2+36M1+8S
	126. 合果木属 Paramichelia Hu	126-1 合果木 Paramichelia baillonii（Pierre）Hu	2n=2x=38	34m+2sm+2st（2SAT）	2A	74	1.74	0.11	57.85	6L+8M2+24M1

续表

植物核型分析（中国园林花卉植物）

科	属	种	染色体数目	核型公式	核型分类	臂指数	最长染色体/最短染色体	臂比>2的染色体比例	核型不对称系数	染色体相对长度组成
XXXVII 锦葵科 Malvaceae	127. 观光木属 Tsoongiodendron Chun	127-1 观光木 Tsoongiodendron odorum Chun	$2n=2x=38$	32m+6sm（2SAT）	2B	76	2.16	0.16	55.93	6L+12M2+12M1+8S
	128. 秋葵属 Abelmoschus Medicus	128-1 秋葵 Abelmoschus esculentus (Linn.) Moench	$2n=40$	20m+14sm+6st	2B	74	2.66	0.35	63.28	8L+4M2+20M1+8S
	129. 蜀葵属 Althaea L.	129-1 蜀葵 Althaea rosea Cav.	$2n=42$	36m+4sm（2SAt）+2st	2A	82	1.59	0.1	59.6	4L+18M2+20M1
	130. 木槿属 Hibiscus L.	130-1 木槿 Hibiscus syriacus L.	$2n=88$	54m+20sm（2SAT）+14st	2B	162	3.66	0.27	61.88	10L+26M2+32M1+20S
	131. 锦葵属 Malva L.	131-1 锦葵 Malva sinensis Caran.	$2n=42$	34m+6sm+2st	2A	82	1.86	0.19	58.89	6L+12M2+24M1
XXXVIII 楝科 Meliaceae	132. 米籽兰属 Aglaia Lour.	132-1 米籽兰 Aglaia odorata Lour.	$2n=84$	36m+16sm+22st+10t	2A	136	1.85	0.48	69.54	4L+34M2+46M1
XXXIX 桑科 Moraceae	133. 无花果属 Ficus L.	133-1 橡皮树 Ficus elastica Roxb. ex Hornem	$2n=40$	8m+6sm（2SAT）+26st（2SAT0	3B	54	2.27	0.8	73.55	6L+6M2+26M1+2S
XL 叶子花科	134. 叶子花属 Bougainvillea Comm. ex Juss.	134-1 叶子花 Bougainvillea spectabilis Willd.	$2n=48$	40m（2SAT）+8sm	1B	96	2.2	0	57.8	2L+24M2+18M1+4S
Nyctaginaceae	135. 紫茉莉属 Mirabilis l.	135-1 紫茉莉 Mirabilis jalapa L.	$2n=58$	38m（4SAT）+18sm+2st	2B	114	2.36	0.1	60.73	8L+16M2+26M1+8S
	136. 莲属 Nelumbo Adans	136-1 荷花 Nelumbo nucifera Gaertn	$2n=2x=16$	8m+8sm	2B	32	2.48	0.38	63.68	2L+6M2+6M1+2S
	137. 萍蓬草属 Nuphar Sibth et Sm.	137-1 萍蓬草 Nuphar pumilum (Timm.) DC.	$2n=2x=34$	30m+4sm	2A	68	1.72	0.06	58.42	4L+10M2+18M1+2S
XLI 睡莲科 Nymphaeaceae	138. 睡莲属 Nymphaea L.	138-1 白睡莲 Nymphaea alba L.	$2n=84$							
		138-2 红睡莲 Nymphaea alba L. var. rubra Lonnr.	$2n=54$							
		138-3 黄花睡莲 Nymphoides mexicana Zucc.	$2n=82$	48m(1SAT)+10m+22st+2t	2A	140	1.87	0.41	65.64	2L+32M2+48M1
		138-4 延药睡莲 Nymphaea stellata Willd.	$2n=76$							
		138-5 睡莲 Nymphaea tetragona Georgi.	$2n=66$							

续表

植物核型分析（中国园林花卉植物）

科	属	种	染色体数目	核型公式	核型分类	臂指数	最长染色体/最短染色体	臂比>2的染色体比例	核型不对称系数	染色体相对长度组成
	139.连翘属 Forsythia Vahl	139-1 连翘 Forsythia suspensa (Thunb.) Vahl	$2n=2x=24$	16m+6sm+2st	2A	46	1.72	0.25	62.13	2L+10M2+10M1+2S
	140.素馨属 Jasminum L.	140-1 迎春花 Jasminum nudiflorum Lindl.	$2n=2x=28$	20m+6sm+2st（2SAT）	2A	54	1.87	0.29	62.96	2L+12M2+14M1
		140-2 茉莉 Jasminum sambac (L.) Ait.	$2n=26$	20m+6sm（2SAT）	2A	52	1.59	0.08	57.93	14M2+10M1+2S
	141.女贞属 Ligustrum L.	141-1 小叶女贞 Ligustrum quihoui Carr.	$2n=46$	28m+10sm（2SAT）+8st	2B	84	2.39	0.39	64.21	4L+18M2+20M1+4S
	142.木犀属 Osmanthus Lour.	142-1 桂花 Osmanthus fragrans Lour.	$2n=46$	28m+12sm（1SAT）+4st（2SAT）+2t	2A	86	1.86	0.26	63.52	2L+22M2+22M1
	143 丁香属 Syringa L.	143-1 暴马丁香 Syringa eticulata var. mandshurica Hara.	$2n=46$	24m+14sm+8st（1SAT）	2B	84	2.98	0.3	62.05	8L+12M2+16M1+10S
		143-2 白丁香 Syringa oblata var. offinis Lingelshe	$2n=36$	14m+16sm+4st+2t	2A	66	1.7	0.44	66.46	4L+12M2+20M1
		143-3 紫丁香 Syringa oblata L.	$2n=46$	42m+4sm	2A	92	1.91	0.04	56.69	8L+8M2+26M1+4S
		143-4 洋丁香 Syringa vulgaris L.	$2n=2x=38$	22m+12sm+2st+2t（2SAT）	2A	72	1.58	0.37	63.76	2L+14M2+22M1
		143-5 滇丁香 Syringa yunnanensis French.	$2n=44$	28m+12sm+4st	2A	84	1.8	0.18	61.42	2L+20M2+22M1
	144.倒挂金钟属 Fuchsia L.	144-1 吊钟海棠 Fuchsia magellanica Lam.	$2n=120$	62m+14sm+36st+8t	2B	196	3.34	0.43	67.6	6L+52M2+54M1+8S
XLIII柳叶菜科 Onagraceae	145.高代花属 Godetia Spach	145-1 古代稀 Godetia amoena Don.	$2n=2x=14$	4m+4sm（2SAT）+6st	2A	22	1.3	0.43	67.36	8M2+6M1
	146.月见草属 Oenothera L.	146-1 月见草 Oenothera Odorata Jacq.	$2n=2x=14$	12m（2SAT）+2sm	2A	28	1.32	0.14	57.15	6M2+8M1
	147.指甲兰属 Aerides Lour.	147-1 脑唇指甲兰 Aerides flabellata Rolfe ex Downie	$2n=2x=38$	34m+4sm	2A	76	1.97	0.11	58.04	4L+12M2+18M1+4S
	148.白芨属 Bletilla Rchb.f	148-1 白芨 Bletilla striata(Thunb.)ex A.Murray Rchb. f.	$2n=2x=32$	18m+6sm+8st	2B	56	3.4	0.44	65.6	6L+12M2+4M1+10S
XLIV兰科 Orchidaceae	149.虾脊兰属 Calanthe R.Br.	149-1 虾脊兰 Calanthe discolor Lindl.	$2n=2x=40$	28m+8sm+2st+2t	2A	76	1.83	0.25	60.91	6L+10M2+20M1+4S
	150.卡特兰属 Cattleya L.	150-1 卡特兰 Cattleya labiata Lindl.	$2n=46$	26m（1SAT）+12sm+8st	2B	84	2.31	0.26	62.52	4L+14M2+26M1+2S
	151.隔距兰属 Cleisostoma Bl.	151-1 大序隔距兰 Cleisostoma paniculatum (Ker-Gawl.) Garay	$2n=2x=38$	16m+10sm+12st	2B	64	2.54	0.47	68.21	8L+10M2+12M1+8S

续表

植物核型分析（中国园林花卉植物）

科	属	种	染色体数目	核型公式	核型分类	臂指数	最长染色体/最短染色体	臂比>2的染色体比例	核型不对称系数	染色体相对长度组成
XLIV 兰科 Orchidaceae	152.舞草属 Codariocalyx Hassk.	152-1 跳舞草 Codariocalyx gnans	$2n=2x=20$	14m+4sm+2st	2B	38	2.67	0.3	60.4	4L+4M2+6M1+6S
	153.贝母兰属 Coelogyne Lindl	153-1 流苏贝母兰 Coelogyne fimbriata Lindl.	$2n=2x=40$	32m+8sm	2B	80	2.04	0.05	58.82	4L+18M2+14M1+4S
		153-2 斑唇贝母兰 Coelogyne fuscescens var. Brunnea (Lindl.) Lindl.	$2n=2x=40$	30m+4sm+6st	2A	74	1.9	0.2	60.95	2L+14M2+22M1+2S
	154.兰属 Cymbidium Sw.	154-1 纹瓣兰 Cymbidium aloifolium (L.) Sw.	$2n=2x=40$	28m+10sm+2st (2SAT)	2A	78	1.44	0.1	58.42	24M2+16M1
		154-2 冬凤兰 Cymbidium dayanum Rchb.f	$2n=2x=40$	34m+6sm	2B	80	2.16	0.05	58.76	2L+24M2+6M1+8S
		154-3 建兰 Cymbidium ensifolium (L.) Sw.	$2n=2x=40$	34m (2SAT) +6sm	2B	80	2.81	0.05	58.27	6L+14M2+14M1+6S
		154-4 蕙兰 Cymbidium faberi Rolfe.var.faberi	$2n=2x=44$	26m+14sm+2st (2SAT) +2t	2A	84	1.9	0.32	62.99	6L+16M2+18M1+4S
		154-5 云南多花兰 Cymbidium floribundum Lindl.	$2n=2x=40$	38m (2SAT) +2sm	2A	80	1.47	0.05	58.14	16M2+24M1
		154-6 春兰 Cymbidium goeringii (Rchb. f.) Rchb. f. var.goeringii	$2n=2x=39$	24m+13sm+2st	2A	76	1.75	0.35	61.13	6L+12M2+19M1+2S
		154-7 虎头兰 Cymbidium hookerianum Rchb. f. Hook.	$2n=2x=40$	26m+14sm	2A	80	1.91	0.2	58.04	6L+12M2+18M1+4S
		154-8 兔耳兰 Cymbidium lancifolium Hook.	$2n=2x=38$	16m+18sm+4st	2B	72	3.02	0.42	63.87	4L+14M2+12M1+8S
		154-9 尖叶吊兰 Cymbidium sp.	$2n=2x=40$	34m+4sm+2st (2SAT)	2B	78	2.84	0.1	58.74	10L+8M2+14M1+8S
		154-10 宽叶虎头兰 Cymbidium sp.	$2n=2x=38$	28m+8sm+2st (2SAT)	2B	74	2.06	0.21	58.45	2L+16M2+12M1+8S
		154-11 大花蕙兰 Cymbidium Sw.cv	$2n=4x=76$	40m (2SAT) +18sm+12st (4SAT) +6t	2B	134	3.3	0.37	64.68	14L+16M2+28M1+18S
	155.石斛属 Dendrobium Sw.	155-1 束花石斛 Dendrobium chrysanthum Lindl.ex Wall.	$2n=4x=76$	46m (1SAT) +16sm (2SAT) +14st	2B	138	2.88	0.34	62.41	22L+14M2+22M1+18S
		155-2 叠鞘石斛 Dendrobium denneanum Kerr	$2n=2x=38$	32m+4sm+2t	2B	74	2.16	0.05	57.06	4L+16M2+12M1+6S
		155-3 石斛 Dendrobium nobile Lindl.	$2n=2x=38$	26m+8sm+4st	2B	72	2.02	0.16	62.94	4L+16M2+14M1+4S

续表

植物核型分析（中国园林花卉植物）

科	属	种	染色体数目	核型公式	核型分类	臂指数	最长染色体/最短染色体	臂比>2的染色体比例	核型不对称系数	染色体相对长度组成
		155-4 石斛属种间杂种 D.denneanum×D.moschatum	2n=44	38m+6sm（1SAT）	2B	88	2.05	0.09	58.5	2L+24M2+16M1+2S
	156.毛兰属 Eria Lindl.	156-1 钝叶毛兰 Eria acervata Lindl.	2n=2x=38	32m+2sm+2st+2t	2B	72	2.08	0.16	59.43	6L+6M2+26M1
	157.斑叶兰属 Goodyera R.Br.	157-1 高斑叶兰 Goodyera procera (Ker-Gawl.) Hook.	2n=2x=38	22m+12sm（2SAT）+4st（2SAT）	2B	72	2.21	0.26	62.21	8L+10M2+14M1+6S
	158.槽舌兰属 Holcoglossum Schltr	158-1 小花槽舌兰 Holcoglossum junceum Tsi.	2n=2x=38	30m+8sm（2SAT）	2B	76	2.69	0.05	59.5	4L+14M2+14M1+6S
	159.沼兰属 Malaxis Soland.ex Sw.	159-1 阔叶沼兰 Malaxis latifolia J. E. Sm.	2n=2x=42	10m+26sm+6st（2SAT）	3B	78	2.12	0.67	68.5	8L+4M2+26M1+4S
	160.兜兰属 Paphiopedilum pfitz.	160-1 硬叶兜兰 Paphiopedilum micranthum T. Tang et F. T. Wang	2n=2x=26	18m+8sm	2B	52	2.72	0.23	59.12	6L+2M2+8M1+10S
	161.钻柱兰属 Pelatantheria Ridl.	161-1 钻柱兰 Pelatantheria rivesii T. Tang et F. T. Wang	2n=2x=38	34m+4sm	2B	76	2.24	0.05	57.96	8L+8M2+18M1+4S
XLIV 兰科 Orchidaceae	162.蝶兰属 Phalaenopsis Bl.	162-1 蝴蝶兰 Phalaenopsis aphrodite Rchb. f.	2n=4x=76	50m+10sm+16st（4SAT）	2B	136	2.48	0.34	63.64	10L+26M2+30M1+10S
	16.3 石仙桃属 Pholidota Lindl.	163-1 节茎石仙桃 Pholidota articulata Lindl	2n=2x=38	32m+4sm+2st	2A	74	1.76	0.05	57.55	2L+14M2+22M1
	164.独蒜兰属 Pleione D.Don.	164-1 独蒜兰 Pleione bulbocodioides (Franch.) Rolfe	2n=2x=41	32m+4sm+5st（1SAT）	2B	77	2.8	0.24	59.52	8L+16M2+8M1+9S
		164-2 黄花独蒜兰 Pleione forrestii Schltr.	2n=2x=44	28m+14sm+2st	2B	86	3.67	0.27	59.13	12L+10M2+10M1+12S
		164-3 台湾独蒜兰 Pleione formosana Hayata	2n=2x=40	20m+12sm+8st	2B	72	2.94	0.5	65.35	10L+12M2+12M1+6S
	165.火焰兰属 Renanthera Lour.	165-1 火焰兰 Renanthera coccinea Lour.	2n=2x=38	26m+8sm+4st（2SAT）	2B	72	2.1	0.26	62.52	4L+10M2+24M1
	166 绶草属 Spiranthes Rich.	166-1 香港绶草 Spiranthes hongkongensis Y.Hu et Barretto	2n=3x=45	18m+18sm+9st（1SAT）	3B	81	2.86	0.53	66.06	6L+12M2+21M1+6S
		166-2 绶草 Spiranthes sinensis (Pers.) Ames	2n=2x=30	8m+22sm（2SAT）	2B	60	2.19	0.47	65.5	4L+14M2+8M1+4S
		166-3 螺旋花苔草 Spiranthes spralis (L.) Chevall	2n=2x=30	14m+16sm	2A	60	1.48	0.4	64.68	12M2+18M1

续表

植物核型分析（中国园林花卉植物）

科	属	种	染色体数目	核型公式	核型分类	臂指数	最长染色体/最短染色体	臂比>2的染色体比例	核型不对称系数	染色体相对长度组成
XLIV 兰科 Orchidaceae	167.白点兰属 Thrixspermum Lour.	167-1 白点兰 Thrixspermum centipeda Lour.	$2n=2x=38$	22m+14sm+2t	2A	74	1.99	0.21	61.23	8L+8M2+16M1+6S
	168.万代兰属 Vanda Joes	168-1 大花万代兰 Vanda coerulea Griff. ex Lindl.	$2n=2x=38$	32m (2SAT) +6sm	2B	76	2.78	0.05	59.8	8L+8M2+12M1+10S
		168-2 琴唇万带兰 Vanda concolor Bl. ex Lindl.	$2n=2x=38$	36m (2SAT) +2sm	1B	76	2.84	0	56.57	8L+10M2+12M1+8S
		168-3 矮万代兰 Vanda pumila Hook. f.	$2n=4x=76$	50m+12sm+8st+6t (2SAT)	2B	138	2.49	0.29	64.28	8L+24M2+32M1+12S
		168-4 英德万代兰 Vanda sp.	$2n=2x=38$	28m+4sm+6st	2B	70	2	0.26	61.16	2L+20M2+12M1+4S
		168-5 纯色万代兰 Vanda subconcolor T. Tang et F. T. Wang	$2n=2x=38$	24m+8sm+6st	2A	70	1.75	0.32	61.59	2L+16M2+18M1+2S
	169.假万代兰属 Vandopsis Pfitz.	169-1 白柊拟万代兰 Vandopsis undulata (Lindl.) J. J. Smith	$2n=2x=38$	24m+8sm+6st	2B	70	2.29	0.37	62.26	6L+12M2+16M1+4S
XLV 罂粟科	170.罂粟属 Papaver L.	170-1 观赏罂粟 Papaver orientale	$2n=42+1B$	12m+30sm+1t	2A	85	1.8	0.5	66.3	22M2+20M1+1S
		170-2 虞美人 Papaver rhoeas L.	$2n=2x=14$	12sm+2st	4A	26	1.46	1	72.25	6M2+8M1
XLVI 胡椒科 Piperaceae	171.豆瓣绿属 Peperomia Ruiz et Pavon	171-1 西瓜皮椒草 Peperomia argyreia E.Morr.C.DC.	$2n=2x=22$	12m+8sm+2st	2A	42	1.83	0.27	63.19	4L+6M2+12M1
XLVII 海桐花科 Pittosporaceae	172.海桐花属 Pittosporum Banks ex Soland	172-1 海桐 Pittosporum tobira（Thunb.）Ait.	$2n=2x=28$	20m+8sm	2B	56	2.21	0.14	59.44	6L+10M2+4M1+8S
XLVIII 悬铃木科	173.悬铃木属 Platanus L.	173-1 法国梧桐 Platanus orientalis L.	$2n=42$	6m+24sm+12st	3A	72	1.86	0.71	70.34	4L+14M2+24M1
XLIX 白花丹科 Plumbaginaceae	174.补血草属 Limonium Mill.	174-1 补血花 Limonium sinuatum (L.) Mill.	$2n=2x=16$	2m+12sm+2st	3A	30	1.66	0.75	69.68	2L+4M2+10M1
L 花荵科 Polemoniaceae	175.天蓝绣球属 Phlox L.	175-1 福禄考 Phlox drummondii Hook.	$2n=2x=14$	4sm+10st (2SAT)	4A	18	1.58	1	78.95	8M2+6M1
LI 马齿苋科 Portulacaceae	176.马齿苋属 Portulaca L.	176-1 半枝莲 Portulaca grandiflora Hook.	$2n=2x=18$	14m+4sm (2SAT)	1B	36	2.21	0	55.58	12M2+4M1+2S
LIII 报春花科 Primulaceae	177.点地梅属 Androsace L.	177-1 点地梅 Androsace umbellata(Lour.)Merr.	$2n=2x=20$	14m+2sm+4st (1SAT)	2A	36	1.38	0.2	61.18	12M2+8M1

续表

植物核型分析（中国园林花卉植物）

科	属	种	染色体数目	核型公式	核型分类	臂指数	最长染色体最短染色体	臂比>2的染色体比例	核型不对称系数	染色体相对长度组成
LIII报春花科 Primulaceae	178.报春花属 primula L.	178-1 报春花 Primula malacoides Franch.	$2n=2x=24$	24m	1A	48	1.52	0	53.86	8M2+16M1
		178-2 海仙报春 Primula poissonii Franch.	$2n=2x=22$	22m	1A	44	1.47	0	56.78	2L+6M2+14M1
	179.飞燕草属 Consolida（DC.）Opiz	179-1 飞燕草 Consolida ajacia Schur.	$2n=2x=26$	18m+8sm	2A	52	1.37	0.23	60.11	16M2+10M1
LIII毛茛科 Ranunculaceae	180.芍药属	180-1 芍药 Paeonia lactiflora Pall.	$2n=2x=10$	8m+2st	2A	18	1.46	0.2	60.28	6M2+4M1
		180-2 牡丹 Paeonia suffruticosa Andr.	$2n=2x=10$	6m+2sm（1SAT）+2st（1SAT）	2A	18	1.31	0.2	61.1	7M2+3M1
	181.毛茛属 Ranunculus L.	181-1 花毛茛 Ranunculus asiaticus L.	$2n=2x=16$	6m+6sm+2st+2t	3A	28	1.49	0.63	66.49	8M2+8M1
	182.木瓜属 Chaenomeles Lindl.	182-1 贴梗海棠 Chaenomeles lagenaria Koidz	$2n=2x=34$	18m+8sm+6st+2t	2A	60	1.83	0.29	66.75	4L+12M2+16M1+2S
	183.棣棠属 Kerria DC.	183-1 重瓣棣棠 Kerria japonica var.pleniflora Witte	$2n=2x=18$	14m+4sm	2A	36	1.24	0.11	57.72	8M2+10M1
	184.苹果属 Malus Mill.	184-1 西府海棠 Malus spectabilis var.rivesii Nash.	$2n=2x=34$	24m+6sm+4st	2A	64	1.69	0.24	63.72	4L+8M2+22M1
		184-2 海棠花 Malus spectabilis Borkh.	$2n=2x=34$	22m+8sm+4st（2SAT）	2A	64	1.84	0.35	62.57	4L+14M2+14M1+2S
LIV蔷薇科 Rosaceae	185.李属 Prunus L.	185-1 红叶李 Prunus Cerasifera Ehrhar f. atropurpurea（Jacq.）Rehd.	$2n=2x=16$	12m+4sm	2B	32	2.14	0.13	59.61	2L+6M2+6M1+2S
		185-2 梅花 Prunus mume Sieb.et Zucc f. duplex Rehd.	$2n=2x=16$	8m+4sm（2SAT）+4st	2B	28	2.13	0.38	65.25	2L+4M2+8M1+2S
		185-3 碧桃 Prunus persica f. duplex Rehd.	$2n=2x=16$	10m+4sm+2st	2B	30	2.21	0.25	61.2	2L+4M2+8M1+2S
		185-4 樱花 Prunus seulata Lindl.	$2n=2x=16$	8m+6sm+2st（2SAT）	2A	30	1.9	0.25	64.03	2L+2M2+12M1
		185-5 榆叶梅 Prunus triloba Lindl.	$2n=8x=64$	32m+14sm+18st	2B	110	2.37	0.41	66.34	6L+16M2+42M1
		185-6 日本樱花 Prunus yedoensis Matsum.	$2n=2x=16$	12m+4sm	1A	32	1.98	0	59.65	2L+2M2+12M1
	186.珍珠梅属 Sorbaria(Ser.)A.Br.	186-1 珍珠梅 Sorbaria kirilowii(Regel)Maxim.	$2n=2x=36$	10m+14sm+12st	3A	60	1.59	0.67	68.98	4L+12M2+20M1
	187.蔷薇属 Rosa L.	187-1 木香花 Rosa banksiae R. br.	$2n=2x=14$	10m+2sm+2st	2A	26	1.68	0.29	60.98	2L+4M2+8M1
		187-2 月季花 Rosa chinensis Jacq.	$2n=4x=28$	20m+8sm	2A	56	1.76	0.14	59.46	2L+8M2+16M1+2S

续表

植物核型分析（中国园林花卉植物）

科	属	种	染色体数目	核型公式	核型分类	臂指数	最长染色体最短染色体	臂比>2的染色体比例	核型不对称系数	染色体相对长度组成
LIV 蔷薇科 Rosaceae	187.蔷薇属 Rosa L.	187-3 丽江蔷薇 Rosa lichiangensis Yu et Ku	$2n=4x=28$	16m（2SAT）+10sm+2st	2A	54	1.82	0.36	62.78	2L+12M2+12M1+2S
		187-4 多花蔷薇 Rosa multiflora Thunb.	$2n=2x=14$	12m+2sm	1A	28	1.59	0	57.76	8M2+4M1+2S
		187-5 十姐妹 Rosa multiflora var. platyphylla Thory	$2n=3x=21$	15m+6sm（3SAT）	2A	28	1.82	0.14	58.17	2L+6M2+4M1+2S
		187-6 香水月季 Rosa odorata（Andr.）Sweet	$2n=2x=14$	8m+4sm+2st（1SAT）	2A	26	1.6	0.14	61.47	6M2+8M1
		187-7 红玫瑰 Rosa rugosa Thunb.	$2n=2x=14$	10m+2sm+2st（1SAT）	2A	26	1.87	0.29	60.64	2L+6M2+4M1+2S
		187-8 圣诞玫瑰 Rose rugosa Thunb.	$2n=36$	18m+8sm（2SAT）+2st	2A	54	1.82	0.29	61.96	2L+8M2+18M1
		187-9 黄刺梅 Rosa xanthina Lindl.	$2n=2x=14$	34m+2sm（2SAT）	1A	72	1.87	0	56.31	4L+16M2+14M1+2S
		187-10 杏花村 Betty prior	$2n=3x=21$	6m+6sm+2st	2A	26	1.7	0.43	63.32	2L+4M2+6M1+2S
		187-11 小姐妹月季 Hybrid polyantha Roses	$2n=3x=21$	15m+3sm+3t（3SAT）	2B	39	2	0.14	60.62	3L+6M2+9M1+3S
	188.绣线菊属 Spiraea L	188-1 绣线菊 Spiraea Salicifolia L.	$2n=2x=36$	12m+9sm	2A	28	1.47	0.14	61.59	6M2+8M1
	189.虎刺属 Damnacanthus Gaertn.f.	189-1 虎刺 Damnacanthus indicus(L.)Gaertn.f.	$2n=38$	24m+6sm+6st	2A	66	1.65	0.22	62.11	4L+10M2+22M1
LV 茜草科 Rubiaceae	190.栀子属 Gardenia Ellis	190-1 栀子 Gardenia jasminoides Ellis	$2n=2x=22$	36m+2sm	1A	76	1.68	0	56.48	2L+18M2+18M1
	191.六月雪属 Serissa Comm.ex Juss.	191-1 六月雪 Serissa foetida Comm.	$2n=2x=18$	18m+4sm（2SAT）	2A	44	1.42	0.09	58.77	10M2+12M1
LVI 杨柳科 Salicaceae	192.杨属 Populus L.	192-1 毛白杨 Populus tomentosa Carr.	$2n=2x=38$	12m+4sm+2st	2A	34	1.89	0.33	62.75	2L+6M2+8M1+2S
	193.柳属 Salix L.	193-1 垂柳 Salix babylonica L.	$2n=4x=76$	26m+6sm+4st+2t	2B	70	2.59	0.26	62.11	4L+10M2+20M1+4S
		193-2 旱柳 Salix matsudana Koidz.	$2n=6x=114$	54m+12sm+10st（2SAT）	2B	142	2.09	0.26	61.84	4L+24M2+46M1+2S
LVII 虎耳草科	194.绣球属 Hydrangea L.	194-1 八仙花 Hydrangea macrophylla Ser.	$2n=2x=20$	60m+22sm+24st+8t	2B	196	2.28	0.39	66	8L+34M2+70M1+2S
				6m+10sm+4st	3A	36	1.48	0.6	67.08	10M2+10M1

续表

植物核型分析（中国园林花卉植物）

科	属	种	染色体数目	核型公式	核型分类	臂指数	最长染色体/最短染色体	臂比>2的染色体比例	核型不对称系数	染色体相对长度组成
LVIII玄参科 Scrophulariaceae	195.虎耳草属 Saxifraga L.	195-1 虎耳草 Saxifraga stolonifera Meerb.	$2n=2x=36$	24m（2SAT）+12sm	2B	72	2.2	0.06	59.61	10L+8M2+8M1+10S
	196.金鱼草属 Antirrhinum L.	196-1 金鱼草 Antirrhinum majus L.	$2n=2x=16$	12m+4sm	2A	32	1.28	0.25	59.13	6M2+10M1
	197.泡桐属 Paulownia Sieb.et Zucc	197-1 毛泡桐 Paulownia tomentosa（Thunb.）Steud	$2n=40$	28m+4sm+8st（2SAT）	2A	72	1.46	0.3	64.31	22M2+18M1
LIX茄科 Solanaceae	198.辣椒属 Capsicum L.	198-1 朝天椒 Capsicum annuum L. var. conoides（Mill.）Irish	$2n=2x=24$	20m+2sm（2SAT）+2st	2A	46	1.78	0.08	55.73	2L+14M2+6M1+2S
		198-2 五色椒 Capsicum frutescens L. var. cerasiforme Bailey	$2n=2x=24$	20m+2sm（2SAT）+2st	2B	46	2.01	0.17	56.51	2L+12M2+6M1+4S
		198-3 番茄椒 Capsicum sp.	$2n=2x=24$	22m+2st	2A	46	1.93	0.08	55.44	2L+14M2+4M1+4S
	199.碧冬茄属 Petunia Juss.	199-1 矮牵牛 Petunia hybrida Vilm.	$2n=2x=14$	14m	1A	28	1.27	0	56.58	6M2+8M1
LX山茶科 Theaceae	200.山茶属 Camellia L.	200-1 金花茶 Camellia chrysantha Tuyama	$2n=2x=30$	24m+6sm	2A	60	1.85	0.07	59.01	4L+8M2+12M1+6S
		200-2 显脉金花茶 Camellia euphlebia Merr.	$2n=2x=30$	14m+16sm（2SAT）	2A	60	1.75	0.27	60.77	2L+14M2+10M1+4S
		200-3 凹脉金花茶 Camellia impressinervis Chang.	$2n=2x=30$	22m+6sm+2t	2A	58	1.62	0.2	61.07	14M2+14M1+2S
		200-4 山茶 Camellia japonica L.	$2n=2x=30$	18m+6sm+6st	2A	54	1.88	0.2	61.54	2L+14M2+10M1+4S
		200-5 野山茶 Camellia yunnanensis Coh.Stuart	$2n=6x=90$	56m（4SAT）+22sm+12st	2A	168	1.8	0.24	62.65	2L+48M2+38M1+2S
LXI旱金莲科 Tropaeolaceae	201.金莲花属 Tropaeolum L.	201-1 金莲花 Tropaeolum Majus L.	$2n=2x=28$	18m+6sm（2SAT）+4st	2A	52	1.95	0.29	61.91	6L+6M2+12M1+4S
LXII马鞭草科 Verbenaceae	202.大青属 Clerodendrum L.	202-1 臭牡丹 Clerodendrum bungei Sterd.	$2n=2x=104$							
	203.马鞭草属 Verbena Limn.	203-1 美女樱 Verbena hybrida Voss.	$2n=2x=10$	8m（2SAT）+2sm	2A	20	1.24	0.2	57.37	4M2+6M1
LXIII葡萄科	204.爬山虎属 Parthenocissus Pl.	204-1 五叶地锦 Parthenocissus quinquefolia（L.）Planch.	$2n=40$	22m+8sm+10st	2A	70	1.87	0.3	65.36	2L+18M2+18M1+2S
LXIV姜科 Zingiberaceae	205.姜花属 Hedychium Koenig	205-1 姜花 Hedychium coronarium Lily	$2n=2x=34$	18m（4SAT）+6sm（2SAT）+8st+2t	2B	58	2.02	0.41	65.83	18M2+14M1+2S

第四节　中国竹类植物基因组染色体研究结果

提要：本书收录了作者对竹亚科（Bambusoideae）33 属 185 种竹子的染色体研究结果，总计有 363 版染色体图。

全世界有 70 多属 1000 余种竹子。我国有 37 属 500 余种。竹子是我国重要速生森林资源植物，对国民经济建设有十分重要的意义。竹子有木本、草木和藤本之分。竹子形态分类主要是依据花的构造，而竹子花期不固定，一般相隔甚长（数年、数十年乃至百年以上），多数竹子一生只开一次花，花后立即死亡，而竹子营养体又有很大的相似性，因此，造成竹子分类的不一致，同时存在几个分类系统。总之，对没有开花的竹子进行分类是植物分类学家长期面临的一个难题。作者首次将染色体资料引入竹子分类中，试图通过核型分析对竹子进行分类。经查明，竹子染色体数目分两大类群，散生竹为 $2n=48$，丛生竹为 $2n=70\pm2$，但是，核型分析却很复杂，表现出显著的多态性，如丛生竹银丝竹（*Bambusa multiplex* cv silvretripe）就有三种细胞型：①$2n=72=34m+16sm+4st+18t$、②$2n=70=36m+16sm+10st+8t(2SAT)$、③$2n=68=38m+16sm+8st+6t(2SAT)$，它们的 N.F.值均为 122，说明发生了 Robertson 易位。着丝粒并合在动物中很普遍，是动物的一种进化机制。在植物中很少见，在丛生竹中为什么普遍存在 Robertson 现象，它与竹子物种形成、起源进化有什么关系，还有待深入研究。竹子是兼性无融合生殖（facultative apomixis）植物，所以，竹子的各种多倍体、非整倍体、混倍体、染色体结构变异及其各种无性生殖"宗""小种"、细胞型被稳定地保存下来，被分类学家们划分为过多的"种"，增大了细胞分类的难度。但有一点是可以肯定的，竹类染色体大数据分析结果表明，不论是散生竹还是丛生竹核型分析均为异源多倍体，按 1945 年 Ammal 认为竹子染色体基数（basic number）$x=12$，散生竹为异源四倍体，丛生竹为异源六倍体。这在植物中整个亚科、整个属均为多倍体的现象是少见的。特别是异源六倍体，整个丛生竹均为异源六倍体，实属罕见。本书收录了我国 114 种丛生竹异源六倍体核型数据，其中西双版纳歪脚龙竹是世界上最大的竹子，一个竹笋就有 26 kg 重。表现出了极好的优势。

六倍体形成有两种途径：①八倍体×四倍体→六倍体，AAAABBBB♀×CCCC♂→F1AABBCC。②四倍体×二倍体→三倍体→加倍→六倍体，AABB♀×CC♂→F1ABC→加倍→AABBCC。第①途径属于远缘杂交，只具有远缘优势。在第②途径中，形成三倍体后，三倍体具有多倍体优势、杂交优势和远缘优势，加倍成异源六倍体后，由于异源六倍体减数分裂正常，将这 3 种优势固定了一起，我们称之为多倍体优势、杂交优势之外的第三种优势。在第三种优势作用下，形成了超级物种，表现出强大的优势。西双版纳歪脚龙竹等六倍体丛生竹类的强大优势，可能就是这样形成的。因此，我们提出"基因组三倍化与杂交相结合，对高等植物进化起了重要作用，具体表现在形成了第三种优势。利用好第三种优势，将会给人类创造出巨大的财富。

2.4.1 部分竹类植物染色体

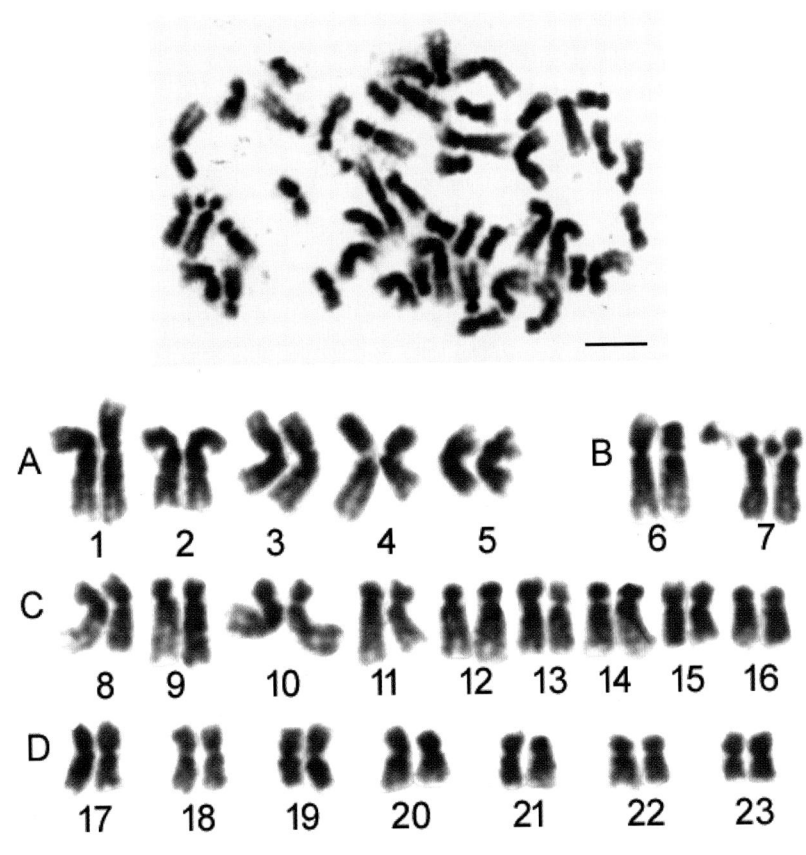

铁竹 *Ferrocalamus strictus* 核型

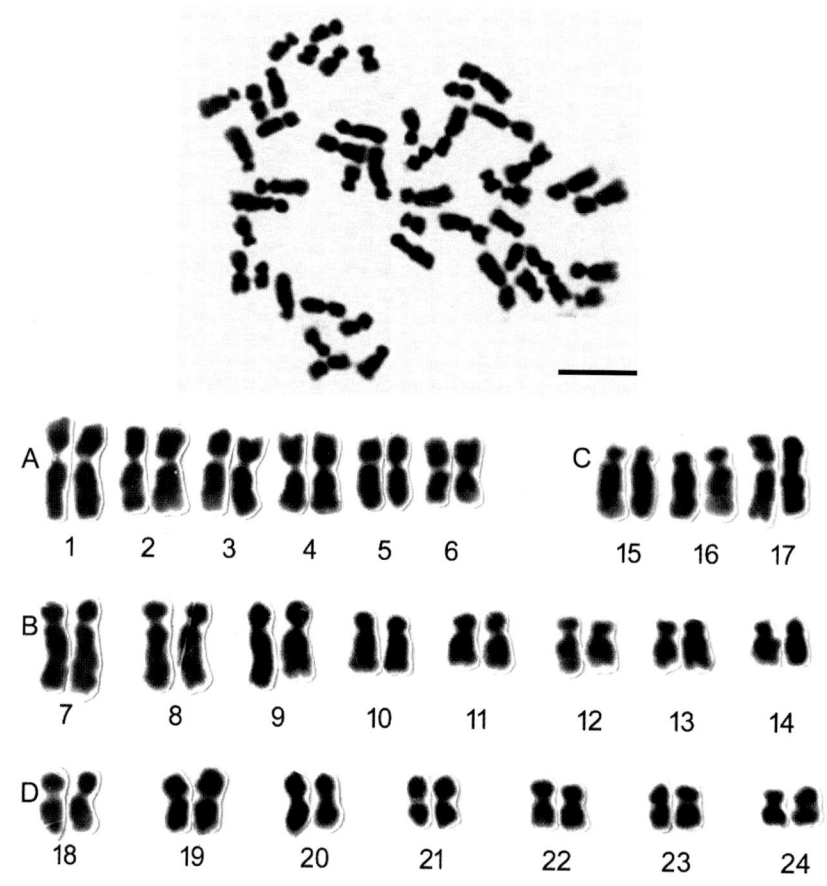

少穗竹 *Oligostachyum sulcatum* 核型

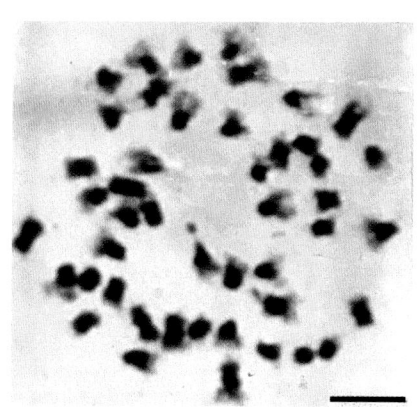

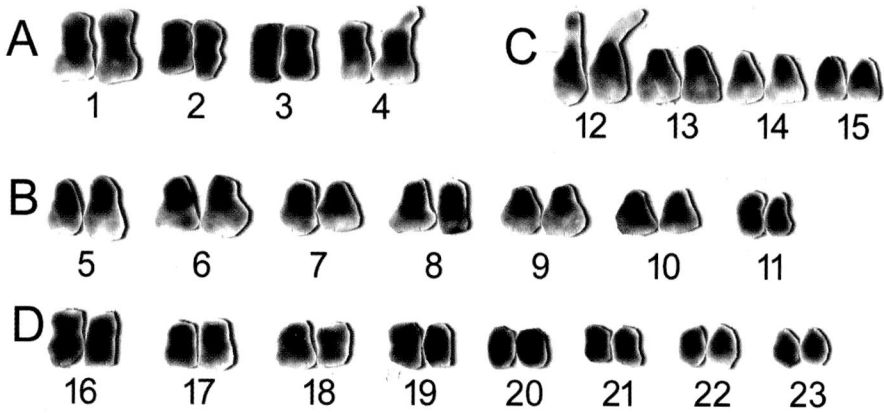

瓜多竹 *Guadua angustifolia* 核型

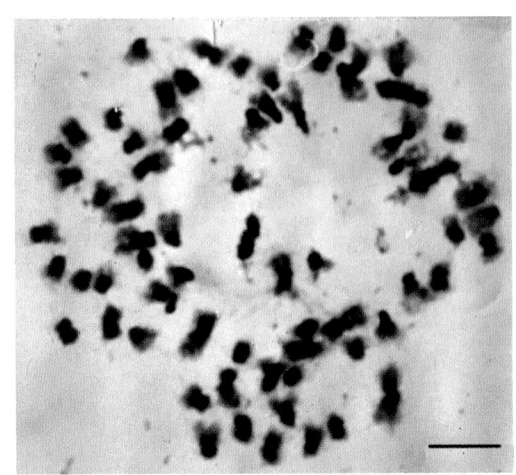
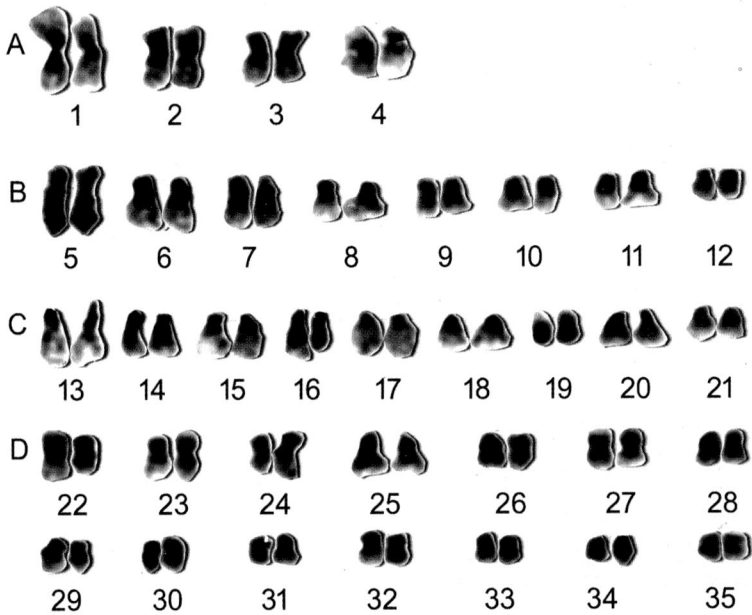

歪脚龙竹 *Dendrocalamus sinicus* 核型

2.4.2 中国竹类植物基因组染色体数据分析

中国竹类植物核型分析

族	属	种	染色体数目	核型公式	核型分类	臂指数	最长染色体/最短染色体	臂比>2的染色体比例	核型不对称系数	染色体相对长度组成
I 梨竹族 Melocanneae	1.梨竹属 *Melocanna* Trin.	1-1 梨竹 *Melocanna baccifera* (Roxb.) Kurz	2n=72	34m+16sm+4st+18t	3B	122	2.8	0.53	68.65	6L+24M2+38M1+4S
		1-2 象鼻竹 *Melocanna* sp.	2n=70	36m+16sm+12t (2SAT)	2B	122	2.8	0.46	66.05	12L+18M2+26M1+14S
	2.思劳竹属 *Schizostachyum* Nees	2-1 沙罗单竹 *Schizostachyum funghomii* McClure	2n=70	36m+16sm+8st+10t	2B	122	3.4	0.49	66.57	10L+22M2+22M1+16S
		2-2 思劳竹 *Schizostachyum pseudolima* McClure	2n=64	34m+16sm+6st+8t	2B	114	2.3	0.44	66.16	12L+18M2+22M1+12S
	3.泡竹属 *Pseudostachyum* Munro	3-1 泡竹 *Pseudostachyum ploymorphum* Munro	2n=70	36m+16sm+10st+8t (2SAT)	2B	122	3	0.49	67.99	16L+8M2+30M1+16S
	4.空竹属 *Cephalostachyum* Munro	4-1 糯竹 *Cephalostachyum pergracile* Munro	2n=68	38m+16sm+4st+10t	2B	122	2.3	0.49	67.01	16L+14M2+30M1+10S
		4-2 金毛空竹 *Cephalostachyum virgatum* (Munro) Kurz	2n=42	18m+18sm+6st	3B	78	2.3	0.41	66.15	18L+8M2+32M1+10S
	5.泰竹属 *Thyrsostachys* Gamble	5-1 大泰竹 *Thyrsostachys oliveri* Gamble	2n=72	42m+16sm+6st+8t	2B	130	2.6	0.57	65.06	12L+8M2+12M1+10S
		5-2 泰竹 *Thyrsostachys siamensis* (Kurz ex Nmunro) Gamble	2n=98	50m+18sm+10st+20t	2B	166	2.9	0.39	65.99	14L+14M2+26M1+18S
	6.梨藤竹属 *Melocalamus* Benth.	6-1 藤竹 *Melocalamus* sp.	2n=70	36m+16sm+4st+14t (2SAT)	2B	122	2.4	0.41	65.89	12L+26M2+54M1+6S
		6-2 霸王藤竹 *Melocalamus* sp.	2n=68	38m+16sm+4st (2SAT)+10t	2B	122	2.5	0.43	66.72	14L+14M2+32M1+10S
	7.单枝竹属 *monocladus* Chia et al.	7-1 单枝竹 *monocladus saxatilis* Chia et al	2n=70	36m+16sm+8st (4SAT)+10t	2B	122	2.8	0.41	65.31	14L+16M2+26M1+12S
	8.新小竹属 *Neomicrocalamus* Keng f.	8-1 新小竹 *Neomicrocalamus prainii* (Gamble) Keng f.	2n=70	36m+16sm+6st+12t (2SAT) +10t	2B	122	2.6	0.43	64.14	12L+12M2+36M1+10S
II 簕竹族 Bambuseae	9.簕竹属 *Bambusa* Retz.corr Schreber	9-1 印度簕竹 *Bambusa arundinacea* (Retz.) Willd.	2n=68	38m+16sm+8st+8t	2B	122	3.9	0.35	62.7	16L+10M2+18M1+24S
		9-1 印度簕竹 *Bambusa arundinacea* (Retz.) Willd.	2n=70	36m+16sm+6st+12t (2SAT)	2B	122	4.2	0.46	68.25	14L+14M2+28M1+14S
			2n=70	36m+16sm+2st+16t (2SAT)	2B	122	3	0.49	65.82	10L+16M2+36M1+8S
			2n=70	36m (2SAT)+16m+2st+16t (2SAT)	2B	122	2.2	0.49	66.29	12L+20M2+28M1+10S
		9-2 簕竹 *Bambusa blumeana* J.A. et J.H. Schult.	2n=70	36m+14sm+4st (2SAT)+10t	2B	122	3.3	0.49	66.8	14L+16M2+24M1+16S
			2n=64	34m+16sm+10st (2SAT) +12t	2B	114	3	0.38	65.33	8L+18M2+36M1+2S
		9-3 小簕竹 *Bambusa flexuosa* Munro	2n=72	36m+16sm+6st+12t	2B	122	2.2	0.44	66	8L+22M2+35M1+6S
		9-4 车筒竹 *Bambusa sinospinosa* McClure	2n=70	36m+16sm+6st (2SAT)+12t	2B	122	2.7	0.46	66.8	14L+14M2+32M1+10S
			2n=70		2B	122	2.6	0.4	66.8	14L+14M2+30M1+12S

续表

中国竹类植物核型分析

族	属	种	染色体数目	核型公式	核型分类	臂指数	最长染色体/最短染色体	臂比>2的染色体比例	核型不对称系数	染色体相对长度组成
		9-5 坭簕竹 Bambusa dissemulator McClure	2n=72	34m+16sm+4st（2SAT）+18t	2B	122	2.6	0.42	68.72	8L+24M2+32M1+8S
		9-6 毛簕竹 Bambusa dissimulator McClure var. hispida Mcclure	2n=68	38m+16sm+4st（2SAT）+10t	2B	122	2.4	0.35	66.15	10L+18M2+32M1+8S
		9-7 木竹 Bambusa rutila McClure	2n=70	36m+16sm+4st+14t	2B	122	2.6	0.49	67.05	12L+14M2+32M1+12S
		9-8 油簕竹 Bambusa lapidea McClure	2n=72	34m+16sm+8st+14t	2B	122	3.5	0.5	67.53	12L+12M2+34M1+14S
		9-9 锦竹 Bambusa subaequalis H. L. Fung et C. Y. Sia	2n=70	36m+16sm+10st（2SAT）+8t	2B	122	2.8	0.49	66.94	14L+12M2+28M1+16S
		9-10 坭簕竹 Bambusa gibba McClure	2n=70	36m+16sm+12st+6t	2B	122	2.6	0.37	64.19	14L+16M2+28M1+12S
		9-11 佛肚竹 Bambusa ventricosa McClure	2n=64	38m+16sm+6st（2SAT）+4t	2B	118	2.83	0.41	65.43	14L+10M2+26M1+12S
II 簕竹族 Bambuseae	9.簕竹属 Bambusa Retz.corr Schreber	9-12 东兴黄竹 Bambusa corniculata Chia et H.L.Fung	2n=68	38m+16sm+2st+12t	2B	122	2.8	0.35	65.51	12L+20M2+24M1+14S
		9-13 吊罗坭竹 Bambusa diaoluoshanensis Chia et H.L.Fung	2n=70	36m+16sm+4st+14t	2B	122	2.6	0.46	68.06	10L+18M2+28M1+14S
		9-14 霞山坭竹 Bambusa xiashanensis Chia et H.L.Fung	2n=70	32m+16sm+12st（2SAT）+10t	2B	122	3.1	0.46	68.62	14L+18M2+28M1+14S
		9-15 牛儿竹 Bambusa prominens H.L.Fung et C.Y.Sia	2n=70	36m+16sm+8st+10t（2SAT）	2B	118	2.7	0.49	67.86	14L+10M2+36M1+10S
		9-16 俯竹 Bambusa nutans Wall. ex Munro	2n=104	64m+18sm+6st+16t	2B	122	2.2	0.49	65.02	10L+18M2+34M1+8S
		9-17 马甲竹 Bambusa tulda Roxb.	2n=104	58m+16sm+14st（2SAT）+14t	2B	186	2.7	0.35	64.83	16L+28M2+48M1+12S
		9-18 大眼竹 Bambusa eutuldoides McClure	2n=70	36m+16sm+6st（2SAT）+12t（2SAT）	2B	180	3.1	0.42	64.11	16L+24M2+58M1+6S
		9-19 银丝大眼竹 Bambusa eutuldoides var. basistriata McClure	2n=68	38m+16sm+6st（2SAT）+8t	2B	122	2.9	0.46	67.06	10L+16M2+34M1+10S
		9-20 青丝黄竹 Bambusa var. viridi-vittata（W. T. Lin）Chia	2n=70	36m+16sm+14st+4t	2B	124	3.3	0.44	65.82	14L+14M2+24M1+16S
			2n=70	36m+16sm+2st+16t	2B	122	2.5	0.43	63.64	12L+16M2+26M1+16S
			2n=70	36m+16sm+8st+10t（2SAT）	2B	122	2.7	0.46	65, 51	8L+18M2+34M1+10S
			2n=70	36m+16sm+8st+10t（2SAT）	2B	122	2.9	0.46	65.63	14L+8M2+36M1+12S

续表

中国竹类植物核型分析

族	属	种	染色体数目	核型公式	核型分类	臂指数	最长染色体/最短染色体	臂比>2的染色体比例	核型不对称系数	染色体相对长度组成
		9-21 撑麻竹 Bambusa pervariabilis McClure	$2n=68$	38m+16sm+4st+10t	2B	122	3.3	0.38	66.47	18L+4M2+26M1+20S
		9-22 花撑蒿竹 Bambusa pervariabilis var.sp.	$2n=70$	36m+16sm+4st+14t	2B	122	2.8	0.46	65.98	10L+20M2+30M1+10S
		9-23 花眉竹 Bambusa longispiculata Gamble ex Brandis	$2n=68$	38m+16sm+6st (2SAT) +8t (2SAT)	2B	122	3.2	0.44	64.85	14L+12M2+28M1+14S
		9-24 青秆竹 Bambusa tuldoides Munro	$2n=70$	36m+16sm+10st (2SAT) +8t	2B	122	3	0.43	66.44	12L+22M2+22M1+14S
		9-25 鼓节竹 Bambusa tuldoides cv.Swollenintemode	$2n=68$	38m+16sm+4st+10t (2SAT)	2B	122	2.7	0.38	65.3	12L+14M2+32M1+10S
		9-26 信宜石竹 Bambusa subtruncata Chia et H.L.Fung	$2n=68$	38m+16sm+4st+10t	2B	122	3, 3	0.38	65.25	10L+20M2+24M1+14S
		9-27 硬头黄竹 Bambusa rigida Keng et Keng f.	$2n=70$	36m+16sm+6st+12t	2B	122	2.1	0.4	66.71	8L+22M2+30M1+10S
		9-28 妈竹 Bambusa boniopsis McClure var.sp	$2n=70$	36m+16sm+10st+8t	2B	122	3	0.43	67.36	12L+20M2+28M1+10S
II 簕竹族 Bambuseae	9.簕竹属 Bambusa Retz.corr Schreber	9-29 花头妈竹 Bambusa boniopsis McCure var.sp.	$2n=70$	36m+16sm+8st+10t (2SAT)	2B	122	3	0.43	66.88	16L+12M2+24M1+18S
		9-30 长枝竹 Bambusa dolichoclada Hayata	$2n=68$	38m+16sm+6st+8t	2B	122	3.2	0.35	65.74	12L+20M2+24M1+12S
		9-31 龙头竹 Bambusa vulgaris Schrader ex Wendl.	$2n=70$	36m+16sm+10st+8t	2B	122	2.6	0.49	67.19	10L+16M2+32M1+12S
		9-32 黄金间碧竹 Bambusa vulgaris Schrader ex Wendland'cv.Vittata'	$2n=70$	40m (2SAT) +12sm+2st+16t	2B	122	2.4	0.43	68.65	14L+22M2+24M1+10S
		9-33 大佛肚竹 Bambusa vulgaris cv. Wamin	$2n=72$	34m+16sm+8st+14t	2B	122	2.3	0.42	66.39	12L+16M2+40M1+4S
		9-34 鱼肚腩竹 Bambusa gibboides W. T. Lin	$2n=68$	38m+16sm+4st+10t (2SAT)	2B	122	2.3	0.29	64.32	10L+18M2+34M1+6S
		9-35 石竹仔 Bambusa piscatorum McClure	$2n=104$	52m+18sm+12st (2SAT) +22t	2B	174	3.1	0.48	67.61	20L+18M2+48M1+18S
		9-36 花竹仔 Bambusa albo-lineata Chia	$2n=72$	34m+16sm+2st+20t (2SAT)	3B	122	2, 4	0.53	69.67	12L+14M2+40M1+6S
		9-37 黄竹仔 Bambusa mutabilis McClure	$2n=68$	38m+16sm+6st (2SAT) +8t	2B	122	3	0.41	66.22	8L+22M2+28M1+10S
		9-38 破篾黄竹 Bambusa contracta Chia et H. L. Fung	$2n=70$	36m+16sm+10st+8t	2B	122	2.9	0.37	64.9	12L+16M2+32M1+10S
			$2n=70$	36m+16sm+6st+12t	2B	122	2.1	0.49	69.66	16L+12M2+36M1+6S

续表

中国竹类植物核型分析

族	属	种	染色体数目	核型公式	核型分类	臂指数	最长染色体/最短染色体	臂比>2的染色体比例	核型不对称系数	染色体相对长度组成
II 簕竹族 Bambuseae	9. 簕竹属 Bambusa Retz.corr Schreber	9-39 孝顺竹 Bambusa multiplex (Lour.) Raeusch. ex Schult.	2n=70	36m+16sm+6st(2SAT)+12t	2B	122	2.4	0.46	65.32	14L+20M2+22M1+14S
		9-40 观音竹 Bambusa multiplex var.rivlereorum R.Maire	2n=68	38m+16sm+8st+6t	2B	122	2.2	0.32	64.95	10L+18M2+40M1
		9-41 黄条竹 Bambusa multiplex cv. Yellowstripe	2n=70	36m+16sm+6st+12t(2SAT)	2B	122	3.1	0.4	66.5	14L+14M2+32M1+10S
		9-42 小琴丝竹 Bambusa multiplex cv. Alphonse-Karr	2n=68	38m+16sm+6st(2SAT)+8t	2B	122	3.6	0.44	63.91	12L+22M2+16M1+18S
		9-43 银丝竹 Bambusa multiplex cv. Silverstrip	2n=70	36m+16sm+10st+8t(2SAT)	2B	122	2.5	0.4	65.75	12L+18M2+30M1+10S
		9-44 垂柳竹 Bambusa multiplex cv. Willowy	2n=70	36m+16sm+4st+14t(2SAT)	2B	122	3.1	0.46	69.04	10L+20M2+32M1+8S
		9-45 凤尾竹 Bambusa multiplex cv.Fernleaf	2n=70	36m+16sm+6st(2SAT)+12t	2B	122	2.8	0.37	65.32	12L+20M2+30M1+8S
		9-46 绵竹 Bambusa intermedia Hsueh Et Yi	2n=70	36m+16sm+12st(2SAT)+6t	2B	122	2.5	0.46	66.61	14L+20M2+20M1+16S
		9-47 甲竹 Bambusa remotiflora Kuntze	2n=72	34m+16sm+6st+16t(2SAT)	3B	122	2.9	0.53	70.2	14L+16M2+34M1+8S
		9-48 油竹 Bambusa surrecta Q. H. Dai	2n=70	36m+16sm+6st+12t	2B	122	2.7	0.43	68.08	16L+10M2+28M1+16S
		9-49 单竹 Bambusa cerosissima McClure	2n=70	36m+16sm+10st+8t	2B	122	3.7	0.49	64.95	14L+10M2+30M1+16S
		9-50 粉单竹 Bambusa chungii McClure	2n=70	36m+16sm+8st+10t(2SAT)	2B	122	2.9	0.46	66.2	14L+16M2+28M1+2S
		9-51 青皮竹 Bambusa textilis McClure	2n=70	36m+16sm+6st+12t	2B	122	2.5	0.31	65.04	14L+14M2+36M1+6S
		9-52 紫竿竹 Bambusa textilis McClure var. textilis cv. Purpurascens	2n=70	36m+16sm+10st+8t	2B	122	3.2	0.4	65.58	10L+20M2+26M1+14S
		9-53 光竿青皮竹 Bambusa textilis var. glabra McClure	2n=68	38m+16sm+4st+10t	2B	122	3.4	0.38	63.3	14L+6M2+40M1+8S
		9-54 崖州竹 Bambusa textilis McClure var. gracilis McClure	2n=70	36m+16sm+6st+12t	2B	122	2.1	0.4	66.71	8L+22M2+30M1+10S
		9-55 料慈竹 Bambusa distegia (Keng et Keng f.) Chia et H.L.Fung	2n=72	36m+14sm+8st+14t(2SAT)	2B	124	3	0.33	65.57	10L+24M2+26M1+10S
		9-56 水单竹 Bambusa papillata Q. H. Dai	2n=70	36m+16sm+6st+12t	2B	122	2.5	0.43	65.76	14L+10M2+32M1+16S
		9-57 藤单竹 Bambusa hainanensis Chia et H. L. Fung	2n=68	34m+16sm+8st(2SAT)+10t	2B	118	3.9	0.47	65.23	14L+10M2+30M1+14S
			2n=70	36m+16sm+8st+10t	2B	122	2.5	0.37	66.31	14L+10M2+38M1+8S

续表

中国竹类植物核型分析

族	属	种	染色体数目	核型公式	核型分类	臂指数	最长染色体/最短染色体	臂比>2的染色体比例	核型不对称系数	染色体相对长度组成
II 簕竹族 Bambuseae	9.簕竹属 Bambusa Retz.corr Schreber	9-58 牛角竹 Bambusa corigera McClure	2n=96	60m+16sm+10st+10t	2B	172	3.7	0.35	63.2	20L+16M2+34M1+26S
		9-59 冲天单竹 Bambusa sp.	2n=72	34m+16sm+6st+16t	2B	122	2.4	0.5	69.4	10L+22M2+32M1+8S
		9-60 溪竹 Bambusa sp.	2n=72	44sm+14sm+4st（2SAT）+10t（2SAT）	2B	130	3.5	0.33	64.25	12L+20M2+24M1+16S
	10.慈竹属 Neosinocalamus Keng f.	10-1 慈竹 Neosinocalamus affinis (RENDLE) Keng f.	2n=68	38m+16sm+4st+10t（2SAT）	2B	122	3.3	0.44	64.3	16L+12M2+24M1+16S
		10-2 孖竹 Neosinocalamus recto-cuneatus W. T. Lin	2n=70	36m+16sm+14st（2SAT）+4t	2B	122	3.2	0.49	66.28	16L+16M2+26M1+12S
	11.绿竹属 Dendrocalamopsis (chia et H.L. Fung) Keng f.	11-1 苦绿竹 Dendrocalamopsis basihirsuta (McClure) Keng f.	2n=70	36m+16sm+2st+16t（2SAT）	2B	122	2.6	0.34	67.55	16L+12M2+28M1+14S
		11-2 大头典竹 Dendrocalamopsis beecheyana (Munro) Keng var. pubescens (P. F. Li) Keng f.	2n=72	34m+16sm+8st+14t	2B	122	3.7	0.5	69.06	12L+16M2+30M1+14S
		11-3 吊丝球竹 Bambusa beecheyana Munro	2n=72	34m+16sm+4st+18t（2SAT）	2B	122	2.1	0.42	67.73	10L+18M2+40M1+4S
		11-4 乌脚绿竹 D.edulis (Odashima) Keng f.	2n=104	60m+18sm+8st+18t	2B	182	3	0.33	64.56	16L+22M2+56M1+10S
		11-5 绿竹 Dendrocalamopsis oldhami (Munro) Keng f.	2n=68	38m+16sm+2st+12t	2B	122	3.6	0.44	65.58	16L+16M2+16M1+20S
III 牡竹族 Dendrocalameae		11-6 孟竹 Dendrocalamopsis bicicatricata (W. T. Lin) Keng f.	2n=68	34m（2SAT）+16sm+4st+14t（2SAT）	2B	118	3.8	0.47	69.2	16L+8M2+26M1+18S
		11-7 花头黄竹 Dendrocalamopsis oldhami f.revoluta. (W.T.Lin et J.Y.Lin) W.T.Lin	2n=70	36m+16sm+8st+10t（2SAT）	2B	122	3.4	0.46	65.92	12L+16M2+28M1+14S
		11-8 细叶麻竹 Dendrocalamopsis sp.	2n=70	36m+16sm+6st+12t	2B	122	2.4	0.43	66.59	12L+14M2+34M1+10S
		11-9 鸡脾竹 Dendrocalamopsis sp.	2n=104	52m+20sm+14st+18t	2B	176	2.7	0.5	65.68	14L+30M2+44M1+16S
		11-10 黄麻竹 Dendrocalamopsis stenoaurita (W. T. Lin) Keng f.	2n=70	36m+16sm+10st+8t	2B	122	2	0.46	65.38	6L+24M2+38M1+2S
		11-11 壮绿竹 Dendrocalamopsis validus Q.H.Dai	2n=70	36m+16sm+4st+14t	2B	122	2.9	0.37	66.7	12L+18M2+30M1+10S
		11-12 吊丝单 Dendrocalamopsis vario-striata (W. T. Lin) Keng f.	2n=104	58m+20sm+10st+16t（2SAT）	2B	182	3.9	0.4	64.38	20L+24M2+30M1+30S
		11-13 大绿竹 Dendrocalamopsis daii Keng f.	2n=70	36m+16sm+4st（2SAT）+14t	2B	122	3.2	0.49	67.78	20L+8M2+30M1+12S

续表

中国竹类植物核型分析

族	属	种	染色体数目	核型公式	核型分类	臂指数	最长染色体/最短染色体	臂比>2的染色体比例	核型不对称系数	染色体相对长度组成
		12-1 野龙竹 Dendrocalamus semiscandens Hsueh et D. Z. Li	$2n=70$	36m+16sm+8st+10t	2B	122	2.7	0.46	65.69	12L+18M2+32M1+8S
		12-2 马来甜龙竹 Dendrocalamus asper (J. A. et J. H. Schult.) Backer ex Heyne	$2n=70$	36m+16sm+10st（2SAT）+8t	2B	122	2.9	0.46	65.63	16L+14M2+24M1+16S
		12-3 椅子竹 Dendrocalamus bambusoides Hsueh et D. Z. Li	$2n=70$	36m+16sm+6st+12t（2SAT）	2B	122	2.8	0.46	66.4	14L+16M2+34M1+6S
		12-4 毛脚龙竹 Dendrocalamopsis barbatus var.intermodiiradicatus Hsueh et D.Z.Li	$2n=68$	38m+16sm+6st+8t（2SAT）	2B	122	2.7	0.41	66.83	14L+18M2+18M1+18S
		12-5 小叶龙竹 Dendrocalamus barbatus Hsueh et D. Z. Li	$2n=68$	38m+16sm+6st+8t	2B	122	3.6	0.41	66.7	10L+16M2+32M1+10S
		12-6 勃氏甜龙竹 Dendrocalamus brandisii (Munro) Kurz	$2n=72$	34m+16sm+4st+18t	3B	122	2.5	0.53	68.99	14L+18M2+30M1+10S
		12-7 美穗龙竹 Dendrocalamus calostachyus (Kurz) Kurz	$2n=68$	38m+16sm+6st+8t	2B	122	3.1	0.41	65.81	14L+18M2+22M1+14S
III 牡竹族 Dendrocalameae	12.牡竹属 Dendrocalamus Nees		$2n=70$	36m+16sm+10st（2SAT）+8t	2B	122	2.3	0.46	67.19	16L+14M2+28M1+12S
		12-8 梁山慈竹 Dendrocalamus farinosus (Keng et Keng f.) Chia et H. L. Fung	$2n=68$	38m+16sm+4st+10t	2B	122	2.7	0.35	64.77	12L+14M2+30M1+12S
		12-9 龙竹 Dendrocalamus giganteus Munro	$2n=70$	36m+16sm+6st+12t	2B	122	2.5	0.49	67.04	12L+14M2+34M1+10S
		12-10 版纳甜龙竹 Dendrocalamus hamiltonii Nees et Arn. ex Munro	$2n=70$	36m+16sm+6st+12t	2B	122	2.6	0.37	65.14	14+20M2+20M1+16S
		12-11 麻竹 Dendrocalamus latiflorus Munro	$2n=64$	36m+16sm+6st（2SAT）+6t	2B	116	3.1	0.41	65.86	14L+10M2+26M1+14S
		12-12 黄竹 Dendrocalamus membranceus Munro	$2n=70$	36m+16sm+8st（2SAT）+10t（2SAT）	2B	122	3.1	0.43	63.77	12L+18M2+26M1+14S
		12-13 吊丝竹 Dendrocalamus minor (McClure) Chia et H. L. Fung	$2n=68$	38m+16sm+4st（2SAT）+10t	2B	122	2.3	0.41	66.84	12L+20M2+24M1+12S
		12-14 花吊丝竹 Dendrocalamus minor var. amoenus (Q. H. Dai et C. F. Huang) Hsueh et D. Z.Li	$2n=68$	38m+16sm+4st（2SAT）+10t	2B	122	3.1	0.41	65.86	12L+18M2+22M1+16S
		12-15 船竹 Dendrocalamopsis ovatus Xia et Chia Yi	$2n=68$	38m+16sm+4st+10t	2B	122	2.5	0.35	65.95	8L+26M2+26M1+8S
		12-16 粉麻竹 Dendrocalamus pulverulentus Chia et But	$2n=68$	38m+16sm+2st+12t	2B	122	2.4	0.44	67.42	10L+22M2+24M1+12S

中国竹类植物核型分析

族	属	种	染色体数目	核型公式	核型分类	臂指数	最长染色体/最短染色体	臂比>2的染色体比例	核型不对称系数	染色体相对长度组成
	12.牡竹属 Dendrocalamus Nees	12-17 龙丹竹 Dendrocalamopsis rongchengensis Yi et C.Y.Sia	2n=70	34m+16sm+8st+12t (2SAT)	2B	124	3.21	0.49	67.73	16L+14M2+20M1+20S
		12-18 歪脚龙竹 Dendrocalamus sinicus Chia et J. L. Sun	2n=68	38m+16sm+2st+12t (2SAT)	2B	122	2.6	0.35	67.68	10L+24M2+22M1+1
			2n=70	36m+16sm+8st+10t (2SAT)	2B	122	3.2	0.43	66.77	10L+16M2+32M1+12S
			2n=68	38m+16sm+4st+10t	2B	122	3.3	0.38	66.14	16L+12M2+24M1+16S
Ⅲ 牡竹族 Dendrocalameae		12-19 融安黄竹 Dendrocalamopsis ronganensis Dai et Huang	2n=72	34m+16sm+4st+18t (2SAT)	2B	122	2.5	0.5	69.2	14L+16M2+34M1+8S
		12-20 白龙巨竹 Dendrocalamopsis sp.	2n=72	34m+16sm+12st (2SAT) +10t	2B	122	2.7	0.5	66.68	12L+18M2+30M1+12S
		12-21 牧竹 Dendrocalamopsis strictus (Roxb.) Nees	2n=68	38m+16sm+6st+8t	2B	122	3	0.44	64.14	12L+12M2+34M1+10S
		13-1 毛筝竹 Gigantochloa levis (Blanco) Merr.	2n=68	38m+16sm+8st+6t	2B	122	2.9	0.44	63.13	14L+16M2+24M1+14S
			2n=72	34m+16sm+8st+14t	3B	122	2.5	0.53	69.48	8L+28M2+28M1+8S
	13.巨竹属 Gigantochloa Kurz ex Munro	13-2 花巨竹 Gigantochloa verticillata (Lour.) Kurz	2n=70	36m+16sm+4st (2SAT) +14t	2B	122	2.9	0.49	69.98	10L+16M2+34M1+10S
		13-3 琴丝滇竹 Gigantochloa rostrata (Willd.) Munro Wong	2n=68	38m+16sm+6st+8t (2SAT)	2B	122	2.4	0.35	62.99	14L+14M2+28M1+12S
		13-4 白毛巨竹 Gigantochloa albociliata (Munro) Kurz	2n=70	34m+18sm+6st+6t	2B	122	3.1	0.49	65.83	14L+18M2+20M1+18S
		13-5 勐遮滇竹 Gigantochloa sp.	2n=70	36m+16sm+6st (4SAT) +12t	2B	122	2.1	0.46	66.52	10L+20M2+30M1+10S
		14-1 黄竿竹 Indosasa levigata Z.P.Wang et G.H.Ye	2n=48	26m+20sm+2st	2A	94	2	0.38	62.51	6L+18M2+14M1+10S
		14-2 摆竹 Indosasa shibataeoides McClure	2n=48	28m+10sm+2st+8t	2B	86	2.1	0.42	66.72	2L+24M2+18M1+4S
Ⅳ 倭竹族 Shibataeeae	14.大节竹属 Indosasa McClure	14-3 中华大节竹 Indosasa sinica C. D. Chu et C. S. Chao	2n=48	22m+24sm+2st	2B	94	3	0.46	64.26	10L+12M2+20M1+6S
		14-4 橄榄竹 Indosasa gigantea Wen	2n=48	22m+20sm (2SAT) +4st+2t	2B	90	2.8	0.5	64.1	10L+12M2+20M1+6S
		14-5 大节竹 Indosasa crassiflora McClure	2n=48	30m+14sm+4st	2B	92	2.6	0.29	61.73	6L+18M2+14M1+10S
		14-6 浦竹仔 Indosasa hispida McClure	2n=48	22m(2SAT)+20sm+4st(SAT)+2t	2B	90	2.9	0.46	64.67	12L+10M2+18M1+8S

续表

中国竹类植物核型分析

族	属	种	染色体数目	核型公式	核型分类	臂指数	最长染色体/最短染色体	臂比>2的染色体比例	核型不对称系数	染色体相对长度组成
	15.唐竹属 Sinobambusa Makino ex Nakai	15-1 唐竹 Sinobambusa tootsik (Sieb.) Makino	2n=48	24m+16sm+6st+2t	2B	88	2.9	0.5	64.97	8L+16M2+12M1+12S
		15-2 月月竹 Sinobambusa sichuanensis Yi	2n=48	32m+14sm+2st	2B	94	2.4	0.33	60.74	10L+8M2+22M1+8S
		15-3 井冈唐竹 S.anaurita Wen	2n=48	26m+16sm+4st+2t	2B	90	2.6	0.38	65.08	8L+14M2+16M1+10S
		15-4 红古唐竹 Sinobambusa rubroligula McClure	2n=48	26m+14sm+6st+2t (2SAT)	2B	88	2.9	0.29	64.29	10L+14M2+10M1+14S
	16.短穗竹属 Brachystachyum Keng	16-1 短穗竹 Brachystachyum densiflorum (Rendle) Keng	2n=48	26m+18sm+2st+2t	2B	92	2.4	0.42	63.5	10L+10M2+14M1+14S
	17.刚竹属 Phyllostachys Sieb.et Zucc.	17-1 刚竹 Phyllostachys sulphurea (Carr.) A.et C.Riv.cv.Viridis	2n=48	28m+8sm+6st (2SAT) +6t	2B	84	3.3	0.42	66.18	12L+10M2+12M1+14S
		17-2 毛环竹 Phyllostachys meyeri McClure	2n=48	24m+20sm (2SAT) +4t	2B	92	2.8	0.42	63.05	10L+12M2+18M1+8S
		17-3 人面竹 Phyllostachys aurea Carr. ex A. et C. Riv.	2n=48	30m+14sm+4st (2SAT)	2B	92	2.4	0.33	61.8	10L+8M2+18M1+12S
		17-4 灰竹 Phyllostachys nuda McClure	2n=48	30m+14sm (2SAT) +2st+2t	2B	92	2.5	0.38	60.8	8L+14M2+20M1+6S
IV倭竹族 Shibataeeae		17-5 石绿竹 Phyllostachys arcana McClure	2n=48	22m+16sm+6st (2SAT) +4t	3B	86	2.4	0.54	65.43	12L+8M2+18M1+10S
		17-6 淡竹 Phyllostachys glauca McClure	2n=48	26m+14sm+4st+4t (2SAT)	2B	88	3.9	0.42	64.76	10L+8M2+18M1+12S
		17-7 黄古竹 Phyllostachys angusta McClure	2n=48	24m+18sm+6st	2B	90	2.6	0.5	62.68	12L+8M2+18M1+10S
		17-8 曲竿竹 Phyllostachys flexuosa (Carr.) A. et C. Riv.	2n=48	28m+14sm+6st	2B	90	3.1	0.42	63.62	12L+8M2+14M1+14S
		17-9 花哺鸡竹 Phyllostachys grabrata S.Y.Chen et C.Y.Yao	2n=48	20m+18sm+6st+4t	3A	86	1.84	0.58	68.39	2L+22M2+22M1+2S
		17-10 角竹 Phyllostachys fimbriligula Wen	2n=48	28m+18sm+2t	2B	94	3.3	0.38	62.55	8L+12M2+20M1+8S
		17-11 乌哺鸡竹 Phyllostachys vivax McClure cv.Vivax	2n=48	28m+12sm+4st (2SAT) +4t	2B	88	3.4	0.42	65.3	10L+12M2+14M1+12S
		17-12 早竹 Phyllostachys praecox C. D. Chu et C. S. Chao	2n=48	32m+10sm+6st (2SAT)	2B	90	3.1	0.33	62.4	10L+12M2+18M1+8S
		17-13 龟甲竹 Phyllostachys heterocycla (Carr.) Mitford	2n=48	26m+14sm+6st (2SAT) +2t	2B	88	3.1	0.46	65.12	14L+10M2+10M1+14S
			2n=48	26m+14sm+4st (2SAT) +4t	2B	88	2.3	0.46	63.17	12L+10M2+14M1+12S

续表

中国竹类植物核型分析

族	属	种	染色体数目	核型公式	核型分类	臂指数	最长染色体/最短染色体	臂比>2的染色体比例	核型不对称系数	染色体相对长度组成
Ⅳ 倭竹族 Shibataeeae	17. 刚竹属 Phyllostachys Sieb. et Zucc.	17-14 毛竹 Phyllostachys heterocycla Mitford cv. Pubescens	2n=48	30m+14sm+4st (2SAT)	2B	92	2.5	0.33	61.57	12L+8M2+20M1+8S
		17-15 圣音毛竹 Phyllostachys heterocycla (Carr.) Mitford cv. Tubaeformis	2n=48	26m+16sm (2SAT) +2st+4t	2B	90	2.6	0.46	63.19	10L+8M2+20M1+10S
		17-16 强竹 Phyllostachys heterocycla (Carr.) Mitford cv. Obliquinoda	2n=48	24m+14sm+2st+8t (2SAT)	2B	86	2.9	0.5	65.13	10L+10M2+16M1+12S
		17-17 花毛竹 Phyllostachys heterocycla (Carr.) Mitford cv. Tao kiang	2n=48	22m+20sm (2SAT) +4st+2t	3B	90	2.8	0.54	64.47	10L+6M2+22M1+10S
		17-18 美竹 Phyllostachys mannii Gamble	2n=48	26m+10sm+4st+8t (2SAT)	2B	84	2.6	0.46	66.09	12L+6M2+18M1+12S
		17-19 京竹 Phyllostachys aureosulcata McClure cv. Pekinensis	2n=48	32m+12sm (4SAT) +2st+2t	2B	92	2.5	0.33	61.87	12L+12M2+12M1+12S
		17-20 金镶玉竹 Phyllostachys aureosulcata McClure cv.Spectabilis	2n=48	24m+16sm+8st (2SAT)	2B	88	2.7	0.46	65.48	12L+6M2+22M1+8S
		17-21 黄竿京竹 Phyllostachys aureosulcata McClure cv. Aureocarlis	2n=48	26m+16sm+4st (2SAT) +2t	2B	90	2.7	0.33	62.19	10L+12M2+16M1+10S
		17-22 紫竹 Phyllostachys nigra (Lodd. ex Lindl.) Munro	2n=48	30m+12sm+4st (2SAT) +2t	2B	90	2.9	0.38	62.67	12L+6M2+22M1+8S
		17-23 毛金竹 Phyllostachys nigra var. henonis (Mitford) Stapf ex Rendle	2n=48	30m+10sm+4st+4t	2B	88	3.1	0.33	64.61	14L+6M2+12M1+16S
		17-24 红壳雷竹 Phyllostachys incarnata Wen	2n=48	30m+10sm+8t (2SAT)	2B	88	2.4	0.38	64.56	10L+10M2+20M1+8S
		17-25 白哺鸡竹 Phyllostachys dulcis McClure	2n=48	36m+8sm+2st+2t	2B	92	3.1	0.25	59.36	10L+10M2+16M1+12S
		17-26 灰水竹 Phyllostachys platyglossa Z. P. Wang et Z. H. Yu	2n=48	26m+16sm+2st+4t	2B	90	3.6	0.42	64.88	10L+10M2+18M1+10S
		17-27 桂竹 Phyllostachys bambusoides Sieb. et Zucc.	2n=48	24m+22sm (2SAT) +2st	2B	94	2.2	0,25	61.55	8L+14M2+16M1+10S
		17-28 斑竹 Phyllostachys bambusoides Sieb. et Zucc. f. lacrima-deae Keng f. et Wen	2n=48	26m+10sm+2st+10t (2SAT)	2B	84	2.8	0.46	67.74	10L+12M2+12M1+14S
		17-29 粉绿竹 Phyllostachys viridi-glaucescens (Carr.) A. et C. Riv	2n=48	26m+16sm+4st+2t	2B	90	2.5	0.46	65.73	12L+8M2+16M1+12S
		17-30 高节竹 Phyllostachys prominens W. Y. Xiong	2n=48	36m+6sm+4st+2t	2A	92	1.93	0.25	60.75	8L+16M2+18M1+6S

续表

中国竹类植物核型分析

族	属	种	染色体数目	核型公式	核型分类	臂指数	最长染色体/最短染色体	臂比>2的染色体比例	核型不对称系数	染色体相对长度组成
	17.刚竹属 Phyllostachys Sieb.et Zucc.	17-31 富阳乌哺鸡竹 Phyllostachys nigella Wen	2n=48	24m+14sm（2SAT）+8st（2SAT）+2t	2B	86	3	0.5	65.87	10L+12M2+14M1+12S
		17-32 毛环水竹 Phyllostachys aurita J.L.Lu	2n=48	28m+10sm+6st（2SAT）+4t	2B	86	3.3	0.38	64.44	10L+10M2+14M1+14S
		17-33 篌竹 Phyllostachys nidularia Munro	2n=48	30m+14sm+2st+2t	2A	92	1.59	0.38	62.92	6L+14M2+26M1+2S
		17-34 黎子竹 Phyllostachys heteroclada Oliver f.purpurata（McClure）Wen	2n=48	28m+10sm+4st（2SAT）+6t	2B	86	2.3	0.38	64.19	14L+6M2+14M1+14S
IV倭竹族 Shibateaeae	18.倭竹属 Shibataea Makino exNakai	18-1 狭叶倭竹 Shibataea lanceifolia C. H. Hu	2n=48	28m+14sm+6st	2B	90	2.2	0.42	63.72	8L+14M2+18M1+8S
		18-2 南平倭竹 Shibataea nanpingensis Q.F.Zheng et K.F.Huang	2n=48	24m+14sm+4st+6t（2SAT）	2B	86	3	0.5	65.11	12L+10M2+18M1+8S
		18-3 鹅毛竹 Shibataea chinensis Nakai	2n=48	28m+12sm+2st+6t（2SAT）	2B	88	2.5	0.38	63.67	8L+14M2+16M1+10S
	19.寒竹属 Chimonobambusa Makino	19-1 寒竹 Chimonobambusa marmorea（Mitford）Makino	2n=48	28m+10sm+2st+8t	2B	86	2.7	0.42	63.87	14L+6M2+18M1+10S
		19-2 方竹 Chimonobambusa quadrangularis（Fenzi）Makino	2n=48	28m+16sm+4st	2B	92	2.4	0.33	61.84	12L+8M2+20M1+8S
		19-3 刺竹子 Chimonobambusa pachystachys Hsueh et Yi	2n=48	26m+16sm+4st+2t	2B	90	2.4	0.42	62.66	4L+18M2+18M1+8S
		19-4 四方竹 Chimonobambusa sp.	2n=48	32m+10sm+4st（2SAT）+2t	2B	90	2.9	0.29	61.74	8L+16M2+16M1+8S
	20.筇竹属 Qiongzhuea Hsueh et Yi	20-1 筇竹 Qiongzhuea tumidinoda Hsueh et Yi	2n=48	26m+16sm+4st+2t	2B	90	2.2	0.38	61.55	14L+6M2+16M1+12S
V香竹族 Chusqueeae	21.镰序竹属 Drepanostachyum Keng f.	21-1 爬竹 Drepanostachyum scandeus（Hsueh et W.D.Li）Keng f. ex Yi	2n=48	34m+8sm+4st（2SAT）+2t	2B	90	2.7	0.29	61.44	8L+16M2+14M1+10S
		21-2 镰序竹 Drepanostachyum sp.	2n=48	24m+16sm+6st（2SAT）+2t	2B	88	2.5	0.5	66.14	6L+16M2+18M1+8S
	22.箭竹属 Fargesia Franch.emend.Yi	22-1 岩斑竹 Fargesia canaliculata Yi.	2n=48	28m+14sm+4st+2t	2B	90	2.2	0.42	64.31	12L+10M2+18M1+8S
		22-2 扫把竹 Fargesia fractiflexa Yi	2n=48	26m+16sm+4st+2t	2B	90	2.2	0.46	65.71	6L+18M2+14M1+10S
		22-3 海南箭竹 Fargesia hainanensis Yi	2n=48	30m+12sm+4st+2t	2B	90	2.1	0.29	62.31	8L+16M2+20M1+4S
		22-4 丰实箭竹 Fargesia ferax（Keng）Yi	2n=48	36m+10sm+2st	2B	94	2.5	0.21	58.71	6L+16M2+20M1+6S
VI北美箭竹族 Arundinarieae	23.玉山竹属 Yushania Keng f.	23-1 百山祖玉山竹 Yushania baishanzuensis Z.P.Wang et G.H.Ye	2n=48	28m+12sm+4st+4t	2B	88	2.9	0.42	62.05	14L+8M2+12M1+14S
	24.酸竹属 Acidosasa C.D.Chu et C.S.Chao	24-1 黄甜竹 Acidosasa edulisWen	2n=48	20m+18sm+6st+4t（2SAT）	2B	86	3.6	0.46	66.35	16L+4M2+12M1+16S

续表

中国竹类植物核型分析

族	属	种	染色体数目	核型公式	核型分类	臂指数	最长染色体/最短染色体	臂比>2的染色体比例	核型不对称系数	染色体相对长度组成
	25.少穗竹属 Oligostachyum Z.P.Wang et G.H.Ye	25-1 少穗竹 Oligostachyum sulcatum Z.P.Wang et G.H.Ye	$2n=48$	26m+16sm+6st（2SAT）	2B	90	2.7	0.38	65.03	10L+12M2+16M1+10S
		25-2 屏南少穗竹 Oligostachyum glabrescens（Wen）Keng f.et Z.P.Wang	$2n=48$	30m+14sm+2st+2t	2B	92	3	0.38	63.39	12L+8M2+14M1+14S
		25-3 胂节少穗竹 Oligostachyum oedogonatum Q.F.Zheng et K.F.Huang	$2n=48$	28m+14sm+2st+4t（2SAT）	2B	90	3.1	0.42	64.83	14L+10M2+10M1+14S
		25-4 四季竹 Oligostachyum lubricum（Wen）Keng f.	$2n=48$	32m+8sm+4st+4t（2SAT）	2B	88	3.8	0.33	63.78	14L+10M2+10M1+14S
		25-5 糙花少穗竹 Oligostachyum scabriflorum Z.P.Wang et G.H.Ye	$2n=48$	30m+12sm+4st+2t	2B	90	2.7	0.38	63	12L+8M2+20M1+8S
	26.大明竹属 Pleioblastus Nakai	26-1 大明竹 Pleioblastus gramineus（Bean）Nakai	$2n=48$	28m+12sm+4st+4t	2B	88	3.5	0.42	63.76	12L+10M2+16M1+10S
		26-2 光箨苦竹 Pleioblastus amarus var. subglabratus S.Y.Chen	$2n=48$	30m+14sm+4st	2B	92	2.03	0.25	62.75	8L+14M2+22M1+4S
		26-3 杭州苦竹 Pleioblastus amarus var. hangzhouensis S. L. Chen et S. Y. Chen	$2n=48$	28m+10sm+6st（2SAT）+4t	2B	86	4	0.42	64.92	12L+12M2+8M1+16S
		26-4 华箨竹 Pleioblastus intermedius S.Y.Chen）	$2n=48$	26m+16sm+4st+2t	2B	90	2.4	0.33	62.36	10L+8M2+22M1+8S
VI北美箭竹族 Arundinarieae		26-5 高舌苦竹 Pleioblastus altiligulatus S.L.Chen et S.Y.Chen	$2n=48$	28m+12sm+6st+2t	2B	88	3.2	0.42	64.79	12L+10M2+16M1+10S
		26-6 宜兴苦竹 Pleioblastus yixingensis S. L. Chen et S. Y. Chen	$2n=48$	28m+16sm+4st+2t	2B	90	2.2	0.42	65.88	10L+14M2+14M1+10S
		26-7 硬头苦竹 Pleioblastus longifimbriatus S.Y.Chen	$2n=48$	30m+10sm+8st	2A	88	1.9	0.33	61.28	10L+6M2+30M1+2S
		26-8 烂头苦竹 Pleioblastus ovatoauritas Wen	$2n=48$	26m+12sm+4st（2SAT）+6t	2B	86	3	0.46	65.44	14L+10M2+10M1+14S
	27.巴山木竹属 Bashania Keng f.et Yi	27-1 巴山木竹 Bashania fargesii（E. G. Camus）Keng f.et Yi	$2n=48$	22m+18sm+6st+2t	2B	88	2.5	0.5	64.79	10L+8M2+18M1+12S
	28.井冈寒竹属 Gelidocalamus Wen	28-1 井冈寒竹 Gelidocalamus stellatus Wen	$2n=48$	18m+24sm+4st+2t	3B	90	3	0.58	65.07	10L+12M2+16M1+10S
	29.矢竹属 Pseudosasa Makino ex Nakai	29-1 矢竹 Pseudosasa japonica（Sieb. et Zucc.）Makino	$2n=48$	28m+16sm+4t	2B	92	2.4	0.42	62.24	8L+14M2+18M1+8S
		29-2 茶竿竹 Pseudosasa amabilis（McClure）Keng f.	$2n=48$	26m+16sm+4st（2SAT）+2t	2B	90	2.6	0.42	63.62	12L+6M2+22M1+8S

续表

中国竹类植物核型分析

族	属	种	染色体数目	核型公式	核型分类	臂指数	最长染色体/最短染色体	臂比>2的染色体比例	核型不对称系数	染色体相对长度组成
		29-3 近实心茶秆竹 Pseudosasa subsolida S.L.Chen et G.Y.Sheng	2n=48	28m+14sm+4st+2t	2B	90	3	0.38	64.02	12L+10M2+14M1+12S
	30.赤竹属 Sasa Makino et Shibata	30-1 铺地竹 Sasa argenteastriatus Camus	2n=48	24m+16sm+8st（2SAT）	2B	88	3	0.5	63.78	10L+12M2+16M1+10S
		30-2 菲黄竹 Sasa auricoma E.G.Camus	2n=48	28m+14sm（2SAT）+6st	2B	90	3.4	0.42	61.96	12L+12M2+10M1+14S
		30-3 华箬竹 Sasa sinica keng	2n=48	24m+18sm+2st（2SAT）+4t	2B	90	2.2	0.5	66.52	10L+8M2+22M1+8S
	31.铁竹属 Ferrocalamus Hsueh et Keng f.	31-1 铁竹 Ferrocalamus strictus Hsueh et Keng f.	2n=46	24m+22sm（2SAT）	2B	92	2.7	0.35	63.2	10L+8M2+20M1+8S
VI北美箭竹族 Arundinarieae	32.箬竹属 Indocalamus Nakai	32-1 髯毛箬竹 Indocalamus barbatus McClure	2n=48	30m+12sm+4st+2t	2B	90	2.1	0.29	62.12	10L+12M2+20M1+6S
		32-2 箬叶竹 Indocalamus longiauritus Hand.-Mazz.	2n=48	30m+12sm+4st+2t	2B	90	2.6	0.38	63.28	8L+18M2+12M1+10S
	33.Guadua 属	33-1 瓜多竹 Guadua angustifolia Stkiata	2n=46	24m（2SAT）+14sm+2t（2SAT）+6t	2A	84	2	0.48	67.44	4L+22M2+14M1+6S
	34.杂交种	34-1 毛箪竹×麻竹 Gigantochloa levis (Blanco) Merr.×Dendrocalamus latiflorus Munro	2n=70	36m+16sm+4st+14t（2SAT）	2B	122	3.5	0.43	66.76	14L+18M2+28M1+10S

第五节 中国药用植物基因组染色体研究结果

提要：目前我国有中药资源 12807 种，其中药用植物 11146 种，种子植物 10188 种，隶属 222 科 1972 属，占药用植物 90% 以上。我们从中选取了 123 科 403 属 563 种药用植物进行了染色体研究。

我国常用药材有 800 多种，道地药材占 80%，道地药材在临床上具有极高的价值和声望。但现代科学还无法阐述其优质形成的原因，更无法进行人工培育。2011 年第 390 次香山会议虽然提出在道地药材中存在道地中药功能基因组或基因，但是，随着时间的演进、自然条件、生态环境的变化和生物有性后代的分离，我国药材的道地性逐渐减少或消失。因此，现在的当务之急是如何解决道地药材的持续利用和新品种培育问题。基于我们多年来对药用植物基因组染色体基础研究，提出建立"基因组三倍化与杂交相结合育种平台"实现对道地药材功能基因组的开发与利用，通过新品种培育将道地药材功能基因完整地保存下来。通过十多年三倍体丹参的培育，证明这是一条有效途径。具体做法是：①将二倍体道地丹参人工诱变成同源四倍体为母本♀；②与二倍体道地丹参为父本♂进行杂交，获得三倍体丹参，形成三倍体后，在"空间"上与其他物种形成了生殖隔离，不产生有性后代，成为永久杂种，这样就将道地丹参基因组完整地固定下来。这是到目前为止，最好的能固定、完整保存基因组的方法。同时，基因组三倍化后还具有多倍体优势和杂交优势，在育种上具有很多优良性状可供选择。另外，大部中药材是以根、茎、叶、花为药材，以种子为药材的只占少数，所以，多倍体育种在药用植物上具有广泛用途。我们在本书中报道有四倍体 41 种，六倍体 20 种，三倍体 10 种，其他倍性 15 种。半夏多倍体复合体就有 4 个倍性，$7\times=91$、$8\times=104$、$9\times=117$、$10\times=130$。这些资料为药用植物多倍体育种提供了借鉴。

2.5.1 部分药用植物染色体

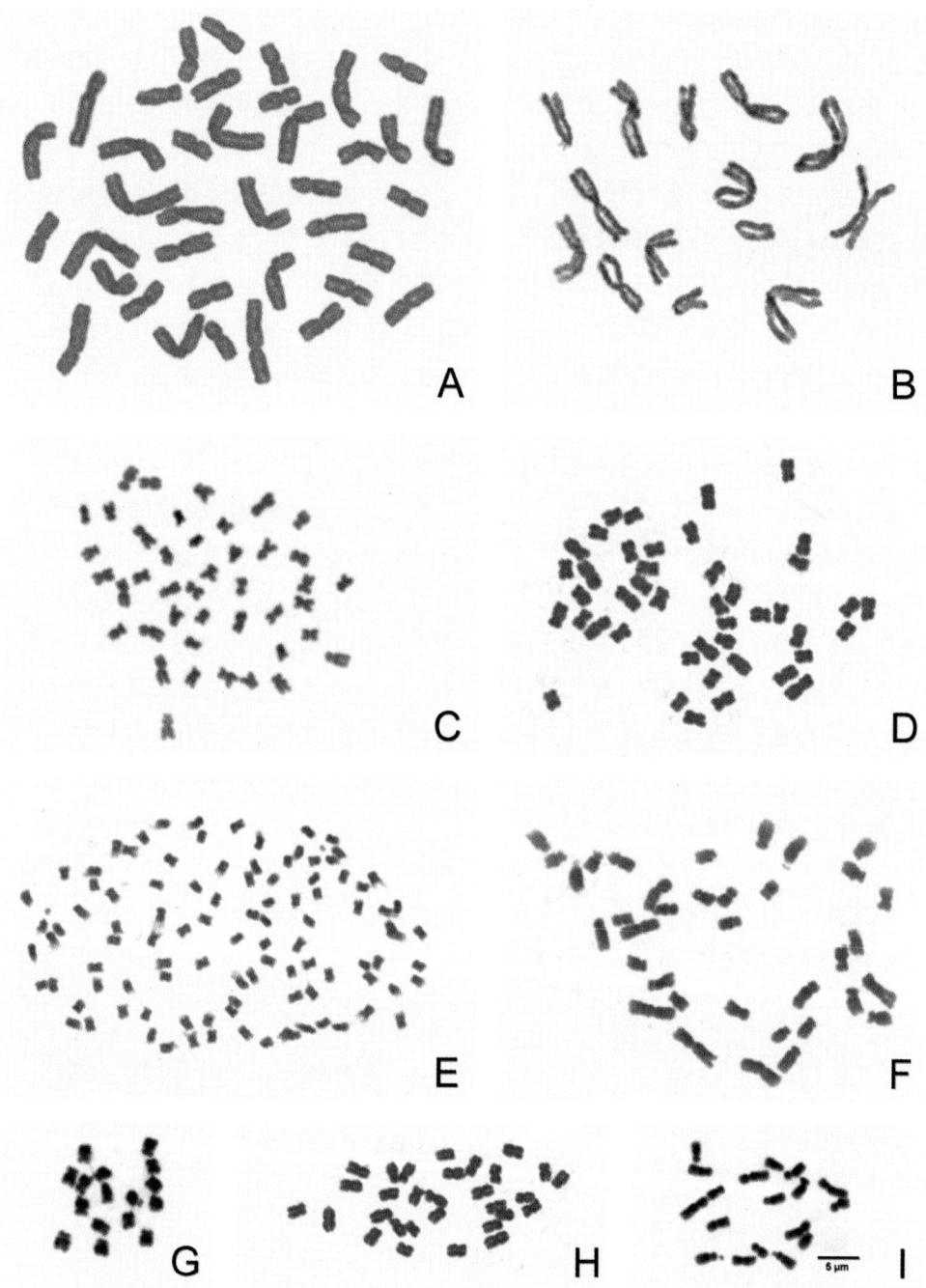

A.观音草 *Peristrophe baphica*，B.泽泻 *Alisma orientale*，C.牛膝 *Achyranthes bidentata*，D.马蓝 *Strobilanthes cusia*，E.杏叶沙参 *Adenophora hunanensis*，F.牛蒡 *Arctium lappa*，G.长春花 *Catharanthus roseus*，H.豨莶 *Siegesbeckia orientalis*，I.金银花 *Lonicera japonica*

2.5.2 中国药用植物基因组染色体数据分析

中药植物核型分析

科	属	种	染色体数目	核型公式	核型分类	臂指数	最长染色体/最短染色体	臂比>2的染色体比例	核型不对称系数	染色体相对长度组成
一、龙舌兰科 Agavaceae	1. 龙舌兰属 Agave L.	1-1 金边龙舌兰 Agave americana L. var. marginata Hort.	2n=120	18m+34sm+40st+28t	3C	174	5.77	0.75	78.23	20L+10M₂+28M₁+62S
二、爵床科 Acanthaceae	2. 穿心莲属 Andrographis Wall.	2-1 穿心莲 Andrographis paniculata（Burm. f.）Nees.	2n=50							
	3. 山一笼鸡属 Gutzlaffia Hance	3-1 山一笼鸡 Gutzlaffia aprica Hance	2n=2x=38	12m+22sm+2st+2T	2A	72	1.52	0.26	65.93	16M₂+22M₁
	4. 观音草属 Peristrophe Ness.	4-1 观音草 Peristrophe baphica（Spreng.）Bremek.	2n=2x=38	26m+12sm	2B	76	2.09	0.05	60.94	8L+8M₂+16M₁+6S
	5. 马蓝属 Strobilanthes Bl.	5-1 马蓝 Strobilanthes cusia（Nees）Ktze.	2n=3x=45	24m+21sm	2A	90	1.61	0.27	63.36	6L+12M₂+27M₁
三、槭树科 Aceraceae	6. 金钱槭属 Dipteronia Oliv.	6-1 云南金钱槭 Dipteronia dyeriana Henry	2n=2x=22	14m+8sm（2SAT）	2A	44	1.84	0.09	60.62	2L+6M₂+14M₁
四、八角枫科 Alangiaceae	7. 八角枫属 Alangium Lam.	7-1 八角枫+（Lour.）Harms	2n=2x=22	2m+10sm+10T	3B	34	2.11	0.64	78.82	2L+6M₂+12M₁+2S
五、泽泻科 Alismataceae	8. 泽泻属 Alisma L.	8-1 泽泻 Alisma orientale（Sam.）Juzepcz.	2n=2x=14	2n=2x=14	2B	24	2.63	0.29	63.42	2L+6M₂+4M₁+2S
	9. 慈姑属 Sagittaria L	9-1 慈姑 Sagittaria sagittifolia L.	2L+6M₂+4M₁+2S 2n=124	2L+6M₂+4M₁+2S	3B	26	3.03	0.82	0.82	2L+8M₂+10M₁+2S
六、苋科 Amaranthaceae	10. 牛膝属 Achyranthes L.	10-1 土牛膝 Achyranthes aspera L.	2n=42	22m+20sm（2SAT）	2A	84	1.8	0.33	63.25	4L+14M₂+24M₁
		10-2 牛膝 A.bidentata Blume	2n=42							
		10-3 红柳叶牛膝 Achyranthes longifolia（Mak.）Mak. f. rubra Ho									
	11. 苋属 Amaranthus L.	11-1 刺苋 Amaranthus spinosus L.	2n=2x=34	10m+8sm+16T	3A	52	1.83	0.59	78.29	2L+14M₂+16M₁+2S
	12. 青葙属 Celosia L.	12-1 青葙子 Celosi argentea L.	2n=72	66m+6sm	1A	144	1.68	0	57.71	2L+30M2+40M1
	13. 杯苋属 Cyathula Bl.	13-1 杯苋 Cyathula prostrata（L.）Blume	2n=32	6m+8sm+2st+16T	3A	46	1.87	0.56	80.13	2L+12M₂+14M₁+4S
七、石蒜科 Amaryllidaceae	14. 仙茅属 Curculigo Gaertn	14-1 仙茅 Curculigo orchioides Gaertn	2n=2x=18	10m+8sm	2A	36	1.79	0.44	63.42	2L+6M₂+10M₁
	15. 石蒜属 Lycoris Herb.	15-1 红花石蒜 Lycoris radiata（L'Her.）Herb.	2n=2x=16	6m+2t+8T	3B	22	2.38	0.63	74.16	6L+2M1+8S
八、漆树科 Anacardiaceae	16. 黄连木属 Pistacia L.	16-1 清香木 Pistacia weinmannifolia J. Poisson ex Franch.	2n=68							

续表

中药植物核型分析

科	属	种	染色体数目	核型公式	核型分类	臂指数	最长染色体/最短染色体	臂比>2的染色体比例	核型不对称系数	染色体相对长度组成
八、漆树科 Anacardiaceae	17. 盐肤木属 Rhus (Tourn.) L.	17-1 盐肤木 Rhus chinensis Mill.	2n=30	14m+12sm(2SAT)+4st	2A	56	1.6	0.33	65.57	$2L+12M_2+16M_1$
	18. 漆树属 Toxicodendron (Tourn.) Mill.	18-1 漆树 Toxicodendron vernicifluum (Stokes) F. A. Barkl.	2n=30	4m+22sm(2SAT)+4st	3A	56	1.58	0.67	68.88	$:2L+12M_2+16M_1$
	19. 水甘草属 Amsonia Walt.	19-1 水甘草 Amsonia sinensis Tsiang et P. T. Li	2n=24							
	20. 罗布麻属 Apocynum L.	20-1 罗布麻 Apocynum venetum L.	2n=24							
	21. 长春花属 Catharanthus G. Don	21-1 长春花 Catharanthus roseus (L.) G. Don	2n=2x=16	4m（1SAT）+12sm	3A	32	1.35	1.35	67.11	$8M_2+8M_1$
九、夹竹桃科 Apocynaceae	22. 萝芙木属 Rauvolfia L.	22-1 萝芙木 R. verticillata (Lour.) Baill	2n=2x=22	2M+2m+10sm+4st+4t (1SAT)	3A	36	1.52	0.64	70.36	$2L+8M_2+12M_1$
		22-2 红果萝芙木 Rauvolfia verticillata (Lour.) Baill. var. rubrocarpa Tsiang	2n=2x=22	12m+8sm+2st	2A	42	1.55	0.27	63.34	$10M_2+12M_1$
	23. 络石属 Trachelospermum Lem.	23-1 络石 Trachelospermum jasminoides (Lindl.) Lem.	2n=2x=20	12m+8sm	2A	40	1.58	0.1	60.58	$10M_2+10M_1$
	24. 盆架树属 Winchia A.DC.	24-1 盆架树 Winchia calophylla A. DC.	2n=42							
十、冬青科 Aquifoliaceae	25. 冬青属 Ilex L.	25-1 枸骨 Ilex cornuta Lindl. ex Paxton	2n=40							
	26. 菖蒲属 Acorus L.	26-1 菖蒲 Acorus calamus L.	2n=44	28m+16sm	2A	88	1.52	0.18	63.04	$2L+18M2+24M1$
		26-2 石菖蒲 Acorus tatarinowii Schott	2n=24	12m+12sm	2B	48	2.03	0.08	61.3	$2L+14M_2+4M_1+4S$
	27. 魔芋属 Amorphophallus Bl. ex Decne	27-1 魔芋 Amorphophallus rivieri Durieu ex Carriere	2n=2x=26	16m+6sm+4st	2B	48	2.3	0.31	62.4	$4L+8M_2+8M_1+6S$
十一、天南星科 Araceae	28. 天南星属 Arisaema Mart.	28-1 东北天南星 Arisaema amurense Maxim	2n=2x=28	14m+10sm+4st(2SAT)	2A	52	1.73	0.21	63.02	$14M_2+12M_1+2S$
		28-2 滇南星 Arisaema austro-yunnanense Buchet	2n=48	30m+18sm	2B	96	2.16	0.21	62.3	$4L+10M_2+34M_1$

第二章 中国植物染色体研究结果 | 197

续表

中药植物核型分析

科	属	种	染色体数目	核型公式	核型分类	臂指数	最长染色体/最短染色体	臂比>2的染色体比例	核型不对称系数	染色体相对长度组成
	28. 天南星属 Arisaema Mart.	28-3 一把伞南星 Arisaema erubescens (Wall.) Schott	$2n=2x=28$	6m+14sm+8st	3A	48	1.65	0.57	68.67	$12M_2+14M_1+2S$
		28-4 异叶天南星 Arisaema heterophyllum Bl	$2n=2x=28$	10m+14sm+4st	3A	52	1.54	0.57	67.74	$12M_2+16M_1$
		28-5 朝鲜天南星 Arisaema peninsulae Nakai	$2n=2x=28$	22m+2sm+2st（1SAT）+2t	2A	52	1.91	1.91	1.91	$16M_2+10M_1+2S$
	29. 千年健属 Homalomena Schott	29-1 千年健 Homalomena occulta (Lour.) Schott	$2n=40$	10m+20sm+10st	3A	70	1.42	0.55	68.62	$20M_2+20M_1$
十一、天南星科 Araceae		30-1 滴水珠 Pinellia cordata N.E.Brown	$2n=2x=26$	18m（2SAT）+8sm	2A	52	1.35	0.15	59.53	$14M_2+12M_1$
	30. 半夏属 Pinellia Ten.	30-2 虎掌 Pinellia pedatisecta Schott	$2n=3x=39$	18m（2SAT）+21sm	2A	78	1.5	0.08	60.81	$18M_2+21M_1$
		30-3 盾叶半夏 Pinellia peltata Pei	$2n=2x=26$	24m（2SAT）+2sm	1A	52	1.34	0	57.22	$14M_2+12M_1$
		30-4 半夏 Pinellia ternata (Thunb.) Breit.	$2n=7x=91$	6m (2SAT) +16sm+4st	3A	48	1.58	0.77	70.33	$16M2+10M1$
		30-5 鹊落萍半夏 Pinellia yaoluopingensis X.H.Guo et X.L.Liu	$2n=2x=26$	54m+37sm	2A	182	1.74	0.22	62.29	$4L+40M_2+47M_1$
	31. 臭菘属 Symplocarpus Salisb. ex Nutt.	31-1 日本臭菘 Symplocarpus nipponica Makino	$2n=30$	20m（2SAT）+6sm	2A	52	1.27	0.08	59.12	$14M_2+12M_1$
	32. 犁头尖属 Typhonium Schott	32-1 独角莲 Typhonium giganteum Engl.	$2n=4x=52$	18m+12sm	2B	60	3.46	0.13	61.44	$4L+12M_2+8M_1+6S$
	33. 五加属 Acanthopanax (Deche. et Planch.) Miq	33-1 刺五加 Acanthopanax senticosus (Rupr. et Maxim.) Harms	$2n=48$	4M+24M+12sm+12st	2A	92	1.69	0.35	63.55	$4L+16M_2+32M_1$
		33-2 三叶五加 Acanthopanax trifoliatus (L.) Merr.	$2n=48$	2n=48	2A	94	1.81	0.42	65.06	$2L+24M_2+18M_1+4S$
十二、五加科 Araliaceae	34. 楤木属 Aralia L.	34-1 楤木 Aralia chinensis L.	$2n=2x=24$	14m+26sm+8sm	2A	88	1.72	0.5	67.17	$26M_2+18M_1+4S$
		34-2 长白楤木 Aralia continentalis Kitag	$2n=4x=48$	20m (2SAT) +4sm	2A	48	1.49	0.17	60.4	$10M2+14M1$
		34-3 辽东楤木 Aralia elata (Miq.) Seem.	$2n=2x=24$	16m+8sm	2A	48	1.49	0.17	61.94	$12M_2+12M_1$
	35. 人参属 Panax L.	35-1 人参 Panax ginseng C. A. Meyer	$2n=48$	18m+26sm+4st	2A	92	1.79	0.5	65.43	$2L+24M_2+20M_1+2S$
		35-2 三七 Panax notoginseng (Burk.) F. H. Chen	$2n=24$	14m+10sm	2A	48	1.57	0.33	62.73	$12M_2+12M_1$

续表

中药植物核型分析

科	属	种	染色体数目	核型公式	核型分类	臂指数	最长染色体/最短染色体	臂比>2的染色体比例	核型不对称系数	染色体相对长度组成
		35-3 西洋参 Panax quinquefolius L.	2n=48	16m+28sm+4st	2A	92	1.72	0.46	66.37	2L+20M$_2$+24M$_1$+2S
十三、马兜铃科 Aristolochiaceae	36. 马兜铃属 Aristolochia L.	36-1 马兜铃 Aristolochia debilis Sieb. et Zucc.	2n=2x=14	2m+12sm	2B	28	2	0.43	66.08	2L+2M$_2$+8M$_1$+2S
	37. 细辛属 Asarum L.	37-1 辽细辛 Asarum heterotropoides Fr. Schmidt var. mandshuricum (Maxim.) Kitag.	2n=2x=26	16m+8sm+2st	2B	50	2.28	0.31	59.79	4L+10M2+8M1+4S
		37-2 汉城细辛 Asarum sieboldii Miq. f. seoulense (Nakai) C.Y.Cheng et C.S.Yang	2n=2x=26	18m+4sm+4st	2B	48	2.33	0.23	59.28	4L+8M$_2$+10M$_1$+4S
	38. 白前属 Cynanchum L	38-1 白薇 Cynanchum atratum Bge.	2n=2x=22							
		38-2 牛皮消 Cynanchum auriculatum Royle ex Wight	2n=2x=22							
		38-3 白前 Cynanchum glaucescens (Decne.) Hand.-Mazz	2n=2x=22	4m+10sm+6st+2T	3A	38	1.5	0.64	72.05	12M$_2$+10M$_1$
十四、萝藦科 Asclepiadaceae		38-4 徐长卿 Cynanchum paniculatum (Bge.) Kitag	2n=2x=22	8m+10sm+4st	3A	40	1.5	0.64	67.31	12M2+10M1
	39. 萝藦属 Metaplexis R. Br.	39-1 萝藦 Metaplexis japonica (Thunb.) Makino	2n=2x=22	12m+10sm (2SAT)	2A	44	1.49	1.49	62.66	14M$_2$+8M$_1$
	40. 杠柳属 Periploca L.	40-1 杠柳 Periploca sepium Bge	2n=2x=22	10m+10sm+2st	2A	42	1.45	0.36	65.98	10M$_2$+12M$_1$
	41. 夜来香属 Telosma Cov.	41-1 夜来香 Telosma cordata (Burm. f.) Merr.	2n=2x=14	12m+2sm (2SAT)	1A	28	1.42	0	57.95	8M$_2$+6M$_1$
十五、凤仙花科 Balsaminaceae	42. 凤仙花属 Impatiens L.	42-1 凤仙花 Impatiens balsamina L.	2n=2x=14	2m+12sm	1A	28	1.78	0	64.99	2L+6M$_2$+4M$_1$+2S
十六、落葵科 Basellaceae	43. 落葵属 Basella L.	43-1 落葵 Basella alba L.	2n=44	12m+26sm(2SAT)+6st	2A	82	1.79	0.45	67.2	2L+24M$_2$+18M$_1$
十七、小檗科 Berberidaceae	44. 淫羊藿属 Epimedium L.	44-1 淫羊藿 Epimedium brevicornum Maxim.	2n=2x=12	4m+6sm+2st (2SAT)	3A	22	1.52	0.67	67.05	8M$_2$+4M$_1$
	45. 鲜黄连属 Jeffersonia Barton	45-1 鲜黄连 Jeffersonia dubia (Maxim.) Benth. et Hook. f	2n=2x=12	6m+6st	2A	18	1.6	0.5	68.36	6M$_2$+6M$_1$
	46. 十大功劳属 Mahonia Nutt.	46-1 十大功劳 Mahonia fortunei (Lindl.) Fedde	2n=2x=28	24m+4sm (4SAT)	2A	56	1.47	0.14	58.33	14M$_2$+14M$_1$
十八、紫葳科 Bignoniaceae	47. 千张纸属 Oroxylum Vent.	47-1 木蝴蝶 Oroxylum indicum (L.) Vent.	2n=2x=28	26m+2sm	1A	56	1.74	0	58	2L+12M$_2$+12M$_1$+2S
十九、红木科 Bixaceae	48. 红木属 Bixa L.	48-1 红木 Bixa orellana L.	2n=2x=14	10m+2sm+2st (2SAT)	2A	26	1.77	0.14	61.01	2L+4M2+6M1+2S

第二章 中国植物染色体研究结果 | 199

续表

中药植物核型分析

科	属	种	染色体数数目	核型公式	核型分类	臂指数	最长染色体/最短染色体	臂比>2的染色体比例	核型不对称系数	染色体相对长度组成
二十、紫草科 Boraginaceae	49. 软紫草属 Arnebia Forsk.	49-1 软紫草 Bixa euchroma (Royle) Johnst.	$2n=23$	11m+8sm+4st (2SAT)	2A	42	1.93	0.42	63.76	$4L+5M_2+12M_1+2S$
	50. 聚合草属 Symphytum L.	50-1 聚合草 Bixa officinale L.	$2n=2x=14$	4m+4sm+6st	3B	22	2.2	0.57	69.15	$2L+2M_2+6M_1+4S$
	51. 山茄子属 Brachybotrys Maxim. ex Oliv.	51-1 山茄子 Brachybotrys paridiformis Maxim. ex Oliv.	$2n=24$	8m+12sm+2st+2T	3A	44	1.67	0.58	68.26	$12M_2+10M_1+2S$
二十一、醉鱼草科 Buddlejaceae	52. 醉鱼草属 Buddleja L.	52-1 密蒙花 Buddleja officinalis Maxim.	$2n=38$							
二十二、仙人掌科 Cactaceae	53. 仙人掌属 Opuntia Mill.	53-1 仙人掌 Opuntia dillenii (Ker. Gawl.) Haw.	$2n=44$							
	54. 沙参属 Adenophora Fisch	54-1 杏叶沙参 Adenophora hunanensis Nannf.	$2n=102$	62m+40sm (2SAT)	2A	204	1.79	0.14	62.06	$2L+46M_2+54M_1$
		54-2 沙参 Adenophora stricta Miq.	$2n=2x=34$	24m+8sm (1SAT) +2st	2A	66	1.28	0.12	59.79	$16M_2+18M_1$
		54-3 轮叶沙参 Adenophora tetraphylla (Thumb.) Fisch	$2n=2x=34$	28m+4sm+2st	2A	66	1.32	0.12	59.11	$12M_2+22M_1$
	55. 牧根草属 Asyneuma Griseb. et Schenck	55-1 球果牧根草 Asyneuma chinense Hong	$2n=24$	22m+2sm	1A	48	1.61	0	55.89	$2L+10M_2+12M_1$
		55-2 牧根草 Asyneuma japonicum (Miq.) Briq	$2n=64$								
二十三、桔梗科 Campanulaceae	56. 风铃草属 Campanula L.	56-1 风铃草 Campanula medium L.	$2n=102$	78m (2SAT) +20sm (4SAT) +4st	2A	200	1.9	0.08	59.79	$4L+44M_2+52M_1+2S$
	57. 金钱豹属 Campanumoea Bl.	57-1 大花金钱豹 Campanumoea javanica Blume	$2n=28$	10m (1SAT)+14sm+4st (2SAT)	2A	52	1.53	0.36	65.62	$14M_2+14M_1$
	58. 党参属 Codonopsis Wall.	58-1 轮叶党参 Codonopsis lanceolata Benth. et Hook	$2n=2x=16$	2m+10sm+4st	2A	30	1.35	0.5	68.17	$8M_2+8M_1$
		58-2 党参 Codonopsis pilosula (Franch.) Nannf	$2n=2x=16$	10m+6sm (2SAT)	2A	32	1.32	0.13	0.13	$8M_2+8M_1$
	59. 半边莲属 Lobelia L.	59-1 半边莲 Lobelia chinensis Lour.	$2n=9x=63$	33m+24sm(2SAT)+6st	2B	120	2.07	0.29	63.85	$3L+27M_2+30M_1+3S$
	60. 桔梗属 Platycodon A. DC.	60-1 桔梗 Platycodon grandiflorus (Jacq.) A. DC.	$2n=2x=18$	14m+4sm (2SAT)	2A	36	1.71	0.11	59.66	$2L+4M_2+12M_1$
二十四、美人蕉科 Cannaceae	61. 美人蕉属 Canna L.	61-1 蕉芋 Canna edulis Ker.-Gawl	$2n=2x=18$	16m+2t	2A	34	1.33	0.11	60.03	$8M_2+10M_1$
		61-2 美人蕉 Canna indica L.	$2n=2x=18$	16m+2sm	2A	36	1.38	0.11	58.58	$10M_2+8M_1$

续表

中药植物核型分析

科	属	种	染色体数目	核型公式	核型分类	臂指数	最长染色体/最短染色体	臂比>2的染色体比例	核型不对称系数	染色体相对长度组成
二十五、大麻科 Cannabaceae	62. 大麻属 Cannabis L.	62-1 大麻 Cannabis sativa L.	$2n=2x=20$	16m+2sm+2st（2SAT）	2A	38	1.43	0.2	60.34	$8M_2+12M_1$
二十六、白花菜科 Capparaceae	63. 白花菜属 Cleome L.	63-1 白花菜 Cleome gynandra L.	$2n=2x=34$	12m+22sm	2A	68	1.53	0.24	63.7	$2L+12M_2+20M_1$
	64. 忍冬属 Lonicera L.	64-1 忍冬 Lonicera japonica Thunb.	$2n=2x=18$	2m+14sm（2SAT）+2st	2A	34	1.61	0.44	66.6	$2L+6M_2+10M_1$
		64-2 金银木 Lonicera maackii (Rupr.) Maxim.	$2n=2x=18$	2m+12sm+4st（1SAT）	3A	32	1.87	0.78	69.71	$4L+2M_2+10M_1+2S$
二十七、忍冬科 Caprifoliaceae	65. 接骨木属 Sambucus L.	65-1 金叶接骨木 Sambucus canadensis L. var. aurea	$2n=36$	12m+8sm+12st+4t（2SAT）		56	1.84	0.5	71.3	$4L+12M_2+16M_1+4S$
	66. 荚蒾属 Viburnum L.	66-1 烟管荚蒾 Viburnum utile Hemsl.	$2n=2x=18$	12m+6sm（2SAT）	2A	36	1.88	0.22	60.22	$4L+2M_2+10M_1+2S$
二十八、番木瓜科 Caricaceae	67. 番木瓜属 Carica L.	67-1 番木瓜 Carica papaya L.	$2n=2x=18$	14m+2sm+2st（2SAT）	2A	34	1.77	0.22	59.3	$10M_2+6M_1+2S$
	68. 石竹属 Dianthus L.	68-1 石竹 Dianthus chinensis L.	$2n=2x=30$	16m+14sm	2A	60	1.67	0.27	62.96	$2L+8M_2+20M_1$
		68-2 瞿麦 Dianthus superbus L.	$2n=2x=30$	18m+12sm	2A	60	1.35	0.27	60.98	$14M2+16M1$
	69 石头花属 Gypsophila L.	69 长蕊石头花 Gypsophila oldhamiana Miq.	$2n=2x=36$							
二十九、石竹科 Caryophyllaceae	70. 剪秋罗属 Lychnis L.	70-1 大花剪秋罗 Lychnis fulgens Fisch	$2n=2x=24$	22m+2sm	1A	48	1.33	0	55.6	$14M_1+10M_1$
	71. 孩儿参属 Pseudostellaria Pax	71-1 太子参 Pseudostellaria heterophylla (Miq.) Pax ex Pax et Hoffm.	$2n=32$	26m+6sm	2A	64	1.5	0.06	58.48	$14M_1+18M_1$
	72. 肥皂草属 Saponaria L.	72-1 肥皂草 Saponaria officinalis L.	$2n=2x=28$	12m+10sm+4st+2t	2A	50	1.62	0.36	66.03	$2L+10M_2+16M_1$
	73. 王不留行属 Vaccaria Medic	73-1 王不留行 Vaccaria segetalis (Neck.) Garcke	$2n=2x=30$	16m(2SAT)+12sm+2st	2A	58	1.4	0.2	62.48	$18M2+12M1$
三十、卫矛科 Celastraceae	74. 南蛇藤属 Celastrus L.	74-1 南蛇藤 Celastrus orbiculatus Thunb.	$2n=46$							
三十一、藜科 Chenopodiaceae	75. 地肤属 Kochia Roth	75-1 地肤 Kochia scoparia (L.) Schrad.	$2n=2x=18$	16m+2sm	2A	36	1.29	0.11	56.16	$8M_2+10M_1$
三十二、金粟兰科 Chloranthaceae	76. 草珊瑚属 Sarcandra Gardner	76-1 草珊瑚 Sarcandra glabra (Thunb.) Nakai	$2n=2x=30$	12m+6sm+12st（2SAT）	3B	48	2.09	0.53	69.33	$6L+6M_2+12M_1+6S$
三十三、鸭跖草科 Commelinaceae	77. 鸭跖草属 Commelina L.	77-1 鸭跖草 Commelina communis L.	$2n=88$	32m+50sm(2SAT)+6st	2B	170	2.46	0.45	65.36	$8L+32M_2+34M_1+14S$

续表

中药植物核型分析

科	属	种	染色体数目	核型公式	核型分类	臂指数	最长染色体/最短染色体	臂比>2的染色体比例	核型不对称系数	染色体相对长度组成
三十四、使君子科 Combretaceae	73. 榄仁树属 Terminalia L.	78-1 诃子 Terminalia chebula Retz.	2n=48	42m+6sm	2A	96	1.73	0.04	57.96	$2L+12M_2+34M_1$
	79. 蓍属 Achillea L.	79-1 蓍草 Achillea alpina L.	2n=72							
	80. 牛蒡属 Arctium L.	80-1 牛蒡 Arctium lappa L.	2n=2x=36	10m+16sm+8st+2t (2SAT)	3B	62	2.34	0.67	69.03	$6L+4M_2+24M_1+2S$
	81. 蒿属 Artemisia L.	81-1 艾蒿 Arctium argyi Levl.et Vant	2n=2x=34	24m+8sm+2st	2A	66	1.96	0.12	60.77	$2L+10M_2+22M_1$
		81-2 菁蒿 Arctium carvifolia Buch.-Ham. ex Roxb.	2n=2x=18	14m+4sm	2A	36	1.37	0.11	59.29	$10M_2+8M_1$
		81-3 牡蒿 Arctium japonica Thunb.	2n=36	28m+4sm+2st+2t	2A	68	1.87	0.17	61.59	$2L+16M_2+18M_1$
	82. 紫菀属 Aster L.	82-1 紫菀 Aster tataricus L. f.	2n=54	48m+6sm	2A	108	1.76	0.11	59.16	$2L+20M_2+28M_1+4S$
	83. 苍术属 Atractylodes DC.	83-1 北苍术 Atractylodes chinensis (DC.) Koidz.	2n=2x=24	6m+14sm+4st	2B	44	2.23	0.42	66.76	$4L+6M_2+10M_1+4S$
		83-2 关苍术 Atractylodes japonica Koidz. ex Kitam	2n=2x=24	12m+12sm	2A	48	1.99	0.25	63.66	$4L+4M_2+14M_1+2S$
		83-3 苍术 Atractylodes lancea (Thunb.) DC.	2n=2x=24	8m+12sm+4st	3A	44	1.92	0.58	68.39	$6L+4M_2+14M_1$
		83-4 白术 Atractylodes macrocephala Koidz.	2n=2x=24	8m+14sm+2st (2SAT)	2A	46	1.93	0.42	66.17	$4L+4M_2+16M_1$
三十五、菊科 Compositae	84. 云木香属 Aucklandia Falc.	84-1 木香 Aucklandia lappa Decne.	2n=2x=36	10m+12sm+12st (2SAT) +2t	3B	58	2.2	0.61	71.14	$6L+6M_2+20M_1+4S$
	85. 鬼针草属 Bidens L.	85-1 鬼针草 Bidens pilosa L.	2n=72	28m+26sm (1SAT) +16st (3SAT) +2T	2A	128	1.8	0.42	67.76	$2L+32M_2+36M_1+2S$
	86. 艾纳香属 Blumea DC.	86-1 香艾草 Bidens balsamifera (L.) DC.	2n=2x=12	2m+10sm	3A	24	1.29	0.83	69.56	$6M_2+6M_1$
	87. 金盏花属 Calendula L.	87-1 金盏花 Calendula officinalis L.	2n=32	8m+16sm+6st (2SAT) +2t (2SAT)	3B	56	2.33	0.63	67.59	$4L+12M_2+12M_1+4S$
	88. 金挖耳属 Carpesium L.	88-1 大花金挖耳 Carpesium macrocephalum Franch et Sav.	2n=40	20m+14sm+6st	2B	74	2.73	0.25	64.83	$4L+10M_2+18M_1+8S$
	89. 红花属 Carthamus L.	89-1 红花 Carthamus tinctorius L.	2n=2x=24	12m+8sm (2SAT) +4st (4SAT)	2A	44	1.43	0.42	66.41	$10M_2+14M_1$
	90. 蓟属 Cirsium Mill.	89-1 大蓟 Cirsium japonicum DC.	2n=2x=34	14m+14sm+6st(2SAT)	2A	62	1.9	0.35	64.51	$4L+8M_2+22M_1$
		90-2 烟管蓟 Cirsium pendulum Fisch. ex.DC.	2n=2x=34	22m+2sm+8st+2t	2B	58	2.25	0.29	64.67	$6L+10M_2+12M_1+6S$

续表

中药植物核型分析

科	属	种	染色体数目	核型公式	核型分类	臂指数	最长染色体/最短染色体	臂比>2的染色体比例	核型不对称系数	染色体相对长度组成
三十五、菊科 Compositae	90. 蓟属 Cirsium Mill.	90-3 小蓟 Cirsium setosum (Willd.) MB.	2n=2x=34	20m+12sm (2SAT) +2t (2SAT)	2A	66	1.95	0.35	62.31	4L+10M₂+18M₁+2S
		90-4 绒背蓟 Cirsium vlassovianum Fisch. ex DC.	2n=2x=28	22m+6sm	2B	56	2.03	0.14	59.73	4L+8M₂+12M₁+4S
	91. 白酒草属 Conyza Less.	91-1 小飞蓬 Conyza canadensis (L.) Cronq.	2n=2x=18							
	92. 秋英属 Cosmos Cav.	92-1 波斯菊 Cosmos bipinnata Cav.	2n=2x=24	16m (2SAT) +6sm (2SAT) +2st (2SAT)	2A	46	1.39	0.25	61.88	2L+10M₂+12M₁
	93. 菊属 Dendranthema (DC.) Des. Moul.	93-1 野菊 Dendranthema indicum (L.) Des Moul.	2n=2x=18	16m+2st	2A	34	1.43	0.11	59.82	10M₂+8M₁
	94. 泽兰属 Eupatorium L.	94-1 佩兰 Eupatorium fortunei Turcz	2n=3x=30	3m+27sm	2A	2A	1.9	0.5	67.04	3L+12M₂+9M₁+6S
	95. 三七草属 Gynura Cass.	95-1 三七草 Gynura segetum (Lour.) Merr.	2n=2x=20	20m	1A	40	1.74	0	57.26	2L+8M₂+8M₁+2S
	96. 泥胡菜属 Hemistepta Bunge	96-1 泥胡菜 Hemistepta lyrata Bunge	2n=28	18m (2SAT) +6sm+4st	2A	52	1.56	0.21	63.22	2L+12M₂+14M₁
	97. 狗娃花属 Heteropappus Less.	97-1 狗娃花 Heteropappus hispidus (Thunb.) Less.	2n=36							
	98. 旋覆花属 Inula L.	98-1 土木香 Inula helenium L.	2n=2x=20	14m+6sm (2SAT)	2A	40	1.81	0.3	62.71	2L+8M₂+8M₁+2S
		98-2 旋覆花 Inula japonica Thunb.	2n=2x=16	12m+2sm (2SAT) +2st	2A	30	1.83	0.25	60.67	4L+12M₁
		98-3 条叶旋覆花 Inula linearifolia Turcz.	2n=2x=16	12m+4sm (2SAT)	2B	32	2.05	0.13	60.56	4L+4M₂+4M₁+4S
	99. 苦荬菜属 Ixeris Cass	99-1 苦荬菜 Ixeris denticulata (Houtt.) Stebb	2n=2x=18	10m+8sm	2A	36	1.52	0.33	62.96	2L+6M₂+10M₁
		99-2 多头苦荬 Ixeris polycephala Cass.	2n=2x=14	8m+6sm	1A	28	1.4	0	61.09	6M₂+8M₁
		99-3 抱茎苦荬菜 Ixeris sonchifolia Hance	2n=48	18m+22sm (6SAT) +4st+4t	2A	88	1.95	0.38	66.25	4L+16M₂+24M₁+4S
	100. 马兰属 Kalimeris Cass.	100-1 马兰 Kalimeris indica (L.) Sch.-Bip	2n=40	18m+14sm+8st (2SAT)	2A	72	1.76	0.3	64.34	2L+14M₂+24M₁
	101. 大丁草属 Leibnitzia Cass.	101-1 大丁草 Leibnitzia anandria (L.) Nakai	2n=2x=46	18m+10sm+18st	3B	74	2.83	0.57	67.82	8L+6M₂+26M₁+6S
	102. 大翅蓟属 Onopordum L.	102-1 大翅蓟 Onopordum acanthium L.	2n=2x=34	22m (2SAT) +12sm	2A	68	1.85	0.12	61.07	2L+14M₂+16M₁+2S

续表

科	属	中药植物核型分析								
		种	染色体数目	核型公式	核型分类	臂指数	最长染色体/最短染色体	臂比>2的染色体比例	核型不对称系数	染色体相对长度组成
三十五、菊科 Compositae	103. 金光菊属 Rudbeckia L.	103-1 黑心菊 Rudbeckia hirta L.	$2n=76$	60m（6SAT）+16sm	2A	152	1.85	0.16	58.8	$2L+36M_2+34M_1+4S$
	104. 千里光属 Senecio L.	104-1 千里光 Senecio scandens Buch.-Ham.ex D.Don	$2n=2x=20$	14m+6sm	2A	40	1.37	0.1	60.62	$12M_2+8M_1$
	105. 豨莶属 Siegesbeckia L.	105-1 豨莶 Siegesbeckia orientalis L.	$2n=30$	22m+8sm	2A	60	1.51	0.07	58.94	$12M_2+18M_1$
	106. 水飞蓟属 Silybum Adans.	106-1 水飞蓟 Silybum marianum (L.) Gaertn.	$2n=2x=34$	22m+6sm+6st（6SAT）	2A	62	1.77	0.18	63.89	$2L+14M_2+18M_1$
	107. 一枝黄花属 Solidago L.	107-1 一枝黄花 Solidago decurrens Lour.	$2n=2x=18$							
	108. 苦苣菜属 Sonchus L.	108-1 苦苣菜 Sonchus oleraceus L.	$2n=2x=18$	4m+14sm	2A	36	1.4	0.33	65.46	$10M_2+8M_1$
	109. 漏芦属 Stemmacantha Cass.	109-1 祁州漏芦 Stemmacantha uniflora (L.) Ditrich	$2n=66$	16m+36sm+14st（2SAT）	3A	118	1.94	0.64	67.85	$4L+28M_2+30M_1+4S$
	110. 山牛蒡属 Synurus Iljin	110-1 山牛蒡 Synurus deltoides (Ait.) Nakai	$2n=2x=34$	10m+12sm+12st（3SAT）	3A	56	1.87	0.59	70.25	$8L+4M_2+20M_1+2S$
	111. 万寿菊属 Tagetes L.	111-1 孔雀草 Tagetes patula L.	$2n=48$	6m（2SAT）+32sm+8st+2t	3B	86	2.32	0.79	71.76	$6L+18M_2+12M_1+12S$
	112. 菊蒿属 Tanacetum L.	112-1 菊蒿 Tanacetum vulgare L.	$2n=2x=18$	12m+4sm+2st	2A	36	1.38	0.11	59.3	$8M_2+10M_1$
	113. 蒲公英属 Taraxacum Weber.	113-1 异苞蒲公英 Taraxacum heterolepis Nakai et Koidz. ex Kitag	$2n=3x=24$	12m+12sm（3SAT）	2A	48	1.81	0.13	63.2	$3L+6M_2+15M_1$
		113-2 蒲公英 Taraxacum mongolicum Hand.-Mazz	$2n=3x=24$	21m（2SAT）+3sm	1A	48	1.59	0	58.13	$3L+9M_2+12M_1$
	114. 款冬属 Tussilago L.	114-1 款冬 Tussilago farfara L.	$2n=60$	30m+22sm+8st	2B	112	2.06	0.27	63.73	$4L+26M_2+26M_1+4S$
	115. 苍耳属 Xanthium L.	115-1 苍耳 Xanthium sibiricum Patr.	$2n=36$	26m+6sm+2st+2t（2SAT）	2A	68	1.54	0.28	62.41	$18M+18M_1$
	116. 白鹤藤属 Argyreia Lour.	116-1 头花银背藤 Argyreia capitata (Vahl) Arn. ex Choisy	$2n=2x=26$	18m（2SAT）+6sm+2st	2A	50	1.72	0.23	61.39	$2L+12M_2+12M_1$
三十六、旋花科 Convolvulaceae	117. 旋花属 Convolvulus L.	117-1 田旋花 Convolvulus arvensis L.	$2n=78$							
	118. 菟丝子属 Convolvulus L.	118-1 菟丝子 Convolvulus chinensis Lam.	$2n=2x=30$	22m+6sm+2st	2A	58	1.83	0.27	59.17	$2L+14M_2+8M_1+6S$

续表

中药植物核型分析

科	属	种	染色体数目	核型公式	核型分类	臂指数	最长染色体/最短染色体	臂比>2的染色体比例	核型不对称系数	染色体相对长度组成
三十六、旋花科 Convolvulaceae	119. 马蹄金属 Dichondra J. R. Forst. et G. Forst.	119-1 马蹄金 Dichondra micrantha Urb.	$2n=2x=16$	10m+6sm	2A	32	1.47	0.13	60.88	$2L+4M_2+10M_1$
三十七、山茱萸科 Cornaceae	120. 梾木属 Cornus L.	120-1 山茱萸 Cornus officinalis Sieb. et Zucc.	$2n=2x=18$	8m+10sm	2A	36	1.51	0.33	64.41	$2L+4M_2+12M_1$
三十八、景天科 Crassulaceae	121. 景天属 Sedum L.	121-1 景天三七 Sedum aizoon L.	$2n=64$	2M+42m+20sm	2A	128	1.52	0.16	60.2	$30M_2+34M_1$
		121-2 费菜 Sedum kamtschaticum Fisch. et C. A. Mey.	$2n=112$							
		121-3 白景天 Sedum pallescens Freyn	$2n=44$							
		121-4 长药景天 Sedum spectabile Boreau	$2n=50$	22m+28sm	2B	100	2.13	0.24	63.01	$4L+16M_2+28M_1+2S$
	122. 南芥属 Arabis L.	122-1 垂果南芥 Arabis pendula L.	$2n=30$	8m+20sm (2SAT) +2st	2A	58	1.59	0.33	65.46	$2L+12M_2+16M_1$
	123. 芸苔属 Brassica L.	123-1 白芥 Brassica alba L.	$2n=2x=24$	14m+8sm+2st (2SAT)	2A	46	1.76	0.17	63.85	$2L+12M_2+8M_1+2S$
		123-2 芥菜 Brassica juncea (L.) Czern. et Coss.	$2n=2x=24$	8m+16sm	2A	48	1.87	0.42	64.27	$2L+10M_2+10M_1+2S$
三十九、十字花科 Cruciferae	124. 荠属 Capsella Medik.	124-1 荠菜 Capsella bursa-pastoris (L.) Medik.	$2n=32$							
	125. 菘蓝属 Isatis L.	125-1 板蓝根 Isatis indigotica Fortune	$2n=2x=14$	10m+4sm	2A	28	1.38	0.14	61.08	$8M2+6M1$
	126. 独行菜属 Lepidium L.	126-1 独行菜 Lepidium apetalum Willd	$2n=2x=28$	6m+16sm+4t (2SAT) +2T	3B	50	2	0.79	70.98	$4L+8M_2+12M_1+4S$
	127. 萝卜属 Raphanus L.	127-1 萝卜 R. sativus L	$2n=2x=18$	12m+6sm (2SAT)	2A	36	1.6	0.11	61.03	$10M_2+4M_1+4S$
四十、葫芦科 Cucurbitaceae	128. 冬瓜属 Benincasa Savi	128-1 冬瓜 Benincasa hispida(Thunb.)Cogn	$2n=2x=24$	12m+12sm (4SAT)	2A	48	1.66	0.25	61.7	$4L+4M_2+16M_1$
	129. 假贝母属 Bolbostemma Franquet	129-1 土贝母 Bolbostemma paniculatum (Maxim.) Franquet	$2n=2x=22$	10m+8sm+4st	2A	40	1.71	0.36	65.26	$2L+10M_2+8M_1+2S$
	130. 葫芦属 Lagenaria Ser.	130-1 葫芦 Lagenaria siceraria (Molina) Standl.	$2n=2x=22$	16m+6sm	2A	44	1.62	0.27	63.09	$2L+12M_2+8M_1$
	131. 苦瓜属 Momordica L.	131-1 木鳖 Momordica cochinchinensis (Lour.) Spreng	$2n=2x=28$	16m+12sm (2SAT)	2A	56	1.37	0.14	61.46	$12M_2+16M_1$

续表

中药植物核型分析

科	属	种	染色体数目	核型公式	核型分类	臂指数	最长染色体/最短染色体	臂比>2的染色体比例	核型不对称系数	染色体相对长度组成
四十、葫芦科 Cucurbitaceae	132. 罗汉果属 Siraitia Merr.	132-1 罗汉果 Siraitia grosvenorii (Swingle) C. Jeffrey ex Lu et Zhi Y. Zhang	$2n=2x=28$	4m+14sm（2SAT）+8st+2t	3A	46	1.92	0.57	70.6	$4L+10M_2+10M_1+4S$
	133. 赤瓟属 Thladiantha Bunge	133-1 赤瓟 Thladiantha dubia Bge.	$2n=2x=18$	4m+4sm+2st+8T	3A	26	1.51	0.67	79.06	$2L+8M_2+8M_1$
	134. 栝楼属 Trichosanthes L.	134-1 蛇瓜 Trichosanthes anguina L.	$2n=2x=22$	16m+6sm	2A	44	1.58	0.18	59.79	$4L+8M_2+10M_1$
		134-2 栝楼 Trichosanthes kirilowii Maxim.	$2n=8x=88$	28m+56sm(1SAT)+4st	2B	172	2.3	0.48	66.27	$8L+26M_2+50M_1+4S$
		134-3 中华栝楼 Trichosanthes rosthornii Harms	$2n=2x=22$	10m+10sm+2t（1SAT）	2A	42	1.87	0.45	64.71	$2L+6M_2+14M_1$
四十一、柏科 Cupressaceae	135. 侧柏属 Platycladus Spach	135-1 侧柏 Platycladus orientalis (L.)Franco	$2n=2x=22$	20m+2sm（2SAT）	2A	44	1.85	0.09	55.71	$12M_2+8M_1+2S$
四十二、莎草科 Cyperaceae	136. 薹草属 Carex L.	136-1 东陵薹草 Carex tangiana Ohwi	$2n=2x=24$							
		137-1 翅茎薯蓣 Dioscorea alata L.	$2n=80$							
四十三、薯蓣科 Dioscoreaceae	137. 薯蓣属 Dioscorea L.	137-2 野山药 Dioscorea doryophora Hance	$2n=122$							
		137-3 穿龙薯蓣 Dioscorea nipponica Makino	$2n=2x=20$	10m+10sm（2SAT）	2A	40	1.26	0.3	61.06	$10M_2+10M_1$
		137-4 薯蓣 Dioscorea opposita Thunb	$2n=122$							
四十四、川续断科 Dipsacaceae	138. 川续断属 Dipsacus L.	138-1 川续断 Dipsacus asperoides C. Y. Cheng et T. M. Ai	$2n=2x=18$	8m+6sm+4t（4SAT）	3A	32	1.41	0.56	70.11	$8M_2+10M_1$
四十五、龙血树科 Dracaenaceae	139. 龙血树属 Dracaena Vand. ex L.	139-1 龙南龙血树 Dracaena angustifolia Roxb.	$2n=40$							
		139-2 海南龙血树 Dracaena cambodiana Pierre ex Gagnep	$2n=40$	24m+14sm+2st	2B	78	2.21	0.25	62.48	$4L+12M_2+20M_1+4S$
四十六、茅膏菜科 Droseraceae	140. 茅膏菜属 Drosera L.	140-1 锦地罗 Drosera burmannii Vahl	$2n=80$	30m+18sm+16st+4t+12T	3A	128	1.71	0.58	73.08	$4L+30M_2+46M_1$
四十七、柿科 Ebenaceae	141. 柿属 Diospyros L.	141-1 君迁子 Diospyros lotus L. var. lotus	$2n=32$							
四十八、胡颓子科 Elaeagnaceae	142. 胡颓子属 Elaeagnus L.	142-1 胡颓子 Elaeagnus pungens Thunb	$2n=2x=28$	8m+14sm+6st	2B	54	3.1	0.43	67.86	$4L+2M_2+20M_1+2S$
四十九、杜鹃花科 Ericaceae	143. 杜鹃花属 Rhododendron L.	143-1 兴安杜鹃 Rhododendron dauricum L.	$2n=2x=26$							
五十、杜仲科 Eucommiaceae	144. 杜仲属 Eucommia Oliv.	144-1 杜仲 Eucommia ulmoides Oliv.	$2n=2x=34$	10m+10sm+4st+2t+8T	3A	54	1.79	0.65	74.87	$4L+8M_2+20M_1+2S$

续表

中药植物核型分析

科	属	种	染色体数目	核型公式	核型分类	臂指数	最长染色体/最短染色体	臂比>2的染色体比例	核型不对称系数	染色体相对长度组成
	145. 巴豆属 Croton L.	145-1 巴豆 Croton tiglium L.	$2n=2x=20$	6m+14sm	2A	40	1.41	0.5	63.91	$10M_2+10M_1$
	146. 大戟属 Euphorbia L.	146-1 续随 Euphorbia lathyris L.	$2n=2x=20$	20m (2SAT)	1A	40	1.74	0	58.02	$4L+2M_2+14M_1$
		146-2 林大戟 Euphorbia lucorum Rupr.	$2n=51$	10m+38sm+3st	2B	99	3.34	0.5	65.98	$6L+14M_2+18M_1+13S$
		146-3 大戟 Euphorbia pekinensis Rupr.	$2n=2x=20$	8m+10sm+2T	2B	38	2.43	0.5	66.11	$4L+6M_2+8M_1+2S$
		146-4 绿玉树 Euphorbia tirucalli L.	$2n=2x=20$	12m+8sm (2SAT)	2A	40	1.27	0.3	61.88	$8M_2+12M_1$
	147. 白饭树属 Flueggea Willd	147-1 一叶萩 Flueggea suffruticosa (Pall.) Baill.	$2n=2x=26$	10m+14sm+2st	2A	50	1.35	0.38	64.47	$12M_2+14M_1$
	148. 麻风树属 Jatropha L.	148-1 佛肚树 Jatropha podagrica Hook	$2n=2x=22$	18m+4st (1SAT)	2A	40	1.8	0.18	59.83	$2L+10M_2+6M_1+4S$
五十一、大戟科 Euphorbiaceae	149. 红雀珊瑚属 Pedilanthus Neck. ex Poit.	149-1 大银龙 Pedilanthus tithymaloides (L.) Poit.	$2n=33$							
	150. 叶下珠属 Phyllanthus L.	150-1 余甘子 Phyllanthus emblica L.	$2n=104$							
		150-2 菁灰叶下珠 Phyllanthus glaucus Wall.	$2n=100$							
		150-3 叶下珠 Phyllanthus urinaria Linn	$2n=2x=26$	10m+14sm+2t (2SAT)	3B	50	2.83	0.54	66.77	$4L+8M_2+10M_1+4S$
	151. 蓖麻属 Ricinus L.	151-1 蓖麻 Ricinus communis L.	$2n=2x=20$	14m (2SAT) +6sm	1A	40	1.34	0	58.35	$8M_2+12M_1$
五十二、紫堇科 Fumariaceae	152. 乌桕属 Sapium P.Browne	152-1 乌桕 Sapium sebiferum (L.) Roxb.	$2n=88$	30m+40sm (2SAT) +10st+8T	2B	158	2.12	0.43	68	$6L+36M_2+38M_1+8S$
	153. 紫金龙属 Dactylicapnos Wall	153-1 紫金龙 Dactylicapnos scandens (D. Don) Hutch.	$2n=2x=16$	14m+2sm	1A	32	1.43	0	60.54	$6M_2+10M_1$
五十三、龙胆科 Gentianaceae	154. 龙胆属 Gentiana L.	154-1 秦艽 Gentiana macrophylla Pall	$2n=52$	32m+18sm+2st	2B	102	2.26	0.19	62.61	$4L+18M_2+26M_1+4S$
五十四、牻牛儿苗科 Geraniaceae	155. 老鹳草属 Geranium L.	155-1 老鹳草 Geranium wilfordii Maxim	$2n=2x=28$	26m+2sm	1A	56	1.48	0	57.22	$2L+12M_2+14M_1$
	156. 天竺葵属 Pelargonium L'Hérit	156-1 香叶天竺葵 Pelargonium graveolens L'Hérit	$2n=30$	4m+22sm (1SAT) +4st (2SAT)	3B	56	2.09	0.73	69.71	$2L+14M_2+12M_1+2S$
五十五、银杏科 Ginkgoaceae	157. 银杏属 Ginkgo L.	157-1 银杏 Ginkgo biloba L.	$2n=2x=24$	4m (2SAT) +8sm (1SAT) +12st	3A	36	1.64	0.83	72.88	$2L+8M_2+14M_1$

续表

中药植物核型分析

科	属	种	染色体数目	核型公式	核型分类	臂指数	最长染色体/最短染色体	臂比>2的染色体比例	核型不对称系数	染色体相对长度组成
五十六、禾本科 Gramineae	158. 芦竹属 Arundo L.	158-1 芦竹 Arundo donax L.	$2n=72$	$40m+26sm+6st$（2SAT）	2B	138	2.01	0.25	62.75	$6L+22M_2+40M_1+4S$
	159. 薏苡属 Coix L.	159-1 薏苡 Coix lacryma-jobi L.	$2n=2x=20$	$12m$（2SAT）$+8sm$	2A	40	1.34	0.2	60.38	$8M_2+12M_1$
		159-2 川谷 Coix lacryma-jobi L. var. mayuen (Roman.) Stapf ex Backer	$2n=2x=20$	$18m+2sm$	1A	40	1.46	0	57.02	$10M_2+10M_1$
	160. 白茅属 Imperata Cyrillo	160-1 白茅 Imperata cylindrica(L.)P. Beauv.	$2n=2x=20$	$14m+2sm+4st$（2SAT）	2A	36	1.49	0.3	64.05	$8M_2+12M_1$
	161. 淡竹叶属 Lophantherum Brongn.	161-1 淡竹叶 Lophantherum gracile Brongn.	$2n=54$	$46m+8sm$	2A	108	1.69	0.11	58.26	$2L+26M_2+26M_1$
	162. 芒属 Miscanthus Andersson	162-1 芒 Miscanthus sinensis Andersson	$2n=2x=18$	$6m+12sm$	3B	36	2.03	0.56	65.85	$4L+4M_2+6M_1+4S$
	163. 芦苇属 Phragmites Adans	163-1 芦苇 Phragmites communis Trin.	$2n=4x=48$, $2n=6x=72$, $2n=8x=96$								
	164. 香根草属 Vetiveria Bory.	164-1 香根草 Vetiveria zizanioides (L.) Nash	$2n=2x=20$	$20m$	1A	40	1.53	0	56.35	$8M_2+12M_1$
五十七、金缕梅科 Hamamelidaceae	165. 枫香树属 Liquidambar L.	165-1 枫香树 Liquidambar formosana Hance	$2n=2x=26$	$6m+16sm+4st$	3B	48	2.78	0.62	66.44	$6L+4M_2+10M_1+6S$
	166. 射干属 Belamcanda Adans.	166-1 射干 Belamcanda chinensis（L.）DC.	$2n=32$	$20m+12sm$	2A	64	1.75	0.19	61.22	$4L+10M_2+16M_1+2S$
五十八、鸢尾科 Iridaceae	167. 番红花属 Crocus L.	167-1 番红花 Crocus sativus L.	$2n=24$	$10m+8sm$（1SAT）$+6st$（4SAT）	2A	42	1.71	0.5	67.09	$10M_2+12M_1+2S$
	168. 鸢尾属 Iris L.	168-1 马蔺 Iris ensata Thunb	$2n=3x=33$	$21m+6sm+3st+3t$	2B	60	2.08	0.27	63.1	$6L+6M_2+18M_1+3S$
		168-2 香根鸢尾 Iris pallida Lamarck	$2n=44$	$22m+22sm$	2A	88	1.91	0.32	63.83	$4L+16M_2+22M_1+2S$
五十九、灯心草科 Juncaceae	169. 灯心草属 Juncus L.	169-1 假灯心草 Juncus setchuensis Buchen. var. effusoides Buchen.	$2n=40$							
六十、唇形科 Labiatae	170. 藿香属 Agastache Clayt.	170-1 藿香 Agastache rugosus（Fisch. et Meyer）Kuntze.	$2n=2x=18$	$2m+14sm+2t$（2SAT）	3B	34	2.53	0.56	70.05	$4L+4M_2+6M_1+4S$
	171. 广防风属 Epimeredi Adans.	171-1 广防风 Epimeredi indica（L.）Rothm.	$2n=2x=34$	$10m$（2SAT）$+24sm$	2A	68	1.71	0.24	65.05	$2L+14M_2+18M_1$

续表

中药植物核型分析

科	属	种	染色体数目	核型公式	核型分类	臂指数	最长染色体/最短染色体	臂比>2的染色体比例	核型不对称系数	染色体相对长度组成
	172. 活血丹属 Glechoma L.	172-1 活血丹 Glechoma longituba (Nakai) Kupr.	$2n=2x=24$	4m+6sm+2st+12T	3A	34	1.81	0.83	82.31	$4L+6M_2+12M_1+2S$
	173. 神香草属 Hyssopus L.	173-1 驱蚊草 Pelargonium×Citrenella	$2n=36$	28m (2SAT) +6sm+2st (2SAT)	2A	70	1.51	0.11	59.73	$18M_2+18M_1$
	174. 夏至草属 Lagopsis Bunge	174-1 夏至草 Lagopsis supine (Steph.) Ik.-Gal.	$2n=2x=22$							
	175. 野芝麻属 Lamium L.	175-1 野芝麻 Lamium barbatum Sieb. et Zucc	$2n=32$	10m+16sm+6st	2A	58	1.65	0.5	66.52	$2L+12M_2+18M_1$
	176. 薰衣草属 Lamium L.	176-1 薰衣草 Lamium angustifolia Mill	$2n=2x=18$	16m (1SAT) +2sm	2A	36	1.77	0.11	58.47	$2L+6M_2+8M_1+2S$
	177. 益母草属 Leonurus L.	177-1 欧益母草 Leonurus cardiaca L.	$2n=2x=18$	2m+12sm+4st	3B	32	2.94	0.89	71.6	$2L+6M_2+6M_1+4S$
		177-2 益母草 Leonurus japonicus Houtt	$2n=2x=20$	6m+10sm+4st	3A	36	1.44	0.6	69.86	$8M_2+12M_1$
	178. 蜜蜂花属 Melissa L.	178-1 蜜蜂花 Melissa axillaris (Benth.) Bakh. f	$2n=2x=34$	12m+20sm(2SAT)+2st	2A	66	1.9	0.47	65.35	$4L+8M_2+20M_1+2S$
六十、唇形科 Labiatae	179. 薄荷属 Mentha L.	179-1 薄荷 Mentha haplocalyx Briq.	$2n=72$	32m+30sm+10st (2SAT)	2A	134	1.8	0.42	66.33	$6L+28M_2+32M_1+6S$
	180. 美国薄荷属 Monarda L.	180-1 美国薄荷 Monarda didyma L.	$2n=2x=34$							
	181. 荆芥属 Monarda L.	181-1 荆芥 Monarda cataria L.	$2n=2x=12$	6m+6sm (2SAT)	2A	24	1.48	0.33	64.23	$4M_2+8M_1$
	182. 罗勒属 Monarda L.	182-1 丁香罗勒 Monarda gratissimum L.	$2n=4x=40$	12m+12sm+8st+2t (2SAT) +6T	3A	64	1.81	0.55	71.75	$2L+16M_2+18M_1+4S$
	183. 紫苏属 Perilla L.	183-1 紫苏 Perilla frutescens (L.) Britt.	$2n=2x=20$	16m+4st (2SAT)	3A	36	1.46	0.8	70.26	$10M_2+10M_1$
		183-2 回回苏 Perilla frutescens (L.) Britton var. crispa (Benth.) Decne.	$2n=2x=34$	10m (2SAT) +20sm (2SAT) +4st (1SAT)	2A	64	1.8	0.47	66.5	$2L+10M_2+22M_1$
	184. 糙苏属 Phlomis L.	184-1 块茎糙苏 Phlomis tuberosa L.	$2n=2x=22$							
	185. 刺蕊草属 Pogostemon Desf.	185-1 广藿香 Pogostemon cablin (Blance) Benth	$2n=64$							
	186. 夏枯草属 Prunella L.	186-1 大花夏枯草 Prunella grandiflora (L.) Jacq.	$2n=2x=28$	22m (1SAT) +6sm	2A	56	1.89	0.14	59.42	$4L+6M_2+16M_1+2S$
		186-2 夏枯草 Prunella vulgaris L.	$2n=2x=28$	10m+14sm+4st	2A	52	1.8	0.29	64	$2L+12M_2+12M_1+2S$

续表

科	属	种	中药植物核型分析							
			染色体数目	核型公式	核型分类	臂指数	最长染色体/最短染色体	臂比>2的染色体比例	核型不对称系数	染色体相对长度组成
六十、唇形科 Labiatae	187. 香茶菜属 Rabdosia (Blume) Hassk.	187-1 细叶香茶菜 Rabdosia ternifolia (D. Don) Hara	$2n=48$							
	188. 鼠尾草属 Salvia L.	188-1 丹参 Salvia miltiorrhiza Bge	$2n=2x=16$	10m+6sm (1SAT)	2A	32	1.64	0.25	63.74	$2L+2M_2+12M_1$
		188-2 荔枝草 Salvia plebeia R.Br.	$2n=2x=16$	2m+12sm+2st	3A	30	1.54	0.75	67.85	$8M_2+8M_1$
	189. 黄芩属 Scutellaria L.	189-1 黄芩 Scutellaria baicalensis Georgi	$2n=2x=18$	12m+6sm (2SAT)	2A	36	1.46	0.33	62.71	$8M_2+10M_1$
		189-2 甘肃黄芩 Scutellaria rehderiana Dieis	$2n=2x=26$	12m+8sm+6st (2SAT)	2B	46	2.06	0.31	64.79	$6L+2M_2+16M_1+2S$
	190. 百里香属 Thymus L.	190-1 百里香 Thymus mongolicus Ronn.	$2n=2x=28$							
六十一、木通科 Lardizabalaceae	191. 木通属 Akebia Decne	191-1 三叶木通 Akebia trifoliata (Thunb.) Koidz.	$2n=64$							
六十二、樟科 Lauraceae	192. 樟属 Cinnamomum Trew Nees & Eberm.	192-1 樟树 Cinnamomum camphora (L.)	$2n=2x=24$	18m (4SAT) +6sm	2A	48	1.91	0.08	59.33	$6L+4M_2+10M_1+4S$
	193. 新木姜子属 Neolitsea Merr.	193-1 勐腊新木姜子 Neolitsea aurata (Hayata) Koidz	$2n=2x=24$	6m+12sm (2SAT) +6st	3A	44	1.99	0.58	69.47	$6L+6M_2+8M_1+4S$
六十三、火筒树科 Leeaceae	194. 火筒树属 Leea L.	194-1 火筒树 Leea macrophylla Roxb. ex Hornem.	$2n=48$	28m+18sm+2st	2A	94	1.86	0.17	63.17	$2L+22M_2+22M_1+2S$
	195. 相思子属 Abrus L.	195-1 广州相思子 Abrus cantoniensis Hance	$2n=2x=22$	18m+2sm+2st	2A	42	1.72	0.09	60.88	$2L+10M_2+8M_1+2S$
	196. 合欢属 Albizzia Durazz	196-1 合欢 Albizzia julibrissin Durazz	$2n=2x=26$	12m+12sm+2st	2A	50	1.37	0.38	64.4	$14M_2+12M_1$
		197-1 华黄芪 Astragalus chinensis L.	$2n=2x=38$	24m+12sm+2st	2B	74	2.01	0.21	61.45	$6L+10M_2+20M_1+2S$
		197-2 扁茎黄芪 Astragalus complanatus R. Br	$2n=2x=16$	2m+6sm+6st+2t (2SAT)	3B	24	2.15	0.63	73.83	$10M_2+4M_1+2S$
六十四、豆科 Leguminosae	197. 黄芪属 Astragalus L.	197-3 膜荚黄芪 Astragalus membranaceus (Fisch.) Bge	$2n=2x=16$	6m (2SAT) +8sm+2st	3B	30	3.21	0.63	68.25	$4L+4M_2+6M_1+2S$
		197-4 蒙古黄芪 Astragalus membranaceus (Fisch.) Bge. var. mongholicus (Bge.) Hsiao	$2n=2x=16$	10m (2SAT) +6sm	1B	32	2.23	0	61.7	$2L+6M_2+6M_1+2S$
		197-5 软毛黄芪 Astragalus mollissimus Gontsch.	$2n=2x=16$	10m+6sm	2A	32	1.26	0.25	62.11	$8M_2+8M_1$
	198. 羊蹄甲属 Bauhinia L.	198-1 多脉叶羊蹄甲 Bauhinia pernervosa L. Chen	$2n=2x=28$	22m+6sm	2A	56	1.44	0.14	60.32	$16M_2+12M_1$

续表

中药植物核型分析

科	属	种	染色体数目	核型公式	核型分类	臂指数	最长染色体/最短染色体	臂比>2的染色体比例	核型不对称系数	染色体相对长度组成
		198-2 红花羊蹄甲 Bauhinia blakeana Dunn	$2n=2x=28$	16m+12sm	2A	56	1.74	0.36	62.89	$2L+10M_2+16M_1$
	199. 木豆属 Cajanus DC.	199-1 木豆 Cajanus cajan (L.) Millisp	$2n=2x=22$	20m+2sm	1A	44	1.46	0	57.9	$14M_2+8M_1$
	200. 刀豆属 Canavalia DC.	200-1 红刀豆 Canavalia maritima (Aubl.) Thou	$2n=2x=22$	10m+12sm	2B	44	2.06	0.18	62.78	$2L+8M_2+10M_1+2S$
	201. 决明属 Cassia L.	201-1 江茳决明 Cassia sophera L.	$2n=2x=14$	4m+10sm	2A	28	1.31	0.14	62.21	$6M_2+8M_1$
		201-2 望江南 Cassia occidentalis L.	$2n=2x=26$	16m+10sm	2A	52	1.52	0.23	61.02	$12M_2+14M_1$
		201-3 决明 Cassia obtusifolia L.	$2n=2x=28$	2m(1SAT)+10sm(2SAT)+12st+4T	3A	42	1.75	0.79	76.49	$2L+14M_2+8M_1+4S$
		201-4 铁刀木 Cassia siamea Lam	$2n=2x=28$	10m+12sm+6st	3A	50	1.69	0.43	65.86	$2L+10M_2+14M_1+2S$
	202. 紫荆属 Cercis L.	202-1 紫荆 Cercis chinensis Bunge	$2n=2x=14$	2m(2SAT)+8sm+4st	3A	24	1.49	0.57	68.7	$10M_2+4M_1$
	203. 蝶豆属 Clitoria L.	203-1 蝶豆 Clitoria ternatea L.	$2n=2x=16$	2m+14sm(2SAT)	2A	32	1.98	0.5	65.99	$6L+2M_2+2M_1+6S$
六十四. 豆科 Leguminosae	204. 猪屎豆属 Crotalaria L.	204-1 假地蓝 Crotalaria ferruginea Grah. ex Benth.	$2n=48$	40m+8sm	2A	96	1.62	0.08	59.26	$2L+20M_2+26M_1$
	205. 山蚂蝗属 Desmodium Desv.	205-1 广东金钱草 Desmodium styracifolium (Osb.) Merr.	$2n=2x=22$	14m+8sm	2A	44	1.76	0.18	60.26	$2L+8M_2+10M_1+2S$
	206. 镶扁豆属 Dolichos L.	206-1 白扁豆 Dolichos lablab L.	$2n=2x=22$	16m+6sm	2A	44	1.87	0.27	62.46	$2L+6M_2+14M_1$
	207. 千斤拔属 Flemingia Roxb.	207-1 大叶千斤拔 Flemingia macrophylla (Willd.) Kuntze ex Merr.	$2n=2x=22$	10m+10sm+2st	2A	42	1.32	0.27	63.21	$8M_2+14M_1$
		207-2 蔓性千斤拔 Flemingia philippinensis Merr. et Rolfe	$2n=2x=22$	14m+8sm	2A	44	1.46	0.18	62.08	$12M_2+10M_1$
		207-3 腺毛千斤拔 Flemingia glutinosa (Prain) Y. T. Wei et S. Lee	$2n=2x=34$	6m+20sm+6st+2t	3A	60	1.96	0.65	70.16	$2L+14M_2+18M_1$
	208. 山羊豆属 Galega L.	208-1 山羊豆 Galega officinalis L.	$2n=2x=16$	6m+10sm	2A	32	1.4	0.38	63.84	$6M_2+10M_1$
	209. 皂荚属 Gleditsia L.	209-1 山皂荚 Gleditsia japonica Miq.	$2n=2x=28$	6m+10sm+4st+2t(2SAT)+6T	3A	44	1.99	0.64	75.77	$4L+10M_2+8M_1+6S$
		209-2 皂荚 Gleditsia sinensis Lam.	$2n=2x=28$	8m+16sm+2st+2T	3A	52	1.74	0.64	67.98	$2L+14M_2+8M_1+4S$

续表

中药植物核型分析

科	属	种	染色体数目	核型公式	核型分类	臂指数	最长染色体/最短染色体	臂比>2的染色体比例	核型不对称系数	染色体相对长度组成
六十四、豆科 Leguminosae	210. 甘草属 Glycyrrhiza L.	210-1 甘草 Glycyrrhiza uralenisis Fisch	$2n=2x=16$	8m+8sm	1A	32	1.23	0	61.77	$10M_2+6M_1$
	211. 木蓝属 Indigofera L.	211-1 尖叶木蓝 Indigofera zollingeriana Miq	$2n=2x=20$	12m+8sm	2A	40	1.62	0.3	63.29	$10M_2+8M_1+2S$
		211-2 马棘 Indigofera pseudotinctoria Matsum.	$2n=2x=16$	14m+2sm	2A	32	1.31	0.13	58.44	$8M_2+8M_1$
		211-3 野青树 Indigofera suffruticosa Mill.	$2n=2x=16$	14m+2sm	1A	32	1.27	0	56.78	$6M_2+10M_1$
	212. 扁豆属 Lablab Adans.	212-1 扁豆 Lablab purpureus (L.) Sweet	$2n=2x=22$	14m+6sm+2st	2A	44	1.83	0.18	61.24	$2L+8M_2+12M_1$
	213. 苜蓿属 Medicago L.	213-1 紫苜蓿 Medicago Sativa L.	$2n=32$	26m+6sm (4SAT)	2A	64	1.48	0.13	59.15	$16M_2+16M_1$
	214. 油麻藤属 Mucuna Adans.	214-1 常春油麻藤 Mucuna sempervirens Hemsl.	$2n=2x=22$	8m+14sm (1SAT)	3B	44	2.3	0.64	64.67	$6L+2M_2+8M_1+6S$
	215. 黄雀儿属 Priotropis Wight et Arn.	215-1 黄雀儿 Priotropis cytisoides (Roxb. ex DC.) Wight et Arn.	$2n=2x=16$	12m+4sm	1A	32	1.49	0	58.94	$8M_2+8M_1$
	216. 补骨脂属 Psoralea L.	216-1 补骨脂 Psoralea corylifolia Franch.	$2n=2x=22$	8m+12sm+2st	3A	42	1.4	0.55	65.81	$8M_2+14M_1$
	217. 葛属 Pueraria DC.	217-1 葛根 Pueraria lobata (Willd.) Ohwi.	$2n=2x=22$	12m+6sm (4SAT) +2st+2T	2A	40	1.78	0.27	63.53	$12M_2+8M_1+2S$
	218. 田菁属 Sesbania Scop	218-1 田菁 Sesbania cannabina (Retz.) Pers.	$2n=3x=24$	15m+6sm+3st	2A	48	1.92	0.25	62.43	$3L+9M_2+9M_1+3S$
	219. 槐属 Sophora L.	219-1 苦豆子 Sophora alopecuroide L.	$2n=36$	20m+12sm (1SAT) +2st+2t	2A	68	1.57	0.28	65.04	$20M_2+16M_1$
		219-2 白刺花 Sophora davidii (Franch.) Skeels	$2n=2x=18$	12m+4sm+2st (2SAT)	2A	34	1.71	0.22	63.14	$2L+6M_2+10M_1$
		219-3 苦参 Sophora flavescens Ait.	$2n=2x=18$	12m+2sm+4st	2A	32	1.57	0.33	61.42	$10M_2+6M_1+2S$
		219-4 越南槐 Sophora tonkinensis Gapnep.	$2n=2x=18$	8m+10sm (1SAT)	2A	36	1.56	0.11	62.87	$8M_2+10M_1$
	220. 葫芦茶属 Tadehagi Ohashi	220-1 葫芦茶 Tadehagi triquetrum (L.) Ohashi	$2n=2x=20$	16m (2SAT) +2sm+2st (2SAT)	2A	38	1.84	0.1	60.66	$2L+8M_2+6M_1+4S$
	221. 野决明属 Thermopsis R. Br.	221-1 小叶野决明 Thermopsis chinensis Benth. ex S. Moore	$2n=2x=18$	4m+10sm+4T	3A	32	1.78	0.78	72.56	$2L+6M_2+10M_1$

续表

中药植物核型分析

科	属	种	染色体数目	核型公式	核型分类	臂指数	最长染色体/最短染色体	臂比>2的染色体比例	核型不对称系数	染色体相对长度组成
六十四、豆科 Leguminosae	222. 车轴草属 Trifolium L.	222-1 白车轴草 Trifolium repens L.	$2n=32$	30m+2sm	1A	64	1.39	0	58.37	$12M_2+20M_1$
	223. 葫芦巴属 Trigonella L.	223-1 葫芦巴 Trigonella foenum-graecum L.	$2n=2x=16$	10sm（2SAT）+6st（2SAT）	3A	26	1.6	0.88	72.48	$8M_2+6M_1+2S$
	224. 紫藤属 Wisteria Nutt	224-1 紫藤 Wisteria sinensis（Sims）Sweet	$2n=4x=32$	24m+8sm	2A	64	1.55	0.06	58.46	$16M_2+16M_1$
	225. 野豌豆属 Vicia L.	225-1 山野豌豆 Vicia amoena Fisch. ex DC.	$2n=2x=12$	2m+8sm+2st	2A	22	1.45	0.5	67.62	$6M_2+6M_1$
		225-2 歪头菜 Vicia unijuga A. Br.	$2n=4x=24$	6m+14sm+4st	2A	44	1.43	0.5	66.2	$10M_2+14M_1$
	226. 肺筋草属 Aletris L.	226-1 粉条儿菜 Aletris spicata（Thunb.）Franch	$2n=2x=18$	14m+2sm+2st	2A	34	1.34	0.22	58.82	$8M_2+10M_1$
	227. 葱属 Allium L.	227-1 野韭菜 Allium ramosum L.	$2n=4x=32$	28m+4sm	2A	64	1.64	0.13	59.01	$14M_2+16M_1+2S$
		227-2 蒜 Allium sativum L.	$2n=2x=16$	10m+2sm+2st（2SAT）+2t（2SAT）	2A	28	1.98	0.25	61.32	$2L+6M_2+4M_1+4S$
		227-3 茖葱 Allium victorialis L.	$2n=4x=32$	26m+4sm+2st	2A	62	1.49	0.13	58.32	$14M_2+18M_1$
	228. 芦荟属 Aloe L.	228-1 油葱 Aloe vera var. chinensis（Haw.）Berg.	$2n=2x=14$	6sm+8st	4B	20	3.41	1	78.75	$8L+6S$
	229. 知母属 Anemarrhena Bunge	229-1 知母 Anemarrhena asphodeloides Bunge	$2n=2x=22$	14m+8sm（2SAT）	2B	44	2.52	0.18	61.35	$6L+10M_1+6S$
六十五、百合科 Liliaceae	230. 天门冬属 Asparagus L.	230-1 天门冬 Asparagus cochinchinensis（Lour.）Merr.	$2n=2x=20$	6m+12sm+2st	3B	38	2.01	0.6	66.29	$4L+8M_2+4M_1+4S$
		230-2 芦笋 Asparagus officinalis L.	$2n=2x=20$	8m+12sm	2B	40	2	0.3	65.48	$4L+8M_2+2M_1+6S$
	231. 吊兰属 Chlorophytum Ker Gawl	231-1 银边吊兰 Chlorophytum capense var. variegatum Hort.	$2n=2x=28$	8m+18sm+2st	2A	54	1.9	0.36	63.38	$4L+8M_2+14M_1+2S$
	232. 万寿竹属 Disporum Salisb.	232-1 金刚草 Disporum ovale Ohwi	$2n=2x=16$	8m+6sm+2st	2B	30	2.06	0.5	66.14	$4L+2M_2+8M_1+2S$
	233. 贝母属 Disporum L.	233-1 伊贝母 Disporum pallidiflora Schrenk	$2n=2x=24$	2m+2sm+8st+12t	3A	28	1.95	0.83	83.24	$4L+4M_2+14M_1+2S$
		233-2 平贝母 Disporum ussuriensis Maxim	$2n=2x=22$	4m+2sm+2st+14t	3B	28	2.48	0.82	80.28	$4L+4M_2+12M_1+2S$
	234. 萱草属 Hemerocallis L.	234-1 黄花菜 Hemerocallis citrina Baroni	$2n=2x=22$	12m+6sm+4st	2A	40	1.77	0.27	63.25	$4L+2M_2+16M_1$
		234-2 大苞萱草 Hemerocallis middendorfii Trautv. et Mey.	$2n=2x=22$	12m+4sm+4st（1SAT）+2t	2A	38	1.9	0.36	66.34	$4L+4M_2+12M_1+2S$

第二章 中国植物染色体研究结果 | 213

续表

中药植物核型分析

科	属	种	染色体数目	核型公式	核型分类	臂指数	最长染色体/最短染色体	臂比>2的染色体比例	核型不对称系数	染色体相对长度组成
	235. 玉簪属 Hosta Tratt.	235-1 东北玉簪 Hosta ensata F. Maekawa	$2n=90$	6m+22sm+40st+22t（1SAT）	3C	118	5.03	0.84	80.97	$16L+8M_2+28M_1+38S$
	235. 百合属 Lilium L.	236-1 卷丹 Lilium lancifolium Thu	$2n=2x=24$	2m+2sm+6st+14t	3A	28	1.94	0.83	82.49	$4L+6M_2+12M_1+2S$
		236-2 轮叶百合 Lilium distichum Nakai	$2n=2x=24$	4m+8st+12t	3B	28	2.11	0.83	81.99	$4L+4M_2+14M_1+2S$
		236-3 松叶百合 Lilium cernuum kom	$2n=2x=24$	4m+6st+14t	3A	28	1.8	0.83	80.66	$4L+4M_2+16M_1$
	237. 沿阶草属 Ophiopogon Ker-Gawl.	237-1 沿阶草 Ophiopogon bodinieri Levl.	$2n=2x=36$	18m+18sm	2B	72	2.42	0.17	61.61	$6L+6M_2+22M_1+2S$
		237-2 麦冬 Ophiopogon japonicus (Thunb.) Ker-Gawl.	$2n=4x=72$	38m+34sm	2B	144	2.77	0.25	62.86	$12L+18M_2+30M_1+12S$
	238. 重楼属 Paris L.	238-1 北重楼 Paris verticillata M.-Bieb.	$2n=4x=20$	12m+4st+4t	2A	32	1.97	0.4	65.55	$4L+4M_2+10M_1+2S$
		238-2 云南重楼 Paris yunnanensis Franch	$2n=2x=10$	6m+4t	2A	16	1.97	0.4	66.7	$2L+2M_2+6M_1$
		238-3 西南重楼 Paris polyphylla Sm. var. thibetica (Franch.) Hara	$2n=2x=10$	6m+4t	2A	16	1.74	0.4	69.66	$2L+2M_2+6M_1$
六十五、百合科 Liliaceae	239. 黄精属 Polygonatum Mill.	239-1 小玉竹 Polygonatum humile Fisch. ex Maxim.	$2n=2x=20$	10m+4sm+6st	2B	34	2.6	0.3	65.79	$6L+4M_2+2M_1+8S$
		239-2 黄精 Polygonatum sibiricum Delar. ex Redoute	$2n=4x=44$								
		239-3 二苞黄精 Polygonatum involucratum (Franch. et Sav.) Maxim.	$2n=2x=18$	6m+6sm+6st	2B	30	2.91	0.44	67.33	$2L+8M_2+4M_1+4S$
		239-4 玉竹 Polygonatum Odoratum (Mill.) Druce	$2n=2x=20$	12m (2SAT) +6sm+2st	2B	38	2.42	0.2	61.31	$4L+4M_2+10M_1+2S$
	240. 绵枣儿属 Scilla L.	240-1 绵枣儿 Scilla scilloides (Lindl.) Druce	$2n=2x=34$	8m+14sm+10st+2t	3B	56	3.97	0.71	71.01	$6L+12M_2+6M_1+10S$
	241. 鹿药属 Smilacina Desf.	241-1 鹿药 Smilacina japonica A. Gray	$2n=36$	12m+14sm（2SAT）+10st	2C	62	5.16	0.5	70.08	$12L+4M_2+20S$
	242. 菝葜属 Smilax L.	242-1 菝葜 Smilax china L.	$2n=128$	4m+50sm+70st+4t（2SAT）	3B	186	2.83	0.88	76.56	$26L+30M_2+38M_1+34S$
	243. 开口箭属 Tupistra Ker Gawl.	243-1 齿瓣开口箭 Tupistra fimbriata Hand.-Mzt	$2n=2x=38$	26m+12sm	2B	76	2.3	0.11	60.09	$6L+8M_2+22M_1+2S$
	244 藜芦属 Veratrum L.	144-1 尖被藜芦 Veratrum oxysepalum Turcz.	$2n=32$	30m+2st（2SAT）	2B	62	2	0.06	54.92	$4L+12M_2+14M_1+2S$
六十六、亚麻科 Linaceae	245. 亚麻属 Linum L.	245-1 宿根亚麻 Linum perenne L.	$2n=2x=36$	34m+2sm（2SAT）	2A	72	1.59	0.06	58.05	$2L+14M_2+20M_1$

续表

中药植物核型分析

科	属	种	染色体数目	核型公式	核型分类	臂指数	最长染色体/最短染色体	臂比>2的染色体比例	核型不对称系数	染色体相对长度组成
六十七、木兰科 Magnoliaceae	246. 八角属 Illicium L.	246-1 八角茴香 Illicium verum Hook. F.	$2n=2x=26$	14m+8sm+4T	2B	48	2.08	0.23	65.5	$2L+10M_2+12M_1+2S$
	247. 五味子属 Schisandra Michx	247-1 五味子 Schisandra chinensis (Turcz.) Baill.	$2n=2x=28$	22m+6sm	2A	56	1.49	0.14	59.79	$2L+12M_2+14M_1$
	248. 秋葵属 Abelmoschus Medic.	248-1 秋葵 Abelmoschus esculentus (L.) Moench	$2n=72$	20m+44sm+8st	2A	136	1.71	0.42	67.29	$4L+30M_2+38M_1$
		248-2 黄蜀葵 Abelmoschus manihot (L.) Medic.	$2n=66$	32m+28sm+4st+2T	2A	126	1.79	0.39	65.17	$6L+22M_2+38M_1$
	249. 苘麻属 Abutilon Mill.	249-1 苘麻 Abutilon theophrasti Medicus	$2n=42$	16m+8sm+8st+2t (2SAT) +8T	3B	66	2.16	0.52	72.51	$10L+4M_2+24M_1+4S$
	250. 蜀葵属 Althaea L.	250-1 蜀葵 Althaea rosea Cav,	$2n=42$	14m (2SAT) +26sm (2SAT) +2st	2B	84	2.07	0.43	65.92	$8L+10M_2+20M_1+4S$
	251. 木槿属 Hibiscus L.	251-1 玫瑰茄 Hibiscus sabdariffa L.	$2n=72$	50m+20sm(2SAT)+2st	2A	144	1.88	0.17	59.22	$4L+30M_2+38M_1$
		251-2 木槿 Hibiscus syriacus L.	$2n=88$	42m+28sm（2SAT）+12st+4t+2T	2B	158	2.94	0.45	66.91	$8L+34M_2+24M_1+22S$
六十八、锦葵科 Malvaceae	252. 锦葵属 Malva L.	252-1 冬葵 Malva verticillata L.	$2n=120$							
	253. 黄花稔属 Sida L.	253-1 桤叶黄花稔 Sida alnifolia L.	$2n=2x=28$	18m+10sm	2A	56	1.53	0.21	61.92	$16M_2+12M_1$
	254. 棉属 Thespesia Sol. ex Correa	254-1 肖槿 Thespesia lampas (Cav.) Dalzell et A. Gibson	$2n=2x=20$	6m (2SAT) +14sm	2B	40	2.22	0.4	66.37	$6L+4M_2+6M_1+4S$
	255. 梵天花属 Urena L.	255-1 梵天花 Urena procumbens L.	$2n=2x=38$	12m(1SAT)+20sm+6st	2A	70	1.96	0.42	66.01	$2L+16M_2+18M_1+2S$
	256. 香椿属 Toona Roem.	256-1 香椿 Toona sinensis (A. Juss.) Roem.	$2n=56$	36m+20sm	2A	112	1.79	0.11	59.91	$4L+22M_2+28M_1+2S$
	257. 麻楝属 Chukrasia A. Juss.	257-1 麻楝 Chukrasia tabularis A. Juss.	$2n=56$							
六十九、楝科 Meliaceae	258. 浆果楝属 Cipadessa Bl.	258-1 浆果楝 Cipadessa cinerascens (Pell.) Hand.-Mazz	$2n=28$	22m+6sm	2A	56	1.51	0.07	60.48	$2L+10M_2+16M_1$
	259. 楝属 Melia L.	259-1 楝 Melia azedarach L.	$2n=2x=28$							
		259-2 川楝 Melia toosendan Sieb. et Zucc.	$2n=2x=28$	20m (2SAT) +8sm	1A	56	1.42	0	61.01	$16M_2+12M_1$

续表

中药植物核型分析

科	属	种	染色体数目	核型公式	核型分类	臂指数	最长染色体/最短染色体	臂比>2的染色体比例	核型不对称系数	染色体相对长度组成
七十、防己科 Menispermaceae	260. 天仙藤属 Fibraurea Lour.	260-1 黄藤 Fibraurea recisa Pierre	$2n=2x=26$							
	261. 防己属 Sinomenium Diels	261-1 防己 Sinomenium acutum（Thunb.）Rehd. et Wils.	$2n=64$							
	262. 千金藤属 Stephania Lour.	262-1 金线吊乌龟 Stephania cepharantha Hayata	$2n=2x=26$	$22m+4sm$	1A	52	1.54	0	58.38	$2L+10M_2+14M_1$
		262-2 汝兰 Stephania sinica Diels	$2n=2x=26$	$22m+4sm$（2SAT）	2A	52	1.44	0.08	58.86	$12M_2+14M_1$
七十一、含羞草科 Mimosaceae	263. 金合欢属 Acacia Mill.	263-1 金合欢 Acacia farnesiana（L.）Willd.	$2n=52$	$24m+14sm+2st+12T$	2B	90	2.33	0.46	69.54	$6L+10M_2+34M_1+2S$
		263-2 儿茶 Acacia catechu（L. f.）Willd.	$2n=2x=26$	$10m+14sm+2t$	2A	50	1.57	0.23	65.32	$2L+8M_2+16M_1$
	264. 海红豆属 Adenanthera L.	264-1 海红豆 Adenanthera pavonina L. var. microsperma（Teijsm. et Binn.）Nielsen	$2n=85$							
	265. 构树属 Broussonetia L'Hér.	265-1 构树 Broussonetia papyrifera（L.）Vent.	$2n=2x=28$	$14m+10sm+4st$	2B	52	2.15	0.36	64.53	$2L+8M_2+18M_1$
七十二、桑科 Moraceae	266. 榕属 Ficus L.	266-1 高山榕 Ficus altissima Blume	$2n=2x=26$							
		266-2 垂叶榕 Ficus benjamina L.	$2n=2x=26$	$10m+6sm$（2SAT）$+8st+2t$	3A	42	1.69	0.54	70.01	$2L+10M_2+14M_1$
		266-3 无花果 Ficus carica L.	$2n=2x=26$	$8m+8sm$（2SAT）	2A	32	1.75	0.5	62.49	$2L+4M_2+10M_1$
	267. 葎草属 Humulus L.	267-1 葎草 Humulus scandens（Lour.）Merr.	$2n=2x=16$	$14m+14sm$	2B	56	2.56	0.21	62.36	$4L+2M_2+22M_1$
	268. 桑属 Morus L.	268-1 桑 Morus alba L.	$2n=2x=28$	$40m+6sm$	2A	92	1.45	0.04	58.21	58.21
七十三、紫金牛科 Myrsinaceae	269. 紫金牛属 Ardisia SW.	269-1 尾叶紫金牛 Ardisia caudata Hemsl.	$2n=2x=46$		2A					$24M_2+22M_1$
		269-2 朱砂根 Ardisia crenata Sims	$2n=2x=26$							
七十四、桃金娘科 Myrtaceae	270. 番石榴属 Psidium L.	270-1 番石榴 Psidium guajava L.	$2n=2x=22$	$6m+16sm$	2A	44	1.58	0.27	64.54	$2L+6M_2+14M_1$
	271. 丁香属 Syringa L.	271-1 羽叶丁香 Syringa pinnatifolia Hemsl.	$2n=46$							
	272. 叶子花属 Bougainvillea Comm. ex Juss.	272-1 三角花 Bougainvillea glabra Choisy	$2n=2x=34$	$24m+10sm$	2A	68	1.4	0.12	58.45	$18M_2+16M_1$
七十五、紫茉莉科 Nyctaginaceae	273. 紫茉莉属 Mirabilis L.	273-1 紫茉莉 Mirabilis jalapa L.	$2n=58$	$28m(5SAT)+24sm+6st$	2B	110	2.16	0.41	64.54	$6L+12M_2+40M_1$

续表

中药植物核型分析

科	属	种	染色体数目	核型公式	核型分类	臂指数	最长染色体/最短染色体	臂比>2的染色体比例	核型不对称系数	染色体相对长度组成
七十六、珙桐科 Nyssaceae	274. 喜树属 Camptotheca Decne	274-1 喜树 Camptotheca acuminata Decne.	$2n=42$							
七十七、木犀科 Oleaceae	275. 连翘属 Forsythia Vahl	275-1 连翘 Forsythia suspensa (Thunb.) Vahl	$2n=2x=28$	12m+12sm+4st(2SAT)	2A	52	1.56	0.36	64.02	$2L+10M_2+16M_1$
	276. 白蜡树属 Fraxinus L.	276-1 白蜡树 Fraxinus chinensis Roxb.	$2n=2x=46$	22m+20sm+4st(4SAT)	2A	88	1.81	0.35	64.16	$4L+18M_2+22M_1+2S$
	277. 女贞属 Fraxinus L.	277-1 小蜡 Fraxinus sinense Lour.	$2n=2x=24$	14m (1SAT) +8sm+2st	2B	46	2	0.33	61.41	$6L+4M_2+12M_1+2S$
七十八、柳叶菜科 Onagraceae	278. 月见草属 Oenothera L.	278-1 月见草 Oenothera odorata Jacq.	$2n=2x=14$	8m (1SAT) +6sm	1A	28	1.47	0	62.05	$8M_2+6M_1$
七十九、兰科 Orchidaceae	279. 白芨属 Bletilla Reichenb. f.	279-1 白芨 Bletilla striata (Thunb.) Reichenb. f.	$2n=32$	14M+16sm+2st	2B	62	2.14	0.44	65.2	$4L+10M_2+10M_1+8S$
	280. 石斛属 Dendrobium Sw.	280-1 霍山石斛 Dendrobium huoshanense C. Z. Tang et S. J. Cheng	$2n=2x=38$	14m+12sm+8st+4T	3B	66	2.08	0.53	69.61	$6L+10M_2+16M_1+6S$
		280-2 铁皮石斛 Dendrobium officinale Kimura et Migo	$2n=4x=76$							
		280-3 金钗石斛 Dendrobium nobile Lindl.	$2n=2x=38$							
	281. 山兰属 Oreorchis Lindl.	281-1 山兰 Oreorchis patens (Lindl.) Lindl	$2n=50$							
	282. 绶草属 Spiranthes L. C. Rich.	282-1 绶草 Spiranthes sinensis (Pers.) Ames	$2n=30$	6sm+22st+2t (2SAT)	4A	36	1.73	1	78.77	$2L+10M_2+18M_1$
八十、棕榈科 Palmae	283. 棕榈属 Trachycarpus Wendl.	283-1 棕榈 Trachycarpus fortunei (Hook.) H. Wendl.	$2n=36$	10m+10sm+16st	3B	56	3.15	0.67	69.81	$6L+8M_2+14M_1+8S$
八十一、白屈菜科	284. 白屈菜属 Chelidonium L.	284-1 白屈菜 Chelidonium majus L.	$2n=11$	1sm+5st+5t (1SAT)	3A	13	1.76	0.91	85.02	$2L+2M_2+7M_1$
	285. 紫金龙属 Dactylicapnos Wall	285-1 紫金龙 Dactylicapnos scandens (D. Don) Hutch	$2n=2x=16$	14m+2sm	1A	32	1.43	0	60.54	$6M_2+10M_1$
八十一、罂粟科 Papaveraceae	286. 荷青花属 Hylomecon Maxim.	286-1 荷青花 Hylomecon japonica (Thunb.) Prantl et Kündig	$2n=2x=24$							
	287. 博落回属 Macleaya R. Br.	287-1 博落回 Macleaya cordata (Wind.) R. Br.	$2n=2x=20$	12m+8sm (1SAT)	2A	40	1.3	0.2	61.2	$8M_2+12M_1$

续表

中药植物核型分析

科	属	种	染色体数数目	核型公式	核型分类	臂指数	最长染色体/最短染色体	臂比>2的染色体比例	核型不对称系数	染色体相对长度组成
八十一、罂粟科 Papaveraceae	288. 罂粟属 Papaver L.	288-1 东方罂粟 Papaver orientale L.	$2n=2x=14$	2m+10sm+2st	3A	26	1.26	0.86	70.69	$6M_2+8M_1$
		288-2 罂粟 Papaver somniferum L.	$2n=2x=14$	2m+6sm+6st (2SAT)	3A	22	1.22	0.71	69.89	$6M_2+8M_1$
八十二、胡麻科 Pedaliaceae	289. 胡麻属 Sesamum L.	289-1 芝麻 Sesamum indicum L.	$2n=2x=26$	8m+12sm+6st (2SAT)	3A	46	1.55	0.54	69.03	$14M_2+12M_1$
八十三、透骨草科 Phrymaceae	290. 透骨草属 Phryma L.	290-1 透骨草 Phryma leptostachya L. var. asiatica Hara	$2n=2x=28$	8m+20sm	2B	56	2.05	0.5	64.92	$4L+8M_2+14M_1+2S$
八十四、商陆科 Phytolaccaceae	291. 商陆属 Phytolacca L.	291-1 商陆 Phytolacca acinosa Roxb.	$2n=36$	14m+10sm+12st (2SAT)	3A	60	1.62	0.56	68.3	$2L+16M_2+18M_1$
		291-2 垂序商陆 Phytolacca americana L.	$2n=36$	20m+16sm	2A	72	1.62	0.17	62.7	$2L+18M_2+16M_1$
八十五、胡椒科 Piperaceae	292. 胡椒属 Piper L.	292-1 荜麦 Piper longum L.	$2n=52$							
八十六、车前草科 Plantaginaceae	293. 车前草属 Plantago L.	293-1 车前 Plantago asiatica L.	$2n=2x=12$	10m+2sm (2SAT)	2A	24	1.38	0.17	60.53	$6M_2+6M_1$
		293-2 平车前 Plantago depressa Willd.	$2n=2x=12$	10m+2sm (2SAT)	2A	24	1.39	0.17	56.05	$8M_2+4M_1$
		293-3 长叶车前 Plantago lanceolata L.	$2n=2x=12$	8m+4sm (2SAT)	1A	24	1.41	0	58.17	$8M_2+4M_1$
		293-4 细叶车前 Plantago lessingii Fisch. et Mey.	$2n=2x=12$	8m+2sm+2t (2SAT)	2A	22	1.48	0.17	62.48	$6M_2+6M_1$
		293-5 大车前 Plantago major L.	$2n=2x=12$	8m+4sm (4SAT)	2A	24	1.34	0.17	60.74	$6M_2+6M_1$
八十七、白花丹科 Plumbaginaceae	294. 白花丹属 Plumbago L.	294-1 白花丹 Plumbago zeylanica L.	$2n=2x=28$	18m (2SAT) +10sm	2A	56	1.52	0.29	63.43	$2L+10M_2+16M_1$
八十八、远志科 Polygalaceae	295. 远志属 Polygala L.	295-1 远志 Polygala tenuifolia Willd.	$2n=2x=38$	28m+8sm+2st (2SAT)	2A	74	1.58	0.11	61.3	$2L+14M_2+22M_1$
八十九、荞麦科	296. 荞麦属 Fagopyrum Mill.	296-1 天荞麦 Fagopyrum cymosum Meissn	$2n=2x=16$	10m+6sm	2A	32	1.27	0.25	61.01	$8M_2+8M_1$
		296-2 荞麦 Fagopyrum esculentum Moench	$2n=2x=16$	16m (2SAT)	1A	32	1.28	0	56.49	$8M_2+8M_1$
		296-3 金荞麦 F. dibotrys (D. Don) Hara	$2n=2x=16$	12m+4sm (4SAT)	1A	32	1.41	0	57.99	$8M_2+8M_1$
八十九、蓼科 Polygonaceae	297. 蓼属 Polygonum L.	297-1 头花蓼 Polygonum capitatum Buch.-Ham. ex D. Don	$2n=2x=22$	18m+4sm	2A	44	1.4	0.18	58.01	$8M_2+14M_1$
		297-2 酸膜叶蓼 Polygonum lapathifolium L.	$2n=32$	6m+16sm+2st+4t (2SAT) +4T	3A	54	1.78	0.75	74.95	$2L+6M_2+24M_1$
		297-3 何首乌 Polygonum multiflorum (L.) Thunb.	$2n=2x=20$	8m+10sm+2st	2A	38	1.85	0.5	65.23	$2L+8M_2+6M_1+4S$

续表

中药植物核型分析

科	属	种	染色体数目	核型公式	核型分类	臂指数	最长染色体/最短染色体	臂比>2的染色体比例	核型不对称系数	染色体相对长度组成
八十九、蓼科 Polygonaceae	297. 蓼属 Polygonum L.	297-4 红蓼 Polygonum orientale L.	$2n=2x=22$	8m+14sm	2A	44	1.3	0.36	65.33	$12M_2+10M_1$
		297-5 戟叶蓼 Polygonum thunbergii Sieb. et Zucc.	$2n=2x=20$							
		297-6 水蓼 Polygonum hydropiper L.	$2n=2x=16$	10m+6sm（2SAT）	2B	32	2.01	0.25	60.84	$4L+10M_1+2S$
	298. 大黄属 Rheum L.	298-1 掌叶大黄 Rheum palmatum L.	$2n=60$	38m+16sm+6st（2SAT）	2A	114	1.59	0.2	62.12	$4L+18M_2+38M_1$
		298-2 药用大黄 Rheum officinale Baill.	$2n=44$	36m（2SAT）+8sm	2A	88	1.41	0.05	60.13	$22M_2+22M_1$
	299. 酸模属 Rumex L.	299-1 皱叶酸模 Rumex crispus L.	$2n=60$							
		299-2 羊蹄 Rumex japonicus Houtt.	$2n=2x=20$	6m+14sm	3A	40	1.41	0.6	65.46	$10M_2+10M_1$
		299-3 巴天酸模 Rumex patientia L.	$2n=2x=34$	4m+20sm+2st+2t（2SAT）+6T	3B	58	2.73	0.76	74.47	$4L+12M_2+16M_1+2S$
		299-4 土大黄 Rumex madaio Mak.	$2n=40$	20m+16sm+4st	2A	76	1.92	0.4	63.12	$4L+16M_2+14M_1+6S$
九十、马齿苋科 Portulacaceae	300. 马齿苋属 Portulaca L.	300-1 马齿苋 Portulaca oleracea L.	$2n=52$							
	301. 土人参属 Talinum Adans.	301-1 土人参 Talinum paniculatum（Jacq.）Gaertn.	$2n=2x=24$	18m+6sm（2SAT）	1A	48	1.72	0	58.56	$2L+8M_2+14M_1$
九十一、报春花科 Primulaceae	302. 珍珠菜属 Lysimachia L.	302-1 过路黄 Lysimachia christinae Hance	$2n=2x=24$	2m+4sm+2st+4t+12T	3A	30	1.99	0.92	86.45	$4L+6M_2+12M_1+2S$
		320-2 珍珠菜 Lysimachia clethroides Duby	$2n=2x=24$	16m+6sm+2st	2A	46	1.21	0.17	60.32	$10M_2+14M_1$
		302-3 黄莲花 Lysimachia davurica Ledeb.	$2n=42$	16m+22sm+4st	2A	80	1.5	0.29	64.63	$24M_2+18M_1$
九十二、石榴科 Punicaceae	303. 石榴属 Punica L.	303-1 石榴 Punica granatum L.	$2n=2x=16$	14m+2sm	1B	32	2.17	0	59.87	$2L+4M_2+8M_1+2S$
九十三、毛茛科 Ranunculaceae	304. 乌头属 Aconitum L.	304-1 乌头 Aconitum carmichaeli Debx	$2n=4x=32$	8m+20sm+4st	3C	60	4.58	0.63	65.6	$8L+16M_1+8S$
		304-2 宽叶蔓乌头 Aconitum sczukinii Turcz.	$2n=2x=16$	2m+6sm+8st	3B	24	2.65	0.88	72.12	$4L+2M_2+8M_1+2S$
	305. 银莲花属 Anemone Linn	305-1 大火草 Anemone tomentosa（Maxim.）Pei	$2n=2x=16$	8m+2sm+2st+4T（1SAT）	2A	26	1.74	0.38	67.33	$2L+6M_2+6M_1+2S$
	306. 耧斗菜属 Aquilegia L.	306-1 耧斗菜 Aquilegia viridiflora Pall.	$2n=2x=14$	10m+2sm+2st（2SAT）	2A	26	1.34	0.14	59.66	$8M_2+6M_1$
	307. 驴蹄草属 Caltha L.	307-1 膜叶驴蹄草 Caltha palustris L. var. membranacea Turcz.	$2n=2x=18$	12m+6sm（2SAT）	1A	36	1.6	0	61.28	$2L+4M_2+12M_1$

续表

中药植物核型分析

科	属	种	染色体数目	核型公式	核型分类	臂指数	最长染色体/最短染色体	臂比>2的染色体比例	核型不对称系数	染色体相对长度组成
	308. 升麻属 Cimicifuga L.	308-1 升麻 Cimicifuga foetida L.	$2n=2x=16$	10m+4sm+2t	2A	30	1.91	0.38	63.17	$4L+4M_2+4M_1+4S$
		308-2 兴安升麻 Cimicifuga dahurica（Turcz.）Maxim.	$2n=4x=32$	20m+8sm+4t	2A	60	1.87	0.38	62.3	$4L+12M_2+12M_1+4S$
	309. 铁线莲属 Clematis L.	309-1 芹叶铁线莲 Clematis aethusaefolia Turcz.	$2n=2x=16$	10m+4sm+2t	2A	30	1.5	0.25	60.29	$10M_2+6M_1$
		309-2 威灵仙 Clematis chinensis Osbeck	$2n=2x=16$	10m+2st+4t（2SAT）	2A	26	1.53	0.38	68.36	$8M_2+8M_1$
		309-3 铁线莲 Clematis florida Thunb	$2n=2x=16$	10m+2st+4t（3SAT）	2A	26	1.51	0.38	66.26	$8M_2+8M_1$
		309-4 大叶铁线莲 Clematis heracleifolia DC.	$2n=2x=16$	10m+2st+4t（4SAT）	2A	26	1.77	0.38	65.93	$2L+8M_2+2M_1+4S$
		309-5 山木通 Clematis finetiana Lévl. et Vant	$2n=2x=16$	6m+4sm+4st+2t（2SAT）	2A	26	1.8	0.38	65.84	$2L+8M_2+2M_1+4S$
九十三、毛茛科 Ranunculaceae	310. 黄连属 Coptis Salisb.	310-1 黄连 Coptis chinensis Franch.	$2n=18$		2A	26	1.56	0.38	69.21	$8M_2+8M_1$
	311. 芍药属 Paeonia L.	311-1 芍药 Paeonia lactiflora Pall.	$2n=2x=10$	6m+2sm+2st	2A	18	1.36	0.2	61.52	$4M_2+6M_1$
		311-2 草芍药 Paeonia obovata Maxim	$2n=2x=10$	6m+2sm+2st+2t（1SAT）	2A	18	1.48	0.4	64.13	$6M_2+4M_1$
		311-3 牡丹 Paeonia suffruticosa Andr.	$2n=2x=10$	6m+2sm+2st	2A	18	1.35	0.2	61.35	$6M_2+4M_1$
	312. 白头翁属 Pulsatilla Adans	312-1 白头翁 Pulsatilla chinensis（Bunge）Regel	$2n=2x=16$	10m+2sm+4st	2A	28	1.8	0.38	62.07	$2L+6M_2+6M_1+2S$
	313. 毛茛属 Pulsatilla L.	313-1 毛茛 Pulsatilla japonicus Thunb.	$2n=28$	14m+12sm+2st	2A	54	1.66	0.36	65.13	$2L+14M_2+12M_1$
	314. 唐松草属 Thalictrum L.	314-1 唐松草 Thalictrum aquilegifolium L. var sibiricum Regel et Tiling	$2n=2x=14$	6m+8sm（2SAT）	2A	28	1.27	0.14	63.78	$6M_2+8M_1$
		314-2 展枝唐松草 Thalictrum squarrosum Steph. ex Wild	$2n=12x=84$	46m+30sm+8st(2SAT)	2A	160	1.63	0.24	63.89	$40M_2+44M_1$
九十四、鼠李科 Rhamnaceae	315. 枣属 Ziziphus Mill.	315-1 酸枣 Ziziphus jujuba var. spinosa（Bunge）Hu H. F. Chow	$2n=2x=24$		2A					
九十五、蔷薇科 Rosaceae	316. 龙牙草属 Agrimonia L.	316-1 仙鹤草 Agrimonia pilosa Ledeb.	$2n=56$	32m+22sm+2st	2A	110	1.89	0.21	61.87	$6L+20M_2+28M_1+2S$
	317. 杏属 Armeniaca Mill	317-1 山杏 Armeniaca sibirica（L.）Lam.	$2n=2x=16$	8m+6sm+2st	2B	30	2.03	0.38	64.08	$2L+6M_2+6M_1+2S$

续表

中药植物核型分析

科	属	种	染色体数目	核型公式	核型分类	臂指数	最长染色体/最短染色体	臂比>2的染色体比例	核型不对称系数	染色体相对长度组成
九十五、蔷薇科 Rosaceae	318. 贴梗海棠属 Chaenomeles Lindl.	318-1 贴皮木瓜 Chaenomeles speciosa (Sweet) Nakai	$2n=2x=34$	20m+6sm+6st+2t	2A	60	1.46	0.41	65.68	$18M_2+16M_1$
	319. 枇杷属 Eriobotrya Lindl.	319-1 枇杷 Eriobotrya japonica (Thunb.) Lindl.	$2n=2x=34$	20m+14sm（1SAT）	2A	68	1.38	0.29	62.03	$14M_2+20M_1$
	320. 蚊子草属 Filipendula L.	320-1 蚊子草 Filipendula palmata (Pall.) Maxim	$2n=56$	44m+10sm+2st	2A	110	1.35	0.14	59.15	$28M_2+28M_1$
	321. 水杨梅属 Geum L.	321-1 水杨梅 Geum aleppicum Jacq.	$2n=42$	22m+18sm+2st	2A	82	1.75	0.29	62.61	$4L+16M_2+18M_1+4S$
	322. 萎陵菜属 Potentilla L.	322-1 萎陵菜 Potentilla chinensis Ser.	$2n=42$	22m+12sm（1SAT）+4st+4t	2A	76	1.51	0.43	66.69	$22M_2+20M_1$
	323. 火棘属 Pyracantha Roem.	323-1 火棘 Pyracantha fortuneana (Maxim.) Li	$2n=2x=34$	14m+14sm+6st	2A	62	1.62	0.47	66.13	$2L+12M_2+20M_1$
	324. 蔷薇属 Rosa L.	324-1 金樱子 Rosa laevigata Michx.	$2n=2x=14$	6m+2sm+6st	3A	22	1.45	0.57	66.96	$6M_2+8M_1$
	325. 地榆属 Sanguisorba L.	325-1 地榆 Sanguisorba officinalis L.	$2n=55$							
	326. 绣线菊属 Spiraea L.	326-1 柳叶绣线菊 Spiraea salicifolia L.	$2n=2x=18$	4m+12sm+2st	3A	34	1.99	0.56	67.27	$4L+4M_2+8M_1+2S$
	327. 栀子属 Gardenia Ellis	327-1 栀子 Gardenia jasminoides Ellis	$2n=2x=22$	12m+10sm（2SAT）	1A	44	1.77	0	61.05	$2L+6M_2+12M_1+2S$
九十六、茜草科 Rubiaceae	328. 耳草属 Hedyotis L.	328-1 红铁扫把 Hedyotis lance Thunb.ex Maxim	$2n=2x=18$	18m	1A	36	1.45	0	60.17	$8M_2+10M_1$
	329. 鸡矢藤属 Paederia L.	329-1 鸡矢藤 Paederia scandens (Lour.) Merr.	$2n=6x=66$	24m+22sm+18st+2t	2B	112	2.22	0.39	67.63	$6L+18M_2+40M_1+2S$
		329-2 云南鸡矢藤 Paederia yunnanensis (H. Lev.) Rehder	$2n=2x=22$	16m+4sm+2st（2SAT）	2A	42	1.46	0.09	60.52	$12M_2+10M_1$
	330. 柑橘属 Citrus L.	330-1 枸橼 Citrus medica L.	$2n=2x=18$	8m+10sm	2A	36	1.53	0.33	64.85	$10M_2+8M_1$
九十七、芸香科 Rutaceae	331. 白藓属 Dictamnus L.	331-1 白藓 Dictamnus dasycarpus Turcz	$2n=36$	10m+20sm+6st	3B	66	2.29	0.67	67.87	$2L+12M_2+18M_1+4S$
	332. 九里香属 Murraya L.	332-1 九里香 Murraya exotica L.	$2n=2x=18$	12m+6sm	2A	36	1.7	0.11	58.83	$10M_2+6M_1+2S$
	333. 黄柏属 Rupr.	333-1 黄柏 Phellodendron amurense Pupr.	$2n=78$	26m+36sm+12st+4t	3B	140	2.2	0.51	67.66	$12L+20M_2+40M_1+6S$

续表

中药植物核型分析

科	属	种	染色体数目	核型公式	核型分类	臂指数	最长染色体/最短染色体	臂比>2的染色体比例	核型不对称系数	染色体相对长度组成
九十七、芸香科 Rutaceae	334. 枳属 Poncirus Raf.	334-1 枳橘 Poncirus trifoliata (L.) Raf.	$2n=2x=18$	2m+12sm (2SAT) +4st	3A	32	1.93	0.56	69.31	$4L+4M_2+8M_1+2S$
九十八、杨柳科 Salicaceae	335. 花椒属 Zanthoxylum L.	335-1 花椒 Zanthoxylum bungeanum Maxim.	$2n=136$							
	336. 柳属 Salix L.	336-1 坡柳 Salix myrtillacea Anderss	$2n=2x=28$	16m+8sm+4st	2A	52	1.88	0.14	64.23	$4L+6M_2+16M_1+2S$
九十九、无患子科 Sapindaceae	337. 倒地铃属 Cardiospermum L.	337-1 倒地铃 Cardiospermum halicacabum L.	$2n=2x=22$	8m+4sm+10st	3A	34	1.68	0.64	69.29	$2L+8M_2+10M_1+2S$
一〇〇、三白草科 Saururaceae	338. 蕺菜属 Houttuynia Thunb.	338-1 鱼腥草 Houttuynia cordata Thunb.	$2n=96$							
	339. 三白草属 Saururus L.	339-1 三白草 Saururus chinensis (Lour.) Baill.	$2n=2x=22$	10m+8sm+2st+2t	2A	40	1.54	0.36	66.5	$12M_2+10M_1$
一〇一、虎耳草科 Saxifragaceae	340. 常山属 Dichroa Lour.	340-1 常山 Dichroa febrifuga Lour.	$2n=20$	6m+12sm+2st	3A	38	1.77	0.6	65.56	$2L+8M_2+8M_1+2S$
	341. 虎耳草属 Saxifraga Tourn.	341-1 虎耳草 Saxifraga stolonifera Meerb.	$2n=2x=36$	16m(1SAT)+16sm+4st	2A	68	1.95	0.44	64.8	$4L+14M_2+14M_1+4S$
	342. 通泉草属 Mazus Lour.	342-1 通泉草 Mazus japonicus (Thunb.) Kuntze	$2n=32$	22m+8sm (2SAT) +2st (2SAT)	2B	62	2.16	0.19	56.6	$4L+6M_2+20M_1+2S$
一〇二、玄参科 Scrophulariaceae	343. 地黄属 Rehmannia Libosch. ex Fisch. et Mey.	343-1 地黄 Rehmannia glutinosa (Gaertn.) Libosch. ex Steud.	$2n=56$	46m+6sm+4st	2A	108	1.56	0.11	61.3	$26M_2+30M_1$
	344. 玄参属 Scrophularia L.	344-1 玄参 Scrophularia ningpoensis Hemsl.	$2n=92$							
一〇三、苦木科 Simarubaceae	345. 臭椿属 Ailanthus Desf.	345-1 臭椿 Ailanthus altissima (Mill.) Swingle	$2n=2x=64$	32m+22sm+4st+6t (1SAT)	2A	118	1.88	0.31	65.6	$4L+26M_2+32M_1+2S$
一〇四、茄科 Solanaceae	346. 颠茄属 Atropa L.	346-1 颠茄 Atropa belladonna L.	$2n=2x=24$	16m+6sm+2st (2SAT)	2A	46	1.65	0.33	64.23	$2L+8M_2+14M_1$
	347. 曼陀罗属 Datura L.	347-1 毛曼陀罗 Datura innoxia Mill.	$2n=2x=24$	18m+6sm	2A	48	1.63	0.08	58.88	$12M_2+12M_1$
		347-2 洋金花 Datura metel L.	$2n=2x=24$	18m+6sm (2SAT)	2B	48	2.14	0.17	61.33	$2L+8M_2+10M_1+4S$
		347-3 曼陀罗 Datura stramonium L.	$2n=2x=24$	18m+6sm	2A	48	1.96	0.17	59.66	$2L+12M_2+6M_1+4S$
		347-4 大花曼陀罗 Datura suaveolens Humb. et Bonpl. Ex Wild	$2n=2x=24$	14m+10sm (2SAT)	2A	48	1.46	0.33	61.2	$14M_2+10M_1$

续表

中药植物核型分析

科	属	种	染色体数目	核型公式	核型分类	臂指数	最长染色体/最短染色体	臂比>2的染色体比例	核型不对称系数	染色体相对长度组成
一〇四、茄科 Solanaceae	348. 天仙子属 Hyoscyamus L.	348-1 莨菪 Hyoscyamus niger L.	$2n=2x=34$	16m+18sm (2SAT)	2A	68	1.5	0.24	63.13	$2L+12M_2+20M_1$
	349. 枸杞属 Lycium L.	349-1 宁夏枸杞 Lycium barbarum L.	$2n=2x=24$	18m+6sm	2A	48	1.56	0.08	58.53	$2L+10M_2+12M_1$
		349-2 枸杞 Lycium chinense Mill.	$2n=2x=24$	18m+6sm	2A	48	1.46	0.08	59.64	$8M_2+16M_1$
	350. 酸浆属 Physalis L.	350-1 酸浆 Physalis alkekengi L.	$2n=2x=20$	2m+8sm+10st (2SAT)	3A	30	1.48	0.7	71.97	$8M_2+12M_1$
		350-2 小酸浆 Physalis minima L.	$2n=2x=20$	4m+12sm+4st (2SAT)	3A	36	1.51	0.8	70.51	$10M_2+10M_1$
	351. 赛莨菪属 Scopolia Jacq	351-1 丽江山莨菪 Scopolia acutangula C. Y. Wu et C. Chen	$2n=2x=20$	18sm+2st (2SAT)	3A	38	1.56	0.8	71.22	$2L+8M_2+10M_1$
	352. 茄属 Solanum L.	352-1 野海茄 Solanum japonense Nakai	$2n=2x=24$	20m (2SAT) +2sm+2st (2SAT)	2A	46	1.87	0.17	61.05	$2L+8M_2+12M_1+2S$
		352-2 白英 Solanum lyratum Thunb.	$2n=2x=24$	16m+8sm	2A	48	1.34	0.17	59.96	$12M_2+12M_1$
		352-3 龙葵 Solanum nigrum L.	$2n=72$	26m+28sm+18st	2A	126	1.66	0.47	68.66	$2L+34M_2+36M_1$
		352-4 野茄树 Solanum verbascifolium L.	$2n=2x=24$	10m+12sm(2SAT)+2st	2A	46	1.48	0.25	64.84	$10M_2+14M_1$
一〇五、百部科 Stemonaceae	353. 百部属 Stemona Lour.	353-1 直立百部 Stemona sessilifolia (Miq.) Miq	$2n=2x=14$	8m+4sm+2st (1SAT)	2A	26	1.75	0.29	61.63	$4L+10M_1$
一〇六、梧桐科 Sterculiaceae	354. 昂天莲属 Ambroma Jacq.	354-1 昂天莲 Ambroma augusta (L.) L. f.	$2n=2x=20$	10m+6sm+4st (2SAT)	2A	36	1.62	0.3	65.6	$4L+6M_2+10M_1$
一〇七、梧麻科 Helicteres L.	355. 山芝麻属 Helicteres L.	355-1 山芝麻 Helicteres angutifolia L.	$2n=2x=18$	16m+2sm (1SAT)	1A	36	1.74	0	59.55	$8M_2+8M_1+2S$
一〇七、柽柳科 Tamaricaceae	356. 柽柳属 Tamarix L.	356-1 柽柳 Tamarix chinensis Lour.	$2n=2x=24$	22sm+2st	3A	46	1.8	0.67	69.34	$2L+8M_2+14M_1$
一〇八、杉科 Taxodiaceae	357. 柳杉属 Cryptomeria D. Don	357-1 柳杉 Cryptomeria fortunei Hooibrenk ex Otto et Dietr.	$2n=2x=22$	20m+2sm	1A	44	1.53	0	57.43	$2L+6M_2+14M_1$
	358. 杉木属 Cunninghamia R. Br	358-1 杉木 Cunninghamia lanceolata (Lamb.) Hook.	$2n=2x=22$							
	359. 红豆杉属 Taxus L.	359-1 东北红豆杉 Taxus cuspidate Siebold et Zucc.	$2n=2x=24$	18m+6sm	2B	48	2.56	0.17	57.03	$2L+12M_2+6M_1+4S$
一〇九、山茶科 Theaceae	360. 山茶属 Camellia L.	360-1 长尾毛蕊茶 Camellia caudata Wall.	$2n=24$							
一一〇、瑞香科 Thymelaeaceae	361. 荛花属 Wikstroemia Endl	361-1 丁哥王 Wikstroemia indica (L.) C. A. Mey.	$2n=27$							

续表

中药植物核型分析

科	属	种	染色体数目	核型公式	核型分类	臂指数	最长染色体/最短染色体	臂比>2的染色体比例	核型不对称系数	染色体相对长度组成
一一、椴树科 Tiliaceae	362. 黄麻属 Corchorus L.	362-1 甜麻 Corchorus aestuans L.	$2n=2x=14$	14m	1A	28	1.35	0	57.5	$6M_2+8M_1$
		362-2 黄麻 Corchorus capsularis L.	$2n=2x=14$	12m+2sm (2SAT)	1A	28	1.38	0	58.32	$4M_2+10M_1$
		362-3 长蒴黄麻 Corchorus olitorius L.	$2n=2x=14$	10m+4sm	1A	28	1.37	0	59.03	$8M_2+6M_1$
一二、旱金莲科 Tropaeolaceae	363. 旱金莲属 Tropaeolum L.	363-1 旱金莲 Tropaeolum majus L.	$2n=2x=28$	14m+14sm (2SAT)	2A	56	1.7	0.36	62.36	$2L+10M_2+16M_1$
一三、榆科 Ulmaceae	364. 榆属 Ulmus L.	364-1 榆树 Ulmus pumila L.	$2n=2x=28$	6m (2SAT) +2sm+20st	3A	36	1.53	0.79	74.92	$2L+12M_2+14M_1$
	365. 莳萝属 Anethum L.	365-1 莳萝 Anethum graveolens L.	$2n=40$	32m+8sm	2A	80	1.56	0.05	58.55	$20M_2+20M_1$
	366. 当归属 Angelica L.	366-1 白芷 Angelica dahurica Benth.et Hook. f. ex Franch. et Sav.	$2n=2x=22$	18m+2sm+2st	2A	42	1.61	0.18	59.45	$2L+8M_2+12M_1$
		366-2 杭白芷 Angelica dahurica var. formosana (Boiss.) Shan et Yuan	$2n=2x=22$	14m+4sm+4st	2A	40	1.43	0.27	62.4	$12M_2+10M_1$
		366-3 当归 Angelica sinensis (Oliv.) Diels	$2n=3x=33$	24m+9sm	2A	66	1.39	0.18	59.93	$15M_2+18M_1$
	367. 芹属 Apium L.	367-1 野芹菜 Apium graveolens L. var. dulce DC.	$2n=2x=20$	8m+2sm+10st	3B	30	2.73	0.6	73.67	$8L+4M_1+8S$
	368. 柴胡属 Bupleurum L.	368-1 北柴胡 Bupleurum chinense DC.	$2n=2x=12$	6m+6sm (1SAT)	2A	24	1.37	0.5	64.52	$6M_2+6M_1$
		368-2 川岛柴胡 Bupleurum falcatum L	$2n=2x=26$	10m+14sm(1SAT)+2st	2B	50	3.76	0.31	64.78	$6L+4M_2+8M_1+8S$
一四、伞形科 Umbelliferae		368-3 大叶柴胡 Bupleurum longeradiatum Turcz	$2n=2x=12$	8m+4sm	2A	24	1.19	0.17	62.89	$8M_2+4M_1$
		368-4 抱茎柴胡 Bupleurum longicaule var. amplexicaule C. Y. Wu	$2n=2x=16$	4m+8sm (2SAT) +4st	2A	28	1.88	0.5	68.84	$2L+4M_2+10M_1$
		368-5 红柴胡 Bupleurum scorzonerifolium Wild.	$2n=2x=12$	8m+4sm	2A	24	1.35	0.17	61.68	$6M_2+6M_1$
		368-6 黑柴胡 Bupleurum smithii Wolff in Act.	$2n=2x=12$	10sm+2st	3A	22	1.99	0.67	70.18	$4L+6M_1+2S$
	369. 葛缕子属 Carum L.	369-1 葛缕子 Carum carvi L.	$2n=2x=20$	10m (1SAT) +4sm+6st	2A	34	1.49	0.5	67.07	$10M_2+10M_1$
	370. 积雪草属 Centella L.	370-1 积雪草 Centella asiatica (L.) Urb.	$2n=2x=18$	6m+12sm (2SAT)	2A	36	1.85	0.44	65.36	$10M_2+6M_1+2S$
	371. 明党参属 Changium H. Wolff	371-1 明党参 Changium smyrnioides Wolff	$2n=2x=16$	4m+12sm	2A	32	1.61	0.25	65.41	$8M_2+6M_1+2S$

续表

中药植物核型分析

科	属	种	染色体数目	核型公式	核型分类	臂指数	最长染色体/最短染色体	臂比>2的染色体比例	核型不对称系数	染色体相对长度组成
一一四、伞形科 Umbelliferae	372. 蛇床属 Cnidium Cusson	372-1 蛇床 Cnidium monnieri (L.) Cusson	$2n=2x=20$	16m+4sm (2SAT)	2A	40	1.93	0.2	60.1	$2L+6M_2+12M_1$
	373. 刺芹属 Eryngium L.	373-1 刺芹 Eryngium foetidum L.	$2n=2x=16$	8m+4sm+4st	2B	28	2.17	0.5	66.84	$4L+2M_2+4M_1+6S$
	374. 茴香属 Foeniculum Mill.	374-1 茴香 Foeniculum vulgare Mill.	$2n=2x=22$	12m+8sm+2st	2A	42	1.33	0.09	61.87	$12M_2+10M_1$
	375. 珊瑚菜属 Glehnia Fr. Schmidt ex Miq.	375-1 珊瑚菜 Glehnia littoralis Fr. Schmidt ex Miq	$2n=2x=22$	18m+4sm	2A	44	1.29	0.18	58.14	$10M_2+12M_1$
	376. 藁本属 Ligusticum L.	376-1 川芎 Ligusticum chuanxiong Hort	$2n=2x=22$	10m+8sm+4st	3A	40	1.41	0.55	64.54	$12M_2+10M_1$
		376-2 辽藁本 Ligusticum jeholense (Nakai et Kitag.) Nakai et Kitag.	$2n=2x=24$	8m+16sm	2A	48	1.55	0.17	64.88	$2L+8M_2+14M_1$
		376-3 藁本 Ligusticum sinense Oliv.	$2n=2x=22$							
	377. 欧防风属 Pastinaca L.	377-1 欧防风 Pastinaca sativa L.	$2n=2x=24$	12m+12sm	2A	48	1.63	0.33	62.08	$10M_2+14M_1$
	378. 欧芹属 Petroselinum Hill	378-1 欧芹 Petroselinum sativum Hoffm.	$2n=2x=22$	18m+2sm+2st (2SAT)	2A	42	1.62	0.09	57.19	$14M_2+6M_1+2S$
	379. 前胡属 Peucedanum L.	379-1 白花前胡 Peucedanum praeruptorum Dunn	$2n=2x=20$	18m+2st (2SAT)	2A	38	1.31	0.1	58.51	$10M_2+10M_1$
	380. 防风属 Saposhnikovia Schischk.	380-1 防风 Saposhnikovia divaricata (Turcz.) Schischk.	$2n=2x=16$	12m+2sm+2st (2SAT)	2A	30	1.38	0.25	61.71	$8M_2+8M_1$
	381. 大叶芹属 Spuriopimpinella Kitag.	381-1 大叶芹 Spuriopimpinella brachycarpa Kitag	$2n=2x=22$	6m+4sm+8st+4t	3A	32	1.58	0.73	70.86	$12M_2+10M_1$
一一五、荨麻科 Urticaceae	382. 苎麻属 Boehmeria Jacq.	382-1 苎麻 Boehmeria nivea (L.) Gaudich.	$2n=2x=28$	6m (1SAT)+4sm+4st+6t+8T	3A	38	1.59	0.71	83.08	$16M_2+10M_1+2S$
	383. 蝎子草属 Girardinia Gaud.	383-1 大蝎子草 Girardinia diversifolia (Link) Friis	$2n=2x=26$	10m+12sm(2SAT)+4st	2A	48	1.66	0.46	66.75	$14M_2+10M_1+2S$
	384. 糯米团属 Gonostegia Turcz	384-1 糯米团 Gonostegia hirta (Bl.) Wedd.	$2n=52$							
	385. 艾麻属 Laportea Gaud.	385-1 珠芽艾麻 Laportea bulbifera (Sieb. et Zucc.) Wedd.	$2n=72$							

第二章 中国植物染色体研究结果 | 225

续表

中药植物核型分析

科	属	种	染色体数数目	核型公式	核型分类	臂指数	最长染色体/最短染色体	臂比>2的染色体比例	核型不对称系数	染色体相对长度组成
	386. 荨麻属 Urtica L.	386-1 荨麻 Urtica fissa E. Pritz	$2n=2x=28$	28T	4A	28	1.89	1	100	$2L+10M_2+14M_1+2S$
一一六、败酱科 Valerianaceae	387. 败酱属 Patrinia Juss	387-1 黄花龙牙 Patrinia scabiosaefolia Fisch. ex Link	$2n=6x=66$	34m+24sm+6st+2t	2B	126	2.07	0.24	64.93	$6L+28M_2+30M_1+2S$
		387-2 白花败酱 Patrinia villosa Juss	$2n=2x=44$	14m+20sm+6st+4t	2A	78	1.78	0.5	68.54	$2L+18M_2+22M_1+2S$
	388. 紫珠属 Callicarpa L.	388-1 紫珠 Callicarpa bodinieri H. Lev.	$2n=40$	20m+14sm+4st(2SAT)+2t	2A	74	1.55	0.35	65.49	$2L+16M_2+22M_1$
	389. 赪桐属 Clerodendrum L.	389-1 赪桐 Clerodendrum japonicum (Tunb.) Sweet	$2n=52$							
		389-2 海州常山 Clerodendrum trichotomum Thunb. ex A. Murray	$2n=104$							
一一七、马鞭草科 Verbenaceae	390. 假连翘属 Duranta L.	390-1 假连翘 Duranta repens L.	$2n=2x=34$	14m(4SAT)+14sm+6st	3A	62	1.64	0.59	67.05	$18M_2+14M_1+2S$
	391. 马鞭草属 Verbena L.	391-1 马鞭草 Verbena officinalis L	$2n=56$	32m+18sm+6st	2A	106	1.93	0.29	64.11	$4L+20M_2+28M_1+4S$
	392. 牡荆属 Vitis L.	392-1 荆条 Vitis negundo L. var. heterophylla (Franch.) Rehd.	$2n=2x=34$	34m	1A	68	1.36	0	55.58	$16M_2+18M_1$
		392-2 单叶蔓荆 Vitis rotundifolia L. f.	$2n=2x=34$	16m+16sm+2st	2B	66	2.11	0.29	64.06	$2L+14M_2+14M_1+4S$
一一八、堇菜科 Violaceae	393. 堇菜属 Viola L	393-1 南山堇菜 Viola chaerophylloides (Regel) W. Bckr.	$2n=54$	34m+18sm+2st	2A	106	1.68	0.11	61.64	$2L+22M_2+30M_1$
		393-2 紫花地丁 Viola yedoensis Makino	$2n=48$	38m+10sm（1SAT）	2A	96	1.53	0.04	60.46	$24M_2+24M_1$
	394. 山姜属 Alpinia L.	394-1 大高良姜 Alpinia galanga Willd.	$2n=48$	22m+18sm+6st+2t	2A	88	1.63	0.46	65.72	$28M_1+18M_1+2S$
		394-2 草豆蔻 Alpinia katsumadai Hayata	$2n=48$	24m+18sm+6st	2A	90	1.78	0.33	64.18	$2L+24M_2+20M_1+2S$
		394-3 益智 Alpinia oxyphylla Miq.	$2n=48$	22m+22sm(2SAT)+4st	2A	92	1.94	0.33	63.93	$2L+22M_2+20M_1+4S$
	395. 豆蔻属 Amomum L.	395-1 砂仁 Amomum villosum Lour.	$2n=48$	6m+20sm+20st+2T	3A	74	1.86	0.75	72.93	$4L+18M_2+24M_1+2S$
		395-2 缩砂仁 Amomum villosum var. xanthioides T. L. Wu et S. J. Senjen	$2n=48$							
一一九、姜科 Zingiberaceae	396. 姜黄属 Curcuma L.	396-1 郁金 Curcuma aromatica Salisb.	$2n=63$							
		396-2 姜黄 Curcuma longa L.	$2n=63$							
		396-3 温郁金 Curcuma wenyujin Y. H. Chen et C. Ling	$2n=64$	26m+32sm+6st	2A	122	1.49	0.25	65.29	$2L+28M_2+34M_1$

续表

中药植物核型分析

科	属	种	染色体数目	核型公式	核型分类	臂指数	最长染色体/最短染色体	臂比>2的染色体比例	核型不对称系数	染色体相对长度组成
一一九、姜科 Zingiberaceae	397. 姜花属 Hedychium Koenig	397-1 姜花 Hedychium coronarium Koen.	$2n=48$							
	398. 山柰属 Kaempferia L.	398-1 山柰 Kaempferia galanga L.	$2n=2x=22$	14m+8sm	1A	44	1.48	0	60.46	$10M_2+12M_1$
	399. 姜属 Zingiber Boehm.	399-1 姜 Zingiber officinale Roscoe	$2n=4x=44$	26m+18sm	2B	88	2	0.05	60.47	$2L+22M_2+16M_1+4S$
一二〇、蒺藜科 Zygophyllaceae	400. 蒺藜属 Tribulus L.	400-1 蒺藜 Tribulus terrestris L.	$2n=36$	14m+18sm(4SAT)+2st+2T	3A	68	1.76	0.61	67.69	$16M_2+18M_1+2S$
一二一、鳞毛蕨科 Dryopteridaceae	401. 贯众属 Cyrtomium Presl	401-1 贯众 Cyrtomium fortunei J. Sm.	$2n=44$	2m+4sm+10st+28t	3A	50	1.49	0.91	86.55	$20M_2+24M_1$
一二二、木贼科 Equisetaceae	402. 木贼属 Equisetum L.	402-1 木贼 Equisetum hiemale L.	$2n=214$							
一二三、裸子蕨科 Hemionitidaceae	403. 凤丫蕨属 Coniogramme Fee	403-1 普通凤丫蕨 Coniogramme sinense Ching	$2n=120$							
一二四、球子蕨科 Onocleaceae	404. 球子蕨属 Onoclea L.	404-1 球子蕨 Onoclea sensibilis L. var. interrupta Maxim.	$2n=74$							

参 考 文 献

陈瑞阳等. 中国主要经济植物基因组染色体图谱. 第一册中国果树及其近缘植物染色体图谱. 北京：万国学术出版社，1994.

陈瑞阳等. 中国主要经济植物基因组染色体图谱. 第二册中国农作物及其野生近缘植物染色体图谱. 北京：科学出版社，2003.

陈瑞阳等. 中国主要经济植物基因组染色体图谱. 第三册中国园林花卉植物染色体图谱. 北京：科学出版社，2003.

陈瑞阳等. 中国主要经济植物基因组染色体图谱. 第四册中国竹类植物染色体图谱. 北京：科学出版社，2003.

陈瑞阳等. 中国主要经济植物基因组染色体图谱. 第五册中国药用植物染色体图谱. 北京：科学出版社，2009.

第三章 探索与发现

科学研究是一个长期积累的过程，有些基础研究需要几十年才能见到结果。所以，把我们一部分研究结果放在这个题目下，供大家参考。有些问题还有很大的发展空间，如植物染色体 G-带、B 染色体、荧光原位杂交、细胞地理学等。银杏性染色体则是我们一个很重要的发现。

第一节 植物 B 染色体

提要：B 染色体又称超数染色体（supernumerary chromosome）、附加染色体（accessory chromosome）或额外染色体（extra chromosome），是在生物界中广泛存在的一种特殊染色体。现已在千余种植物和近三百种动物中被发现。B 染色体与物种中正常染色体 A 染色体不同：独立于整倍体基因组之外，减数分裂时不与 A 染色体发生联会和配对，细胞分裂后期不分离，非孟德尔遗传，富含异染色质，不含对宿主主要性状有影响的基因等。

B 染色体 DNA 的分子组成既与 A 染色体极为相似，具有 A 染色体 DNA 分子组成的一般特征：富含重复序列和转座成分，染色体三大功能组件 DNA 高度同源；又与之相区别，含有 B 染色体特异的 DNA 序列，这些 DNA 序列可以为探讨 B 染色体的起源和进化提供有价值的信息。本章以含 B 染色体和不含 B 染色体的黑麦和玉米为材料，利用显微切割技术、LA-PCR 技术、AFLP-银染技术、RAPD 分子标记技术以及染色体荧光原位杂交技术等从不同层次和不同方面，对黑麦和玉米 B 染色体的 DNA 分子组成等进行了研究。

3.1.1 B 染色体分子生物学

B 染色体是独立存在于物种染色体组之外的一种特殊染色体，又被称为超数染色体（Supernumerary chromosome）、附加染色体（aeeessory chromosome）或额外染色体（extra chromosome）（Beukboom1994）。迄今已在 1000 多种植物和 300 多种动物中发现了 B 染色体。一般认为 B 染色体具有以下特征（王玉元，1997）：（1）与常染色体（A 染色体）形态不同，在绝大多数含 B 染色体的物种中，B 染色体小于 A 染色体，其组成绝大部分为异染色质；（2）B 染色体遗传不遵循孟德尔遗传法则，B 染色体在减数分裂时不分离，具有一定的积累和消减机制，例如在被检测的 1306 个玉米花粉粒中 B 染色体着丝粒不分离的频率约为 56.6%不分离主要发生在减数第二次分裂，在减数第一次分裂中也偶有发生；(3) B 染色体有时在有丝分裂后期不分离，在含 B 染色体的物种个体中，不同的体细胞所含有的 B 染色体数目会有差异；(4) B 染色体遗传活力是惰性的，其上不携带与主要性状相关的基因；(5) B 染色体的存在对物种的影响通常表现为中性，在个别物种中会在适应性及生长等方面产生影响，这种影响会随物种个体携带 B 染色体数目的增加而上升，例如在玉米中含 B 染色体的花粉会优先与卵细胞受精；Polwman（1994）对生长在河岸边的细香葱（*Allium schoenoprasum*）进行研究，发现在干旱情况下物种个体所含 B 染色体的频率与该物种种子的发芽率呈正相关，能增强种子萌发时抗干旱的能力。

众所周知，DNA 分子自身结构特点及其表达产物对染色体及其行为会产生决定性的影响，B 染色体也不例外，现代分子生物学的发展使了解 B 染色体的 DNA 结构特征成为可能。目前应用于 B 染色体研究的分子生物学技术手段主要包括：Southern 杂交、荧光原位杂交、减法杂交和染色体显微分离扩增技术。由于实验材料的限制，有关 B 染色体分子生物学的研究主要以禾本科的黑麦（*Secale cereale*）、玉米（*Zea mays*），菊科短毛菊属的（*Brachycome dichromosomatica*），两栖类的蛙类（*Leiopelma hochstetter*），昆虫中的果蝇（*Drosophila subsilvestter*）及寄生黄蜂（*Nasonia vitripennis*）的研究最为深入。综合近年有关 B 染色体分子生物学研究报道可以发现，B 染色体 DNA 在组成上既与 A 染色体 DNA 类似，又有其独特的序列和特征。

3.1.1.1 A、B 染色体共有序列

早在 70 年代，Timmis 等（1975）和 Flavell 等（1975）对含 B 染色体和不含 B 染色体的两种黑麦基因组进行 DNA 复性动力学研究，就曾发现二者之间在 DNA 组成方面无显著的差异，这也说明 B 染色体的 DNA 分子组成与常染色体的 DNA 分子组成相类似。减数分裂时 B 染色体因不与 A 染色体配对，可以很容易的识别和分离。Sandery（1991）等最早将染色体微切方法应用于 B 染色体 DNA 序列的研究。他们将减数分裂时不配对的 B 染色体进行原位裂解和提抽，并进行了微克隆。Southern 杂交分析显示，克隆文库中的克隆与含 B 染色体和不含 B 染色体的两种黑麦总体 DNA 都有杂交信号；因 B 染色体分离数目少，克隆文库不能覆盖整条 B 染色体，因而没有发现 B 染色体特异序列，但该研究结果有力地支持了 B 染色体与 A 染色体在 DNA 组成上十分相似的看法。Houben（1996）等利用显微切割技术对黑麦 B 染色体的近端部位置进行了切割和 DOP-PCR（Degenerate Oligonucleotide Primer-Polymerase Chain Reaction）扩增，扩增产物与不含 B 染色体的黑麦总体 DNA 也有很强的杂交信号，同样验证了黑麦 B 染色体与 A 染色体的 DNA 组成具有相似性。其他植物中的 B 染色体，如玉米（Stark，1996）等，都与黑麦 B 染色体一样，有着与各自物种 A 染色体 DNA 组成类似的特征。在动物中，利用 AP-PCR（Arbitrarily Primed PCR）技术，Tanic（2000）等对同品系黄颈鼠（*Apodemus flavicolls*）中不含 B 染色体和含 1B、2B 和 3B 染色体的个体基因组的总体特征进行了初步分析，虽然从 DNA 定量方面可以区分基因组是否含有 B 染色体，但是 B 染色体的有无并不改变基因组的总体特征。在 B 染色体功能组件，如着丝粒、端粒和复制起始点等方面，郭歌等（1998）以寡核苷酸（CCCTAAA）3 为引物，从 B 染色体端部微切片

段的 PCR 产物中扩增出了黑麦 B 染色体端粒相关序列。经原位杂交，该序列被定位到黑麦所有染色体的端部，同时它还与玉米端粒相关序列高度同源。张荣信等（1999）将黑麦 B 染色体着丝粒微切产物进行 LA-PCR（Linker Adapter PCR）扩增并筛选到一个高度重复序列 pRBC6。该序列在玉米、水稻（*Oriza sativa*）、顶芒山羊草（*Aegilops comosa*）、拟南芥以及酵母（*Saccharomyces cerevisiae*）中找到了与之高度同源的序列。两色蜀黍（*Sorghum bicolor*）BAC（Bacterial Atificial Chromosome）克隆中的亚克隆中为 745 bp 大小的重复序列，其在禾本科多种植物中保守，并通过原位杂交定位于植物的染色体着丝粒区，在黑麦 B 染色体着丝粒区也有杂交信号。这些研究进一步表明植物 B 染色体在功能组件构成方面与 A 染色体也十分相似。

DNA 甲基化被认为是细胞调控基因活性的重要机制之一，在很多已知 B 染色体序列中都发生了甲基化。Neves（1992）等利用 5-aczacytidine 处理黑麦根尖，使其在 DNA 合成时掺入异常的胞嘧啶，发生类似甲基化的修饰，处理后的根尖细胞在有丝分裂后期染色体不能正常分离，其行为与 B 染色体在有丝分裂和减数分裂时不分离的行为十分类似。这一研究从一个侧面说明 B 染色体的细胞学行为与其 DNA 分子的修饰，特别是 DNA 甲基化密切相关。冷泉港实验室对拟南芥 4 号染色体上的一个 0.5~0.7 Mb 的异染色质区域进行了深入的研究（The Cold Spring Harbor Laboratory，2000），他们发现该区域主要由一些高度重复序列转座子（Transposon）和反转座子（Retrotransposon）)组成，基因被重复序列高度分割，密度也非常低，整个区域 DNA 序列的甲基化程度都非常高。传统的细胞化学方法早已证实，B 染色体主要是由高度凝集的异染色质组成，不难想像，B 染色体 DNA 也具有与异染色质类区似的特征。

3.1.1.2　B 染色体特异序列

植物 B 染色体 DNA 特异序列最早是从黑麦中获得的。Sandery（1990）等用多种内切酶对含 B 染色体和不含 B 染色体的黑麦总 DNA 进行酶切分析，在 *Dra* I、*Eco*R I 和 *Bam*H I 酶切的总 DNA 产物中都发现了一个仅在含 B 染色体的基因组中出现的 DNA 片段。Southern 杂交分析显示三者可相互产生杂交信号，但杂交模式有所不同。这表明在黑麦 B 染色体上存在一个具有 *Dar* I、*Eco*R I 和 *Bam*H I 酶切位点的高拷贝重复序列家族，并被命名为 Dl100。将 Dl100 与含 B 染色体的黑麦种群及黑麦近源种 *Secale vavilavii* 进行杂交，结果显示 Dl100 出现在黑麦与 *S.vavilavii* 在进化上相互分离之后，但又早于黑麦含 B 染色体种群的出现。随后，Blunden（1993）等利用相似的方法又分离到另外一个 B 染色体特异重复序列家族——E3900。E3900 仅出现在黑麦中，禾本科的其他作物中不含这一序列。Houben（1996）等利用显微切割技术对黑麦 B 染色体的近端部位置进行了微切和 DOP-PCR 扩增，在扩增产物中检测到黑麦 B 染色体特异序列 Dl100 和 E3900 序列。当以 Dl100 序列特异引物对 DOP-PCR 产物进行扩增时，还发现了 Dl100 类似序列，这些序列与黑麦 A 染色体 DNA 也有杂交信号，揭示 B 染色体上特异序列可能从 A 染色体演化而来。E3900 和 Dl100 性质类似，只是具有不同的重复单位，它们紧密邻接，都定位于 B 染色体长臂的端部，E3900 则更靠近端粒部位。在 E3900 的一个亚克隆 pRAB2100 中存在一个 A、B 染色体共有的中度重复序列，该序列还同时存在于小麦和大麦基因组中。pRAB2100 的端部序列还与玉米的 Copia-类似反转座子 PREM2 和 Ropie 的多嘌呤结合位点高度同源。E3900 的第 2853～3384 位点与反转座子 Ty3/gypsy 家族中的 crwydryn 具有很高的同源性；该区域还包含编码部分 gag 蛋白的开放读码框（Open reading frame，ORF）。而在对 Dll00 单位重复序列的分析中发现，其序列中包含一个微小反向重复可转座因子（miniature inverted-repeated transposable element，MITE）Tnr1。除此之外，在 E3900 和 Dll00 中都还含有与 Tnr 类似的序列（Langdon，2000）。

Lin（1997）等利用 RAPD（Random Amplification Plymorphic DNA）技术，从含 A、B 重组染色体的 1B 和 Bl 玉米株系中筛选出 B 染色体特异性标记：TB-4Sa、TB-1La、TB-3Sb、UBC313、UBC345、UBC349 和 UBC426。1B 染色体为 B 染色体和 1 号染色体之间的重组体，即具有 1 号染色体的着丝粒和 B 染色体的长臂；Bl 染色体则具有 B 染色体的着丝粒和 1 号染色体的长臂。由于不同的突变株系中 B 染色体长臂与 1 号染色体发生交换位置不同，可以通过检测特异标记在 1B 和 Bl 染色体上出现的频率来确定标记在 B 染色体上的位置。利用这种方法上述标记被定位于 B 染色体。Alfenito（1993）在对玉米的研究中，发现一系列与玉米及多种植物着丝粒序列同源性很高的 B 染色体特异序列。最近的研究表明，玉米 B 染色体着丝粒特异序列与玉米 4 号染色体着丝粒序列最为密切。

John 等（1991）最早报道了利用减法杂交获得的一个短毛菊属 *Brachycome dichromosomatica* B 染色体特异重复序列 Bd-49，其重复单位为 176 bp，Leach（1995）等对 Bd-49 重复序列又做了更深入的研究。Bd-49 序列位于 *Brachycome dichromosomatica* B 染色体着丝粒部位，经甲基化分析后发现 B 染色体上的这一重复序列是高度甲基化的，其甲基化点为 ^{m}CG 和 ^{m}CCG。Bd-49 重复序列在短毛菊属的其他物种中也有出现，降低杂交的严紧度和延长杂交时间，在 *Brachycome dichromosomatica* 的 A 染色体上也能检测到与 Bd-49 类似的重复序列。Franks（1996）等又以 Bd-49 序列为探针，对全基因组文库进行了筛选，从 9000 个噬菌体克隆中仅筛选到一个含 Bd-49 序列的克隆 λBd-49，在随后对 λBd-49 亚克隆的分析中，共找到 6 个含 Bd-49 的亚克隆，部分克隆序列富含 AT，并分布于整个基因组，其中 Bd-49-6f 与莴苣（*Lactuca sativa*）基因组中的一个反转座子序列有 70% 的同源性，与 *Cladosporum fulvum* 的一个反转座子的同源性为 65%。Jamilean 在纤细还阳参（*Crepis capillaris*）中发现一个名为 pCc21 的 DNA 片段。序列分析显示 pCc21 序列中包含一段 54 bp 富含 CA 的序列：GATCTAACACACGCAGTACACACACCAAACACACACAGACACACATAGAAAACGT。这一 CA 富含区分布于整个基因组中，在 A、B 染色体之间的差异则表现为该序列拷贝数的差异。在寄生黄蜂（*Nasonia vitripennis*）中存在一种特殊的 B 染色体，即 PSR（Paternal Sex Ratio）染色体。*N. vitripennis* 是以单双倍体来决定个体的性别，即受精卵发育成为雌性，未受精卵发育成雄性，但是当雌性与含有 PSR 染色体的雄性交配后，其后代全部都为雄性个体。细胞学的观察和研究发现在受精卵第一次有丝分裂时，PSR 染色体可以使来自父方的常染色体高度凝集并在随后的分裂过程中丢失。Nur（1988）等构建了含有雄性 PSR 染色体的噬菌体文库，在 6000 多个克隆中找到 36 个只与含 PSR 染色体个体有杂交信号的克隆，并最终筛选出四个 PSR 染色体特异的高拷贝重复序列：PSR2、PSR18、PSR22、PSR1O5。

PSR2 重复序列在不同克隆间具有 92%～99%的同源性，并含有一个 171 bp 长的重复单位；PSR18 的重复单位为 154 bp，它和 PSR2 序列之间具有 73%的同源性；PSR105 为 PSR18 的一个亚家族，具有 214 bp 长的重复单位；PSR22 重复序列中则包含一个 183 bp 长的重复单位。这些序列在 PSR 染色体中的含量可高达 30%，在 *N.vitripennis* 的近源种 *N. Longicornis* 和 *N.Giraulit* 中也发现了它们的同源序列。这几个 PSR 特异重复序列中都有两个短的保守序列，即保守区 I：AAAGTCNGACTT 和保守区 II：TTTTATAAAA，对于这两个保守区的功能尚不清楚。根据 PSR 染色体在有丝分裂过程中的反式作用，这两个保守区可能参与结合某些维持常染色体正常结构的蛋白，或者它们编码某些导致常染色体凝聚的蛋白，致使染色体凝聚。Gutknecht（1995）等利用基因组 DNA 酶切差异显示的方法在果蝇中发现一个新的 B 染色体特异的高度重复序列，命名为 PsSP216。PsSP216 具有一般高度重复序列的特征：以 216 bp 为单位高度重复，不同的重复序列单位之间同源性在 93%以上。原位杂交时，该序列与整条 B 染色体都有非常强烈的信号，在 A 染色体着丝粒部位也有信号。PsSP216 高度重复序列可能携带了某些与染色体结构特别是着丝粒结构域有关的信息。

生长在新西兰的一种蛙类 *Leiopelma hochstetter* 性别决定为 ZW/ZZ 型，该物种中的 B 染色体与 W 染色体具有非常类似的特征，已发现的 B 染色体特异序列：34L4605、34L4477、34 L7416、34A16649、35A2393、35A2433 和 35A18317 同样也出现在 W 染色体上（Sharbel，1998）。*Leiopelma hochstetter* 中，B 染色体的产生很可能是单价 W 染色体重组和重复序列快速扩增的结果，B 染色体的进化很可能与雌性决定 W 染色体上的特异 DNA 有关。

现代分子生物学的技术手段揭示了 B 染色体 DNA 的一般特征，使有谜一般身世的 B 染色体不再不可测。在 B 染色体分子生物学的相关报道中都清楚的显示出 B 染色体与 A 染色体 DNA 组成的相似性，从分子水平上为 B 染色体来源于 A 染色体提供了强有力的证据。B 染色体一旦在物种中产生，会沿着其独特的进化路线前进，并具有种种特殊的性状和行为。就目前的研究来看，在动物特别是昆虫和两栖类中 B 染色体具有与性染色体类似的特征，而在植物中通常不存在性染色体；似乎在动物和植物中 B 染色体的产生途径会有所区别。虽然已经在不同物种中发现了一些 B 染色体特异的序列，但这些序列基本上都是高度重复序列，也只定位于 B 染色体的着丝粒和端粒部位，就认识 B 染色体的本来面目而言，未知还远远大于已知。真正了解 B 染色体的起源和进化，不仅可以揭示遗传物质及其载体的演变途径，还可以开发 B 染色体在染色体工程中的巨大潜能。

（作者：祁仲夏、陈瑞阳。原载：细胞生物学杂志，2002，24（4）199-202）

3.1.2 黑麦 B 染色体显微切割和微克隆

1928 年 Randolph 在玉米中发现一种较少的染色体，其形态特征与常染色体不同。为了与常染色体区别，将其命名为 B 染色体。此后，陆续在 1300 多种植物和近 500 种动物的自然群体中发现（Jones，1993）。B 染色体被认为是可有可无的染色体，其形态特征、生物学效应和生物行为在许多物种中进行了详细的研究（Jones，1982）。但是，对 B 染色体 DNA 的具体特性却知道的很少。随着生物学的发展，对 B 染色体的研究也不断深入。在 B 染色体的分子组成、特异性序列等方面也有了深入的报道。

B 染色体分子组成的早期研究主要基于对含有 B 染色体和不含有 B 染色体的总体 DNA 进行比较，证明 B 染色体与 A 染色体 DNA 没有明显的区别（Rimpau，1976）。此后，随原位杂交技术及分子生物学技术的发展，发现许多物种 B 染色体的大部分 DNA 相似或相同于 A 染色体 DNA（Tsujimoto，1992；Cuagrado，1994；McQuade，1994；Jamilena，1995；Franks，1996）。由于 B 染色体 DNA 与 A 染色体 DNA 有较高的同源性，给分离 B 染色体 DNA 带来很大的困难。显微切割技术的发展给 B 染色体分子生物学研究带来了曙光。Sandery 等在 1991 年最早报道了应用微切微克隆技术分离到一个黑麦 B 染色体和 A 染色体都普遍存在的序列。此后，McQuade（1994）和 Jamilena 等（1995）也先后将微切微克隆技术应用于 B 染色体的研究。此技术的应用，加速了 B 染色体分子生物学研究进程。

本文采用微切微克隆技术构建了黑麦 B 染色体的部分 DNA 文库，为研究黑麦 B 染色体的分子组成，筛选 B 染色体特异性 DNA 序列以及筛选 B 染色体的关键性序列（自主复制序列，着丝粒和端粒序列等）奠定了基础。

3.1.2.1 材料与方法

1. 材料

（1）植物材料：黑麦（*Secale cereale*），JNK 品种，含 0～4 条 B 染色体，由英国威尔士大学 R. N. Jones 教授惠赠。

（2）接头和引物：引物参照邓汉湘（1992）报道合成。将合成的 24 nt 和 10 nt 寡核苷酸序列等摩尔混合，58℃退火 1h 制成接头。以 24 nt 寡核苷酸序列为引物。

↓ *Eco*R I
接头：5′-CGGGAAT TCTGGC TCTGCGACATG-3′
3′- C TGTACCT AG-5′

2. 方法

（1）染色体标本制备及显微分离：参照文献（宋文芹，1996）方法进行。

（2）PCR 扩增前处理：取一个 0.5 mL Ependorf 离心管作阴性对照。将收集 B 染色体的 Ependorf 管稍离心后，向上述两管中分别加入 2 μL*Sau*3AI 1×Multi-buffer（Promega），再加入 0.5 μL（10 mg/mL）蛋白酶 K 溶液，37℃消化 3 h，75℃灭活 15 min。

向上述反应液中加入 2 μL（20 μmol/L）接头，0.6 μL T$_4$DNA 连接酶缓冲液，1 μL（6 U/μL）T$_4$DNA 连接酶，14℃连接 16 h。

（3）PCR 扩增：向连接反应液中加入 10 μL10×TaqDNA 聚合酶缓冲液，6 μL MgCl$_2$（25 mmol/L），10 μL 10×dNTP（2.5 mmoL/L each），10 μL 引物（2mmol/L），5U TaqDNA 聚合酶，补双蒸水至 100 μL。首先在 70℃下延伸 10 min，补平接头。随后按 95℃变性 1 min，56℃退火 1 min，72℃延伸 1 min，运行 30 个循环，然后 72℃延伸 5 min。取 1 μL 上述 PCR 产物为模板，100 μL 体系，按上述反应条件进行 20 个循环的次级 PCR。

（4）PCR 产物 Southern 杂交分析：黑麦种子萌发后，取其根尖制片，鉴定得到 0B（不含 B 染色体）黑麦及+B（含 B 染色体）黑麦种子，继续培养，取其黄化苗，用 CTAB 法（Marray，1980）提取 0B、+ B 黑麦总 DNA。将所提总 DNA 经 *Eco*R I 消化后，与 PCR 产物同时用 1.2% 的琼脂糖凝胶电泳。转移在硝酸纤维素膜上进行杂交（萨姆布鲁克，1989）。

将次级 PCR 产物经酚/氯仿抽提，乙醇沉淀后用随机引物法标记地高辛探针。标记方法，杂交过程及杂交信号检测按宝灵曼提供的地高辛标记检测试剂盒说明书进行。

（5）PCR 产物的克隆：将次级 PCR 产物纯化后，用 *Eco*R I 酶切，低熔点胶回收，取 200 ng 回收产物连接到去磷酸化的 pUC19 载体中。取 1/20 连接产物转化大肠杆菌 DH5α，在含有 X-gal、IPTG 和氨苄青霉素的 SOB 固体培养基上 37℃培养过夜，选取白色菌落。

（6）重组质粒筛选及酶切鉴定：将所得白色菌落，用 Kieser（1984）法粗筛重组质粒。1.2%琼脂糖凝胶电泳，选取落后于空载质粒的重组质粒。以碱法提取质粒，经 *Eco*R I 酶切，琼脂糖电泳，检测到 200～550 bp 酶切外源片段。

3.1.2.2　结果与讨论

1. B 染色体的识别及显微分离

本实验所用材料二倍体黑麦（$2n$=14+0～4B）JNK 品种中，14 条 A 染色体均为中部或近中部着丝点. 而所含 0～4 条 B 染色体，大小占 A 染色体的 1/2 左右，为近端部着丝点，从形态上很容易识别和进行显微分离。

2. LA-PCR 扩增

本文所采用的 LA-PCR 方法，是对模板 DNA 进行酶切后，在酶切片段两端连上接头，然后引物与接头序列匹配进行扩增。从 PCR 产物电泳结果看，阴性对照未出现杂交信号，说明操作过程中没有同源 DNA 污染。从电泳结果可见得到了 0.2～2 kb 的扩增片段，呈涂片状，主要集中在 0.2～0.8 kb 之间，无特异带。从整个 PCR 过程和引物特点分析，不难想到 PCR 产物成分多样，包括单拷贝序列和重复序列。

3. Southern 杂交分析

从 Southern 杂交分析结果来看，阴性对照及 λDNA/*Eco*R I -*Hin*d III 标记并未出现杂交信号。B 染色体 PCR 产物及 0B、+ B 黑麦总体 DNA 均出现信号。0B、+ B 黑麦总体 DNA 信号无明显区别。这与前人研究的结果 B 染色体与 A 染色体的 DNA 没有明显的区别吻合（Rimpau，1976；Tsujimoto，1992；Cuagrado，1994；McQuade，1994；Jamilena，1995；Franks，1996）。在微切过程中，黑麦 B 染色体与 A 染色体有明显区别，极易识别，证明分离的 B 染色体 PCR 产物来源于黑麦 B 染色体。

4. PCR 产物克隆

为进一步分析 PCR 产物的成分，我们取连接产物的 1/20 进行克隆，获得 5000 个白色菌落。经 Kieser 法初筛，结果 80%白色菌落为重组克隆。进一步酶切鉴定，外源片段约为 200～550 bp。PCR 产物在 0.2～2 kb 之间，而克隆得到的外源片段在 200～550 bp。这可能与 PCR 及克隆过程中小片段优先有关。阮国庆（1996）等用此方法克隆得到 100～

600 bp 的单拷贝片段，平均为 350bp，与本文结果基本相同。

在引物上含有 *Eco*R I 酶切位点，便于克隆。但这种单一位点克隆，容易产生假阳性。虽然经过去磷酸化，但去磷酸化后连接效率降低。如果去磷酸化不彻底，仍能产生假阳性。本实验中的假阳性可能与去磷酸化不彻底有关。

本实验中的 B 染色体 PCR 产物如全部用于克隆，可以得到高覆盖率的 B 染色体 DNA 文库，关于微切微克隆方法构建 DNA 文库的覆盖程度，目前尚未见文献报道。此方法得到的外源片段均较小，1991 年 Hadano 用 SUP-PCR（Single Unique Primer PCR）方法构建了人 4 号染色体短臂的特异性文库，克隆片段平均为 350bp，经筛选鉴定 40% 为单拷贝克隆。1995 年 Jamilena 等人用 DOP-PCR（Degenerate Oligonucleotide-Primed PCR）技术对微切的 *Crepiscapillaris* 的 20 条 B 染色体构建了专一性 DNA 文库。1996 年 Houben 等人又将微切 DOP-PCR 技术成功地用于黑麦 B 染色体末端异染色区域研究，检测到 B 染色体专一序列，以上资料为本实验的进一步深入提供了可靠的依据。本文所得大量克隆经进一步 Southern 杂交及原位杂交分析，可用于分析 B 染色体的关键序列以及某些特异性序列，以丰富 B 染色体特异性探针。这对于研究 B 染色体的分子结构和 B 染色体的起源以及构建 B 染色体人工载体均具有重要意义。

（作者：张荣信、宋文芹、李秀兰、陈瑞阳。南开大学学报，1999，32（2）103-106）

3.1.3 玉米 B 染色体特异 RAPDs 标记的分离与染色体定位

B chromosomes, also named supernumerary chromosomes, derived from standard chromosome (A chromosome) set. Once they originated, B chromosomes are in a herited in a non-Mendelian way and express various independent behavior. They do not pair with A chromosomes, exhibit meiotic and mitotic instability, nondisjunction and have little or no phenotypic effects on the host (Beukeboom 1994, Jones 1982). Since B chromosome DNA is quite identical to corresponding sequence on the A chromosome complement, B chromosome specific DNA sequence has been the attractive subject of B chromosome research. So far only a few of B chromosome specific DNA sequences have been identified (Langdon 2000, Page 2001). So more B chromosome specific sequences are required for explanation of the origin and behavior of B chromosome. In this article we described the isolation and characterization of a B chromosome specific DNA sequence from maize genome RAPD analysis.

The seeds of black Mexican sweet maize with B chromosome B542A (+B) and without B chromosome B542B (0B) were obtained from Maize Genetic COOP Stock Center, Illinois, USA. They were cultured in sterile Linsmaier-Skoog (LS) media in the dark for about 15 days. Genomic DNA was isolated based on the procedure described by Murray and Thompson (1980). Genomic DNA of ten 0B and ten +B individuals were mixed respectively for the further RAPD analysis. RAPD analysis was performed according to Williams et al (1990). The amplification process was 1 min at 94℃, 1 min at 36℃, and 1 min at 72℃ for 40 cycles

(Lin 1997). Specific band was cut with a sterile knife and transferred into an eppendorf tube containing 100 μL TE buffer. DNA was released by heating in 97℃ for 10 min. One μL of the solution could be used for 20 μL reaction volume. Amplification was the same as in RAPD analysis. The DNA fragment was sequenced in Sangon (Shanghai). Sequence analysis was performed in BioEdit.

Chromosome spreads were made according to Chen et al (1979). FISH was carried out as described in Nonradioactive in situ Hybridization Application Manual (Second Edition, Roche). Probe was labeled as DIG (digoxigenin- dUTP) by random-primed DNA synthesis.

Total 360 RAPD primers were explored. One + B genome specific RAPD marker was obtained from the amplification of primer S480. S480 sequence was 5′- GACC2CTAGTC- 3′. The marker was named as B480, and it appeared consistently in the repeat RAPD analysis (Fig. 3.1).

Corn B chromosomes were little telomeric chromosomes, which were quite different from its A chromosomes in morphology (Fig. 3.2C). The result of FISH showed veryclearly that B480 was located on centromeric domains of B chromosomes (Fig. 3.2A). At the same time, strong signal was also observed in nucleolus (Fig. 3.2B). Furthermore, very weak signal could be observed on the A chromosomes.

The sequence of B480 enriched the base A and T. They occupy 64 % of the total base composition. Genbank database search illustrated that B480 was homologous to corn ARS1 (Autonomously Replicating Sequence) (Genbank Accession No. X12753) and ARS2 (X12754) (Berlani 1988) (Fig. 3.3). The sequence, 5′- GCATATTTTATT2TATTTTTGAC- 3′ from site 340 to 365 in B480, was conserved in the genomes of human (Homo sapiens), Caenorhabditis elegans, Arabidopsis thalian and other species. To date, only ARS in Saccharomyces cerevisiae and Drosophila melanogaster are clarified (Gilbert 2001). So further analysis of B480 will not only be very helpful for explanation of B chromosome special characters and its evolution path, but also widen our knowledge of ARS in plants, which could be very valuable for construction of plant artificial chromosome (PAC).

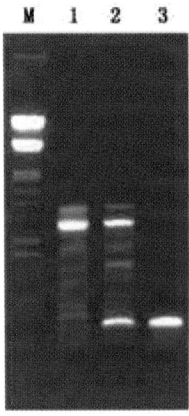

Fig. 3.1　RAPD result using primer S480 and the reamplification of differential band B480. Lane M, size marker (λDNA/ EcoR I + Hind III); Lane 1, 0B genome; Lane 2, +B genome; Lane 3, B480.

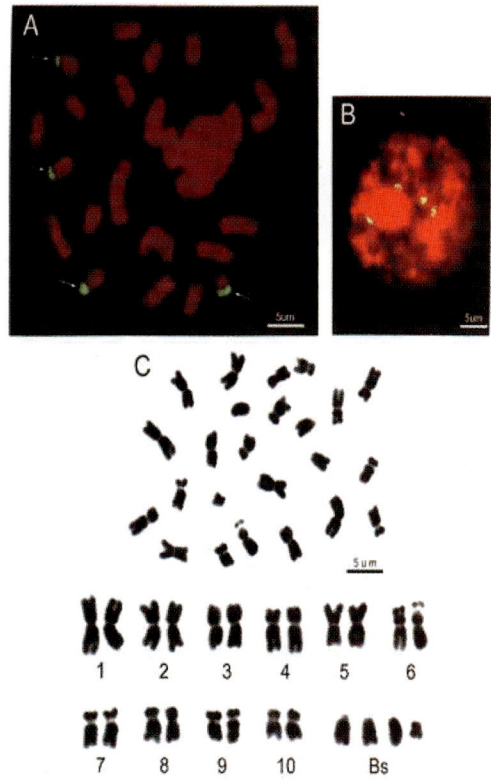

Fig. 3.2 FISH analysis with labeled B480 and the karyotype of corn（Zea mays）with 4 B chromosomes. bar = 5 μm. The distribution of B480 on the metaphase chromosomes. B. The distribution of B480 in the nucleolus. C. The karyotype of corn（$2n = 20+ 4B$）.

（Authors：Qi ZX，Li XL，Chen CB，Song WQ，Chen RY.
Published In：Acta Botanica Sinica，2002，44（4）499-501

3.1.4 黑麦 B 染色体端粒相关序列的克隆

B 染色体是广泛存在于动、植物中的一类附加于常染色体组之外的额外染色体，它是造成动、植物染色体数目多态性的主要原因之一。目前关于 B 染色体已有较深入的细胞遗传学研究（Jones，1982）。但由于 B 染色体与 A 染色体在 DNA GC 百分含量、重复序列比例及复杂度等方面的相似性给 B 染色体 DNA 的分离分析工作带来极大困难（Timmis，1975），以至于 B 染色体 DNA 的分子组成及起源进化问题仍是一难解之谜。端粒是由染色体末端 DNA 序列和其相关蛋白构成的特殊染色体末端结构。它对维持染色体末端稳定性，保证染色体末端稳定复制有重要意义。此外，端粒还可能参与间期细胞核三维结构的建成，可能会影响基因表达或染色体的行为（Wang，1993）。端粒 DNA 包括端粒简单重复序列及端粒相关序列两部分，目前还没有关于 B 染色体端粒 DNA 的研究报道。本文试图通过染色体显微切割和 PCR 技术分离扩增黑麦 B 染色体端粒相关序列，以期对黑麦 B 染色体末端 DNA 组织结构进行探索性研究，这对进一步搞清 B 染色体端粒结构具有重要意义。

3.1.4.1 材料和方法

1. 材料

黑麦（*Secale cereale* L.）JNK 居群，由英国威尔士大学 R. N. Jones 博士惠赠。引物（CC-CTAAA）3 由赛百盛公司合成。

2. 方法

（1）黑麦染色体标本制备及显微切割参照宋文芹等（1996）的方法进行。

（2）染色体端部显微切割片段的 PCR 扩增。①染色体去蛋白：显微切割染色体片段中加入 3 μL 蛋白酶 K 处理液（1× *Sau*3A I 缓冲液，0.1 g/L 蛋白酶 K），45℃保温 5 h 后，75℃ 10 min 灭活蛋白酶 K。② *Sau*3A I 酶切显微切割染色体片段：上述体系中补加 2 μL 反应液成为 5 μL 酶切体系（0.1 U *Sau*3A I，1× *Sau*3A I 缓冲液，37℃保温 2 h 后，70℃ 20 min 灭活内切酶。③ 染色体 DNA 酶切片段的随机连接：上述体系中补加 5 μL 反应液成为 10 μL 连接体系（1× T4 连接酶缓冲液，0.5U T4 连接酶）14℃连接 10 h 后转入下述 PCR 反应体系。④初级 PCR 扩增：20 μL 体系包含 10 μL 连接混合物、1× PCR 缓冲液、5% PEG 8000、0.1% NP-40、0.4 μmol/L 引物、200 μmol/L 4× dNTP、1 U Taq DNA 聚合酶。92℃预变性 5 min 后，加入 1 U Taq DNA 聚合酶，92℃ 1.5 min，55℃ 5 min，72℃ 2 min，循环 20 次。取 3 μL 初级 PCR 扩增产物扩大为 30 μL 二级 PCR 反应体系，其中包含：3 μL 初级 PCR 产物、1× PCR 缓冲液、0.1% NP-40、0.2 μmol/L 引物、200 μmol/L 4× dNTP、1.5 mmol/L Mg^{2+}、1.5 U Taq DNA 聚合酶。92℃ 1.5 min，58℃ 1 min，72℃ 2 min，循环 30 次。最后 72℃延伸 8 min。⑤ 取二级扩增产物 10 μL，10g/L 琼脂糖凝胶电泳检测。

（3）染色体原位杂交。①探针标记：B 染色体短臂端部显微切割片段 PCR 扩增产物以随机引物法进行 Bio-11-dUTP 标记。②染色体制片：去壁低渗法制片，自然干燥，−20℃保存备用。③杂交：制好的染色体标本于 30%~100%乙醇中脱水，各 5 min，风干。每张载片加 200 μL RNaseA（10 g/L），37℃保温 40 min。2× SSC 冲洗 3 次每次 3 min，70%~100%乙醇脱水 5 min。每张载片加 30 μL 杂交液（50%甲酰胺，2×SSC，5×Denharts，10%硫酸葡聚糖，500 mg/L 变性鲑精 DNA，探针 4 μL）37℃杂交过夜。2×SSC、0.1% SDS，37℃洗片 3 次，每次 30 min，0.1×SSC、0.1% SDS，37℃洗片 3 次，每次 30 min；2×SSC 洗片 2 次，每次 10 min，1× PBS 吐温-20 室温洗片 5min。5% BSA、1×PBS 吐温-20 封阻 10 min 后，排干，加入 50 μL Avidin-HRPO（40 mg/L），37℃保温 60min。每片加 200 μL DAB（0.5 g/L），4℃暗处理 20 min 后，DAB/H$_2$O$_2$（40：1）4℃暗显色 20 min，1×PBS 冲洗，5% Giemsa 染色 4 min。水冲片，自然干燥，镜检。

（4）PCR 产物的克隆。B 染色体短臂端部显微切割片段 PCR 扩增产物中加入 1U Klenow 酶，37℃保温 30 min 后，酚、氯仿抽提纯化，无水乙醇沉淀。将沉淀溶于 18 μL 双蒸水中，加入 2 μL 绿豆芽核酸酶 10×缓冲液、绿豆芽核酸酶 1 U，37℃保温 15 min 后，酚、氯仿抽提纯化，无水乙醇沉淀 DNA，溶于 20 μL TE 中。取其中 3 μL 连入 pUC19

载体中，转化 DH5α，X-gal/IPTG 筛选白色菌落，*Eco*RⅠ/*Bam*HⅠ双酶切鉴定重组克隆。

（5）序列分析。重组克隆子 pp3 利用 M13 正向引物进行测序。

3.1.4.2 结果和讨论

1. B 染色体形态识别及显微切割

B 染色体短臂端部片段的显微切割，二倍体黑麦常染色体组有 14 条 A 染色体，均为中部或近中部着丝点染色体。实验用材料 JNK 中一般含有 0~4 条 B 染色体，不同的个体中 B 染色体的数目不同，其中以具有两个 B 染色体的个体居多。B 染色体在形态上与 A 染色体差异很大，为近端部着丝点染色体，大约为 A 染色体的 1/2，与 A 染色体容易区别。本实验中共切取了 10 个 B 染色体短臂端部片段，同时切取 10 个 A 染色体短臂端部片段作为 PCR 扩增反应的正对照。

2. B 染色体短臂端部显微切割片段的 PCR 扩增

参照 Burr（1992）及洪德军和陈受宜（1995）以总体 DNA 为基础的 TAS 序列 PCR 扩增法，建立了用于扩增单条染色体端粒相关序列的二级单引物 PCR 扩增法。端粒简单重复序列（CCCTAAA）$_n$ 中不含有 *Alu*Ⅰ、*Ava*Ⅱ、*Fok*Ⅰ、*Hae*Ⅲ、*Rsa*Ⅰ、*Sau*3AⅠ等酶切位点，而其亚端部重复序列对这些酶是敏感的，因而用这些酶对染色体 DNA 进行酶切，可以产生一端为端粒简单重复序列、另一端为酶切末端的 DNA 片段。将这样的酶切片段随机连接，其两端的端粒简单重复序列就可以为 PCR 反应提供引物结合位点，从而实现对其中端粒相关序列的扩增。按本文设计的 PCR 扩增体系将显微切割的 B 染色体短臂端部片段扩增后，得到分子量为 350 bp、550 bp 的两条清晰扩增带。作为阳性对照的 A 染色体也得到 350 bp、550 bp 的两条清晰扩增带。Burr（1992）从玉米总体 DNA 中扩增克隆的 TAS 序列长度分别为 382 bp、311 bp、369bp；洪德军（1995）扩增的水稻 TAS 序列分子量为 503 bp，由此可见我们建立的显微切割-PCR 扩增体系扩增的 TAS 序列的分子量与 Burr 等及洪德军和陈受宜从总体 DNA 中扩增的 TAS 序列的分子量相当，说明我们显微切割-PCR 扩增体系中染色体标本制备及蛋白酶 K、*Sau*3AⅠ、T4 连接酶对显微切割染色体片段的处理没有对 PCR 扩增造成不利影响。

黑麦二倍体基因组 DNA 含量为 5~10 pg（Flavell，1975），单条 B 染色体约占其中的 5.5%，我们的显微切割方法平均可以切取单条 B 染色体全长的 1/5~1/10，按此量估算 10 个显微切割片段的 DNA 含量约为 0.4 pg。每次二级 PCR 扩增 30 μL 体系中包含 400~800 ng DNA，此量已足够用于一般的探针标记、分子克隆等实验，所以我们建立的显微切割-PCR 体系是高效、灵敏的。

3. 染色体原位杂交

将 B 染色体端部显微切割片段 PCR 扩增产物作为探针进行染色体原位杂交实验，结果显示 B 染色体短臂末端呈现清晰的杂交信号，长臂末端没有杂交信号，杂交信号集中在染色体端部较大的区域，在染色体内部没有散布的杂交信号，且端部的杂交信号一直延伸到染色体的最末端，说明 PCR 产物是位于 B 染色体短臂端部的重复序列。A 染色体末端也显示了清晰的杂交信号，且信号的强度及大小在不同 A 染色体间存在差异，说明 B 染色体亚端部区域在分子结构上与 A 染色体有一定的同源性。从杂交结果可以看出具随体的（1R）染色体在其随体处杂交信号较强，且杂交信号覆盖了其核仁组织区（NOR），但是黑麦 A 染色体的 210 bp、480 bp、610 bp、630 bp 四种特征性亚端部重复序列家族与 NOR 区域均不能杂交（Flavell，1982），这暗示我们的 PCR 产物中可能存在与 A 染色体特征性亚端部重复序列家族不同的 DNA 组分。

4. PCR 产物的克隆及序列分析

取 30 μL 二级 PCR 扩增产物纯化后，连入 pUC19 载体克隆。经 X-gal/IPTG 筛选得到 73 个白色菌落。取其中 45 个经快速提取质粒鉴定，得到 8 个有外源插入片段的重组子。经 *Eco*R I、*Bam*H I 双酶切鉴定其外源插入片段分子量均为 550 bp，与 PCR 产物中的第二条带的分子量相当，并对其中一个重组子 pp3 进行了序列分析。根据 Genbank 提供的同源性检索信息，此序列与玉米亚端部序列克隆子 pBF266 的 123-279 碱基区域的同源性高达 92%。就目前资料检索的情况来看，黑麦、玉米端粒相关序列之间存在高度同源性还未见报道。端粒是染色体末端 DNA 的特殊组织形式，其独特的功能有独特的分子结构基础（Blackburn，1991）。我们的实验结果暗示在黑麦 B 染色体端粒相关序列中也存在特殊 DNA 结构。在 PCR 产物克隆过程中，尽管我们多方尝试仍无法得到包含 PCR 扩增产物中 350 bp 主带的转化子。Sandery（1990）也发现利用 pUC18 载体无法得到黑麦 B 染色体末端异染色质区域 D1100 重复序列家族 1.1 kb *Dra* I 及 *Eco*R I 酶切片段的克隆子，作者指出可能是 D1100 家族的特殊分子结构阻碍了其克隆，这与我们的实验结果相符。那么这条 350 bp 扩增带中隐含什么样的与端粒结构相关的序列信息呢？这一问题的解决有待于通过进一步的克隆及序列分析等工作去研究。

5. 高密度 RFLP 图谱中的应用

RFLP 是当前作物遗传图谱构建的主要遗传标记。端粒是染色体物理图谱和遗传图谱的末端，是 RFLP 图谱的重要组成部分。但由于端粒主要由重复序列构成，端粒简单重复序列同源性较高，所以染色体末端缺乏 RFLP 选择标记。因而克隆端粒相关序列在这一点上就显得格外重要了。Kilian 和 Kleinhofs（1992）利用盒式 PCR 扩增方法克隆了大麦端粒相关序列，并利用其中两个克隆子 pAKp3 和 pAKp6 作了 RFLP 分析，证实其显示的多态性符合孟德尔遗传规律，并指出 RFLP 技术可以用于染色体末端 TAS 序列的作图工作，李平（1996）等成功地将两个 TAS 克隆子 TelA2、TelC1 分别定位于水稻第三和第九染色体的末端。本文建立的显微切割-PCR 扩增方法在此意义上的重要性在于，它不同于上述方法中 TAS 序列的扩增和克隆是以总体 DNA 为基础的，它的扩增和克隆是针对单条染色体进行的，因而更具有选择性。按照本文的实验设计，只要染色体端部具有拟南芥端粒简单重复序列（CCCTAAA）$_n$ 的同源序列，即可利用本文建立的显微切割-PCR 方法得到针对某一染色体的 TAS 序列克隆。玉米（Burr，1992）、水稻（Hong，1995）、人、大麦（Trude，1991）中具有此端粒简单重复序列（Biessmann，1994），且端粒简单重复序列具有高度种间同源性，据此我们可以推测：本文建立的显微切割-PCR 扩增法可以用于玉米、水稻、大麦和人单条染色体 TAS 序列的克隆，并在其他植物中也有利用的可能，进而为作物高密度遗传图谱的构建提供有价值的端部探针。

（作者：郭歌、陈成彬、李秀兰、宋文芹、陈瑞阳。原载：植物学报，1998，40（12）1123-1128）

3.1.5 黑麦 A，B 染色体着丝粒区同源性的荧光原位杂交分析

B 染色体虽然已在 1300 多种植物和近 500 种动物的自然群体中被发现（Jones，1993），但关于 B 染色体的起源及遗传功能至今还是一个不解之迷，B 染色体是一种自然的附加染色体（accessory chromosome），它的显著特征是：一般比 A 染色体小；异染色质化；减数分裂时与 A 染色体不发生联会和交换；在遗传上是惰性的；为非孟德尔式遗传；并且不带有重要基因，基本没有功能等（Jones，1982）。基于这些特殊性质，B 染色体引起许多生物学家的关注，其中关于 B 染色体着丝粒、端粒和复制起点的分子结构成为研究的焦点。一旦搞清 B 染色体这些重要元件的 DNA 序列，B 染色体可能成为构建人工染色体载体的理想对象，它独立于整倍体基因组之外，可在高等真核生物转基因中发挥重要作用。

本文在黑麦 B 染色体端粒相关序列研究的基础上（郭歌等，1998），对黑麦 A、B 染色体着丝粒区同源性进行了荧光原位杂交分析，发现黑麦 B 染色体着丝粒区与 A 染色体着丝粒区具有高度同源性。

3.1.5.1 材料与方法

1. 材料

黑麦（*Secale cereale* L.）JNK 居群，含 0～4 条 B 染色体，由英国威尔士大学 R.N. Jones 教授惠赠。

2. 方法

（1）接头和引物。引物合成及接头制备参照夏家辉等人（夏家辉等，1994）的方法并加以改进。

（2）B 染色体着丝粒区探针池的获得。① 染色体标本制备。参照宋文芹（1998）等人的方法制备。②显微切割。按李秀兰（1998）等人的方法，将染色体制备在 50 mm×24 mm 的盖玻片上，利用普通光学显微镜，通过自制的玻璃台架在 100 倍油镜下用显微操作器操纵细玻璃针进行切割，④ LA-PCR 扩增。LA-PCR 参照夏家辉（1994）等人报道加以改进。将空白对照及所切 4 条 B 染色体着丝粒片段经蛋白酶 K 处理，*Sau*3A I 酶切，连接接头后补平。按 95℃ 1 min，56℃ 1 min，72℃ 1 min，运行 30 个循环，最后 72℃ 延伸 10 min，进行初级 PCR 扩增。取 1 μL 上述 PCR 产物为模板，100 μL 体系，相同反应条件进行 20 个循环的次级 PCR。

（3）③PCR 产物的荧光原位杂交（FISH）分析。参照宋文芹等人（1998）方法制备染色体标本，然后将标本用 100 μg/mL 的 RNase（2×SSC：0.3 mol/L 氯化钠，0.03 mol/L 柠檬酸钠）37℃处理 1 h，2×SSC 洗 15 min。乙醇系列（70%→85%→100%）脱水，各 5 min，干燥后，70%甲酰胺 70℃变性 2 min。冷乙醇系列（70%→85%→100%）脱水，各 5 min。干燥后加 10 μL 杂交液（50%甲酰胺，10%硫酸葡聚糖，2×SSC，5 μg 鲑鱼精 DNA，1×Denhardts，25 ng 地高辛标记微切 PCR 产物探针），盖上玻片，37℃杂交过夜，2×SSC，0.1% SDS 37℃洗 15 min，0.1×SSC，0.1% SDS 37℃洗 15 min。室温下 2×SSC 洗 15min。1×TNT（100mmol/L Tris-HCl pH=7.5，150mmol/L NaCl，0.2% Tween-20）

37℃洗 5 min。加入用 1% Blocking buffer 按 1：50 稀释的 Anti-DIG-Rhodamin，37℃保温 30 min。1×TNT 中 45℃洗 15 min。100 μL DAPI 染色 15 min。在 Option 荧光显微镜下观察。

3.1.5.2 结果分析

1. B 染色体的识别和着丝粒区的显微切割

在供试二倍体黑麦 JNK 居群中，14 条 A 染色体均为中部或近中部着丝点染色体。而所含 B 染色体，均为近端部着丝点染色体，大小约占 A 染色体的 1/2，从形态上很容易识别，切割并回收了 4 条 B 染色体的着丝粒区片段，每个片段大小约为 0.3 μm。

2. LA-PCR 扩增

将回收的 4 条 B 染色体的着丝粒区片段进行 PCR 扩增，取 10 μL 次级 PCR 产物在琼脂糖凝胶上电泳。从电泳结果看，阴性对照泳道未出现溴化乙锭染色信号。说明没有污染存在.扩增片段在 0.1~2 kb 之间，呈涂片状，主要集中在 0.2~0.8 kb。

3. PCR 产物的荧光原位杂交分析

从荧光原位杂交结果可以看出，2 条 B 染色体着丝粒区均出现明显的红色杂交信号。证明 PCR 产物确实来源于 B 染色体着丝粒区。同时在 A 染色体着丝粒区也有相同强度的红色杂交信号，说明两者有较高的同源性，染色体的其他部位无散布杂交信号。

3.1.5.3 结果讨论

B 染色体独立于 A 染色体组之外；具有不稳定性和多态性；不产生明显的表型影响；被认为"可有可无"（Jones，1982），由于这些因素的影响，B 染色体的研究进展一直缓慢，而且多集中在细胞水平分子生物学兴起后，人们开始在分子水平上究 B 染色体，但由于 A，B 染色体 DNA 在 GC 含量、重复序列比例及复杂程度等方面相似性给 B 染色体的分离分析带来了很大的困难（Rimpau，1975；Timmis，1975）。1981 年显微切割微克隆技术的诞生（Scalenghe，1981）给分子细胞遗传学带来了新的革命，它将细胞学技术和分子生物学技术结合起来，为特定染色体或染色体特定区域研究提供了新手段，也对认识 B 染色体的分子组成起到了重大的推动作用，1991 年 Sandery 等人将显微切割微克隆技术应用于黑麦 B 染色体研究，分离到 1 个 A、B 染色体都普遍存在的序列，1996 年 Houben 等人对微切割的 20 条黑麦 B 染色体长臂末端片段及剩余部分建立 2 个探针池，杂交分析表明，多为 A、B 染色体共有序列，且在长臂末存在其专一序列。本实验应用显微切割技术建立黑麦 B 染色体着丝粒区探针池，并进行了染色体荧光原位杂交分析，证明黑麦 A、B 染色体着丝粒区 DNA 具有高度同源性，为 B 染色体起源于 A 染色体提供了一个佐证，这与人研究 A、B 染色体 DNA 有很大的相似性（Rimpau，1975；Timmis，1975），和同源性（Sandery，1991；Houben，1996）结果一致。

关于 B 染色体起源，虽然普遍认为起源于某条 A 染色体（Jones，1982；Markarian，1958；Battaglia，1964，郑国锠 1975），但其起源及进化途径至今尚不清楚，"关节断离"假说（Markarian，1958）（和"次溢痕区断裂"假说（Battaglia，1964）都认为次缢痕区比较脆弱，容易断裂或"关节断离"，断离产生的末端片段进化为 B 染色体，但此断离片段并不具有着丝粒，它是如何获得着丝粒活性的无法解释，Jones 和 Rees（1982）认为中着丝粒和亚端部着丝粒 A 染色体的不同断裂形式及着丝粒融合产生了中部、亚中部和端部着丝粒的 B 染色体。而郑国锠（1975）等人提出的细胞融合学说则认为，B 染色体是不同细胞的 A 染色体在细胞融合穿壁时产生的带着丝粒的片段。究竟 B 染色体起源那种途径，目前尚无定论。但从我们的实验结果看，A、B 染色体着丝粒区高度同源，B 染色体应更倾向于起源带着丝粒的片段，而不是起源于次缢痕区断裂和"关节断离"假说所阐述的无着丝粒的片段。

（作者：张荣信、陈成彬、李秀兰、宋文芹、陈瑞阳。原载：科学通报 1999，44（5）520-524）

3.1.6 黑麦 B 染色体 DNA 组成特征

In the history of evolution chromosome genetics does not always obey Mendelian rules. Polymorphisms frequently occur when housekeeping genes are passed on from one generation to another. B chromosomes（Bs）are among these kinds of polymorphisms. They are also called supernumerary, accessory or extra chromosomes, and they are additional and optional extras to the standard A chromosome set（As）. They are found in thousands of plants and animals（Beukeboom, 1994）. They are often morphologically distinct from the As, being smaller and more highly heterochromatic. B chromosomes do not pair with A chromosomes, and they are inherited in a non-Mendelian way, exhibiting meiotic and mitotic instability and nondisjunction（Muntzing, 1974）. Bs occur in many species of higher plants and animals over a wide range of geographical distribution（Jones and Rees, 1982）. They are a common cause of numerical chromosome polymorphism in these species. Because a B chromosome can co-exist with an A chromosome, they are thought of as parasitic chromosomes making use of cellular machinery required for normal chromosome function（Puertas 2002; Jones and Houben, 2003）. The effects and behavior of Bs have been well studied in many species, but little is known about their origin and evolutionary pathways. Their origin has puzzled scientists for a long time in the research field of chromosome biology, and nobody has yet found the answers to this enigma. Earlier efforts have been made to isolate B-specific DNA sequences（Sandery *et al.* 1990; Alfenito and Birchler 1993; Blunden *et al.* 1993; Houben *et al.* 1996; Stark *et al.* 1996; Cheng and Lin 2003, 2004）, but the results have been limited. It is believed that the origin of Bs may have multiple-pathways（Jones and Houben 2003）. The commonly accepted opinion is that they are derived from A chromosomes, because most of their DNA sequences are similar to those of the As（Chilton and Mccarthy 1973; Alfenito and Birchler 1993; Jamilena *et al.* 1995; Stark *et al.* 1996; Cheng and Lin 2004）. In the present study, using microdissection and microcloning, we constructed a DNA library of the whole of the B chromosome of rye（*Secale cereale* L., $2n = 2x = 14$）, screened out new rye B-specific sequences from this pool, blasted these with othe rsequences and analyzed the possible pathway of the origin of the rye B chromosome.

3.1.6.1 Results

1. B chromosomes contain specific and common repeats

B chromosomes are about half the length of A chromosomes, and subtelocentric, whereas A chromosomes are metacentric or sub metacentric. It is therefore easy to distinguish them under a microscope. It was demonstrated that a manually prepared glass needle with a tip of 1–2 μm was suitable for microdissection and for the collection of chromosomes from metaphase spreads (Zhou et al. 1999). Preparing slides of spread chromosomes under conditions of aseptic manipulation, suitable specimens were selected. With this method, 10 B chromosomes were collected, as shown in Fig. 3.3A (before isolation) and Fig. 3.3B (after isolation). After the first round of polymerase chain reaction (PCR) amplification the signals detected through the agarose gel electrophoresis were faint. The signals became obvious after the second round of amplification, and yielded products ranging in sizes from 100 bp to several kilo-base pairs.

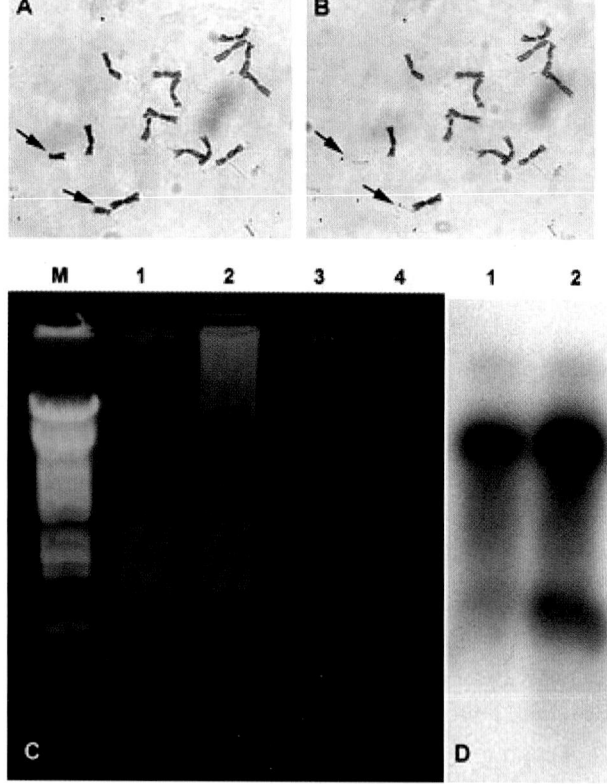

Fig. 3.3 Microdissection and PCR amplification of rye B chromosomes.
(A) Before isolation of microdissection of B chromosomes (arrow). (B) After isolation of microdissection of B chromosomes (arrow)(C)Linkage adapter-polymerase chain reaction(LA-PCR)amplification of B chromosome after microdissection, M. λDNA digested with EcoR I /HindIII. 1. First round PCR products; 2. Second round PCR products; 3. Negative control of first round PCR; 4. Negative control of second round PCR products. (D) Southern blot using second round PCR products as a probe. 1. 0B DNA; 2. + B DNA.

As a negative control, for monitoring the possible contamination with DNA, a sample containing no template DNA was set up during all stages of the microdissection and amplification. None of the products, either in the first round or in the second round, was amplified from the negative controls (Fig. 3.3C).

In order to confirm the reliability of the linkage adapter (LA) -PCR products, and its distribution in the rye genome, Southern-blot analysis was carried out using the second round PCR products as a probe, since the DNA content in the first round of PCR products was quite low. Genomic DNA with and without B chromosomes was digested using HaeIII overnight for the Southern blot. The results showed that the signals of the second round PCR products were obvious and similar, but not the same for 0B and +B genomic DNA (Fig. 3.3D). This confirms the reliability of the method of preparing chromosome samples, microdissections, and LA-PCR. It also indicates that the DNA constitution of B chromosomes is similar, but not identical to that of A chromosomes, indicating that the Bs contain both specific and common repeats.

2. Bs contain a few low copy specific sequences

The second round LA-PCR products from the 10 microdissected Bs were cloned into the pAMP-1 vector using a non-T4 ligase plasmid construction method (Fig. 3.4). More than 97% of clones in this small B chromosome library were positive. A total of 200 white colonies were screened through PCR amplification (Fig. 3.5). The length of the inserts ranged from 100 bp to 2 000 bp, and approximately 98% of clones contained inserts. Compared with the traditional T4 ligase system, the ligation method used here had a higher efficiency due to the 12 base-long stick ends. More than 95% of clones were positive. An estimation of the nature of these inserts was determined by dot-blot hybridization with genomic DNA with (+B) or without (0B) as a probe. The probes hybridize with most of the clones. There were four kinds of signals: (i) 0B No/+BNo (0B/+B Yes/No: with/without signals): clones were very low copies or false positive; (ii) 0B Yes/+B Yes: common sequences of both A and B chromosomes, which were the highest number of all the four kinds of hybridization signals; (iii) 0B No/+B Yes: B-specific sequences; and (iv) 0B Yes/+B No: A chromosome specific sequences. This was contradictory in that the inserts were amplifications from microdissection of Bs. A possible reason is that when the two kinds of genomic DNA were labeled as probes, the labeling could not cover the whole genome. Some of the sequences were not labeled. The result 0B Yes/+BNo occurred if the unlabeled sequences were in the +B genome. This small B chromosome library of high quality laid a foundation for the isolation of B chromosome specific sequences.

R56 is a B chromosome centromeric associated sequence with high copy, while there are other sequences with low, moderate or high repeats.

Although many positive clones were obtained, further hybridization was needed to confirm whether these clones were derived from B chromosomes. Southern blots were made using PCR products of positive clones as probes. The results showed that few clones were derived from Bs. Two of them, R56 (Fig. 3.5B) and R111 (Fig. 3.6A), had signals of moderately repeat sequences, which existed mainly on +B genomic DNA. Others, R53 (moderate repeat sequence, Fig. 3.6B), R0 and R200 (highly repeated sequences, Fig. 3.6C and Fig. 3.6D respectively), showed signals on both +B and 0B genomic DNA.

To determine further whether these sequences were specific to B, and on which part of the chromosome they were located, we examined the 0B and +B genomes with fluorescence in situ hybridization (FISH) experiments, with R53, R56, R111, R0, and R200. The results showed that R56 is a centromeric associated sequence (Fig. 3.5C). Sequencing of R56 was carried out by Sangon Company (Shanghai, China), showing its length at 61 bp. This short sequence showed a strong FISH signal on both the B chromosome and the interphase nucleus, but not on the A chromosome, implying that it is a B-specific sequence with high copy. No FISH signal was detected for any other sequences, indicating that these sequences are either low copy or below the level of detection by FISH signal.

(A) Screening for positive clones using polymerase chain reaction (PCR). The first lane from left is λDNA digested with *Eco*R I /*Hin*d III as a marker. (B) Southern blot analysis for R56, 1. 0B DNA; 2. +B DNA. (C) Fluorescence in situ hybridization (FISH) result for R56.

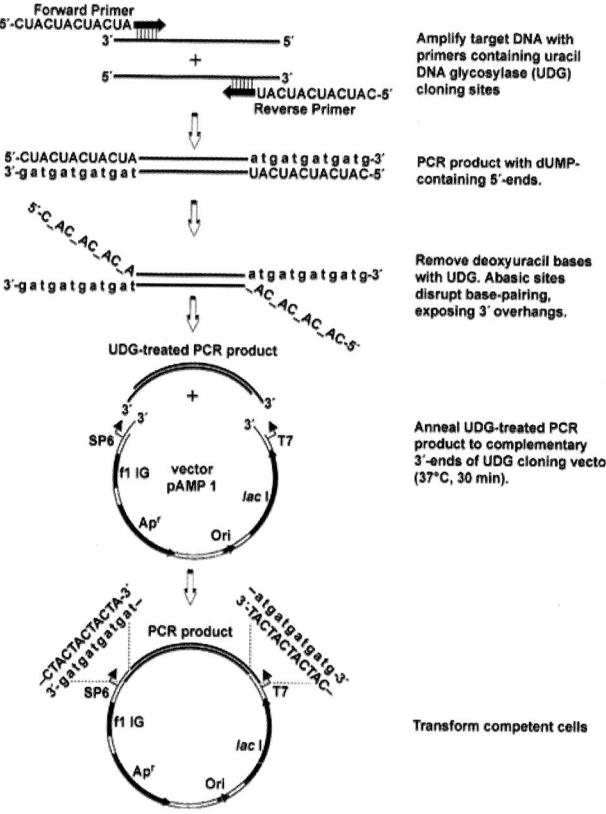

Fig. 3.4　Non-T4 ligase plasmid construction

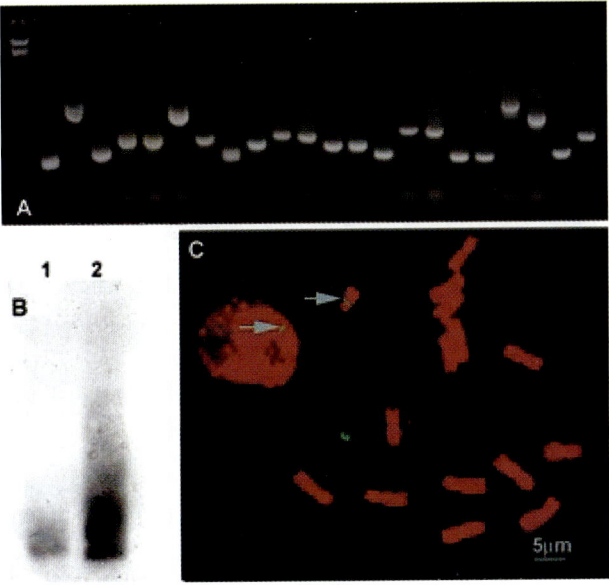

Fig. 3.5 B-specific DNA sequence

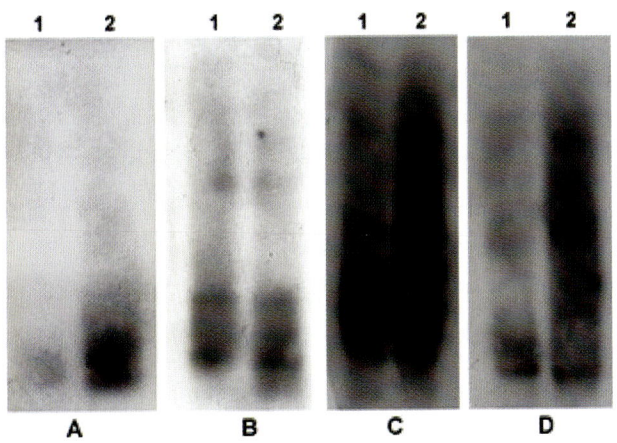

Fig. 3.6 Southern blot analysis for R111 (A), R53 (B), R0 (C), R200 (D).1. 0B DNA; 2. +B DNA

3.1.6.2 Discussion

We have demonstrated the presence of B-specific DNA markers by direct cloning of LA-PCR products from a microdissected chromosome library of rye Bs. Our library was constructed with microdissection and microcloning, and gave inserts ranging in length from 100 bp to 2000 bp. By examining electrophoretograms, Southern blots and FISH signals, we determined that at least one repetitive sequence is associated with the B chromosome centromere region. Although several repetitive DNA families of rye B-specific sequences have been found (Sandery *et al.* 1990, 1991; Blunden *et al.* 1993), most of the characterized inserts were low-copy sequences, suggesting that LA-PCR amplification selects against repetitive sequences. The preferential selection for low-copy sequences by degenerate oligonucleotid primer (DOP) -PCR was also noticed in the ampli-fication of flow-sorted chromosomes of tomato (Arumuganathan et al. 1994), and wheat (Liu et al. 1997), and in microdissected B-chromosomes of Crepis capillaris (Jamilena et al. 1995). This feature of LA-PCR and DOP-PCR could be of significance for the enrichment of molecular markers in a given chromosome. In the present study we report the isolation and characterization of new DNA sequences of rye Bs. The southern hybridization experiments with LA-PCR products showed that the signals of the PCR products were obvious and similar, but not the same, for both of 0B and + B genomic DNA. This indicated the presence of B-specific sequences. The hybridization experiments also demonstrated that most of the specific and common sequences were low copies, matching the preferential character of LA-PCR and DOP-PCR amplification. Southern blots of positive clones of PCR products showed that R56 and R111 are present in moderately repeated numbers, which exist mainly on +B genomic DNA, while R53 showed signals on both +B and 0B genomic DNA in moderately repeated numbers. The other two, R0 and R200, showed signals on both + B and 0B genomic DNA in high repeat numbers. Furthermore the FISH results showed that R56 is present in high copy numbers in the Bs. There are other high copy, tandemly repeated B-specific sequences that have been reported in rye (Sandery et al. 1990; Blunden et al. 1993; Houben et al. 1996), indicating that high copy sequences could be one of the characters of B chromosomes.

Sequencing data for these sequences indicate that R56 is 61 bp in size. Blasting results in the National Center for Biotechnology Information (NCBI) GenBank database illustrated that no homologous sequences were found, suggesting this sequence may be a new one for rye Bs. The FISH result also showed that R56 is located at the B chromosome centromere region, indicating that it is a B centromere-associated sequence. Sandery et al. (1990) reported a high copy, tandem repeated B chromosome-specific sequence D1100 in rye, which was associated with telomere repeated sequences of the A chromosomes. Chromosome FISH results of Zhang et al. (1999), using LA-PCR products from microdissection of rye B chromosome centromere as a probe, showed that both A and B chromosome centromere regions had the same level of hybridization signals, indicating the homologous origin of the A and B chromosomes. It is a general consideration that the B chromosome is derived from A chromosomes, but how this happened remains unclear. The above data support the hypothesis that Bs could originate from the breakage of A chromosomes centromeres and the refusion of centromeres in certain cases (Jones and Rees, 1982). However, our FISH results showed that the B chromosome centromere region of rye contains a certain specific sequence, implying a possible different organization from that of the A chromosomes. This result supports another hypothesis: namely, that Bs may come from the breakage of an A chromosome secondary constriction, rather than a centromere, because this part is more vulnerable and easily broken (Markarian and Schulz 1958; Battaglia 1964). The question of how B chromosomes gained new centromere sequences, rearrangement of chromosomes and restructuring of the genome, may be critical in obtaining new sequences that have the potential to form Bs (Jones and Houben 2003). No FISH signals were detected from R53, R111, R0, R200, although Southern blots indicated that these sequences were with higher copy numbers than R56. A possible explanation is that these sequences were not highly repeated enough to enable FISH detection. Another reason may be the chromosome rearrangements blocking the signals, and rendering them below the level of detection of the fluorescence microscope. Researchers have focused on the isolation of B chromosome specific sequences for a long time, but thus far, results have been limited. This implies the similarity of DNA sequences on both A and B chromosomes. Homologous sequences could be found on A chromosomes even for B-specific sequences, since under low-stringency conditions of Southern blot or FISH, these specific sequences can appear common of both As and Bs (Page et al. 2001). It is very difficult to isolate B-specific sequences, because of the high similarity of DNA constitution on As and Bs. However, these isolations continue, since B-specific sequences can not only reveal the origin and evolution of Bs, but also provide clues for the specific behavior of B chromosomes. By combining genetic and molecular biology approaches, we hope eventually to reveal the enigmatic origin and function of the B chromosome.

3.1.6.3 Materials and Methods

1. Plant material

The initial source of rye material was obtained from seeds presented by Professor RN Jones, University College of Wales Aberystwyth, UK. The seeds were reproduced for two to three further generations in the field. Cytogenetic p reparation and microdissection Seeds were germinated on moist filter paper in Petri dishes at 25℃. Root tips (1–2 cm) were treated in 1.25 mmol/L Hydroxyurea (Hu) for 18 h, double distilled water for 3 h and 4 μmol/L amiprophos-methyl (APM) for 2.5 h. Chromosome spreads were prepared according to Chen et al. (1979). In brief, root tip cells were digested in an enzyme mixture including 2.5%(w/v) cellulase and pectinase for 1 h, double distilled water for 20 min and Carnoy's fixation (3∶1, methanol: acetic acid, v/v). Finally, root tips were squashed on a slide and dried with hot air. The B chromosomes were microdissected and collected using a glass needle with a tip of 1–2 μm under the microscope at 1 000× magnification. The tips of the needles were then broken into a 0.5 mL Eppendorf tubes containing 20 μL of proteinase K solution (10 mg/mL), 2 μL1 × Multi-buffer (Promega, Madison, WI, USA), followed by digestion for 4 h at 37℃. After maintaining at 75℃ for 15 min, 0.2 U EcoR I (Promega) was added to digest at 37℃ for 4 h, 75℃ for 15 min.

2. Amplification by EcoR I LA-PCR

Two microliters of EcoR I linker adaptors (20 μmol/L), with the sequences 5'-CTCGTAGACTGCGTACC-3'and 3' -CATCTGACGCATGGTTAA-5', 0.6 μL10 ×T4 DNA ligasebuffer (Promega), and 1 μLT4 DNA ligase (Promega) were added into the tube. The ligation between the adaptor and digested chromosomal DNA was carried out at 14℃ for 16 h. T4 DNA ligase was then inactivated at 70℃ for 20 min. The first round of PCR was carried out in the same tube by adding 5 μL 10 × PCR buffer, 3 μL MgCl2 (25 mmol/L), 5 μL dNTPs (2 mmol/L), 5 μL primer 1 (2 μmol/L, 5'--CUACUACUACUAGA CTGCGTACCAA TTC-3'), 5 μL primer 2 (2 μmol/L, 5'-CAUCAUCAUCAUGACTGCGTACCAATTC-3'), 5U Taq DNA polymerase, and double distilled water to 50 μL. A first extension at 72℃ was carried out in order to patch the linker. After denaturation at 96℃ for 5 min, amplification was performed with 35 cycles at 96℃, 1 min; 56℃, 1.2 min; 72℃, 1.3 min, followed by a final extension at 72℃ for 15 min. The second round of PCR was carried out using 2 μL of the first-round products as a template. The method was the same as described above, except for 20 instead of 30 cycles of amplification and no patching for the linker. Library construction After purification of the PCR products with a DNA Purification Kit (TaKaRa, Dalian, Liaoning, China), 0.5 μL purified DNA was ligated into the pAMP1 vector, in a 20 μL ligation volume at 37℃ for 30 min. The 20 μL ligation volume contained 0.5 μL purified DNA (50 ng/μL), 2 μL pAMP1 vector (25 ng/μL), 1 μL uracil DNA Glycosylase (UDG) (1 U/μL), 1.5 μL deionized water and 15 μL 1 × renaturation buffer (20 mmol/L Tris Cl, 50 mmol/L KCl, 1.5 mmol/L $MgCl_2$, pH 8.4). Then 1 μL of the ligation mixture was used for transformation of DH5α competent Escherichia coli cells by the Mg^{2+} transformation method (Ausubel et al. 1992). The transformed cells were selected on a plate containing ampicillin (100 g/L), isopropyl-1-thio-galactopyranoside, and 5-bromo-4-chloro-3-indolyl-galactopyranoside.

3. Analysis of plasmid clones

Recombinant plasmids (white colonies) were isolated by alkaline lysis. Inserts were released by PCR amplification. The insert size was estimated in 1.4% agarose gel using known DNA markers. To determine the positive nature of the clones, digoxygenin-dUTP (DIG)-labeled genomic rye DNA was hybridized to Genescreen Plus membrane (NEN) containing recombinant clones by dot blotting.

4. Southern hybridization

Genomic rye DNA was isolated from leaf tissue using the method of cetyltrimethyl ammonium bromide (CTAB) and was digested with Hae III (Promega) at 37℃ overnight. DNA fragments were resolved in 0.7% agarose gels in 0.5 × TBE buffer (1 × means 89 mmol/L Tris, 89 mmol/L boric acid, 2 mmol/L ethylene diamine tetra acetic acid (EDTA), pH 8.0) without ethidium bromide (EB) to avoid background of Southern blot signals, and subsequently transferred to Genescreen Plus membrane (NEN). For hybridization, DNA probes were labeled with DIG (Roche, Penzberg, Bayern, Germany) by random-primed DNA synthesis. Hybridization was carried out overnight at 42℃ in DIG Easy Hyb (Roche). Blots were washed successively in 2 × standard saline citrate (SSC), 0.1% sodium dodecyl sul phate (SDS) at room temperature (1 × SSC is 0.15 mol/L NaCl and 0.015 mol/L sodium citrate); 0.5 × SSC, 0.1% SDS at 68℃ twice. The hybridization signal was detected by enzyme immunoassay with chemiluminescent substrate for alkaline phos-phatase（CSPD）（Roche）. The membrane was exposed to X-ray film for appropriately 1–2 min.

5. Sequencing and sequence analysis

Sequencing was carried out by Sangon Company. Analysis of sequences similarity was performed with BioEdit sequence alignment editor and Vector NTI suite 7.0（InforMax，Frederick，MD，USA）.

6. Fluorescence in situ hybridization

Fluorescence in situ hybridization was carried out as described in the Nonradioactive In Situ Hybridization Application Manual（2nd edn，Roche）. Briefly, the probe was labeled as for Southern hybridization. Hybridization buffer contained 50% deionized formamide，2 × SSC，50 mmol/L sodium phosphate（pH 7.0），5% dextran sulfate，and 3 ng/ μL probe. The buffer was denatured at 98℃ for 10 min before being used. Slides with metaphase spread chromosomes were treated with 70% deionized formamide in 2 × SSC at 68℃ for 2 min，then 15 μL denatured hybridization buffer was applied onto the slides and incubated at 80℃ for 5 min and at 37℃ overnight. The slides were then washed with 2 × SSC，30% deionized formamide at 37℃ for 5 min and 2 × SSC at 37℃ for 5 min twice. A signal was detected with anti-DIG-fluorescent-conjugate. Metaphase spreads were counterstained with 10 ng/mL propidium iodle（PI）. The epifluorescence signal was viewed directly in a fluorescence microscope（Olympus BX51，Japan）. Photographs were taken with a digital camera.

（Authors：Hong Long，Zhong-Xia Qi，Xiao-Ming Sun，Cheng-Bin Chen，
Xiu-Lan Li，Wen-Qin Song and Rui-Yang Chen. Published。In：Joumal
of integrative plant biology，2008，50（2）183-189）

3.1.7 植物 B 染色体序列分离及关键功能组件的研究

B 染色体是独立存在于物种常染色体之外的一种特殊染色体，又被称为超数染色体、附加染色体或额外染色体。尽管从 20 世纪 30 年代发现 B 染色体以后，对 B 染色体的研究一直没有间断，但是由于 B 染色体比较特殊的性质和特征，对其 DNA 组成以及起源、进化等问题一直悬而未解。随着现代科学技术和知识水平的不断提高，有着许许多多未解之谜的 B 染色体引起了学者们广泛的研究兴趣，相信有关 B 染色体的"谜底"会被一一解开。

3.1.7.1 高质量黑麦单条 B 染色体--DNA 文库构建及特异片段的筛选鉴定

1. 染色体标本的制备与识别

黑麦（*Secale cereale* L.）是研究 B 染色体行为机制的理想试材，因此,本研究利用由英国威尔士大学 R.N.Jones 教授惠赠的黑麦 JNK 居群（2*n*=14+0~6B）为研究材料，采用本实验室在国内首次建立的植物染色体去壁低渗制备方法，获得了黑麦大量中期染色体

分裂相，在显微镜下可以清晰的辨别出 B 染色体。黑麦 B 染色体大小约为常染色体的一半，且为端着丝粒染色体。另外我们观察到 B 染色体的发生呈多态性，同一种群内，不同个体间 B 染色体的数目也有差异，即使在 B 染色体发生率很高的种群中也可能有。在我们使用的材料中有些个体不含 B 染色体，B 染色体一般为 1～5 个，其中以 2 个为多数（图 3.7），高质量染色体标本的制备为后续工作奠定了坚实的基础。同时研究中还发现黑麦细胞内除可有可无的 B 染色体外，其常染色体也存在数目的变化，如在研究中我们已检测到 $2n=2x=15$ 及 $2n=2x=16$ 的细胞类型（图 3.8、图 3.9）。通过核型分析与 SSR

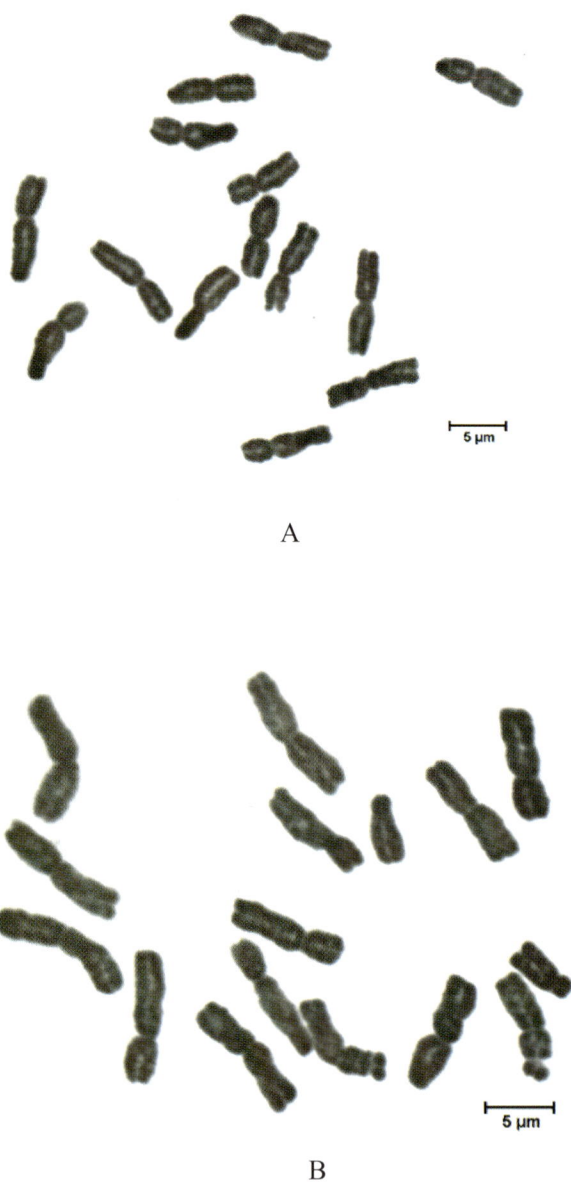

图 3.7　黑麦 0B 和+B 染色体
A：0B 染色体 $2n=14$，B：+B 染色体 $2n=14+2B$

探针杂交分析表明 2n=15 类型是多了一条 5R 染色体。2n=16 的是多了一对大的染色体。并且还发现即使是同在一个细胞内的 B 染色体也可存在一定的差异,如在个别黑麦单株细胞内除检测到正常的端着丝粒类型的 B 染色体外,还发现一种新出现的小的中着丝点类型 B 染色体,该类型 B 染色体与已报道的黑麦 B 染色体相比明显偏小(图 3.10),性状观察表明,其对黑麦生长及发育性状也未有影响,因此可以确定是一类新形成的 B 染色体。该种类型 B 染色体的发现对于探究 B 染色体的起源及进化机制具有十分重要的意义。

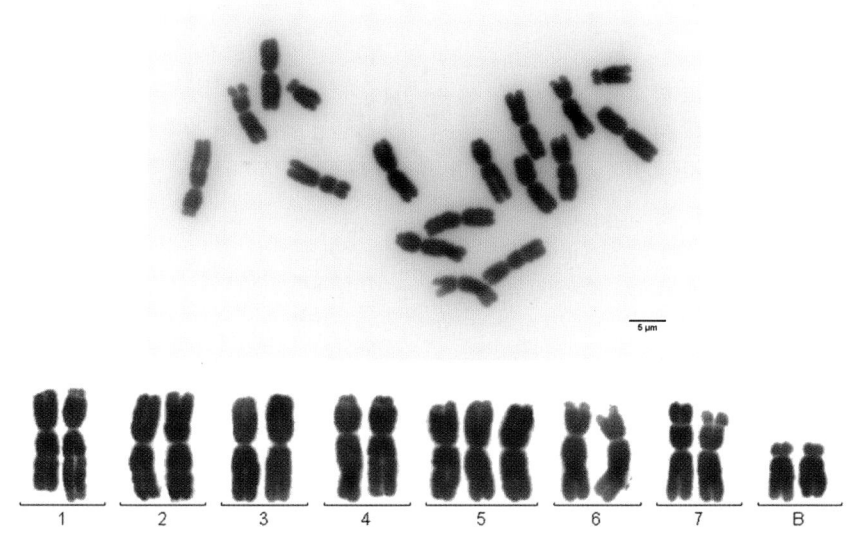

图 3.8　黑麦 2n=15+2B 核型分析(第 5 对三条染色体)

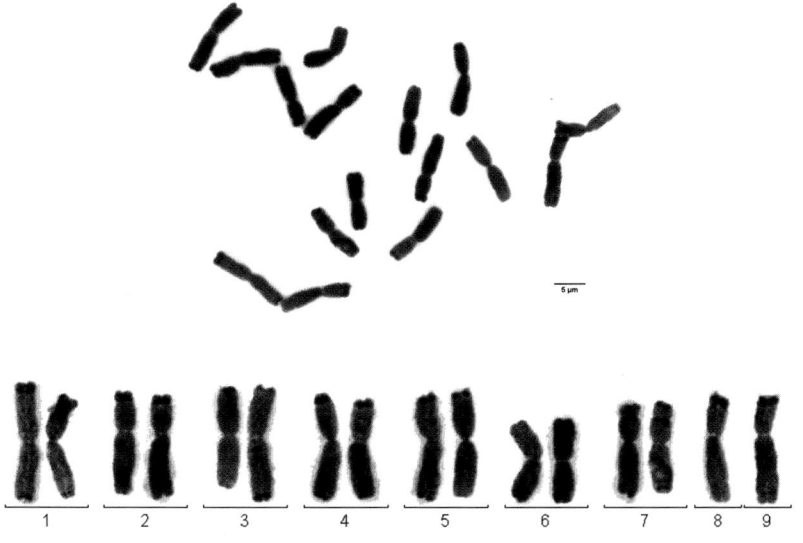

图 3.9　黑麦 2n=16 四体黑麦(根据染色体的带型进行配对排列)

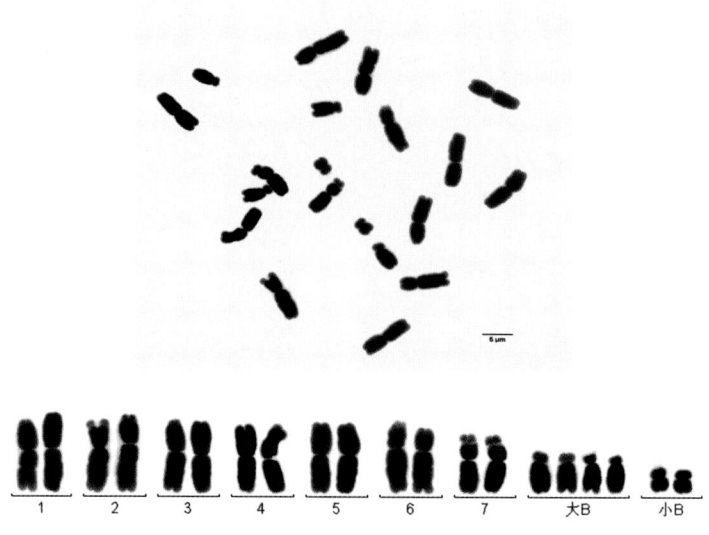

图 3.10 黑麦 2n=14+4B+2 小 B 核型

2. 黑麦 B 染色体的显微分离

在严格无菌的条件下制备染色体标本,并挑选出适合分离的染色体标本(图 3.11)。黑麦核型中 A、B 染色体形态区别明显,在 40×目镜准确识别 B 染色体后,将直径 1 μm 左右的微细玻璃针仔细的移动至标本侧下方,利用针尖小心地铲起染色体,待标本附着于针尖后将之与针尖一起断于离心管中,如此往复收集 B 染色体约 500 条/次,为后续试验提供充足的单条染色体。如图所示,为显微分离前(图 3.11A)和显微分离后(图 3.11B)染色体。

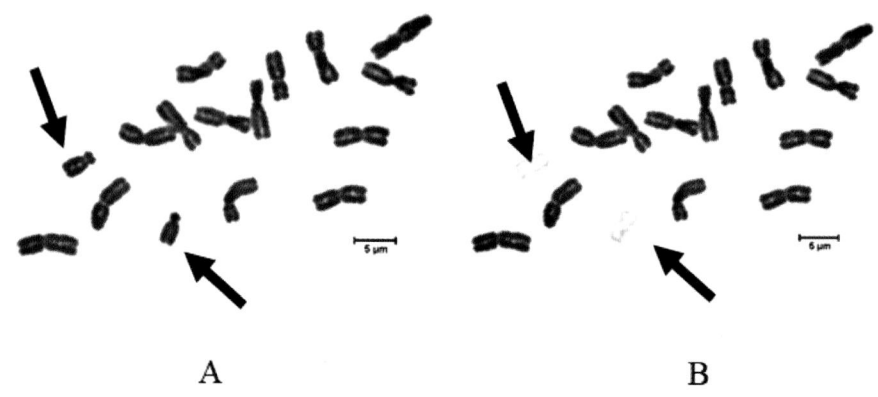

图 3.11 微切前后黑麦染色体
A:微切前,B:微切后的染色体,箭头示微切前后的染色体

3. 黑麦 B 染色体的 LA-PCR 扩增

黑麦的 1C DNA 含量为 9.5 pg,平均一条染色体的 DNA 含量为 1.2~1.4 pg。黑麦 B 染色体的大小约相当于 A 染色体大小的二分之一,其 DNA 含量约为 0.6~0.7 pg。因此,显微分离 B 染色体 500 条,约 350 pg。将上述微切产物经蛋白酶 K 消化、Sau3A I 酶切

及 LA 接头连接后采用 LA-PCR 法对单染色体微切产物进行初级和次级 PCR 扩增。在这个过程中为避免外源 DNA 的污染，操作都严格在无菌条件下进行，并在实验的各个环节设立严格不含染色体 DNA 的阴性对照和以 1 pg 基因组 DNA 为模版的阳性对照。首轮扩增后，电泳检测信号十分微弱，扩增产物大小从 100 bp 至几 kb 不等；经二次扩增后，电泳信号明显增强，扩增产物大小较初级 PCR 相对变大，从 200 bp 至几 kb 不等，负对照未检测到任何扩增信号，表明在整个微分离和 PCR 扩增过程中，无外源 DNA 污染（图 3.1）。但在后期的文库构建及克隆、测序过程中发现采用文献报道的由 24 mer 寡核苷酸片段碱基和 10 mer 寡核苷酸片段碱基组成的接头进行酶切产物连接获得的阳性克隆相对较少，而且最为严重的是引物自连现象十分普遍，对后续获得大量有效克隆和测序造成很大困难。为此，在反复试验的基础上，我们对 LA-PCR 接头进行了改进，将原来的接头修改成由 23 mer 和 19 mer 组成：

5'—GATCCTGAGCTCGAATTCGACCC—3'
5'—GGGTCGAATTCGAGCTCAG—3'

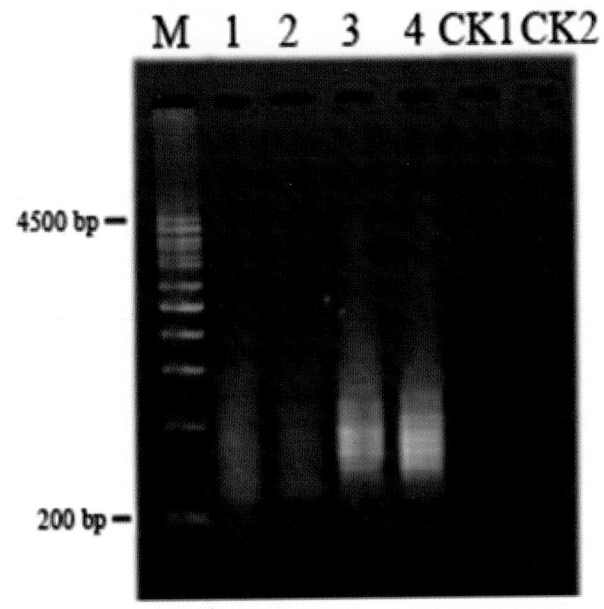

图 3.12 黑麦 B 染色体 DNA LA-PCR 扩增电泳结果
M：DNA 分子量标准；1，2：初级 PCR 扩增产物；3，4：次级 PCR 扩增产物；
CK1：初级 PCR 负对照；CK2：次级 PCR 负对照

研究表明，这两个序列组成的接头更加稳固，没有大的空档，能有效的避免大量接头自连，提高接头与染色体酶切片段的连接效率。同时防止平端连接，同时只将 23 mer 5'端磷酸化，使接头和 DNA 片段具有相同的黏性末端，连接上的接头的 DNA 片段可以通过与接头互补的引物进行扩增，无论是编码区还是非编码区都可以得到较好的扩增，构建的 DNA 文库相对较为完整。

4. LA-PCR 扩增产物 Southern 杂交分析

为了进一步验证 LA-PCR 产物的可靠性及在基因组中的分布,进行 Southern 印迹分析。分别将 10 μg 含 B 染色体和不含 B 染色体黑麦基因组 DNA 用限制性内切酶 *Eco*R I 消化,将+B 和 0B 染色体黑麦基因组酶解产物、B 染色体初级 PCR 扩增产物和次级 PCR 扩增产物经 1.2%琼脂糖凝胶电泳分离后转膜,与 DIG-dUTP 标记的黑麦(含 B 染色体)基因组 DNA 探针进行 Southern 杂交,结果(图 3.13)表明,黑麦基因组 DNA、B 染色体 LA-PCR 初级扩增产物和次级扩增产物均出现明显的呈涂片状的杂交信号,而阴性对照则无任何杂交信号,表明扩增产物确实来自黑麦基因组,没有外源 DNA 污染(图 3.13)。

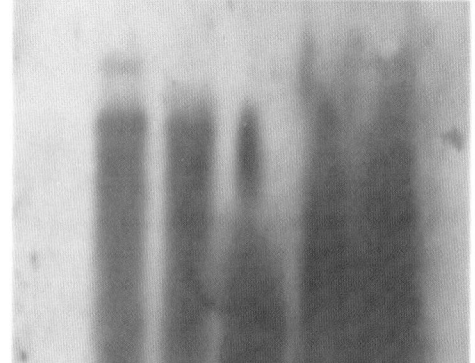

图 3.13 黑麦 B 染色体 LA-PCR 扩增产物 Southern 杂交检测
1:阴性对照; 2:不含 B 染色体黑麦基因组 DNA 酶切产物; 3:含 B 染色体黑麦基因组 DNA 酶切产物;
4:B 染色体 LA-PCR 初级扩增产物; 5,6:B 染色体 LA-PCR 次级扩增产物

5. LA-PCR 扩增产物的荧光原位杂交

由于 Southern 杂交的结果仅能证明扩增产物来源于黑麦基因组,但没有提供直接的证据证实扩增产物是否来自于分离的黑麦 B 染色体。为此,本研究用 DIG-dUTP 标记的初级和次级 LA-PCR 产物为探针,在 JNK 黑麦根尖细胞有丝分裂中期染色体上进行荧光原位杂交定位分析,如图 3.14A 所示,分裂中的两个 B 染色体长臂末端上均匀出现强烈的红色(初级)特异的杂交信号,图 3.14B 是次级产物与黑麦杂交的结果,从图 3.14B 中可以看出。次级产物信号均匀分布在除末端以外的整条染色体上,色彩均匀,未见背景干扰。信号明显强于初级 PCR 扩增产物。实验结果证实扩增产物来自于被分离的 B 染色体,而不是来自基因组的其他染色体。同时杂交结果显示其他常染色体着丝粒区域和长、短臂中间也可检测到明显的杂交信号,暗示这些部位可能存在与 B 染色体同源的序列。

表 3.1　黑麦 B 染色体微克隆文库的质量分析

文库参数	数值	文库参数	数值
克隆数目	33400	重复序列比例	67%~80%
插入片段长度范围	200~2300 bp	单/低拷贝序列比例	15%~30%
插入片段主要长度范围	200~800 bp	空载率	2%
插入片段平均长度	700 bp	文库冗余度	0~5%

（2）阳性克隆序列分析与荧光原位杂交定位

从文库中随机选取经 PCR 初筛为阳性的克隆 3550 个进行测序。测序结果表明，1417 个序列（约占测序序列的 40%）与已报道的黑麦 BS 特异序列 E3900 家族和 D1100 家族相关。其中 E3900 家族重复区序列同源和部分同源 793 个，与 D1100 家族重复单元相关序列 624 个，与小麦家族 BAC 克隆、磷脂酰丝氨酸脱羧酶、线粒体 DNA 具有同源性的序列为 223 个；与黑麦 BAC 克隆重复序列、黑麦转座酶假基因、微卫星序列相关的 293 个；与拟南芥端粒、黑长臂猿 BAC 克隆、小家鼠 BAC 克隆相关的序列 28 个；与酵母菌等相关序列为 31 个，其余序列经序列比对未发现同源序列，推测可能为新报道序列。随后对包括 E3900 家族和 D1100 家族在内的部分序列进行了 FISH 定位分析，以便进一步确定哪些序列为黑麦 B 染色体特异序列，哪些序列为黑麦 A、B 染色体共有序列及序列在染色体上的分布特征。结果如下：

①黑麦 B 染色体长臂末端特异序列的筛选、鉴定

D1100 和 E3900 作为已证实的黑麦 B 染色体特异序列，其在细胞染色体上的定位信息对后续其他 B 染色体特异序列的定位具有重要的参考价值。因此首先根据测序及序列比对结果，分别随机挑选含 D1100 和 E3900 序列的克隆并标记为探针进行 FISH 定位分析，杂交结果显示 D1100 和 E3900 主要集中在 B 染色体长臂末端（见图 3.17G 和 I），但当降低杂交严谨度后在 A 染色体短臂中间靠近着丝粒的位置也可检测到散布的杂交信号。从杂交结果可以看出 D1100 与 E3900 紧密连接，并有部分信号重叠在一起，但 E3900 更靠近 B 染色体长臂的末端部分，这与 Houben 等人的研究结果一致。值得注意的是，在所构建的 B 染色体文库中，与 D1100 和 E3900 高度同源或部分同源的克隆占很大比例，这一方面可能与相关序列的优先扩增相关，而更可能是与黑麦 B 染色体本身 DNA 序列组成特征密切相关，D1100 和 E3900 类型的重复序列可能是黑麦 B 染色体 DNA 的重要组成部分。

随后根据测序信息及序列比对结果，选取 200 个阳性克隆做进一步的 FISH 定位分析，以确定所选序列是否为黑麦 B 染色体所特有，或表现出与 A 染色体明显不同。FISH 定位结果显示，RB1、RB2、RB136、RB6、RB316、RB21、RB40、RB150、RB316 均特异的被定位于黑麦 B 染色体的长臂末端（图 3.17），为黑麦 B 染色体特异序列。序列分析显示：

RB1 长度为 312 bp，序列富含 AT（52.9%），序列比对显示其与 Thinopyrum bessarabicum 的一个 RAPD marker 序列高度同源（99%）并与小麦 Ty3-gypsy 反转座子蛋白具有较高相似性（图 3.17A）。

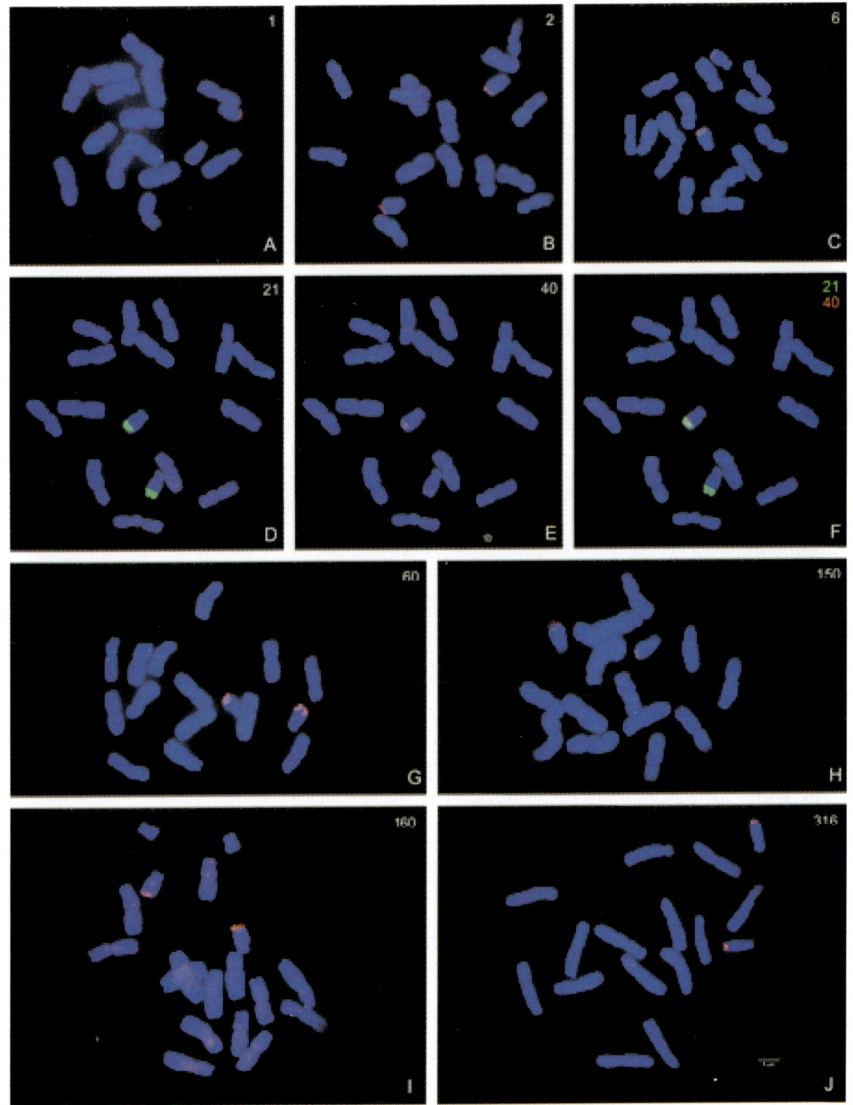

图 3.17　黑麦 Bs 单条 DNA 文库克隆 RB1、2、6、21、40、60、150、160、316 在黑麦染色体上的 FISH 定位结果（上述克隆全部定位在 Bs 长臂末端）

RB2 长度为 1059 bp，序列富含 GC（51%），比对发现其与小麦一个 BAC 克隆的部分序列高度同源（93%），并与水稻的一个假定蛋白高度相似（图 3.17B）。

RB6 长度为 881 bp，序列组成分析显示其 AT、GC 含量基本一致，与 Triticum monococcum 中编码精氨基琥珀酸裂合酶的基因具有一定相似性（48%）（图 3.17C）。

RB 150 长度为 454 bp，富含 AT 序列（54.7%），与已报道的黑麦 L155-20H-1 重复序列高度相似（图 3.17H）。

RB 316 序列长度为 831 bp，富含 AT（53.4%），与黑麦重复单元 E3900 具有 55% 的同源性（图 3.17J）。

RB21 序列长度为 731 bp，AT 含量 56.1%，分析表明其与小麦 BAC1551N13 部分序列、肿胀小麦 BAC 391M13 和 BAC 1144M20 部分序列、小麦 BAC198E19 质体乙酰辅酶 A 羧化酶（ACC-1）基因部分序列高度同源。FISH 定位在 BS 长臂末端，为 Bs 特异序列，信号占长臂的 1/3，非常强烈（图 3.17D）。

RB40 序列长度为 567 bp，AT 含量为 52%，经比对未发现与之同源序列，为首次报道，特异定位于 Bs 的长臂末端，信号专一，在其他染色体上未发现信号（图 3.17E）。

RB160 序列长度为 759 bp，AT 含量 52.7%，与黑麦重复单元 E3900 家族和 gag 假基因完整序列高度同源，是 E3900 序列的一部分，FISH 定位在 Bs 长臂末端，与报道的 E3900 FISH 定位一致，比 D1100 更靠末端（图 3.17I）。

RB60 序列长度为 819 bp，AT 碱基含量 58.4%，与黑麦重复单元 D1100 家族部分序列、小麦 3B 染色体特异 BAC 文库 ctg1030b 中部分序列高度同源，FISH 定位于 Bs 长臂末端，在 E3900 上端，靠近 E3900 信号，部分信号与 E3900 重叠（图 3.17G）。

RB136 长度为 848 bp，AT 碱基含量 58%，与黑麦重复单元 D1100 家族部分序列、小麦 3B 染色体特异 BAC 文库 ctg 部分序列高度同源。FISH 定位在 Bs 长臂末端，与 D1100 信号基本一致（图未显示）。

②黑麦 B 染色体短臂特异序列的筛选、鉴定

依托黑麦 B 染色体文库克隆信息，通过 FISH 定位筛选到两个 B 染色体短臂特异序列 RB165 和 RB306，其只在短臂上有信号，其他染色体上均没有杂交信号出现。

RB165 长度为 589 bp，AT 碱基含量为 56.2%，序列比对未发现与之同源序列，为首次报道。FISH 定位结果显示，RB165 特异性的定位于黑麦 B 染色体短臂靠近着丝粒位置，信号呈散布形式存在（图 3.18A），可以确定该序列为黑麦 B 染色体短臂特异序列。

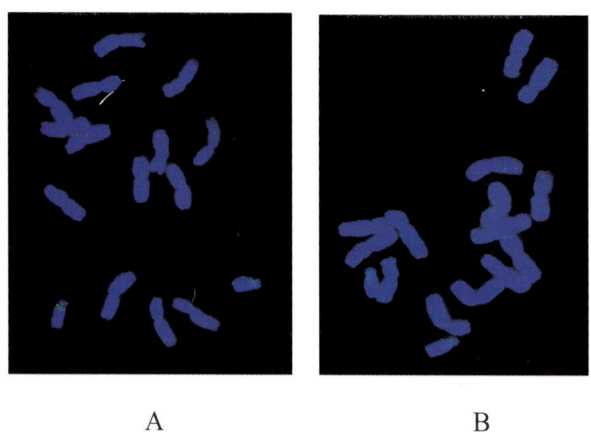

A   B

图 3.18　黑麦 B 染色体 DNA 文库克隆 RB165 与 306 FISH 定位结果
A：RB165 序列，B：RB306 序列，绿色信号为 BS 短臂特异信号

RB306 长度为 589 bp，AT 碱基含量为 56.7%，序列比对未发现与之同源序列，为首次报道。FISH 定位表明，该序列特异性定位于 B 染色体短臂，信号亦成弥散状分布（图 3.18B），在其他常染色体上未见杂交信号。因此，该序列鉴定为黑麦 B 染色体短臂特异序列。

③黑麦 A、B 染色体共有着丝粒序列的筛选、鉴定

着丝粒是染色体上比较宽阔的区域，位于主缢痕部位与着丝点并列，由高度浓缩的重复 DNA 序列组成，是构成人工 B 染色体载体的重要元件。目前在植物中已经发现了几个着丝粒相关序列，如 pSau3A9\CCS1 等，并先后将这些序列定位在不同的植物上。依据着丝粒序列的组成特征，为筛选获得黑麦 B 染色体着丝粒特异序列，为构建人工 B 染色体提供必需的功能组件。研究中从已构建的黑麦 B 染色体文库中，根据斑点杂交筛选的结果，挑选出杂交信号强烈的潜在重复序列进行测序及 FISH 定位。通过对 89 个克隆的测序及 FISH 定位分析，证实其中 5 个序列 RB183、RB72、RB99、RB88 和 RB113 特异的定位于 B 染色体着丝粒部位，同时在 A 染色体的着丝粒部位也可检测到特异的杂交信号（图 3.19）。进一步分析表明，RB183 长度为 1017bp，AT 碱基含量 52.9%，与已报道的 Triticum monococcum subs o.aeoioooides clone BAC TbBACs.comol 具有较高同源（83%）。虽然杂交信号为 A、B 染色体共有，但在 B 染色体着丝粒部位的杂交信号明显比相应 A 染色体强（图 3.19C）。RB72 的杂交信号模式特征与 RB183 相似，也是 B 染色体上杂交信号明显强于 A 染色体（图 3.19A）。结果暗示 RB183 和 RB72 在黑麦 B 染色体着丝粒部位可能具有更高的拷贝数。RB88 序列长 1061bp，AT 含量 51.7%，比对发现与水稻假定反转座子蛋白具较高同源性（图 3.19B）。RB99 长度为 367 bp，与玉米着丝粒 CentA 具较高同源性（70%），根据序列推测其可能为黑麦着丝粒区的类反转座子序列（图 3.19E）。与张荣信（1999）得到的着丝粒序列 pRBC6 高度同源。RB113，789 bp，AT 碱基含量 53.9%，序列比对分析，与小麦 BAC 文库 TaaCsp3BFhA_0100L17 部分序列同源，一致性达 91%，与小麦 3B 染色体特异文库 ctg 系列部分序列一致性达 91%，FISH 定位在黑麦每一条染色体着丝粒的位置，B 染色体信号强，几乎覆盖整个短臂（图 3.20）。

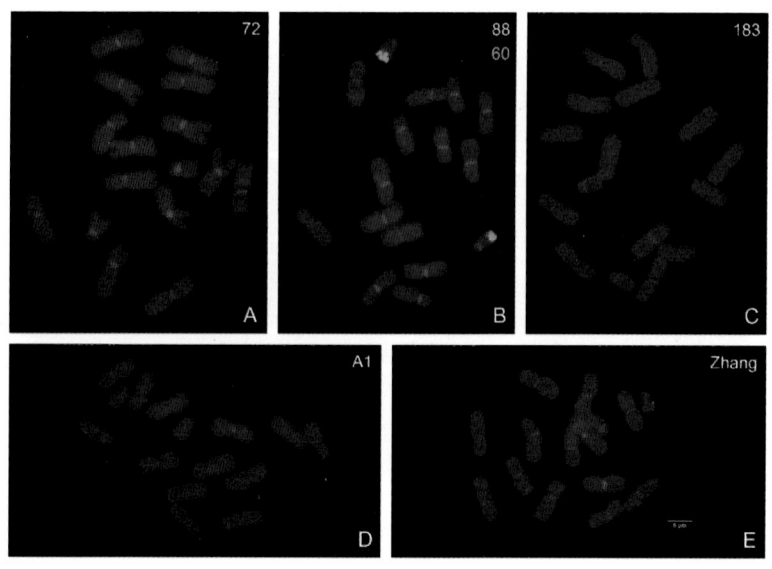

图 3.19 黑麦 B 染色体 DNA 文库克隆 RB72、88、183、99、A1（AFLP 筛选）在黑麦染色体上的定位

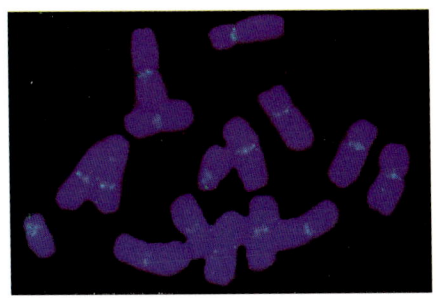

图 3.20　黑麦 B 染色体 DNA 文库克隆 RB113 序列在黑麦染色体上定位的结果

④黑麦 A、B 染色体共有中、高度重复序列的筛选、鉴定

大规模测序已表明，重复序列在物种基因组中占有很大比例，并已有报道证实，这些序列并不都是可有可无的冗余序列，而很可能在维持物种染色体的稳定性，提供表观遗传调控信息方面具有重要作用。前期研究已证实，黑麦 B 染色体可能含有更加丰富的 DNA 重复序列，因此揭示 B 染色体重复序列的组成特征，对于探究 B 染色体 DNA 序列组成特征及其演化和可能的功能具有重要意义。为此，对通过斑点杂交实验及测序证实为黑麦 B 染色体来源的重复序列进行了 FISH 杂交定位分析。结果表明，克隆 RB3、RB127、RB87、RB289、RB318、RB191、RB118、RB220、RB190、RB195 均被成功定位在黑麦染色体上，除个别染色体外，各探针序列基本弥散分布整个基因组中，但它们之间所显示的杂交信号有很大区别的，在 B 染色体与 A 染色体之间信号也存在一些差异。

RB3 序列长度为 728 bp，AT 含量 52.7%，是一个中度重复序列，与肿胀小麦亚种硬粒小麦 BAC221H19 部分序列、小麦 cDNA 克隆 SET4_J21 部分序列、小麦 3B 染色体特异文库部分序列高度同源。从图 3.21 FISH 定位结果可以看出，该序列与其他信号分布位置不同，其主要分布在染色体长、短臂两端，信号比较集中，在 B 染色体长臂末端有较强弥散信号，短臂上未发现信号。在 A 染色体的长、短臂末端信号较强，长、短臂略有区别，在染色体的其他部位未发现有信号（图 3.22A3、图 3.22A6）。

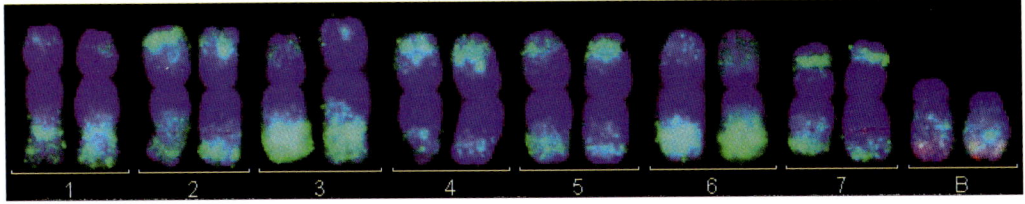

图 3.21　RB3 序列 FISH 定位所显示的信号

RB289 序列长度为 755 bp，AT 含量为 52.9%，分析表明其与肿胀小麦亚种硬粒小麦 Copia 反转座子部分序列、单粒小麦 BAC453N11 部分序列高度同源。FISH 杂交结果表明，该信号在染色体的位置与 RB3 序列很相似，信号强度比 RB3 弱，B 染色体信号分布在长臂末端。常染色体信号分布在长、短臂的两端，中间靠近着丝粒部分没有信号分布（图 3.22B2、图 3.22B3）。

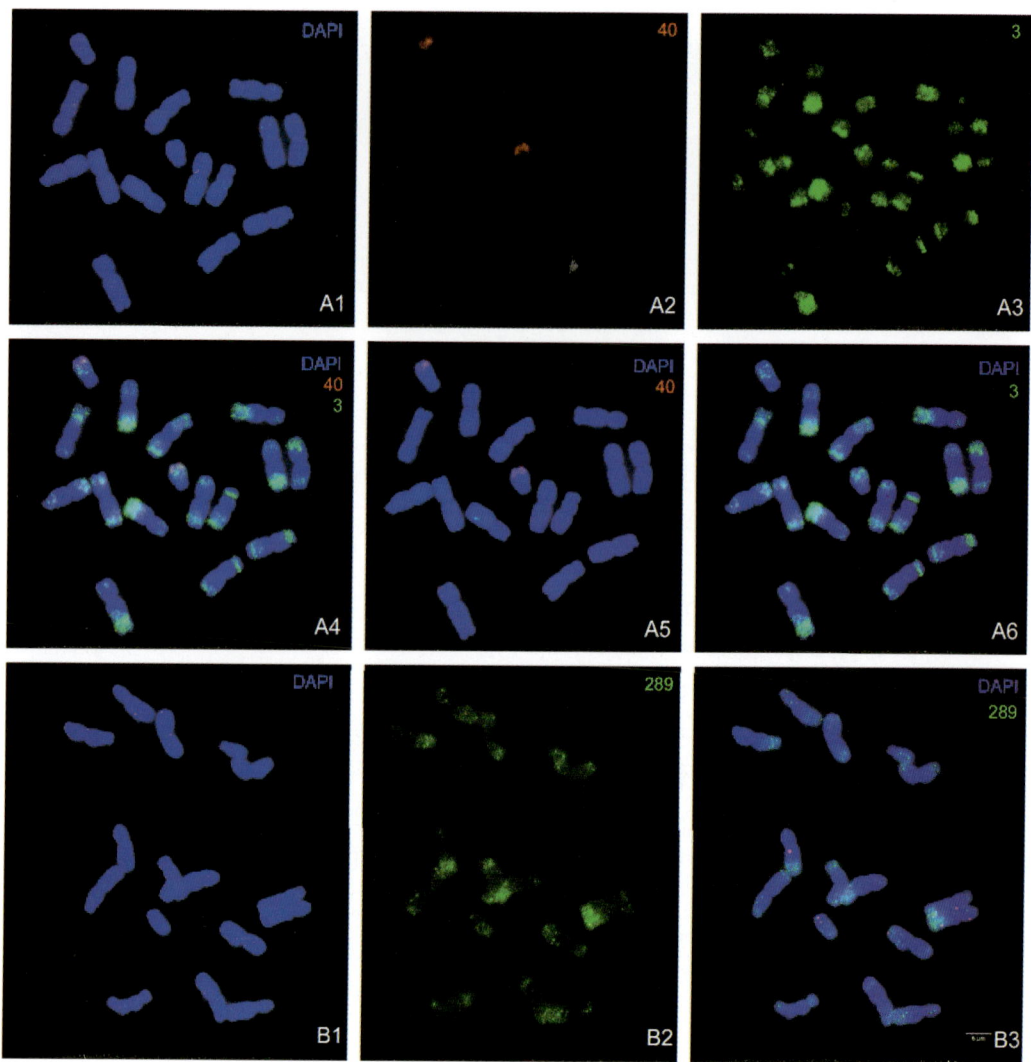

图 3.22　黑麦 B 染色体 DNA 文库克隆 R3、R40、R289 在黑麦染色体上 FISH 定位结果
A1：DAPI 染色，A2：RB40 信号图像，A3：RAB3 信号图像，A4：RAB3 与 RB40 合成图像，A5：A1 与 A2 合成图像，A6：A1 与 A3 的合成图像，B1：DAPI 染色，B2：RAB 289 信号图像，B3：B1 与 B2 的合成图像

RB127 序列长度为 476 bp，富含 AT 碱基序列（57.6%）与 *Triticum turqidum* subsp.dicoccoidesdones M13 和 BAC1 有 31%同源，序列在染色体上分布不均匀，有的染色体上很少，两条 B 染色体基本相同（图 3.23C）。

RB87 序列长度为 553 bp，富含 GC 碱基（53.3%），是一个中度重复序列，序列比对与黑麦 Revolver 转座酶假基因系列高度同源，一致性达 92%，与小麦 3B 染色体特异 BAC 文库部分序列同源，一致性达 76% FISH 杂交结果显示，该序列信号主要分布在黑麦染色体着丝粒附近，长短臂的两端没有信号。在两条 B 染色体分布特征有所不同，其中一条 B 染色体杂交信号强，呈弥散型分布于整条染色体，而另外一条 B 染色体信号较弱，在长臂末端没有信号（图 3.23A）。

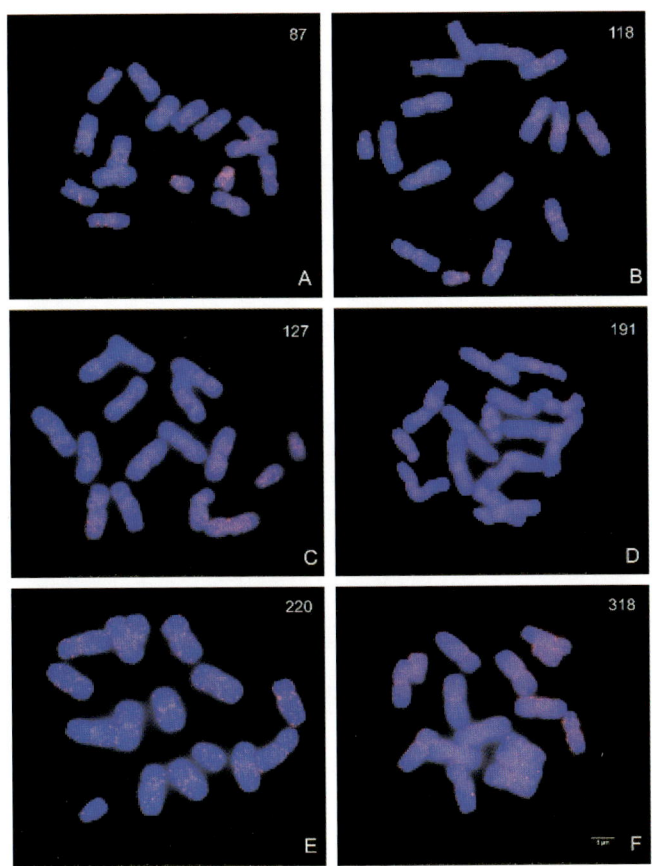

图 3.23 黑麦 B 染色体 DNA 文库克隆 RB87、118、127、191、220、318 在黑麦染色体上 FISH 定位的结果

RB118 序列长度为 500 bp，AT 碱基含量 53.8%，为 *Secale cereale* Revolver-1 pseudogene for transposAse，FISH 杂交结果显示，该序列主要分布在染色体着丝粒附近，其中一条 B 染色体杂交信号较强，分布在整条染色体上，而另一条杂交信号稍弱，并没有覆盖整个染色体（图 3.23B）。

RB318 序列长度为 867 bp，AT 碱基含量为 51.3%，与黑麦 Revolver-2 pseudoqene for transposAse 99%同源，是一个散布的高度重复序列。信号弥散分布在黑麦每条染色体上，在 B 染色体上具有较强的杂交信号，两条 B 染色体没有什么区别（图 3.23F）。

RB191 长度为 756bp，AT 含量 52.2%，与 *Triticum aestivum* cDNA clone SET4J21、*Triticum turqidum* sunsp，durum clone BAC 221H19 genomic sequen 及 *Triticum urartu* clone BAC 292N12 genomic sequence 高度同源（均为 98%以上）。FISH 定位结果表明，该序列为散布的中度重复序列，除染色体末端外信号均有分布。B 染色体长臂末端未显示信号，其他部位信号比 A 染色体强（图 3.23D）。

RB220 序列长度为 452 bp，AT 含量 55.3%，与黑麦 607_IV-G6 克隆微卫星序列、黑麦克隆 543_II-C12 微卫星序列、小麦 3BS 特异 BAC 文库 ctg0079b 部分序列高度同源。FISH 定位表明，该序列为散布的重复序列，点状信号分布在黑麦染色体长、短臂着丝粒附近，一条 B 染色体信号较多，一条信号较少。(图 3.23E)。

RB195 序列长度为 582 bp，AT 含量较高为 60.5%，经序列分析未发现同源序列。经 FISH 定位信号散布于每一条染色体上，是一个中度重复序列（图未显示）。

RB190 序列长度为 572 bp，AT 含量为 52.3%，与一粒小麦磷脂酰丝氨酸脱羧酶 ZCCT2、ZCCT1、SNF2P 基因，假定转座酶基因、P450 型蛋白基因部分序列高度同源；部分序列与小麦 3B 染色体特异文库部分序列高度同源。蛋白比对与水稻 OSIGBa0147J19.11、水稻假定 Ty3-gypsy 反转座子蛋白同源，但同源性较低。FISH 定位结果表明，该序列是一个中度重复序列，定位在黑麦每一条染色体的全长上，A、B 染色体没有区别（图 3.24B）。

1CTATTTTGGGTTTTCGGTTTTGGGATTCTGGCTTTTCTATTTGGGGTTTTGGGTTTA
AGGTCTAAGGTGTA

71GGTTTAGGTTCTAGTGTCTATAAGTTAGGGCCGGGATTAAGCTTAGGGTGTACTA
TTTAGGGTTTAGGGT

141 TTAGGGTTTAGGGTT

图 3.24　RB534 克隆的序列分析，划线部分为拟南芥型端粒序列的重复单元

⑤黑麦 B 染色体端粒序列的筛选、鉴定

端粒位于染色体的端部对于维持染色体结构的完整性具有重要意义，端粒序列对于构建 B 染色体人工载体也是必需元件之一。为此，在筛选获得黑麦 B 染色体长臂末端特异序列基础上，进一步对黑麦 B 染色体特异端粒序列进行了筛查。结果显示其中一个克隆 RB534 与拟南芥型端粒重复序列密切相关（图 3.24），具有拟南芥型端粒的序列特征。FISH 定位结果显示，该 RB534 在黑麦 B 染色体和 A 染色体端部均显示出较强的杂交信号（图 3.25）。可以确定该序列为黑麦端粒序列的组成部分。

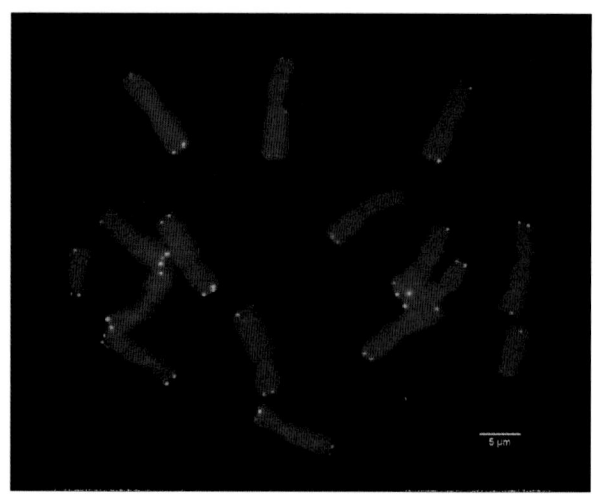

图 3.25　黑麦 B 染色体 DNA 文库 RB534 克隆在黑麦染色体上 FISH 定位的结果

通过上述的研究可以发现，B染色体在DNA组成上既有与A染色体相似之处，又有其独特的序列特征，即一类为A、B染色体共有序列，如RB3、87、118、127、72、183等，它们中的部分已发生了特异的点突变，产生了一些差异；另一类是B染色体所特有的序列，仅在B染色体中出现，如RB1、2、6、40、150等。但是发现的序列中A、B染色体具有的共同序列多，B染色体特异序列发现的少。通过对B染色体文库进行测序分析及FISH定位，了解了部分克隆在染色体上分布的信息（表3.2），但目前还没有筛选到B染色体特异的着丝粒序列，是否B染色体与A染色体着丝粒序列是同样序列，目前还是个未知数，还需要进一步的研究，总之黑麦B染色体特异序列的筛选对于构建人工B染色体载体是十分重要的。

表 3.2 黑麦 B 染色体文库部分克隆序列分析及定位在染色体上的信息

序号和大小	AT 含量（%）	同源比对	FISH 定位
RB160，759 bp	52.7	与黑麦重复单元 E3900 家族和 gag 假基因完整序列高度同源	Bs 长臂末端
RB60，819 bp	58.4	与黑麦重复单元 D1100 家族、黑麦非编码 RNA D1100mRBNA 部分序列、小麦 3B 染色体特异 BAC 文库 ctg1030b 中某些序列高度同源	Bs 长臂末端
RB183，1017 bp	52.9	与小麦野生亚种拟山羊草小麦（野生一粒小麦）TbBAC30 克隆部分序列，黑麦反转座子 Bilby（黑麦着丝粒 Ty1-copia 反转座子家族）部分序列高度同源	黑麦着丝粒的位置，Bs 着丝粒及短臂
RB6，881 bp	48.2	与单粒小麦和肿胀小麦某 BAC 克隆中部分序列 70% 同源，	Bs 特异序列，定位在 Bs 长臂末端
RB220，452 bp	55.3	与黑麦 607_IV-G6 克隆微卫星序列、黑麦克隆 543_II-C12 微卫星序列、小麦 3BS 特异 BAC 文库 ctg0079b 部分序列高度同源	散布的重复序列，定位黑麦染色体长短臂着丝粒附近
RB318，867 bp	51.3	与诸多黑麦转座酶假基因高度同源	散布的高度重复序列，黑麦每一条染色体上
RB87，553 bp	46.7	与黑麦 RevolveR 转座酶假基因系列（92%）、小麦 3B 染色体特异 BAC 文库部分序列高度同源（76%）	散布的重复序列，定位黑麦染色体长短臂着丝粒两侧
RB316，831 bp	53.4	由 E3900 尾部序列+头部序列+新序列组成	Bs 长臂
RB127，476 bp	57.6	与肿胀小麦 BAC 391M13 and BAC1144M20 部分序列、小麦 3B 染色体特异文库部分序列同源，覆盖率较低，同源性不高	散布的高度重复序列，黑麦每一条染色体上
RB150，494 bp	54.7	15~249bp 序列与黑麦属诸多亚种中 gyspy 类型反转座子部分序列高度同源，剩余序列为发现明显同源性蛋白比为大麦 Sukkula-1 多聚蛋白，一致性达 78%	Bs 特异序列，定位在 Bs 长臂末端
RB118，500 bp	53.8	与黑麦 RevolveR 转座酶假基因系列（80%）、小麦 3B 染色体特异文库部分序列高度同源（79%）	散布的高度重复序列，黑麦每一条染色体上
RB1，312 bp	52.9	与百萨偃麦草 RAPD 标记 DNA 高度同源，一致性为 77%	Bs 特异序列，在 Bs 长臂末端
RB2，1059 bp	49.9	与小麦 BAC 文库 TaaCsp3BFhA_0100L17 部分序列、黑麦 JNK 2F2R-1500 亚区基因组部分序列、黑麦 E3900 重复单元和 gag 假基因部分序列高度同源	Bs 特异序列，在 Bs 长臂末端
RB88，1061 bp	51.7	与小麦 BAC 文库部分序列、小麦 3B 染色体特异文库部分序列、黑麦微卫星克隆 362_IV-H7 部分序列高度同源	黑麦每一条染色体着丝粒

续表

序号和大小	AT 含量（%）	同源比对	FISH 定位
RB191，756 bp	52.2	与小麦 cDNA 克隆 SET4_J21 部分序列、肿胀小麦 BAC 221H19 部分序列、黑麦 RevolveR 转座酶部分序列高度同源	散布的高度重复序列，黑麦每一条染色体上
RB21，731 bp	56.1	与小麦 BAC1551N13 部分序列、小麦 3B 染色体特异 BAC 文库 ctg0079b 一系列部分序列、肿胀小麦 BAC 391M13 and BAC 1144M20 部分序列、小麦 BAC198E19 质体乙酰辅酶 A 羧化酶（ACC-1）基因部分序列高度同源	Bs 特异序列，在 Bs 长臂末端
RB289，755 bp	52.5	与肿胀小麦亚种硬粒小麦 Copia 反转座子部分序列、单粒小麦 BAC453N11 部分序列高度同源	散布的高度重复序列，在部分黑麦长短臂末端
RB72，846 bp	55	与肿胀小麦亚种硬粒小麦 BAC219E24 部分序列、小麦 BAC 41C8 部分序列、小麦 3B 染色体特异文库系列部分序列高度同源，与水稻 OSIGBa0161P06.1、水稻反转座子 Ty3-gypsy 假定蛋白高度同源	着丝粒序列，定位在黑麦每一条染色体着丝粒
RB99，367 bp	49.1	与玉米着丝粒序列 CentA 高度同源（356 bp70%同源），与水稻（358bp 68%同源），种茅山羊草（315bp 60%同源），拟南芥（179bp 69%同源）	着丝粒序列，定位在黑麦每一条染色体着丝粒
RB40，567 bp	52	无同源	Bs 特异序列，在 Bs 长臂末端
RB3，728 bp	52.7	与肿胀小麦亚种硬粒小麦 BAC221H19 部分序列、小麦 cDNA 克隆 SET4_J21 部分序列、小麦 3B 染色体特异文库系列部分序列高度同源	重复序列，定位在常染色体每一条染色体长短臂两端，Bs 长臂末端
RB113，789 bp	53.9	与小麦 BAC 文库 TaaCsp3BFhA_0100L17、小麦 3B 染色体特异文库 ctg 系列部分序列	黑麦每一条染色体着丝粒
RB136，848 bp	58	与黑麦重复单元 D1100 家族部分序列、黑麦非编码 RBNAD1100mRBNA 部分序列、小麦 3B 染色体特异 BAC 文库 ctg 系列部分序列高度同源	Bs 长臂末端
RB165，589 bp	56.2	蛋白比对与高粱假定蛋白 SORBIDRAFT 系列同源，但同源性较低；	BS 短臂
RB190，572 bp	52.3	与单粒小麦磷脂酰丝氨酸脱羧酶 ZCCT2、ZCCT1、SNF2P 基因，假定转座酶基因、P450 型蛋白基因部分序列高度同源；部分序列与小麦 3B 染色体特异文库部分序列高度同源。蛋白比对与水稻 OSIGBa0147J19.11、水稻假定 Ty3-gypsy 反转座子蛋白同源，但同源性较低	散布重复序列黑麦每一条染色体上
RB306，589 bp	56.7	蛋白比对与高粱假定 SORBIDRAFT 系列蛋白有同源性，一致性较低	BsS 短臂
RB195，582 bp	60.5	无同源	散布重复序列，黑麦每一条染色体上

3.1.7.2 黑麦 0B/+B 基因组 ISSR 分析

在对黑麦 0B 和+B 基因组的 ISSR 分析中，经过 36 对引物的筛选，在近 1500 条清晰可辨的条带中仅有 16 条明显差异带产生，其中+B 基因组特异带有 12 条，0B 基因组特异带 4 条。经回收扩增后在含 B 的黑麦基因组中只获得了 8 个特异带（图 3.26）。

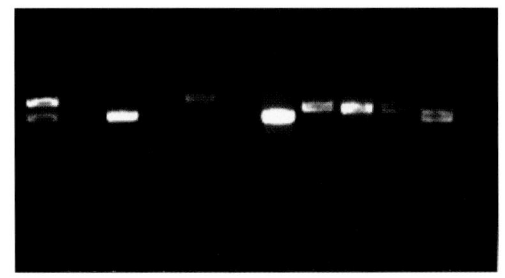

图 3.26　经回收扩增后得到的 B 染色体 ISSR 特异条带

将这些差异带扩增产物分别与含 B 和不含 B 基因组标记的探针进行斑点杂交，只有 I3、I4、I7、I8 号在含 B 和不含 B 基因组探针中都有杂交结果，但信号较弱，其他均无信号显示。其中，I4、I-7、I8 号均为含 B 基因组特异带，I-3 为不含 B 基因组特异带，点杂结果显示在两者中都有信号，说明，这 4 个片段为高/中拷贝重复序列，并且进一步说明 B 染色体与 A 染色体有很高同源性；其他未显示信号，分析可能为低拷贝片段。进一步将这些差异片段测序和 Blasta 分析，结果见表 3.3。

表 3.3　ISSR 分子标记克隆的序列分析及 FISH 定位信息

序号	序列大小	AT 碱基含量	同源比对结果	染色体定位
I1	957 bp	50.7%	高粱假定蛋白 m RNA，末段一段序列同源 9e-66，75%；高粱假定蛋白 SORBIDRAFT_10g024663，64%	无信号
I2	280 bp	62%	与水稻基因组 1 号染色体 PAC 克隆 P0784G04 中的一段序列同源达 86%，4e-69，蛋白比对与水稻假定蛋白 OsI_01941 同源达 92%	散布点状信号
I3	579 bp	55.3%	与小麦 3B 染色体特异 BAC 文库 ctg0464b 部分序列高度同源，一致性达 92%	散布的重复序列全部黑麦染色体，Bs 信号很少
I4	391 bp	50.4%	与小麦 3B 染色体特异 BAC 文库 ctg 系列部分序列同源，同源性达 80%	散布的重复序列全部黑麦染色体，Bs 上有信号
I5	413 bp	54.96%	无同源性，新序列	无信号
I6	582 bp	62.7%	无同源性，新序列	无信号
I7	406 bp	54.9%	序列比对无明显同源性	散布的重复序列，全部黑麦染色体上呈现点状信号
I8	380 bp	53.4%	与小麦 BAC 克隆 122F14 质体乙酰辅酶 A 羧化酶基因（ACC-1）部分序列、肿胀小麦重叠克隆 BAC 326E2 and BAC 354M17 部分序列、小麦 P450 基因、假定锚蛋白基因、TAC TaBx3B-TaBx4B 克隆部分序列高度同源	散布的重复序列，全部黑麦染色体上呈现点状信号

为了进一步分析这些片段在染色体上的分布，将获得的差异片段进行 FISH 定位，结果表明：利用 ISSR 获得差异片段均为 A、B 染色体共有重复序列，但相对于 A 染色体来说，B 染色体信号较弱（图 3.27）。

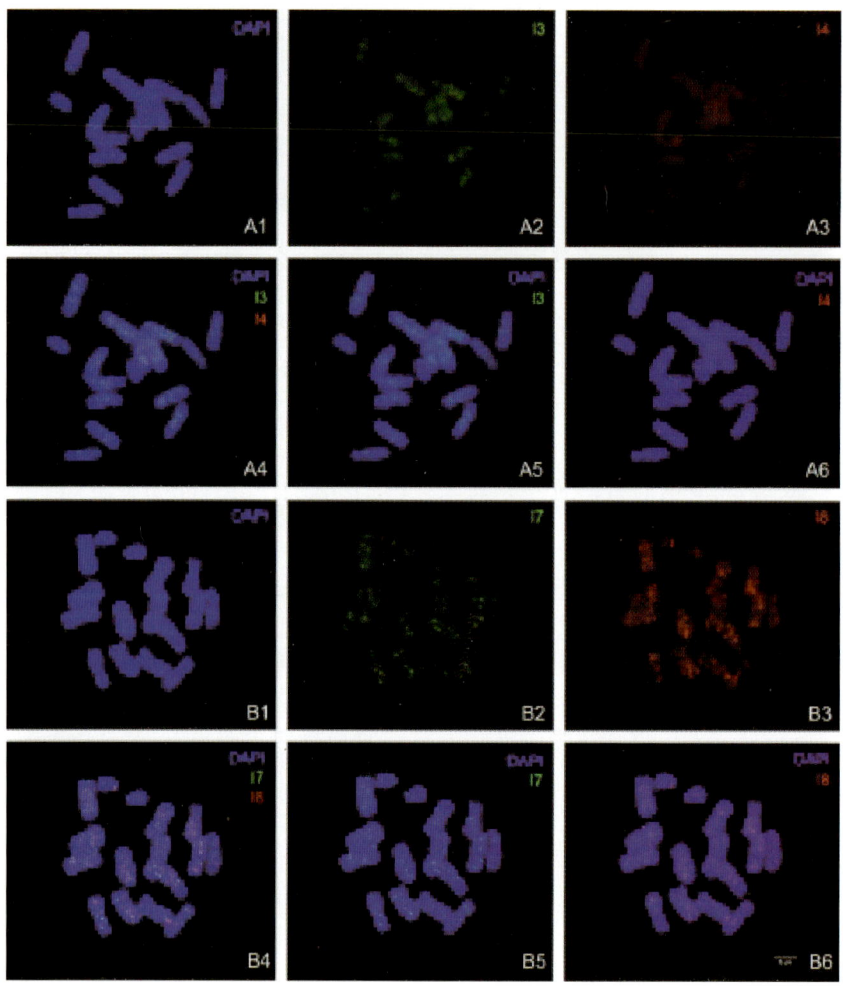

图 3.27 黑麦 0B 与+B 基因组 DNA ISSR 差异分子标记 I2、I3、I4、I7、I8 原位杂交结果
A1：DAPI 染色，A2：I-3 信号图像，A3：I-4 信号图像，A4：A1+A2+A3 合成图像，A5：A1+A2 合成图像，A6：A1+A3 合成图像，B1：DAPI 染色，B2：I-7 信号图像，B3：I-8 信号图像，B4：B1+B2+B3 合成图像，B5：为 B1+B2 合成图像，B6：B1+ B3 合成图像

3.1.7.3 黑麦单条染色体的比较荧光原位杂交分析

B 染色体的起源及演化一致颇具争议，至今仍没有确凿的证据对其做出明确的解释。而阐释 B 染色体的起源及演化特征对于了解 B 染色体的功能及其行为特征，进而对 B 染色体进行开发、利用具有举足轻重的重要性。关于黑麦 B 染色体的起源，大多数研究者认为其来源于 A 染色体，是 A 染色体演化的产物，但 B 染色体到底来源于 A 组染色体的哪一条？还是哪几条？其是否真的来源于被多数研究者所接受的 7R 染色体？目前仍有很大的疑问。为此，在获得高质量黑麦染色体制片的基础上，我们对黑麦 14 条常染色体及 B 染色体分别进行了显微分离及 LA-PCR 扩增（图 3.28、图 3.29），并将每一条染色体扩增产物进行标记，进行单染色体的比较荧光原位杂交分析，以确定黑麦 B 染色体到底与哪条或哪几条 A 染色体更为相似，通过分析可以非常直观的揭示 B 染色体

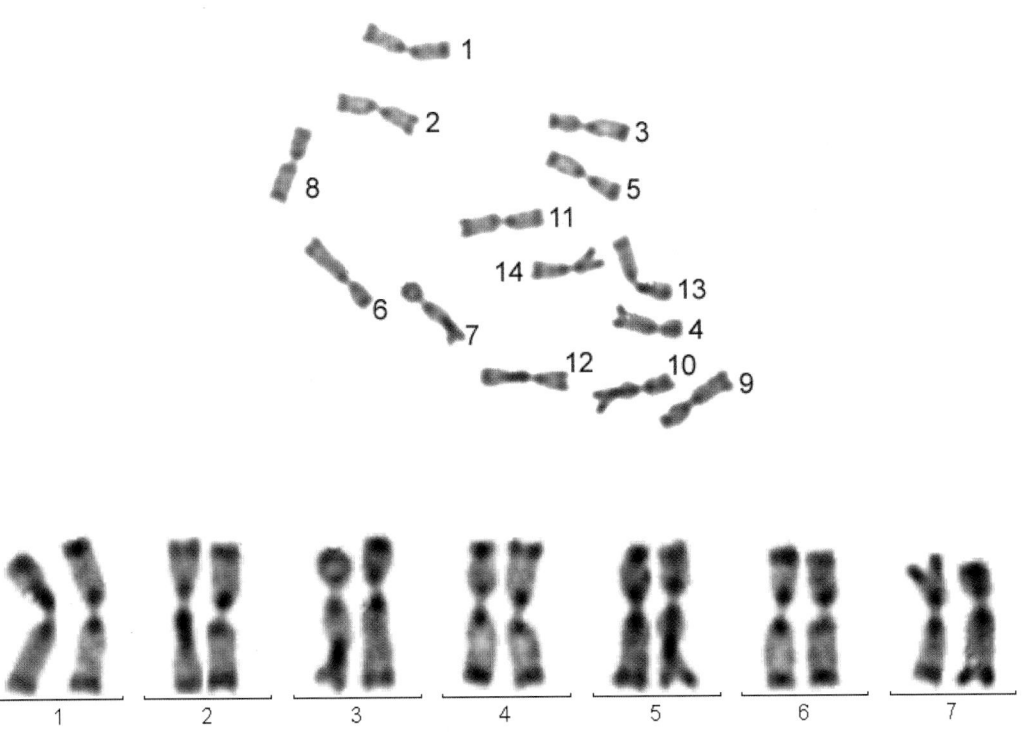

图 3.28 黑麦 14 条染色体显微分离图像及核型

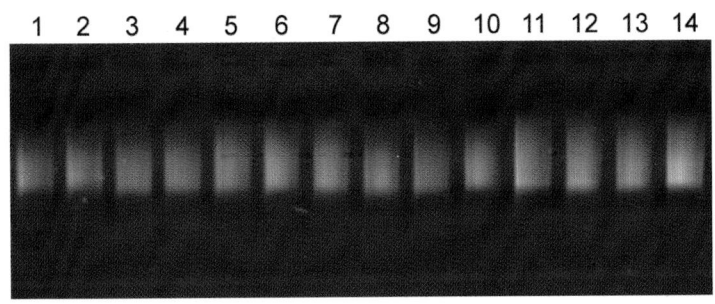

图 3.29 分离 14 条黑麦 A 染色体 LA-PCR 扩增的电泳图

与其他 A 染色体在 DNA 序列组成上的关系,进而可以揭示 B 染色体的起源及演化特征。

1. 以 1R～7R 扩增产物为探针的比较荧光原位杂交结果

以 1R 染色体微切扩增产物为探针进行 FISH 定位,杂交信号除在 1R 染色体上呈均匀弥散分布外,在其他 6 对 A 染色体上也呈弥散状分布,但信号分布并不均匀。在 B 染色体信号虽也呈弥散状分布,但可以发现在 B 染色体长臂末端和着丝粒位置并未有明显的杂交信号(图 3.30B)。该 FISH 杂交结果表明,1R 染色体虽为常染色体但其 DNA 序列组成也以中、高度重复序列占优势,并且重复序列组成与其他常染色体的组成可能差异不大,致使 7R 染色体微切扩增产物含有丰富的重复 DNA 序列,并在随后杂交中,在其他 A 染色体也可检测到较强的弥散型杂交信号。但可以发现 1R 染色体与自身及其他常染色体杂交信号模式（信号强度、分布特征）又有明显不同，暗示不同 A 染色体间存在自身特异的基因或 DNA 序列。B 染色体本身具有丰富的中、高度重复序列，因此以 1R 染色体微切扩增产物为探针在两条 B 染色体上都检测到较强的弥散型杂交信号，但杂交模式的不同也显而易见（图 3.30C）。因此，推测 1R 染色体与 B 染色体可能关系较远。

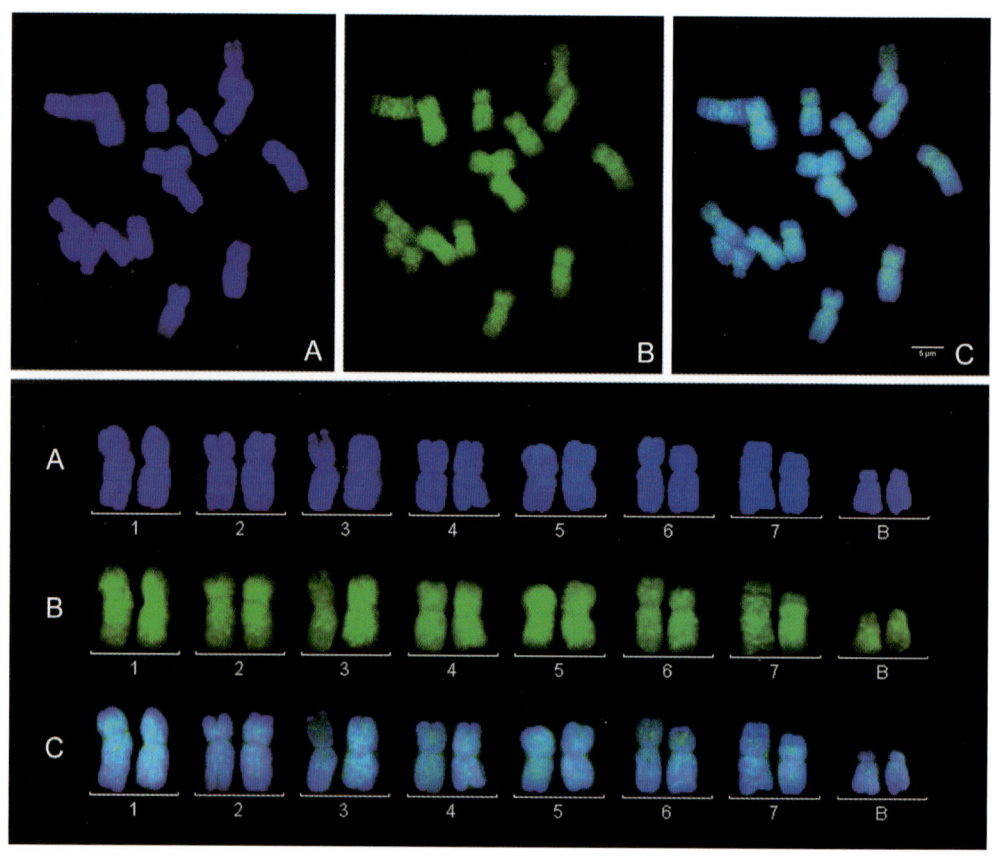

图 3.30　以 1R 染色体微切产物为探针与黑麦荧光原位杂交的结果
A：DAPI 染色的染色体，B：信号图像，C：合成图像，绿色为 1R 号染色体 FISH 信号

以 2R 染色体微切扩增产物为探针进行 FISH 定位，定位分析结果与以 1R 染色体为探针的杂交模式基本一致，信号在 2R 染色体及其他常染色体上也呈弥散状分布，但染色体端部信号更加强烈（图 3.31B），与 Giemsa C 带的类型基本相似（图 3.31），除 6R 短臂末端外，所有染色体短臂都有染色深、大而明显的末端带，但没有着丝点带，在 7R 染色体上有明显的核仁缢痕带。这暗示 2R 染色体异染色质非常丰富。与之相似，在 Bs 上也可检测到弥散的杂交信号，但与其他 A 染色体不同，杂交信号在 B 染色体长、短臂端部并没有大而深的末端带，因此，B 染色体起源于 2R 的可能性很小。

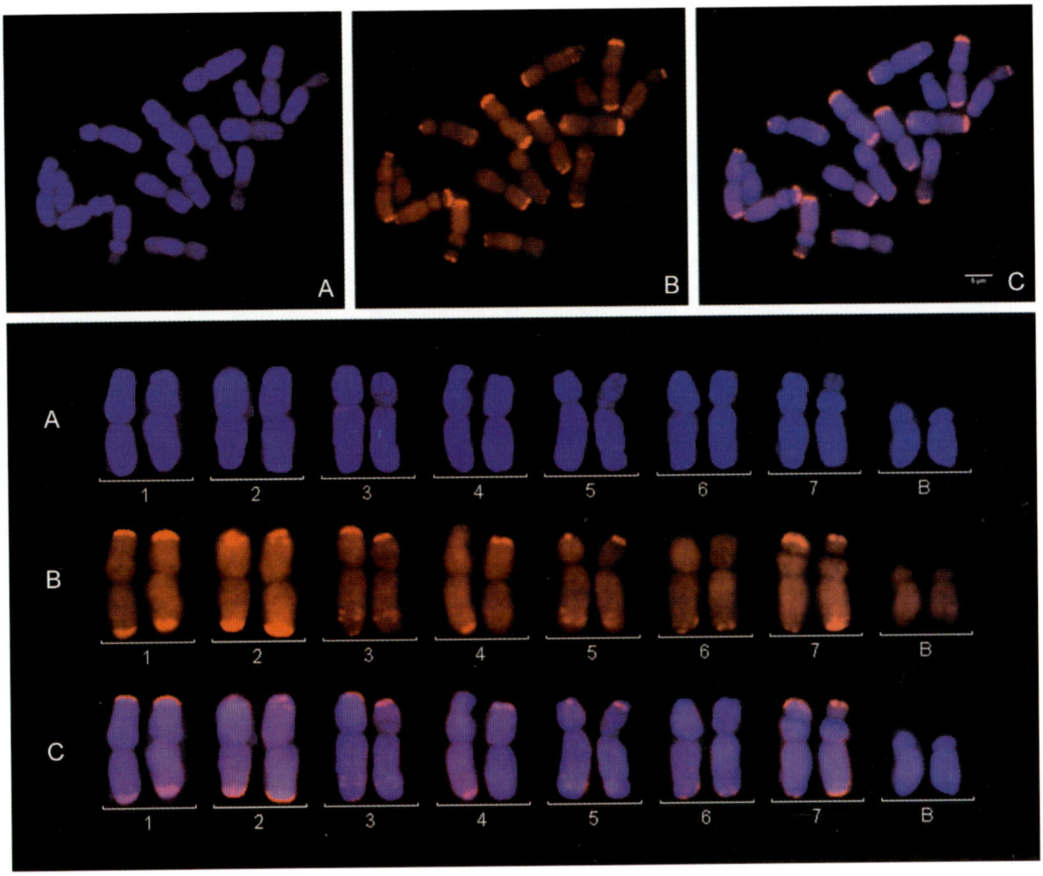

图 3.31 以 2R 号染色体 PCR 产物为探针与黑麦荧光原位杂交的结果
A：DAPI 染色的染色体，B：信号图像，C：合成图像，红色为 2R 号染色体 FISH 信号

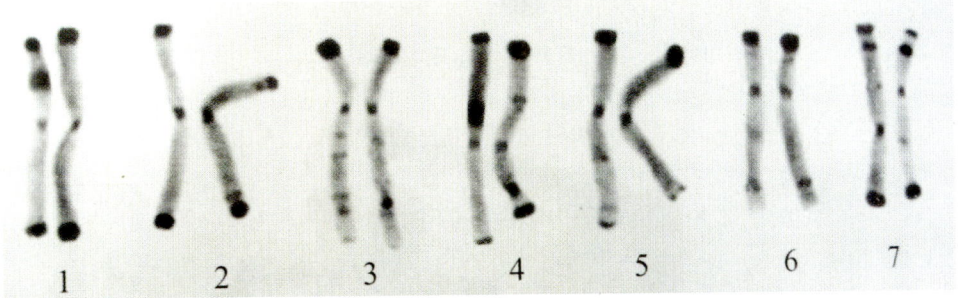

图 3.32 黑麦染色体 Giemsa C 带核型

以 3R 染色体微切扩增产物为探针进行 FISH 定位，杂交信号在 3R 及其他 A 染色体上也呈弥散状分布，之间差异并不显著，表明 3R 染色体上一部分序列组成与其他 A 染色体十分接近（图 3.33B 和图 3.33C）。但与两个 B 染色体的杂交结果显示，虽也可在 B 染色体上检测到弥散的信号分布，但在其长、短臂末端均未检测到杂交信号，暗示 3R 染色体与 B 染色体的亲缘关系更远一些。

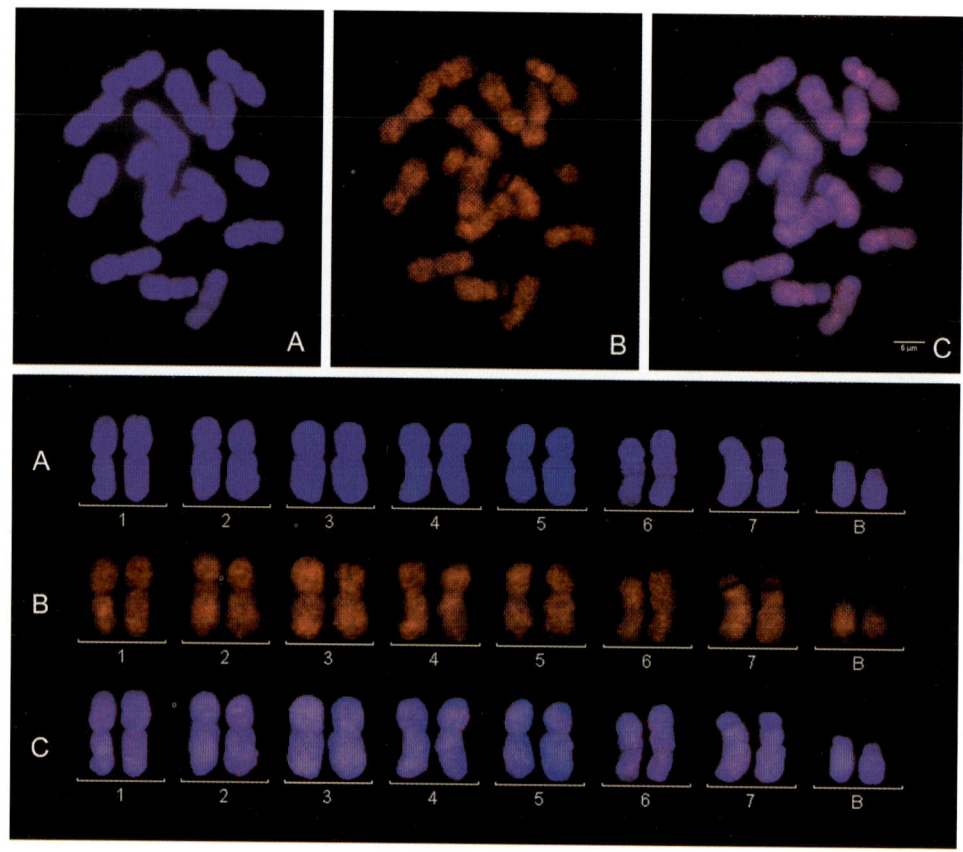

图 3.33 以微切 3R 染色体 PCR 产物为探针与黑麦荧光原位杂交的结果

以 4R 染色体微切扩增产物为探针进行 FISH 定位，杂交信号在常染色体之间基本一致，呈均匀弥散状分布。但在 B 染色体中，只在其中一个小 B 染色体长臂上端具有较强的杂交信号，其他位置信号都较弱（图 3.34B 和图 3.34C）。因此，可以排除 B 染色体来源于 4R 染色体。

以 5R 染色体微切扩增产物为探针进行 FISH 定位，定位分析结果显示，5R 染色体自身杂交信号较为强烈，在染色体上分布均匀。其他 A 染色体上杂交信号虽也呈弥散状分布，但不同 A 染色体之间仍有一定区别，差异主要体现在染色体端部（图 3.35B 和图 3.35C）。分析发现 5R 染色体信号与 7R 染色体信号相当，5R 与 7R 是否有一定的关联还要进一步研究。而在 B 染色体上，尽管在其长臂末端未检测到信号，但其他部位信号分布均匀，强烈。因此，从现有的杂交结果显示，B 染色体与 5R 染色体十分相似，其是否起源于 5R，还有待于进一步试验证实。

以 6R 染色体微切扩增产物为探针进行 FISH 定位，杂交信号显示，在所有黑麦常染色体的末端杂交信号都较弱，分布也不均匀，暗示 6R 染色体在序列组成上可能比较特殊，其在 Bs 上的杂交信号也较弱（图 3.36B 和图 3.36C）。因此，推测 6R 染色体与 B 染色体关系不大。

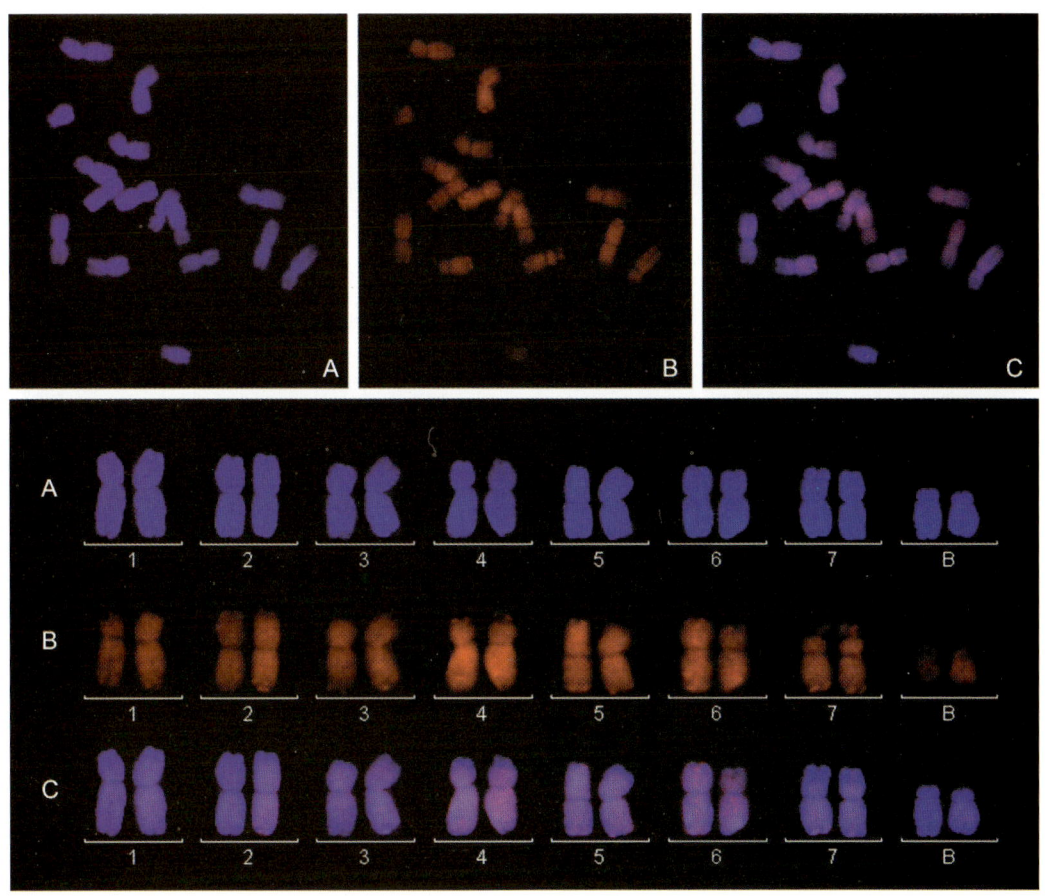

图 3.34 以微切 4R 号染色体 PCR 产物为探针与黑麦荧光原位杂交的结果
A：DAPI 染色的染色体，B：信号图像，C：合成图像，红色为 4R 号染色体 FISH 信号

以 7R 染色体微切扩增产物为探针进行 FISH 定位，7R 染色体具有随体，杂交结果显示，杂交信号虽在各条 A 染色体上均有分布，但分布模式及信号强度在各条 A 染色体间有较大差异（图 3.37B 和图 3.37C）。暗示 7R 染色体 DNA 序列组成可能也比较特殊。7R 与 B 染色体的杂交结果显示，两条 B 染色体一条信号比较强而另外一条稍弱，且在两条染色体长臂末端均未检测到杂交信号。7R 染色体与 B 染色体杂交信号强度相当，因此，推测 7R 染色体与 B 染色体关系较近。假设 B 染色体来源 7R 染色体随体丢失后的那一段，那么 B 染色体应该与 7R 有很多相同的序列，7R 与 B 染色体 AFLP 结果表明，它们之间亲缘关系较近，如果起源于 7R 染色体，B 染色体与 7R 不同源的长臂末端序列又是如何获得等等，因此，还缺少足够的实验证据证明 B 染色体起源于 7R 染色体。

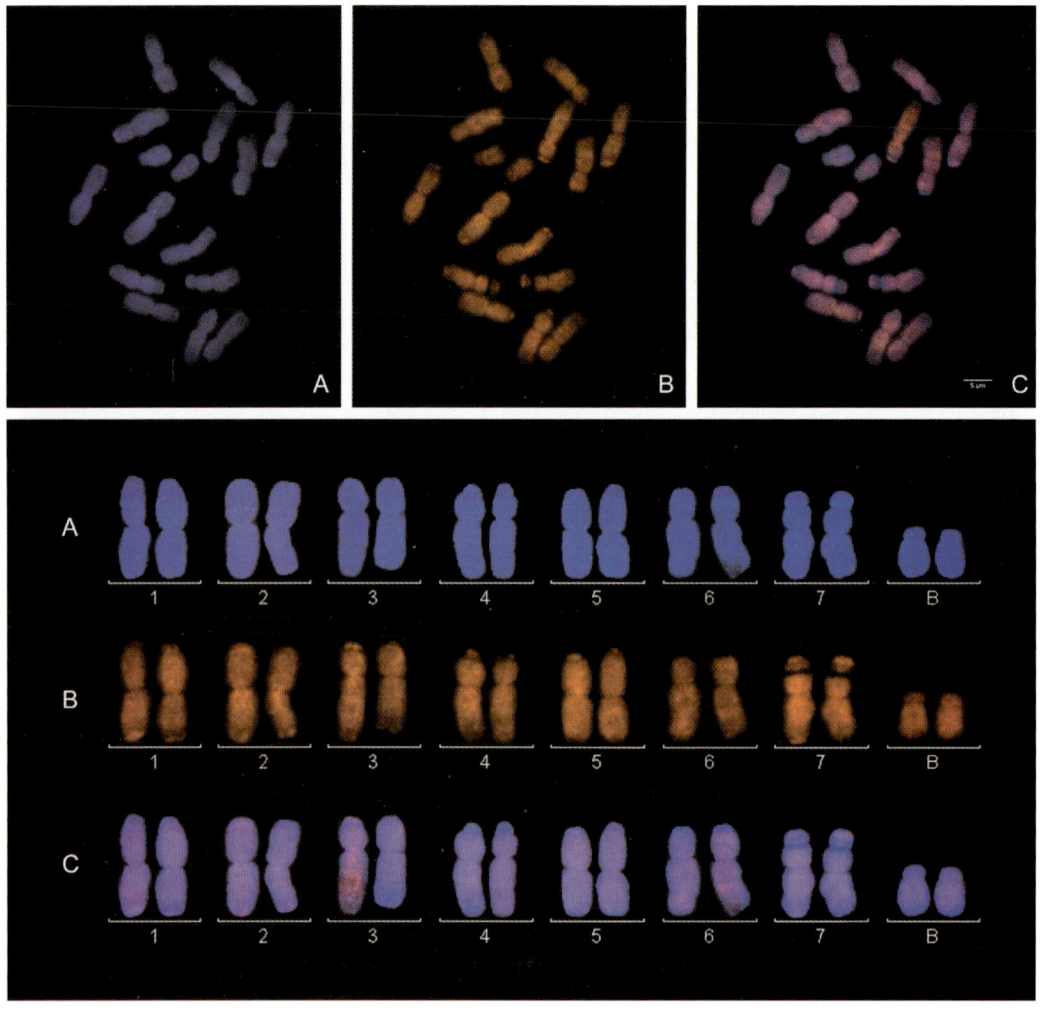

图 3.35 以 5R 号染色体 PCR 产物为探针与黑麦荧光原位杂交的结果
A：DAPI 染色的染色体，B：信号图像，C：合成图像，红色为 5R 号染色体 FISH 信号

通过上述的结果分析，我们初步认为 B 染色体来源于 A 染色体，除 B 染色体长臂末端外其他的部位基本都能与常染色体显示相似的杂交信号，同源性高的位置信号强一些。说明了 A、B 染色体的 DNA 序列之间具有较高水平的相似性。利用 B 染色体 DNA 序列作探针与含 B 的染色体杂交，结果表明 B 染色体几乎全部都显示了很强的杂交信号，其他常染色体着丝粒周围信号最强，推测黑麦中的 B 染色体是否可能起源于 A 染色体的近着丝粒区？而且目前的分子生物学研究证实在许多物种中 B 染色体含有来源于一条 A 染色体或者不同 A 染色体中大多数非编码重复序列或者可移动元件，我们利用微分离技术分离了 14 条常染色体，PCR 扩增后标记探针，逐一将每条 A 染色体都与 B 染色体杂交，发现在 B 染色体上并没有出现与 A 染色体一样的信号，说明 B 染色体形成后异染色质化迅速的形成了与 A 染色体不同的特殊序列，从 E3900 的片段杂交情况来看也可以说明，E3900 是 B 染色体长臂末端发现的一个重复序列家族，用 E3900 作探针与黑麦基因组杂交后，B 染色体末端信号比较集中，可能发生了兆碱基数量级的扩

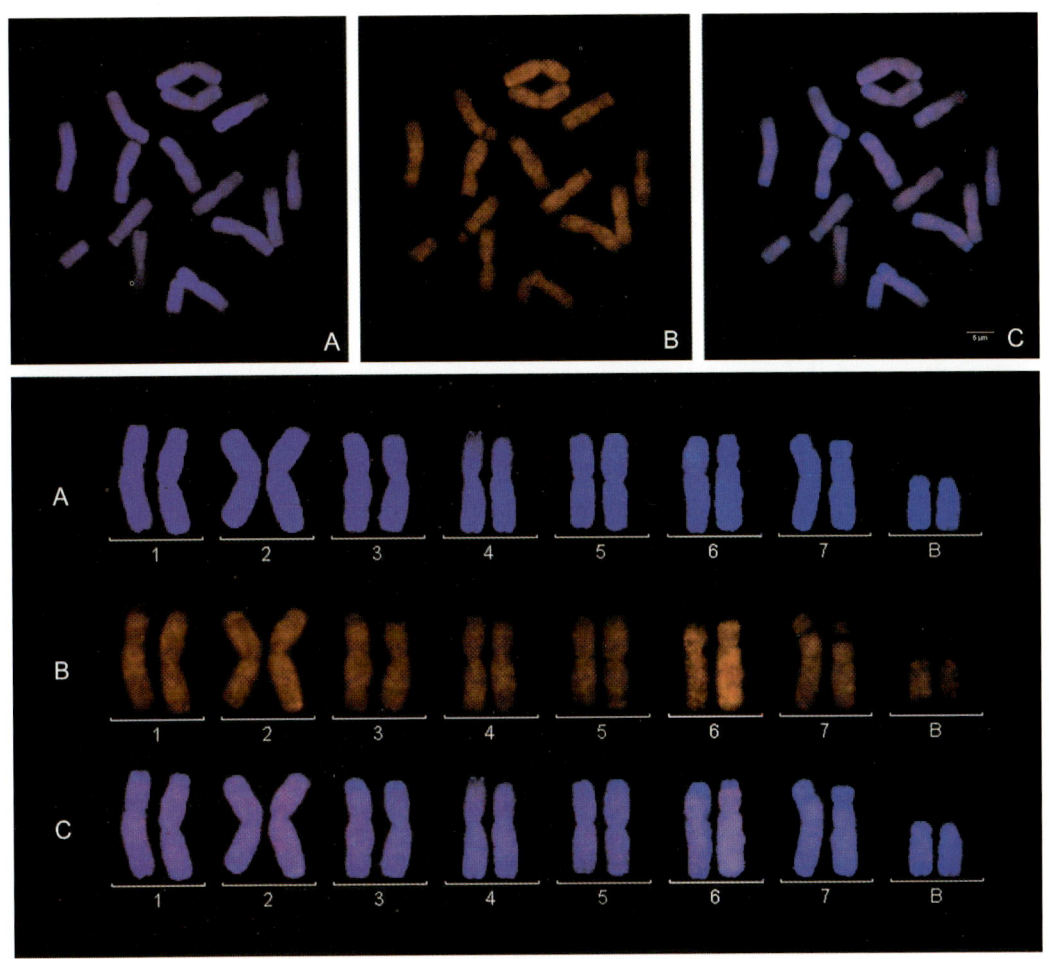

图 3.36 以微切 6R 号染色体 PCR 产物为探针与黑麦荧光原位杂交的结果
A：DAPI 染色的染色体，B：信号图像，C：合成图像，红色为 6R 号染色体 FISH 信号

增，以致形成强烈的信号，B 染色体的其他部位也有杂交信号出现。A 染色体在着丝粒周围出现比较强的杂交信号，这些结果进一步暗示 B 染色体起源于 A 染色体。还有一种推测就是 B 染色体来源于黑麦的具随体染色体，通过我们的实验证明，如果是来自于随体染色体那么利用 18S、5S 杂交会在 NOR 区产生杂交信号，实验证明在 B 染色体上没有 NOR 序列（图 3.38）。从我们筛选获得 RB306 和 165 两个序列来看，他们是特异定位于 Bs 着丝粒短臂周围，在其他常染色体该位置上并没有信号的出现，这说明 B 染色体着丝粒到短臂这些部位与 A 染色体还有差别。另外，通过对大量黑麦染色体的分析，我们发现了很多株 $2n=15$ 的染色体和 16 的染色体，15 的染色体为两条 B 染色体，16 的没有 B 染色体，更为巧合的是 15 条染色体中 R5 为三条，那么怎么形成 3 个 R5 染色体，是否与 B 染色体有关。Jones 和 Rees 总结 B 染色体的起源机制中，其中之一就是 B 染色体起源三体片段，在三体大麦自交产生的三体个体（$2n=2x+1=15$）后代中有 14 条染色体（正常二倍体染色体数目）和一个具着丝粒的小片段，在进一步的自交中又产生了 14 条染色体和 2 个具着丝粒的片段。这两个具着丝粒的片段以高频率传递到后代，

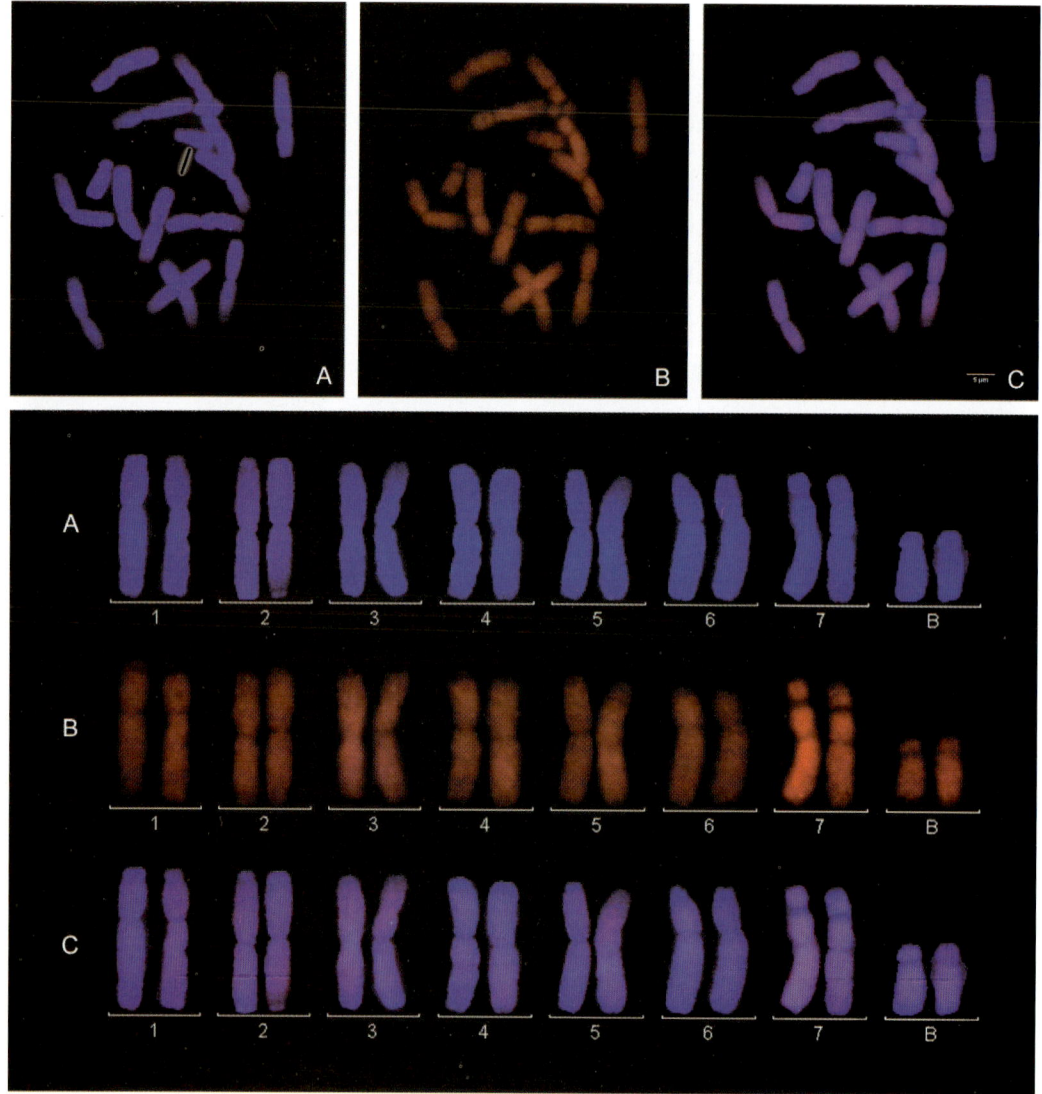

图 3.37　以微切 7R 号染色体 PCR 产物为探针与黑麦荧光原位杂交的结果
A：DAPI 染色的染色体，B：信号图像，C：合成图像，红色为 7R 号染色体 FISH 信号

对大麦的表型几乎无影响，这些片段可能就是一些物种中 B 染色体来源。还有一种可能就是两条 B 染色体融合形成一条 A 染色体，以至于形成 15 和 16 条的染色体。这些非整倍体染色体的形成可能更为复杂，还有待于进一步实验佐证。另外在实验中还发现一些染色体结构异常，如染色体的断片、双着丝点染色体等（图 3.39），这些结果说明在 JNK 黑麦群体中染色体受外界条件的刺激后染色体发生断裂，有了断裂点后染色体可能发生融合，图 3.39 是发生断裂后形成的双着丝粒的染色体（红箭头），旁边是断下来的段片（黄箭头）。但是 Bs 不会是这样简单的断裂方式形成的。另外还有一个有趣的问题，我们分析的大量测序片段中，发现从 B 染色体文库中得到的序列很多与小麦 B 组 3 号染色体相关，是否 B 染色体起源于种间小麦属 B 组的某个染色体，相关的研究正在进行中。

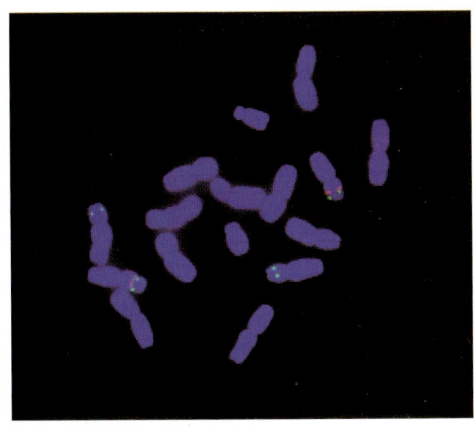

图 3.38　以 18S rDNA 和 5S rDNA 为探针与黑麦+B 染色体 FISH 结果
红色信号为 18S rDNA，绿色信号为 5S rDNA

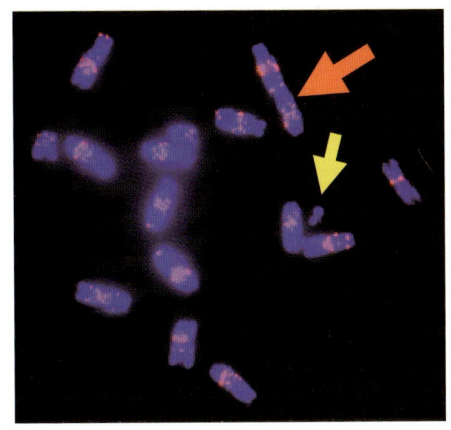

图 3.39　黑麦染色体结构变异
红色箭头示双着丝点染色体；黄箭头示断片

3.1.7.4　黑麦小 B 染色体形成的可能机制

以往的研究表明，在某些物种中甚至同时存在有不同类型的 Bs，例如在 *B. dichromosomatica* 中，可以同时观测到大的 B 染色体和微小 B 染色体。袋娃 *Gastrotheca espeletia* 含有三种类型的 B 染色体，这三类 B 染色体在大小和形态上都不相同。蝗虫 *Eyprepocnemis plorans* 中也含有不同大小的 B 染色体。但是 Muntzing 等人在黑麦研究中显示，不同区域黑麦种群中 B 染色体在根尖细胞有丝分裂中期时的形态都十分类似，Lima-de-Faria（1963）观察了多个黑麦种群减数分裂粗线期的 B 染色体，发现它们的形态也相当一致。根据 Muntzing 和 Lima-de-Faria（1952）的描述，黑麦中的 B 染色体只有 A 染色体约一半大小，并且是近端部着丝粒染色体，没有发现其他类型的 B 染色体。但是我们连续三年都对经染色体鉴定含 B 染色体的黑麦种子进行保种繁殖。而在随后的连续追踪研究中发现，在一些含 B 染色体种子的繁殖后代中新出现了一对小的 B 染色体。染色体观察显示，这对小 B 染色体多数情况下可与大 B 染色体共存于同一细胞。

这对小 B 染色体是如何形成的？是否和已经存在的 B 染色体有共同的形成机制？带着这些问题，我们对黑麦中这类新发现的小 B 染色体的形成机制进行了一系列探究。首先利用显微分离技术分离了 10 条小 B 染色体，随后利用其 LA-PCR 扩增产物进行比较原位杂交分析。结果表明，小 B 染色体微切扩增产物与 A 组染色体及大 B 染色体均有杂交信号，其中杂交信号在 A 组染色体上主要分布于着丝粒附近，在大 B 染色体上信号在其短臂及着丝粒附近均有较强的均匀分布，而其长臂末端未有信号（图 3.40）。由此推测小 B 染色体是否来源于大 Bs 的长臂末端缺失？随后利用已报道的大 B 染色体长臂末端特异序列 D1100 和 E3900 与同时含有小 B 染色体和大 B 染色体的细胞进行 FISH 分析。结果显示，D1100 和 E3900 杂交信号只出现在大 B 染色体的长臂末端，而在小 B 染色体上未检测到杂交信号（图 3.41，图 3.42）。由此确定小 B 染色体缺失了大 B 染色体的长臂末端。为进一步证实这个结论，利用已筛选获得的着丝粒特异探针 RB183 与小 B 染色体进行 FISH 杂交。结果表明，小 B 染色体与大 B 染色体一致，在着丝粒处显示出比 A 染色体更加强烈的杂交信号（图 3.43）。随后以简单重复序列（ACC）5 为探针的 FISH 结果也证实小 B 染色体与大 B 染色体信号分布模式一致（图 3.44）。一系列结果均表明，小 B 染色体和大 B 染色体有较高的相似性，其不同可能就在于小 B 染色体缺失了大 B 染色体的长臂末端，这极有可能是这些新形成的小 B 染色体的主要形成机制。而对于已经存在的 B 染色体其形成机制是否也是如此，还有待于深入探究。

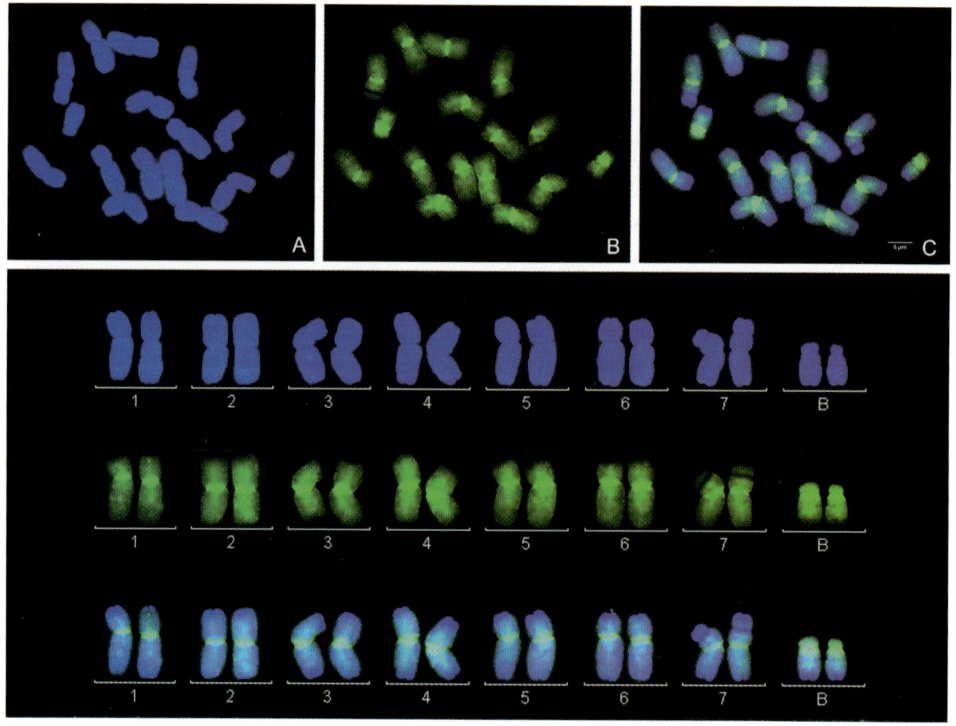

图 3.40　以微切小 B 染色体 PCR 产物为探针与含有大 B 染色体荧光原位杂交的结果
A：DAPI 染色，B：信号图像，C：合成图像

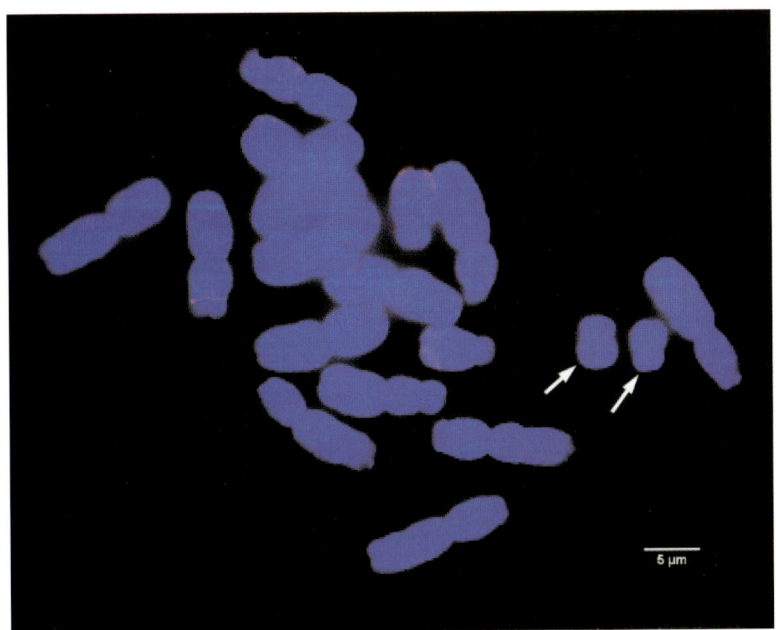

图 3.41 以 D1100 为探针与含小 B 染色体 FISH 的结果

箭头示小 B 染色体；红色信号为大 B 染色体末端特异信号，小 B 染色体末端未见信号

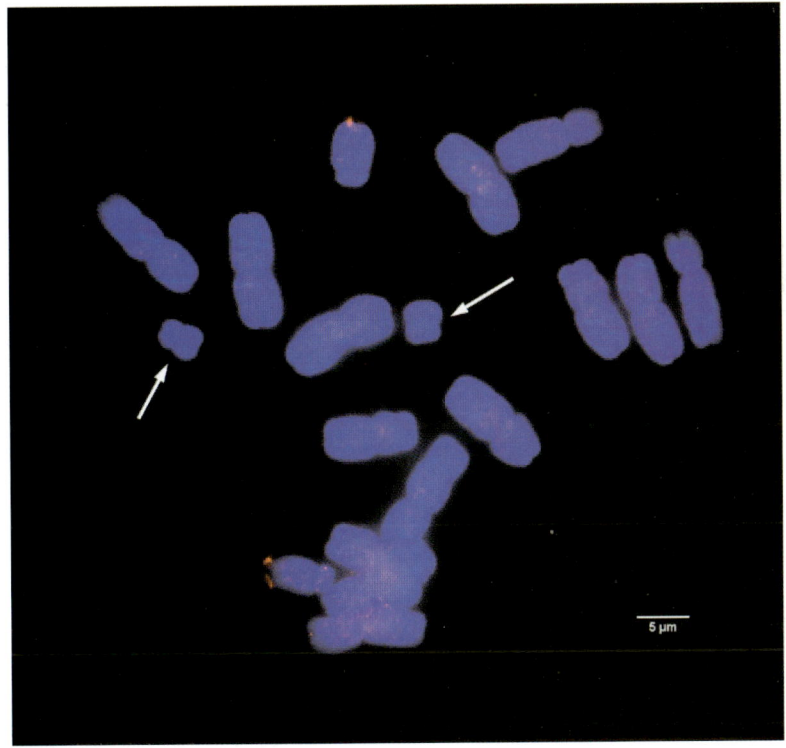

图 3.42 以 E3900 为探针与含小 B 染色体 FISH 结果

箭头示小 B 染色体；红色信号为大 B 染色体末端特异信号，小 B 染色体末端未见信号

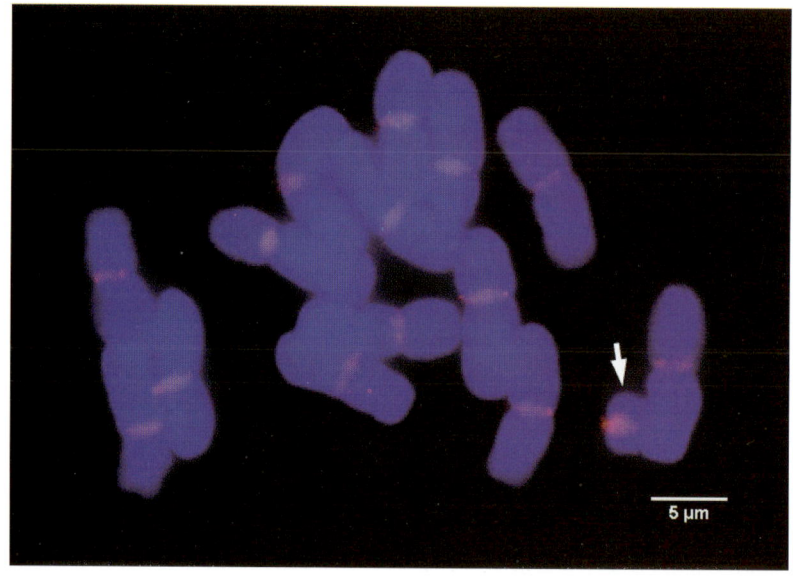

图 3.43　以 RB183 为探针与含小 B 的染色体 FISH 结果
箭头示小 B 染色体

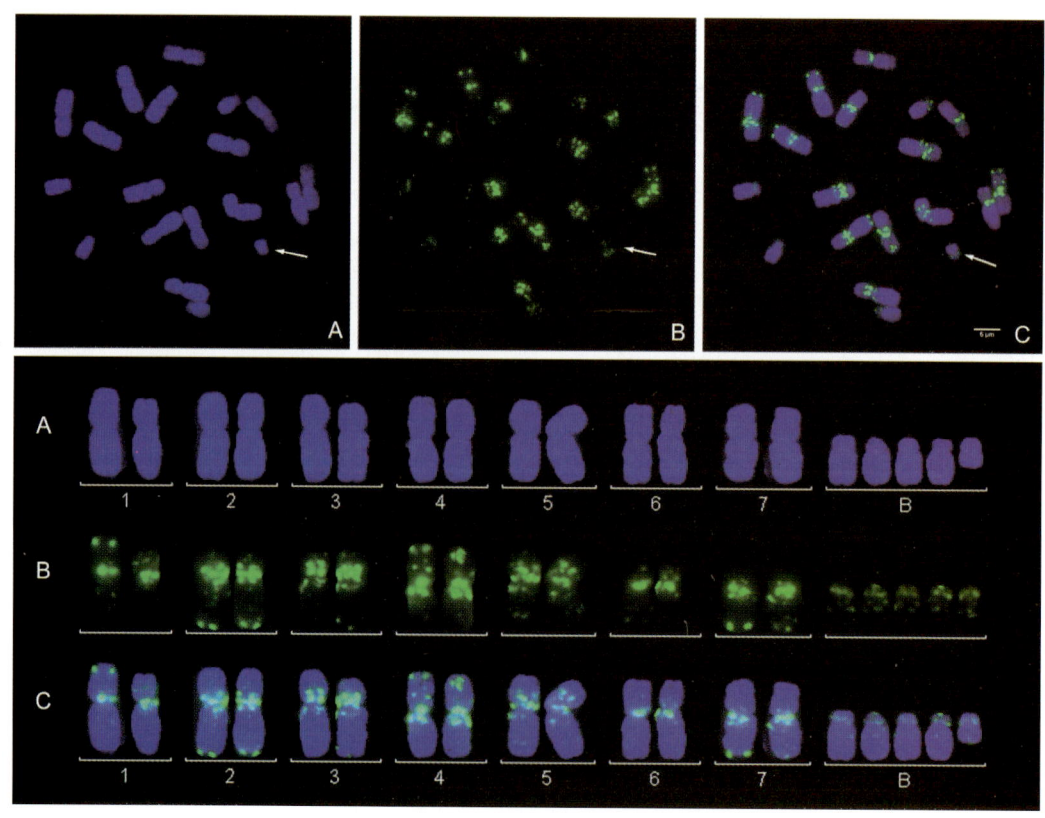

图 3.44　以简单重复序列 AAC 为探针与黑麦 14+ 4B+1Bs FISH 结果
A：DAPI 图像，B：信号图像 C：合成图像

3.1.7.5 黑麦 A、B 染色体简单重复序列组成分析

大规模基因组测序已证实,重复 DNA 序列在物种基因组内占有很高比例,黑麦也不例外,已有报道表明黑麦基因组内重复序列含量高达 92%。我们前期研究结果也证实黑麦基因组内含有丰富的重复 DNA 序列。重复序列主要包括三类:(1)简单重复序列(Simple Sequence Repeats,SSR)或微卫星,如 $(CAG)_n$;(2)复杂重复序列(Complex Repeats),如转座元件(Tarnsposable Elements,TE);(3)数学上定义的重复序列(Mahtematieally Defined Repeats,MDR)。在生物学上研究的主要是前两种重复序列。与以低拷贝或单拷贝形式存在的编码基因或调控序列不同,已有研究表明,这些以中、高拷贝存在的大量重复序列,在物种基因组内并非完全可有可无,而是在维持物种染色体稳定性、促进基因组进化和提供表观遗传调控信息等方面具有重要作用。因此探究物种中每一条染色体重复 DNA 序列的组成,对于从细胞遗传学角度揭示每一条染色体的行为及组成特征,进而揭示整个染色体组的起源及演化具有重要参考价值。黑麦基因组内重复序列含量丰富,为了进一步研究黑麦中每一条染色体中重复 DNA 序列的组成特征,特别是 B 染色体重复 DNA 序列组成特征,以便寻求 A 染色体、B 染色体 DNA 序列组成上的异同。根据以往文献报道,选择了 14 个具有代表性的 SSR 序列进行 FISH 分析。结果表明在这 14 个 SSR 序列中,11 个可在 A 染色体、B 染色体上检测到杂交信号。

具体杂交结果如下:

以(AAG)5 为探针,FISH 杂交结果显示该序列在 A 组染色体上呈弥散型分布,个别染色体杂交信号很强。在 B 染色体上只在长臂末端有较强点状信号分布(图 3.46D2、D3,图 3.48A)。

以(GGC)5 为探针,FISH 杂交结果显示该序列在 A 组染色体和 B 染色体上均呈现弥散型分布,表明该序列为 A 染色体、B 染色体可能具有相似的组成和分布(图 3.45A2、A4,图 3.49B)。

以(ACG)5 为探针,杂交信号主要存在于 A 染色体着丝粒和次缢痕位置,而在 B 染色体主要存在于长、短臂末端(图 3.45D2、D4,图 3.48E)。

以(AAC)5 为探针,杂交信号主要存在于 A 染色体着丝粒长短臂附近、末端,B 染色体主要分布于短臂上,在长臂末端也有少量信号(图 3.45B3、B5,图 3.49F)

以(CAC)5 为探针,杂交信号遍及整个 A 染色体,其中着丝粒部位信号比较强,而在 B 染色体上以点状分布于长、短臂着丝粒附近(图 3.45C3、C5,图 3.49E)。

以(CAG)5 为探针,杂交信号在 A 染色体着丝粒周围比较强,个别 A 染色体末端也有杂交信号,在 B 染色体上信号覆盖整个短臂,在长臂靠近着丝粒附近也有弱的点状信号分布(图 3.45D3、D5,图 3.48D)。

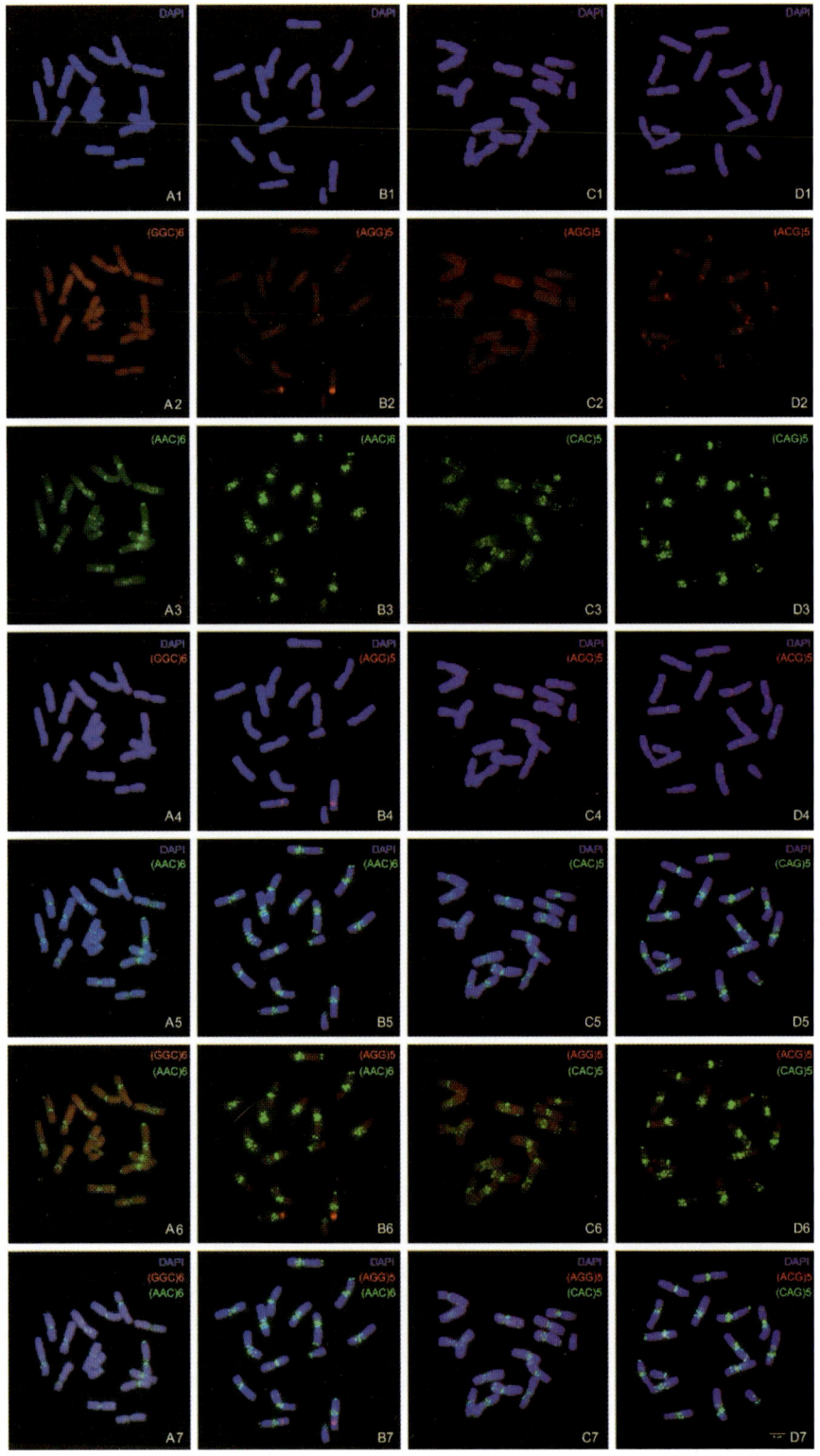

图 3.45　SSR 重复序列 GGC、AGG、AGC、CAC、CAG、ACG FISH 结果

以（AG）8为探针，杂交信号呈弥散状分布于整个A染色体，在B染色体上的分布与As相似（图3.46A2、A3，图3.48F）。

以（AAT）5为探针，杂交信号其中分布于着丝粒附近，且在A、B染色体之间没有显著区别（图3.46C2、C3，图3.48B）。

以（AGG）5为探针，该序列为Bs特有序列，只在B染色体长臂末端检测到信号，经过改进杂交条件后A染色体上只有2R、3R一条短臂上4R一条长臂上有杂交信号（图3.46B2、B4，图3.49C、D）。

以（CA）8为探针，杂交信号呈弥散型分布于着丝粒附近，A染色体、B染色体间信号分布没有显著不同（图3.47A3、A6，图3.48C）。

以（GATA）4为探针，杂交信号主要分布于A染色体长短臂着丝粒附近，在B染色体上信号主要存在于短臂，长臂信号较弱（图3.47B3、B6图3.49A）。

以（GACA）4、（CAT）5、（ACT）5为探针，在A、B染色体上都未检测到杂交信号，表明此种类型的重复序列单元在黑麦基因组含量较低。

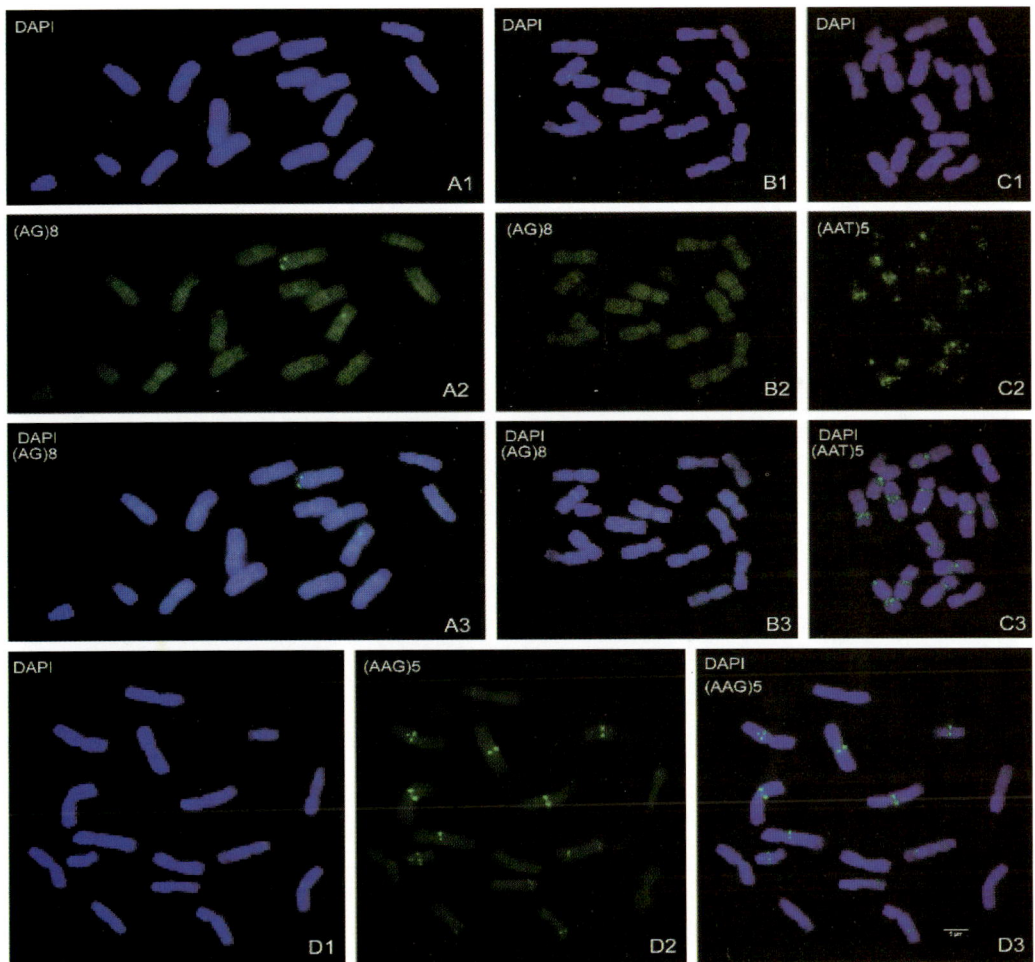

图3.46　SSR重复序列AG、AAt、AAG FISH杂交结果

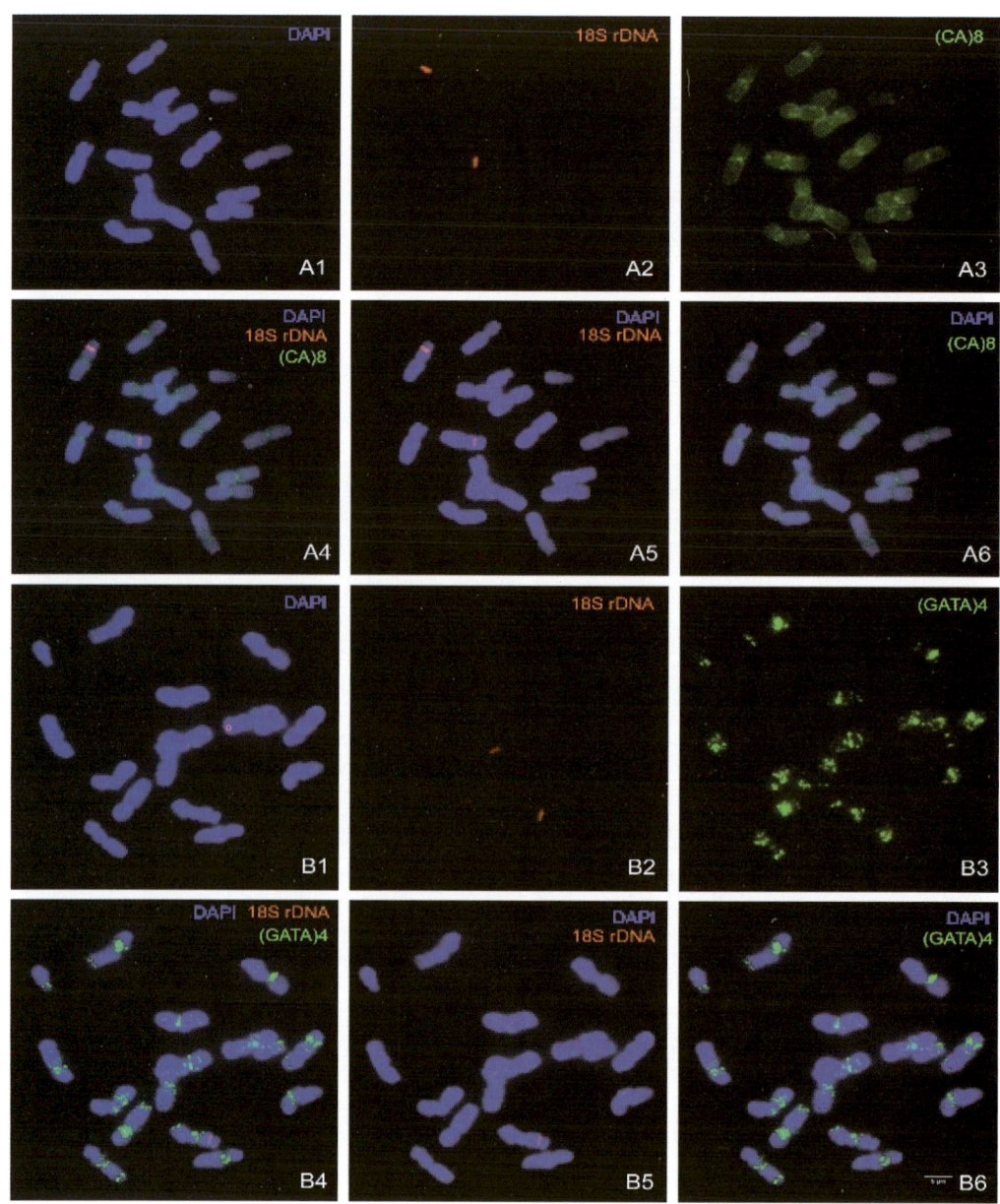

图 3.47　SSR 重复序列 CA+18S rDNA、GATA+18S rDNA FISH 结果

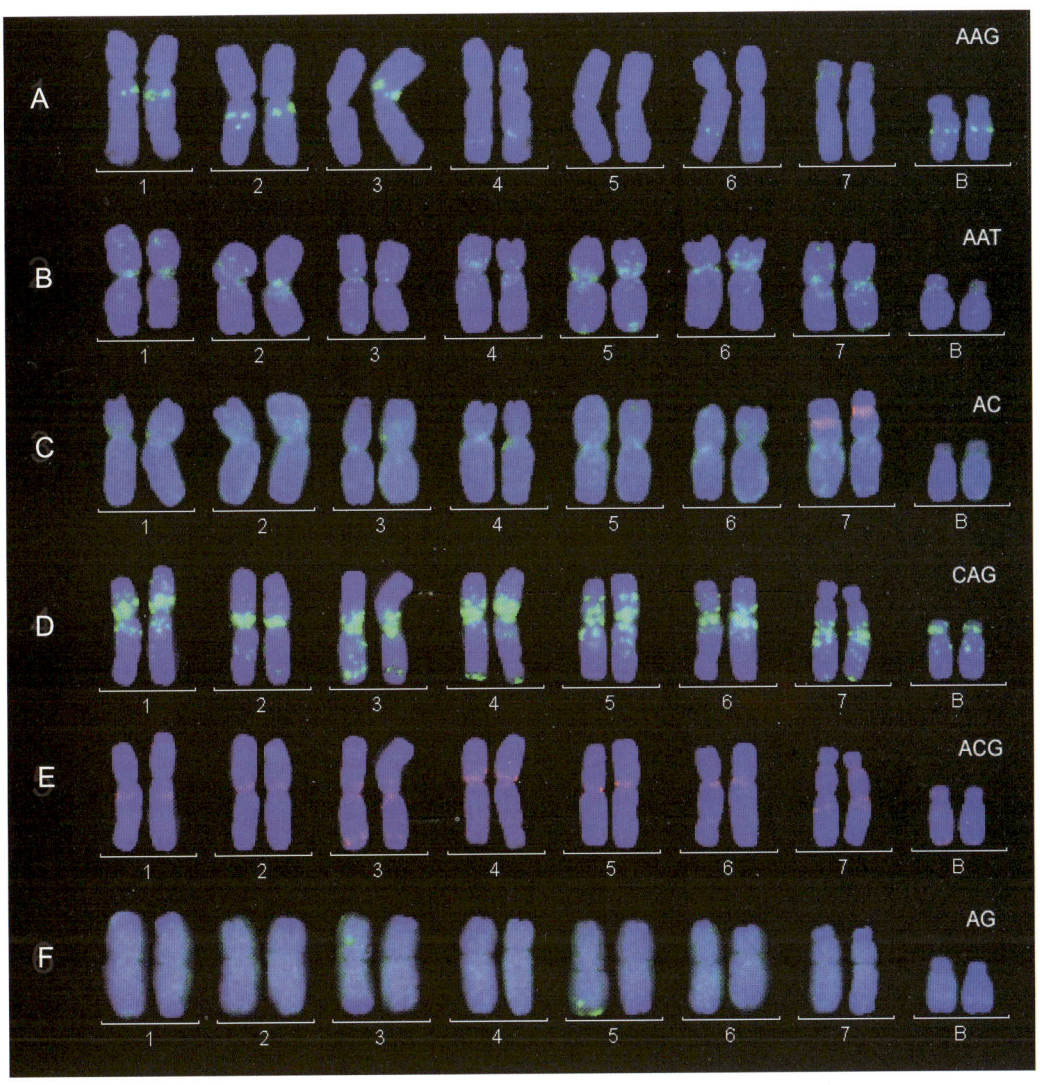

图 3.48　黑麦+Bs SSR 重复序列 AAG、AAT、AC、CAG、AGG、AG FISH 核型分析

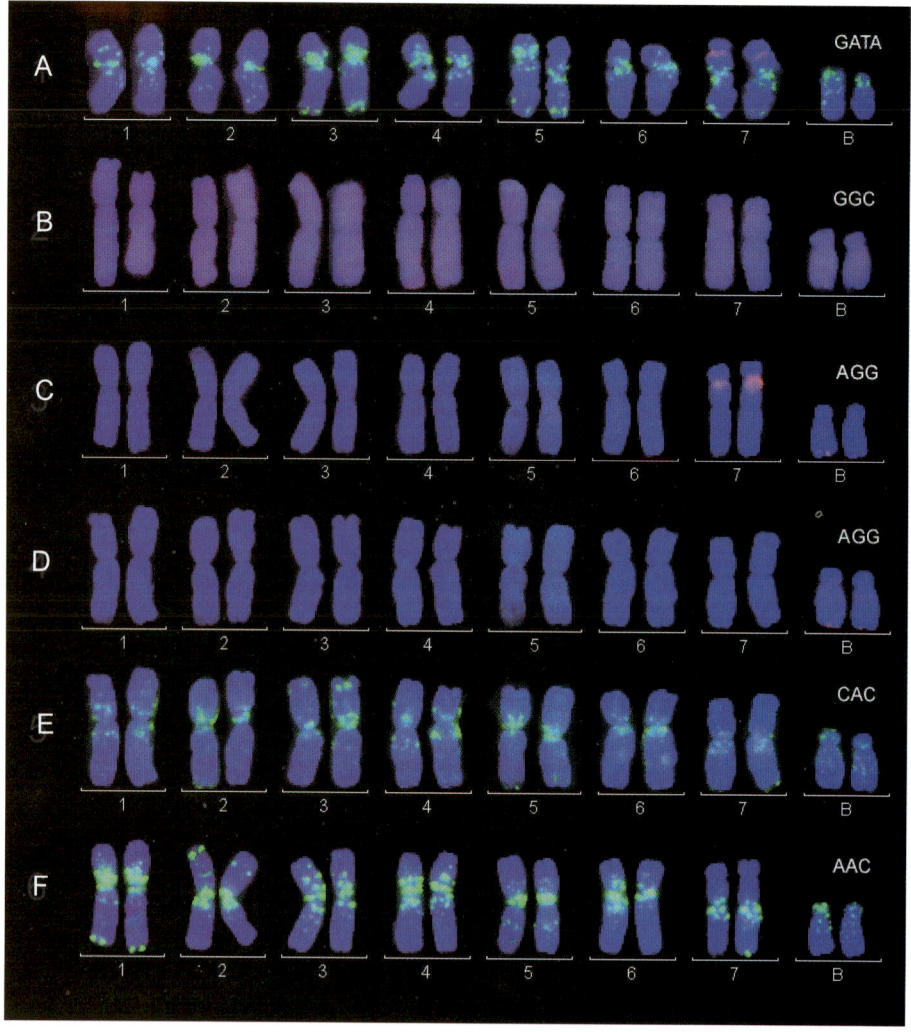

图 3.49 黑麦+Bs SSR 重复序列 GATA、GGC、AGG、CAA、AAC FISH 核型分析

上述结果表明，每个 SSR 序列所产生的带型还是比较稳定的，基本上每对染色体所产生代的数量和位置也基本上一致，个别有些变异可能与片段的缺失有关。另外不同 SSR 重复序列较强的信号大部分都主要集中在黑麦 A 染色体着丝粒附近，B 染色体主要集中在短臂，长臂信号比较弱，只有 AGG 序列在 B 染色体长臂末端有特异信号，长臂信号相对于短臂信号比较弱。这说明 A 染色体和 B 染色体在简单重复序列组成上有一定的差别，总体来讲 B 染色体分布的信号比较少，而且信号相对于 A 染色体来说信号也比较弱，这可能与它本身高度异染色质化有一定的关系。还需要进一步扩大 SSR 序列的筛选以便能够找到更好的解释。

3.1.7.6 B 染色体三大功能组件的研究

关于 B 染色体是如何起源的，众说纷纭。Dhar 等人认为，B 染色体来源于 A 染色体或性染色体，是 A 染色体片段脱落或性染色体变异而来。Cabrero 等则认为，它的起源是多层次的,既有可能是种内起源,也可能是种间起源。我们的实验结果支持 Dhar 的观点,认为 B 染色体起源于 A 染色体产生的片段和变异。通过 FISH,将黑麦 B 染色体的扩增次级产物杂交到黑麦染色体上,在黑麦的每条染色体上都看到了大量的杂交信号,说明了 A、B 染色体的 DNA 序列之间具有高水平的相似性。尤其是着丝粒附近信号几乎和 B 染色体的强度一致,因此推测 B 染色体可能起源于 A 染色体的近着丝粒区域。另外,我们在 B 染色体短臂上发现的 RB165 和 RB306 新序列,在 B 染色体长臂末端发现了除 D1100 和 E3900 家族外的 RB1、2、6、150、21、40 六个序列是新发现的 B 染色体长臂末端特异序列,上述这些序列在 A 染色体上没有一点信号,那么这些特异序列如何起源的?是否与在进化过程中某一个特殊的事件作用于 B 染色体使 B 染色体产生了变异,使某个片段得到了一系列扩增后发生了重组,从而在进化成与 A 染色体不一样的序列,而后 B 染色体沿着其自身独特的进化路线前进,并具有种种特殊的性状和生物学行为形成了 B 染色体特有的功能。那么到底 B 染色体有着那些功能这些一直是个谜,本研究拟进行黑麦+B 的转录组测序,寻找 B 染色体的功能序列,为进一步研究黑麦 B 染色体的功能提供更进一步的数据。

目前有关 B 染色体关键功能组件的筛选已经获得了较大的进展,在植物方面研究的比较清楚的是端粒序列,本实验室从 1998 年就开始对 B 染色体端粒进行研究(郭歌等 1998)以寡核苷酸(CCCTAAA)3 为引物,从微切片段的 PCR 产物中扩增出了黑麦 B 染色体端粒相关序列。经原位杂交,该序列被定位到黑麦所有染色体的端部,同时它还与玉米端粒相关序列高度同源。本研究从 B 染色体文库中筛选获得 RB534 序列,这个序列与拟南芥序列类似,具有拟南芥型端粒的序列特征。FISH 定位结果显示,该 RB534 在黑麦 B 染色体和 A 染色体端部均显示出较强的杂交信号。可以确定该序列为黑麦端粒序列的组成部分。张荣信等(1999)对黑麦 B 染色体着丝粒进行了微切割,并从扩增(LA-PCR)产物中克隆了 B 染色体着丝粒区序列 pRBC6。该序列在玉米、水稻、种毛山羊草(*Aegilops comosa*)拟南芥以及酵母中都找到了与之高度同源的序列。本项研究又从 B 染色体文库中筛选获得 5 个 B 染色体和 A 染色体共有的着丝粒序列 RB72、RB88、RB183、RB113、A1,并将这些序列 FISH 定位在 A、B 染色体着丝粒上。Jiang 等(1996)从两色蜀黍(*Sorghum bicolor*)BAC 克隆的亚克隆中发现一个 745 bp 大小的重复序列 p*Sau*3A9,该序列在禾本科多种植物中保守,并通过原位杂交定位于染色体着丝粒区,在黑麦 B 染色体着丝粒区也有杂交信号。目前有关着丝粒的研究获得了相当丰富的数据。利用抗 CEN P2A 类似物抗体的免疫共沉淀技术,拟南芥染色体着丝粒的 178 bp 卫星重复序列被鉴定和克隆,目前水稻和玉米的着丝粒序列虽然尚未完全测定,但这将有助于着丝粒序列的比较研究。玉米功能性着丝粒序列是由 156 bp 卫星序列组成,水稻第 4 号染色体着丝粒的 1116 Mb 重复群序列表明,该着丝粒区域有基本单元为 155、164 bp 的 CentO 卫星 DNA 片段。有关 B 染色体着丝粒的研究结果发现,玉米 B 染色体包含 114 kb 串联重复序列,该序列与玉米异染色质钮和着丝粒 CEN 4 的部分区域有同源性,利用玉米截短的 B 染色体也取得了重要的进展。Kaszas 等(1998)分析了 25 个具有不同着丝粒长度的玉米 B 染色体遗传行为,发现着丝粒区域大小与减数分裂传递率有直接的关系。如果 B 染色体着丝粒区域小于 1 000 kb,或者缺失了中间 370 kb 片段的染色体,在减数分裂过程中，染色体具有很高的丢失频率。但是所有含有不完整着丝粒的染色体能够稳定地进行多次有丝分裂。然而在玉米 B 染色体中，着丝粒相关卫星序列组成的区域大小与玉米 B 染色体减数分裂传递效率也明显相关，从而暗示这种选择阈值维持了 B 染色体中卫星序列组成的最小范围。Phelps-Durr 等（2004）研究了玉米 B 染色体着丝粒大小和减数分裂遗传的关系，结果表明玉米 B 染色体着丝粒区域约 9000 kb，其中含有 1.4 kb 的重复序列，着丝粒的缩短与减数分裂的遗传稳定性相关，最小可遗传的大小为 110 kb。利用 CENH3 蛋白抗体对一系列的玉米 B 染色体着丝粒错分裂后代的研究发现，与 CENH3 蛋白互作的 700 kb 的着丝粒区域是由 CRM、一部分 B 染色体特异的重复序列 ZmBs 及少量的 CentC 序列共同组成；并且不同 B 染色体的着丝粒所结合的 CENH3 蛋白数量高度可变，而着丝粒功能的发挥又与 CENH3 蛋白数量密切相关。韩方普等（Han et al., 2007）将玉米 BS 的着丝粒易位到反向重复的玉米 9 号染色体短臂上，进行染色体类型的断裂-融合-桥（breakage-fusion-bridge）循环研究，筛选到不同类型的微小染色体，发现植物染色体的着丝粒的失活受表观遗传控制，利用免疫染色结合荧光原位杂交技术，以 CENH3 抗体对自交后代中具有双着丝粒的 5 个稳定遗传的微小染色体进行功能研究发现只有 1 个着丝粒具有功能活性，完整的 B 染色体着丝粒可以易位到第 9 号染色体上。着丝粒研究方面虽然已经取得了一定的进展，但着丝粒的情况比较复杂，目前尚不清楚着丝粒的精确 DNA 序列。此外，具有着丝粒功能的 DNA 序列难以在体外进行操作，基于这两点原因，着丝粒结构已成为当前构建人工染色体研究中的一大难点和热点。随着几种模式植物基因组测序的完成，组成着丝粒的 DNA 序列已逐渐被人们所认识，利用功能性着丝粒 DNA 序列来构建植物人工染色体，虽然很少有成功的报道，但是这必将有助于植物着丝粒功能的研究及其关键性元件的分离和鉴定。

有关复制起始点的研究本实验室祁仲夏（2002）等人利用 RAPD 分子标记技术，对玉米 0B 和+B 基因组 DNA 进行分析，发现 B480 和 B1320 两个标记，并将这两个标记定位在 B 染色体着丝粒部位，他们的序列特征较特殊，碱基 A/T 含量高达 67%，与玉米自主复制起始序列 ARS1 和 ARS2 同源性较高，可能为复制起始点序列，同时也在玉米、烟草和番茄等植物中发现这些 DNA 部分序列可以在酵母中自主复制，虽然不清楚这些序列在各自的基因组中所起的作用，但他们在基因组中的频率却与基因组中应具有的复制起始序列拷贝相当，Berlani 等人研究显示，这些序列在基因组中的分布是相当分散的，而不是串联重复排列，对发现的这些序列详细研究表明他们分属两个重复序列家族 ARS1 和 ARS2，他们都含有高的 AT，约占 65%~70%，嘌呤和嘧啶集中分布，这两个序列都含有酵母 ARS 中 ACS 相似序列。人类人工染色体构建成功后，Harringten（1997）等认为不需要了解这些元件的组成成分就可以构建人工染色体，他们把人的端粒 DNA、人的基因组 DNA 和人工合成的 α-卫星 DNA 混合起来转入人的肿瘤细胞中，产生人工合成的人类染色体，他们的依据是裸露的 DNA 在体内可以非同源连接，端粒 DNA 能产生端粒，从目前的报道看，与 DNA 复制有关的特殊序列是客观存在的，这些序列在不同位点的数目和分布会有很大的变化，任何 DNA 序列都可以完成 DNA 起始复制功能，由此可见 DNA 复制起始点并不集中在某些位点，这些研究结果为人工染色体的组装提供了理论依据。

构建植物人工 B 染色体最关键的核心技术一是如何获得端粒（telomere，TEL），着丝粒（centromere，CEN）和自主复制序列，二是如何组装成人工 B 染色体载体。目前人工染色体的组装有两种方法可以借鉴，一是组装法（即 bottom-up 法），用组成染色体的基本元件来组装一个新的染色体；二是截短法（即 top-down 法），从一个正常的人类染色体开始，逐步将其截短，最后得到一个很短的染色体，即微小染色体（minichromosome）。人类人工染色体的构建就是将上述三种功能组件按恰当的顺序排列在一起，当新生的染色体片段具有这三个功能组件且三者位置合适时，他们便可以在细胞分裂中稳定存在，形成一条独立的染色体，利用此种方式形成的染色体与原染色体相比，着丝粒 DNA 同源性比较低，所以不会与原染色体配对，其行为会与正常的 A 染色体有很大的区别。从细菌质粒的比较中我们发现 B 染色体具有很多与质粒类似的特征，独立于细菌基因组之外，在细菌中，质粒是可有可无的，且不影响细菌的正常生理活动；它可以携带额外基因，使细菌获得一些有益的性状，同样 Bs 也具有发展成为高等生物大尺度基因载体的潜能。特别是高等生物的多数性状多由多基因控制，因此，当通过载体导入一组基因时，B 染色体的优越性就凸现出来，它可以携带上百兆的 DNA 序列。以 B 染色体构建人工载体的另一个优点在于 B 染色体在减数分裂时不与 A 染色体配对，保证了构建载体的正常传递。

关于植物人工染色体的构建已经取得重要进展。Yu 等（Yu *et al.*，2006，2007）利用截短法通过插入端粒重复（telomeric repeats）序列将 A 染色体和 B 染色体截短（truncation）构建了玉米人工染色体载体，这种方法借鉴人类人工染色体的构建方法，1 个 2.6 kb 的拟南芥端粒重复序列用于根癌农杆菌介导或基因枪粒子轰击转化玉米幼嫩胚胎，由于截短的位置是随机的，需要用 FISH 来鉴定截短的染色体，但截短的频率较低。由于 Bs 的遗传惰性，截短后的 B 染色体比 A 染色体存活率高。几乎与此同时，Carlson 等（2007）借鉴人类人工染色体的构建方法，首次利用组装法在体内（in vivo）从头合成（de novo）了玉米微小染色体（minichromosome），该载体的选择标记基因含有 7～190 kb 的玉米基因组 DNA 片段，包括着丝粒卫星序列和反转座子序列以及其他重复序列，通过基因枪粒子轰击转化玉米胚胎组织中，镜检结果显示有独立于常染色体之外的荧光信号，有丝分裂和减数分裂可以稳定遗传。然而，这样构建的人工染色体载体中新着丝粒由于表观遗传而造成的可能失活以及减数分裂稳定性等问题还没有解决，形成有功能活性着丝粒所必需的 CENH3 蛋白也没有 FISH 鉴定，微小染色体的遗传机制尚不明确（Houben *et al.*，2008）。

因此，运用分子生物学技术对 B 染色体的结构和作用机制进行深入研究，有着诱人的前景。如进一步开发 B 染色体在染色体工程中的巨大潜能，利用 B 染色体对作物进行育种改良，使其更适应环境的生长等。

（作者：宋文芹、陈成彬、王春国、赵换、张力鹏。原载：2008 年国家自然科学基金总结）

第二节 植物染色体 G-带研究

提要：所有高等动物染色体上都有显示 G-带的报道，但植物染色体没有 G-带，这是难以想象的。1982 年 Drewry 以红松（*Pinus resinosa*）为材料发表了 G-带论文。虽然该论文的 G-带不明显，但启发很大，使我们对植物染色体 G-带产生极大兴趣。因此除了建立植物染色体 G-带方法，我们还在红松、重楼、百合、蚕豆、一粒小麦、小麦、玉米、洋葱、银杏等多种植物上获得了 G-带。所获得的植物染色体 G-带受到国内外高度重视，但至今该技术没有取得突破，没有得到好的 G-带带型。可惜的是，我们自己也没有进一步研究，此外正遇上分子生物学时代，人们对细胞生物学染色体研究处于低潮。

分析这几种植物染色体 G-带之所以会成功，可能有几种原因，①选择的植物都是比较大的染色体，而且是处在早中期的染色体，②采用 WDH 法制备染色体标本，用蒸气法代替火焰干燥法，③植物染色体 G-带与人类和动物染色体 G-带的表现可能不一样，还有待进一步深入研究。

3.2.1 利用胰蛋白酶和尿素诱导植物染色体 G-带

Since the beginning of 1970s, the G-banding technique for studying human and other animal chromosome has been well established. But despite much effort in many laboratories, attempts to reveal G-bands in plants had not been successful for a long time. More than a decade ago, we used, in preparing plant chromosomes, a method similar to that for preparing animal chromosome(Chen, 1979). Only in recent years, some achivements in G-banding were reported in several species of plants（Chen, 1986, 1987, Drewry, 1982, 1988, Zhan, 1987, Zhang, 1987, Zhu, 1986）. However, only a few of these reports showed bands that appeared to be comparable in quality to those of animals; many of them could only be regarded as G-band equivalents instead of real ones. As a result some questions arised: Does the G-banding cast a general significance that applies to plants as well as to animals? And if so, how can we manage to reveal G-bands in highly condensed plant chromosomes? Thus, a new method of general applicability, was developed, later called TUG method. This procedure used a double treatment with trypsin and urea（Chen, 1988）. Following this method, G-bands have been clearly revealed in 16 species of plants. The G-bands of some plants with large chromosomes, for example, *Lilium davidii*, are even clearer than those of some animals reported lately. In the present paper, details of the TUG method and its application to karyotype analysis of *Lilium davidii*, *Vicia faba*, *Hordeum vulgate*, *Ginkgo biloba*, and *Triticum monococcum* are reported, and the mechanism of G-banding and some factors involved in G-banding treatment are discussed.

3.2.1.1 Materials and Methods Meterials

Five species of plants (Table 3.4)

Chromosomes were prepared with the methods of wall-degradation, hypotonic treatment and steam-drying (Chen RY, 1979, 1982, 1986)

1. Chromosome preparation

1) Roots from germinating seeds growing at 25℃, bulbs were cultured in moist wood scraps till roots were 1-3 cm long. 2) Roots tips were excised and treated with 0.02% colchicine solution for 1-2 h. 3) Excising the good root tips and immersing them in 0.075 M KCl solution for 30 min at 20-25℃ for pre-hypotonic treatment. 4) They were fixed in methanol: glacial acetic acid 3∶1 fixation solution for 30 min and then washed with distilled water. 5) Treated with 2.5% enzyme mixture 2.5 g cellulase ("Onozuka" R-10) and 2.5 g pectinase(form aspergilus niger, from Serva, NY)in 100 ml distilled water] at 25℃ for about 40-60 min. Washed with distilled water or 0.075 M KCl twice and incubated at 25℃ for 15 min. 6) Freshly prepared methanol: acetic acid (3∶1) was added to the material. 7) Put 1-2 root tips on the slide which was kept in frozen distilled water. Added a (few) drop (s) of fixation solution to them.Tore the material with a pair of forceps and removed the supernatant. 8) Two drops of fixation solution were added onto the slide, which were dried with steam from the undersurface of the slide and then placed in a 60-80℃ oven for 2 h. The slide was stored at room temperature till use for banding 1-2 d later. G-banding procedures。

Table 3.4 Source and number of chromosomes in 5 species of higher plants used for the TUG method

Taxon	Source	Chromosome number (2n)
Lilium davidii Duchartre	From Prof. Zheng Guochang, Lanzhou University	24
Vicia faba L.	Market in Tianjin	12
Hordeum vulgare cv.Tianjin L.	Nankai University, Tianjin	14
Ginkgo biloba L.	Hebei Province Institute of Pomology, Changli, Hebei	24
Triticum monococcum L.	Chinese Acad Agri Sci, Beijing	14

2. Bands methods

(1) TUG method

Treated the chromosome samples with 0.01% trypsin (Difcö) in 0.02% EDTA-CMF solution at 4-10℃ for a few sec. Washed with 0.85% NaCl. Then treated with 5.3 Murea (1 part of 8 Murea mixed with 2 parts of phosphate buffer, pH 7.4) at 4-10℃ for a few sec. Washed with 0.85% NaCl and then stained with Giemsa solution (Giemsa: phosphate buffer=1∶30, pH 7.4) for 5 min.

(2) Trypsin method

Treated the chromosome samples with 0.01% trypsin as mentioned above for a few seconds. Washed with 0.85% NaCl. Then stained with Giemsa solution.

(3) Urea method

Treated the chromosome samples with 5.3 Murea (1 part of 8 Murea mixed with 2 parts of phosphate buffer, pH 7.4) at 20-25℃ for a few seconds. Washed with 0.85% NaCl. Then stained with Giemsa solution mentioned above.

3.2.1.2 Results

1. G-bands observation on chromosomes of several plant species

By using the TG, UG and TUG methods mentiond above, we revealed G-bands on chromosomes of the following species: *Hordeum vulgare*, *Vicia faba*, *Ginkgo biloba*, *Triticam monococcum* and *Lilium davidii*. Among the 3 methods, TUG method, i.e.treating chromosomes successively with both trypsin and urea, produced the clearest bands and results were regularly reproducible. In Hordeum vulgare, G-bands were distributed along the whole length of its chromosomes (Fig. 3.50). In contrast, the C-band appeared only nearer the centromere. Fig. 3.51 showed the G-band of *Vicia faba*. There were (more than) 20 odd bands on its M chromosome. In *Triticium monococcum* and *Ginkgo biloba*, each chromosome also showed clear G-bands (Fig. 3.52 and Fig. 3.53).

Fig. 3.50　G-bands in chromosomes of *Hordeum vulgare* cv.Tianjin L.

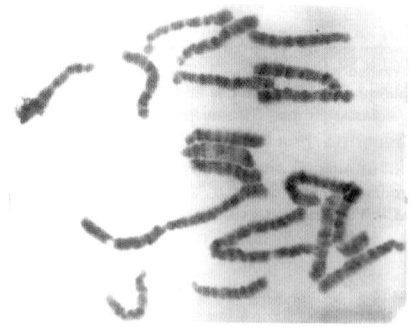

Fig. 3.51　G-bands in chromosomes of *Vicia faba* L.

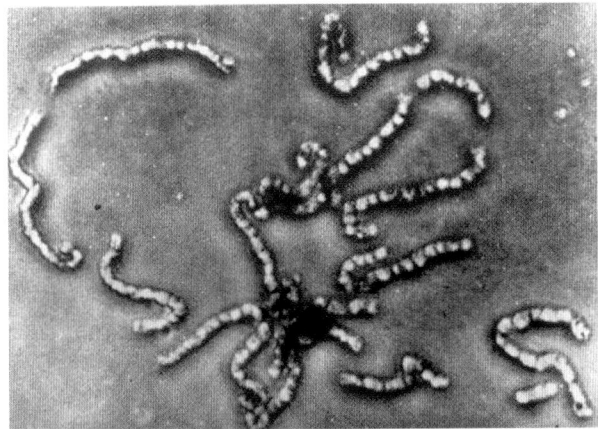

Fig. 3.52　G-bands in chromosomes of *Triticum monococcum* L.（photographed from UNIVAR transmitted-light interference contrast microscope）

2. Comparison of G-bands on homologous chromosome

To determine the correspondence between G-bands of homologous chromosomes, cells at 3 stages of metaphase (pro-metaphase, early-metaphase and mid-metaphase) from *Lilium davidii* were adopted for examination. Fig. 3.54A, B and C demonstrated that all 3 substages revealed clear G-bands. These were the clearest G-bands so far obtained in plants, and were very similar to those of animal chromosomes. Bands of pro-metaphase chromosomes are easily distinguishable. Furthermore, in each cell, bands of the 2 homologous chromosomes were identical in number, relative width and location which will be very useful in pairing homologous chromosomes. Correlation between band number and cell division stages It was found that the number of G-bands was closely related to the division stages in plant chromosomes. Fig. 3.55 showed the chromosome III of *Lilium davidii*. There were 32 bands at late prophase, 21 at pro-metaphase, 17 at early metaphase and 12 at mid-metaphase. These

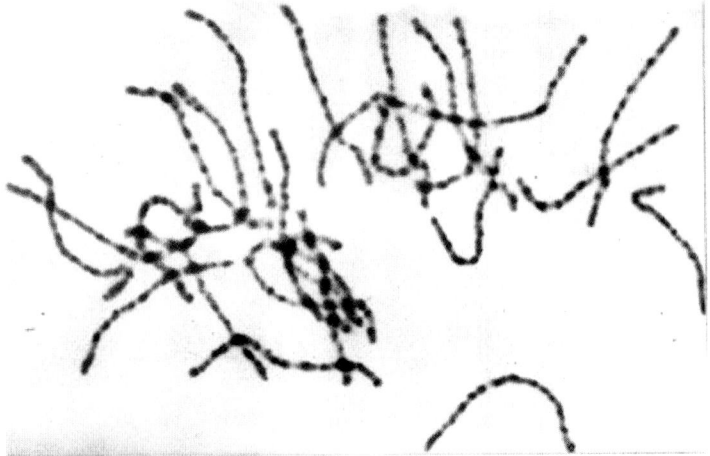

Fig. 3.53　G-bands in chromosomes of *Ginkgo biloba* L.

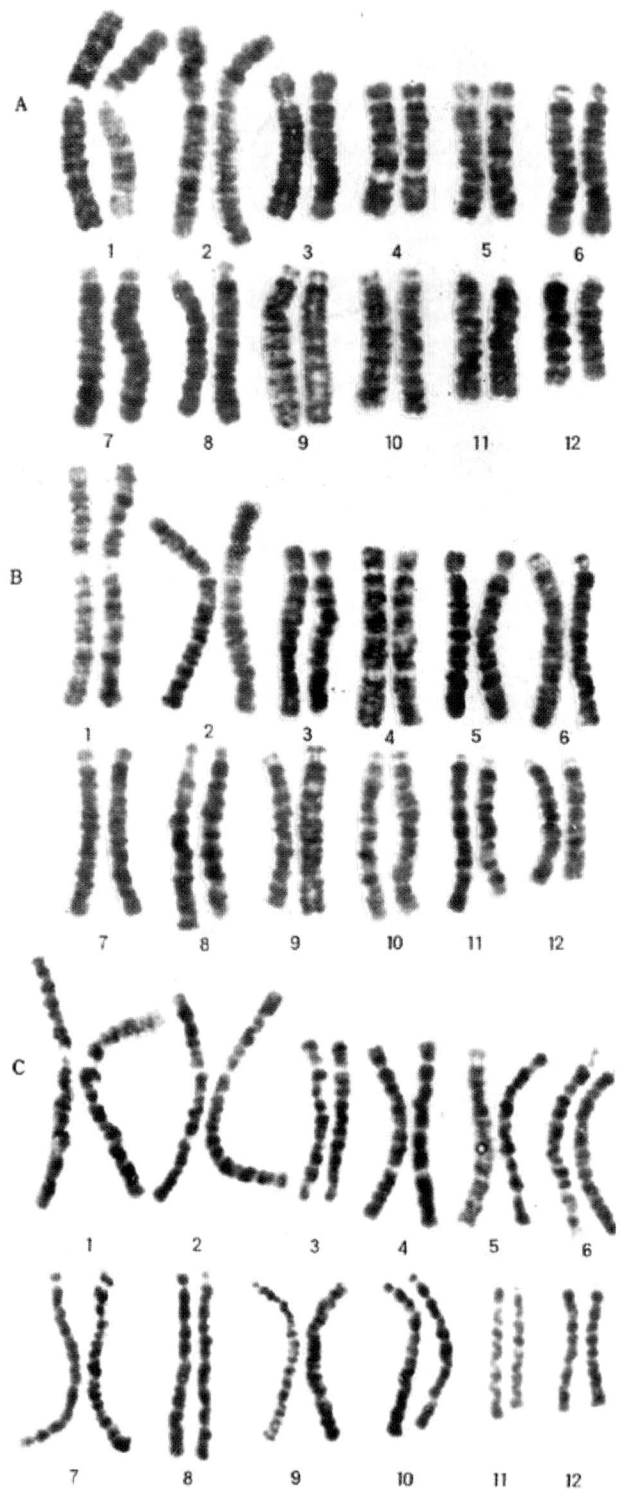

Fig. 3.54 G-banded karyotype of *Lilium davidii*. In 3 stages（A.mid-metaphase： B.early-etaphase：C.Pro-metaphase）.The homologous chromosomes show similar G-banded characteristics

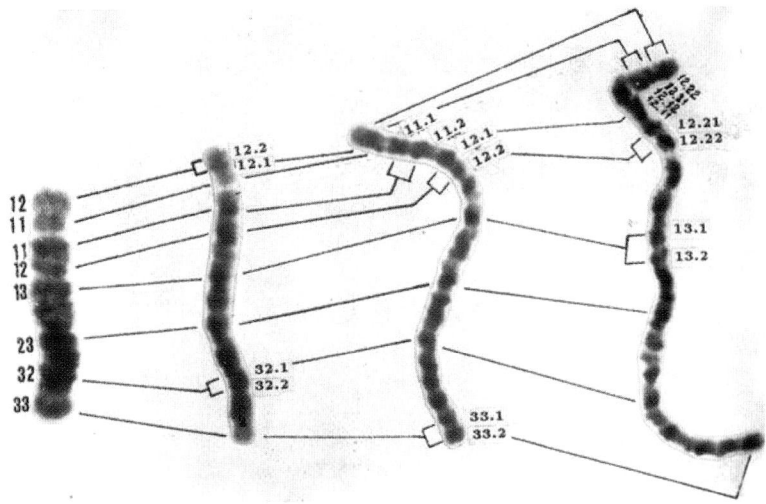

Fig. 3.55 Gbands showing in chromosome III from different stages of *lilium davidii*. The band number increases with mid-metaphase to late-prophase

results were very similar to those found in the high resolution G-banding study of *Muntiak muntiak*（Rohmd, 1982）and humans（Yunis, 1978）. Thus it can be said that there were many similarities between G-bands of plant chromosomes and those of animal chromosomes. They both exhibited dark and light bands and their number varied with the cell division stages. From prophase to metaphase，bands darkened and their number decreased as a result of because of the union of sub-bands，indicating that a band in metaphase may include several sub-bands in prophase. Hence the most favorable stage for G-banding lies somewhere between prophase and early metaphase，when the chromosome structure is most relaxed, rather than in mid metaphase.

3.2.1.3 Discussion

1. Relationship between G-banding and the method used in preparing plant chromosome samples

Recently，some assumptions have been proposed to explain why plant chromosomes do not show G-bands. Greilhuber（1977）pointed out that if the crux of the problem does not lie in the method of preparing the plant chromosome samples，then it is plausible to assume that G-bands do not exist at all. From our work，it becomes evident that the main trouble is that suitable means for preparing the chromosome samples to reveal the G-bands had not yet been devised. It has been proposed that the procedures in preparing chromosome samples for C-banding should not be copied for G-banding procedures（Chen，1986），because degradation with HCl is unfavorable to the induction of G-band（Cominges，1973）and the flame-drying makes chromosomes insensitive to the band-revealing treatments. So the traditional HCl hydrolyzing squashing method was abandoned and cell wall-degradation hypotonic treatment method was adopted instead（Chen，1979）. A mixture of cellulase and pectinase was used to degradate the cell wall. Then chromosome samples were made by means similar to that used in human samples. Slides were dried with steam (60-80℃), which can assure heating the slides evenly. The enzyme degradation and hypotonic treatment are the crucial steps in the procedures. The concentration of enzyme mixture higher than 5%, especially a high concentration of cellulase, will interfere with the revealing of G-bands. On the other hand, hypotonic treatment with KCl is absolutely necessary as using pure water would being about untoward influences.

2. Peculiarities of plant chromosome and the G-banding methods

Drewry (Drewry, 1982) first revealed G-bands and C-bands in *Pinus resinosa* using trypsin. Chen *et al.* (1987) reported G-bands in several species of plants revealed by the samemethod. It has been proved that the trypsin method can widely be applied in animal and plant chromosome G-banding. However, we found that the time of trypsin digestion in plant and human chromosomes should be notably different, i. e. much shorter for palnts material. This illustrates that plant and human chromosomes are distinct in structure, and this distinction probably involves not only DNA, but also the type and content of chromosomal protein. In generall, DNA of plant chromosome are tightly coiled and highly condensed, so the proteins covering chromosome may be less than that of animal chromosomes, requiring a much shorter digestion time. In contrast, the degree of DNA condensation of human chromosomes is relatively low, and there may be more associated proteins on their surface; hence a longer digestion time is needed. According to Mckay (1973), the bands were resulted from differences in condensation of the chromatin and were intrinsic properties that could be revealed by banding treatments, rather than artifacts produced during the course of manipulation. As a result, we believed that there is a close relationship between the proteins covering the surface of chromosomes and G-band revealing. The TUG methods recommended eliminate proteins mildly so that much clearer G-bands may therely be obtained. It is accepted that more in depth study should be programmed to further improve this method.

(Authors: Chen Ruiyang, Song Wenqin, Li Xiulan,
An Zhuping. Published in: Cell Research 1994, 4, 79-87)

3.2.2 百合（*Lilium davidii* L.）G-带核型分析

The G-banded karyotype of *Lilium davidii* revealed the simultaneous demonstration of G- and C-bands of chromosomes in *Pinus resinosa*, there have been many reports about G-banding technique in plants (Chen *et al.*, 1986, 1987; Zhang &Yang 1986; Zhu *et al.*, 1986; Zhan *et al.*, 1987; Song *et al.*, 1987). But few papers about G-banded karyotype in plants have been published so far. The genus *Lilium* ($2n=24$) is noted for its large chromosomes, but it is impossible to identify individual chromosomes on the basis of morphological characteristics such as size and arm ratio sinece most of chromosomes are subterminal and of similar size. In addition, the similarity of chromosomal morphology among *Lilium* species is remarkable. As a result, resolution of conventional karyotype comparison between Lilium species was limited (Li *et al.*, 1984). Presented in the present paper is G-banded karyotype analysis on *Lilium davidii*, which involved double-treatment with trypsin and urea for identification of individual chromosomes and comparison between the G-bandes. The C-banded karyotype of *Lilium davidii* is also given for comparison.

3.2.2.1 Materials and Methods

The bulbs of *Lilium davidii* were kindly provided by Prof. Zheng Guochang, Departmentof Biology, Lanzhou university.

1. Chromosome preparation

(1) Bulbs of *Lilium davidii* were cultured in moist wood scraps till roots were 3-5 cm long.

(2) Root tips were excised and treated with 0.02% colchicine solution for 1-2 hours.

(3) They were immersed in 0.075 M KCl at 20-25℃ for 30 minutes for pre-hypotonic treatment.

(4) They were fixed in 3∶1 (methanol∶glacial acid) fixation solution for more than 30 minutes and then washed with distilled water.

(5) Treated with 2.5% enzyme mixture [2.5 g cellulose ("Onozuka"R-10, from kinki Yakult MFG.Co., Ltd., Japan) and 2.5 g pectinase (from Asoergillus niger, from Serva, New York) in 100ml distilled water] at 25℃ for about 60 minutes. Washed with 0.075 M KCl twice and incubated at 25℃ for 30 minutes.

(6) One or two root tips were placed on each slide that has been kept in ice water, adding a drop of fixation solution. They were torn apart with a pair of forceps.

(7) Two drops of fixation solution were added on the slide, which were dried with steam from the lower side of the slide and then placed in 60-80℃ oven for several hours. The slide was stored at room temperature till it was used to reveal G-bands 1-2 days later.

2. G-banding procedure

The chromosome samples were put in to 0.85% NaCl solution, and then treated with 0.01% trypsin (Difco, in 0.02% EDTA-CMF solution) at 4-10% for 20-40 seconds. They were washed with 0.85% NaCl, and treated with 5.3 Murea (2 parts of 8 Murea in 0.02% EDTA and one part of 1/15 M phosphate buffer adjusted to pH 7.4) at 4-10℃ for 40-60 seconds. Again they were washed with 0.85% NaCl, and stained with 1∶40 Giemsa∶phosphate buffer (pH 7.4) for 10 minutes. Finally they were washed with flowing water and dried on air.

3. C-banding procedure

The chromosome samples were treated with the BSG method (Vosa, 1974).

Free hand measurement of the location, number and width of bands followed Yunis (1978).

To construct the ideogram, photographs of cell with four division stages showing clear bands were used: late prophase (LP), prometaphase (PM). Early metaphase (EM) and mid-metaphase (MM). Images of chromosomes were cut out along their outline and then four homologous chromosomes of each stage were chosen from the same or different cells to show the banding characteristics of homologous chromosomes. Arms of the chromosomes of *Lilium davidii* each were divided into several zones basically in accordance with the standards used for studying human chromosomes.

3.2.2.2 Results

1. C-banded karyotype of *Lilium davidii*

For comparison with the G-bands, the C-banded karyotype of *Lilium davidii* is presented in Fig. 3.56-3.67. It can be seen that most chromosomes reveal C-bands only at the centromere though the chromosomes 4, 5 and 7 each have an intercalary band in the lang arm. The total number of bands are very small. This result is very similar to that reported about *Lilium lancifolium*.

2. The G-banding characteristics of 12 paris of homologous of *Lilium davidii*

All the 12 pairs of homologous chromosomes show clearly abundant G-bands. And the band number decreases from late prophase to mid-metaphase. For instance, the chromosome 2 shows 42 bands at late prophase, 26 at prometaphase, 20 at early metaphase and 15 at mid-metaphase. The phenomenon is similar to that obtained in high resolution study of human chromosomes.

Figures 3.56-3.67 are the G-banded karyotype of Lilium davidii. The chromosomes are arranged in decreasing order of the length of their short arms. The chromosomes 1 and 2, the langest chromosomes of all, are the only two pairs which are metacentric. In the chromosome 1 (Fig. 3.56), there is a non-G-stained zone adjacent to each centromere at the same position as the dark C-band demonstrated in the C-banded preparation; in the chromosome 2 (Fig. 3.57), only a light-stained C-band is found at each centromere where a dark G-band shows clearly in the G-banded preparation. The chromosome 3 (Fig. 3.58) is subterminal. Nine G-bands are found in each of their long arms at mid-metaphase, which one dark C-band is displayed at each centromere. The chromosome 4 (Fig. 3.59) hans G-bands distributed along each long arm at metaohase. One C-band is markedly situated in the long arm containing a secondary constriction and a non-stained zone is exactly located at the same position in the G-banded preparation. The chromosome 5 (Fig. 3.60) has eight G-bands in each long arm at mid-metaphase and one light-stained C-band at the centromere. It is interesting that a deep-stained C-band could be observed in the long arm of one chromosome of the pair but not in the other. The chromosome 6 (Fig. 3.61) has eight G-bands in each long and a non-stained zone in each short arm in G –banded karyotype, while a dark C-band is found at the same position. The chromosome 7 (Fig. 3.62) has ten G-bands in each long arm. The similar phenomenon as in the chromosome 5 also occurs in this homologous pair. The difference between them is that the dark C-bad in the chromosome 7 is nearer to centromere than in the chromosome 5. The chromosomes 8 (Fig. 3.63) and 9 (Fig. 3.64) have similar G-banding patterns with about ten G-bands in each long arm. The chromosomes 10 (Fig. 3.65), 11 (Fig. 3.66) and 12 (Fig. 3.67) also have similar G-banding patterns. Ten G-bands are found in their long arm at mid-metaphase.

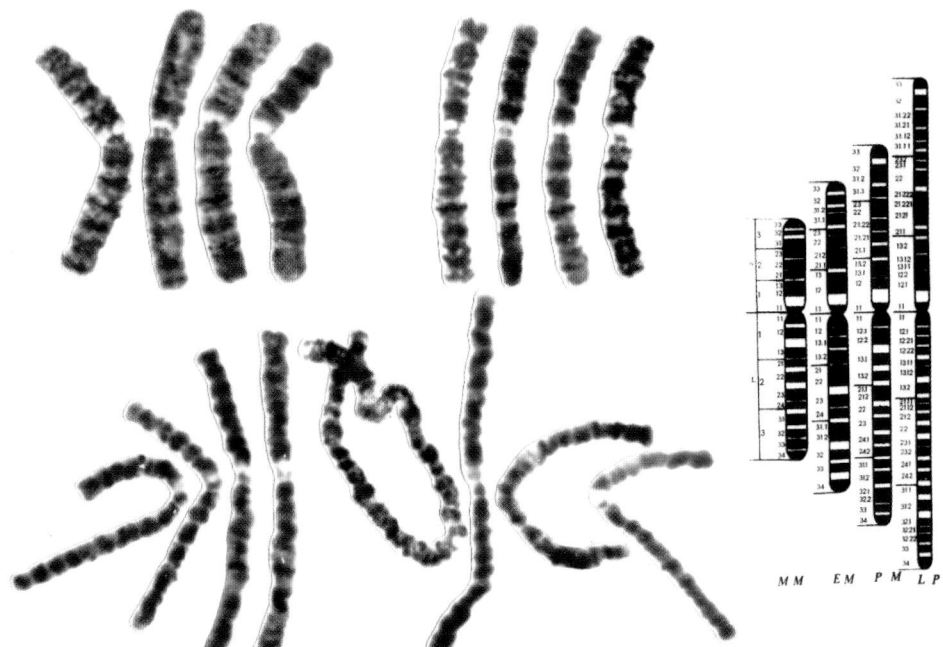

Fig. 3.56　G-banding pattern of chromosomes 1 of *Lilium davidii* L., showing 41 bands at late prophase (LP), 29 at pro-metaphase (PM), 22 at early metaphase (EM) and 19 at mid-metaphase (MM). A notable non-G-stained zone is adjacent to the centromere

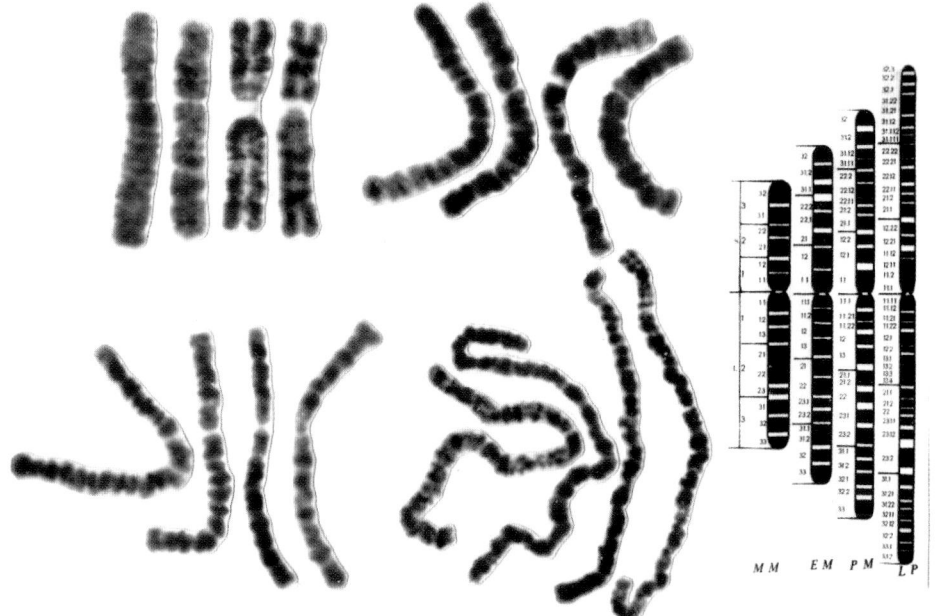

Fig. 3.57　G-banding pattern of chromosomes 2 of *L. davidii*, showing 42 bands at LP, 26 at PM, 20 at EM and 15 at MM

Of all the chromosomes from 7 to 12, there is only one C-band situated at the centromere. The chromosomes could be identified by their G-banding pattern and their morphological characteristics with high resolution.

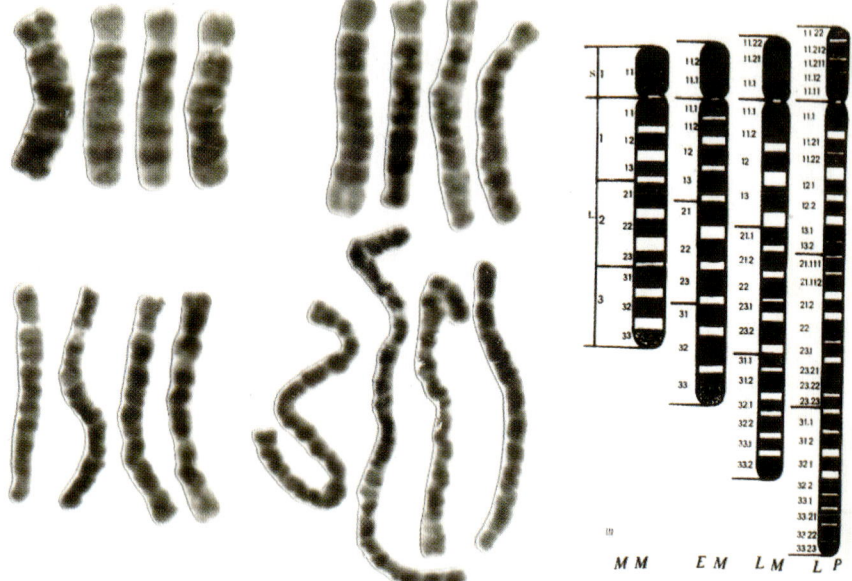

Fig. 3.58 G-banding pattern of chromosomes 3 of *L. davidii*, showing 28 bands at LP, 18 at PM, 12 at EM and 9 at MM

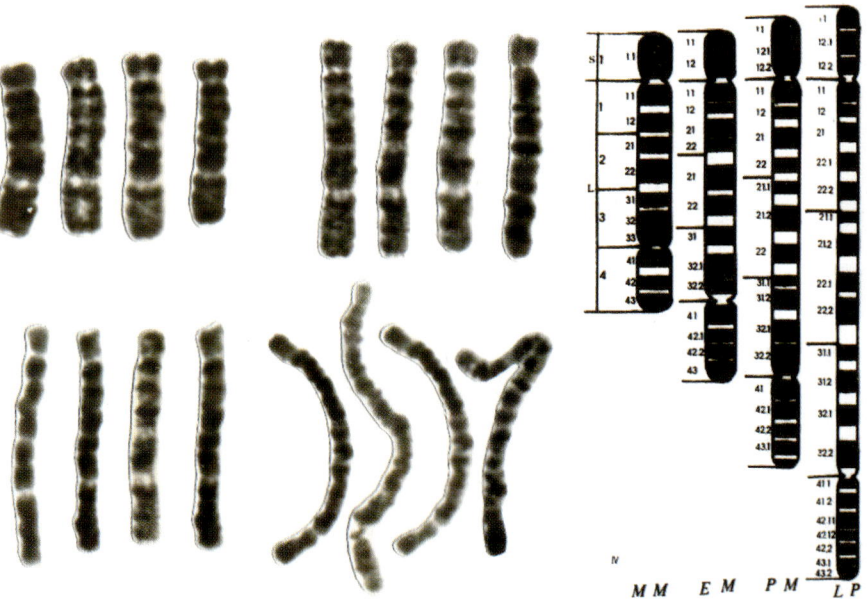

Fig. 3.59 G-banding pattern of chromosomes 4 of *L. davidii*, showing 23 bands at LP, 19 at PM, 15 at EM and 9 at MM. A non-G-stained zone is located at the secondary constriction in the long arm, where a dark C-band occurs

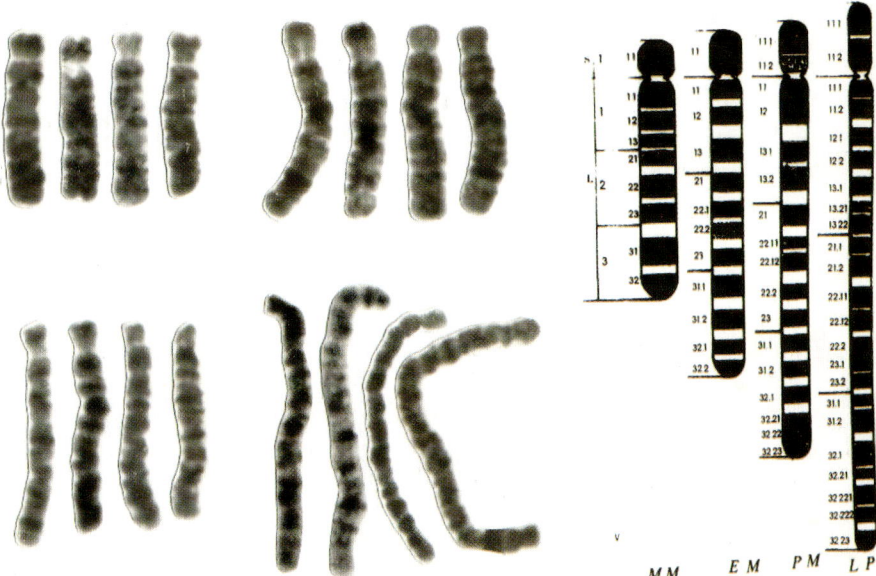

Fig. 3.60　G-banding pattern of chromosomes 5 of *L. davidii*, showing 23 bands at LP, 17 at PM, 12 at EM and 8 at MM

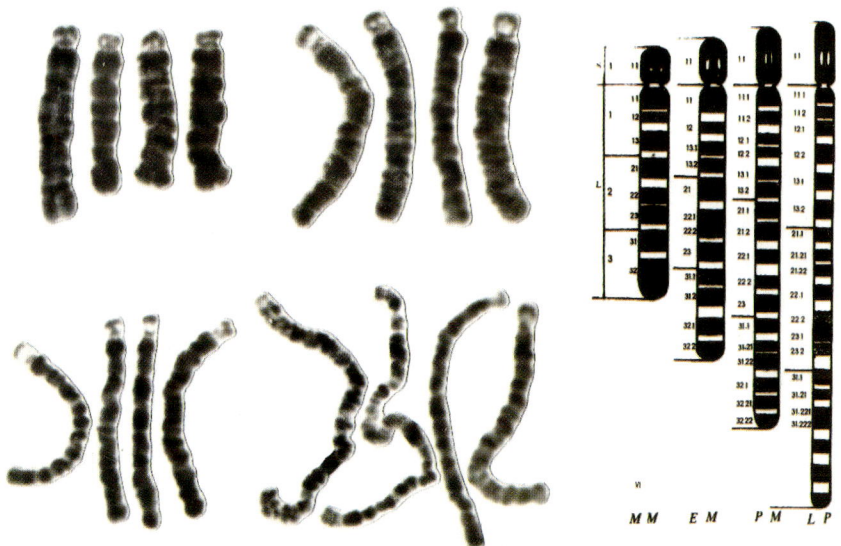

Fig. 3.61　G-banding pattern of chromosomes 6 of *L. davidii*, showing 21 bands at LP, 18 at PM, 13 at EM and 8 at MM. There is a non-stined zone in each of the pair's short arm, while a dark C-band is found at this position

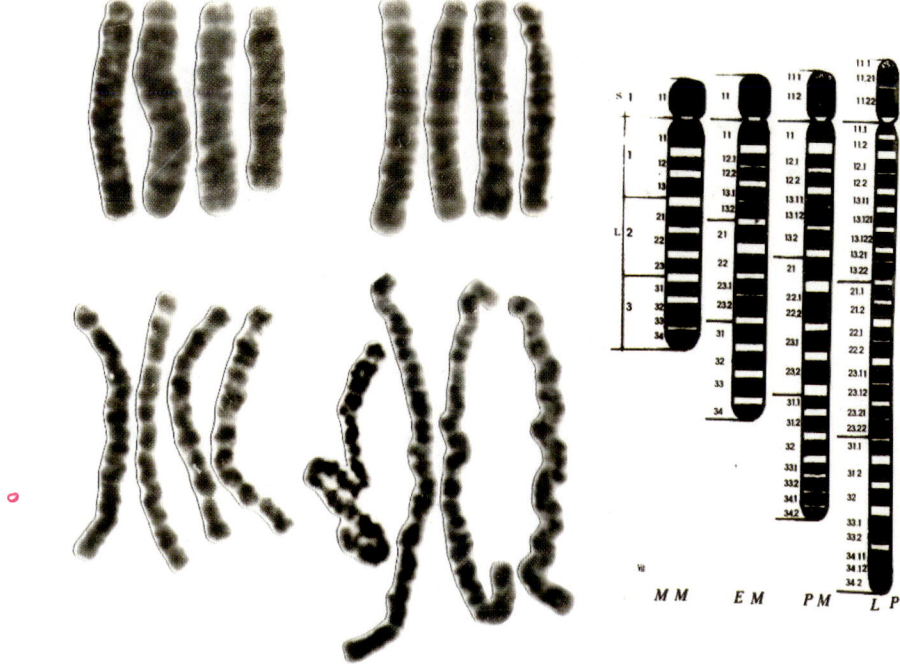

Fig. 3.62　G-banding pattern of chromosomes 7 of *L. davidii*, showing 26 bands at LP, 20 at PM, 14 at EM and 10 at MM

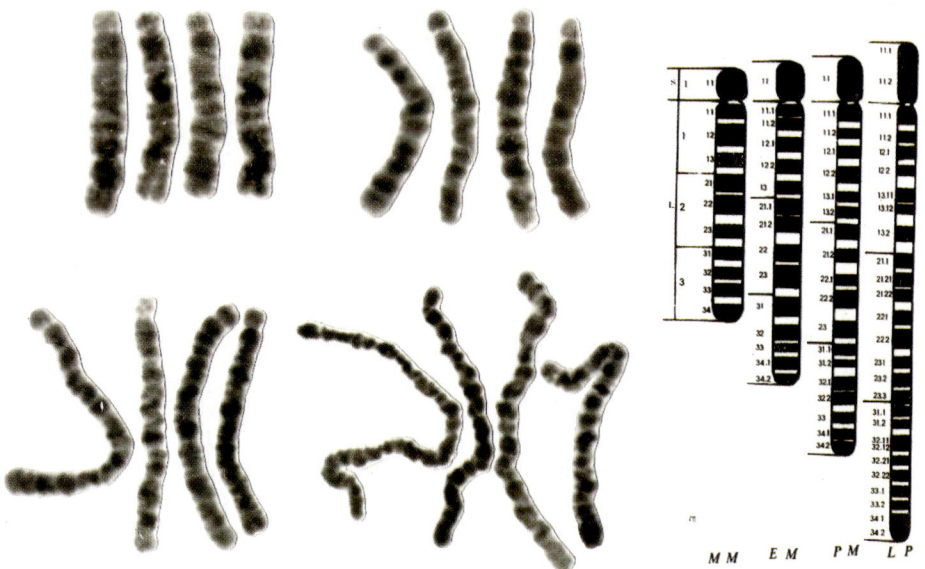

Fig. 3.63　G-banding pattern of chromosomes 8 of *L. davidii*, showing 27 bands at LP, 19 at PM, 15 at EM and 11 at MM

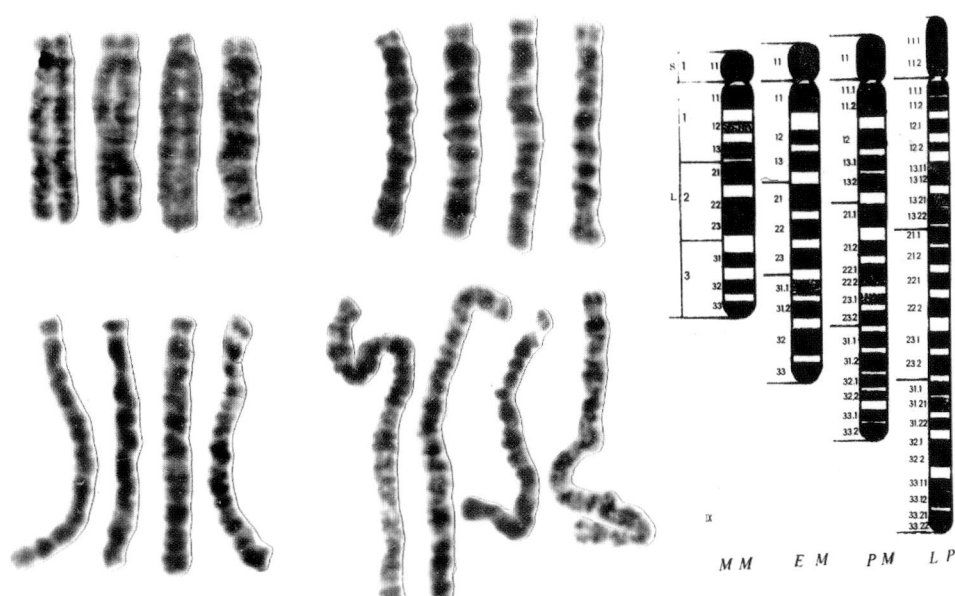

Fig. 3.64 G-banding pattern of chromosomes 9 of *L. davidii*, showing 25 bands at LP, 18 at PM, 11 at EM and 10 at MM

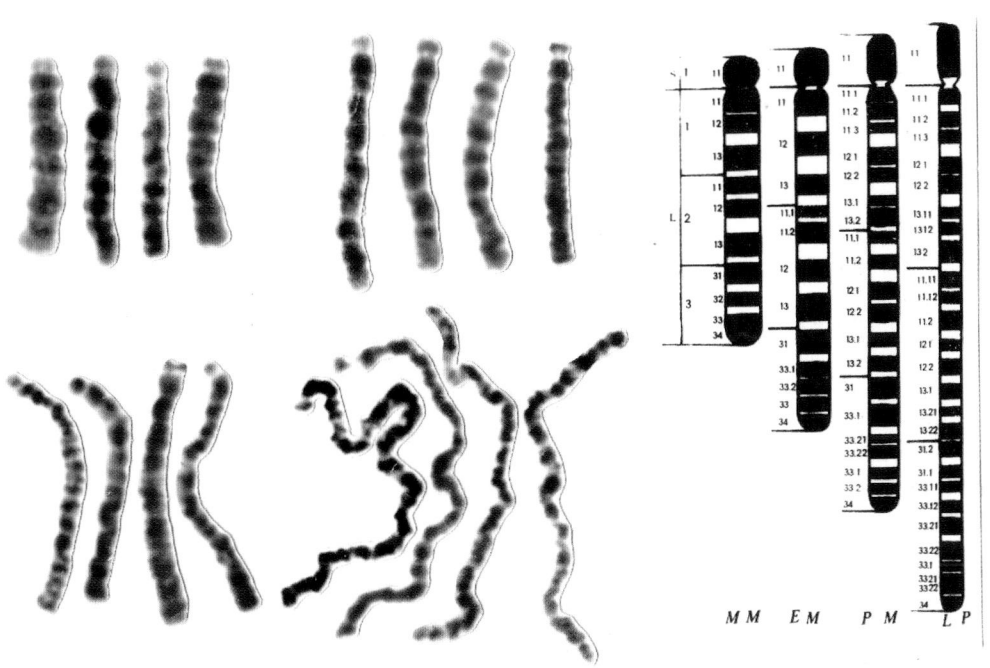

Fig. 3.65 G-banding pattern of chromosomes 10 of *L. davidii*, showing 26 bands at LP, 21 at PM, 13 at EM and 10 at MM

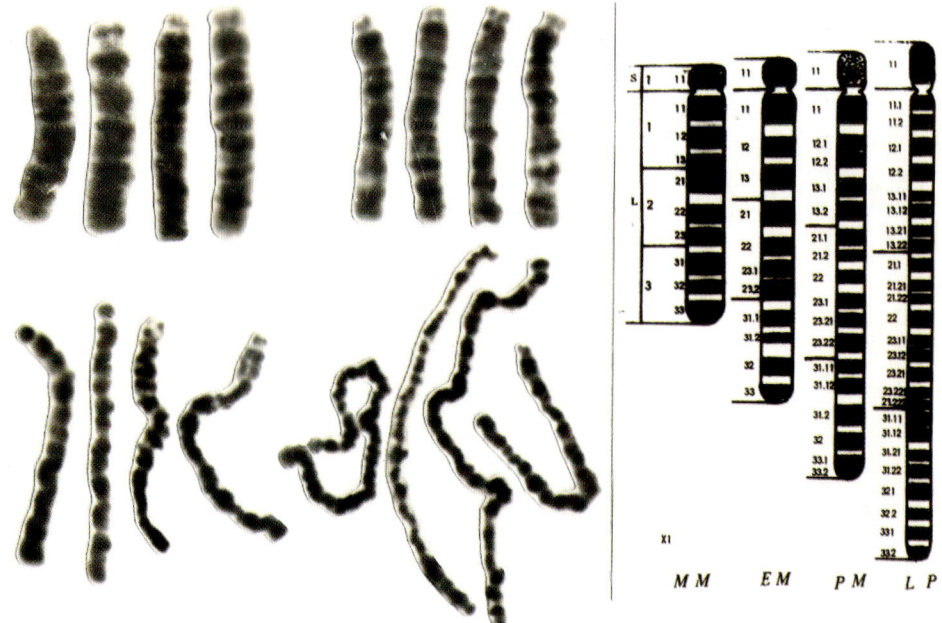

Fig. 3.66 G-banding pattern of chromosomes 11 of *L. davidii*, showing 26 bands at LP, 18 at PM, 12 at EM and 10 at MM

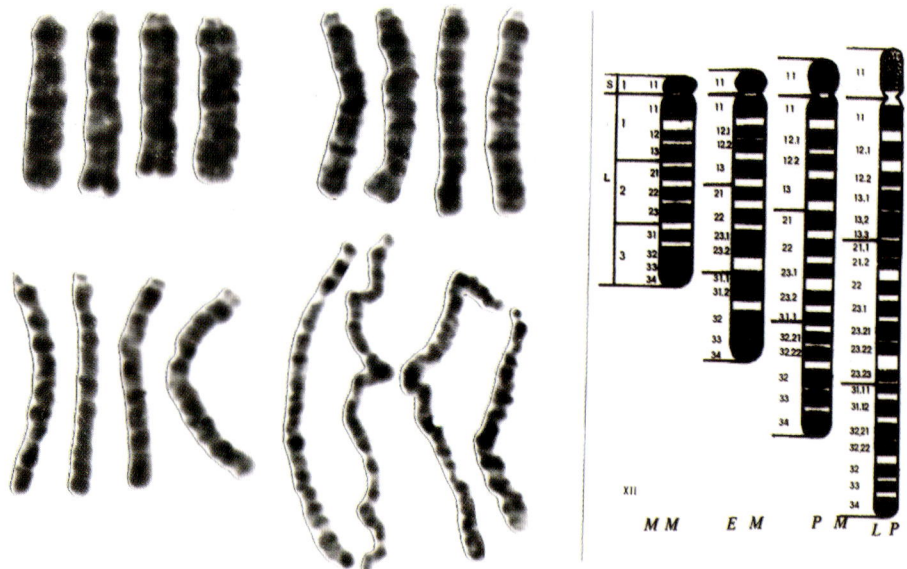

Fig. 3.67 G-banding pattern of chromosomes 12 of *L. davidii*, showing 20 bands at LP, 15 at PM, 14 at EM and 10 at MM

3.2.2.3 Discussion

It can be seen in the present work that there exists a close correspondence betwee G-and C-bands of *Lilium davidii*. At the position, were a dark C-band is shown, a notable non-G-stained zone is obtained. It might be inferred that the G-bands of *Lilium davidii* correspomd to euchromatin, while the C-bands to heterochromatin. Moreover, the G-banding pattern obtained in the present work is very similar to the Q-banding pattern reported by Kongsuwan and Smyth (1977) and Holm (1976) in terms of the location, width and number of bands. The G-bands are narrow and distributed along almost the entire length of the chromosome arms.

For more than 10 years, hypotheses have been proposed in order to explain the failure in demonstrating G-bands in plant chromosomes. For instance, Greilhuber (1977) suggested that if it is not because G-bands do not exist in plants al all, then it is because the method for preparing plant chromosome samples is not suitable. Nagl (1976) pointed out that the failure in demonstrating G-bands in plant chromosomes is due to the highly condensed state of chromosomes and the high proportion of repeated DNA sequences. Our work work shows that the main reason is the lack of a suitable method for preparing chromosome samples and revealing the bands. An important finding is that the methods for preparing chromosome samples in C-banding should be avoided in G-banding, for the degradation with HCl is unfavourable to the induction of G-bands (Comings *et al.*, 1973) and the flame-drying makes chromosomes insensitive to the band-showing treatment. Therefore, the traditional HCl hydrolyzing-squashing method was substituted by wall degradation hypotony (Chen *et al.*, 1979) with the satisfactory results.

(Authors: Chen Ruiyang, An Zhuping, Song Wenqin, Li Xiulan, and Zong Wenxing, Kenji Taniguchi, Hong Deyuan. Published in: Cathaya, 1992, 4: 9-20.)

3.2.3 植物染色体 G-带的深入研究

Since 1970's, the chromosomal banding techniques have made a rapid development in the field of chromosomal study of human and many other animals (Yunis, 1978; Comings, 1973; Rohme, 1982).They are of great significance in the research and utilization of cytogenetics. But the development of chromosome banding techniques in plants has been over ten years behind that in animals. So far, G-bands have been induced in Pinus resinosa(Drewry 1982), *Lilium davidii*, *Pinus armandii*, *Paris polyphylla* (Chen, 1986), *Hordeum vulgare* (Zhu, 1986), *Secale cereale* (Zhang, 1986), *Zea mays* (Zhan, 1987), *Vicia faba*, *Triticum monoccum* and *Ae. squarrosa* (Chen, 1987). There are also many problems in plant chromosome banding, which have not so far been completely solved, of them are universalism and repetition, experimental procedure suitable for more condensed chromosomes, distinguish long chromosomes and pairing homologous chromosomes.

In recent years, the authors have obtained distinct G-bands in many plants and some problems about G-banding technique and mechanism of plant chromosomes have been studied.

3.2.3.1 Materials and Methods

Used as experimental materials were *Lilium davidii*(2n=24), *Hordeum vulgare*(2n=14), *Sereale*(2n=14), *Ginkgo biloba*(2n=24), *Triticum boeoticum*(2n=14), *Aegilops squarrosa*(2n=14), *Vicia faba*(2n=12), *Apium graveolens* var.*clulce*(2n=22), *Plantago asiatical*(2n=12) and so on.

Chromosome were prepared with the methods of wall-degradation hypotony and dteam-drying.

The banding treatments are as follows:

(1) The trypsin method

Chromosome preparations were treated with 0.01% trypsin(made up with Diföcs 250:1, 0.02% EDTA-CMF solution) for several Seconds at 4℃, swashed with 0.85% NaCl and stained with 2.5% Giemsa [phosphate buffer (pH=7.4): Giemsa 40:1].

(2) The urea method

The preparation were treated for 2-4 minutes with a mixture of 2:1 8 Murea (made up with 0.02% EDTA): 1/15 M phosphate buffer (pH=7.4) at room temperature (20-25℃) swashed with 0.85% NaCl and stained with 2.5% Giemsa for 5-10 minutes.

(3) The trypsin-urea method

The preparations were treated in the same way as the method A →swashed with 0.85% NaCl →1:28 Murea: phosphate buffer (pH=7.4) for several seconds at 4-10℃ → swashed with 0.85% NaCl → Giemsa staining.

3.2.3.2 Results and Discussion

1. The universalism of plant chromosome G-banding

It is important to find out whether G-bands can be obtained in all of plants. For this purpose, the authors selected 19 plant materials randomly, including 2 species of gymnosperms, 5 species of dicotyledons and 12 species of monocotyledons. The experimental results demonstrate that G-bands can be shown in all these plant materials in cluding chromosomes of *Vicia faba*, *Secale cereal*, *Triticum boeoticum*, *Taestivum*, *Allium cepa*, *Aegilops squarrosa*, *Hordeum vulgare*, *Apium graveolens*, *Plantago asiatica* and *Ixeris denticalata* though the bands of latter are fewer.

2. The comparison of G-banding distribution in homologous chromosomes

268 cells from *Lilium davidii* at different division phases were chosen for examination. All of these chromosomes showed distinct G-bands and the number of bands between homologous chromosomes from different cells were basically the same. This is very useful to the match of homologous chromosomes.

3. Experimental procedure of G-banding for plant chromosomes

In 1982, Drewry used Trypsin-Giemsa method to induce G-bands and C-bands simultaneously in Pinus resinosa.We also obtained G-bands in *Lilium davidii*, *Pinus armandii* and *Paris polyphylla* in 1985 with this method (Chen, 1986). This indicates that the trypsin-Giemsa method is suitable for the study of plant chromosome G-banding technique. So far, G-bands have also been reported in Celery (Murata, 1984). Zhu (1986), Zhang (1986) and Zhan (1987) reported G-bands in *Hordeum vulgare* and *Secale cereal* by using AMD before marking chromosome preparations. Recently we reported G-bands in *Allium cepa*, *Hordeum vulgare*, *Triticum boeoticum*, *Aegilops squarrosa*, *Vicia faba*, and some other plant materials with the urea method (Chen, 1987). All of these results show that the G-banding techniques which have been used in animal chromosomes are also appropriate for plant chromosome. However, the time needed for showing the bands is much longer because of the more compact structure of plant chromosomes.

It is now known that both trypsin and usea are denaturants of protein, and they mainly denature histone. But the action of trypsin requires the optimal substrate concentration, temperature, pH, ionic strength, etc. The band-showing time is as short as several seconds. If the time is prolonged, the edge of the chromosomes will become blurred or they will be digested completely. The effect of urea is much milder than that of trypsin. The band-showing time is as long as several minutes, but the quality of bands induced by urea method are not as good as that by trypsin method. Taking advantages of trypsin and urea, we combine these two methods together. The bands obtained with trypsin-urea method are very clear. It is easily controlled and the reproducibility is high. We conclude that it is an effective method to study G-banding in plant chromosomes at present.

4. Preparation making and repetition of G-bands

The showing of bands is closely related to the way making chromosome samples. The degradation with HCl is unfavourable to the induction of G-bands (Comings, 1973). The flame-drying will make chromosomes insensitive to the band-showing treatment. The time of band-showing will be much longer. So it should avoid using the methods which cause C-bands in G-banding technique in plant chromosomes. The authors used steam-drying method (80-90℃) instead of flamed-drying (Chen, 1986). This proves to be a useful improvement.

The influence of wall-degradation hypotony is also obvious. If the concentration of enzyme mixture is higher than 5%, especially the high concentration of pectolase will affect the showing of G-bands. Another factor is that the hypotony of KCl is absolutely necessary to get G-bands, which agrees with Kurata (1987).

5. Relationship between the number of bands and the division phase of chromosomes

It was found that the number of G-bands is closely related to the division phase of chromosomes. The reason why plant chromosomes are so difficult to show G-bands at metaphase is the high degree of contraction. If the nuclear DNA content and the total length of the metaphase chromosomes of human is defined as 1, it will be 3.5-16.5 in plants. The distance between chromosomes is 0.6 μm at prometaphase in human chromosomes, while this distance is only 0.1 μm or less in plant chromosomes. It will cause the fusion of adjacent G-bands, soit is not resolvzble at the light microscopical level. In one cell of *Lilium davidii*, these are 14 bands at metaphase, 16 bands at earlymetaphase, 23 bands at lateprophase and 41 bands at prophase. In one cell of *Triticum boeoticum*, there are 88 bands at earlymetaphese and 237 bands at alteprophase. These results are very similar to those obtained in *Muntiacus muntiak*(Rohme, 1982)and humans(Yunis, 1978). So, the suitable phase of showing G-bands is not metaphase but lateprophase and earlymetaphase. Lateprophase is suitable for small chromosomes and earlymetaphase for big chromosomes.

（Authors：Chen Ruiyang. Published in：Proc.Sino-Jpn.Symposium Pl.Chromos. Plant Chromosome Research 1987，187-193.）

3.2.4 植物染色体高分辨 G-带的研究

自七十年代以来，染色体分带技术在人类和其他高等动物染色体研究中都取得了很大的进展，尤其是近年 Yunis 等人（1978）提出的高分辨 G-带技术，为细胞遗传学的发展和应用开创了一个崭新的时期。

然而，植物染色体分带的研究发展却极为缓慢，一直停留在 C-带和 N-带的水平上，而应用价值最大的 G-带技术，至今还未完全突破。张自立 1981 年，1985 年曾先后使用胰酶法和荧光显带技术，进行了植物染色体 G 显带的尝试，但都未能获得 G-带，得到的荧光带亦与 C-带基本相同。近年一些学者使用 AMD 诱导植物染色体 G-带，但至今尚未获得良好的结果。Murata（1984）对芹菜染色体进行了显带研究，但也只得到了类似 G-带。

植物染色体为什么难显 G-带，Grieilhuber（1977）曾认为植物染色体比起人的染色体要致密得多，在高度致密的染色体上 G-带无法显现。Grilelhuber 的观点影响较大，被许多人引用，认为植物界可能不表现 G-带。

1982 年 Drewyrt（1982）采用胰酶-Giemsa 法在红松染色体上同时诱导出 G-带和 C-带，尽管 G-带不十分清楚，但它告知人们，植物染色体和动物染色体一样，也可以显示 G-带，从而否定了 Grieilhuber 植物染色体不显 G-带的主张。

作者（1986）采用胰酶法，在 *Lilium dvidii*，*Pinus armardii* 和 *Pairs polyphylla* 伸长的染色体上，首次诱导出清晰的 G-带。两年来，作者在原有工作的基础上，又首次运用改良的尿素法在多种植物材料上获得了 G-带，进一步证明，植物染色体 G-带具有普遍性，这对植物细胞生物学和细胞遗传学的发展将具有重要意义。

3.2.4.1 材料和方法

（1）材料

野生一拉小麦（*Triticum bocoticum*）、节节麦（*Aogilops squarrosa*）、玉米（*Zea mays*）、蚕豆（*Vicia Faba*）、大麦（*Hordeum vulgare*）、洋葱（*Allium cepa*）、吊兰（*Chorophytum cepense*）、川百合（*Lilium davidii*）。

（2）标本制备

采用去壁低渗、蒸气干燥法制备染色体标本。

① 尿素法 0.02% EDTA 配制 8 mol/L 尿素，使用时与 pH 7.4 1/15 mol/L 磷酸缓冲液按 2:1 混合，室温下处理 2~4 min，然后以 pH 7.4 磷酸缓冲液与 Giemsa 40:1 染色）。

② 胰酶法 0.01%胰酶（Difocs 250:1 0.02% EDTA-CMF 溶液配制）4℃左右处理数秒，Giemsa 染色。

（3）染色体 G-带分析方法

① 使用 Reicherf 产 Univar 扫描显微分光光度计（波长 400~700nm，扫描步距 0.05 μm）对显微镜下染色体 G-带进行了光谱分析，测量带纹的数目、宽度、面积和积分光密度，绘制带纹分布曲线图（图 3.68）。

② 染色体 G-带显微照片使用 Opton-03 型显微光度计及其图像分析系统进行带纹测量，绘制带纹分布曲线（图 3.68）所示（此项工作由复旦大学协助进行）。

3.2.4.2 结果与讨论

作者成功地使用尿素法在野生一粒小麦、节节麦、大麦、玉米、蚕豆、洋葱和吊兰，川百合等植物染色体上诱导出了 G-带（图 3.69）。与以往的 C-带不同，G-带的带纹数目多，分布在染色体的全长上，并且每条染色体均可显带，带的数目、位置、宽窄、浓淡各不相同。前期染色体带呈颗粒状，中期呈明显的横带状，与哺乳动物染色体的 G-带很相似。

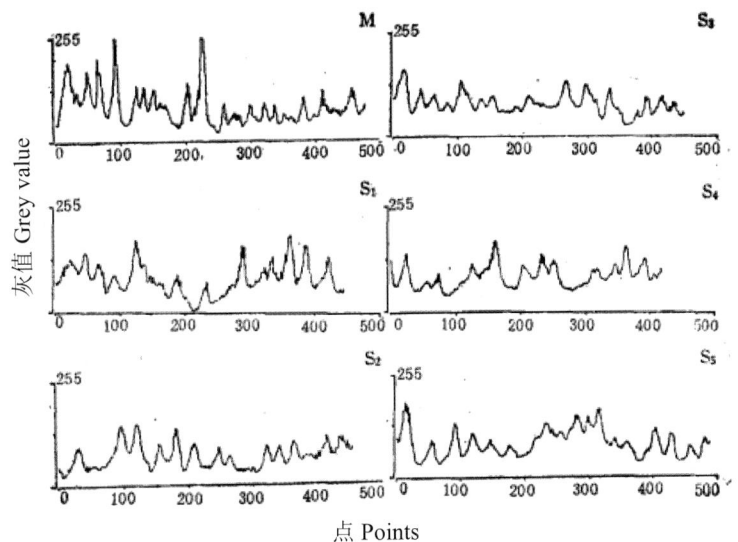

图 3.68 蚕豆染色体高分辨 G-带带纹分布曲线图 M_1，S_1，S_2，S_3，S_4，S_5 染色体 G-带见图版 II-4

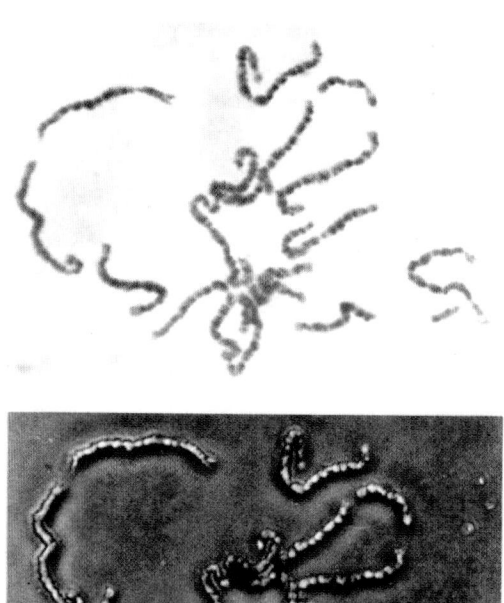

图 3.69 野生一粒小麦染色体高分辨 G-带
上：透射光，下：微分干涉

同时作者观察到随细胞分裂时期向前推移，带纹数目逐渐增多。我们使用 Uinvar 型扫描显微光度计测得川百合前期染色体带纹数目是中期的 2.92 倍，野生一粒早中期细胞有 8 条带，晚前期为 237 条带，即中期一条深带，到前期变成 2.67 条亚带，这与 Rohme 在赤虎中、Yuins 在人类中观察到的结果是很接近的。

1971 年巴黎会议确定，人类中期单倍染色组 G-带的数目为 321 条，1978 年 Yunis 利用氮甲蝶呤阻断法，将人类中期染色体 G-带的数目提高到 395 条，晚前期 1250 条，为中期染色体带纹数目的 3.16 倍，这被称为染色体分带的第二世代或高分辨 G-带技术。目前植物染色体 G-带也可接近这个水平，图 3.70 显示的是蚕豆不同时期的染色体 G-带，因此我们认为，植物染色体 G-带也可达到高分辨显带水平，而进入染色体分带的第二时代，为植物染色体亚单位的识别以及基因定位的研究展现了新的前景。

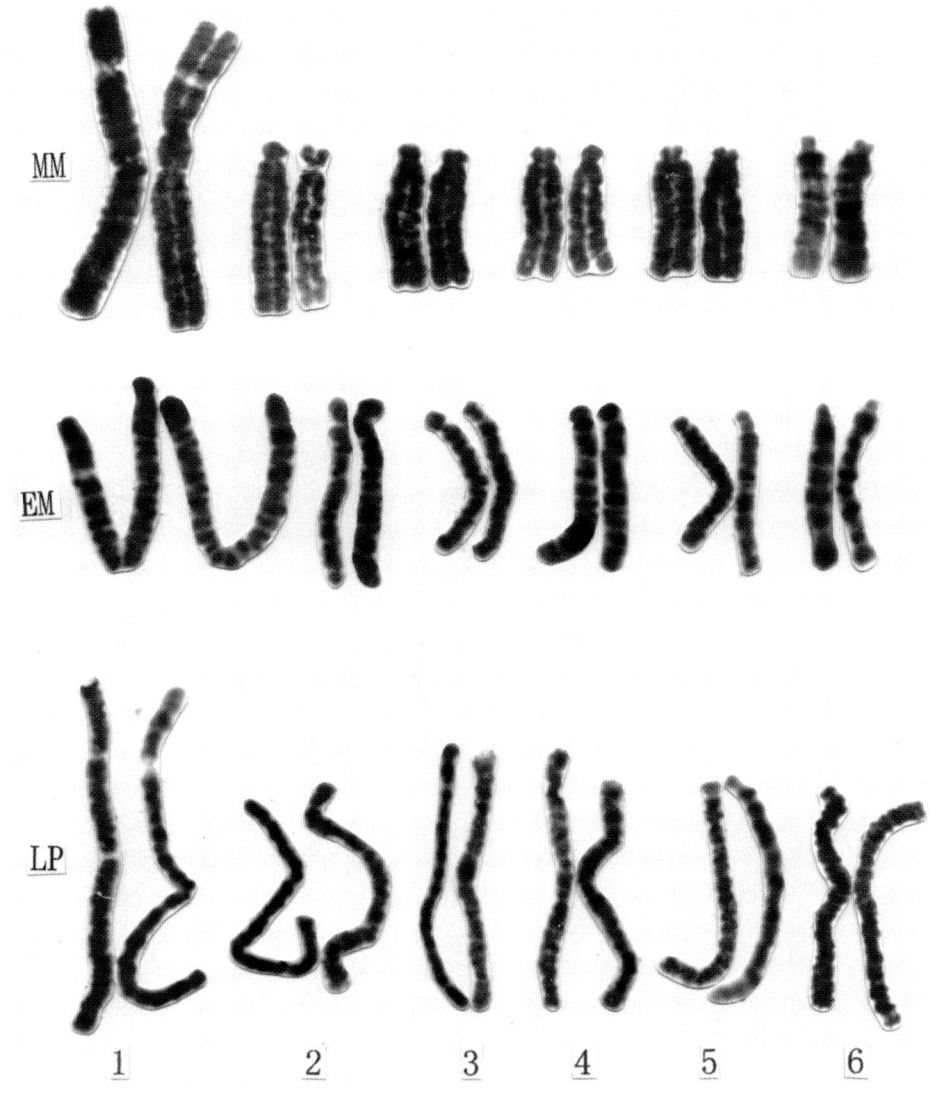

图 3.70 蚕豆不同时期的 G-带核型

因此我们认为尿素法对于高度致密的植物染色体 G 显带研究更适宜。

植物染色体之所以难显 G-带，是因为其螺旋化程度太高，结构过于致密，如果把人类有丝分裂染色体浓缩度视为 1，则植物染色体浓缩度为 3.5～16.5，人类有丝分裂早中期染色体染色粒间的距离为 0.6 μm，而植物染色体中染色粒间距离只有 0.1 μm，甚至更小，因而使相邻的带纹互相重叠连成一片，以致在光镜下无法分辨。但是与动物相比植物染色体停留在前期的时间比较长，如洋葱 71 min，蚕豆 64 min，而一般动物只有 2～3 min，利用这一特点，缩短前处理时间，降低前处理药品的浓度，可以得到较多的伸长了的染色体，诱导出清晰的 G-带。

胰酶和尿素都是蛋白质的变性剂，主要使组蛋白变性，但胰酶的作用，需要其最适的底物浓度、温度、声、离子强度等多种条件，作用时间快，显带处理仅数秒，时间稍长，染色体边缘就会发毛，甚至完全消化掉，相对来说尿素的作用要温和得多，显带时间长达数分钟，时间及温度上的微小差异对带纹质量影响不大，所以便于掌握，重复性高，具有更高的应用染色体标本制备与 G-带的显示有密切的关系，通常的 HCl 处理和火焰干燥法都可使部分 DNA 变性，有选择地抽提蛋白质，而常染色质区不论丢失的时间和数量都先于异染色质区，因此染色体标本制备过程，对 G-带的影响要大于对 C-带的影响，因此在植物染色体 G-带研究中，要尽量避免使用 C-带标本制备方法。

（作者：陈瑞阳、安祝平、宋文芹、李秀兰、苏健英、郑坚瑜。原载：科学通报，1987，14，1096-1098）

3.2.5 黑麦 G-带和 C-带核型分析

Secale cereale is a traditional material for genetic studies. In the early years when Giemsa C-banding technique was developed, researched, researchers obtained typical C-banding pattern on chromosomes of rye (Gill, 1974; Chen et al., 1979). Vosa (1974) analysed the karyotype pf rye by Giemsa and flouoresence method. Clesr and large telomeric bands were found to be anotable characteristics of the species, and this has been used to clarify the relationship between C-bands and heterochromatin (Weimarck 1975). The C-bands of each chromosome are so marked and unique that we can identify individual chromosome easily. But, too detect translocation and determine the alien additional lines, more sensitive method must be developed.

Giemsa G-banding technique is a valuable method in studying human and animal chromosomes, but its application to plant chromosomes is very difficult. In his paper "Why plant chromosomes do not show G-bands?" Greilhuber (1977) pointed out that if it is not because there is no G-band on plan chromosomes at all, then the suitability of the methods used should be questioned. In 1982, Drewry reported for the first time that G-bands were demonstrated on chromosomes of *Pinus resinosa*, and from then on, there have been several reports on G-banding in plants (Chen et al., 1985). But few of them are about G-banded karyptype. In the present paper, we studied the G-banded karyotype of rye for the first time and compared it with the C-banded one, trying to find a way to study rye chromosomes more precisely.

The rye seed tested were supplied by the Chinese Acad. Agri. Sci. The seeds were germinated on moist filter paper at 23-25℃ (36 hours). Tender root tips were excised and treated with 0.02% colchicine solution. Chromosome samples wer prepared following the Wall Degradation Hypotonic Method (Chen et al., 1982). C-bands were obtained following the BSG method (Vosa, 1974) and G-bands following the TUG method (Chen et al., 1988).

3.2.5.1 Results

Fig. 3.71 shows the G-banded karyotype of rye together with the C-banded one Eight chromosomes were metacentric, six were submetacentric. There were were large and notable telomeric C-bands at the ends of both short and long arms of No.1, 2 and 7 chromosomes, while bands were found only at the end of short arms of No.3, 4, 5 and 6 chromosomes. Centromeric and intermediate C-bands were relatively narrow. This result is in agreement with those reported by others. Unlike C-bands, the G-bands of rye were distributed along the whole length of chromosomes, and both dark and light, both wide and narrow bands could be observed. Homologous chromosomes very easy. No.1 and 2 chromosomes were two pairs of chromosomes with similar banding pattern. Five G-bands were found in their short arms and six in their long arms, and one narrow C-band was shown at the centromere.

No.3 chromosomes were a pair of submetacentric chromosomes. A large C-band was seen at each end of the short arms and several narrower C-bands were distributed at the centromere and along the long arm. In G-band preaprations, four bands were distributed along the short arm and eight along the long arm, but the end of the short arm which showed a large C-band was non-staining zone.

No.4 chromosomes were a pair of submetacentric chromosomes with a large C-band at each end of their short arms and some narrower C-bands the centromere and on the long arms. In G-band preparations, four band were shown in each short arm and six in the long arm.

No.5 chromosomes were a pair of metacentric chromosomes. A large C-band was found at the end of the short arm. Two narrow C-bands could be seen at the centromere and the end of the lang arm, respectively. Ten G-bands were evenly distributed along each chromosome.

No.6 chromosomes were a pair of submetacentric chromosomes. There was a large C-band at the end of the short arm of each chromosome, corresponding to a non-G-staining zone. The short and lang arm of each chromosome could be treated to show five and seven G-bands respectively.

No.7 chromosomes were also a pair of submetacentric chromosomes, but each bore a satellite, and thus could be identified most easily. Two notable C-bands were found at both ends of the short and long arms, and the secondary constriction also showed a notable dark C-band. In G-band preparations, three bands were situated in the short arm while five in the long arm. And opposite to that found in C-band pattern, the secondary constriction showed a non-G-staining zone.

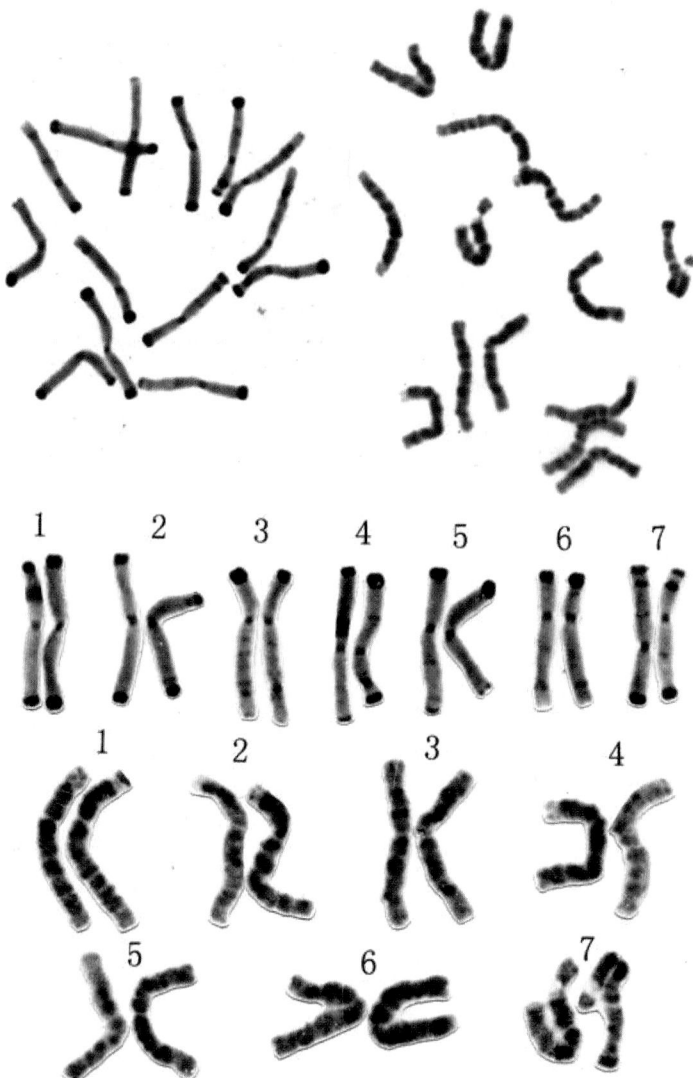

Fig. 3.71 The G-and C-banded karyotype of *Secale cereale*

3.2.5.2 Discussion

Rye is a traditional material so frequently used in genetic study that there have been a great amount of papers dealing with it. Earlier studies on rye chromosomes were based on their morphological characteristics such as chromosome number, length and centromere position. Later, the identification of chromosomal translocations and determination of homology in additional lines were carried out by applying Giemsa C-banding technique (Darvey, 1975; Singh, 1976, 1977). Drewry (1982) reported the G-bands in *Pinus resinosa*. This is the first report of the G-banding pattern in plants. More recently, Drewry (1987) reported the G-banded karyotype of *Pinus resinosa*; Chen (1988) obatinred G-bands in several in several species of higher plants such as *Lilium davidii*, *Triticum monoccucum*,

Ginkgo biloba, *Vicia faba* etc., Wang (1988) demonstrated G-bands in chromosomes of *Vicia hajastana*. In the present study, G-banding technique is employed to identify the individual chromosomes of rye and the G-banded karyotype is constructed for the first time. It can be seen that there exists a close correspondence between G-and C-bands of rye. At most positions which showed dark C-bands, notable non-G-staining zones were obtained. This led us to the belief that the G-bands of rye correspond to euchromatin while the C-bands to heterochromatin.

An important factor that adds ton the difficulty of G-banding in plant chromosomes is that swelling of cells in hypotonic solution is prevented by the cell walls. And acid hydrolysis is commonly used to allow squashing of plant cells for chromosome spreading. However this technique prevents G-banding. So the Wall Degradation and Hypotony Method was applied and a good result has been obtained. We hope that greater headway on G-banding of plant chromosomes should be achieved by development of this method.

(Authors: Li Xiu-lan, An Zhu-ping, Song Wen-qin, Chen Rui-yang. Published in Proc.Sec.Sino.Jpn.Symposium Pl.Chromos.1992 plant Chromosome Research p129-132.)

第三节 染色体分离及应用

提要：90 年代初本实验室建立了染色体分离与 PCR 扩增的方法，特别是实现了水稻 24 条染色体的分离与扩增，微切达到了 0.2 μm。这项技术对于当时的植物基因组计划的实施上，具有很大的应用潜力。并得到了上海细胞生物学研究所王亚辉教授、中国科学院许智宏院士、北京大学翟中和院士、东北师大郝水院士、武汉大学杨弘远院士、兰州大学郑国锠院士、中科院植物所洪德元院士、上海植物生理研究所洪孟民院士等 8 位专家的推荐我们参加国家"95 计划"和"水稻基因组计划"，虽然最终没有能实现，但得到了 1995 年国家科委特别支持经费 15 万元。在这些经费的支持下，我们的染色体分离工作得到了很大的发展。本节主要介绍这部分的研究工作。

3.3.1 黑麦染色体的显微分离与 PCR 扩增

自 Scalenghe 等（1981）首次将染色体显微切割与微克隆技术在果蝇唾腺染色体上取得成功后，这项技术很快被应用到小鼠和人的染色体研究上（Senger，1990；Fisher，1985）并取得了很多有价值的结果。目前，该技术已在甜菜（Jung，1992）、大麦（Schondelmaier，1993）、小麦（Diego，1993）、黑麦（Sanders，1991）、蚕豆（宋文芹等，1996）等植物染色体显微切割和克隆方面得到应用。由于人类基因组、水稻基因组计划的启动和与人类疾病及各种经济性状有关的基因的克隆、基因连锁图、RFLP 图的绘制等，都是以染色体为背景，而染色体 DNA 的显微切割和微克隆技术，在基因分离、定位、特异性探针筛选、染色体原位杂交等方面具有广泛的应用前景和潜力，因此，该技术倍受细胞生物学家和遗传学家青睐。我们选择了在遗传研究上具有重要意义的黑麦为材料，对植物染色体显微分离与 PCR 体外扩增条件进行了综合研究，建立了一套从植物染色体标本制备到单条染色体显微分离和 PCR 扩增的方法，对单条染色体 DNA 文库构建及特异性探针的筛选具有重要参考价值。

3.3.1.1 材料和方法

1. 材料

黑麦（*Secale cereale* L.）种子由英国 Wales 大学 Neil Jones 教授惠赠。

PCR 试剂购自华美生物工程公司北京分公司。

引物设计参照 Hadano 等（1991）报道的序列，由香港大学动物学系孙梅博士惠赠。

2. 方法

（1）黑麦染色体标本制备　于 25℃下培养黑麦种子，12 h 后加入 1.25 μmol/L 羟基脲（hydroxyurea）处理 18h，将羟基脲用蒸馏水洗 3 次后水培 4 h，然后加入 4 μmol/LAPM（amiprophosmethyl）溶液在室温下处理 4 h，切下根尖用 0.075 mol/L KCl 进行前低渗 30 min，70%乙醇固定，涂片于 24 mm×50 mm 的玻片上，空气干燥，3% Giemsa 染色 30 min，镜检后-20℃保存备用。

（2）显微切割　在显微镜下挑选分散好的黑麦染色体，用自制的 1 μm 左右的玻璃针在显微操作器下进行分离，将分离的单条染色体直接放入微量离心管中。

（3）PCR 扩增　参照 Hadano 等（1991）的方法略加修改。将装有黑麦染色体的微量离心管进行离心，加入 8.5 μL 的纯水、1.5 μL（0.01 mg/L）蛋白酶 K，37℃消化过夜。然后进行 PCR 扩增：①向含有 DNA 的水溶液中加入 10×PCR buffer、4×dNTPs（2.5 mmol/L）、镁离子（25 mmol/L）、人工合成的非特异性引物（14 μmol/L，TAGATCTGATATCT-GAATTCCC）、Taq DNA 聚合酶 1.5 U。补水至 30 μL，按以下条件进行低强度 6 个循环：90℃ 5 min，22℃ 120 min，50℃ 20 min（缓慢升温 1℃/min）。②上述循环结束后再补加 30 μL PCR 混合液（10× PCR buffer，4× dNTPs（2.5 mmol/L）、镁离子（25 mmol/L）、Taq DNA 聚合酶 1 U）进行 11 次高强度 PCR 循环，92℃ 1.5 min，50℃ 1.5 min，72℃ 3min。③从（2）步 PCR 产物中取出 10 μL，加入 50 μL PCR 混合液（10× PCR buffer，4×dNTPs（2.5 mmol/L），引物（14 μmol/L），1.5 U Taq DNA 聚合酶）作第三次 PCR 扩增，92℃ 1.5 min，55℃ 1 min，72℃ 3 min，35 次循环后于 72℃延伸 10 min。

（4）琼脂糖凝胶电泳　取 PCR 产物 10 μL，2%琼脂糖凝胶电泳，254 nm 紫外灯下观察并照相。

（5）Southern 杂交黑麦基因组 DNA 提取　按常规方法，探针标记采用随机引物方法，探针检测按地高辛标记检测说明进行。Southern 印迹转移采用虹吸法，将 DNA 转移到尼龙膜上，68℃杂交过夜，经过洗膜后，加显影液暗处显色。

3.3.1.2 结果和讨论

利用羟基脲和 APM 双阻断处理同步化方法获得的黑麦染色体，由 14 条 A 染色体和 4 条 B 染色体组成。选定分离的细胞后，先进行核型分析，确定每条染色体的编号。为防止同源染色体产生变异或出现人为配对错误，不进行同源染色体混合。根据染色体编号，逐一进行单条染色体分离、收集，并拍照记录，按上述方法可以准确地获得黑麦基因组的每一条染色体。收集的每条染色体经蛋白酶 K 处理后，进行 3 次级联 PCR 扩增，模板 DNA 由纳克（ng）级可以扩增到微克（μg）级。单一引物法 PCR 扩增后的产物经 2%琼脂糖凝胶电泳检测，可以明显看到扩增产物的分子量都集中在 200~20000 bp 之间，呈弥散性分布，这同本实验采用非特异性引物扩增的特点相符。PCR 产物经地高辛标记后与黑麦基因组 DNA 进行 Southern 杂交，结果表明 PCR 产物确与黑麦基因组同源，证明了染色体显微切割和 SUP-PCR 扩增技术的可靠性。

以往用于植物染色体显微切割的 PCR 扩增方法均为特异性引物法，这种方法在扩增前需要对染色体 DNA 进行酶切、连接等步骤。采用 Hadano 等（1991）报道的 SUP-PCR DNA 扩增方法，与其他随机引物法所不同的是，其引物不含随机序列，它是通过扩增条件来实现非特异性扩增的，在 PCR 循环中进行长时间低温复性，让引物 3′端充分同微量模板结合，使其各段序列都有得到扩增的可能，缓慢升温至延伸温度，非特异性结合不会因温度骤升而脱开，经高强度循环后，DNA 被大量扩增。实验结果表明，微切一段（10^{-15}g）染色体就可满足扩增和克隆的需要，这一点优于其他非特异性扩增法。为了进一步完善染色体的分离和扩增技术，在实验中我们对植物染色体的显微分离和 PCR 扩增的程序进行了探索和改进，使该方法更加完善和系统化。Brown 和 Green-field（1987）研究证明，显微切割与微克隆法获得的克隆数过多地低于期望值（＜1%），主要原因是固定液中酸诱导了 DNA 脱嘌呤，并估算出克隆到的 DNA 每 100 bp 即有一个碱基缺失，从而对这一方法所构建的 DNA 文库能否确切反映基因组内容提出质疑。因此，作者用 70%乙醇取代了传统的甲醇：冰醋酸（3：1）固定液，使酸对 DNA 的破坏作用降低到最小程度。另外，我们发现纤维素酶和果胶酶中可能含有核酸降解酶类，对 DNA 有部分降解作用。因此在对材料酶解时应采用轻度酶解。同时在染色体显微切割和扩增过程中，应尽量减少中间环节，不用油室和微小的收集液滴，将粘有染色体片段的显微切割针尖直接断在纯水中，经蛋白酶 K 处理后，进行 PCR 扩增，省掉酶切、连接等中间环节，避免了染色体丢失，减少了污染，这对高灵敏度的 PCR 扩增是至关重要的。1983~1986 年，美国的两个实验室（National Laboratories at Livermove and Los Alamos）用流式细胞分类器和分子生物学技术建立了人类 24 种染色体特异性的基因文库。流式细胞分类器分离染色体主要靠染色体的 DNA 含量和着丝粒指数。对一些 DNA 含量和染色体大小相近的就难以分辨，这就不可避免地出现染色体之间的污染，污染率可达 10%~50%（邓汉湘等，1992）。我们采用显微切割技术分离植物的单条染色体，解决了流式细胞分类器的污染问题，而且也可缩短建立特异性 DNA 文库的时间。本实验从切割染色体到完成扩增、Southern 杂交和建立单条染色体的探针库仅需 5 d 的时间。随着基因组分析的不断深入，基因在染色体上的位置信息变得越来越重要，目前应用 RFLP

分子标记遗传作图已在水稻、大麦和小麦等多种作物上进行，黑麦 1R 染色体的 RFLP 遗传连锁图也已建立，分子标记的位点非常集中。仅在黑麦 1R 染色体上就具有多种抗性基因。因此显微分离能够帮助我们直接地根据研究需要切割任意的整条染色体和染色体区，制备大量的特异性探针，提供特异性的区带遗传标记。如果该方法能和物理图谱、遗传图谱结合，将会缩小基因的筛选范围，提前进入功能基因的分离。

（作者：宋文芹、李秀兰、许文胜、陈瑞阳。原载：植物学报 1998，40（2）158-162.）

3.3.2 黑麦 1R 染色体的微切微克隆研究

染色体微切微克隆技术，最早由 Scalenghe 等（1981）提出，并首次在果蝇唾腺多线 X 染色体上取得成功。由于该技术操作复杂，不容易掌握，因此该技术在当时没有得到推广和应用。1989 年 Lùdecke 等（1989）将 PCR 技术应用到染色体显微切割中来，使收集的染色体从 100～200 条减少到数个染色体 DNA 片段。随后，该项技术被广泛应用于人类与动物的染色体上。随着核酸分子标记技术的发展，不少研究者利用此项技术制备整条染色体绘画探针和构建探针池，尤其在与人类遗传疾病有关的特异染色体区带的微切微克隆方面有多篇报道（邓汉湘等，1991，1992，1994，Guan，1994）。由于植物染色体同步化和染色体制片比较困难，从而限制了染色体微切微克隆技术在植物染色体上的应用。目前只有在甜菜中与抗线虫有关的染色体（Jung，1992）、大麦 1HS 染色体（Schondelmaier，1993）、小麦染色体（Albani，1993）及黑麦 B 染色体（Houben，1996）上有报道。我国目前在蚕豆染色体方面开展了研究（宋文芹等，1996）。黑麦 1R 染色体上具有许多抗病基因，例如在育种上具有重要意义的抗霉粉病基因 Pm8、抗秆锈病基因 Sr31、抗叶锈病基因 Lr26、抗黄锈病基因 Yr9 和抗绿病菌基因 Gb 等（Schlegel，1986），但与这些基因紧密连锁的分子标记很少。人们期待通过筛选与目的基因紧密连锁的分子标记，以便在 1R 染色体上定位和克隆出这些基因。本文以黑麦 1R 染色体为材料，显微分离出完整的 1R 染色体并进行 PCR 扩增与微克隆，初步构建 1R 染色体探针库，建立了一种简捷的植物染色体微切微克隆方法。为进一步筛选 1R 染色体上控制重要农艺性状基因的分子标记或单拷贝探针（SCP）和增加 1R 染色体 RFLP 图谱密度奠定了基础。

3.3.2.1 材料和方法

1. 染色体标本的制备和染色体显微分离

黑麦（*Secale cereale* L.）1R 染色体的标本制备参考宋文芹等报道的方法，略加修改。种子经 70%酒精消毒后，在 25℃条件下培养。待根尖长到 0.1 cm 时，用 Hu（羟基脲）处理 18 h，水培养 4 h，最后用 APM（甲基氨草磷）处理 4 h，70%酒精固定备用。将染色体标本置于载物台上，用自制的玻璃针，通过显微操作器的螺旋杆，对准待分离的染色体，在显微镜目镜视野的监视下进行显微切取，小心地放入 Eppendorf 管中。

2. 1R 染色体的体外扩增

PCR 方法参照邓汉湘等（1992）的报道，略加修改。

(1) 1R 染色体 PCR 前的预处理　将 1R 染色体收集到 Eppendorf 管中，加入 2 μL Sau 3A I、1× Multibuffer（Promega）和 0.5 μL 10 g/L 蛋白酶 K 溶液。同时取一干净的空 Eppendorf 管做同样操作，作为阴性对照。在 37℃下消化 4 h，78℃保温 15 min，灭活蛋白酶 K。接着加入 0.3 μL 10×缓冲液 B 和 0.2 μL Sau3A I（共 2 U，Promega）。37℃下酶切 4 h，78℃保温 15 min，灭活 Sau3A I。然后加入预先制备好的 20 μmol/L 接头 2 μL，0.6 μL T4DNA 连接酶 10×缓冲液和 1 μL T4DNA 连接酶（共 6 Weiss units，New England）。在 12～14℃下，连接 14～16 h。接头的制备是用 DNA 合成仪合成的 24 碱基（引物）、10 碱基两个寡聚核苷酸片段，按摩尔比 1∶1 混合，58℃下退火 1 h。

　　　　↑*Eco*R I
5′CGG<u>GAATTC</u>TGGCTCTGCGACATG 3′——Primer（24 base）
3′CTGTACCTAG 5′

(2) Cohesive adapters single primer PCR（CASP-PCR）

完成连接反应后，按下列步骤向连接混合液中加入 PCR 反应液（10 μL 10× PCR 缓冲液，6 μL MgCl$_2$（25 mmol/L），10 μL dNTPs（2 mmol/L），10 μL 引物（2 μmol/L），5 U Taq DNA 聚合酶（Promega），用无菌双蒸水补齐至 100 μL）。在 70℃下延伸 10 min，补平接头 3′末端的 18 碱基。随后在 96℃下变性 1 min，56℃下退火 1.2 min，72℃下延伸 1.3 min，共进行 35 个循环。最后在 72℃下延伸 5 min。取 2 μL 初级 PCR 产物作为模板，再次扩增 20 个循环，作为次级 PCR 产物。

(3) 探针的标记与纯化

次级 PCR 产物经酚、氯仿抽提，乙醇沉淀，三蒸水溶解后，用随机引物法标记探针。具体过程参见 Direction of DIG DNA Labeling and Detection Kit（Boehringer Mannhein）。

(4) Southern 印迹杂交及杂交信号检测　Southern 印迹参见《分子克隆实验指南》（Sambrook，1989）。杂交过程及杂交信号的检测参见 Direction of DIG DNA Labeling and Detection Kit（Boehringer Mannhein）。

(5) 克隆和重组子的鉴定

次级 PCR 产物纯化后用 *Eco*R I 酶切，低熔点胶回收，连到 100 ng 预先用 CIP（牛小肠碱性磷酸酶）去磷酸化的 pUC19 质粒上（先经 *Eco*R I 酶切），取出 1/10 体积的连接物转化感受态菌 *E.coli*（DH5α）。在含氨苄青霉素、IPTG 和 X-gal 的 LB 平板上涂布。37℃倒置培养过夜。随机挑取白色重组菌落于少量 LB 液过夜培养，煮沸 5 min，离心，取 5 μL 上清液进行 PCR 扩增。两引物为 pUC19 多克隆位点两侧的序列：5′ACAGGAAACAGC-TATGACCA3′和 5′CGTTGTAAAACGACGGCCAG3′（Kao 1991）。电泳检测插入片段大小。

3.3.2.2　实验结果

1. 黑麦 1R 同源染色体的显微分离

黑麦 1R 染色体是次中部着丝点具随体的染色体。由于具有随体，因此很容易识别。染色体分离所使用的是黑麦一个完整的体细胞染色体（2n=14 + 2B）。

2. 1R 染色体的 PCR 扩增产物分析

2%琼脂糖凝胶电泳结果显示，初级和次级扩增产物分布均匀，片段大小在 250～2000 bp，主要在 250～700 bp 以内，并未发现有特异带。另外，取 2 μL 初级产物进行次级 PCR 扩增，可获得大量 1R 染色体的扩增产物，保证了足够量探针的获得，也为将来的克隆研究提供充足的外源 DNA。

3. Southern 印迹杂交

在以 1R 染色体 PCR 产物为探针的 Southern 杂交图上，初级和次级 PCR 产物及黑麦基因组 DNA 上都有明亮的杂交信号，而初级和次级 PCR 负对照未检测到杂交信号。此外，我们在以 DIG 标记的黑麦基因组 DNA 为探针的 Southern 杂交时，也得到了类似的结果。这表明 PCR 产物与黑麦基因组 DNA 之间具有同源性。

4. 克隆

用 1/10 连接物转化 *E.coli*（DH5α），获得 10 000 多个重组菌落。随机挑出 100 个克隆子进行分析。电泳结果显示克隆子的 PCR 扩增片段长度不到 650 bp。而克隆进的外源片段的实际长度为 PCR 扩增片段减去 108 bp（40 bp 的两引物序列长度和 68 bp 的 pUC19 多克隆位点序列长度）。所以，插入片段实际长度大约为 250 bp 到 500 bp。

3.3.2.3 讨论

本文以黑麦 1R 染色体为材料，将 CASP-PCR 方法引入植物染色体微切微克隆的研究，并结合本实验室的条件对该方法进行了改进。经典的微切微克隆将微切到的染色体或特异片段收集于油室进行处理（Lüdecke，1989；邓汉湘等，1992）。而我们对显微分离到的染色体处理时不用建立油室，将分离到的染色体直接收集于微量离心管中，避免了油室所要求的一系列复杂显微操作。并且对收集到的染色体 DNA 进行蛋白酶 K 消化、*Sau*3A Ⅰ 酶切和连接反应，由原来油室中的纳升级操作变为微升级操作，简化了操作程序，使得该技术在常规实验室条件下都可进行。另外，从显微分离到 PCR 整个实验流程，都始终在一个 Eppendorf 管中进行，减少反应混合液的转移次数，操作方便，减少了污染。

前人在进行染色体微切微克隆时，常常微切染色体多达几十条甚至上百条，或是微切染色体特异区域几十段（Lüdecke，1989；邓汉湘等，1991，1992）。而本文只用 2 条 1R 染色体为起始 DNA 模板进行体外扩增就获得了足够量的扩增产物。从而把底物 DNA 的量降低到 2 条 1R 染色体。而在染色体标本的同一个分裂相细胞中便可分离得到 2 条同源染色体。这不但减少显微操作工作量，而且避免了从不同细胞中显微分离染色体时由于识别错误所带来的污染。

从扩增产物的电泳图可见，Eppendorf 管中显微分离到的染色体确实得到了扩增。由于黑麦 1R 染色体为次中部着丝点具随体染色体，在形态上与其他染色体有明显差别，保证识别上不会出错，保证 Eppendorf 管中收集的染色体肯定只有 2 条 1R 染色体，这可初步推断 PCR 产物来自 1R 染色体。

为进一步验证扩增产物是否真正来源于微切染色体，我们用 DIG-11-dUTP 分别标记 PCR 产物和黑麦基因组 DNA，作为探针与扩增产物和总体 DNA 进行 Southern 杂交。

扩增产物与总体 DNA 泳道上杂交信号显著，对照则未检测到杂交信号。两次杂交实验都证明了 PCR 产物与黑麦染色体的同源性，表明 PCR 产物来自显微分离收集到的 2 条 1R 染色体，也表明该方法的可靠性及其在植物染色体上的适用性。

正如预期结果那样，扩增产物电泳图中未检测到特异带，表明 CASP-PCR 不会优先扩增某特异结构的 DNA 序列。Southern 印迹杂交结果也没有特异带信号出现。而在研究植物染色体 DNA 中应用较多的 DOP-PCR 则会优先扩增低拷贝序列（Jamilena, 1995），所扩增的产物不能代表整条染色体，也就不能制备用于整条染色体绘画的探针池。虽然 CASP-PCR 产物为"涂片状"，但是否扩增出整条染色体序列，则有待进一步的研究来证明。

接头中含有的限制性内切酶酶切位点提高了克隆效率。我们随机挑出 100 个克隆子进行分析。由此分析出：有的菌落质粒空载，不含插入片段。在所分析的 100 个菌落中空载体约占 5%。随机挑出的 100 个克隆子的插入片段大小比较接近，在 250 bp 到 500 bp 之间。看来没克隆到大片段或者大片段的克隆效率极低。这可能是由于小片段含量比大片段多许多，在与载体连接时竞争载体所致。克隆所得的这种长度的 DNA 片段适合作为 DNA 分子标记物。从这些克隆中筛出含单拷贝序列的克隆，制成 SCP 池，结合 RFLP、CISS（染色体原位抑制性）杂交技术，可以筛选与特定基因连锁的分子标记，促进 STS（sequence-tagged site）的筛选、目的基因的定位和克隆，还将为其他植物材料染色体特异区域分子水平的研究提供有效方法。

在目前所构建的较详细的黑麦 1R 染色体的 RFLP 图谱（Wang, 1991）中，只有 16 个分子标记，其中的 12 个位点都集中于着丝粒附近的 15 cM 的区域上，而且着丝粒在图谱中的具体位置也没有确定下来。在标记这 16 个基因位点时，所用的探针均来自小麦的 cDNA 或 gDNA 克隆。这就迫切需要筛选出能用作分子标记又来自黑麦 1R 染色体本身的 SCP。此外，在目前所构建的黑麦 1R 染色体的 RFLP 图谱中，长、短臂远端区没有分子标记的区域还很大，所以有待分离更多的探针来增加图谱的密度。从传统的 cDNA 或基因组文库中筛选分子标记耗资大、需时长，而应用染色体显微切割技术从特定染色体区域 DNA 文库中获得 SCP，是最直接和快速的方法。Schondelmaier（1993）分离了大麦第 1 号染色体的短臂（1HS），并成功地将筛选到的 SCP 用于 RFLP 分析和遗传作图上，证实了用显微切割分离 SCP 的实用价值。本文已显微分离出黑麦 1R 染色体，并初步构建了该条染色体的探针库。可以利用筛选到的黑麦 1R 染色体 SCP 进行 RFLP 图谱分析，可以将 SCP 标记在 RFLP 图谱上，从而增大 1R 染色体的遗传图谱的密度。另外又可通过 CISS 杂交将 SCP 定位在 1R 染色体上。这为研究单条染色体遗传图谱与细胞学图谱（即染色体分带带型图）之间的对应关系奠定了基础。而高密度的 1R 染色体遗传图谱的获得，为通过重组技术，充分利用 1R 染色体上的抗病等优良性状基因，提供了可靠的遗传背景和理论基础，在育种上具有重要意义。

（作者：江赐忠、宋文芹、李秀兰、陈瑞阳。原载：植物学报 1998, 40 (11) 988-993.）

3.3.3 水稻染色体的显微分离与克隆

遗传学家们用了 50 多年时间，发现了将近 350 个水稻基因，由分子标记已定位的控制水稻性状的主基因已超过 30 个（郑康乐等，1997）。美国康乃尔大学绘制了含有 726 个分子标记的遗传连锁图（Causse，1994）。日本水稻基因组构建了具有 2276 个分子标记的高密度遗传图（Nagamira，1997）。中国水稻基因组已初步完成了水稻基因组物理图谱的构建和利用双倍分离群体构建了水稻的遗传连锁图（Hong，1997；徐吉臣等，1994），这些工作大大推进了水稻基因组研究的进展。但是，由于水稻染色体识别与分离的困难，中国和日本水稻基因组计划采用以水稻基因组总 DNA 建立基因文库的方案，而与人类基因组采取的从分离纯化单条染色体建库的策略有相当大的差距。以总 DNA 建库的方法虽然前期工作容易获得大量克隆片段，读出水稻基因全部核苷酸顺序，但后期工作，即确定某一片段在染色体上的位置及克隆功能基因，会遇到很大困难，由于其文库容量多达数百万个重组子，加上不同克隆片段之间可能出现部分同源性，有不少辛苦的筛选工作会没有结果。

水稻基因组由 12 条染色体组成，预测 DNA 总长度约为 4.3×10^8 bp（郑康乐，1992），平均每条染色体的总长度约为 35.8 Mb，与人类基因组总长度 3.6×10^9 bp 相比，不论总长度或是平均每条染色体长度，都小了近 10 倍，水稻基因组是高等植物中比较小的基因组之一。在读出和读懂 DNA 核苷酸顺序上应比人类基因组容易得多，但是以总 DNA 建库的方案，反而加大了它的难度，单靠现有的 2276 个（日本）和 565 个（中国）分子标记，将数万个克隆定位在 12 条染色体上还存在诸多不足，因此构建水稻基因组单条染色体和特定区 DNA 文库，筛选每条染色体特异性分子标记，对推进我国水稻基因组计划的进程具有重要理论意义和应用价值。

本文在对水稻染色体识别的基础上，对水稻染色体进行了分离、PCR 扩增和克隆，为筛选出水稻每条染色体和特定区分子标记打下了基础。

3.3.3.1 材料与方法

1. 水稻种类及染色体标本制备

供试水稻（*Oryza sativa*）为广陆矮 4 号品种，由复旦大学遗传研究所杨金水教授提供萌动的水稻种子，经羟基脲（HU）和 Amirophos-methyl（APM）双阻断同步化处理，根尖用 70% 乙醇固定（0~4℃保存），用蒸气干燥法制片，Giemsa 染色，无菌蒸馏水冲洗，镜检后迅速进行染色体显微分离。

2. 染色体显微分离

待确定分离的细胞后，首先进行核型分析，确定染色体序号，再根据编号进行分离。显微分离的仪器为经作者改装的安装在防震台上的液压传动式显微操作器，采用硅化的直径 1 μm 左右的微细玻璃针，分离的染色体直接放在微量离心管底部，-20℃ 或液氮中保存备用。具体操作方法见参考文献（李秀兰等，1998）。

3. PCR 扩增

SUP-PCR 扩增参照参考文献（Hadano *et al*.，1991）的方法并加以改进.向含有染色体的微离心管中加入 10 μL 纯水，1.5 μL（μg/μL）蛋白酶 K，3 μL 10×Taq 缓冲液，0.5 μL Mg^{2+}（25 mmol/L），纯水 5 μL 至总体积为 20 μL，50℃消化 2 h，94℃保温 5 min。

（1）第一步 PCR 扩增：向消化反应体系中补加 10×PCR 缓冲液 1 μL，Mg^{2+}（25 mmol）1.2 μL，4×dNTPs（2.5 mmol）1.6 μL，BVE22cc 引物（14 μmol，TAGATCTGATATCTGAATTCCC）1.5 μL，PEG-6000（60%）4 μl，吐温—202 μL，95℃变性 5 min 后，加入 1 U Taq DNA 聚合酶，90℃ 5 min→22℃ 120 min→22℃梯度升温到 50℃（1℃/min），50℃保温 20 min，进行 6 次循环。

（2）第二步 PCR 扩增：向第一步 PCR 反应体系中补加 10×PCR 缓冲液 1 μL，Mg^{2+}（25 mmol）0.6 μL，4×dNTPs（2.5 mmol）0.8 μL，Taq DNA 聚合酶 0.5 μL，H$_2$O 7 μL，92℃ 1.5 min→50℃ 1 min→72℃ 2 min，11 个循环。

（3）第三步 PCR 扩增：取 5 μL 第二次扩增产物到一新的微量离心管中，加入 10×PCR 缓冲液 3 μL，Mg^{2+}（25 mmol）3 μL，4×dNTPs（2.5 mmol）2.4 μL，BVE22cc 引物（14 μmol）2.4 μL，Taq DNA 聚合酶 2 U，纯水补足体系 30 μL，92℃ 1.5 min→55℃ 1 min→55℃ 1 min→72℃ 2 min，循环 35 圈后，72℃延伸 5 min。

4. SUP-PCR 产物的鉴定

将第三次 PCR 产物、水稻基因组 DNA 于 1.5%琼脂糖凝胶电泳，转膜后用地高辛标记的水稻基因组 DNA 为探针，进行 Southern 杂交。具体过程参见 Boehinger Mannheim 地高辛随机引物标记试剂盒说明书。

5. SIP-PCR 产物的克隆与重组子的鉴定

（1）PCR 产物的纯化、酶切和回收：将三次 SUP-PCR 产物经酚、氯仿抽提、乙醇沉淀，三蒸水溶解后，经 *Eco*R I 酶切，低熔点胶回收。

（2）克隆载体的制备：将质粒 pUC19 经 *Eco*R I 酶切、异戊醇、氯仿抽提，乙醇沉淀，回溶后用牛小肠碱性磷酸酶去磷酸化。

（3）PCR 产物的连接、转化：取酶切回收的 PCR 产物 200 ng，去磷酸化的 pUC19 载体 200 ng，45℃水浴 5 min，加入 3U T4 连接酶，1 μL 的 10×T4 DNA 连接酶缓冲液，10 μL 连接体系，12℃连接 16 h，取 5 μL 连接产物转化大肠杆菌 DH5α，在含有 X-gal、IPTG 和氨苄青霉素的 LB 固体培养基上培养过夜，选取白色菌落。

（4）重组子的筛选和酶切鉴定：将转化所得白色菌落用 Kieser 法筛选重组质粒. 琼脂糖凝胶电泳，选取落后于空载质粒的重组质粒，以碱法提取质粒（卢圣栋，1993），经 *Eco*R I 酶切、琼脂糖凝胶电泳，在紫外检测仪下筛选外源插入片段，照相。

3.3.3.2 结果与讨论

1. 水稻染色体的识别

水稻基因组有 12 对同源染色体（2n=24），染色体较小，前中期染色体长度分布在 3~6 μm 范围，各相邻染色体之间差异也不显著，因此在水稻细胞学研究中，染色体识别比较困难，对于染色体编号有较多争议，致使历史上很多研究者的结果不能统一（Kurata，1978，1981；Khush，1984）。1990 年第二届国际水稻遗传学会上统一了水稻染色体及连锁群的编号（Khush，1990）。作者测定了 100 个中期细胞染色体（如表 3.5

所示），并与 Kurata 以 20 个水稻前中期细胞计算的结果相似（苍田のり，1982）。水稻 1~12 染色体的相对长度、着丝点和臂比都比较稳定，但由于各相邻染色体之间差异较小，除 4 号（K4）和 9 号（K10）外，其他染色体很难凭目测识别，必须进行核型分析，综合考虑染色体参数后，才能准确识别。应用酶解、去壁、火焰干燥法制备水稻前中期染色体常显示有深染色区，对水稻染色体的识别很有帮助（陈瑞阳等，1980）。

表 3.5 从 100 个细胞计算出的前中期染色体的特征
Table 3.5 The premetaphase chromosomes characters calculated from 100 cells

序号 number	相对长度平均值 Relative length average values	范围 range	臂比平均值 Arm ratio average values	范围 range	类型 type
1 （K1）	12.42	9.96~14.80	1.48	1.04~1.92	m
2 （K2）	10.49	8.83~13.06	1.46	1.03~2.07	m
3 （K3）	10.06	7.91~11.66	1.28	1.06~1.75	m
4 （K4）	8.98	7.69~11.66	3.57	1.88~3.97	st
5 （K9）	7.20	5.92~8.02	1.76	1.16~2.26	sm
6 （K6）	8.29	6.73~9.66	1.26	1.04~1.57	m
7 （K11）	6.20	4.66~7.29	1.42	1.04~1.72	m
8 （K7）	7.69	6.01~9.06	1.58	1.08~2.08	m
9 （K10）	6.29	4.14~7.39	3.28	1.37~4.47	st
10 （K12）	6.04	4.44~7.20	1.98	1.14~1.98	sm
11 （K8）	7.35	5.50~8.56	1.42	1.02~2.14	m
12 （K5）	8.42	7.14~9.75	1.73	1.10~2.53	sm

具随体染色体（satellite chromosome）。

2. 水稻染色体的显微分离

首先选取染色体分散良好的细胞，显微拍照，进行核型分析，确定染色体序号，再根据染色体序号进行显微分离。当前用于染色体显微分离和切割的方法主要有两种，一是利用安装在倒置显微镜上的显微操作仪进行（Senger，1990）。这种方法虽然较易操作，但由于倒置显微镜的物镜一般只有 40×，从而限制了倒置显微镜在染色体微切中的应用，特别是像水稻 3 μm 以下小染色体的分离和染色体特异区的显微切割，显得尤其困难。另一种方法是利用激光显微镜进行染色体切割和分离（Fukui，1992），该方法虽然提高了染色体切割的精度，但是方法复杂，需要将染色体制在一种尼龙膜上，不需要的染色体部分用激光烧掉，万一去除不干净很容易造成污染，切割的染色体（片段）连膜一起脱下进行收集，而且激光显微镜价格昂贵，一般实验室难以实现。为解决以上矛盾，提高分辨率和放大倍数，使染色体识别和显微切割同步进行，作者使用普通明视野电光源显微镜，利用改装的安装在防震台上的液压传动式显微操作仪，在 100× 油镜下进行染色体分离和切割获得成功（李秀兰等，1998）。这不仅解决了染色体识别、分离与切割同步进行，而且提高了染色体微切的精度，使微切的染色体片段缩小到 0.2~1.0 μm，基本达到了激光显微切割的水平。作者实验初期阶段将水稻 1 号染色体切割成 5 个特定区，每个片段大约 7 Mb，根据对人类、水稻和小麦染色体长度的估算，1 Mb 相当于遗传图谱的 1 cM，而 1～3 cM 被认为是遗传图的有效距离。应用作者建立的显微切割技术，获得 1～3 cM 的染色体片段，建立水稻染色体特定区 DNA 文库是完全可能的，它将是筛选与特定基因紧密连锁的标记，构建高密度遗传连锁图的有效方法。

3. 水稻染色体的 PCR 扩增

从水稻广陆矮 4 号一个完整细胞 24 条染色体和第一号染色体的 5 个特异区 SUP-PCR 扩增后的结果和从琼脂糖凝胶电泳检测的结果可以看出，DNA 片段长度在 80～600 bp 之间，表明应用染色体显微分离、直接进行 PCR 体外扩增，可以建立水稻每条染色体 DNA 文库，为进一步筛选水稻基因组各单条染色体特异性探针群奠定了基础。

当前用于显微切割染色体采用的扩增方法主要有两类，即特异性引物随机扩增法和非特异引物扩增法。前者在扩增前需对微切 DNA 进行酶切和连接，通过小体积显微操作完成，相比而言，非特异引物扩增法更为直接。它只需 20～24 bp 的寡核苷酸分子为引物，通过数次循环采用非严格的反应条件，就可使模板 DNA 各段序列得到扩增。本实验使用 SUP-PCR 方法是一种高灵敏度扩增任意痕量未知 DNA 序列的方法，与其他随机引物法不同的是其引物不含随机序列，它是通过在初级扩增体系中加入适量的表面活性分子聚乙二醇-6000，使引物和模板有效碰撞的机会增多，然后通过低温长时间复性，使引物充分和模板结合，再继续缓慢升温至延伸温度，经过数次高严格的循环，体系由原来的皮克（pg）级扩增微克（μg）级，实现了对微切一个片段（3～4 Mb 大小）进行直接扩增和克隆，极大提高了工作效率。

4. SUP-PCR 产物的鉴定

以水稻基因组 DNA 为探针的 Southern 杂交结果显示，对应水稻基因组 DNA 和水稻染色体 SUP-PCR 产物的位置检测到明显的杂交信号，证明 SUP-PCR 产物确来源于水稻染色体 DNA。

5. PCR 产物的克隆

将 PCR 的 *Eco*R I 消化产物与 pUC19 质粒连接起来，取一条染色体 1/5 的连接产物（相当于 10 μg DNA，一条染色体扩增一次可获得 50 μg DNA）转染 *E. coli* DH5α，获得了大量的蓝、白色菌落。我们共分析了其中的 13 个白色菌落，其插入片段的大小平均为 80～500 bp。这种长度的片段极适合分子标记的筛选。

染色体微切、微克隆技术在动、植物基因组研究方面有着重要作用，利用该技术可以建立特定染色体或染色体区带的 DNA 文库。从中筛选大量探针，提供高密度遗传图及精细物理图构建所需的位标。作为分子生物学和遗传学研究的辅助手段，微切、微克隆技术必将发挥更大的作用。

（作者：宋文芹、李秀兰、祁仲夏、梁思源、陈瑞阳。
原载：南开大学学报（自然科学）2000, 33（3）83-88）

3.3.4 水稻第 9 号染色体 DNA 文库的构建

自 1988 年美国康乃尔大学 McCouch 等（1998）绘制出第一张较完整的水稻 RFLP 遗传连锁图以来，随着日本、美国、韩国、中国等国家水稻基因组研究计划的开展，水稻遗传图上遗传标记的数量在迅速地增加，至今，日本发表的水稻遗传图上已含有 2276 个遗传标记。然而，无论美国还是日本，遗传图上的标记绝大部分都是来自于 RFLP 标记，它们未必是真正的单拷贝序列，不能排除遗传图上许多标记的同源序列可能存在于同一染色体或不同染色体的多个位点上，因此，这些标记并非均匀分布于水稻的 12 条染色体上（Nagamira，1997）。而且 RFLP 遗传连锁图上的遗传距离与实际物理距离并不一致，往往造成基因组某些区域分子标记较少，影响进一步构建完整的物理图谱及基因定位克隆。通过染色体显微分离构建基因组特定区域的 DNA 文库，筛选分子标记，可加大此特定区域 RFLP 图谱的密度，克服分子标记分布不均的问题。该方法已成功应用于少数几种植物的染色体专一性 DNA 文库的构建（Sandery，1991；Jung，1992；Schondelmaier，1993；Chen，1995）。由于水稻染色体小，分离困难，这方面的工作国内外尚未见报道。本文显微分离了水稻单条第 9 号（K10）染色体，并构建了该染色体的 DNA 文库，现将结果报道如下。

3.3.4.1 材料与方法

1. 材料

供试水稻品种为中国水稻基因组计划规定品种广陆矮 4 号，由复旦大学杨金水教授提供。

2. 方法

（1）染色体标本制备和染色体显微分离：水稻种子萌发后经羟基脲（HU）和 Amiprophosmethyl（Amp）双阻断同步化处理，根尖用 70%乙醇固定，蒸气干燥法制片，Gimesa 染色。染色体分类根据 1990 年第二届国际水稻遗传学会议规定标准，参照 Kurata（1978）K1～K12 染色体编号。利用液压传动式显微操作仪（日产 NARISHIGE），在 100×油镜下用自制玻璃针分离水稻单条 K10（按国际标准为第 9 号，本文中称 K10）染色体，将分离的染色体直接放于微量离心管底部。

（2）单条 K10 染色体的 SUP-PC 扩增：SUP-PCR 的扩增方法参见 Hadano（1990）的方法，并加以改进。

（3）SUP-PCR 产物的鉴定：为了确证 SUP-PCR 产物是否来自于所分离的水稻 K10 染色体，将单条 K10 染色体及一条随机分离的水稻染色体 Kn（非 K10 染色体）同时进行 SUP-PCR 扩增，将第三次 PCR 产物于 1.5%琼脂糖凝胶电泳，转膜后分别用水稻基因组 DNA 及 K10 染色体 PCR 产物以地高辛标记探针进行 Southern 杂交。具体过程参见 Boehinger Mannheim 地高辛随机引物标记试剂盒说明书。

（4）SUP-PCR 产物的克隆与克隆子的鉴定：将 K10 染色体第三次 SUP-PCR 产物经酚、酚/氯仿抽提，乙醇沉淀，溶于 H_2O 中，经 *Eco*R I 酶切后，于 1.5%琼脂糖凝胶电泳，冻融法回收凝胶中酶切后的 PCR 产物，连接到预先用 CIP 去磷酸化的 pUC19 质粒上，取出连接物转化感受态大肠杆菌 DH5α，在含氨苄青霉素、IPTG、X-gal 的 LB 平板上涂布。37℃倒置过夜，挑取白色菌落，接种于 LB 培养基后加入 15%体积的甘油冻存。取少量菌液沸水浴 5 min，离心取上清液为模板，以 pUC19 多克隆位点两侧的序列为引物进行 PCR 扩增。两个引物的序列为 5′-ACAGGAAACAGCTATGACCA-3′和 5′-CGTTGTAAAAC-GACGGCCAG-3′（Hadano，1991）。扩增产物以 1.5%琼脂糖凝胶进行电泳。

（5）基因组 Southern 杂交分析：将 PCR 扩增出的插入序列以随机引物法标记探针进行基因组 Southern 杂交分析。用 *Eco*R I 酶切水稻基因组 DNA 10 μg，于 0.8%琼脂糖凝胶电泳，电场强度 2V/cm 持续 6h，转移至硝酸纤维素膜，晾干，80℃烤 1 h，取 5 mL 预热的预杂交液（含 5× SSC，0.1%肉桂酸钠，0.02% SDS，1%封阻剂，50%去离子甲酰胺），放入杂交膜，42℃预杂交 2 h 以上，另取 5 mL 42℃预热的杂交液，按 5～10ng/mL 加入探针标记液，入杂交膜，37℃杂交 16 h，2× SSC 溶液（含 0.1% SDS）室温洗膜 10 min，两次，0.5×SSC 溶液（含 0.1% SDS），50℃洗膜 20 min. 按照 Boehinger Mannheim 地高辛随机引物标记试剂盒说明书进行杂交信号检测。

3.3.4.2 结果与讨论

1. 水稻 K10 染色体的显微分离与 SUP-PCR 扩增

水稻 K10 染色体是唯一具随体的水稻染色体，容易识别，我们在 100×油镜下分离了该染色体。经 SUP-PCR 扩增后产物呈涂片状。扩增产物的大小集中在 120～500 bp。以水稻基因组 DNA 为探针的 Southern 杂交结果显示，对 K10 和 Kn 染色体 SUP-PCR 产物的位置检测到明显的杂交信号，而负对照的位置则未检测到杂交信号；以 K10 染色体 SUP-PCR 产物标记探针和以 Kn 染色体 SUP-PCR 产物封阻进行 Southern 杂交时，结果只在对应 K10 染色体 SUP-PCR 产物的位置呈现强烈的杂交信号。以上结果表明水稻 K10 染色体确被特异性的扩增。

2. pUC19 文库中重组子的分析

应用连接物转化大肠杆菌 DH5α 共获得近 9×10^4 个转化子，其中 4.5×10^4 个为白色菌落。我们共分析了其中 100 个克隆子（白色菌落），插入片段的长度为 PCR 扩增出的片段长度减去 108 bp（两个引物总长度为 40 bp，多克隆位点长度为 68 bp），插入片段的长度范围为 80～500 bp，平均 210 bp，其中 40%超过 200 bp。在所分析的 100 个重组子中只有 4%因质粒空载或不含质粒未扩增出插入片段。有的重组经 PCR 扩增后两个扩增片段，推测是由于两个连在一起的菌落被同时挑取造成的。为了估计此文库的丰富度，我们将 100 个重组子的插入片段点于硝酸纤维素膜上，以随机选取的三个重组子用地高辛标记探针进行点杂交，结果未检测到杂交信号，表明此文库中重复的克隆子是比较少的。我们用基因组 Southern 杂交对此文库中的克隆子进行了初步分析，随机选取 9 个克隆子的插入序列标记探针进行基因组 Southern 杂交分析，其中 2 个克隆子检测到单拷贝序列；3 个杂交信号较弱；4 个克隆子检测到重复序列。水稻是禾谷类植物分子遗传学研究中的模式植物，它的基因组很小（C=0.6 pg）（McCouch，1998）平均每条染色体只有 50 fg，我们以单条水稻 K10 染色体为模板进行扩增克隆，即获得了 4.5×10^4 个克隆子，此效率（900 个克隆子/fg）明显高于甜菜（Jung，1992）、大麦（Schondelmaier，

1993）的染色体特异性文库构建中的克隆效率。利用基因组 Southern 杂交已从本文库中筛选到单拷贝序列，进一步对文库中大量克隆的分析工作正在进行，本文库的构建为获取水稻 K10 染色体上的分子标记及研究该染色体特异性 DNA 序列结构提供了新的途径。

（作者：梁思源，宋文芹，李秀兰，毛英伟，陈瑞阳。

原载：南开大学学报（自然科学）1999，32（2）107-110）

3.3.5 水稻 4 号染色体的分离与 DNA 文库的构建

Rice is not only one of the most important food crops in the world，but also a model plant among the cereals for molecular genetics studies. It has 24 chromosomes and the smallest genome（4×10^8 bp）of any monocot. Some of important genes，for example，Ph（phenol staining），Gm-2（Gall-midge resistance），Xa-1，2（Bacterial blight resistance）are located on chromosome 4 of rice[1].Most important ofall，sequencing of rice chromosome 4 will be carried out by the Rice Genome Project（RGP）in China.

With the development of the Human Genome Project（HGP）in recent years，analysis of restriction fragment length polymorphisms（RFLP）Has become a powerful tool gene tagging，gene manipulation and physical mapping；these methods are also widely used in rice genome research. McCouch *et al.* constructed the first RFLP map of rice in 1988（McCouch，1988）. Causse *et al.* enriched their molecular linkage map to 726 markers based on an interspecific backcross population. But there were only 84 markers mapped on chromosome 4. Another linkage map of rice which was derived from a cross between Nipponbare（a *Japonica variety*）and Kasalath（an *Indica variety*）was constructed by Kurata *et al.* in Japan，and included a total of 1285 RFLP and random amplified polymorphism DNA（RAPD）markers. Using these probes they also constructed a physical map by YAV clones which has been arranged over 52% of the genome length. Their physical map of the rice chromosome 4 contained 114 DNA markers and 258 individual YACs which formed 17 contigs and covered 48% of the chromosome length（about 16.9Mb）（Kurata，1997；Koike，1997）. In China hong utilized bacterial artificial chromosomes（BACs）to build a large-scale conting map of rice including 631 contigs which covered 92% of the genome，based on fingerprinting and hybridization（Hong，1997，1998）.

However，most of these markers from RFLP maps in U.S.A. and Japan came from randomly selected genomic clones，cDNA clones and a few markers from other plants，such as oat，barley and maize. These random markers are distributed unevenly along the chromosomes. Many gaps in rice maps still need to be filled. In addition many of these clones are multi-copy probes，and few STS（Sequence-Tagged Site）markers are mapped. Because the genetic distance between two markers doesn't match the physical distance，we need more probes to saturate the map in order to reduce the errors. Cloning by microdissection and microcloning has proved to be an extremely useful technique to obtain more chromosome-specific molecular markers (Fiedler, 1991; Jung, 1992). Thus, our chromosome-specific library of rice can provide a large number of additional chromosomal assigned DNA markers which will facilitate mao-based cloning and construction of ahigh resolution physical map of rice chromosome 4.

3.3.5.1 Materials and Methods

1. Materials

Orysa sativa L. guangluai No.4 was kindly provided by Prof. Yang Jinshui (Fudan University). Eight RFLP markers, R2373, R1854, R288, C1100, C1016, C891, C335 and C445 were kindly provided by Prof.Sasaki, the head of Rice Genome Project (RGP) in Japan.

2. Preparation of metaphase chromosome spreades and microdissection

The procedures for preparation of metaphase cell and isolation of rice chromosome 4 were essentially the same as those described by Song et al (1996).

3. Pretreatment paior to LA-PCR

Two rice chromosome 4 homologous from one cell were collected in a 0.5-ml Eppendof tube by microdissection. 2 μl of *Sau 3A* I 1×multi-buffer (Promega) and 0.5 μl of 10mg/ml proteinase K solution were added. A clean tube without chromosome DNA was used as a negative control.Both were digested at 37℃ for 4 h, then incubated at 78℃ for 30 min to inactivate proteinase K. Next 0.3 μl of 10 ×buffer B and 0.2 μl of 10 units/ μl *Sau 3A* I (Promega) were added. The chromosomal DNA was digested at 37℃ for 4 h, followed by incubation at 78℃ for 15 min to inactivate *Sau 3A* I.

The adapter was prepared by mixing two synthetic oligonucleotides of which the 5' end phosphorylated in an equimolar ratio: 24-mer sequence

(1): 5'-CGGGAATTCTGGCTCTGCGACATG-3', 10mer sequence (2): 3'-CTGTAC CTAG-5' then annealing at 58℃ for 1 h. The adapter was as followed: 5'-CGG*GAATTC* TGGCTCTGCGACATG-3'

↑ *Eco*R I   3'-CTGTACCTAG-5'

The DNA fragments cut by *Sau 3A* I were ligated with 0.6 μl of T4 DNA ligation 10×buffer, 1 μl of 6 units/ μl, T4 ligase (Promega) at 12-14℃ for 14-16 h.

4. LA-PCR

After ligation, Pcr reaction mixture (100 μl) was added [1×PCR reaction buffer, 2mM MgCl$_2$, 0.2 mM each dNTP, 5 units Tag polymerase (Goldworld), 0.2 μM primer]. The 24-mer sequence was used as the primer. The following PVR conditions were used: first at 72℃ for 10 min to fill the recessed 3' end of the adapter, then 35 cycles at 96℃ for 1 min, 54℃ for 1 min, 72℃ for 1.5 min, finally an extension at 72℃ for 5 min. 1 μl of primary product was amplified in a secondary PCR reaction identical to that described above ithout the filling reaction.

5. Probe preparation

Rice genomic DNA was extracted by the CTAB method.The purified PCR products and rice genomic DNA were labeled by the random priming method, according to the Direction of DIG DNA Labeling and Detection Kit (Boehring Mannhein).

6. Southern blot hybridization

First, labeled total rice DNA was hybridized to a filter containing the secondary PCR products and rice genomic DNA. Second, the purified secondary PCR products were labeled as probes to hybridize with eight RFLP markers of rice chromosome 4: R 2373, R1854, C1100, C1016, C891, C445, C335, R288. The method was described in Molecular Cloning (Sambrook, 1982).

7. Cloning into pUC19

After PCR, the amplified products were purified by phenol/chloroform ex traction, digested with *Eco*R I, and ligated to dephosphorylated pUC19 vector with T4 ligase. One-fifth of the ligation mixture was used in transformation of the highly efficient *E.coli* DH5α competent cell. One fifth of the transformed mixture was plated on to LB medium with ampicillin and IPTG-X-gal.

8. Characterization of microclones

After transformstion, ehite colonies were isolated for initial characterization. Rice total DNA was labeled with α-^{32}P-dCTP and hybridized to filters containing recombinant microclones, as described by Kao *et al.*(1982). The single copy probe C335 and rice genomic DNA as controls. The size of inserts contained in the microclones was estimated by PCR in which specific primers flanking the cloning sites of pUC 19 were used to amplify the insert sequence in each clone. Two 20-mers were used as the primers: 5'-ACAGGAAACAAGC TATGACCA-3' and 5'-CGTTGTAAAACGACGGCCAG-3'.

3.3.5.2　Results

Rice chromosome 4 was isolated from one metaphase cell (Fig. 3.72) using a micromanipulator combined with a Nikon microscope and a 1 μm tip-glass needle. Rice chromosome 4 is easy to identify, as is submetacentric and the 4th longest chromosome.

1. Amplification of microdissected chromosome 4 using *Sau 3A* I linker adapter

From the 1.2% agarose gel electrophpresis pattern(Fig. 3.73), the primary and secondary PCR products were smears of fragments ranging from 250 bp-2000 bp, and there was no detectable amplification of extraneous DNA in negative control. Appearance of specific bands was observed, which suggests no preferential amplification of a subset of DNA sequences.When stained with ethidium bromide, the amount of amplified DNA in the secondary reaction is estimated in the μg range, which ensures enough material for probes.

2. Characterization of PCR products by southern blot hybridization

Southern blot analysis showed strong hybridization signals between rice genomic probe DNA and PCR products in lane 3, whereas no signal is seen in the negative control in lane 4; this confirmed that rice DNA sequences have been amplified(Fig. 3.74). Additionally, pooled

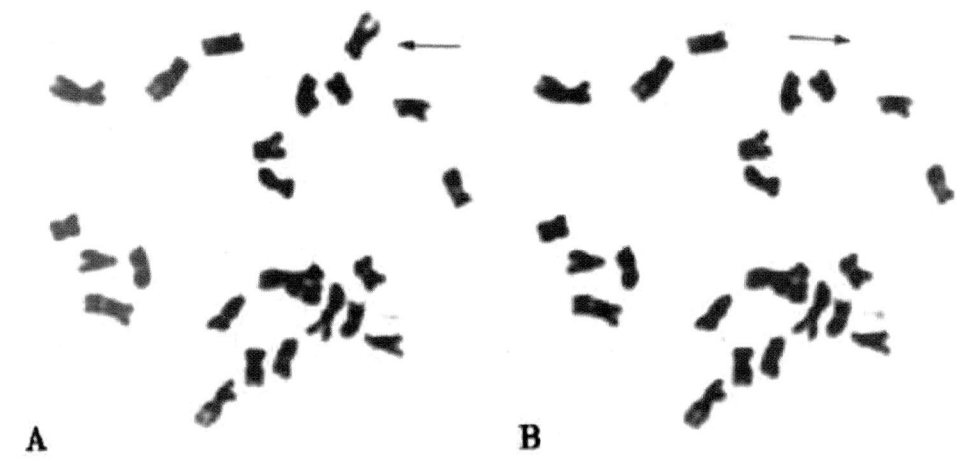

Fig. 3.72　Microdissection of rice chromosome 4
A. Before microdissection of rice chromosome, arrow indicates chromosome 4; B. After microdissection of rice chromosome 4

PCR products were used as probes to hybridize with eight RFLP markers of rice chromosome 4: R2373, R1854, C1100, C1016, C891, C445, C335, R288. As shown in two recent works, these eight DNA markers were selected specially for their even distribution along the linkage map of rice chromosome 4. Amoog these markers C445 and R2373 were lacated at the two ends of thelinkage map and R2373, R1854, C1100, C1016, C891, R288 were used to construct a rice physical map. Furthermore, C1100, C1016, C891, C445, C335 were selected from a rice cDNA library, and parts of their sequences were released on Gene Bank. So these eight probes are site-specific uniquences. All of these markers except for R288 demonstrated intensive hybridization signals with amplified products. These results gave further evidence that these PCR products were derived from rice chromosome 4.

Fig. 3.73　Amplified products of LA-PCR
1. Primary PCR products
2. 2Secondary PCR products
3. λ/Hind III +EcoR I
4. Negative control

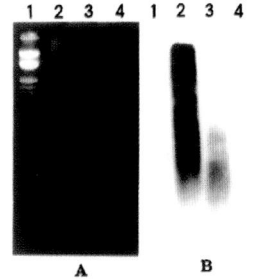

Fig. 3.74　Southern blot hybrisization analysis between rice total DNA and PCR products
A.Before hybridization
B.After hybridization 1. λ/Hind III+EcoR I, 2.Rice genomic DNA, 3.Secondary PCR products, 4.Negative control

3. Characterization of microclones

Analysis of 500 clones with colony hybridization revealed that about 42% of them hybridized either strongly (highly repetitive sequences) or less strongly (moderately repetitive or slightly repetitive sequences) and 58% showed no detectable hybridization (containing unique or very low-copy sequences) (data was not shown). We selected 66 microclones randomly from the recombinant clones to determine their insert size, which ranged from 140 to 500 bp with a mean length of 300 bp (Fig. 3.75).

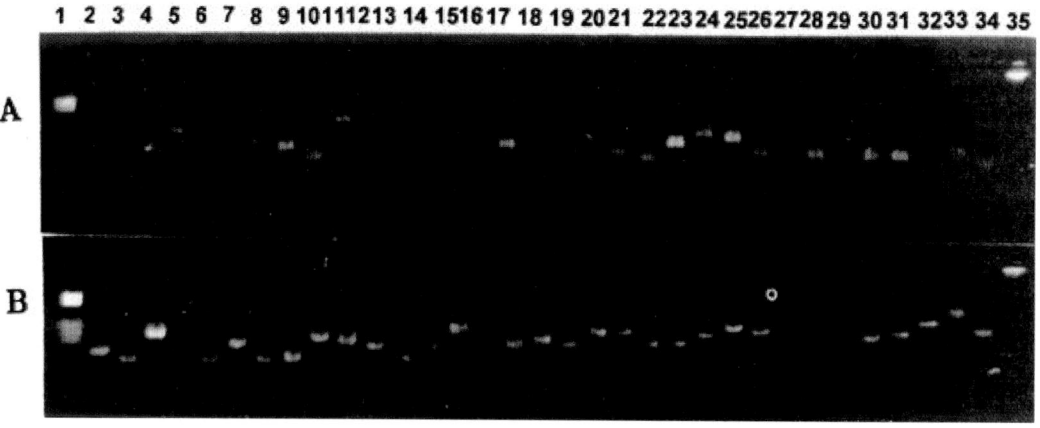

Fig. 3.75　Insert of recombined plasmids. A and B represent different sets of clones
1: pBR322/*Hae* III; 2-34: Inserts of microclones; 35: λ/*Hind* III+*Eco*R I

3.3.5.3　Discussion

Because of the lack of special reagents such as human-mouse hybrid cell lines, it is difficult to verify chromosome specificity of PCR products in plant. Southern blot hybridization with total genomic DNA can only prove their origin from certain plants but not accurately from specific chromosomes(Zhou, 1998). Furthermore, some researchers using FISH found that the hybridization signals were dispersed on each chromosome due to repetitive sequences in higher plants (Hernould, 1997; Vega, 1994).In contrast it is quick and simple to identify where the PCR products come from using specific DNA markers, which have been mapped on designated chromosomes. Cheng *et al.* (1998). succeeded in proving that their PCR products were derived from rice chromosome 5 using STS and microsatellite markers.In prder to verify to verify the chromosome-specific origin of our library, firstly Southern blot analysis with genomic DNA was used to prove our products really contained rice sequences.Then, we carried out hybridization with eight RFLP markers that were site-specific single copy probes. The results indicated that our library not only originated from rice chromosome 4 but also covered a fairly high proparttion of it. The fluorescence in situ hybridization (FISH) using the secondary PCR products of rice chromosome 4 as probe was also carried out and the same result was gotten as Hernould *et al.* (1997), that hybridization signals were distributed on every chromosome of rice including chromosome 4 (data not shown). Maybe the dispersion of in ditu hybridization signals was a characterization of higher plants because of high proportion of reprtitive sequences. However, this is still a problem deserving of further discussion.Individual and multiple chromosome specific unique probe fluorescent images of high quality by FISH are certainly good means to double check the target chromosome, which we would like to perform in the future.

It is relatively simple and efficient to isolate a large number of microclones containing unique sequences by colony hybridization. 58% of the clones in our library can be expected to contain single or low-copy sequences which is similar to the content of such sequence in the rice genome (50%). This result demonstrates that LA-PCR had no preferences for repetitive sequences. Thus, thousands of microclones with unique sequences can be conveniently isolated and used in screening other rice genomic libraries for thecorresponding genes.

Even though RFLP linkage maps of rice have been completed in Japan and America, and the construction of a rice physical map is being carried out, there are still many gaps in the maps. A saturation map of every small genomic region is indispensable for map-based cloning of genes of interest and filling up gaps in a physical map. Currently there are 163,215 ESTs (Expressed Sequence Tags) and about 30000 STS markers located on human chromosome (Schuler, 1996). In the rice genome, another 900 RFLP markers were added to the linkage map after the Japanese RGP published their physical map (Kurata, 1997). However, compared with the progress of the Human Genome Project, DNA markers to the rice linkage map have not progressed so far. Thus, it is necessary to construct DNA libraries of each chromosome and to isolate large numbers of unique sequence probes from defined chromosomes. The probes from our library can be used to directly create "Sequence-Tagged Sites" (STS) for rice chromosome 4. In addition, The probes can be conveniently used to screen YAC libraries with large inserts. If rice chromosome 4 contains 35.4 Mb, it appears feasible to use unique-sequence microclones to isolate 70-100 corresponding YACs with average insert size of 500 kb, to construct contigs for entire chromosome. Our library can provide enough genomic landmarks for high-resolution physical map of the entire rice chromosome 4. The probes can also be used to screen various cDNA libraries for isolating expressed sequences form rice chromosome 4. Most of cDNA sequences have functional significance and can be regarded as candidate genes. Finally, our results will provide a great deal of important molecular information, such as telomere and centromere sequences. These sequences will facilitate locating these sites on physical map and research on the function of plant chromosome.

(Authors: Mao Yingwei, Si Yuan Liang, Wenqin Song, Xiulan Li, Ruiyang Chen. Published in: Cell research, 1998, 8, 285-293)

3.3.6 蚕豆大 M 染色体长臂端部的显微切割与 PCR 扩增

染色体微切与微克隆技术由于可以在分子水平上对特定染色体区域进行基因定位和结构研究，因此，当 Scalenghe（1981）首次在果蝇唾腺染色体上取得成功后，很快将这种技术应用于小鼠、人的染色体上。并利用该技术对染色体特定区域从分子水平上进行详细研究，并得到很多有价值的结果。我国对人类染色体微切与微克隆也已有报道，如：夏家辉（1994）等成功地构建了人类 7 号染色体专特性探针池和 14 个染色体区带特异性探针池。

但植物染色体的显微切割体外扩增与微克隆技术由于染色体同步化和制片困难与动物相比进展缓慢，目前只有在甜菜中与抗线虫有关染色体及 Schondelmaier（1993）在大麦 IHS 染色体、Albani（1993）在小麦染色体上作过报道。我国则尚未开展这方面的研究。

本文作者利用自制的微切割装置对蚕豆大 M 染色体端部 2 μm 片段做了显微切割和采用非特异性 DNA 引物对切割片段进行了 PCR 扩增，并用分子杂交技术证明该方法的可靠性。试图为进一步研究高等植物染色体结构积累基础资料。

3.3.6.1 材料和方法

1. 材料和标本制备

本实验使用的材料为普通蚕豆（*Vicia faba*），采用了改进的染色体制片方法，即蚕豆根尖经低浓度（0.002%）秋水仙水溶液处理后，直接用 70%乙醇固定 2 h 以上，然后涂于 50 mm×24 mm 的盖玻片上，空气干燥，5%的 Giemsa 染色 5～10 min，镜检后进行显微切割。

2. 显微切割与 PCR 扩增

将硅化过的直径为 1 μm 左右的玻璃针在显微操作仪下切取 20 条蚕豆大 M 染色体长臂末端 2 μm（约 0.3 pg DNA）片段，放入微量离心管中，离心后加 20 μL 纯水，4℃过夜。向含有微切片段的 20 μL 纯水中加入 1.5 μL/ μL 蛋白酶 K，3 μL10×Taq buffer，0.5 μL Mg^{2+}（25 mmol/L），纯水 5 μL，至总体积为 30 μL，37℃消化 1 h，70℃保温 15 min，灭活后加入人工合成的非特异性引物 5'-TTGCGGCCGCATrNNNNTC-3'和其他 PCR 试剂 2 μL Taq 10×buffer，4 μL dNTP（2.5 mmol/L），0.4 μL Mg^{2+}（25 mmol/L），2UTaq 酶，补水至 50 μL，按以下条件进行 PCR 体外扩增。前 10 个循环：94℃ 1.3 min→28℃ 2 min→57℃ 2 min；其余 20 个循环：94℃1 min→43℃ 2 min→68℃ 3 min；循环结束后，68℃保温 7min，然后取初级 PCR 产物 5 μL 进行第二次 PCR，反应体系组分如下：10×buffer 5 μL，Mg^{2+}（25 mmol/L）1 μL，引物（0.3 μg/ μL）5 μL，Taq 酶 4U，4×dNTP（各 2.5 mmol）5 μL，加水至 50 μL，石蜡油覆盖后，进行 30 个循环：94℃ 1 min→43℃ 2 min→68℃ 3 min；取次级 PCR 扩增产物进行 3%琼脂糖凝胶电泳分析。

3. 生物素斑点杂交

用随机引物法标记第二次 PCR 产物，取蚕豆基因组 DNA 10 μg，点于尼龙膜上，以 Bio-11-dNTP 标记第二次 PCR 产物作为探针，42℃下杂交 16 h，BSA 封阻，Avidin-HRPO 亲和，DAB，H$_2$O$_2$ 显色，Giemsa 染色。

3.3.6.2 实验结果

从微切后的图中可以明显看出，切割后的染色体长臂短了约 1/5（2 μm）。相当于约

0.3 pg DNA，总共切割 20 个大 M 染色体长臂端部片段，DNA 总量约 6 pg。取显微切割所得大 M 染色体长臂端部 DNA 片段，经 PCR 初级、次级扩增后，其 DNA 含量由微切所得的 6 pg 扩增到 150 μg，其总量足以进行电泳检测、克隆等实验。微切片段经初级、次级两次 PCR 扩增后，PCR 产物经 3%琼脂糖电泳检测结果可以明显看到，经初级、次级两次 PCR 扩增的产物中均可检测到 PCR 产物，尤其在约 2000 bp 附近可看到 1 条明显带纹，这条带纹经 PCR 次级扩增后更为清晰、明显，表明经次级扩增后 DNA 产物明显增多。用生物素标记的第 2 次 PCR 产物作探针，与蚕豆基地组 DNA 杂交，蚕豆基因组 DNA 上有强烈的杂交信号，而作为对照的 PUC19 质粒 DNA 上无杂交信号，此信号表明探针同蚕豆基因组有很高的同源性，从而证明在实验中染色体微切和 PCR 扩增技术的可靠性。

（作者：宋文芹，崔香芹，许文胜，李秀兰，彭永康，陈瑞阳。原载：科学通报，1996，41（4）361-363）

3.3.7 黑杨 1 号染色体的显微分离、克隆和定位

Forest trees are the dominant life form in many ecosystems. They provide structural and functional habitat for two-thirds of the Earth's terrestrial species and contain greater than 90% of all terrestrial biomass（BRADSHAW *et al.*，2000；TAYLOR，2002）. Forests cover about 3.8 billion ha., or 30% of the global land surface. Managed and unmanaged forests throughout the world provide recreational and environmental benefits such as carbon sequestration，renewable energy supplies，watershed protection，improved air quality，biodiversity and habitat for endangered species（Proposal to sequence Populus Genome white paper, http: //genome.jgi-psf.org/Poptr1/Poptr1.info.html）.

The necessity for model species of plants is well recognized and in this role，*Arabidopsis thaliana* has gained a supreme acceptance amongst plant scientist. While many aspects of tree biology are common to all plants，and hence can be studied in very tractable model species such as Arabidopsis thaliana，some unique facets of tree anatomy and physiology must be investigated in trees themselves. The genus Populus has been adopted as a model for forest tree genetics. Populus，one of only two genera in the family Salicaceae，first occurred in the fossil record ca. 60 MYBP. The genus Populus is especially well suited to serve as the model genome for trees because of the following reasons：1）a small genome size– the haploid genome size is ca. 480 ± 20Mbp，2）rapid juvenile growth，3）ease of clonal propagation and 4）high-throughput transformation and in vitro propagation（BRADSHAW *et al.*，2000；TAYLOR，2002；BRUNNER *et al.*，2004）.

A chromosome microdissection and microcloning technology was developed in 1981（SCALENGHE *et al.*，1981）. Subsequently，it has evolved into an efficient tool for generating chromosome specific DNA libraries of many species（THALHAMMER *et al.*，2004）. This procedure could also be used for the generation of chromosome specific molecular markers， for isolation of chromosome specific sequences using specific PCR primers or for the indirect physical mapping of low/single copy sequences. With the improvements in the techniques of microdissection and the polymerase chain reaction(PCR), this strategy was simplified largely. Construction of chromosome specific libraries is a potential strategy for the construction of high-density genetic linkage maps of individual chromosomes and the comprehensive analysis of genomes in forest trees. 'Chromosome painting' refers to the hybridization of fluorescently labeled chromosome specific, composite probe pools to cytological preparations. Chromosome painting allows the visualization of individual chromosomes in metaphase or interphase cells and the identification of both numerical and structural chromosomal aberrations with high sensitivity and specificity (RIED et al., 1998). These techniques would also facilitate genomics research in forest trees. However, the application of these techniques to forest trees has seen little reports, largely due to the difficulty of chromosome preparation.

The present study was performed to establish a method for single chromosome microdissection, cloning and painting in forest plants using poplar (*Populus tremula*) as a model. A single chromosome 1 was microdissected from the metaphase spreads of poplar root-tip cells with fine glass needles. The dissected chromosomes were amplified *in vitro* by the *Sau*3A linker adaptor-mediated PCR technique. A microclone library of chromosome 1 was constructed. The amplification products were used as a complex probe mixture and hybridized on the metaphase spreads of poplar.

3.3.7.1 Materials and Methods

1. Plant materials

Populus tremula was used as the experimental material in this study. The branches were collected from the plantation in The Research Institute of Forestry, The Chinese Academy of Forestry.

2. Preparation of mitotic chromosomes

Chromosome samples were prepared by wall degradation hypotonic method according to CHEN et al. (1979) with minor modifications. In brief, root tips were removed and immersed successively in saturated paradichlorobenzene for 3 h, in double-distilled water for 30min, mixed with 2.5% (w/v) cellulase and pectinase (Sigma, Germany) for 30 min, rinsed in double-distilled water for 15 min, and finally fixed in 70% ethanol for 5 min. Put on the fixed material on the sterile coverslip (22 × 60-mm) and placed drop of 70% ethanol to them, using a microscope slide as a carrier to stabilize the coverslip. Tore the material and removed the supernatant. Added two drops of 70% ethanol on the slide, and then dried with hot air. Twenty slides were produced and counted per tree. Metaphase spreads for FISH were prepared on common microscope slide. The quality of chromosome spreading was evaluated under a microscope (80i, Nikon, Japan). Slides with good quality were used for karyotype analysis, chromosome microdissection and FISH.

3. Karyotype analysis and identification of chromosome 1

Karyotype analysis was carried out using photographs of clear and well-spreaded metaphase chromosomes according to LI and CHEN (1985). Chromosome 1, the largest chromosome, was identified under a microscope with 1000× magnification (oil objective) based on the result of the karyotype analysis.

4. Microdissection of chromosome 1

Each air-dried chromosome specimen was immediately used for microdissection. In the chromosome complements prepared by enzymatic maceration, chromosome 1 was unambiguously identified by its size and shape (the largest chromosome). The target chromosome was isolated using an traditional light microscope (BH-2, Olympus, Janpan) equipped with a micromanipulator (MMO-203, Narishige, Japan) and transferred by a fine glass needle pulled by PC-10 puller (Narishige, Japan) into a 0.5 ml tube according to Li et al. (1998).

5. Amplification by *Sau*3A linker adaptor-mediated PCR amplification (LA-PCR)

The *Sau*3A linker adaptors, with the sequences 5'-CGGGAATTCTGGCTCTGCGACATG-3' and 5'-GATCCATGTC-3' were prepared as described by DENG et al. (1992). Isolated chromosomal DNA was treated in 10 μl of 50 ng/μl proteinase K (Merck, Germany) solution at 37℃ for 2h. The proteinase K was then inactivated at 65℃ for 20 min. The chromosomal DNA was digested by *Sau*3A (0.002U in 1 × T4 ligase buffer, Takara, Japan) at 37℃ for 2h. The *Sau*3A was inactivated at 65℃ for 20 min, and 20 uM of prepared *Sau*3A linker adaptors and 1U of T4 ligase (Takara, Japan) were added. The ligation between the adaptors and digested chromosomal DNA was performed at 16℃ for 16 h. The first round of PCR was carried out in the same tube by adding 10 μl of 10 × Taq buffer, 6 μl of 25 mM $MgCl_2$, 2 μl of 10mM dNTPs, 6 μl of 10 μM 24-mer primer, 2.5U of Taq DNA polymerase (Takara, Japan), and distilled water to 100 μl. PCR amplifications were performed in a thermal cycler (MG 5331, Eppendorf, Germany) using the following programme: after denaturation at 94℃ for 5 min, amplification was performed with 35 cycles of 1 min at 94℃, 1 min at 56℃, and 2.5 min at 72℃, followed by a final extension at 72℃ for 10 min. The second round of PCR was carried out using 2 μl of the firstround products as template. The method was the same as described above, except that 20 cycles of amplification were carried out. To monitor possible extraneous DNA contamination, we maintained a negative control (no template DNA) and a positive control (1 pg genomic DNA as template) throughout the whole process.

6. Southern blot hybridization analysis

The poplar genomic DNA was isolated from leaf tissue using CTAB method according to MURRAY and THOMPSON (1980). DNA molecular weight was checked for quality and quantity by agarose gel (0.8%) electrophoresis and fluorometry (ND-1000, NanoDrop, America). Appropriate amounts of *Eco*R I (Takara, Japan) digested genomic DNA, and the two rounds PCR products from chromosome and controls were separated by electrophoresis (0.8% agarose) and transferred onto nylon membranes (Pall, America). After the poplar genomic probes had been labeled with digoxygenin (DIG) -11-dUTP (Roche, Germany), Southern hybridization and detection were performed following the instructions of the Roche

DIG High Prime DNA Labeling and Detection Starter Kit (Roche, Germany).

7. Construction and characterization of a single chromosome DNA library

The second LA-PCR products from the isolated poplar chromosome 1 were purified using a DNA purification kit (Sangon, China). A 1 μl aliquot of the purified DNA (about 100 ng) was ligated into T-easy vector (Sangon, China) in a 10 μl reaction volume at 16℃ for 16 hours. A 1 μl aliquot of the ligation mixture was used to transform *Escherichia coli* strain DH5a competent cells by heat shock. After co-cultivation in luria broth (LB) medium at 37℃ for 1 h, one-tenth (v/v) of the cell solution was plated onto LB plates containing 100 ng/μl of ampicillin, 50 ng/μl of X-Gal and 50 ng/μl IPTG for selection of white colonies. 160 randomly selected recombinant plasmids (white colonies) were isolated by alkaline lysis; the insert DNA was amplified using M13 forward/reverse primer. The sizes of the inserts were estimated by extrapolating to molecular weight standards on a 1.5% agarose gel. To determine the low/unique or high copy (repetitive sequence) nature of the inserts, DIGlabeled genomic DNA of poplar was hybridized to nylon membranes (Pall, America) containing recombinant clones by dot blotting.

8. Fluorescence in situ hybridization (FISH)

The second round PCR products were labelled with DIG-dUTP (Roche, Germany) by randomly primed DNA synthesis, being used as a complex probe mixture for FISH on the metaphase spreads of poplar. FISH was carried out as described by QI *et al.*(2002). The slides were examined with a Nikon 80i fluorescence microscope (Nikon, Japan). FISH images of suitable metaphases were acquired and stored with a cooled charge-couple device (Spot Diagnostic, America) equipped with the version 4.0.8 Spot Rtke soft (Spot Diagnostic, America).

3.3.7.2 Results

1. Chromosome preparation and karyotype analysis

Using root-tip as source material and wall degradation hypotonic method (CHEN *et al.*, 1979), we succeeded in preparing good-quality slides of chromosomes in *Populus tremula*. As shown in Fig. 3.76A, chromosomes were spreaded evenly on the slide with a low background. The somatic chromosome number was $2n = 2x = 38$. Twenty slides with good quality were used for karyotype analysis. The result was shown in Fig. 3.76B.

2. Individual chromosome 1 identification and microdissection

The precise identification of microdissected chromosomes was an indispensable prerequisite for the reliability of results obtained in this study. According to the established standard karyotype, chromosome 1, the largest submetacentric chromosome, could be identified at prometaphase or metaphase. After being identified and marked, the target chromosome was successfully isolated by a fine glass needle using an improved chromosome microdissection method in our lab (Li *et al.*, 1998) (Fig. 3.77).

3. *Sau*3A linker adaptor-mediated PCR (LA-PCR) amplification of chromosome 1 DNA

In this study, a single chromosome was collected separately into a tube and was used for *Sau*3A LA-PCR. After two sucessive rounds of amplification, the microdissected chromosomes all yielded products that were observed as bright smear electrophoretic bands of DNA with a range from 200 bp to 2,500 bp on a 1% agarose gel, with most fragments being concentrated between 300 bp and 1,200 bp (Fig. 3.78, lane 2). At the same time, the positive control with 1pg of poplar genomic DNA as template for PCR generated products with sizes between 200 bp to 4,000 bp (Fig. 3.78, lane 3).

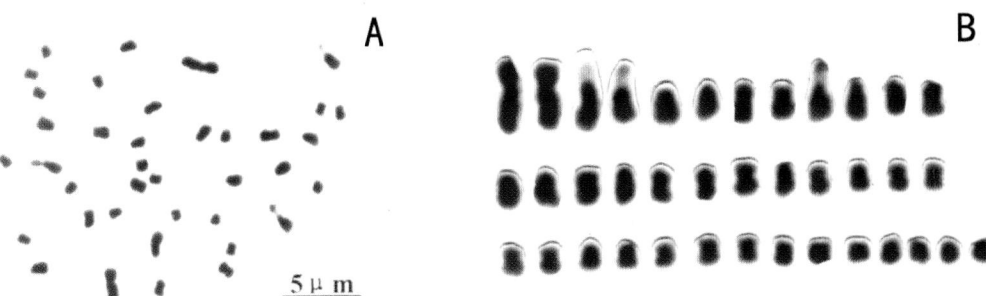

Fig. 3.76　Image of *Populus tremula* somatic metaphase chromosomes (A) and the karyotype (B)

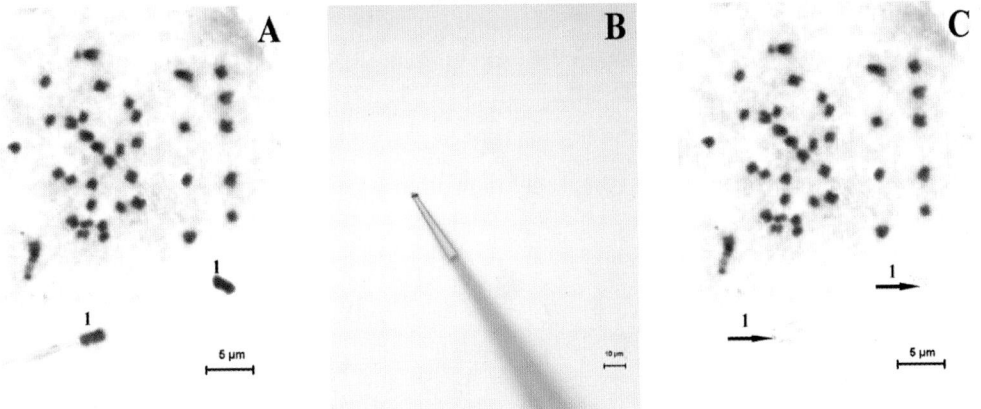

Fig. 3.77　Procedure of isolation of an individual chromosome 1 in poplar by micromanipulator
A: A Mitotic metaphase image before microdissection of the chromosome 1 (arrow), B: The target chromosome adhering to the tip of a glass needle, C: The individual chromosome 1 1 in A (arrow) was removed from the cell, respectively

As a negative control for monitoring possible contamination with DNA, a sample without template DNA was set up during all stages of the microdissection and amplification procedures. None of the products was amplified from the negative control (Fig. 3.78, lane 1). The results indicate that DNA from chromosome 1 was amplified successfully without contamination of exogenous DNA. Southern hybridization with DIG-labeled genomic DNA confirmed that the products were amplified from the poplar genome (Fig. 3.78).

4. Construction of the chromosome 1 specific DNA library

The second round LA-PCR products from the single chromosomes 1 were used to prepare ligation mixtures for transformation. Approximately 3×10^5 recombinant clones (white colonies) were obtained. 160 randomly selected white colonies were selected for

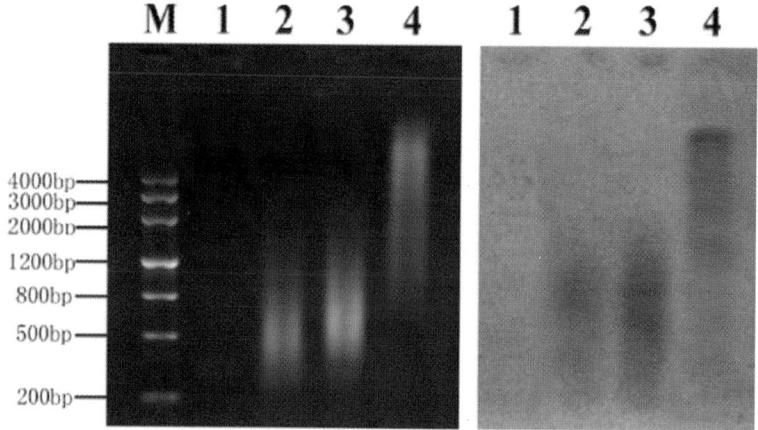

Fig. 3.78 Linker adaptor-mediated polymerase chain reaction (LA-PCR) amplification with microdissected chromosomes 1 and southern blot hybridization analysis. Left: The LA-PCR products of microdissected chromosomes: the DNA molecularweight mark lane M, the negative control (lane 1), the single chromosome 1 as DNA template (lane 2), the positive control (lane 3), EcoR I -digested genomic poplar DNA (lane, 4). Right: Southern blot hybridization of the PCR products with DIG labeled genomic poplar DNA

further analysis The length of the inserts ranged mainly from 230 to 2,200 bp, with an average of 800 bp, as estimated by 1.5% agarose gel electrophoresis (Fig. 3.79). The colony microarray hybridization analysis were carried out to estimate the copy number of the inserts. of the 160 clones examined, 40% represented low/unique copy sequences, while 60% of the clones were medium or highly repetitive sequences (Fig. 3.80).

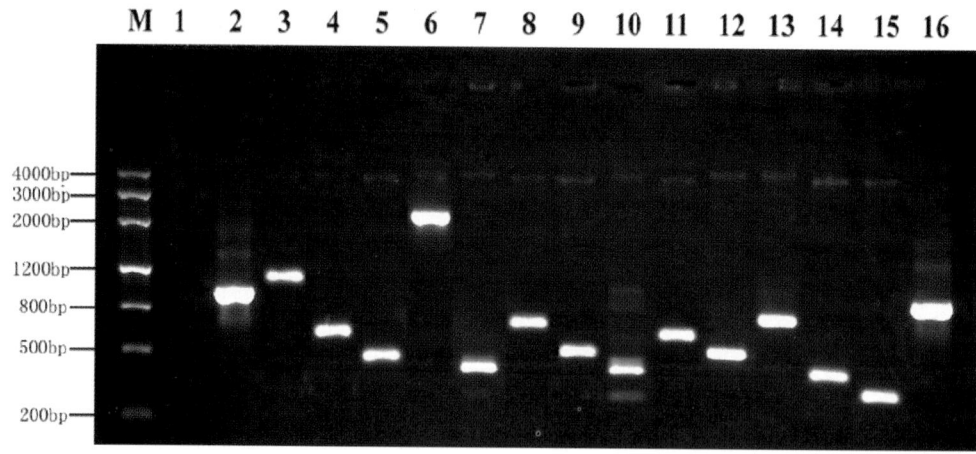

Fig. 3.79 Partly electrophoresis result of amplified cloning fragments by PCR using M13 forward reverse primers

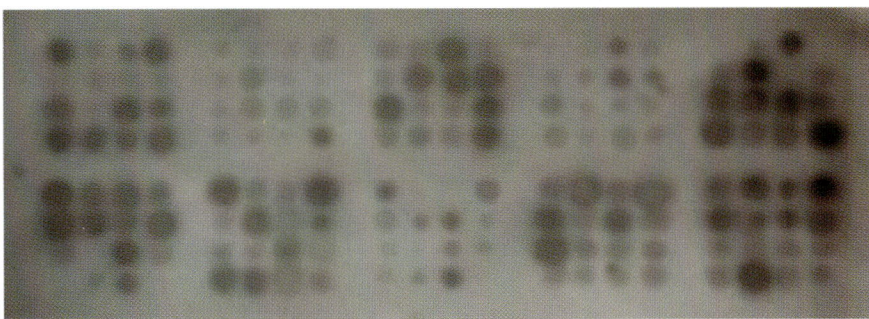

Fig. 3.80 Result of bacterial colonies hybridization. The 160 randomly selected recombinant clones were positioned on a membrane in a grid microarray. The colony microarray was hybridized with a DIG-labeled genomic poplar DNA probe

5. Chromosome painting of chromosome 1

The LA-PCR products originating from individual microdissected chromosome 1 were labeled with DIGdUTP, and hybridized to mitotic metaphase spreads in the absence of a competitor. Signals were mainly observed uniformly in all chromosomes 1 (Fig. 3.81). At the same time, some signals were also observed on the terminal and centromeric regions of other chromosomes (Fig. 3.81). Then, unlabeled shared total DNA of poplar was added as a competitor. FISH experiments, which included different probe/competitor combinations, were performed. In all cases, the pattern of signals distributing had not greatly change, although weaker than without pre-hybridization with competitor DNA (even competitor DNA in 100-fold and excess for a 2 h prehybridization) (data not shown).

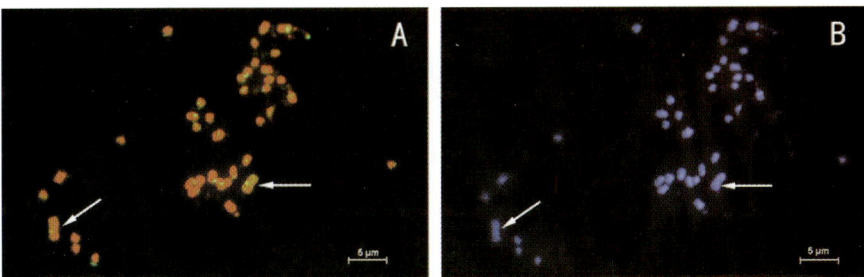

Fig. 3.81 Result of chromosome painting using microdissection amplification products as prob
A: Metaphase spread of poplar hybridized with labeled LA-PCR products of chromosome 1 without blocking DNA. B: The same chromosomes as shown in A after staining with 4', 6-diamidino-2-phenylindole. Arrows indicate the chromosome 1

3.3.7.3 Discussion

1. Preparation of mitotic chromosomes

A good-quality chromosome spread was essential in this method and was technically demanding to perform. This step may be a bottleneck hindering this type of research in poplar because they usually have thick cell walls and were highly lignified, making chromosome spreading difficult. Using root-tip as source material and the wall degradation hypotonic method according to CHEN *et al.*(1979)with minor modifications, we succeeded in preparing high quality slides of chromosomes of *Populus tremula*.

The period of chromosome fixation in acetic acid should be as short as possible, because acetic acid damages DNA by depurination. An extended period of fixation in acetic acid could result in relatively short(50–100 bp long)chromosomal DNA fragments(Houben *et al.*, 2002). Therefore, we avoided fixing root-tip in fixation solution (methanol: acetic acid = 3: 1) and used 70% ethanol for short time fixation in chromosome preparation. The chromosome spreads were then prepared in 70% ethanol instead of 45% acetic acid. These steps avoided acid depurination and mechanical damage of chromosomal DNA.

2. Identification and microdissection of target chromosomes

Correct identification of the target chromosomes is also an important step required for single chromosome microdissection. Generally, this is achieved by choosing plants with target chromosomes bearing prominent morphological features, such as the largest or smallest chromosomes (Fukui *et al.*, 1992; Chen and Armstrong, 1995; Stein *et al.*, 1998; Huang *et al.*, 2004), or by using specific plant species, for example, monosomic lines (Jung *et al.*, 1992; Vega *et al.*, 1994), or tritelosomic lines (Schondelmaier *et al.*, 1993). In *Populus tremula*, the chromosome 1, the largest submetacentric chromosome, could be unambiguously identified at prometaphase or metaphase. Only metaphases with unequivocally identifiable and spatially well-separated chromosomes were used for microdissection.

However, because of the small sizes of chromosomes in poplar, it is difficult to identify the target chromosome correctly under a 40 x objective. In this paper, we used a technique that was improved by our lab.(Li *et al.*, 1998)for the precise chromosome microdissection by ordinary light microscope. It had not only solved the difficulty of chromosome microdissection by oil objective, but achieved the microdissection on high magnifying multiple(magnification 1,000x~1,500x), and greatly improved the accuracy of chromosome microdissection.

Generally, the sterility requirements to prevent contaminations have to be the same in PCR as for tissue culture. Exceptional care must be taken in preparing reagents for chromosome microdissection to avoid contaminant DNA being amplified (Houben *et al.*, 2002). Each working solution is prepared, divided into singleuse aliquots, tested, and used for microdissection only if satisfactory. When preparing solutions, wear gloves and, to the extent it is possible, conduct all manipulations in a laminar flow hood. Irradiate plastic and glassware, buffers and stock solutions (except nucleotides, primers, linker-adaptors and enzymes) with UV light (260 nm wavelength) for 12 hours.

3. Construction of chromosome specific DNA library

In this paper, We have successfully constructed a plasmid library from a single microdissected poplar chromosome 1 by used LA-PCR. In previous studies, PCR-mediated cloning was usually performed based on DNA templates provided by several isolated target chromosomes (Houben *et al.*, 2002). However, microdissection of specific chromosomes as small as those in poplar is complicated and elaborate work, requiring skill and experience, hence it is impracticable to isolate many copies of the same chromosome from different metaphases. To tackle this problem, we tried to acquire DNA fragments by conducting PCR based on a single copy of a target chromosome template. Conditions for amplification of chromosomal DNA were optimized to detect and amplify DNA between 50–100 femtogram. Our results indicate it is feasible to amplify DNA fragments from a single chromosome with small size for constructing the chromosome specific DNA library.

To date, the cloning efficiency of PCR products from microdissected plant chromosomes has ranged from 2×10^4 (JUNG et al., 1992) to 5×10^5 (Chen and Arnstrong, 1995), and then to 4×10^6 (Stein et al., 1998). All members of the *Populus* genus have a genome contained 19 nearly identical, metacentric chromosomes, a nuclear content of 2C= 1.2 pg and the haploid genome size of poplar is only ca. 480 Mbp (Bradshaw et al., 2000; Taylor, 2002). Since chromosome 1 covers about 5% of the genome per haploid (Zhang et al., 2005), it is deduced to contain 0.06 pg or approximately 24 Mbp. The microclone library of chromosome 1, in this study, included 3×10^5 clones, with an average insert size of 800 bp. For a 99% probability that every sequence of chromosome 1 is represented in at least one recombinant clone, 1.4×10^5 clones of chromosome 1 are needed (Sambrook et al., 1989). So this library maybe can represent the whole of chromosome 1 of poplar, in the main. However, when linker adapter is used for fragment of one restriction enzyme, theoretically there is a possibility of causing bias. Especially the case of highly repetitive sequences the bias became larger. So, till now, none of the chromosome microclone libraries reported could reach a 100% coverage rate. The way to obtain a genomic library that covers a whole target chromosome would be to employ different combinations of restriction enzymes, adaptors or primers to generate libraries with different complements of DNA sequences (Jung et al., 1992). The quality of the chromosome library, on the other hand, is influenced by the enrichment of unique/low copy or chromosome specific sequences. The poplar nuclear genome contains approximately 60%–80% repetitive DNA sequences (Flavell et al., 1974, 1980; Houben et al., 2002). In the library of chromosome 1, the frequency of repetitive inserts was 60%, while that of low/single copy sequences was 40%. Similar results were obtained by Chen and Armstrong (1995), Liu et al. (1997), Stein et al. (1998) and Zhou et al. (1999).

4. Feasibility of chromosome painting

The concept of chromosome painting was first introduced in 1988 (Lichter et al., 1988; Pinkel et al., 1988). Chromosome painting has over the last few years become an established procedure in laboratories working with mammalian chromosomes (Antonacci et al., 1995). To ensure specific hybridization to related chromosome segments, repetitive sequences need to be excluded from the hybridization process by, for example, pre-hybridization with a large excess of unlabelled total genomic DNA or the fast-reassociating (Cot-1) fraction of genomic DNA (Houben et al., 2002).

However, chromosome painting in plants is relatively underdeveloped, although this technique would be useful for evolutionary studies and plant breeding. A comprehensive study by Fuchs *et al.* (1996) outlines past endeavors to paint chromosomes from various plant species with large genomes (2C DNA content of 11 to 40 pg, such as barley, wheat, *Vicia faba*, *Picea abies*). No specific painting of the chromosomes was obtained although a number of different approaches, including pre-hybridization with a large excess of total unlabeled genomic DNA, were tested.

Successful chromosome painting in plants with relatively large genomes has been reported for B chromosomes of *Secale cereale* (Houben et al., 1996), *Brachycome dichromosomatica* (Houben *et al.*, 1997) and the Y chromosome of *Rumex acetosa* (Shibata *et al.*, 1999). However, the painting of the described B or Y chromosome was possible because of enrichment for chromosome specific repetitive sequences, rather than the chromosome specific low- and single-copy sequences which are responsible for painting mammalian chromosomes (Houben *et al.*, 2002).

These results were similar to our present study. Hybridization with or without competition by unlabeled genomic DNA showed signals on all the metaphase chromosomes. Our painting pattern indicated that the amplification products of chromosome probably failed to specially label chromosome 1 owing to the presence of a large number of chromosome 1 nonspecific repetitive sequences, which were dispersed at high frequency in the genome. This conclusion accorded with component of chromosome 1 specific DNA library, constructing in this study, in which frequency of repetitive inserts was 60%, while that of low/single copy sequences was 40%. So, we could consider that Chromosome 1 contain large amounts of chromosome specific, repetitive DNA sequences and dispersed repetitive DNA sequences, common to all chromosomes.

Without doubt, with the method described here, the high resolution obtainable by the microdissection of poplar chromosomes could provide a valuable tool for constructing a genetic map, the physical mapping of chromosomes and the isolation of useful genes, as well as improving genetic mapping and comparative genomic researches in poplar.

(Authors: Y.Zhang, S.G.Zhang, L.W.Qi, B.Liu, J.M.Gao, C.B.Chen, X.L.Li, and W.Q.Song.Published in: Silvae Genetica 2005, 54 (4) 145-256)

3.3.8 银杏W染色体显微分离及荧光原位杂交分析显示性染色体不同杂交带型

Ginkgo biloba is a dioecious gymnosperm species with both male and female plants having $2n=2x=24$ chromosomes, consisting of four metacentrics and twenty subtelocentrics. Several studies on the sex determination system and sex chromosomes of ginkgo have appeared since the 1950s. Newcomer(1954)and Chen *et al.*(1987)both reported the ZW-type sex chromosome system might be present in ginkgo. Chen *et al.* (1987) found that the size of the Ag-NOR on chromosome 1 from females is different while it is the same from the male.

So far, however, no further study on the chromosome 1 has been reported.

The direct strategy for isolating sequences from chromosomes of interest is to separate them by a flowsorting procedure, or by microdissection. A chromosome microdissection technology was developed in 1981 (Scalenghe et al., 1981). Subsequently, it has evolved into an efficient tool for generating chromosome specific DNA libraries of many species (Ponelies et al., 1997; Thalhammer et al., 2004).

Chromosome painting refers to the hybridization of fluorescently-labeled, chromosome specific, composite probe pools to cytological and structural chromosomal aberrations with high sensitivity and specificity (Ried et al., 1998). The concept of chromosome painting was first introduced in 1988 (Lichter et al., 1988; Pinkel et al., 1988). It has over the last few years become an established procedure in laboratories working with mammalian chromosomes (Antonacci et al., 1995). In plants, however, chromosome painting is relatively underdeveloped. In plants, to ensure specifi chybridization to related chromosome segments, repetitive sequences need to be excluded from the hybridization process by, for example, blocking with a large excess of unlabeled total genomic DNA or the Cot-1 fraction of genomic DNA (Houben et al., 2002). Painting of sex chromosomes has been performed in *Rumex acetosa* by Shibata et al. (1999) and in *Silene latifolia* by Hobza et al. (2004). Hobza et al. (2004) used a modified FAST-FISH protocol based on a short hybridization time combined with a low concentration of probe and succesfully distinguished the sex chromosomes by differential labelling patterns.

Here, by applying microdissection and painting the W chromosome, we found a different labelling region, meaning a different sequence structure, there on the sex chromosome.

3.3.8.1 Materials and Methods

Plant materials and chromosome preparation Root tips and tender buds of 15 male and 12 female ginkgo plants were used in this study. Slides were prepared by cell wall degradation hypotonic method according to Chen et al. (1979) with minor modifications. In brief, root tips were removed and immersed successively in saturated para-dichlorobenzene for 3 h, rinsed in double-distilled water for 30 min, treated with 2.5% (w/v) cellulose and pecitinase (Sigma, Germany) for 30 min, rinsed in double-distilled water for 15 min, and fi nally fixed in 70% ethanol for 5 min. The fixed material was put on the sterile coverslip (22×60-mm) with a drop of 70% ethanol, using a microscope slide as a carrier to stabilize the coverslip. The material was torn into fine pieces and the debris was removed. Two drops of 70% ethanol were added on the slide and then dried with hot air. Metaphase spreads for FISH were prepared on a common microscope slide.

1. Microdissection and LA-PCR amplification

Each air-dried chromosome specimen was Immediately used for microdissection. The target chromosome, the W chromosome, was isolated using an traditional light microscope (BH-2, Olympus, Japan) equipped with a micromanipulator (MMO-203, Narishige, Japan) and transferred by a fine glass needle pulled by an OC-10 puller (Narishige, Japan) into a 0.5 ml tube according to Li *et al.* (1998).

The *Sau*3A linker adaptors, with the sequences 5'-CGGGAATTCTGGCTCTGCGACATG-3' and 5'-GATCCATGTC-3' were prepared as described by Deng *et al.* (1992). Isolated chromosomal DNA was treated in 10 μl of 50 ng/μl proteinase K (Merck, Germany) solution at 37℃ for 2 h. The proteinase K was then inactivated at 65℃ for 20 min. The chromosomal DNA was digested by *Sau*3A (0.002U in 1×T4 ligase buffer, Takara, Japan) at 37℃ for 2 h. The *Sau*3A was inactivated at 65℃ for 20 min, and 20 μM of prepared *Sau*3A linker adaptors and 1U of T4 ligase (Takara, Japan) were added. The ligation between the adaptors and digested chromosomal DNA was performed at 16℃ for 16 h. The first round of PCR was carried out in the same tube by adding 10 μl of 10×Taq buffer, 6 μl of 25 mM $MgCl_2$, 2 μl of 10 mM dNTPs, 6 μl of 10 μM 24-mer primer, 2.5 U of Taq DNA polymerase (Takara, Japan), and distilled water to 100 μl. PCR amplifications were performed using the following programme: after denaturation at 94℃ for 5min, amplification was performed with 35 cycles of 1 min at 94℃, 1 min at 56℃, and 2.5 min at 72℃, followed by a final extension at 72℃ for 10 min. The second round of PCR was carried out using 2 μl of the first round products as template. The method was the same as described above, except that 20 cycles of amplification were carried out. To monitor possible extraneous DNA contamination, we maintained a negative control (no template DNA) and a positive control (1 pg genomic DNA as template) throughout the whole process.

2. Southern blot hybridization analysis

The ginkgo genomic DNA was isolated from leaf tissue using the CTAB method according to Murray and Thompson (1980). DNA molecular weight was checked for quality and quantity by agarose gel (0.8%) electrophoresis and fluorometry (ND-1000, NanoDrop, America). Appropriate amounts of *Sau*3A digested genomic DNA of female and male plants, and the second rounds PCR products from chromosome and controls were transferred onto nylon membranes (Pall, American) after 0.8% agarose gel electrophosis and hybridization with DIG-labeled ginkgo genomic DNA at 42℃ for 16 h. The membranes were washed with 0.2×SSC containing 0.1% SDS at 65℃ for 30 min. Labeling and detection were performed following the instruction of the Roche DIG High Prime DNA Labeling and Detection Starter Kit II (Roche, Germany).

3. Fluorescence in situ hybridization (FISH)

FISH was carried out as described by Qi *et al.* (2002) with minor modifications. The second round PCR products were labeled with DIG-dUTP (Roche, Germany) by randomly-primed DNA synthesis. Hybridization buffer contained 50% deionized formamide; 2×SSC; 50 Mm sodium phosphate, pH 7.0; 5% dextran sulfate; 10 ng μl-1 probe and 200 ng μl-1 unlabelled total genomic DNA. The probe and the unlabelled total genomic DNA were mixed and denatured at 94℃ for 10 min before being used. Slides with metaphase spreads were treated with 70% deionized formamide in 2×SSC at 70℃ for 2 min. Denatured hybridization buffer (10 μl) was then applied to the slides, which were incubated at 37℃ for 2 h. Finally, the slides were washed with 30% deionised formamide in 2×SSC at 37℃ for 5 min and twice in 2×SSC at 37℃ for 5 min. Metaphase spreads were counterstained with 100 ng ml-1 4', 6-diamidino-2-phenylindole (DAPI). Fluorescence signal was detected using anti-DIG-fl uorescent-conjugate (Roche). The hybridization signals were visualized and recorded using Nikon 80i fluorescent microscope and a cooled CCD camera and were then processed using Adobe Photoshop.

3.3.8.2 Results

Individual W chromosome microdissection Using root tips and tender buds as source material an cell wall degradation hypotonic method, we succeeded in preparing good-quality slides of chromosomes in ginkgo.As shown in Fig. 3.82A, chromosomes were spread evenly on the slide with a low background. The somatic chromosome number was $2n=2x=24$, consisting of fourmetacentrics and twenty subtelocentrics. Chromosome 1, which harbors a satellite, was the biggest chromosome. In male, the satellites of chromosome 1 are homoeomorphic, while in female they are heteromorphic, and one is apparently bigger than the other (Fig. 3.82A). The chromosome 1 with the bigger satellite was microdissected from the metaphase spreads of female ginkgo root-tip cells with fine glass needle controlled by a micromanipulator (Fig. 3.82B) and then used for two rounds of LA-PCR amplification. The southern blot analyses of the second round PCR products confirmed that the products were amplified from the ginkgo genome (Fig. 3.83).

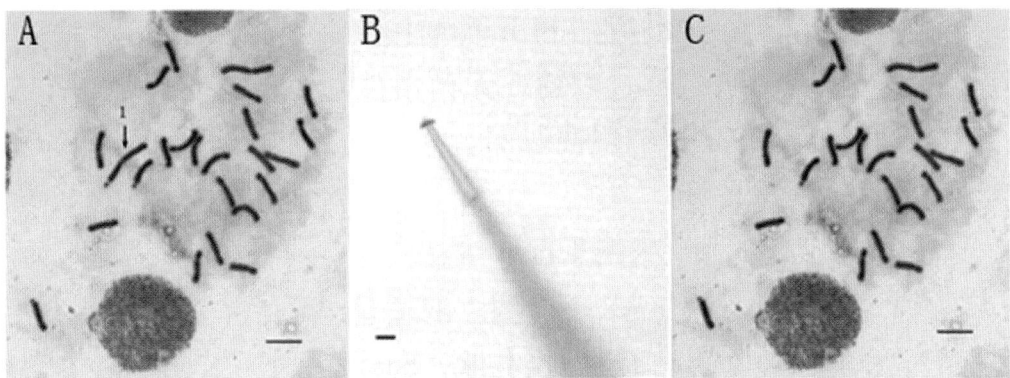

Fig. 3.82 Isolation of an individual chromosome 1 in ginkgo by micromanipulator
A: Mitotic metaphase spread of ginkgo before microdissection. A suitable chromosome 1 with big satellite was found under the inverted microscope (arrow); B: The target chromosome adhering to the tip of a glass needle; C: The target chromosome was removed from the slide. Bar=5, 10, 5 μm, respectively

Blocking with twenty-fold excess of unlabeled genomic DNA, the DIG-labeled second-round LA-PCR products originating from individual microdissected chromosome 1 were hybridized to mitotic metaphase spreads. Signals were mainly observed along the entire

W chromosome while along about 1/2 length at the end of the long arm of the Z chromosome, the signals were weaker than at other parts of the Z chromosome. At the same time, some signals were also observed on the terminal, centromeric, or other regions of other chromosomes (Fig. 3.84).

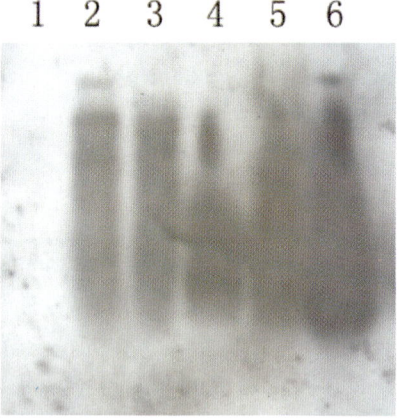

Fig. 3.83   Southern blot hybridization analysis of second-round LA-PCR products with DIG-labeled genomic ginkgo DNA. Lane 1 was the negative control (no DNA template in LA-PCR), lane 2-3 were the Sau3A digested total genomic ginkgo DNA of female and male respectively, lane 4-5 were the LA-PCR products, and lane 6 was the positive control (genomic DNA as template in LA-PCR).

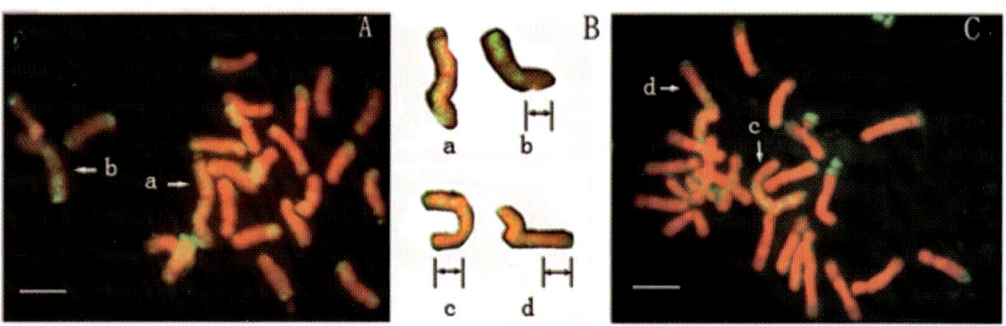

Fig. 3.84   Chromosome painting patterns on female (A) and male (C) metaphase chromosomes of ginkgo using DIG-labeled secondround LA-PCR products. chromosome marked as "a" was chromosome 1 with big satellite while "b, c, d" were chromosome 1 with small satellite. FISH signals on the segments marked with arrows in B were apparently weaker than other segments of chromosome 1.Bar=5 μm. Chromosome painting of W chromosome

3.3.8.3   Discussion

In plants, no specific painting of the chromosomes was obtained although a number of different approaches, including pre-hybridization with a large excess of total unlabelled genomic DNA, were tested. Several experiments with DOP-PCR amplified probes from microdissected chromosomes or chromosome regions hybridized to metaphase chromosome complements, unequivocally revealed dispersed hybridization signals on all chromosomes in the case of *Vicia faba*, *Hordeum vulgare*, *Triticum aestivum*, *Picea abies* and *Petunia hybrida* (Fuchs et al., 1996). In our study, we modified the FISH procedure including changing the hybridization time and the amount of DNA probe or using unlabelled genomic DNA for blocking (data not shown). The results were still similar to the cases described above that all chromosomes harbor signals. However a clear difference emerged between the intensity of the signal on the chromosome of probe origin and that of the other chromosomes. Such uniform hybridization patterns were found irrespective of the blocking conditions and the concentrations of genomic DNA used in these experiments. The number of repetitive sequences in plants is too large for efficient blocking by conventional prehybridization procedures. In the nuclear genome of higher plants, especially in the repetitive sequence, most cytosine are methylated. The unmethylated loci maily disperse in the low and single copy sequence regions. Methylation sensitive restriction enzyme cannot recognize the methylated cytosine loci. If it could, then low and single copy sequences would be relatively rich in the products. Many researchers used the methylation sensitive restriction enzyme such as *Sau*3A to digest the DNA. The LA-PCR method also can concentrate the low copy sequences, and it can amplify longer fragments than DOP-PCR.

In plants, chromosome painting is possible only on specialized chromosomes, such as the B and Y chromosome or on chromosomal regions that contain specific highly repetitive sequences. Painting of sex chromosomes has been successfully performed in *Rumex acetosa* by Shibata et al. (1999) and in *Silene latifolia* by Hobza et al. (2004). In *Rumex acetosa*, strong signals were aberved on the Y1 and Y2 chromosomes, and weak signals were also observed on the X chromsome and autosomes. In *Silen latifolia*, Hobza et al. (2004) found that the X chromosome probe revealed a clear signal on the entire X chromosomes while the signal on the other chromosomes, including the Y, was of lower intensity. Similar results were obtained with the Y chromosome probe when the Y chromosome was strongly labeled. The different labeling patterns of sex chromosomes showed that the composition of the DNA sequences in the X and Y chromosome differs. In our study, different labelling patterns have been shown on the sex chromosomes of *Ginkgo biloba*. Along about 1/2 length at the end of the long arm of the Z chromosome, the signals were weaker than at other parts of chromosome1. This indicated that the composition of DNA sequences in the W chromosome and the Z chromosome might differ, especially in the parts where the FISH signal intensity differed. We believed this part on W chromosome became specialized as a result of the accumulation of chromosome-specific repetitive sequences in the process of sex chromosome evolution. It would harbor sex-specific sequences and might be a non-recombining region. Further study, such as construction of a W chromosomespecific genomic library and isolation of female specifi cDNA markers would reveal more information about the evolution and divergence of the sex chromosomes in *Ginkgo biloba*.

(Authors: T.Y. Lan, R.Y. Chen, X.L.Li, F.P. Dong and W.Q. Song. Published in: Botanical Studies (2008) 49: 33-37.)

3.3.9 通过微切、扩增建立植物（*Vicia faba*）染色体区段特异性基因文库研究初探

Positional cloning by microdissection and microcloning has been proved to be an extreemly useful technique. Chromosome or subcromosome libraries can provide both high-density RFLP markers for genetic linkage map construction and specific probes for genomic library Screening.

The microdissection of metaphase chromosomes with glass needles (Lo *et al.*, 1989) or laser micro-beams (Hadano *et al.*, 1991; Scalenghe *et al.*, 1981) and subsequent amplification and is a technique which was just developed in the last five years.

Sine 1981, after the first successful example of microdissection and direct cloning by manipulation of polytene chromosome in *Drosophila* (Lo *et al.*, 1989), the cloning of su bpicogram quantities of DNA had been hampered by the difficulties in getting sufficient chromosome materials, in micromanipulation and the relatively low cloning efficiencey. Usually at least 100 chromosomes or segments were needed to be collected per cloning cycle (Edstrom *et al.*, 1987).

The introduction of Polymerase Chain Reaction (PCR) into the microcloning protocol had greatly improved the eifficiency of this technique (Lo *et al.*, 1989, Saunders *et al.*, 1989). To date, several ways of sequence independent amplification of microdissected DNA are available, which could be sorted into two types: 1) Methods to use specific primers to amplify microdissected DNA, which Has been priorly ligated into vector or adaptors providing primer-binding sequenee. The linkers including pUC/M13 sequencing plasmid (Lo *et al.*, 1989; Mackinnon *et al.*, 1990), Mbo 1 linker-adaptor (Bohlander *et al.*, 1992; Johnson *et al.*, 1990; Kao, 1987; Yu, 1992) and *Sau*3A primerlinker (Jinno *et al.*, 1992; Deng *et al.*, 1992) Random amplification with partially random sequence as a primer (Wensley *et al.*, 1990, Guan *et al.*, 1992). However, both of the approaches have been almost exclusively employed in human and ani-animal chromosomes.

Only till the last three years, had the application of microcloning in plants been reported. A direct cloning of microdissected B-chromosome of rye obtained only 20 clones (Sandery *et al.*, 1991). The vector-mediated PCR method was used to construct microdissection libraries of *Beta Patellaris* chromosome conferring nematode resistanee (Jung *et al.*, 1992), and of the short arm from Chromosome 1H of barley (Schondelmaier *et al.*, 1993) Two to five pieces of chromosome arms of wheat cells were amplified in a simplified adaptor-mediated PcR reaction.

Single unique primer-PCR (SUP-PCR) proved to be one of the most effective methods Designed to amplify a single micordisseeted segment of human chromosome with primer BVE22cc (Hadano *et al.*, 1991; Yokoi *et al.*, 1994). The present paper applied it to clon a single plant chromosome specific region-the Nueleolar organizer region (NOR) of chromosome 1 from field bean (*Vicia faba*). With a simplified procedure. Not only the use ofmicromanipulation was eliminated, but the quantity of the PCR products of one-microdissected materials was increased as com-paerd to that by the original method. It could facilitate the study of chromosome microdisseetion and microcloning in plant.

3.3.9.1 Materials and Methods

1. Preparation of metaphase chromosome spreads

Seedlings of *V. faba* with 2-cm-long side roots were incubated successively in 1.25 μmol/L hydroxyurea, double distilled H_2O and 4 μmol/L APM (Amiprophosmethyl) at 25℃ as described By Schübert *et al.* (Schubert *et al.*, 1992) to get metaphase cells. After being fixed in 70% ethanol for above 24 hours at -20℃, the harvested root tips were squashed in dividually and quickly on cold clean wet eover slip in a drop of 3:1 methanol-actic acid solution. Air dried, stained with 1:10 Giemsa solution (in phosphate buffer, pH 6.8) for 5min.

2. Microdissection

Siliconized rods with an external diameter of 1.0mm were pulled on a pipet puller toform microneedle with a tip of 1-2 μm, and bended to an angle of 120°, UV treated for 20 min before use.

The cover slip carrying chromosomes was put upside down and fixed on a glass chamber. Microdissection was performed under a Nikon mieroscope of 40×magnification using microneedles controlled through a micro-manipulator. The tip of the needle sticking a single NOR segment was broken off into the bottom of a 0.5ml microtube, and added 10 μl microtube, super pure water, stored at 4℃ overnight. Then DNA-free proteinase K was added to a final concentration of 1.0 $\mu g \mu l^{-1}$, incubated at 37℃ for 2 h to deproteinize completely. Inactivation of proteinase K was at 75℃ for 15 min.

3. SUP-PCR amplification of the microdissected DNA

All the steps, as well as the sequence of the primer, followed essentially the previous deseription by Hanado *et al.* (Hadano *et al.*, 1992), except for a magnified reaction system.

The fisrt step of PCR: samples ready for amplification were added with the first PCR MixtureHadano *et al.*, 1992) to a final volume or 20 μl covered with paraffin oi1. Six rounds of low stringeney cycles were performed: 5 min at 90℃ and 1h at 24℃, then the temperature was increased gradually to 50℃ and was kept for 20 min.

The second step: added the second PCR mixture to a 60 μl final volume, 92℃ for 2 min, 50℃ for 1 min, 72℃ for 4 min. Twelve cycles were performed.

The third step: 10 μl of the porducts were transferred to a new microtube containing 50 μl of the third PCR mixture, oi1 covered. Thirty-two cycles were carried out as followed: 1.5 min at 92℃, 55℃ for 1.5 min, 72℃, 3 min. 10 μl of the products were electrophoresed on 1.5% agarose gel. Another 2 μl from the products weretaken out to detected DNA concentrations by their absorbing values at 254 nm UV light with the produets of negative control as background.

4. Characterization of the amplified chromosome DNA by southern hybridization

To confirm the bean origin of the amplified microdissected DNA, 1/6 of the final products in one tube, and of the negative control, were separated on a 1.5% agarose gel then transferred to a nylon membrane (Hybrida) using SSC solution. The membrane was prehybridized for 1 h at 68℃ in a solution containing 5×SSC, 1% Blocking reagent (Boehringer Mannheim), 0.1% N-lauroylsarosine (Sigma), 0.2% SDS, Hybridization with total genomic DNA of *Vicia faba* was carried out at 68℃ overnight. The probes were labelled with digoxigenin by Random Priming Method. Signal detection followed the instructions in the specification enclosed in the 'DIG labelling and detection Kit' (Boehringer Mannheim), with alkaline phosphatase as a detecting enzyme.

5. Cloning microamplified DNA into pUC18

One-twelfth of the microproduct was purified with phenol/chloroform extraction, and ethanol precipitation, followed by *Eco*R I digestion to create cloing sites. The DNA was ligated into pUC18 with T4 ligase. One-fifth of the ligation mixture was used to transform E coli JM109 competent cells prepared with $CaCl_2$ method. One-fifth of the transformed mixture was plated on to LB medium with ampicillin and IPTG-X-Gal.

The recombinant plasmids were isolated by alkaline lysis, subjected to electrophoresis (0.7% agarose gel).

3.3.9.2 Results

1. Microdissection

Root tips were uded to provide sufficient metaphase chromosomes after treatment of hydroxyures combining with APM. The highly effective synchronization method could create an MI of as high as 50% in Vicia faba (Schubert et al., 1992). In its karyotype ($2n=12$) (Li, 1989) the chromosome 1 is easy tobe distinguished by its reatively large size and NOR structure. A single segment was microdissected by siliconized glass needle.

2. Direct amplification of the microdissected DNA

A little amount of genomic DNA of Vicia faba was amplified as a positive control, and on DNA was added in the negative control. The product of genomic DNA has a larger mean size than that of microdissected DNA, reflecting some mechanical or biochemical damage of the chromosomal DNA during the spread preparation and the microdissection.UV absorbing value showed that that approximately 60 μg DNA was obtained in eve 3-step PCR tube, so totally about 360 μg of DNA could be the amplified amount of one microdissected segment. All the products appeared an additional smear in the gel, sizing below 110 bp, which had been decreasing greatly after a few days of storage at 4℃. This problem was ignored as small primer polymers forming during amplification. Also the result of Southern hybridization was consistent with this possibility.

3. Characterization of the PCR product by southern genomic hybridization

The strong signal produced by hybridization between genomic probes and the microproducts confirmed that the DNA obtained from the amplification did derive from the microdissected chromosome of *Vicaa faba*. The small molecules in products both of microdissected DNA and of the negative control got no signals detectable.

4. Cloning of the amplifled DNA from microdissection

One a hundred fiftieth of the transformed mixture was plated on LB-Amp-X-Gal plate, resulting in 43 white clones. The gel analysis recombinant plasmids were from about 2.94 to 3.6 kb, indicating that the sizes of the inserts were between 0.25 and 0.9 kb. Detailed characterization and identification of the microclones were not carried out.

3.3.9.3 Discussion

In order to establish a general a way to construct chromosome specific DNA library, Vicia faba was selected as the materials for its easily obtained karyotype and well-recognized chromosomes. That made the microdissection easily performed.

To avoid the acid-induced depurination of chromosomal DNA, we employed the 70% ethanol as fix solution which has been used in chromosome preparation for microdissection in mammalian cell (Edstrom et al., 1987), as a substitution for 3: 1 methanol: acedic acid. The root tips after fixation were squashed directly on coverslip in a drop of 3: 1 solution. So the time of the acid action was reduced to as short as several seconds.

Sup-PCR with BVE22cc primer was designed to amplify any unknown DNA seguences with a high sensitivity. Less than 10 fg of DNA template could be amplified effectively (Hadano et al., 1991). It has been used successfully in microdissection and microcloning of human chromosomes 4p 50-100% and Xqter regions (Yokoi et al., 1994). BVE22cc primer is a 22 mer sequence, with 3 restriction sites for EcoR I, BgI II and EcoR V respectively. It is different from the other random primers used in sequence-independent amplification by its defined sequence. The low stringent conditions in SUP-PCR cycles and the existence of polyethylene glycol (PEG) make it possible to prime DNA sequences nonspecifically. Usually a single piece of chromosome region is sufficient for the following amplification. A lot of work in microdissection was saved.

Here we applied SUP-PCR method from animal materials to amplify a defined region pf plant chromosome. But the present protocol is not all the same as the original one. Instead of 1 μl collecting volume and 3 μl initial amplification system adopted in the previous description (Hadano et al., 1991; Yokoi et al., 1994), we collected the chromosome DNA directly in microtube, and the initial volume of PCR was 20 μl. In the third-step cycles, 1/6 of the second PCR products were used in a final amplifying system of 60 μl. So all the 3-step products of one piece of NOR region from Vicia faba could amount to 360 μg in combination. In Hadano's report, the chromosome fragments were collected into a 3 μl PCR system in the first cycle, and 9 μl in the second. Both of them require micromanipulation, which is time-consuming and strenuous. In our work the micromanipulation was eliminated by adopting the magnified system. Gel analysis and Southern hybridization indicated that the simplified SUP-PCR method worked well in amplifying the microdissected chromosome of Vicia faba.

An important factor that would cause a great consequences in the experiment is extraneous DNA contamination. The following. The following precautions must be taken: all the solutions should be prepared under sterile conditions in a laminar flow cabinet using reagents and equipments stringently autoclaved (123℃, 90 min). New micropipette tips are consistently used (autoclaved). The microneedles and all the other equipments should be UV treated 30 min before use.Genomic DNA from Vicia faba was used as probe to identify the microamplified DNA. The remarkable blotting signal showed that SUP-PCR did originate from the DNA of *Vicia faba*. Using microdissected DNA as probe is not feasible for 2 reasons below:

1) Usually the DNA amount of microdissected section from a chromosome is at fg (10^{-15} g) level, though chromosomes of vicia faba is morphologically larger in this work, the greatest amount of the microdissected DNA cannot exceed 1 pg (10^{-12} g). But it's very hard to manage the probe labelling reaction when the template DNA is fewer than 1 ng (10^{-9} g).

2) The possible contaminated DNA in the microdissecting process could also produce hybriding signal, which cannot be the persuasive proof of the homogenesis. The best way of DNA characterization is *in situ* hybridization. But it's hard to control in plant cells and needs more work, which is not a good step in a simple and rapid laboratorial method comparing to the regularly used Southern blotting.

In the cloning of the microproducts, the smear of unsteady DNAs appeared in the gel analysis was neglected for several reasons: 1) they were decompositing automatically, 2) the possible primer polymers would be digested apart by *Eco*R I, 3) the sizes of the small DNA moleculars were not continuous with that of the microamplified DNA, so even if they were successfully cloned, it is still easy to identify.

The sizes of the liner molecules of the recombinant plasmids ranging around 2.94-3.59 kb, the second bands from front end in all the recombinant plasmids) were roughly calculated according to the relationship between base-pair-number longarithm and DNA migrating distance on the gel under special conditions (concentration of the gel, temperature and time of electrophoresis, etc.) The size of pUP18 is 2.69 kb, so the inserts range from o.25 to 0.9 kb, as consistent with microamlified products. For the information about the NOR region of *Vicai faba* was not our focus study, detailed characterization and identification of the obtained clones and further completed cloning work of DNA from this region were not carried out.

In brief, the method reported here represents a simple and efficient protocol for the construction of subchromosome DNA libraries of specific region in plant, involving microdissection of specific chromosome region, direct amplification with single unique primer, and characterization by Southern hybridization as cloning into plasmid vector. This reproducible technique has also been used in microcloning the telomere of B-chromosome in rye and in amplification of every individual chromosome of rice as well.

（作者：翟晓玲、陈瑞阳。原载：热带亚热带植物学报 1997, 5（3）60-68）

3.3.10 黑杨1号染色体抗病基因同源序列的克隆与鉴定

Plants utilize a variety of strategies to withstand attack by the huge assortment of pathogens found in their habitats. The existence of resistance genes (R-genes) that are able to detect the presence of specific pathogen races by recognizing ligands encoded by the avirulence genes (Av r) of pathogens is one of the most effective and predominant strategies (Richter and Ronald, 2000). Recently, over 30 disease R-genes have been cloned from a wide range of plant species; they confer resistance to various plant pathogens, including viruses, bacteria, oomycetes, fungi, nematodes, and insects (Cooley et al., 2000; Ferrier-Cana et al., 2003; Liu et al., 2003). R-genes are grouped into five classes based on the structure of their protein products (Dangl and Jones, 2001). Most R-genes encode proteins containing a nucleotide-binding site (NBS) domain and a stretch of leucine-rich repeats (LRRs) at the C-terminus. Q3 Based on motifs located on the N-terminal of the NBS domain, the NBS-LRR superfamily can be further subdivided into the TIR (Drosophila Toll and human inter-leukin receptor-like) subfamily (e.g., the tobacco gene N) and the non-TIR subfamily (e.g., the Arabidopsis gene RPS2), members of which typically contain N-terminal coiled-coil (CC) domains (Meyers et al., 1999; Pan et al., 2000b; Cannon et al., 2002). The other four classes consist of (1) cytoplasmic serine/threonine Q4kinases (exemplified by the gene Pto in tomato), (2) extracellular LRRs anchored to a transmembrane domain (for instance, the tomato gene Cf-9), (3) receptor-like kinases (RLKs) with extracellular LRRs and an intracellular serine/threonine kinase domain (e.g., the rice gene Xa21 and the Arabidopsis gene FLS2), and (4) the HM1 gene encoding atoxin reductase (Di Gaspero and Cipriani, 2003).

Genes encoding NBS-LRR proteins are found in dicots and monocots, although there are some differences in gene structure. NBS-LRR proteins that have an N-terminal TIR domain have not been found in monocot genomes, whereas NBS-LRR proteins that have an N-terminal CC domain are present in both dicots and monocots (Pan et al., 2000b). Conserved motifs in NBS domain have been used to isolate resistance gene analogs (RGAs) in a broad range of plant species, such as soybean (Kanazin et al., 1996; Penuela et al., 2002), potato (Leister et al., 1996), *Arabidopsis thaliana* (Aarts et al., 1998; Meyers et al., 1999, 2003), maize (Collins et al., 1998), rice and barley (Leister et al., 1998), lettuce (Meyers et al., 1998), common bean (Rivkin et al., 1999), grapevine (Di Gaspero and Cipriani 2003), apple (Baldi et al., 2004), pine (Liu et al., 2005 Q5), peach (Decroocq et al., 2005; Lalli et al., 2005), and apricot (Soriano et al., 2005). These RGAs are useful in physical mapping and as gene candidates in positional cloning. However, R-gene families are large multigene families and are abundant in plants. For ex-ample, RGAs constitute ca. 1% of the genome in Arabidopsis (ca. 200 RGAs) (Meyers et al., 1999, 2003). Recent analysis of the rice genome identified ca. 500 non-TIR/NBS/LRR-genes that account for 1% or more of the entire rice genome (Monosi *et al.*, 2004; Zhou *et al.*, 2004). To date, RGA amplification is usually performed based on the whole genome, resulting in a very large number of different RGAs. This makes the identifica-tion of any one specific RGA (for example, those closely linked to a given R-gene or candidates of a target R-gene) very time-consuming and laborious.

The chromosome microdissection and microcloning tech-nology was developed in 1981 (Scalenghe *et al.*, 1981). Subse-quently, it has evolved into an efficient tool for generating chromosome-specific DNA libraries of many species (Thal-hammer *et al.*, 2004). This procedure could also be used for the generation of chromosome-specific molecular markers for isolation of chromosome-specific sequences using specific PCR primers or for the indirect physical mapping of low/single copy sequences (Houben *et al.*, 2002).

The necessity for model species of plants is well recognized, and in this role, A. thaliana has gained a supreme acceptance among plant scientist. While many aspects of tree biology are common to all plants, and hence can be studied in very tractable model species such as A. thaliana, some unique facets of tree anatomy and physiology must be investigated in trees themselves. The genus Populus has been adopted as a model for forest-tree genetics and has been proposed for whole-genome sequencing for the following reasons: (1) small ge-nome size (the haploid genome size is ca. 480620 Mbp), (2) rapid juvenile growth, (3) ease of clonal propagation, and (4) high-throughput transformation and in vitro propagation (Bradshaw *et al.*, 2000; Taylor 2002; Brunner *et al.*, 2004).

Information on the genetic and physical locations of markers related to economi cally important genesis useful for map-based cloning and marker- assisted plant bree ding. We have dem on-strated that the combined tec hnology of chromosome microdissection and linker adaptor-mediated PCR (LA-PCR) is suitable for constructing chromosome-specific DNA libraries in forest trees possessing small chromosomes, such as poplar (Zhang *et al.*, 2005). In this study, using poplar as a model, we devel -oped a technique that combines chromosome microdissection and homologous sequence amplification in order to acquire RGAs from a single chromosome. As ingle chromosome 1 was microdis sected from the metaphases preads of poplar (*Populus tremula*) root-tip cells with fine glass needles. A number of RGAs were amplified with the DNA fragments of *P. tremula* chromosomes 1 as a template, using consensus-degenerate hybrid oligonuc leotide primers based on conserved mot ifs o f the NBS domain. This article reports the sequence characterization and diver sity analysis of these RGAs as well as their relati on ships with the NBS sequences of know n R- ge nes fr om o ther plant species.

3.3.10.1 Material and Methods

1. Plant material

Populus tremula branches were collected from the p lantation in the Reseach Institute of Forestry, C hinese Academy of Forestry.

2. Chromosome preparation, microdissection, and DNA amplification from single chromosomes

Root tips were used for chromosome preparation and microdissection. The method of Sau3A LA-PCR was used to acquire DNA fragments from single chromosomes. This procedure is described in detail in Zhang et al. (2005), with the exception that only one round of LA-PCR was carried out in the present study.

3. Primer design and PCR amplification of RGA sequences

A pair of degenerate primers was designed based on two highly conserved motifs in the NBS of the products of the tobacco N, Arabidopsis RPS2 and RPS5, and flax L6 and rice Xa1 genes, using a new primer design strategy for PCR amplification of distantly related gene sequences based on consensus degenerate hybrid oligonucleotide primers (CODEHOPs) according to Rose et al. (2003). The primer NBS-F: 59 -ATGCCAGGA (A/G) TNG GNAA (A/ G) ACN AC-39 was based on the conse- rved P-loop motifs M GG (P/V) GK TT of the N BS domain, while NBS-R: 5 9-ACCTCAAGAGCNA (A/G) NGGNA (A/G) -NCC-3 9 was developed from the conserved hydrophobic domain with the GLPLAL motif of the NBS domain. The "N"represents A/G/C/T.

For amplification of NBS from Populus tremula using these degenerate primers, PCR was performed with 2mL of the first-round products of LA-PCR as template. Each 25 mL PCR reaction volume consisted of 2.5 mL of 10× Taq buffer (Takara, Japan), 0.5 mL of 10 mM dNTPs, 1 mL of 10 mM μm each degenerate primer, 0.5 U of Taq DNA polymerase (Takara, Japan). Three control reactions were performed: a negative control (no template DNA, CK1), a control with only 2 mL of 10-mM NBS-F primer (CK2), and a control with only 2 mL of 10-mM NBS-R primer (CK3). PCR amplifications were performed in a thermal cycler (MG 5331, Eppendorf, Germany) using the following procedures: after denaturation at 94℃ for 2 min, amplification was performed for 35 cycles at 94℃ for 0.5 min, at 58℃ for 1 min, and at 72℃ for 1 min, and a final step of elongation at 72℃ for 10 min.

4. Cloning and sequencing of amplified RGAs

PCR products were separated on a 2% agarose gel, and DNA fragments of the appropriate size were extracted from the gel. The target bands were ligated to the pUCmT–easy vector (Sangon, China) and then transferred into Escherichia coli strain DH5α competent cells by heat shock (Sambrook et al., 1989). Individual clones were distinguished by 4-base restriction enzyme digestion. Clones showing the same restriction patterns were presumed to be identical and therefore sorted into the same class. One to four clones from each defined class were sequenced at the Sangon company using the PRISM Ready Reaction DyeDeoxy Termination cycle sequencing kit with an ABI-3700 DNA Sequencer (Applied Biosystems, United States). The nucleotide sequences of P.tremula RGAs have been deposited in the GenBank database under the accession numbers DQ104366–DQ104388.

5. Multiple Alignments and Tree Construction of RGAs

Nucleotide and amino acid sequences derived from positive clones were searched for their similarity to cloned R-genes and their products using the BLAST algorithm (Alt-schul et al., 1997) in GenBank, which was accessed via the NCBI Web site (http://www.ncbi.nlm.nih.gov/). Phylogenetic analyses were performed on the putative poplar RGA nucleotide sequences as well as on the deduced protein sequences. Searches for introns were done using the Net Plant Gene program (Hebsgaard et al., 1996). Predicted amino acid sequences were generated using the translate tool on the ExPASy proteome server (http://www.expasy.org).

Pairwise compari sons and multiple alignments were performed on nucleotide sequences and deduced amino acid sequences using the Align program (Myers and Miller, 1988) and Clustal W, version 1.8 (Thompson et al., 1997), with default costs ettings for opening and extending gaps.A dissimilarity matrix and a phylogenetic tree were obtained from this alignment following then eigh borjoining method (Saitou and Nei, 1987) with the Poisson correction distance (Nei and Kumar, 2002). The reliability of the tree was establi shed by conducting 1000 neighbor-joining boot-strap samplings teps (Felsenstein, 1985). The phylogenetic tree was rooted using an AfsR gene (P25941) from Streptomyces coelicolor as an outgroup; nonplant proteins Apaf -1 and CED-4 were not used in the phylogenetic analysis because they are more distantly related to plant NBS-encoding R proteins than the S. coelicolor sequence (Meyers et al., 2003). T he RGA s equences in the TIR subgroup were aligned using the Clustal W program, and the multiple sequence alignment editor program GeneDoc (Nicholas et al., 1997) was used to generate a consensus sequence from the subgroup alignments.

NBS sequences of the following R-genes were used in our analysis: N (U15605), L6 (U27081), M (U73916), Prf (U65391), RPM1 (X87851), RPS2 (U12860), RPS5 (AF074916), RPP1(AF0 98962), RPP5(U97106), RPP8(AF089710), I2C-1(AF004878), Mi1-2 (AF039682), RP1-D (AF107294), Pi-B (AB013448), Dm3 (AF113948), Xa-1 (AB002266), GPA2 (AF195939), Bs2 (AF202179), and Cre3 (AAC05834).

3.3.10.2 Results

1. Amplification of DNA fragments from poplar chromosome 1

Using root tip as source material, we succeeded in preparing good-quality slides of chromosomes in Populus tremula. Chromosome 1, the largest submetacentric chromosome, could be identified at prometaphase or metaphase. After being identified and marked, the target chromosome was successfully isolated by a fine glass needle (Fig. 3.85). After one round of amplification, the PCR products were observed as a faint smear of DNA with a range of 200–1500 bp on a 1% agarose gel (data not shown). None of the products was amplified from the negative control (no template DNA). The results indicate that DNA from chromosome 1 was amplified successfully without contamination of exogenous DNA. Southern hybridization with DIG-labeled genomic DNA confirmed that the products were amplified from the P. tremula genome (data not shown).

2. Identification of RGAs from poplar chromosome 1 by PCR with degenerate primers

Only a clear band of ca. 520 bp was detected in the amplification product with primers NBS-F and NBS-R, using the first-round products of LA-PCR or genomic DNA as template.

No band was detected in control by using only primer NBS-F or only primer NBS-R (Fig. 3.86). The 520-bp band was excised and cloned, and 161 individual clones were obtained and classified into 12 classes by digestion with 4-base restriction enzymes. One to four clones of every class were selected and sequenced.

In total, 30 clones were sequenced (25 RGA clones from clones from chromosome 1 and 5 RGA clones from genomic DNA). Seven presented no resemblance to NBS domains of R-genes, while 23 appeared related to the NBS of known R-genes. Analysis of RGA sequences revealed the presence of stop codon and/or frame shift mutations in 5 of 18 RGA clones from chromosome 1 and in 2 of 5 RGA clones from genomic DNA, while others contained an uninterrupted ORFQ9, indicating that they might be pseudogenes. Their deduced amino acid sequences showed not only the two motifs targeted by the primers used but also the internal motifs characteristic of the NBS-LRR gene class. No putative splicing sites were detected in these clones.

3. Diversity analysis of the RGA clones from chromosome 1 and genomic DNA

Diversity analysis of 23 NBS-RGAs, isolated from chromo-some 1 (P1-RGA1 to P1-RGA18) and total geomic DNA (P-RGA1 to P-RGA5) in P. tremula, was conducted using the neighbor-joining method, and the phylogenetic tree was con-structed (Fig. 3.87). The RGA sequences were grouped into eight clusters or families (A–H), all of which were highly sup-ported by bootstrap values. A parsimony analysis produced a similar tree (data not shown), confirming the robustness of the tree. Three families (B, C, and H) consisted of only one RGA, while other families were each composed of several mem-bers. Identity scores for members of a given family ranged between 70% and 99%. Inter family identities ranged from 50% to 60%, depending on the families compared and the RGAs considered, except P-RGA2. The highest levels of interfamily identity were observed between members of the families D, E, and F. Sixteen RGAs (P1-RGA1 to P1-RGA18, except P1-RGA7 and P1-RGA9) from chromosome 1 were highly similar among themselves (percent identities >80%) and thus formed a "compact" group (D, E, F), while the P1-RGA7 and P-RGA1 from genomic DNA were classified into family A, and P1-RGA9 falls into family H alone. At the same time, the five RGAs (P-RGA1 to P-RGA5) amplified from genomic DNA exhibited quite diverse and generally low similarities among themselves (percent identities ranging from 10% to 65%) and were classified into four distinct groups: A, B, C, and G. The similarities of RGAs from the two kinds of sources were generally low (percent identities < 30%).

4. Comparative analysis of NBS domain RGAs from poplar with other cloned R-genes

Searches of GenBank using BLAST algorithms (BLASTN and BLASTP) demonstrated that the P. tremula RGAs shared significant homologies with well-characterized R-genes from other plants, as indicated by the E values ranging from 2e-13 to 6e-32. For instance, the P1-RGA13 clone shared 42% identity and 60% similarity with the NBS region of N. The lowest BLAST score, on the other hand, was observed between P1-RGA6 and M (32% identity and 53% similarity). BLASTP searches also detected four homologous sequence from

P. tremula (AJ490333), *Populus balsamifera* (AJ490334), *Populus deltoides* (AJ416708), and (*Populus tomentosa* × *Populus bolleana*) × *P. tomentosa* (DQ018370).

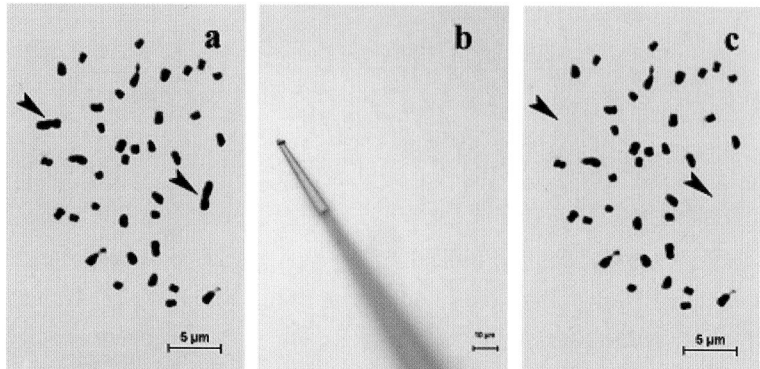

Fig. 3.85 Procedure of isolation of an individual chromosome 1 in *Populus tremula* by micromanipulator
a, Mitotic metaphase image before microdissection of the chromosome 1 (arrow). b, Target chromosome adhering to the tip of a glass needle. c, Individual chromosome 1 in a (arrow) removed from the cell. Scale bar ¼5, 10, 5 mm, respectively

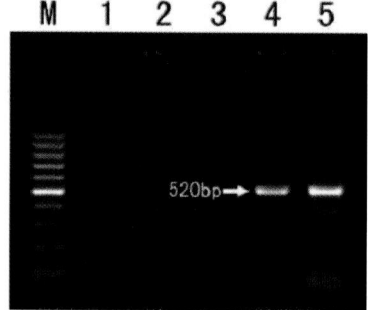

Fig. 3.86 Resistance gene analogs obtained from Populus tremula chromosomes 1: DNA molecular-weight mark (lane M; 100-bp ladder); negative control without DNA template using NBS-F and NBS-R as primer (lane 1); NBS-F alone as primer and LA-PCR products as template (lane 2); NBS-R alone as primer and LA-PCR products as template (lane 3); NBS-F and NBS-R as primer and LA-PCR products as template (lane 4); NBS-F and NBS-R as primer and genomic DNA as template (lane 5)

The amino acid sequences deduced from the *P. tremula* RGAs and from NBS-LRR R-genes from other plants enabled us to construct a phylogeny. The phylogenetic analysis was based on the NBS region, using the R-genes and four homologous RGAs sequence from other Populus spp. The neighbor-joining phylogenetic tree was constructed from the amino acid alignment of NBS domains of these R-genes and *P. tremula* RGAs (Fig. 3.88). The tree had long branch lengths and closely clustered nodes, reflecting a high level of sequence divergence. The distinction between the eight *P. tremula* RGA families was clearly confirmed and was consistent with the nucleotide analysis observed in the relationships between *P. tremula* RGA families (Fig. 3.87).

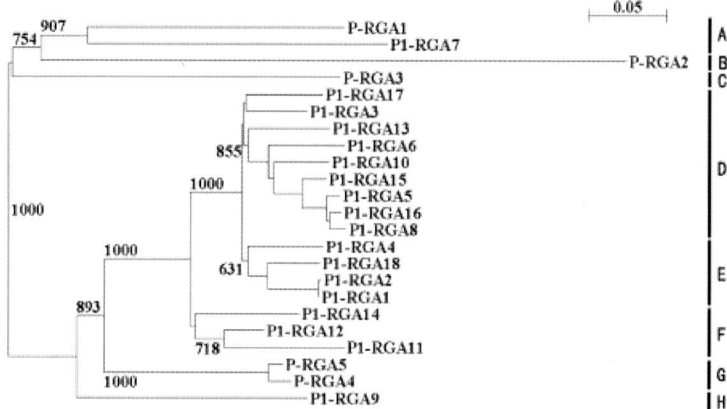

Fig. 3.87 Phylogenetic tree for resistance gene analog (RGA) nucleotide sequences isolated from the *Populus tremula*, P1-RGA1 to P1-RGA18 (from chromosome 1) and P-RGA1 to P-RGA5 (from genomic DNA), based on the neighbor-joining method (Saitou and Nei 1987). Numbers (%) on the main branches represented bootstrap values (for 1000 iterations). Eight RGA families (A –H) were obtained using a 70% identity threshold value. The scale bar shows a distance equal to 5% nucleotide diversity.

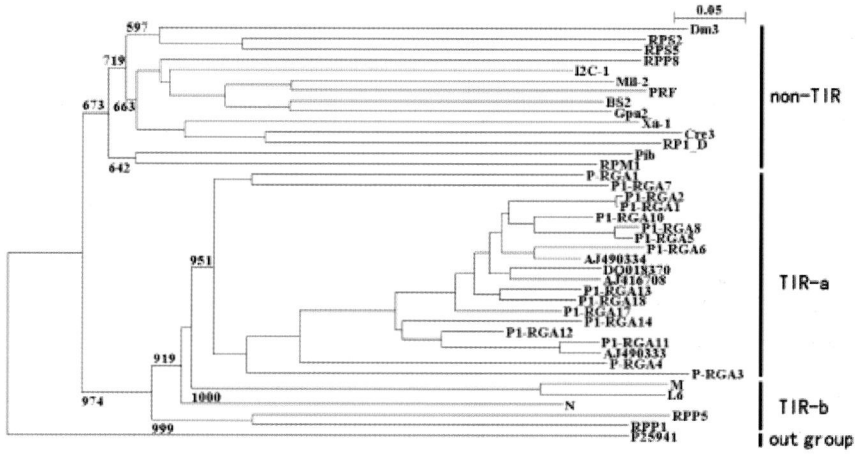

Fig. 3.88 Phylogenetic tree based on the alignment of the consensus amino acid sequences of *Populus tremula* resistance gene analog (RGA) and the nucleotide-binding site domain of 19 known R-genes and four RGA sequences from other Populus spp. using the neighbor-joining method (Saitou and Nei 1987). Kimura's correction was applied. Numbers (%) on the main branches represent bootstrap values (for 1000 iterations). The scale bar in the bottom left corner displays a distance corresponding to 5% amino acid substitutions per site

When the Streptomyces coelicolor AfsR gene (P25941) was used as an outgroup, the phylogenetic tree showed that it could be split into two major branches that were strongly supported on the basis of the bootstrap analysis and of corresponding to the non-TIR and TIR subclasses (Fig. 3.88). All *P. tremula* RGAs identified in this study were present in one branch containing the N, L6, M, RPP1, and RPP5 genes. According to the previously defined distinction between the TIR class and the non-TIR class, all isolated *P. tremula* RGAs seemed to belong to the TIR-NBS-LRR subfamily of R-genes. Among the TIR group, two subclusters emerged, 16 *P. tremula* sequences joining the TIR-a and the other R-genes joining the TIR-b. Bootstrap values for the TIR-a and TIR-b subgroups were, respectively, 951 and 1000.

This classification was further supported by the presence of internal conserved motifs in the NBS domain. Alignments of deduced amino acid sequences demonstrated that the *P. tremula* RGAs also contained the conserved motifs in NBS domains, including the P-loop (Kinase-1), RNBS-A-TIR, Kinase-2, RNBS-B/Kinase-3, RNBS-C, and GLPL signatures (Fig. 3.89), while RNBS-A-TIR has been found only in the TIR-NBS-LRR proteins. For example, a phenylalanine (F) was located 15 residues downstream of the P-loop motif, and the tryptophan residue (W) characteristic of non-TIR class found at the final position of the Kinase-2 motif was not present in any of the 16 *P. tremula* RGA sequences analyzed (Fig. 3.89).

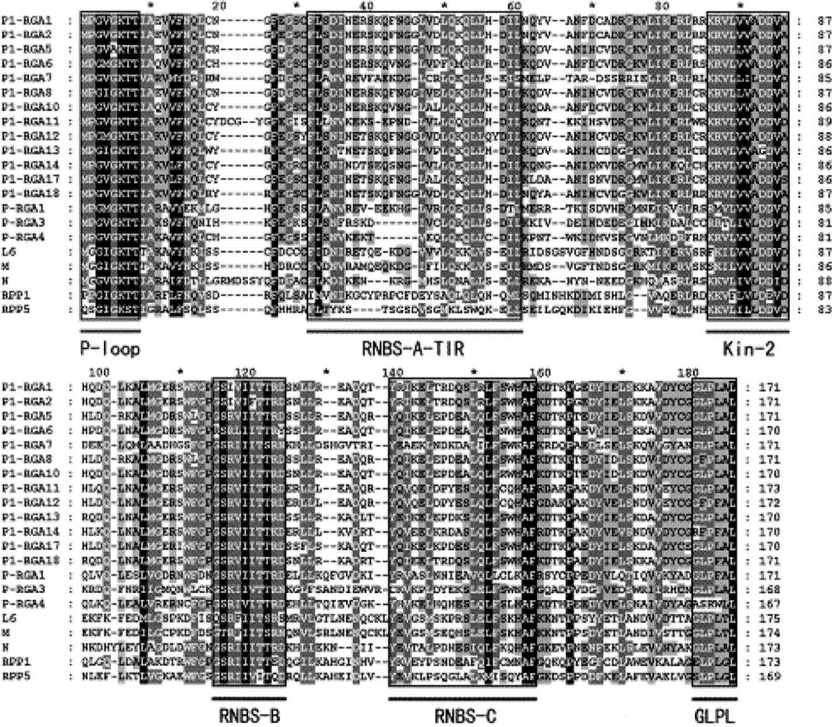

Fig. 3.89 Alignment of the translated *Populus tremula* resistance gene analog consensus sequences, along with the nucleotide-binding site domain of known R-genes from other species, using CLUSTAL W, version 1.8 (Thompson et al. 1997). Consensus residues above 60% are shown by gray shading. Gaps are indicated by dots. The six previously reported conserved motifs of the NBS domain (P-loop, NBS-A-TIR, Kinase-2, NBS-B, NBS-C, and GLPL) are boxed in black lines (Meyers et al. 1999; Cannon et al. 2002). EMBL/DDJB/GenBank Nucleotide Sequence Databases, accession numbers N (U15605), L6 (U27081), (U73916), RPP1 (AF098962), and RPP5 (U97106)

3.3.10.3 Discussion

1. Using chromosome microdissection technique in RGA amplification from single chromosomes

In this study, we develop a method for amplifying RGAs from single chromosomes. Using chromosome microdissection and LA-PCR techniques, DNA fragments can be easily acquired from special single chromosomes and be used as the template for RGA amplification. Correct identification of the target chromosomes is an important step required for single chromosome microdissection. Generally, this can be achieved by choosing plants with target chromosomes bearing unique morphological features (Zhang *et al.* 2005). But in some model organisms that have a previous map available, the products of the LA-PCR can also serve as a template for chromosome identification; for example, Liu *et al.* (2004) designed 18 pairs of primers of chromosome 4 based on the rice map to amplify the chromosome-4 specific sequences from the 27 single chromosome amplification products. In Populus tremula, chromosome 1, the largest submetacentric chromosome, can be unambiguously identified at prometaphase or metaphase. Only metaphases with unequivocally identifiable and spatially well-separated chromosomes were used for microdissection. The products of LA-PCR from single chromosomes not only can serve as templates for RGA amplification but also can be used for chromosome identification as well as for other research purposes, such as homologous amplification of other genes and the construction of single chromosomal DNA libraries (Huang *et al.*, 2004; Zhang *et al.*, 2005).

Howeve r, the procedure of L A-PCR is complicated, invo lving the preparation of linker adaptors and primers, the digestion of single chromosomes, the ligation of chromosomal DNA fragments with linker adaptors, and soon (Huang *et al.*, 2004; Zhang *et al.*, 2005). In a ddition, the digestion time should be strictly controlled. The sizes of the chromosomal DNA fragments obtained by LA-PCR are crucial for RGA amplification (Huang *et al.*, 2004). Since the predicted sizes of RGAs amplified with primers NBS- Fand NBS- Rareca. 520 bp, only those LA-PCR products with a size greater than 520 bp can potentially serve as templates for RGA amplification. Therefore, single chromosomal DNA should be partially rather than completely digested with *Sau*3A. In this study, in which single chromosomes were digested at 37℃ for 2 h, the first-round PCR products were observed as faint-smear electrophoretic bands of DNA with a range from 200 bp to 1500 bp on a 1% agarose gel (Zhang *et al.*, 2005). This size range proved to be suitable for RGA amplification.

2. RGA amplification in poplar using degenerate primers

Using a pair of CODEHOPs matching conserved motifs from NBS domains of NBS-LRR R-genes from various species, DNA amplification and extensive cloning from poplar were performed. The results presented here showed the great potential of heterologous PCR approaches for the cloning and study of R-genes in a perennial plant such as the poplar. As previously suggested (Rose *et al.*, 2003), the use of CODEHOPs PCR strategy with low degeneracy rather than a normal degenerate primer with greater degeneracy seems to be more efficient for isolating highly divergent NBS sequences. When the normal designed degenerate primers from soybean (Kanazin *et al.*, 1996), potato (Leister *et al.*, 1996), and sunflower (Gentzbittel *et al.*, 1998) were used, major amplification products were in the 520 bp size range, as expected, but PCR products with unexpected sizes (> 600 bp or < 500 bp long) were also observed. When sequenced, these latter fragments were found to have no motifs characteristic of cloned R-genes. Some of them showed identity to retrotransposon-like elements, as determined through BLAST searches of GenBank(data not shown). These results indicated that the CODEHOPs PCR strategy was effective in increasing the efficiency of RGA cloning.

The PCR-derived poplar NBS sequences were identified as RGAs or portions of R-genes based on the following attributes. First, these sequences contained uninterrupted ORFs. Second, the appearance of stop codons in a series of closely related but nonidentical cones was evidence that these amplifications were not the result of random sequencing or PCR amplification errors. Third, mutations that might have been introduced during PCR amplification were identified by comparison for related sequences obtained from eight independent PCR amplifications. The error rate was estimated to be 0.05% and appeared negligible for our analyses. Fourth, all sequences contained the conserved motifs characteristic of NBS R-genes(Meyers *et al.*, 1999; Pan *et al.*, 2000b). Fifth, the poplar RGAs shared significant homologies with well-characterized R-genes from other plants.

3. Diversity of RGAs in poplar

The reported ratio of non-TIR to TIR-type NBS sequences in the complete Arabidopsis genome is 1:3 (Meyers *et al.*, 1999). However, this ratio can vary among different species. Thus, Kanazin *et al.*(1996)in soybean, Tian *et al.*(2004)in sugar beet, and Martinez Zamora *et al.* (2004) in strawberry identified mainly TIR-subfamily sequences, but Donald *et al.* (2002), using in grapevine a similar PCR strategy with degenerate primers, found a high proportion of non-TIR: TIR sequences (6:1). In this study, the NBS encoding RGAs from P. tremula showed considerable sequence variation; however, most of them were identified showing strong sequence similarity with almost all known TIR type R-genes. It was noteworthy that no non-TIR-type NBS sequences were isolated from either the *P. tremula* chromosome 1 or the genome. The failure to detect non-TIR encoding sequences in *P. tremula* suggests that this R-gene family is absent or has diverged beyond recognition in poplar. Similarly, non-TIR-type RGA se-quences are absent in apricot genomes(Soriano *et al.*, 2005). Although the PCR degenerate primers used in this study were designed from the conserved motifs of the NBS domain across the non-TIR: TIR subfamilies, slight differences in sequence might explain why non-TIR-type sequences were not amplified(Soriano *et al.*, 2005). In fact, Baldi *et al.* (2004) found mainly non-TIR-class RGAs in apple when the LDD-AS primer located at the end of the Kinase-2 motif was used. However, with primers designed from the P-loop and from the hydrophobic motif, the result was the opposite, and most RGAs obtained were the TIR type. In conclusion, the absence of *P. tremula* non-TIR-class RGAs, which were also found in other species, may reflect an unequal distribution of TIR/non-TIR sequences or may be due to a bias in the PCR strategy used (Lalli *et al.*, 2005; Soriano *et al.*, 2005). For these reasons, there was the probability that the primers we used in this study preferentially amplify TIR sequences. But the fact that most RGAs were amplified from single chromosomes and only a small number of RGAs were amplified from genomic DNA may be one of the important reasons for absence of non-TIR RGAs. At this point, it was not possible to determine if this was due to the presence of only the TIR class of RGA Q10 in the *P. tremula* chromosome 1 or to a bias in the amplification of RGAs due to limitations of primer design.

The compara tive anal ysis of the diversity of RGAs amplified from a single chromosome 1 and total genomic DNA showed obvious differences. Most of the RGAs a mplified from chromosome 1 were distributed in a few fam ilies. The results p rovidedus with some interesting insights into the numbers and organization of Q11R-genes on the single chromosome 1. In most plant species studied to date, R-genes are known to cluster to regions on chromosomes that represent loci involved in specific pathogen recogntion and ultimately resistance to that particular pathogen (Michelmore and Meyers, 1998). Genomic sequencing of *Arabidopsis* has revealed that NBS-encoding sequences tend to be clustered in the genome, and numerous R-genes have beenm apped to the clusters of NBS-encoding sequences on Arabidopsis chromosomes IV and V (Michelmore, 2000). This clustering appears to occur also in poplar. In this study, all 18 RGAs from chromosome 1 were highly similar to each other (percent identity > 80%), suggesting possible geneduplication during evolution. In contrast, all of the RGAs from the genomic DNA were much more diverse in their amino acid sequences.

4. Variation and evolution of NBS-encoding sequences in poplar

R-genes of plants, both dicot and monocot, comprise a significant portion of the genome, represented by ca. 200 genes in the entire genome of *Arabidopsis thaliana* (Meyers *et al.*, 2003). Recent analysis of the rice genome identified ca. 500 non-TIR/NBS/LRR genes that account for 1% or more of the entire rice genome(Monosi *et al.*, 2004; Zhou *et al.*, 2004). Using a heterologous PCR approach, up to 23 different RGAs were identified from a single chromosome 1 and genomic DNA, indicating the presence of a large number of R-genes in the poplar genome. *Populus tremula* NBS sequences appear to be grouped into distinct families. While the RGAs belonging to the same family showed high levels of identity, sequence variation between RGA families was very obvious, suggesting independent evolution of these different families.

Molecular data are increasingly consistent with the hypothesis that plants have R-genes arrayed in complex clusters (Michelmore and Meyers, 1998; Noel *et al.*, 1999; Noir *et al.*, 2001). According to several articles that mapped RGA, these genes were distributed in the genome by clusters, and some clusters have been identified as regions where resistance to a pathogen was previously mapped (Decroocq *et al.*, 2005). Q12 Decroocq *et al.* (2005) reported that the colocalization of three analogs of virus R-genes with two distinct genomic regions linked to Plum pox virus resistance in *Prunus davidiana*. Within a cluster of RGAs, there may be more than one gene conferring resistance to different isolates of a specific pathogen or to biologically diverse pathogen taxa (Cooley *et al.*, 2000). Several studies have shown that within a disease-resistance cluster, genes are often found that show high levels of simi-larity to R-genes but do not encode a functional product (Ori *et al.*, 1997; Song *et al.*, 1997). Other data revealed that clusters of R-genes may contain sequences related in function but not in sequence (Dixon *et al.*, 1996). Indeed, clustering of R-genes and homologous sequences may facilitate the generation of diversity and new resistance specificities.

The evolution of R-genes remains largely unexplored, but useful information has recently been gained from molecular genetic analyses. Several genetic mechanisms, including point mutation, recombination, unequal crossing over, and gene conversion, have been proposed to account for the evolution of R-genes (Michelmore and Meyers, 1998; Noel *et al.*, 1999). These analyses refer to crop and modern plants with short life cycles, and so far, few data have been obtained from perennial plants such as poplar. In our study, the results show that single nucleotide differences, insertions, deletions, and duplications were all found in P. tremula RGAs. In a similar study in tomato, only 10% of cloned RGAs turned out to be pseudogenes (Pan *et al.*, 2000a), a much lower fraction than reported even for other plants. This may reflect a difference in RGA organization between genomes of annuals and perennials. Pseudogenes are nearly as abundant as functional genes in the human genome (Harrison *et al.*, 2002). Probably serving as reservoirs of potential variation, pseudogenes are believed to be advantageous because they may allow for recombination and gene conversion between alleles or paralogs of functional R-genes (Michelmore and Meyers, 1998). For long-lived species such as poplar, with a long life span, the generation of new resistances via somatic mutations may be particularly advantageous (Michelmore and Meyers, 1998). Comparisons of variability in the R-gene domains that determine pathogen specificity among long-lived individuals need to be performed.

5. Functional inferences from NBS domain evolution

Plant R-genes are assumed to encode proteins that recognize specific pathogen-derived avirulence proteins and initiate signal transduction pathways leading to complex defense responses (Grant and Mansfield 1999). In these signaling processes, NBS domains play an important role. Diversity among NBS sequences is believed to be critical for the specificity of interaction either with a pathogen elicitor or with proteins downstream in the various signaling pathways that have been identified by genetic analysis (Aarts *et al.*, 1998). Other studies using precise domain swapping have shown a number of LRRs to be essential for the interaction with avirulent factors (Van der Hoorn *et al.*, 2001; Wulff *et al.*, 2001). Based on NBS sequences, the cloned RGAs in the *P. tremula* can be grouped into eight classes (Fig. 3.87). The concurrent maintenance of different families suggests that the availability of R-genes containing various NBS domains might contribute in part to the evolutionary fitness of poplar.

Sequence analysis of different loss-of-function alleles in the R-genes RPM1, RPS2, and

Prf demonstrated the functional requirements underlying the conservation of the consensus motifs characterizing the NBS region of plant R-genes (Pan *et al.*, 2000b). Specific signatures were revealed by comparing *P. tremula* RGAs with the different closely related R-genes isolated from monocot and/or dicot species that are distantly related to poplar. Within families characterized by NBS sequences of hypothetically common origin, consensus motifs appeared extended, and short additional conserved blocks were observed. Since these genes or RGAs result from an ancient divergence, this conservation suggests a functional family specificity. It is possible that some of the differences between the NBS sequences encoded by R-genes influence the specificity of interactions either with a pathogen elicitor or with proteins downstream in the signal transduction cascade (Noir *et al.*, 2001).

Further genetic and biochemical analyses are required in order to better define the structural and functional roles of particular motifs and the diversity within the NBS domain. Elucidating the evolution of R-genes is of particular importance for the understanding of how plants maintain and adapt their defenses to pathogens (Pan *et al.*, 2000b). Since highly contrasting situations hold for different plants, investigations of various species, including tree species, appear necessary in addition to studies on model plants such as Arabidopsis and rice. In this endeavor, poplar could constitute an attractive model for perennial plants.

(Authors: Yong Zhang, Shougong Zhang, Liwang Qi, Bo Liu, Jianming Gao, Xiaoqiang Chen, Chengbin Chen,, Xiulan Li, Wenqin Song..Published in: Int.J.Plant Sci. 2006, 167 (3): 402-413)

第四节 荧光原位杂交与基因定位

提要：荧光原位杂交（Fluorescence In Situ Hybridization，FISH）技术和荧光分带（fluorochrome banding）技术可以使染色体呈现特定的杂交信号和带型，为识别和分析染色体提供标记。由于这些标记都是来源于染色体特定位点区域客观的 DNA 序列或结构组成特点，使其比来源于染色体形态的测量数据更为准确、可靠。因此，这种加入了特定位点信息的染色体研究，已经广泛应用于细胞遗传学领域，使核型分析技术进入了新的时代。

3.4.1 荧光原位杂交技术的研究进展

荧光原位杂交技术（Fluorescence In Situ Hybridization，FISH）是一种遗传学实验技术，其基本原理是：荧光素直接与寡聚核苷酸结合或者采用间接法以生物素、地高辛等标记，获得寡聚核苷酸探针，探针与变性后的染色体、细胞或者组织中的核酸按照碱基互补配对原则进行杂交，经变性—退火—复性—洗涤形成靶 DNA 与核酸探针的杂交复合体，经直接检测或通过免疫荧光系统检测，最后在荧光显微镜下观察或显影，对待测 DNA 进行定位、定性或相对定量分析。简单地说，染色体原位杂交技术就是在核酸分子碱基互补配对原则的基础上，利用放射性或非放射性物质标记外源核苷酸制备探针，探针与变性处理后的染色体单链 DNA 杂交，利用添加不同滤光片的荧光显微镜分别观察染色体和杂交信号并进行高质量拍照，再通过图像处理软件进行图片叠加，将探针核酸序列在染色体上的位置显示出来。

随着应用的推广和技术方法的更新，FISH 技术也在逐步完善。从同时杂交的探针数目来看，从单色 FISH 发展到多色 FISH，从一次杂交一个探针到同时杂交显示不同杂交信号颜色的多个探针；从杂交对象来看，从早期的中期染色体玻片标本到减数分裂时期的粗线期染色体玻片标本，甚至是伸展 DNA 纤维玻片标本。根据使用探针类型的不同，FISH 还从早期的重复序列探针杂交，衍生出了基因组原位杂交（Genome In Situ Hybrieization，GISH）、细菌人工染色荧光原位杂交（BAC-FISH）和酵母人工染色体荧光原位杂交（YAC-FISH）等技术（Shi 等，1998；Zhang 等，2004；Hao 等，2006）。近年的研究结果表明，FISH 研究中已经很少使用同位素探针，同时，随着冷 CCD 拍照系统的改进及电脑软件的发展与运用，FISH 技术的灵敏性和分辨率都得到了显著提高。下面就对该技术的发展历程中的重要改进相关事件进行详细综述。

3.4.1.1 多色荧光原位杂交与染色体条形码

多色 FISH（Multicolor-FISH，McFISH），是指不同的 DNA 探针用不同的半抗原标记，杂交后再通过不同的荧光素进行检测，从而在荧光显微镜不同滤色片下显示出多种颜色信号，进而对不同的靶 DNA 探针同时进行定位和分析。该方法可以同时确定多个基因在同一染色体组中的位置及排列次序。在一定的实验基础上，利用多色 FISH 可以准确地确定多个彼此邻近或密切连锁的基因在染色体上的物理次序。为了同时检测 rDNA 在染色体组中的分布位点情况，很多研究通过 McFISH 的方法，同时将 45S rDNA 和 5S rDNA 探针定位于染色体组中，这有助于研究二者的相互位置关系（Kim et al., 2005b；Choi et al., 2008；Witkowska et al., 2009）。

随着 McFISH 的发展，在人类染色体的 McFISH 研究过程中，又衍生出彩色显带 FISH。显带 FISH 是在 FISH 技术的基础上，制作出全染色体带型图，获得多色染色体条码。与传统的明暗相间的带型不同，所得带型图是五颜六色的带型，结果使染色体排序更客观、更易自动化（Müller et al., 1997）。Müller 等通过一次荧光原位杂交，将人类所有染色体通过一组 DNA 探针区分开来。其理论基础是人类和灵长类染色体上的非重复序列具有很高的同源性，但由于染色体进化方向差异，这些同源序列在人类和灵长类染色体上的排列存在较大差异。多色染色体条形码的基本原理是：首先，运用体细胞杂交技术，建立含有供试物种染色体随机片段的杂交细胞库；其次，将这些杂交细胞库随机分成几部分，经 Alu-PCR 扩增出其中的人类 DNA 片段，再用不同的荧光分子标记，最终制备成混合探针池。所以，该研究首先将杂交细胞库的不同部分分别标记为红色和绿色荧光探针。在荧光显微镜下检测时，染色体上与不同 DNA 探针杂交的区域分别显示红色荧光或绿色荧光，与两种 DNA 探针都杂交上的部位可以用黄色显示，而且，根据红色和绿色荧光各自的强弱不同，重叠后显示的颜色也不尽相同。杂交后，人的 24 条染色体中每一条染色体都显示出独特的信号序列，形成各自独特的彩色带型，这一结果被称为"染色体条码"。经计算机密度计法分析，该杂交模式在各种标记和杂交实验中具有很好的重复性。因此，染色体条码在诊断细胞遗传学、基因组研究和快速鉴定染色体及染色体重排等研究中具有重要意义。另外，彩色显带 FISH 还可以很好的检测染色体内的异常结构（赵萌等，2002）。

3.4.1.2　基因组原位杂交（GISH）

基因组原位杂交（Genomic In Situ Hybrieization，GISH）技术是 20 世纪 80 年代末 90 年代初发展起来的一种原位杂交技术（Pinkel et al.，1986），主要是利用一个物种的基因组 DNA 作探针对其他物种染色体进行荧光原位杂交，其杂交信号检出设计专门用于检出重复 DNA 序列，因此可以直观地显示不同物种间共同的重复序列在染色体上的分布情况（Takahashi et al.，1997）。最初应用于动物方面的研究，但很快在植物中得到推广应用。用于检测植物杂交后代是否存在或渗入的外源 DNA 情况，后来逐渐成为一种基因组进化分析的新手段，用于研究不同种间的物种进化和亲缘关系研究，是研究不同种基因组间重复序列的变异性和保守性的一种简便方法。它用来自一个物种的总基因组 DNA 作为探针，以适当浓度的另一物种总基因组 DNA 进行封阻，在靶染色体上进行原位杂交。相对于标记 DNA 探针，高浓度封阻 DNA 优先与靶染色体上共有保守序列杂交，剩下的外源种特异性序列与探针杂交。在已有的报道中表明，GISH 能清晰地显示小麦基因组中的外源附加系、代换系及异位系等。任南等（1997）用生物素标记水稻基因组 DNA 作为探针，在玉米有丝分裂中期染色体上进行基因组荧光原位杂交，发现许多杂交带，且杂交区域与非杂交区域间隔排列，其中杂交带主要分布在染色体臂的中部。宁顺斌（2000）分别用水稻基因组 DNA 探针对玉米染色体进行杂交，以及反过来用玉米基因组 DNA 探针对水稻染色体进行杂交，发现水稻基因组 DNA 探针杂交玉米染色体时，所有玉米染色体上都显示出多个间隔的杂交带区；而用玉米基因组 DNA 探针杂交水稻染色体时，所有水稻染色体上都显示连续的杂交信号。这些结果表明玉米和水稻基因组之间具有高度同源性。

GISH 技术还是跟踪检测杂种的真实性及其后代外源染色体或染色体片段的存在与否的一个快速、灵敏和准确的手段（Melo et al.，2015）。它能对外源染色体或片段计数，显示它们的大小、位点和形态以及与寄主染色体的重组情况。该技术在染色体较小的水稻中成功地鉴定了栽培稻与药用野生稻 F1 杂种的真实性及其回交后代中的异源附加系。因此，无需经过基因表达产物分析来推断基因转移与否，就能检测出整合于染色体的外源染色质。例如，用不同的半抗原标记来自两个不同的基因组的总 DNA，含等量两种标记 DNA 的探针杂交液加热变性后，经预退火反应，将杂交混合液加到其中一个物种的染色体制片上，杂交后再用不同的荧光素进行检测，可以用来分析两个物种间相同的和不同的重复 DNA 顺序在其中一个物种的染色体上的分布特征，这类应用在番茄、油棕、水稻、黑麦等物种及其杂种后代的分析中均有报道（Escalante et al.，1998；Madon et al.，1999；Hassani et al.，2008；覃瑞等，2009；Książczyk et al.，2010；Xu et al.，2016）。

3.4.1.3 荧光原位杂交技术及 rDNA-FISH

1969 年，Gall 和 Pardue 用放射性同位素标记爪蟾的 rDNA，并将该 rDNA 与爪蟾染色体标本进行杂交，成功将 rDNA 定位于染色体的核仁组织区，该实验的实施标志着染色体荧光原位杂交技术的初步建立。在早期的染色体原位杂交实验中，所用探针均为放射性同位素标记，这类探针的优点是检测灵敏度很高，所以在原位杂交试验建立初期很受欢迎。但随着原位杂交技术的不断发展，放射性同位素探针的缺点越来越明显。首先，试验产生的放射性废液对实验人员和环境均具有不可避免的危害作用；其次，试验过程烦琐，试验周期长；第三，放射性同位素探针不易保存，只能现做现用，且染色体杂交后颜色深，影响实验精确度。因此，在这种情况下，急需探索新的探针标记物质。

至 1982 年，Langer-Safer 等首次使用缺刻平移法标记用不具放射性的生物素进行探针标记，所得探针杂交信号较稳定，标志着染色体原位杂交技术取得了突破性进展。与传统的同位素标记探针进行的原位杂交技术相比，用生物素标记探针具有以下四个优点：(1) 试验周期时间明显缩短；(2) 生物素所标记探针的化学性质稳定，可进行重复使用长达数月；(3) 生物素标记探针的原位杂交在信号检测时，信噪比明显提高，背景噪音明显减低；(4) 通常情况下，所得杂交结果的清晰度或分辨率会高于同位素标记的探针（Langer-Safer et al., 1982）。该探针的成功应用标志着非放射性原位杂交技术（Non isotopie in situ hybridization）的建立，这是染色体原位杂交在方法上取得的一个重大进展。非放射性原位杂交，有许多优点，不仅标记的探针稳定，非特异性杂交信号污染少，检测方法简便快速，而且能通过多级免疫酶联反应进行多次信号放大，显著提高信号检出率（Lichte et al., 1990）。因此，该方法一经公布便被广泛应用（Singer et al., 1987；Kennedy et al., 1989；Zhuang et al., 1994）。即使在近些年的研究中，也不乏生物素标记探针的应用（Cooper, 2015）。

在荧光素物质出现后，Pinkel 等首先开始使用荧光素标记探针，杂交信号可以直接在荧光显微镜下进行观察（Pinkel et al., 1986），此即荧光原位杂交技术（Fluorescence In Situ Hybridization, FISH）。与其他原位杂交技术相比，FISH 具有许多独特的优点。最主要的是，该技术方法程序简单，灵敏度高，对比明显，图像反差大。目前，该种探针标记方法已广泛应用于基因定位、细胞遗传图谱构建、转基因后代的细胞学鉴定等多方面的研究中（Trask et al., 1993；Zhang et al., 1999；Wang et al., 2006；Di et al., 2010；Ibáñez et al., 2015）。

核糖体 DNA 是染色体分析中常用的细胞遗传学标记物。rDNAs 位点数量和分布的研究为阐明基因组的组织和分布特征提供了重要的信息。Kong 等（2009）研究了 18S-25S rDNA 在羽扇豆属四个种之间的分布特征，并对其进化关系进行了讨论。在多倍体基因组的物种及其杂交后代的鉴别中，rDNA 位点数目及分布也提供了重要依据（Linares et al., 1996；Snowdon et al., 1997）。在物种进化和种间关系的研究中，rDNA 分布特征也具有重要意义。Mantovani 等（2005）研究指出，在粗鳍丽脂鲤的四个类群中，5S rDNA 位点较为保守，45S rDNA 位点更具有可变性。Zhang 等（2016b）在黄瓜属 rDNA 分布特征研究中指出，45S 和 5S rDNA 的进化模式截然不同，5S rDNA 具有多倍体化的倾向，而 45S rDNA 呈现出位点数量增加的趋势。Qi 等（2015）通过对菊科 12 个近缘种的 45S 和 5S rDNA 杂交位点差异分析，为菊科的系统发生和进化研究提供了理论依据。2013 年之前的报道中，研究多用重组质粒或者 PCR 扩增制备探针的方法制备 rDNA 探针（Linares et al., 1996；Snowdon et al., 1997；Gu et al., 2003；Noleto et al., 2007；Ying et al., 2011），2013 年，陈成彬等（2013a，2013b）报道了一种新的 rDNA 探针制备方法，即在寡核苷酸探针合成时进行荧光基团修饰，不用再进行荧光探针的标记，与之前的标记技术相比，具有方法简单且成本低廉的优点，rDNA 定位研究的试验成本和周期都大大缩减。

3.4.1.4 FISH 分辨率和基因定位

因为检测信号分布太少，低拷贝或单拷贝的 DNA 序列直接定位是很难的。因而可以通过间接法利用该基因所在的 BAC 克隆的荧光原位杂交实现（Sazanov et al., 2010）。水稻 BAC 文库和其他文库相比，具有显著优势，因此被广泛应用于水稻基因组研究。Yan 等（Yan et al., 1998）用两个生物素标记的水稻 BAC 克隆与水稻染色体进行了原位杂交。这两个克隆与稻瘟病抗性基因、青叶蝉抗性基因和东格鲁球状病毒抗性基因等 3 个基因紧密连锁。这两个克隆定位于 4 号染色体的长臂和短臂，FL 值分别是 40%和 100%。信号检出率达到 46.8%和 59.2%。信号位置与水稻分子图谱完全一致。研究结果显示了用原位杂交在染色体上定位 BAC 克隆的优势，有助于应用 BAC 文库对低拷贝或单拷贝基因进行染色体定位。通过将基因序列直接标记为探针与染色体进行荧光原位杂交，可以直接完成基因在染色体上的定位。但由于相对于 BAC 序列，基因序列较短，可供检测的荧光信号不够强，用常规的荧光原位杂交技术直接检测时通常难以检测到目标信号。为此，在很多报道中采取了不同了改进或增强单拷贝基因探针信号的方法，以达到对低拷贝短基因片段进行定位的目的。1992，年 Lemieux 等（1992）采用改进的荧光原位杂交方法，通过使用碱性（pH11）对苯二胺抗褪色溶液，得到了不需要额外处理或任何扩增系统的染色体条带，可以直接在人类染色体上检测小至 500 bp 的单拷贝基因。由于条带的产生与 5-溴代脱氧尿嘧啶的结合时间有关，可以同时观察到 R-带或 G-带、结构异染色质的成分或染色体不对称性，因此该技术可以得到高分辨率的染色体图谱。相对于动物，单拷贝基因在植物染色体上的检测更为困难，为了提高单拷贝基因在植物荧光原位杂交中的检出率，Wang 等（2006）在研究中尝试在玉米粗线染色体上检测单拷贝基因。结果表明，由于粗线期染色体较为舒展，变相拉长了目标基因探针的检测范围，因此该方法可以检测到的探针倡导缩短至 3.1 kb，达到了检测单拷贝基因的技术水平。

通常情况下，中期染色体的分辨率经被评估大约为 3 Mb，这个数量的 DNA 相当于染色体 G-带分析中一个较小的带，这使得中期染色体非常适合于大片段 DNA 细胞遗传学定位。如何提高 FISH 技术的分辨率是一个重要课题。而 FISH 的分辨率取决于载体 DNA 的浓缩程度。想要在中期染色体上检测 DNA 序列相近的探针顺序几乎是不可能的（Raap et al., 1998），但是更放松的状态下，染色质间期核的分辨率大大增加，FISH 分辨大约率为 100 kb（Trask et al., 1989；Lawrence et al., 1990；Trask et al., 1991）。同时，在 100 kb-1 Mb 的范围内，FISH 信号之间的位置关系与他们在基因组上排列的顺序呈现线性相关的关系（Van et al., 1992）。

特别是在多色荧光原位杂交模式中，染色体间期相的 FISH 为 100kb-1Mb 片段的定位和排序提供了有力的工具。但是为了进一步提高 FISH 分辨率，除了提高硬件的检测能力以外，更加舒展和解凝状态的染色体也很受研究者欢迎。Moers 等（1995）通过近场扫描光学显微镜研究表明，提高显微镜的分辨率可以获得更好的 FISH 分辨率。然而，使用去浓缩的染色质和常规的荧光显微镜相结合是更为实用（Raap et al., 1996）。Brandriff 等（1991）使用仓鼠卵子与人类精子融合后得到的原核，染色质高度去浓缩，FISH 分辨率达到了约 50 kb。Wiegant 等（1992）、Parra 和 Windle（1993）分别采用不同技术方法，将哺乳动物细胞核中的裸 DNA 纤维固定到玻片上，此类染色体上进行的荧光原位杂交统称为 DNA 纤维荧光原位杂交技术（Fiber FISH）。早期的 Fiber FISH 报道表明 1 kb 的线性 DNA 长度为 0.34 μm，其光学分辨率为 0.2~0.3 μm，观察对象的不同放大倍数和不同型号和质量的数码相机均会影响观察结果。Florijn 等（1995）也报道了在荧光显微镜下可以清楚的观察到带有 1~3 kb 插入片段的质粒探针在 DNA 纤维上的杂交信号。

DNA 纤维 FISH 的关键就在于如何制备高质量的线性 DNA 纤维。理想地来说，制备的 DNA 长度应与完全自然伸展的 DNA 纤维相近，并且断裂点应尽量少。与其他载体上的 FISH 相比，纤维 FISH 在 DNA 作图方面，主要优点很明显：

1. 它的分辨率大大提高，分辨率为 1-2 kb，灵敏度可达 200 bp（Florijn et al., 1996），与常规分子生物学的限制性酶切图谱相差无几（Zhong et al., 1996），能满足图位克隆的要求（Heiskanen et al., 1994）。

2. 材料要求不高。各种方式包括用限制性酶消化制备的线性 DNA 分子均可以满足试验要求，因而对材料的要求也不高，不像中期染色体 FISH，要取分裂生长旺盛的分生组织为材料；并且植物中期染色体的制备需要很好的实验基础和丰富的实验经验。

3. 在 10~1000 kb 以上的线性 DNA 分子伸展后每 μm 长度约为 2.3 kb，因而可把长度结果（μm 级）直接转换为 DNA 片段大小（kb），大大加速了物理图谱的构建进程（Weier et al., 1995）。但是，DNA 纤维 FISH 也有缺点，比如在 DNA 纤维制备过程中，不可避免的会受到局部物理损伤，或者被没有完全去除的蛋白覆盖，进而影响与探针的杂交。

研究表明，除了改变用于杂交的染色体的状态（采用非中期的浓缩程度较低的染色体），对于可供检测的荧光信号不够强的较短基因序列来说，采取荧光信号放大系统将信号进行级联放大后再行检测，同样可以大大提高单拷贝基因在中期染色体上的检出率。同时，生物样品的种类、制备方法、原位靶 DNA 和显微镜硬件配置对检测灵敏度都有很大的影响。这里主要介绍下酪胺荧光信号放大系统（tyramide signal amplification，TSA）。TSA 在免疫细胞化学研究中具有重要的里程碑意义。Kersten（Kerstens et al., 1995）和 Raap（Raap et al., 1995）分别报道了 TSA 对 FISH 杂交信号强度的大幅增加的效果。同时，一些免疫细胞化学研究也表明 TSA 具有很强的信号放大潜力（Berghorn et al., 1994；Merz et al., 1995；Hunyady et al., 1996；Schöfer et al., 1997）。现在的问题是，能否用 TSA 检测出在传统 FISH 试验中检测不到的短的特异 DNA 序列和低丰度的 mRNA 靶点？Zubáčová 等（2011）的研究有力的证实了 TSA 系统在这方面的潜力。

阴道毛滴虫基因组的高度重复性和各种基因家族的大量扩张，更增加了其单拷贝基因定位的难度。但 Zubáčová 等利用 TSA 系统，成功的提高了传统 FISH 的灵敏度，完成了单个拷贝基因在阴道毛滴虫中期染色体上的定位，结果使两个编码丝氨酸棕榈酰转移酶和色氨酸酶的单拷贝基因（分别为 1419 bp 和 1377 bp）分别定位于染色体 I 和 II，为这两对染色体提供了特异的染色体标记物，也为单拷贝基因的 FISH 定位提供了有力的技术保障。FISH 对单拷贝基因定位的研究，同样也可以用来对转基因后插入位点进行检测。曹莹莹等（2013）用荧光原位杂交的方法，以正常羊及正常人染色体标本为对照，通过生物素标记的探针与转基因羊染色体标本进行杂交，经信号放大及 DAPI 染色后发现转基因羊在不同染色体上各有一对明显的荧光信号，而对照组无信号。得出了 hEPO 基因在转基因山羊染色体上的整合是单位点整合，且分别整合在不同的染色体上的结论。吴庆洲等（2010）应用荧光原位杂交技术，检测到外源 HBV 基因以单位点形式稳定地整合到转基因小鼠的染色体上。金危危等（2002）也利用荧光原位杂交技术在供试 8 个转基因水稻株系的染色体上检出了转入的外源基因 barnase-ps1 片段的整合位点。Ibáñez 等（2015）通过荧光原位杂交的方法，对小鼠转基因精子进行了筛查，检测单个精子中的整合转基因，并对含转基因精子的比例进行了评估。

3.4.1.5 重复荧光原位杂交技术

1995 年，Wang 等介绍了一种新的荧光原位杂交方法，即重复荧光原位杂交（rehybridization，ReFISH）。研究用生物素标记的特异性 DNA 探针与已经进行过荧光原位杂交的中期染色体再次杂交，使得不同的 DNA 探针检测相同的中期染色体成为可能（Wang et al.，1995）。这一研究结果具有重要意义，特别是对于稀有材料的 FISH 杂交开辟了新的领域。同时，多次杂交结果在同一组染色体上体现出来，也为研究不同染色体的特别标记提供了新的思路。2002 年，Müller 等也提出了 ReFISH 的概念，描述了同一样品与不同 DNA 探针重复杂交，杂交一轮之后，洗去已结合的 DNA 探针，再进行新一轮杂交的过程。如此循环，结果可以在同一组染色体上同时显示出来多次杂交所用 DNA 探针的全部分布情况。该研究结果显示对同一染色体标本的重复杂交可多达四次，Kitayama 等所实施的重复原位杂交试验也验证了这一结果，也成功完成了四次重复杂交（Kitayama et al.，2010）。随着染色体制片的改进和制片处理方法的改进，染色体重复杂交次数也有了很大提高。ReFISH 不仅能大大提高玻片标本利用率，特别是对难以获得的稀有染色体标本具有重要意义（Kitayama et al.，2010），另外一个重要意义是通过不同批次探针标记在同一染色体组上的分布，获得不同染色体之间的标记信息的差异，为不同染色体的识别提供准确的标记，这对染色体短小和不同染色体间形态结构比较相似的物种来说尤为重要。在重复杂交次数方面，报道在同一染色体玻片标本上完成四次重复荧光原位杂交的例子较多（Müller et al.，2002；Shearer et al.，2014），杂交次数最多报道来自 Tran 等（2016），重复杂交次数可高达八次，染色体能经受的连续杂交次数多少与染色体玻片标本的质量和操作人员的熟练程度密切相关。

（作者：赵瑞红，导师 宋文芹。原载：2018 南开大学博士论文）

3.4.2 菠菜 rDNA 及端粒多色荧光原位杂交分析

随着分子细胞遗传学技术的进步,传统的核型分析技术已经逐渐显露出弊端,通过荧光原位杂交(FISH)技术对染色体进行分子标记,可以为核型分析提供有效的细胞学标记,弥补传统核型分析的不足。多色荧光原位杂交技术是 FISH 技术的一个重要分支,利用多色荧光原位杂交技术可以同时将多种不同标记的探针定位到染色体上,目前利用多色荧光原位杂交进行分子核型分析已经成功地在酸模(*Silene latifolia*)和松属中应用(刘博等,2005;Bie Shu *et al.*,2004)。核糖体 RNA 基因(rDNA)是高度保守的重复序列家族,拥有几百乃至上千个拷贝,成簇分布于一对或多对染色体上(Bie Shu *et al.*,2004)。对 rDNA 在染色体上定位的研究既可以为分子核型分析提供标记,同时也可以为核型进化、系统发育提供重要信息,因而是目前研究最为广泛的染色体分子标记之一。端粒是染色体两个端部特化结构,通常由串联重复序列 DNA 组成,伸展到染色体的 3′端。它是一种在植物界包括被子植物,裸子植物和蕨类植物中广泛存在的重复序列(Rogers *et al.*,1987;Beech *et al.*,1993),其主要分布于染色体的末端或其他部位,是染色体分子核型分析的重要标记之一。

菠菜是藜科菠菜属植物。传统核型分析表明菠菜共有 $2n=2x=12$ 条染色体,包括 1 条中着丝粒染色体,2 条近中着丝粒染色体,2 条近端着丝粒染色体和 1 条端着丝粒染色体,其中随体分别位于 5 号和 6 号染色体。本文借助 rDNA 和端粒序列在菠菜染色体上的定位,纠正了随体的数目,是对菠菜传统核型分析的重要补充,同时也为菠菜的分子核型研究提供了证据。

3.4.2.1 材料和方法

1. 实验材料

本实验取菠菜的嫩芽和根尖为实验材料,菠菜采自于天津市蓟县。

2. 染色体标本的制备

染色体标本的制备参照陈瑞阳等(陈瑞阳等,1979)的去壁低渗法。

3. 探针标记及荧光原位杂交分析

制备探针的 25S rDNA,5S rDNA 和端粒序列均来源于拟南芥。25S rDNA 和 5S rDNA 用生物素切刻平移试剂盒(Roche)标记,端粒序列和 5S rDNA 用地高辛切刻平移试剂盒(Roche)标记。进行多色荧光原位杂交时,探针 25S rDNA、端粒序列、地高辛标记的 5S rDNA 与生物素标记的 5S rDNA 按 2∶2∶1∶1 的比例混合。荧光原位杂交及信号检测按 Qi 等(2002)的方法进行。

3.4.2.2 结果

1. 多色荧光原位杂交结果

菠菜中期染色体 25S rDNA、5S rDNA 以及端粒的多色荧光原位杂交结果图 3.90A。图中绿色信号为 25S rDNA 杂交位点,红色信号为端粒序列杂交位点,黄色信号为 5S rDNA 杂交位点。3 种杂交信号清晰且表现稳定,就信号强度而言,25S rDNA 的信号最强，其次是端粒序列，5S rDNA 的杂交信号较前两者弱。图 3.90E，F 分别为 25S rDNA、5S rDNA 单色荧光原位杂交结果，验证了 25S rDNA 杂交位点为 6 个，5S rDNA 杂交位点为 4 个。

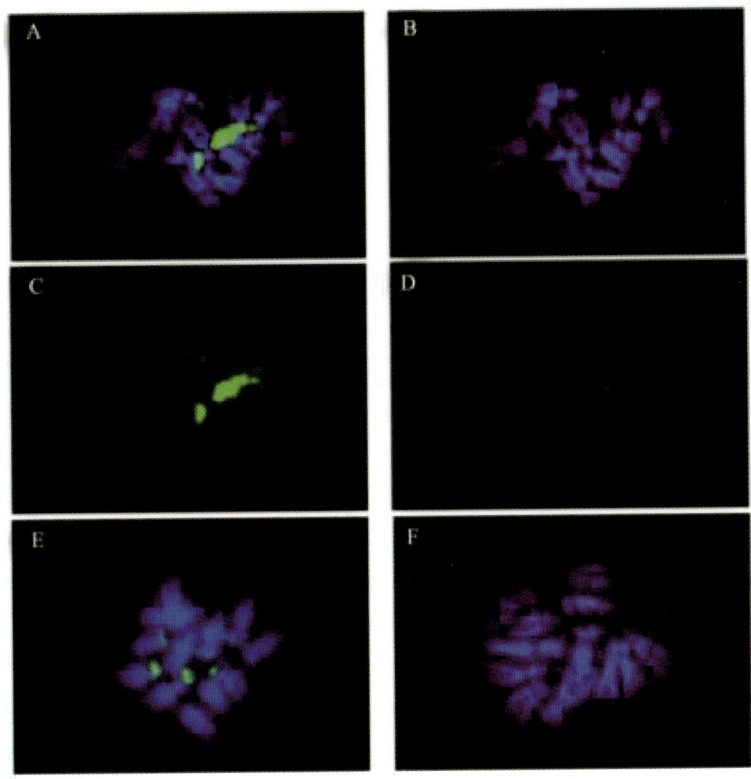

图 3.90　25S rDNA、5S rDNA 及端粒 DNA 与菠菜有丝分裂中期染色体的多色荧光原位杂交结果
A：25S rDNA、5S rDNA 及端粒 DNA 杂交结果，为 B、C 和 D 的叠加图；B：DAPI 染色图；C：生物素标记荧光信号；D：地高辛标记荧光信号；E：生物素标记 25S rDNA 单色杂交结果；F：地高辛标记 5S rDNA 单色杂交结果

Fig. 3.90　Multicolor fluorescence in situ hybridization with 25S rDNA, 5S rDNA and telomere DNA on metaphase chromosomes of spinach
A: FISH image using 25S rDNA, 5S rDNA and telomere DNA as the probes, merged by B, C and D; B: DAPI-stained chromosomes; C: detection of the biotin-labeled signals; D: detection of the DIG-labeled signals; E: FISH image using biotin-labelled 25S rDNA as the probe. F: FISH imageusing DIG-labelled 5S rDNA as the probe

2. 核型分析

将菠菜中期染色体多色荧光原位杂交结果进行同源染色体配对及排序，结果见图 3.91。25S rDNA 杂交信号为 6 个，分别位于 3 号，5 号以及 6 号染色体的次缢痕及随体部位，其中位于 3 号染色体的杂交信号较其余信号小且强度弱；5S rDNA 杂交信号为 4 个，分别位于 3 号染色体和 5 号染色体长臂，信号强度和大小没有明显差异；端粒序列杂交信号位于 6 号染色体端部以及其余染色体的端部和着丝粒部位，其中 6 号染色体端部的杂交信号较其余信号略强。

图 3.91　菠菜中期染色体核型图
Fig. 3.91　The karyotype of spinach

3.4.2.3　讨论

1. rDNA 在菠菜染色体上的定位研究

高等植物 rDNA 是编码各种 rRNA 前体的基因，包括 45S rDNA 和 5S rDNA。45S rDNA 由 18S、5.8S 和 25S rDNA 构成，目前已有大量的研究证明，在大多数物种中其主要定位在核仁组织区（NOR）即染色体的次缢痕部位（徐延浩等，2007；刘博等，2005）。25S rDNA 是 45S rDNA 的一部分，其定位位点与 45S rDNA 一致。从染色体结构来讲，NOR 的数目应该与随体数目相对应，因此，25S rDNA 的杂交位点数目即是染色体随体的数目。菠菜传统核型分析表明其随体个数为 2 对，分别位于 5 号和 6 号染色体，而通过 25S rDNA-FISH 结果表明随体个数为 3 对，分别位于 3 号、5 号和 6 号染色体。位于 3 号染色体上的 25S rDNA 杂交信号比其余的信号要弱且小。FISH 是一种半定量技术，尽管其结果不能直接显示基因的实际拷贝数，杂交信号的大小和强度却间接反映了基因拷贝数的多少，因此可以判断位于 3 号染色体上的 25S rDNA 的拷贝数较少。传统的核型分析主要依靠肉眼判断而不借助分子标记，当染色体上的随体太小，次缢痕不明显的时候，很容易将其判断为没有随体，因而造成了误差。诸如此类的情况在棉花等植物染色体中也有发生（Bie *et al.*，2004）。菠菜的 5S rDNA 定位于 3 号和 5 号染色体长臂，均与 25S rDNA 位于同一染色体。5S rDNA 在染色体上的位置随物种不同而不同。在高等植物中，包括被子植物和裸子植物，5S rDNA 的位置并不恒定，目前还没有发现其位于 NOR 区的物种；在部分苔藓类植物和原核细胞中，5S 则定位于 NOR 区（Srivastava *et al.*，1991；Sone *et al.*，1999）。

2. 端粒序列在菠菜染色体上的定位研究

端粒是一种广泛分布于动植物中的重复序列，它是染色体重要的功能元件，主要作用是维持染色体的完整性和个体性。一个基因组内的所有端粒都是由相同的重复序列组成，但不同物种的端粒的重复序列是不完全相同的。哺乳类和其他脊椎动物端粒的重复序列中的保守序列是 TTAGGG，串联重复 500~3 000 次，序列长度在 2 kb 到 20 kb 之间不等。植物界中大多数植物都有端粒序列，包括蕨类植物，如溪苔（*Pellia epiphylla*），和裸子植物，如油果樟（*Zamia furfuracea*）和赤松（*Pinus sylvestris*）（Fuchs 等 1995）。目前发现的植物端粒的重复序列中的保守序列是 AAATGGG，这个序列是首先在拟南芥中发现，因而也称为拟南芥型端粒序列。不过有的植物，如洋葱（*Alliun cepa*），在其染色体端部则没有端粒序列，取而代之的是其他的一些串联重复序列。在对端粒的染色体定位研究发现，在一些动物中，人的端粒序列不仅定位于染色体端部，在中间和近端区域也有发现（Meyne et al., 1990）。在植物中，端粒序列也不仅仅定位于染色体端部，如在蚕豆（Vicia faba）和苏铁（Cycas revoluta）中，端粒可以定位于着丝粒部位（Shubert, 1992, 1995; Hizume et al., 1998），在赤松中则可以定位于 interstitial 区域，而在凤仙花（Tradescantia commelinoides）中，则定位于端部的异染色质区（Cox et al., 1993）。从分子水平上研究发现，拟南芥的着丝粒 DNA 和 interstitial DNA 中确实含有端粒序列（Richards et al., 1991; Regad et al., 1994），因而可以很好的解释端粒序列在这些区域的定位。

本研究采用的端粒序列为拟南芥型端粒序列，即（AAATGGG）$_n$，杂交信号位于 6 号染色体端部以及其余染色体的端部和着丝粒部位，即在菠菜染色体上，端粒主要定位于端着丝粒染色体及近中着丝粒染色体的两端和着丝粒部位。6 号染色体着丝粒端的信号比其他的端粒杂交信号要强，主要原因可能是由于这个部位是端粒和着丝粒重叠的部位，因此其端粒序列的拷贝数要比其他信号部位多。

（作者：兰添颖、刘博、董凤平、陈瑞阳、李秀兰、陈成彬。原载：遗传，2007，29（11）1405-1408）

3.4.3 豆科三属八种植物的核型及 rDNA 定位研究

在植物基因组中研究最广泛的遗传单元之一是核糖体 RNA 基因（rDNA）以及它们在染色体上的定位。rDNA 是高度保守的重复序列家族，拥有几百乃至几千个拷贝，成簇分布于一对或多对染色体上（Pedersen and Linde-Laursen, 1994）。通过荧光原位杂交（FISH）技术将 rDNA 在染色体上进行定位，可以为核型分析提供有效的细胞学标记，特别对于染色体较小且形态相近的物种，这种染色体标记更是核型分析的重要工具。如 Zoldos 等（1999）在染色体难以区分的栎属（Quercus）11 个种中定位了 rDNA，使 24 条染色体中的 6 条得已明确的分辨。rDNA 虽然在序列上高度保守，在染色体上的物理位置具有固定性，但在物种进化过程中并非是一成不变的，即使在亲缘关系很近的种中甚至种内，rDNA 位点在染色体上的位置、数目和拷贝数也可能有差异。Naganowska 和 Zieliska（2002）通过观察羽扇豆属（Lupinus）5 个种的 18S-25S rDNA 位点的数目和大小分析种间差异；Torrell 等（2003）在蒿属（Artemisia）7 个种中，通过 rDNA-FISH 结合荧光分带分析，确定了其种属间的分类关系；Thomas 等（2001）检测了瑞士黑麦草（Lolium rigidum）8 个基因型中 rDNA 位点的数目和位置，以确定染色体的结构变异；Galasso 等（1998）发现豇豆（Vigna unguiculata）在人工种植过程中，其 rDNA 位点的数目有所增加；Taketa 等（1999）通过对大麦属（Hordeum）9 个野生种及细胞型的研究，发现 rDNA 的定位为研究该属的核型进化、系统发育提供了非常有价值的信息；Hanson 等（1996）、Singh 等（2001）、Gu 和 Xiao（2003）通过 rDNA 的定位分别为棉属（Gossypium）、大豆属（Glycine）和山茶属（Camellia）的多倍体的起源提供线索叶或常绿乔、灌木。约 70 余种，广泛分布热带至温带地区，我国有 21 种，14 变。豆科槐属（Sophora L.）为落种，2 变型。国槐（S.japonica L.）是优良的行道树和蜜源植物，其两个变型五叶槐（*S.japonica* L.f.*oligophylla* Franch.）和龙爪槐（*S.japonica* L.f.*pendula* Loud.）及该属另两个种黄金槐（*S.xanthantha* C.Y.Ma.）和红花槐（*S.rubriflora* Tsoong.）是重要的园林绿化树种；刺槐属（*Robinia* L.）的刺槐（*R.pseudoacacia* L.）、毛洋槐（*R.hispida* L.）和紫穗槐属（*Amorpha* L.）的紫穗槐（*A.fruticosa* L.），是绿化和固沙造林树种（陈德昭等，1994）。对于这些树种，陈瑞阳等（2003）已对其进行了核型分析，但分子细胞遗传学领域的研究尚属空白。我们将 45S rDNA 作为探针在这 8 个树种的中期染色体上进行原位杂交定位，为更精细的核型分析提供染色体标记，同时研究该基因在槐属、刺槐属和紫穗槐属基因组中的分布、特点及种间差异，为研究物种分化过程中染色体的形态变化提供依据。

3.4.3.1 材料与方法

1. 材料

国槐（凭证标本号：Chen Ry 2003-123）、五叶槐（Chen Ry 2003-124）、龙爪槐（Chen Ry 2003-125）、黄金槐（Chen Ry 2003-126）、红花槐（Chen Ry 2003-61）、刺槐（Chen Ry 2003-121）、毛洋槐（Chen Ry 2003-122）采自南开大学校园。紫穗槐（Chen Ry 2003-136）种子采自河北省乐亭。标本存放于南开大学生命科学学院标本室。

2. 染色体标本的制备

除紫穗槐取种子根尖外，其他 7 种材料均是在早春时取幼芽。材料用饱和对二氯苯水溶液预处理 3 h，之后用固定液（甲醇：冰醋酸=3：1）固定。染色体标本的制备采用陈瑞阳等（1979）的去壁低渗法。选取染色体形态和分散较好的标本用于荧光原位杂交。

3. 荧光原位杂交

含有 rDNA 序列的 pBluescript 由武汉大学宋运淳教授惠赠。rDNA 片段来自番茄（Lycopersicon pennellii LA716），长 9.1 kb，包含 18S -5.8S-28S 编码区和非转录间隔序列（简称 45S rDNA）（Arumuganathan *et al.*，1994）。通过随机引物法用 DIG（Digoxigenin-dUTP，Roche）标记 45S rDNA 作为探针。杂交及杂交后信号的检测按 Qi 等（2002）的方法进行。

3.4.3.2 结果

1. 核型分析结果

我们将上述 8 种植物进行了核型分析（陈瑞阳等，2003），结果见表 3.6。
rDNA 位点：数字示具 45S rDNA 位点的染色体编号，*示随体染色体。

表 3.6 槐属、刺槐属和紫穗槐属 8 种植物核型主要特征及 45S rDNA 位点的比较
Table 1 The karyolypic data and 45S rDNA loci of the 8 species in this study

种类 Taxa	核型公式 Karyolype formula	平均臂比 A.A.R	最长/最短 Li/St	臂比>2 的比率 P.C.A%	核型不对称系数 Ax K%	类型 Type	rDNA 位点 rDNA bei
国槐 *S. Japanica*	2*n*=4*x*=28=18*m*+10*xn*	1.678	2.04	29	61.70	2B	2,7
五叶槐 *S. Japanica oligophylla*	2*n*=4*x*=28=14*m*+10*xn*+4*xl*	1.913	2.38	29	63.48	2B	2,4

龙爪槐 S. japonica pendula	$2n=4x=28=18m+8xn+2xl$	1.717	2.34	36	61.38	2B	2,7
黄金槐 S. xanthoanina	$2n=4x=28=14m+12xn+2xl$	1.962	1.97	21	62.38	2A	1,3
红花槐 S. rubriflora	$2n=3x=21=6m+6xn+6xl+3l$	3.631	1.72	57	69.97	3A	5*
刺槐 R. Pseudoacocia	$2n=2x=22=4m+8xn+10xl$	3.562	2.38	82	71.67	3B	6*,9*
毛洋槐 R. hispida	$2n=2x=30=10m+12xn+6xl+2l$	2.770	2.06	60	66.41	3B	2,6,11*,13*
紫穗槐 A. fruticosa	$2n=2x=40=32m+8xn$	1.427	1.70	0	58.27	1A	8,10*,13

A.A.R：Average arm ratio；Lt：Longest arm；St：Shortest arm；P.C.A.：Percentage of chromosome with arm ratio>2.

2. 45S rDNA 在中期染色体上的定位

8 个种的中期染色体与 45S rDNA 的荧光原位杂交结果见图 3.92，同源染色体配对及排序结果见图 3.93；45S rDNA 位点所在染色体编号见表 3.6。在所研究的 8 个种中，未见单条染色体上具多个 45S rDNA 位点的现象。具体观察结果分列如下。

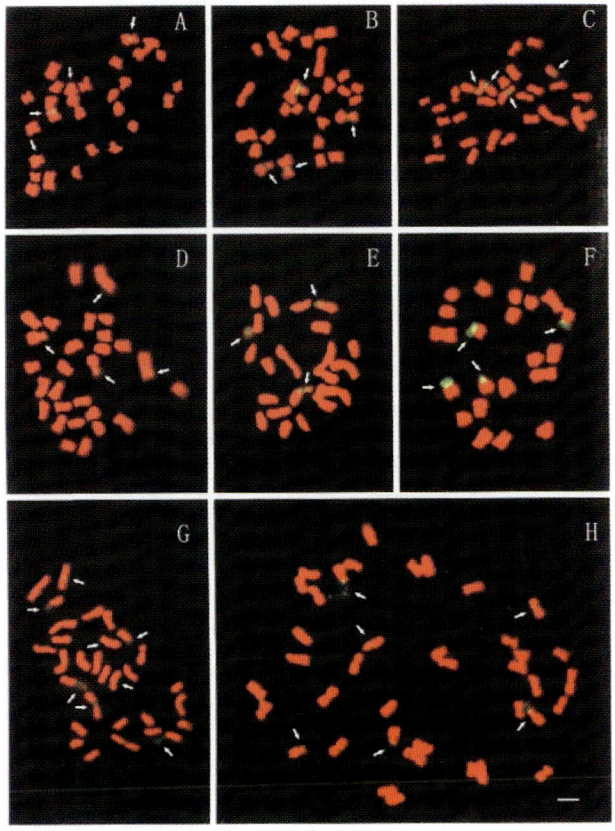

图 3.92 45S rDNA 与 8 种植物中期染色体的荧光原位杂交结果

Fig. 3.92 Fluorescence in situ hybridization with 45S rDNA on metaphase chromosomes of the eight species
A. *S.joponica*；B. *S.japonica* f.oligophylla；C. *S.japonica* f.pendula；D. *S.xanthantha*；E. *S.rubriflora*；F. *R.pseudoacacia*；G. *R.hispida*；H. *A.fruticosa*. bar =5 μm

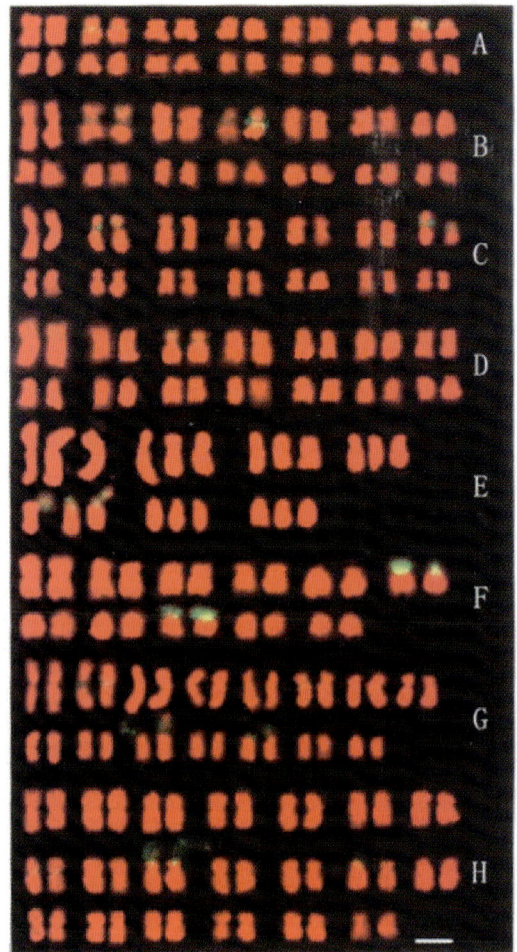

图 3.93 8 种植物中期染色体核型图
Fig. 3.93 The karyotypes of the eight species
A. *S.joponica*；B. *S.japonica f.oligophylla*；C. *S.japonica f.pendula*；D. *S.xanthantha*；E. *S.rubriflora*；F. *R.pseudoacacia*；
G. *R.hispida*；H. *A.fruticosa.* bar =5 μm

国槐：一对信号位于 2 号近中部着丝粒染色体的着丝粒区域，信号大小略有差异。另一对位于 7 号中部着丝粒染色体的着丝粒区域，比前一对信号更大；其中一个信号的强度远超过另一个信号，几乎覆盖了整个短臂。

五叶槐：一对强度相同的信号位于 2 号中部着丝粒染色体的着丝粒及靠近短臂的区域，但其中一个较大，覆盖了整个短臂。另一对位于 4 号近中部着丝粒染色体的着丝粒周围，两个信号的大小和强度均有明显差异。

龙爪槐：两对信号分别位于 2 号和 7 号近中部着丝粒染色体的着丝粒及靠近短臂区域，强度差异不大，但都有一个信号覆盖整个短臂。

黄金槐：信号强度在槐属 5 种植物中最弱。位于 1 号中部着丝粒染色体短臂上的信号几乎不可见，另一对位于 3 号近中部着丝粒染色体着丝粒周围的信号强度略大，其中一个覆盖整个短臂。

红花槐：第5组3条随体染色体的其中两个随体大而明显，另一条的随体很小，附于染色体端部。3个信号位于次缢痕及整个随体，强度较大。

刺槐：信号强度在所有材料中最高，但两对信号内也同样存在明显的强度差异。它们分别位于6号和9号近端部着丝粒染色体的随体上。

毛洋槐：两对较小的信号位于2号中部着丝粒染色体和6号近中部着丝粒染色体的着丝粒区域；另两对较大的信号位于11号近端部着丝粒染色体和13号端部着丝粒染色体的次缢痕并包围整个随体。

紫穗槐：一对很大的信号位于10号近中部着丝粒染色体的随体上，并在染色体的端部各有一处信号强度明显增强；另有两对较小的信号位于8号中部着丝粒染色体和13号近中部着丝粒染色体的着丝粒。

3.4.3.3 讨论

1. 结合 rDNA 位点的核型特征分析与比较

按 Stebbins（1971）的核型分类标准，具1A核型的紫穗槐较为原始，具3B核型的刺槐属较为进化，槐属居于中间。槐属中，国槐及其变型五叶槐和龙爪槐的核型属于比较对称而原始的2B类型，三者的核型数据比较接近（表3.6），且都在2号染色体上有一对 rDNA 位点，表明两个变型相对于原种并未在染色体水平上发生明显变化。尤其是龙爪槐，与国槐尤为接近，甚至它们的45S rDNA 在基因组中的位置完全一致。黄金槐的核型是比国槐更为对称的2A类型，而红花槐则是略进化的3A类型，二者虽然染色体长度差异小于国槐，但平均臂比却明显超过国槐（表3.6），说明染色体两臂的不对称性显著提高。在本研究中，国槐等4个四倍体种均未观察到随体的存在，rDNA 杂交信号仅位于着丝粒至短臂区域（图 3.92A～D），即中部 NOR。陈瑞阳等（2003）认为国槐中4号染色体其中一条的短臂距离长臂较远（图片未显示，形态类似本文龙爪槐中7号染色体，见图3.93C），将这条短臂认定为随体，4号为随体染色体。根据 rDNA 杂交结果和对多个细胞的染色体形态的观察，我们认为还是将原定为随体的部分重新认定为短臂较为合理，rDNA 的杂交信号所处位置应为着丝粒而非次缢痕。由于被处理为随体的部分不计入染色体长度，校正后的核型中染色体的排列顺序也相应改变，原定为4号的染色体在本文中列为2号染色体。由此可以看出，rDNA 在染色体上的定位可以对传统的核型分析进行校正，尤其对于染色体较小且形态特征不明显的材料，可以部分的提高同源染色体配对的准确性。国槐、五叶槐、龙爪槐和黄金槐均为四倍体，均检测出4个45S rDNA 位点（图3.92A～D）；在三倍体红花槐中，检测出3个位点（图3.92E）。在槐属中 rDNA 位点的数目与倍性成正相关。另一方面，具 rDNA 位点的两对染色体上的信号基本呈现出共同的特点：一大一小、一强一弱，大而强的信号大都几乎覆盖整个短臂，小而弱的信号则多集中于着丝粒周围。由此我们推测，45S rDNA 在染色体上的这种分布特点可能普遍存在于槐属各个种甚至其二倍体祖先种中，但有待将 rDNA 定位于更多槐属植物加以证明。

2. 45S rDNA 在各个种基因组中位置、大小的比较和进化趋势

在本文所研究的植物中（红花槐除外），杂交信号在不同位点表现出大小和强度明显的不同。这种现象在其他物种中也存在（Hanson et al., 1996; Vanzela et al., 2002; Muravenko et al., 2003）。在我们的研究结果中，这种差异甚至表现在同源染色体上的信号之间。Zoldos 等（1999）在栎属中发现同源位点间信号大小和强度的差异，其中一个信号甚至无法稳定地出现。但这种现象并不多见。尽管原位杂交结果不能直接显示基因的实际拷贝数，杂交信号的大小和强度却间接反映了基因拷贝数的多少，因此 FISH 被认为是一种半定量技术（Maluszynska and Heslop-Harrison, 1993）。拷贝数与基因的扩增、缺失和不等交换有关（Arnheim et al., 1980），因此这些很可能也是造成信号差异的主要原因。由 45S rDNA 构成的核仁组织区（NOR）在形态上一般表现为染色体的次缢痕，与随体相连。在本文研究结果中，所有随体染色体均呈现明显的杂交信号，信号不仅位于次缢痕，还包围了整个随体（图 3.92：E～H），可见这 3 属植物的随体主要由多拷贝的核糖体 RNA 基因组成。在毛洋槐和紫穗槐中，杂交信号还位于着丝粒至短臂等区域。在黄瓜（*Cucumis sativus* L. cv. Winter Long, Koo et al., 2002）和本文中槐属的 4 个四倍体种，更是出现在了所有 45S rDNA 位点全部位于着丝粒周围的情况。多种植物的 45S rDNA 物理定位也发现该基因除了在次缢痕及随体位置外，在其他染色体上还有数目不等的位点分布（Pedersen and Linde-Laursen, 1994; Hajdera et al., 2003; Raina et al., 2001），这些位点的杂交信号大多小于次缢痕，很多不具转录活性，它们的起源尚不清楚（Taketa et al., 1999）。45S rDNA 位点多位于着丝粒和短臂，位于长臂的情况较为少见。Lima-De-Faria（1976）曾分析了 700 多个种的 NOR，发现 87% 的种中 NOR 位于染色体的短臂上。Lim 等（2001）猜测 rDNA 位点在染色体上的这种位置特点可能与其功能有关，是某种分子或物理上的制约作用使染色体臂可以与核仁相联。rDNA 位点作为一种有效的染色体标记，可以为研究基因组在分子和染色体水平上的进化提供线索（Taketa et al., 1999），可以有效地反映种、属间的分化程度（Thomas et al., 2001）。在槐属的 4 个四倍体种中，45S rDNA 位点的数目完全一致，在基因组中的分布也比较相近，表明他们在染色体结构水平上没有较大的分化。

（作者：刘博、陈成彬、李秀兰、齐力旺、韩素英。原载：云南植物研究，2005，27（3）：261-268）

3.4.4　30 种植物 45S rDNA 在中期染色体上的物理定位

Ribosomal RNA genes（rDNA）are divided into two types：18S-5.8S-25S rDNA（45S rDNA）and 5S rDNA. 5S rDNA is located independently from 45S rDNA in higher eukaryotes（Sone et al., 1999）. Genomic regions coding for rRNA are organised as arrays of tandem repeats where coding regions are separated by untranscribed spacers（Federoff, 1979）. A satellite is a short segment of a chromosome separated from the restby a constriction typically associated with the formation of a nucleolus. 45S rDNA sequences are markers for nucleolar organiser regions（NORs）, and NORs are usually detected cytologically as visible secondary constrictions associated with a nucleolus and a distal satellite. However, the number of 45S rDNA loci may not be consistent with the number of satellites in some plants and NORs may also be located outside the secondary constrictions (Pedersen and Linde-Laursen, 1994; Raina et al., 2001; Hajdera et al., 2003).

Due to the high degree of sequence conservation between rDNAs, DNA sequence probes from one species can be used to identify similar units in most other eukaryotic species. In addition, rDNA units assume specific positions on chromosomes (Maluszynska and Heslop-Harrison, 1993)and so can be used to investigate species that are otherwise difficult to karyotype, thereby providing markers for morphologically indistinct chromosomes (Armstrong et al., 1998; Hasterok et al., 2001).

Various classical and novel cytogenetic techniques, including silver staining(Ag-NOR), in situ hybridization, and primed in situ DNA labelling (PRINS; Kubaláková et al., 1997), among others, have been developed for physical mapping of 45S rDNA. Fluorescence in situhybridisation (FISH) developed by Bauman et al. (1980) and Mukai et al. (1990) is an efficient molecular cytogenetic method to identify the number and physical location of repeated DNA sequences in genomes (Schwarzacher et al., 1994). Thus, 45S rDNA has been located in many important plants. For example, a 45S rDNA probe was used to identify chromosome 6 in maize as a specific cytogenetic DNA marker (Sadder and Weber, 2001). Detailed karyotypes of *Lilium longiflorum* and *L. rubellum* were constructed with the assistance of 45S rDNA as probes (Lim et al., 2001). 45S rDNA sequences were also located on pachytene complements of selected *Brassica* spp. in order to construct comparative genetic and physical maps in the family Brassicaceae (Ziolkowski and Sadowski, 2002). 45S rDNA is also valuable for analysis of polyploid species and in phylogenetic studies. For example, it was used in studies on lupins that are considered to be of polyploid origin. Some interspecific variation was observed in the number and size of the 18S-25S rDNA loci in five species of Lupinus (Naganowska and Zieli′nska, 2002). In cowpea, cytological changes that occurred during the domestication and evolution of this crop were examined by observing the number of active rDNA clusters (Galasso et al., 1998). In a word, the rDNA-FISH technique has become an excellent tool for identifying and comparing chromosomal organisation.

The aim of this study was to provide chromosome 'landmarks' in a wide range of plant species and to describe the various distribution patterns of 45S rDNA in higher plants by comparing the number and positions of 45S rDNA loci.

3.4.4.1 Materials and Methods

1. Plant materials

The 30 species investigated in this study are listed in Table 3.7. Seeds were germinated in Petri dishes at 25℃. Root tips, tender leaves and buds were used for chromosome analysis.

Chromosome preparation For FISH analyses, slides were prepared according to Chen et al. (1979) with minor modifications. In brief, plant tissues were immersed successively in saturated p-dichlorobenzene for 3 h, in double-distilled water (DDW) for 30 min, mixed with

2.5% (w/v) cellulase and pectinase for 40 min, rinsed in DDW for 15 min, and finally fixed in fixative [methanol: acetic acid (3:1 v/v)] for 30 min. Fixed material was placed on a slide and a drop fixation solution added. The fixed tissue was torn and the supernatant removed. Two drops of fixation solution were added to the slide, which was then baked over an alcohol burner.

2. Probe preparation

Cloned 45S rDNA was obtained from Professor Song Yunchun, Wuhan University, China.The 9.1 k bp-long 45S rDNA sequence from tomato (Lycopersicon pennellii LA716; Arumuganathan et al., 1994) was labelled with digoxigenin (DIG) -dUTP (Roche) by randomly-primed DNA synthesis.

Fluorscence in situ hybridization (FISH) FISH was carried out as described by Qi et al. (2002). Hybridisation buffer contained 50% (v/v) deionized formamide, $2 \times$ SSC, 50 mM sodium phosphate, pH 7.0, 5% (w/v) dextran sulphate and 3 ng·μl^{-1} 45S rDNA probe. The probe was denatured at 94℃ for 10 min before being used. Slides with metaphase chromosome spreads were treated with 70% (v/v) deionized formamide in $2 \times$ SSC at 70℃ for 2 min. Denatured hybridisation buffer (10 μl) was then applied to the slides, which were incubated at 37℃ overnight. Finally, the slides were washed with $2 \times$ SSC, 30% deionized formamide at 37℃ for 5 min, and twice in $2 \times$ SSC at 37℃ for 5 min. 45S rDNA were detected using an FITC-labelled anti-DIG antiserum. Metaphase spreads were counter-stained with 100 ng·ml^{-1} 4′, 6-diamidino-2-phenylindole (DAPI) and preparations examined with an OLYMPUS BX 51 Microscope. Images were taken with a cooled CCD camera.

3.4.4.2 Results and Discussion

To confirm the existence of various distribution patterns of 45S rDNA loci on chromosomes, we examined 30 taxonomically diverse plant species belonging to 18 families by FISH. The numbers and positions of all visible rDNA loci on the chromosomes of the 30 species are listed in Table 3.7. For comparison, the number of satellites is also included. FISH results from 15 selected species are shown in Figure 1A-O. Four patterns emerged.

The first pattern observed was that all 45S rDNA loci were detected in the secondary constrictions (SC) and satellites of all satellite-chromosomes (SAT-chromosomes). The number of the 45S rDNA loci was equal to the number of satellites. Fourteen species were included in this pattern (Table 3.7; Fig. 3.94A-E).

Table 3.7 Species investigated for the presence and number of 45S rDNA Loci and satellites

family	species	chromosome number	45S rDNA Loci number of				satellites origin
			SC	SAT	T	PC	
Adoxaceae	Viburnum sargentii*	2n+18	2	-	-	8	Research Institute of Forestry
Apiaceae	Sanicula lamelligera*	2n=20	2	-	-	2	Research Institute of Forestry
Asceraceae	Silybum marianum	2n=34	2	-	-	2	Campus of Nankai University
Berberidaceae	Berberis thunbergii*	2n=28	2	-	-	2	Campus of Nankai University
Bignoniaceae	Catalpa speciosa*	2n=40	2	2	-	2	Campus of Nankai University

family	species	chromosome number	SC	SAT	T	PC	45S rDNA Loci number of satellites origin
Caprifoliaceae	Kolkwitzia amabilis*	2n=32	2	2	-	2	Research Institute of Forestry
Chenopodiaceae	Beta vulgris (diploid)	2n=2x=18	2	-	-	2	Sugar beet Research Institute
	Beta vulgris (triploid)	2n=3x=27	3	-	-	3	Sugar beet Research Institute
	Beta vulgris (tetraploid)	2n=4x=36	4	-	-	4	Sugar beet Research Institute
Cruciferae	Brassica campestris spp.chinensis	2n=20	2	-	-	2	Tianjin Vegetabie Research Institute
Cucubicaceae	Cucumis sativus*	2n=14	-	-	6	-	Tianjin Cucumber research Institute
Lamiaceae	Amorpha fruticosa	2n=16		3	2		Greenhouse of Nankai University
Leguminosae	Amorpha fruricosa	2n=40	2	2	2	2	North of Hebei Province, China
	Gleditsia japonica	2n=28	4	-		4	Campus of Nankai University
	Robinia pseudoacacia*	2n=22	4	-		4	Campus of Nankai University
	Robinia hispida	2n=30	4		4	4	Campus of Nankai University
	Sophora japonica	2n=4x=28	-	-	4	-	Campus of Nankai University
	S.japonica f.oligophylla	2n=4x=28	-	-	4	-	Campus of Nankai University
	S. japonica f.pendula	2n=4x=28	-	-	4	-	Campus of Nankai University
	S.xanthaiha	2n=4x=28	-	-	4	-	Campus of Nankai University
	S. rubriflora	2n=3x=21	3	-	-	3	Campus of Nankai University
Moraceae	Ficus carica*	2n=26	4	-	-	4	Campus of Nankai University
Oleaceae	Syringa oblata	2n=46	2	-	-	2	Campus of Nankai University
Plantaginaceae	Plantago major*	2n=12	4	-	-	4	Campus of Nankai University
Rosaceae	Kerria japonica f.pleniflora*	2n=18	-	10	-	-	Research Institute of Forestry
	Malus spectabilis	2n=34	2	2	-	2	Campus of Nankai University
Salicaceae	P.ussuriensis	2n=2x=38	4	-	-	4	Research Institute of Forestry
	P.pseudo-simonii	2n=2x=38	4	-	-	4	Research Institute of Forestry
	Populus nigra	2n=3x=57	6	-	-	6	Research Institute of Forestry
Sapindaceae	Acer buergerianum*	2n=26	6	4	-	6	Campus of Nankai University
	Lilchi chinensis*	2n=30	2	4	-	2	Guangdong Academy of Agricultural Sciences
Ulmaceae	Ulmus pumila*	2n=26	-	3	2	-	Campus of Nankai University

*Denote species for which FISH results with the 45S rDNA probe are shown in Fig. 3.94.
*The number of 45S rDNA Loci on different regions of the chromosomes.
SC=secondary constriction, SAT=satellite, T=terminal region, PC=peri-centromeric region.

Seven species were included in a second pattern. Here, besides SCs and satellites of all SAT chromosomes, 45S rDNA loci were also located in the terminal and/or peri-centromeric regions of other chromosomes (Table 3.7; Fig. 3.94F-J). However, the intensity of FISH signals on these other chromosomes was lower than that on SAT-chromosomes in some species (Fig. 3.94F, H and J).The additional rDNA sites, and the differences in size and intensity of the signals, may originate from events such as interspecific hybridisation, partial or total deletion of copies of rDNA or, in the polyploid species, by loss of rDNA loci (Vanzela et al., 2002). Galasso et al. (1998) found that during the domestication and evolution of *Vigna unguiculata* populations, 45S rDNA sites increased in number. They also proposed that the increased 45S rDNA clusters might be associated with diversification and the selection pressures on this crop. However, it is also possible that some rDNA sites might have been lost during evolution, or translocated and fused with other rDNA sequences (Thomas *et al.*, 1997; Snowdon *et al.*, 1997).

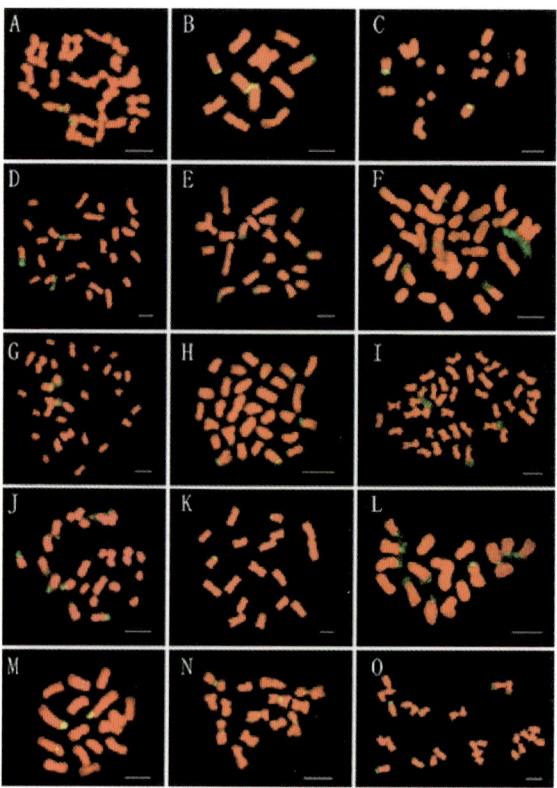

Fig. 3.94 Panels A–O Mitotic metaphase plates of the 15 selected plant species (see Table 3.7) after fluorescence *in situ* hybridisation (FISH) with a 45S rDNA probe. Chromosomes were counter-stained with DAPI, and the 45S rDNA probe was detected by an FITC (yellow green)-labelled anti-DIG antiserum. Panel A, *B. thunbergii* ($2n = 28$); B, *P. major* ($2n = 12$); C, *S. lamelligera* ($2n = 20$); D, *F. carica* ($2n = 26$); E, *R. pseudoacacia* ($2n = 22$); F, *R. hispida* ($2n = 30$); G, *K. amabilis* ($2n = 32$); H, *L. chinensis* ($2n = 30$); I, *C. speciosa* ($2n = 40$); J, *A. buergerianum* ($2n = 26$); K, *V. sargentii* ($2n = 18$); L, *K. japonica* f. *pleniflora* ($2n = 18$); M, *S. miltiorrhiza* ($2n = 16$); N, *C. sativus* ($2n = 14$); O, *U. pumila* ($2n = 26$). Bars = 5 μm

Among the 30 species examined, only *Viburnum sargentii* showed a third pattern. Four pairs of SAT- chromosomes were found in this species, but only one pair gave FISH signals in their satellites (Fig. 3.94K).

It is possible that copies of rDNA repeats present in the other three pairs were too few to be detected (Hanson *et al.*, 1996; Chen, 1999).

The fourth pattern was shown by the remaining eight species. In these species, there were no visible satellites. All 45S rDNA loci were located in the terminal and peri-centromeric regions of chromosomes (Fig. 3.94L-O). Interestingly, in *Salvia miltiorrhiza* (Fig. 3.94M) and *Ulmus pumila* (Fig. 3.94O) five FISH signals were detected, thus one of the loci could not be paired. This may be explained if the other locus possessed too few copies of the rDNA repeat so that it could not be detected, like those in *V. sargentii*.

Lima de Faria (1976) analysed the NORs in over 700 species and reported that, in 87% of cases, the nucleolus was located on the short arm of the chromosome. Lim *et al.* (2001) speculated that such striking conservation in karyotype morphology suggests some molecular or physical constraint exists for chromosome arms to associate with the nucleolus. Surprisingly, in both *U. pumila* and *Catalpa speciose* (Fig. 3.94I), which gave the second pattern described above, one pair of rDNA loci was located in the terminal regions of the long arms of an acrocentric chromosome and a submetacentric chromosome, respectively. This may be due to chromosomal rearrangements that involved moving the NORs to termial regions of the long arm. It is still not known whether such long-arm sites of 45S rDNA have become silenced.

In *Cucumis sativus*, the peri-centromeric regions of six chromosomes exhibited rDNA signals (Fig. 3.94N) that agreed with previous reports (Hoshi *et al.*, 1999; Koo *et al.*, 2002). Chen *et al.* (1999) reported four pairs of 45S rDNA signals. This discrepancy may be due to variability in the rDNA loci, as found in the genus *Oryza*. Varieties found in temperate regions have a single rDNA locus, while those from tropical and sub-tropical regions have two rDNA loci. It is likely that the number of rDNA loci was reduced under selection pressures such as low temperature.

Due to the four distribution patterns of rDNA loci on the chromosomes, rDNA sequences provide excellent cytological markers for karyotype analysis. Knowledge of the relative physical locations and number of rDNA loci is important and useful, both for the construction of physical maps of chromosomes and for phylogenetic studies (Schrader *et al.*, 2000). To our knowledge, many of the species in this study have not previously been subjected to molecular cytological investigation. Physical mapping of 45S rDNA in metaphase chromosomes has been carried out for the first time in all of them, except *C. sativus* (Chen *et al.*, 1999; Hoshi *et al.*, 1999; Koo *et al.*, 2002) and diploid *Beta vulgaris* (Schmidt *et al.*, 1997).

We are grateful to Professor Yunchun Song (Wuhan University, China) for providing the 45S rDNA clone.

（Authors：B. Liu，C. B. Chen，X. L. Li，R.Y. Chen，and W. Q. Song.
Published in：Journal of Horticultural Science & Biotechnology（2005）80（3）287-290）

3.4.5　rDNA 序列在多种蔬果类植物染色体上的定位

蔬果类植物在世界范围内广泛分布，具有巨大的经济价值和营养价值，诸多优势与人们的生活息息相关，应用 rDNA 探针进行荧光原位杂交定位方面的研究在蔬果类植物中还较少，除了一些零星报道之外（刁英等，2005；郝剑瑾等，2009；兰添颖等，2007；李琦等，2007；李艳燕，2008），对于我国栽培蔬菜和水果植物的核型还没有比较系统的研究。本实验作者利用改进的 rDNA 序列荧光原位杂交方法，将 18S、5S rDNA 序列直接在 5'端进行荧光修饰，并利用双色荧光原位杂交技术对 30 种蔬果类植物不同属间、种间中期染色体上进行 18S-5S rDNA 的物理定位，进而了解 rDNA 序列在这些蔬果类植物基因组中位点数目和分布特点，同时证明该方法的可行性，使 rDNA-FISH 杂交具有省时、经济、快速、信号强等特点，便于推广和应用。

3.4.5.1 材料与方法

1. 材料

以兰州百合、芹菜、萝卜等 26 种蔬菜和西瓜、甜瓜 4 种水果共 9 科 21 属 30 种植物为实验材料，名称见表 3.8，凭证标本保存在天津市农业科学研究院科润蔬菜研究所。

表 3.8　30 种蔬果类植物的染色体数目和 18S-5S rDNA FISH 杂交结果

材料名称	染色体数目	18S rDNA（绿色）位点数目	5S rDNA（红色）位点数目
Ⅰ. 百合科 Liliaceae			
兰州百合 *LiLum davidi* var. *unicdor* Duchartre	2*n*=24	8	2
蒜 *Allium.satvium* L.	2*n*=16	4	4
葱 *Allium fistulosum* L.	2*n*=16	2	2
Ⅱ. 十字花科 Brassicaceae			
旱萝卜 *Raphanus sativus* L.ssp. hanluobu	2*n*=18	2	4
长型水萝 *Raphanus sativus* L.ssp shuiluobo	2*n*=18	2	4
樱桃水萝卜 *Raphanus sativus* L. var.*radculus pers*	2*n*=18	2	4
白菜型油菜 *Brassica campestris* L.	2*n*=20	2	6
花椰菜 *B. oleracea* L. Var. *botrytis* L.	2*n*=18	4	2
乌塌菜 *B.campestris* L.ssp.chinensis L.	2*n*=20	2	6
芥蓝 *B. alboglabra* Baile	2*n*=18	2	2
雪里红 *B. juncea* Coss. Var.*crispifolia* Bailey	2*n*=36	4	10
红菜苔 *B. compestris* L. var.*purpurea* Bailey	2*n*=18	2	8
Ⅲ. 菊科 Asteraceae			
莴苣 *Lactuca sativa* L. Var. *capitata* L.	2*n*=18	2	2
茼蒿 *Chrysanthemum coronarium* L.	2*n*=18	2	2
Ⅳ. 葫芦科 Cucurbitaceae			
丝瓜 *Luffa.acutangula*（Linn.）Roxb.	2*n*=26	2	2
苦瓜 *Momordica charantia* L.	2*n*=22	4	2
西瓜 *Citrullus lanatus* L.	2*n*=22	4	2
白薄皮甜瓜 *Cucumis melo* L.ssp.baopi	2*n*=24	2	2
白花皮甜瓜 *Cucumis melo* L.ssp. huapi	2*n*=24	2	2

续表

材料名称	染色体数目	18S rDNA（绿色）位点数目	5S rDNA（红色）位点数目
厚皮甜瓜 *Cucumis melo* L.houpi	2*n*=24	2	2
黄瓜 *Cucumis sativus* L.	2*n*=14	6	2
V．茄科 *Solanaceae*			
辣椒 *Capsicum annuum* L.	2*n*=24	4	2
茄子 *Solanum melongena* L.	2*n*=24	2	2
VI．伞形科 *Umbelliferae*			
芹菜 *Apium graveolens* L.	2*n*=22	2	2
茴香 *Foeniculum vulgare* Mill.	2*n*=22	2	2
VII．豆科 *Leguminosae sp.*			
菜豆 *Phaseolus vulgaris* L.	2*n*=22	2	4
豌豆 *Pisum sativum* L.	2*n*=14	2	6
蚕豆 *Vicia faba* L.	2*n*=12	2	2
VIII．藜科 *Chenopodiaceae*			
菠菜 *Spinacia oleracea* L.	2*n*=12	4	4
IX．旋花科 *Convolvulaceae*			
空心菜 *Ipomoea aquatica* Forsskal	2*n*=30	4	2

2．实验方法

（1）探针的合成及染色体标本制备

染色体标本的制备采用去壁低渗法（陈瑞阳等，1979）选择分散良好的染色体用荧光显微镜标尺记下坐标的位置或用记号笔在反面标记染色体的位置，将染色体标本放在-20℃冰箱中保存待用。

（2）18S rDNA 和 5S rDNA 探针制备

从拟南芥 18S 和 5S rDNA 序列中选取一段重复序列，直接由上海生工合成探针，18S rDNA 在 5'端进行 FAM 绿色荧光标记；5S rDNA 探针 5'端进行 TAMRA 红色荧光标记。

（3）染色体荧光原位（FISH）杂交方法

A：染色体标本预处理

①在荧光显微镜上记下已标记分裂相的坐标，并用玻璃刀在载玻片背面标记分裂相的位置。

②将载玻片在 45%醋酸中浸泡 5 min 褪色，空气干燥；

③多聚甲醛固定 10 min，70%、85%、100%系列乙醇脱水各 5 min，风干；

④在标记处加 30 μL 70%去离子甲酰胺/2×SSC，盖上盖玻片，于 PCR 仪或烘箱中 70℃处理 2~4 min。

⑤去掉盖片，-20℃的冷乙醇系列（70%，85%，100%）脱水，每级 3 min，空气干燥。

B：杂交

①用 2×SSC 稀释探针至 5ng/μL。
②每张标本加 10 μL 18S 和 5S 混合探针，并盖 18 mm×18 mm 的盖玻片。
③于湿盒中 37℃杂交 2 h 以上或过夜。

C：杂交后洗脱
①在 4×SSC，0.2% Tween 20 中室温下避光洗涤 10 min。
②蒸馏水冲洗片刻，空气干燥（避光）。

D：杂交信号检测
①滴加 5 μL 含有 DAPI 的防荧光淬灭剂，盖上盖玻片。
②荧光显微镜观察，并用冷 CCD 拍照。
③荧光显微镜观察，在紫外光激发下可观察到蓝色的分裂相，在绿色激发光激发下可观察到红色的 5S 杂交信号。在蓝色光激发下可观察到绿色的 18S 荧光信号。

E：图像采集与处理
Nikon 80i 荧光显微镜观察杂交信号，冷 CCD 进行图像采集，Spot Rtke 4.1 软件进行图像合成。

3.4.5.2 结果及分析

1. 百合科三种植物 18S-5S rDNA FISH 定位

兰州百合（*LiLum davidi* var. *unicdor* Duchartre）（图 3.95-1）具有 8 个 18S rDNA 位点（绿色）和两个 5S rDNA 位点（红色），18S rDNA 位点均位于一对染色体的副缢痕位置，有三对染色体副缢痕上显示了清楚的信号，这些信号位于三对染色体长臂上，与核型分析的随体位置一致，这与李懋学等人 1984 年报道的结果吻合。

蒜（*Allium.satvium* L.）（图 3.95-2）18S rDNA 在两对染色体上有信号，证明有两对随体染色体，他们均位于着丝粒附近，为中间随体。5S rDNA 信号位于一对染色体上，其中一对信号位于染色体的短臂上，能看到 4～5 个信号位点，一对信号位于长臂靠近着丝粒附近。

葱（*Allium fistulosum* L.）（图 3.95-3）18S rDNA 位于一对染色体上，为一对随体染色体，5S rDNA 具有 2 个位点均位于一对染色体短臂中间部位。

2. 十字花科七种植物 18S-5S rDNA-FISH 定位

旱萝卜（*Raphanus sativus* L.ssp. hanluobu）（图 3.95-4）、水萝卜（*Raphanus sativus* L.ssp shuiluobo）（图 3.95-5）和樱桃萝卜（*Raphanus sativus* L. var.*radculus pers*）（图 3.95-6）三个品种 18S rDNA 都定位在一对染色体上，5S rDNA 定位在两对染色体上，其中有一对是与 18S rDNA 在同一对染色体上，18S rDNA 信号在前，5S rDNA 信号在后，位于随体染色体上，信号也比较强，另一对染色体上的 5S rDNA 信号在染色体着丝粒部位。

白菜型油菜（*Brassica campestris* L.）（图 3.95-7）随体比较大，这与陈瑞阳等（2003）报道的结果一致，属于大随体，18S rDNA 位点有两个，位于一对染色体的短臂靠近着丝粒的部位，这与核型分析的结果相吻合。5S rDNA 位于三对染色体上，其中一对与 18S rDNA 位于同一条染色体上，5S 位点在 18S rDNA 后面更靠近着丝粒。

花椰菜（*B. oleracea* L. Var. *botrytis* L.）（图 3.95-8）18S rDNA 位于两对染色体上，

其中一对信号比较强，一对信号比较弱。5S rDNA 定位到一对染色体的长臂靠近着丝粒位置上，信号强度比 18S rDNA 亮。

乌塌菜（*B.campestris* L.ssp.chinensis L.）（图 3.95-9）18S rDNA 位点位于一对具随体染色体上，两个信号有些差别，一个信号靠近短臂末端，一个信号靠近短臂着丝粒附近，信号强度比较弱。5S rDNA 信号位于三对染色体上，其中一对与 18S 在同一条染色体上，一对位于染色体长臂上，位于小染色体短臂上的信号稍弱。

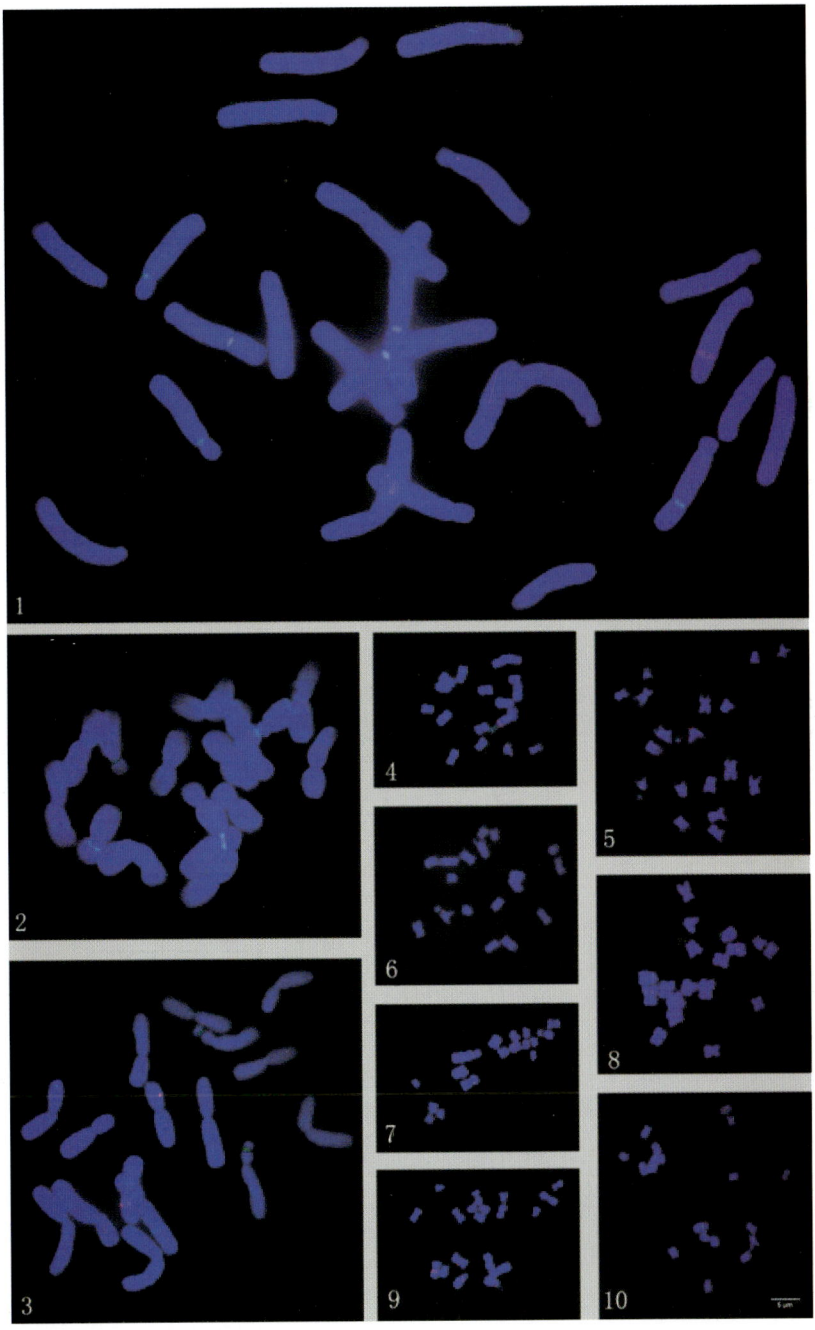

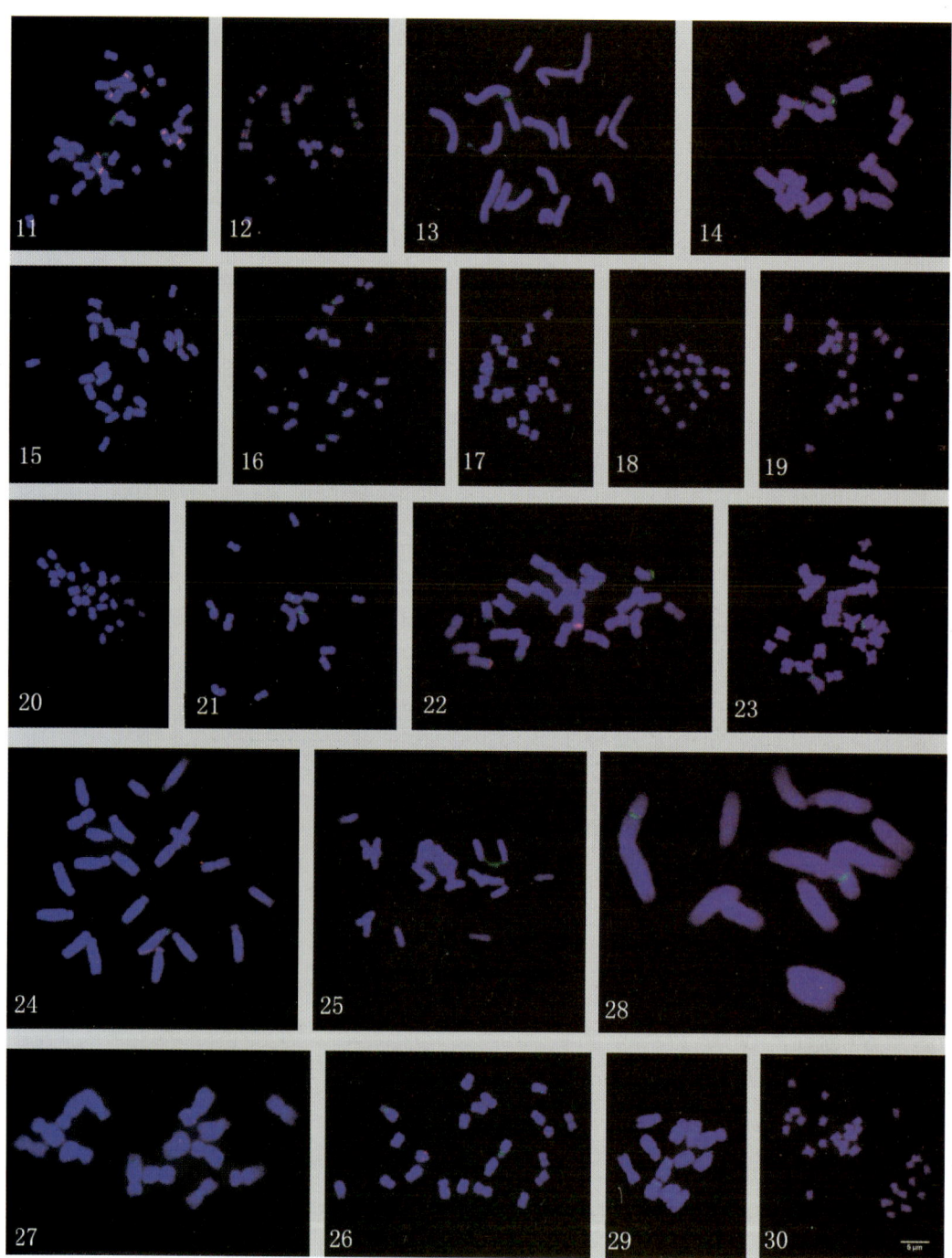

图 3.95　18S-5S rDNA 混合探针对 30 种蔬果植物染色体 FISH 杂交结果
（18S rDNA 为绿色信号，5S rDNA 为红色信号）

芥蓝（*B. alboglabra* Baile）（图 3.95-10）18S rDNA 杂交信号有一对，位于染色体短臂靠近着丝粒，一条染色体的信号几乎覆盖整个着丝粒，一条更靠近短臂末端。5S rDNA 杂交信号位于染色体长臂靠近着丝粒处，两个信号呈现出点状形态，信号强度基本一致。

雪里蕻（*B. juncea* Coss. Var.*crispifolia* Bailey）（图 3.95-11）18S rDNA 在两对染色体上有信号，2 对均在染色体短臂末端，一对信号覆盖整个随体，另外一对比较小，5S rDNA 位于 5 对染色体上，5S rDNA 信号都比较强，其中两对与 18S rDNA 在同一条染色体短臂上，其中一对信号非常强。

红菜苔（*B. compestris* L. var.*purpurea* Bailey）（图 3.95-12）18S rDNA 位点位于一对小染色体上，信号整个覆盖染色体的短臂，信号强度比较强。5S rDNA 信号位于四对染色体上，其中一对与 18S rDNA 在同一条染色体上，一对位于一对小染色体短臂末端，一对位于长臂上，一对位于着丝粒附近，信号都比较强。

3. 菊科两种植物的 18S-5S rDNA-FISH 定位

莴苣（*Lactuca sativa* L. Var. *capitata* L.）（图 3.95-13）18S-5S rDNA 信号均位于一对具随体染色体上，18S rDNA 信号位于副缢痕与核仁组成区（NOR），覆盖整个随体。5S rDNA 信号位于短臂靠近着丝粒附近，信号强度比较亮。

茼蒿（*Chrysanthemum coronarium* L.）（图 3.95-14）18S-5S rDNA 信号都位于一对染色体上，18S rDNA 位于短臂末端，5S rDNA 位于着丝粒附近，两个探针信号强度相当。这个结果与李真等（2011）利用 45S rDNA 为探针在中亚苦蒿上的杂交结果一致。

4. 葫芦科七种植物 18S-5S rDNA-FISH 定位

丝瓜（*Luffa.acutangula*（Linn.）Roxb.）（图 3.95-15）具有一对 18S rDNA 信号，信号弥散几乎覆盖这对染色体的短臂，一对 5S rDNA 信号位于短臂中间部位，本研究得试验结果与徐延浩（2007）报道的丝瓜具有 5 对 45S rDNA 位点结果不一致，5S rDNA 一致，这也许与 18S rDNA 与 45S rDNA 在基因组中的大小有一定关系，也许与品种之间的差异相关，还有待于进一步利用 45S rDNA 作探针进一步验证。

苦瓜（*Momordica charantia* L.）（图 3.95-16）18S rDNA 杂交信号检出分别位于 2 对染色体短臂末端，信号强弱没有明显区别，5S rDNA 杂交信号位于一条小染色体短臂上，信号在两条染色体上没有区别。与李琦（2007）报道的结果一致。

西瓜（*Citrullus lanatus* L.）（图 3.95-17）18S rDNA 信号有 4 个，分布在两对染色体上，一对 5S rDNA 信号位于一对具随体染色体上，与 18S rDNA 信号位于同一对染色体上，紧挨着 18S rDNA 的信号。与李琦等（2007）的报道不同，这可能与所用的西瓜品种有关。

白色薄皮甜瓜（*Cucumis melo* L.ssp.baopi）（图 3.95-18）、花薄皮甜瓜（*Cucumis melo* L.ssp. huapi）（图 3.95-19）、厚皮甜瓜（*Cucumis melo* L.houpi）（图 3.95-20）虽然表型特征不同，但所具有的染色体数目、18S-5S rDNA 的位点数目均相同，18S rDNA 位于一对随体染色体上，信号比较强，几乎覆盖了整个短臂。5S rDNA 位于另外一对染色体的短臂末端。

黄瓜（*Cucumis sativus* L.）（图 3.95-21）18S rDNA 信号全部位于 6 条染色体的中部着丝粒位置。5S rDNA FISH 信号位于一对染色体的短臂上，一个信号强一个信号稍弱一些，也许与 5S rDNA 在染色体上的拷贝数有关。

5. 茄科两种植物 18S-5S rDNA-FISH 定位

辣椒（*Capsicum annuum* L.）（图 3.95-22）18S rDNA 信号分布在两对染色体短臂末端，相对 5S rDNA，18S rDNA 信号比较弱。从辣椒核型分析的角度来看，与李懋学等报道的辣椒具一对随体染色体有差别，与陈瑞阳等 2003 报道的一致。

茄子（*Solanum melongena* L.）（图 3.95-23）18S rDNA 定位在一对染色体短臂副缢痕的位置，信号比较大，几乎覆盖整个短臂，一对较弱的 5S rDNA 信号定位在一对染色体长臂末端，信号呈点状。

6. 伞形科两种植物 18S-5S rDNA-FISH 定位

芹菜（*Apium graveolens* L.）（图 3.95-24）18S rDNA 位于一对具随体染色体上，信号在短臂的末端，两条染色体上的信号强度很均匀。5S rDNA 信号位于一对中着丝点染色体上，位于短臂的末端。

茴香（*Foeniculum vulgare* Mill.）（图 3.95-25）18S rDNA 特异地定位在一对具随体染色体，形成了随体相接的形态，整个信号覆盖了随体部位。5S rDNA 位于一对染色体长臂靠近着丝粒的位置。信号强度大小一致。

7. 豆科三种植物 18S-5S rDNA-FISH 定位

菜豆（*Phaseolus vulgaris* L.）（图 3.95-26）18S rDNA 特异地定位在一对随体染色体上，信号强度覆盖整个短臂。两对染色体具有 5S rDNA 信号，一对信号位于长臂信号弱一些，一对位于着丝粒附近信号相对强一些。

豌豆（*Pisum sativum* L.）（图 3.95-27）在一对染色体长臂副缢痕处显示 18S rDNA 信号，表明豌豆的随体在长臂，这与陈瑞阳等 2003 报道豌豆具随体染色体在第五号染色体的长臂结果一致。

蚕豆（*Vicia faba* L.）（图 3.95-28）只在一对大染色体副缢痕处显示 18S 和 5S rDNA 信号，5S rDNA 信号呈点状，靠近短臂与 18S rDNA 信号紧密相连，18S rDNA 信号在副缢痕处比 5S rDNA 信号强。

8. 藜科菠菜 18S-5S rDNA-FISH 定位

菠菜（*Spinacia oleracea* L.）（图 3.95-29）18S rDNA 在两对染色体次缢痕及随体部位显示了杂交信号，一对信号较强，一对信号较弱。结果与兰添颖等报道（2007）结果不一致，这可能与 18S rDNA 的分布有关或者与品种的性别有关。5S rDNA 信号定位于两对染色体长臂末端，信号强度、大小没有区别。

9. 旋花科空心菜 18S-5S rDNA-FISH 定位

空心菜（*Ipomoea aquatica* Forsskal）（图 3.95-30）18S rDNA 位于两对染色体上，与吴建桥等人 45S rDNA 定位结果一致都是位于两对染色体上。5S rDNA 位于一对染色体的长臂上，5S rDNA 信号比 18S rDNA 信号强。

3.4.5.3 讨论

以往的研究都是利用 45S rDNA 或 25S rDNA，在植物上很少使用 18S rDNA，本研究首次应用 18S-5S rDNA 探针，并且首次利用直接合成探针，在探针的 5'端直接修饰 FAM 或 TAMRA，免去了烦琐的探针标记，节省了时间、节省了经费，而且信号强度等基本上与 45S rDNA 探针效果相当，从我们的试验结果看基本上达到了预期的效果。像葫芦科植物（苦瓜和西瓜）、茄科植物（辣椒和茄子）、十字花科植物（白菜型油菜、乌塌菜、旱萝卜、樱桃水萝卜、长型水萝卜、芥兰、花椰菜、红菜苔），他们的染色体数目分别相同且形态差异较小，仅根据常规核型分析结果很难区分各个品种，而 18S rDNA、5S rDNA 在这些植物上的染色体位点数和分布模式为准确识别这些物种提供了标识，更重要的是，本研究所采用的 18S rDNA 探针在其他作物中通用性很强，虽然与 45S rDNA 杂交所获得的信号在一些物种上有一些差异，但也许是和所用品种有一定的关系，也可能与 18S rDNA 所包含的序列有一定关系。还需要进一步改进探针序列的选择，增加信号强度，使一些较弱的信号能够观察到。

目前已有大量的研究证明 18S-5.8S-25S rDNA 在大多数物种中其主要定位在核仁组织区（NOR）即染色体的次缢痕部位。18S rDNA 是 45S rDNA 的一部分，其定位位点应与 45S rDNA 一致。从染色体结构来讲，NOR 的数目应该与随体数目相对应，因此，18S rDNA 的杂交位点数目即是染色体随体的数目。本实验采用双色荧光原位杂交技术在 30 种蔬果类植物不同属间、种间中期染色体上进行了 rDNA 的物理定位，所研究的 30 种蔬果植物 18S rDNA 的杂交信号从 2~8 个不等，其中兰州百合最多为 8 个杂交信号，黄瓜 6 个杂交信号，蒜、西瓜、苦瓜、花椰菜、雪里蕻、辣椒、空心菜和菠菜为 4 个杂交信号，其余均为 2 个杂交信号。除菠菜、丝瓜和白菜型油菜与 25S rDNA 和 45S rDNA 信号位点不一致外，其余均与已报道的一致，这可能与所取的品种来源相关。或许是由于 45S rDNA 与 18S rDNA 他们本身的结构所造成。在高等植物中，45S 包括 18S-5.8S-28S rDNA，由许多的 45S rDNA 重复单位所组成，一个 45S rDNA 重复单位包括转录区和非转录区，它们串联在一起形成多个重复序列，45S rDNA 重复单位长度为 7.8~18.5 Kb，而 18S rDNA 只是其中的一部分结构，18S rDNA 重复单位长度约 1.8 Kb，比 45S rDNA 小得多，由于重复的次数少，可能在杂交的过程中所获得信号比较弱，不能够显示出来，可能是造成这种差异的原因，还有待进一步研究。

在高等植物中，包括被子植物和裸子植物，5S rDNA 的位置并不恒定，位点 2~10 个不等，最多的是雪里蕻 10 个信号，红苔菜 8 个信号，白菜型油菜、乌塌菜和豌豆为 6 个杂交信号，菠菜、蒜、旱萝卜、樱桃萝卜、水萝卜、菜豆为 4 个杂交信号。由于 5S rDNA 和其他 rDNA 在染色体的分布彼此独立，有些物种分布在相同染色体上，如本试验中的蚕豆、大蒜、西瓜、萝卜（3 种）、白菜型油菜、乌塌菜、红苔菜、雪里蕻、莴苣、茼蒿、菠菜等物种都有一对染色体 18S rDNA 与 5S rDNA 在同一染色体上。但没有发现 5S rDNA 与 18S rDNA 在相同的位点上，虽然 5S rDNA 在功能上也与核仁有关，但它并不位于核仁，5S rDNA 的结构由编码区和间隔区组成，编码区的长度和序列物种间变化较大。

另外在我们所观察的这 30 种蔬果植物中都发现了 rDNA 信号强弱在不同的染色体上有变化，这种变化在有的植物染色体中明显。如雪里蕻、红苔菜等，这些信号的强弱有些可能是由于染色体所处的空间位置导致信号大小不一样，但是大多数信号强弱的变化应该是由于 rDNA 序列的在不同染色体拷贝数不同而造成，如果染色体上 rDNA 拷贝数越多，相应显示的信号越强，反之则越弱。也有学者认为这种信号微弱的区域是曾经存在于着丝粒附近的 45S rDNA 区域在大规模的染色体重排中移向染色体末端后所剩的残余（Lim *et al.*，2005）。

（作者：牛凯、陈成彬、刘慧静、古瑜、王春国、孙德岭、宋文芹。
原载：南开大学学报，2017，50（5）67-75）

3.4.6　rDNA 和端粒重复序列在七种杨属植物中的多色荧光原位杂交定位

The black cottonwood tree（*Populus trichocarpa*）as amodel forwoody plant study has been completely sequenced. Thewhole-genome shotgun sequences were assumed into 2 447 major scaffolds estimated to（485±10）Mb（±SD），and 385 Mb of them were located on the linage map（Tuskan *et al.*, 2006). However, the draft was not enough to characterize the actual chromosomes in many other species contained in the *Populus*. The work at the cytogenetic level in the *Populus* is poorly defined，especially when compared to other models such as *Arabidopsis*，Rice，and sorghum（Islam-Faridi *et al.*, 2009）The identification of homologous pairs in ordinary cytogenetic observation is blocked by the small and similar chromosomes at the morphological level.

FISH signals derived from a single repetitive DNA probe or a cocktail containing several probes can provide a hybridization pattern that allowed identification of the mostly chromosomes. The application of the multicolor FISH analysis methods to distinguish the chromosomes has achieved such as the *Spinacia oleracea*，*Larix principis-rupprechtii* and *Silene latifolia*（Lan *et al.*, 2007; Lengerova *et al.*, 2004; Liu *et al.*, 2007）. Repetitive DNA sequence can generate unique FISH patterns on individual chromosomes for karyotype analysis. The rDNA probes，which contained 18S-5. 8S-25S rDNA（or a part of them，25S rDNA）and 5S rDNA，were widely used to anchor cytogenetics maps. The 18S-5. 8S-25S rDNA and the 5S rDNA probes hybridized to *Picea abies*，*Pinus sylvestris* and *Larix deciduas* show different number and locus，coupled with chromosomes measuring data were used to identify each chromosome（Lubaretz *et al.*，1996）. Although the encoding regions of 18S-5. 8S-25S rDNA and 5S rDNA are conservative，the rDNA loci and number are not unchangeable. Variable number of ribosomal DNA loci was detected among the several *Populus* species（Prado *et al.*, 1996）. In this study，an efficient technique of multicolor FISH was used to analyze the seven *Populus* species which belong to five *Populus* sections，which enabled the application of the probes labeled with two fluorescence to map simultaneously three different DNA sequences along the metaphases chromosomes. Comparison of karyotype constructed with these landmarks produced by the multicolor FISH among the seven *Populus* species was carried out.

3.4.6.1　Materials and Methods

1. Plant materials and chromosome preparation

Seven *Populus* species belong to five *Populus* sections were collected（Table 3.9）. And two of them were triploids. The cuttings were grown in water culture. The root tips, about 1 cm long, were collected and treated by a saturated aqueous solution of α-bromonapthalene and paracide for 3 hours at room temperature and then fixed in Carnoy's fixative (ethanol: acetic acid= 3:1) for 1 h. The fixed roots were spread on the slide with water film by fine pointed for ceps and air-dried.

Table 3.9 The name and origin of experimental materials

Section	Species	Voucher	Origin
Leuce	*P. tomentosa* (2n=2x=38) *P. tomentosa* (2n=3x=57)	L.W.Qi	Datong, Shanxi
Aigeiros	*P. nigra* var. *thevestina* (2n=2x=38) *P. euramericana* cv. Wuhei-1 (2n=3x=57)	L.W.Qi R.Y.Chen	Datong, Shanxi Wuqing, Tianjin
Tacamahaca	*P. pseudo-simonii* (2n=2x=38)	L.W.Qi	Dalian, Liaoning
Leucoides	*P. lasiocarpa* (2n=2x=38)	L.W.Qi	Jianshi, Hubei
Turanga	*P. euphratica* (2n=2x=38)	L.W.Qi	Datong, Shanxi

2. Probes preparation and labeling

The 2.3 kb subclone of the 25S rDNA coding region of *Arabidopsis thaliana* (Unfried and Gruendler, 1990) was labeled with DIG-16-dUTP (Roche) and Biotin-11-dUTP (Roche) at ratio 1:1 by randomly primed DNA synthesis according to manufacturer's instructions. The 5S rDNA was amplified by the primers (F: 5'-AAGACTAGTAGCCTTCGGCCTC-3' R: 5'-GTCGACACCTAATTCCACCAA-G-3'), and the PCR products were labeled by DIG-16-dUTP. The *Arabidopsis*-type telomere repeat sequences (TTTAGGG)$_n$ were got by the twice PCR method. The first PCR without DNA templates was carried out by primers (TTTAGGG)$_5$, (CCCTAAA)$_5$ and then the PCR products as the templates proceeded the second PCR without primers. The telomere repeatswere labeled by Biotin-11-dUTP. Three probes were mixed at ratio 1:2:1 (25S:5S:telomere).

3. In situ hybridization and signals detection

The *in situ* hybridization proceeded as described previously (Liu *et al.*, 2005). The slides were treated at 70℃ for 2 min in 70% for mamide, 2×SSC to denature the chromosome DNA. Probes were denatured at 94℃ for 10 min in the 2×SSC, 5% dextran sulfate, and 50% formamide. Anti-DIG-fluorescent-conjugate (FITC, Roche) and Streptavidin-Cy3 conjugate (Sigma) enlarge the signals. The chromosomes were counterstained with DAPI (4', 6-diamidino-2-phenylindole, Sigma) and covered by Vectashield mounting medium (Vector Laboratory). Nikon 80i fluorescence microscope and a cooled CCD camera (SPOT, Diagnostic Instruments) were used to detecte the hybridization signals and then the version 4.0.8 of the Spot soft ware (Diagnostic Instruments) exported the stacked images.

3.4.6.2 Results

The chromosomes spread on the slide well, and were free of cell walls and nuclear membranes to low the background. The chromosome number of each species was carefully counted.

1. The rDNA sites and telomere repeat sequence

The 25S rDNA probe marked the 18S-28S rDNA sites on the chromosome. And two 18S-28S rDNA sites in one set of genome were identified in the *Populus* except only one in *P.euphratica* (white arrows, Fig. 3.96E). Two pairs, four sites were found in diploid *Populus* species ($2n=2x$): *P.tomentosa*($2x$), *P.nigra* var. *Thevestina*, *P.pseudo-simonii* and *P.lasiocarp* (white arrows, Fig. 3.96A, B, C, D), while six sites were found in triploid ($2n=3x$) *P.tomentosa* ($3x$), *P.euramericana* cv. Wuhei-1 (white arrows, Fig. 3.96F, G). A single 5S rDNA locus was observed in one set genome of all seven *Populus* species (grey arrows, Fig. 3.96A, B, C, D, E, F, G). The *Arabidopsis*-type telomere repeat sequenceswere located on the end ofmost chromosomes in the seven *Populus* species (Fig. 3.96A, B, C, D, E, F, G).

2. Karyotyping the seven *Populus* species

Multicolor FISH was used to identify and characterize individual chromosomes of the *Populus* species (Fig. 3.96). The length of short and long arms was measured according to the telomere repeat sequences sites, and arm ratio (length of long arm /length of shortarm) and relative length were estimated for each chromosome. Complementary chromosomes were arranged in order of decreasing length. The slightly stronger 18S-28S rDNA site was on the 2nd to 5th chromosome in six *Populus* species, and the other one was mainly on the 7th and 8th chromosome (white stars, Fig. 3.97A, B, C, D, E, F, G). The 5S rDNA was located on the 11th to 15th chromosome (grey stars, Fig. 3.97A, B, C, D, E, F, G). The 18S-28S rDNA sites and 5S rDNA sites were all identified on the end of chromosomes (Table 3.10).

Table 3.10　The location of the rDNA sites in seven Populus species

Species	25S sites	5S sites
P. tomentosa $2n=2x=38$	4, 8	1 2
P. nigra var. *thevestina* $2n=2x=38$	3, 7	1 1
P. pseudo-simonii $2n=2x=38$	5, 7	1 4
P. lasiocarpa $2n=2x=38$	2, 8	1 2
P. euphratica $2n=2x=38$	8	1 2
P. tomentosa $2n=3x=57$	5, 7	9
P. euramericana cv. Wuhei-1 $2n=3x=57$	3, 7	1 5

3.4.6.3　Discussion

The number of 18S-28S rDNA in our research were similar to the report of Prado (1996), and the same hybridization result was also gained in the research of Islam-Fardi (2009), where they all found two 18S-28S rDNA sites in *P.nigra*, *P.deltoidsand*, *P.trichocarpa* (one major and one minor). However, as were ported previously (Dong et al., 2007), only one 18S-28S rDNA site was observed in *P.euphratica*. It probably coursed by the high salt and high temperature selection pressures, which differ from the other *Populus* species, forced *P.euphratica* to evolve in dependently early in the speciation and lose the major sites. The major 18S-28S rDNA site was on the 2nd to 5th chromosome in six *Populus* species, and the minor one was on the 7th and 8th chromosome including *P.euphratica*. The 18S-28S rDNA sequence of *P. trichocarpa* from NCBI was found only one in the 10th chromosome in the *Populus trichocarpa* database (build 1.2) by using the blast, and the other

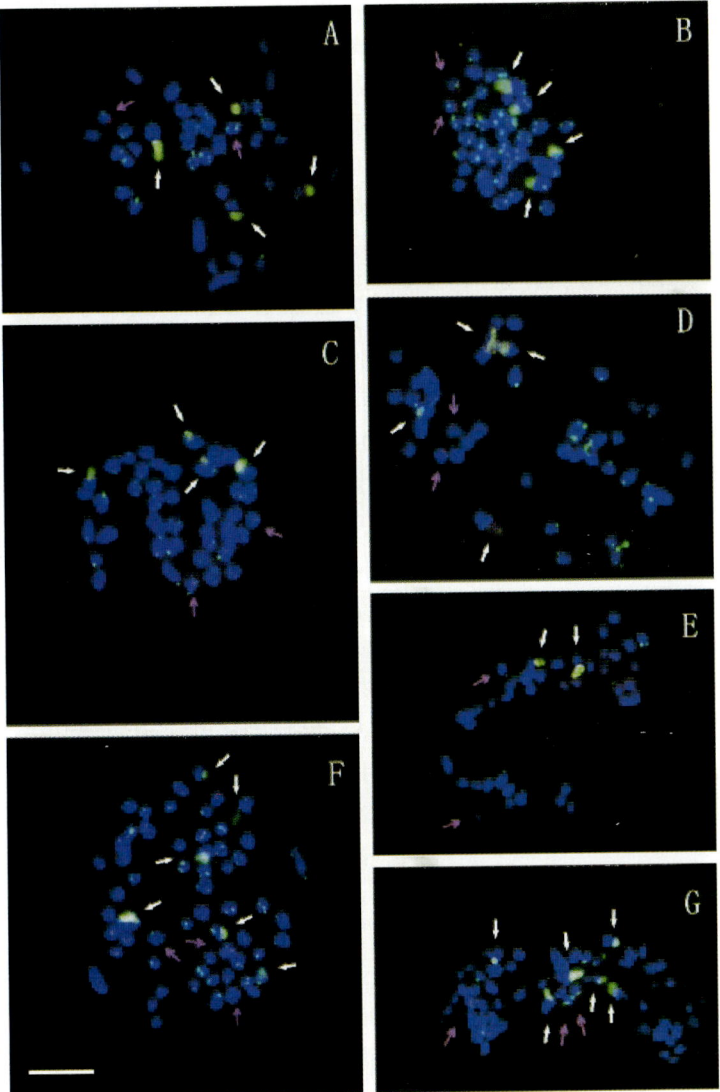

Fig.3.96 Multicolor fluorescence in situ hybridization with 25S rDNA, 5S rDNA and telomere on metaphase chromosomes of *Populus*

A: *P. tomentosa* (2n= 2x= 38); B: *P. nigra* var. *thevestina*; C: *P.pseudo-simonii*; D: *P. lasiocarpa*; E: *P.euphratica*; F: *P. tomentosa* (2n= 3x= 57), (bar= 5 μm); G: *P. euramericana* cv. Wuhei-1 (bar= 5 μm)

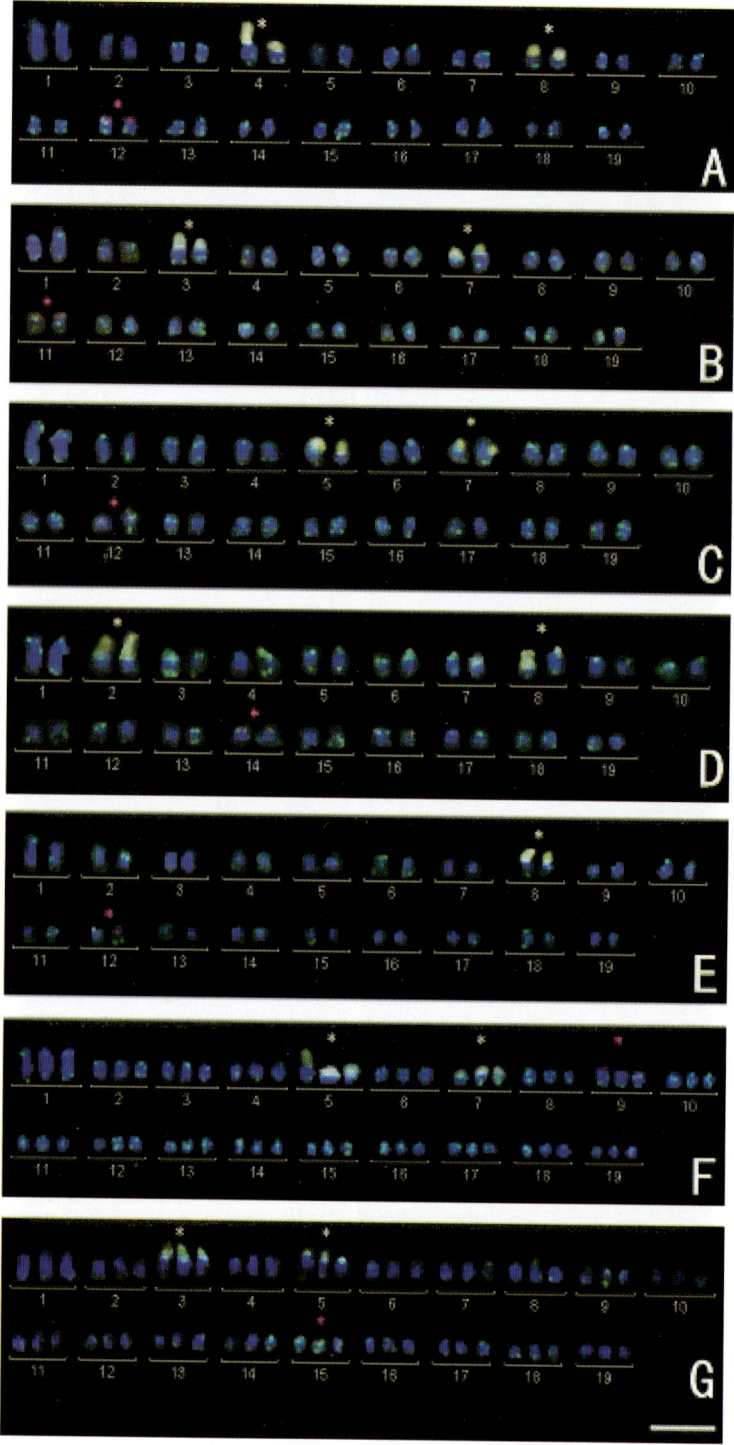

Fig. 3.97 Multicolor fluorescence in situ hybridization with 25S rDNA, 5S rDNA and telomere on metaphase chromosomes of *Populus*

A: P. tomentosa ($2n= 2x= 38$); B: P. nigra var.thevestina; C: P. pseudo-simonii; D: P. lasiocarpa; E: P. euphratica; F: P. tomentosa ($2n= 3x= 57$); G: P. euramericana cv. Wuhei-1 (bar= 5 μm) blast consequences were located on the not placed contigs. Probably, the 18S-28S rDNA sites among the seven *Populus* species were not same to each other as result of partial chromosome's in adequate dispersion.

However, these observations can still raise the conclusion that the major 18S-28S rDNA site was on the former chromosome than the minor one. And the translocations of the 18S-28S rDNA site in one of the seven *Populus* species were possible. Subsequent studies to confirm the translocations were needed in these species. The satellites were colocalized with the 25S rDNA in all seven *Populus* species. Identifying the satellites by using FISH was easier than in opticalm icroscope due to the higher sensitivity. The 5S rDNA was identified only one and the same results were found in *P. trichocarpa* and *P. nigra* (Islam-Faridi et al., 2009; Ribeiro et al., 2008) The 5S rDNA sequence of *P. trichocarpa* from NCBI was not found in any chromosome in the *Populus* trichocarpa database (build 1.2) by using the blast. The karyotype showed that 5s rDNA was on the 9th to 15th chromosome. In Islam's research, they considered that 5S rDNA was likewise identified on the 3rd chromosome. But from this result one can infer that 5S rDNA site is more likely on the latter chromosome.

According to the karyotyping classification (Stebbins, 1971), the genus *Populus* belongs to the 2B type: the slightly asymmetric chromosome. And as result of the morphologically small confusing chromosomes, chromosomes identification under optical microscope is quiet difficult. Here, we demonstrated a multi-color FISH which was employed two rDNA probes and *Arabidopsis*-type telomeric repeats sequences to discriminate the partial poplar's chromosomes and tried to establish the molecular karyotype of the seven *Populus* species. Karyotyping the seven *Populus* species showed the difference among the *Populus* sections.

Kulak used similar method that the 5S and 25S rDNA were used to identify the chromosomes of *Brassica* amphidiploids (Kulak et al., 2002). However, the rDNA location and karyotype can be applied to identify the partial chromosomes, but more landmarks were needed due to much more chromosomes in the *Populus*. In addition, the specific centromere probe is needed to measure the length of the long and short arm more exactly. These experiments will show the more difference and relationship among the *Populus* species.

In this study, the combinatorial labeling of probes method in one FISH experiment, coupled with the chromosomes measuring data was used to discern more chromosome targets and compare the location of the rDNA locus in the seven *Populus* species. Summarization of the results in this investigation could provide the more karyotype analysis formation to the cytogenetic research of the *Populus*.

(Authors: Hu Baoquan, Dong Fengping, Wang Chunguo, Qi Liwang, Song Wenqin, Chen Chengbin.
Published in: Acta Scientiarum Naturalium Universitatis Nankaiensis. 2012. 45 (1) 58-64)

3.4.7　25S rDNA 在杨属植物染色体上的定位

核糖体 RNA 基因（rDNA）是植物基因组中研究最广泛的遗传单元之一。rDNA 是拷贝数很高（500~40000bp）的串联重复序列，成簇分布于一个或数个位点（Long and Dawid, 1980; Peders-en and Linde-laursen, 1994）。25S rDNA 是 45S rD-NA 基因重复单位中的一员，位于核仁组织区（NOR）上。由于 rDNA 序列在真核生物中高度保守并且拷贝数高，应用荧光原位杂交（FISH）技术可以很方便的进行观察和定位。对于具有较多小染色体和形态对称染色体的植物物种，25S rDNA 是进行染色体组型分析的非常好的细胞学标记（Schrader et al., 2000）。rDNA 序列虽然高度保守，但在其数目和染色体上的物理位置在进化过程中并非一成不变，即使在同一属或亚属的近缘种中也存在着差异。杨属（Populus）植物具有基因组较小、物种丰富、生长迅速、再生能力强等特性而成为木本植物研究中模式物种的首选（Bradshaw and Stettler, 1995; Taylor, 2002）。在我国，杨树分布极为广泛，是最重要的用材树种之一，并在城市绿化中具有防沙滞尘功能，是园林绿化的首选树种。国际上较公认的杨属植物约含 30 个生物种（Eckenwalder, 1996），分为五组（五派）：白杨组（Section Leuce）、黑杨组（Section Aigeiros）、青杨组（Section Tacamahaca）、大叶杨组（Section Leucoides）和胡杨组（Section Turanga）。其中，在白杨组和黑杨组中都发现了天然三倍体植株（陈成彬等，2004）。对于杨属植物，陈成彬等（2005）及齐力旺等（2004）已对其进行了核型分析，但是杨树的染色体较小数量较多，传统的核型分析无法准确的显示染色体 NOR 区的数量并对其进行准确的定位，本文以 25S rDNA 作为探针在杨属植物 5 个组 7 个种（含品种）的中期染色体上进行原位杂交定位，并对 25S rDNA 在杨属植物染色体上的数量、分布及种间差异，与常规的核型分析进行了比较，为研究杨属植物进化过程中染色体 NOR 区变化及其相连染色体的变异提供依据。

3.4.7.1 材料和方法

1. 材料

杨属五组中的二倍体（$2n=2x=38$）代表种和白杨组和黑杨组中天然三倍体（$2n=3x=57$）共 7 种植物，名称及来源见表 3.11，凭证标本存南开大学生命科学学院标本室。

表 3.11 供试材料的名称及来源
Table 3.11 The name and origin of experimental materials

组 Section	种名 Species	凭证标本 Voucher	来源 Origin
白杨组 Leuce	毛白杨 P. tomentosa（$2n=2x=38$） 毛白杨 P. tomentosa（$2n=3x=57$）	齐力旺 L.W.Qi 03-4	山西大同 Datong, Shanxi
黑杨组 Aigeiros	箭杆杨 P. nigra var. thevestina（$2n=2x=38$） 武黑 1 号 P. euramericana cv. Wuhei-1（$2n=3x=57$）	齐力旺 L.W.Qi 03-97 陈瑞阳 R.Y.Chen 03-1	山西大同 Datong, Shanxi 天津武清 Wuqing, Tianjin
青杨组 Tacamahaca	小青杨 P. pseudo-simonii（$2n=2x=38$）	齐力旺 L.W.Qi 03-51	辽宁大连 Dalian, Liaoning
大叶杨组 Leucoides	大叶杨 P. lasiocarpa（$2n=2x=38$）	齐力旺 L.W.Qi 03-47	湖北建始 Jianshi, Hubei
胡杨组 Turanga	胡杨 P. euphratica（$2n=2x=38$）	齐力旺 L.W.Qi 03-132	山西大同 Datong, Shanxi

2. 杨树品种和染色体标本的制备

7 种材料均取自杨树植物的幼芽。材料用含有饱和 α-溴代萘与饱和对二氯苯的水溶液 25℃预处理 3 h，之后用固定液（甲醇∶冰醋酸=3∶1）固定。染色体标本的制备采用陈瑞阳等（1979）的去壁低渗法。选取染色体形态和分散较好的标本用于荧光原位杂交。

3. 荧光原位杂交

25S rDNA 片段来自拟南芥 25S rDNA 的一个 2.3 kb 的亚克隆（Unfried and Gruendler，1990）。通过随机引物法用 DIG（Digoxigenin-dUTP，Roche）标记 25S rDNA 作为探针。杂交及杂交后信号的检测按 Qi 等（2002）的方法进行。

3.4.7.2 结果

7 个种的中期染色体 25S rDNA 的荧光原位杂交及核型分析见图 3.98 和图 3.99。杨属植物除胡杨组外，白杨组、黑杨组、青杨组和大叶杨组的二倍体种都具有 4 个杂交信号，白杨组和黑杨组的天然三倍体都具有 6 个杂交信号，所有杂交信号均位于染色体短臂的端部，信号的强弱相近，而大小不同。

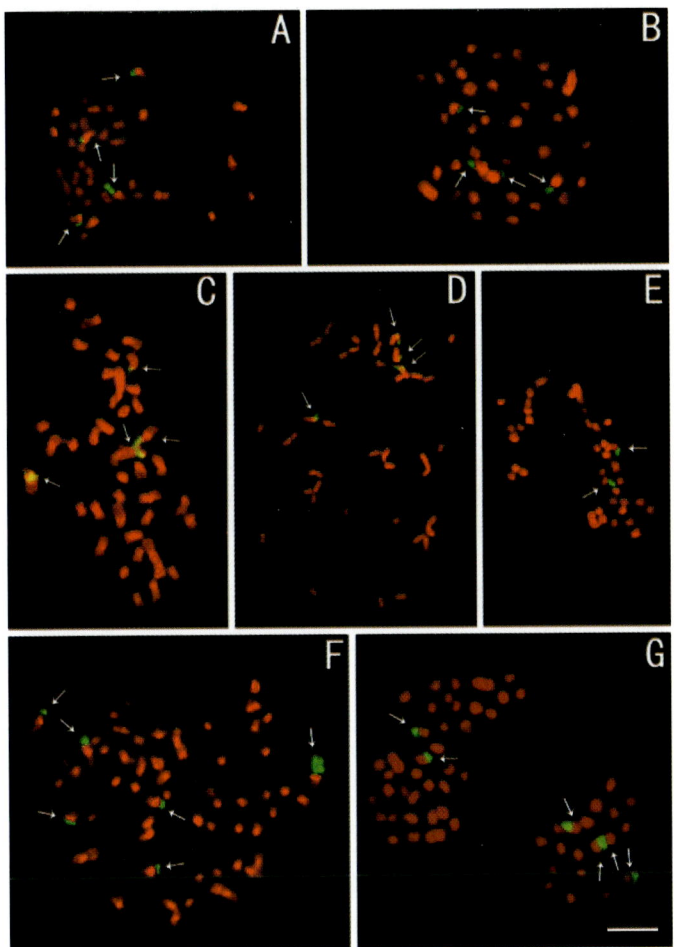

图 3.98 25S rDNA 与 7 种植物中期染色体的荧光原位杂交结果

Fig. 3.98 Fluorescence in situ hybridization with 25S rDNA on metaphase chromosomes of the seven species.
A. *P.tomentosa*（2n=2x=38）；B. *P.nigra* var. *thevestina*；C. *P.pseudo-simonii*；D. *P.lasiocarpa*；E. *P.euphratica*；F. *P.tomentosa*（2n=3x=57）；G. *P.euramericana* cv. Wuhei-1（bar=5 μm）

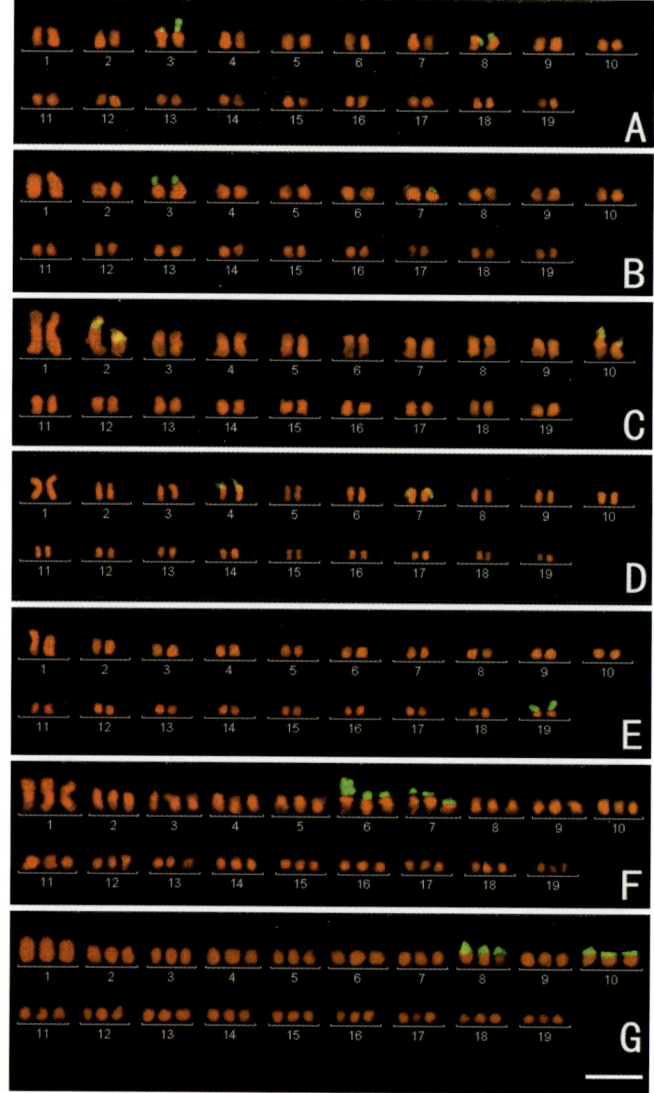

图 3.99　7 种杨属植物中期染色体核型图

Fig. 3.99　The karyotypes of the seven species

A. *P.tomentosa*（2*n*=2*x*=38）；B. *P.nigra* var. *thevestina*；C. *P.pseudo-simonii*；D. *P.lasiocarpa*；E. *P.euphratica*；F. *P.tomentosa*（2*n*=3*x*=57）；G. *P.euramericana* cv. Wuhei-1（bar=5 μm）

3.4.7.3　讨论

核糖体 RNA 基因（rDNA）所定位的染色体是物种染色体组中的一个重要成员。由于 25S rDNA 一般都定位在染色体的随体位置，在染色体组研究及核型分析中携带随体的染色体一般最易与其他染色体相区别，并且由于不同类型品种之间，随体染色体数目可能存在一定的差异。因此，研究随体染色体的数目及其在染色体组中的编号位置对于研究杨属植物的起源分化规律具有一定指导意义。

1. 结合 rDNA 位点的核型分析和与传统核型分析的比较

根据对杨属植物的常规核型分析，发现杨属五组之间核型基本结构十分相近，按 Stebbins（1971）的核型分类标准，杨属植物均为 2B 类型，为较不对称的类型，即由对称核型向不对称核型发展。常规核型分析依据光镜下观察染色体标本，确定染色体的类型、随体数目并进行配对组合。但是这种方法有时存在一定的不准确性，尤其对于那些随体并不明显的染色体配对，会产生一些困难。对具有形态相近小染色体的杨属植物，在应用 FISH 技术对 rDNA 重复序列进行染色体定位分析后，发现其 25S rDNA 位点数目与常规核型分析得出的随体数目并不完全相同。

毛白杨二倍体：2 对杂交信号位于 3 号和 8 号中部着丝粒染色体的短臂端部。而毛白杨核型公式为：1M +27m +2sm +6st +2t（齐力旺等，2005），没有观察到随体，25S rDNA 的荧光原位杂交后发现在 3 号染色体上有一对较明显的随体存在，其中一个随体较大，形成了一对异型随体。第 8 号染色体上的随体杂交信号相对 3 号染色体上的信号较小（图 3.99A）。

箭杆杨：核型结果分析表明箭杆杨核型公式为：3M +29m（2SAT）+5sm +1st（陈成彬等，2005），只观察到一对染色体具随体。而 25S rDNA 的荧光原位杂交结果显示，有 2 对染色体带有荧光杂交信号分别位于 3 号和 7 号中部着丝粒染色体的短臂端部。3 号染色体随体相对 7 号 rDNA 荧光原位杂交信号较大（图 3.99B）。

小青杨：荧光原位杂交信号位于 2 号近中部着丝粒染色体和 11 号近端部着丝粒染色体的短臂端部。小青杨核型公式：27m+6sm（1SAT）+4st（2SAT）+1t，观察到 3 条具随体染色体。荧光原位杂交结果清楚的显示出 25S rDNA 定位于 2 号染色体和 11 号染色体上，其中 2 号染色体的信号较大和较强（图 3.99C）。

大叶杨：荧光原位杂交信号位于 4 号和 7 号近端部着丝粒染色体的短臂端部。核型分析结果显示，大叶杨核型公式为：2M +22m+8sm+6st，未发现具有随体染色体，荧光原位杂交结果显示在大叶杨 4 号染色体近端部着丝粒处具有一对较明显的随体。相对较小的 25S rDNA 信号定位于 7 号近端部着丝粒染色体上（图 3.99D）。

胡杨：只有 1 对较大的荧光原位杂交信号位于最小的 19 号近端部着丝粒染色体的短臂端部，与胡杨核型分析结果：2M +23m +3sm +10st（2SAT）相一致。胡杨的随体大小要超过其所定位的染色体的大小（图 3.99E）。

毛白杨三倍体：25S rDNA 荧光原位杂交信号位于 6 号和 7 号近中部着丝粒染色体的短臂端部。毛白杨三倍体常规核型公式为：1M +31m+10sm+15st（陈成彬等，2004），没有观察到随体染色体，荧光原位杂交结果显示 6 号染色体具有较明显的随体，7 号染色体的随体大小和信号强度都稍小于 6 号染色体（图 3.99F）。

武黑 1 号：25S rDNA 荧光杂交信号位于 8 号中部着丝粒染色体和 10 号近中部着丝粒染色体的短臂端部。武黑 1 号核型公式为：2M +38m（4SAT）+14sm（2SAT）+3st，其显示的随体数目与荧光原位杂交的 25S rDNA 位点数目一致（图 3.99G）。

本研究结果清楚的显示了杨属 5 个组的二倍体和天然三倍体的 25S rDNA 的分布情况，与 PI 复染中所显示的随体所在位置一致（未显示），有些随体由于染色体高度折叠压缩的原因而没有被观察到。rDNA 探针的染色体荧光原位杂交与核型分析相结合，可对核型分析结果进行校正,提高同源染色体配对的准确性。

2. 杨属植物 25S rDNA 位点数目和位置变化的比较

植物属内 rDNA 位点数目和位置的变化已经在很多植物中观察到,例如在豆科(Leguminosae)的三个属(刘博等,2005),菜豆属 *Phaseolus*(Ga-lasso *et al.*,1995;Pedrosa *et al.*,2003),碧冬茄属 *Petunia*(Benabdelmouna *et al.*,1997),棉属 *Gossypium*(Ji *et al.*,1999),栎属 *Quercus*(Zoldos *et al.*,1999),豇豆属 *Vigna*(Moscone *et al.*,1999)的属内不同种或野生种与栽培种之间 rDNA 数目和位置都存在着差异。对杨属植物 5 个组的二倍体和三倍体的 25S rDNA 的数目和形态进行比较,发现毛白杨二倍体、箭杆杨、小青杨、大叶杨、毛白杨三倍体和武黑 1 号具有相似的 25S rDNA 分布特征,都具有 2 组可配对的 25S rDNA 位点(二倍体为 4 个,三倍体为 6 个),信号在形态大小上表现为其中一组要大于另一组,只是信号在染色体组中的排列位置不同;而胡杨则只有 1 对大的 25S rDNA 位点,并位于最小的 19 号近端部着丝粒染色体上,而其他六种杨属植物,携带较大杂交信号的一组同源染色体较大,排列顺序比较靠前。与其他组杨属植物不同的是,毛白杨的二倍体和三倍体以及胡杨中,较大的随体本身还具有一个缢痕(图 3.99A、E、F),称为衔接随体(洪德元,1990)。

从胡杨的分子核型分析中可以得出,胡杨组具有较多的 st、t 染色体,核型不对称系数比黑杨组、青杨组、大叶杨组高,为 63.03,胡杨的 25S rDNA 的数目和位置也与其他组差异较大。而白杨组具有更高的核型不对称系数为 65.97(陈成彬,2005),但其 25S rDNA 的数目和位置变化不大。

据 1935 年在新疆发现的化石推断,胡杨距今大约有 300~600 万年的历史,是杨属中最古老的一种(王世绩,1996)。干旱高盐碱的特殊生态环境保证了胡杨组与杨属其他植物发生分歧后的生殖隔离,也保证了其基因组的原始性。胡杨组与其他组杨属植物相比,虽然保持了核型的相似性,但其具有较高的核型不对称系数,根据 Stebbins(1971)对植物界核型进化的一般规律是由对称型向不对称型发展,核型越不对称越进化的阐述,可见胡杨属于较进化的类型,同时 25S rDNA 数目和位置的差异也显示在胡杨的进化过程中发生了明显的染色体变化。综合以上研究结果,说明胡杨自身进化的动力较高,发生了更多的染色体事件,这种随机事件发生频率高是与环境压力相适应的(Ayala *et al.*,2000)。染色体事件频率提高推动变异的产生、加速进化,使胡杨更加适应极端的生态环境,甚至产生了储水储碱等特异表型。

综上所述,笔者认为在杨属植物进化的过程中,胡杨组与其他杨属植物在 rDNA 数目和位置上存在着显著差异,这种染色体组型的变异可为研究杨属的起源进化,以及准确的判断胡杨在进化过程中的分歧时间提供了线索。在白杨组和黑杨组的天然三倍体中发现带有 25S rDNA 杂交信号的 3 个同源染色体的形态和信号的大小都存在着一定的差异,这显示其可能是异源三倍体,但这个推断还需要进一步的试验证据。

(作者:董凤平、韩素英、张守攻、齐力旺、刘博、李秀兰、陈成彬。

原载:云南植物研究,2007,29(4):423~428)

3.4.8 利用组合标记探针对华北落叶松进行多色荧光原位杂交和详细的核型分析

The chromosomes ($2n = 2x = 24$) of *Larix principis-rupprechtii* are composed of six pairs of large metacentrics and six pairs of medium-sized submetacentrics. The identification of homologous pairs is hampered by their high degree of similarity at the morphological level in each group. As one of the most extensively used methods in molecular cytogenetics producing chromosome landmarks, fluorescence in situ hybridization (FISH) has significantly facilitated karyotype construction, especially in species with morphologically similar chromosomes. This study developed a simple but effective use of combinatorial labeling probes to distinguish chromosomes of *Larix principis-rupprechtii* by multicolor FISH. Three highly repetitive sequences in *Larix* were selected: 25S rDNA hybridized at all of the secondary constrictions of two pairs of metacentrics and the largest pair of submetacentrics; 5S rDNA hybridized at subtelomeric sites of one pair of metacentrics that also harboured 25S rDNA on different arms; LPD family sequences are tandem repeats hybridized at proximal regions of 22 chromosomes. The three different probes were labeled with only two different labels, hybridized to metaphase chromosomes of *Larix principis-rupprechtii*, simultaneously visualized, and unequivocally distinguished in a single FISH experiment. These multicolor FISH marks largely improved the karyotype analysis of *Larix principis-rupprechtii*.

Larix principis-rupprechtii is a perennial deciduous arbor mainly planted in north of China. The diploid chromosome number ($2n = 2x = 24$) is very conserved in the family of Pinaceae, with little karyotype differentiation between species (Friesen *et al.*, 2001). The chromosomes of *Larix principis-rupprechtii* constitute a bimodal karyotype in which they are divided into two groups, a group of twelve large metacentrics and the other group of twelve medium-sized submetacentrics (Zhang *et al.*, 1985). However, in each group the identification of homologous pairs is hampered by similar length and slight differences at the morphological level. On the other hand, chromosome landmarks produced by fluorescence in situ hybridization (FISH) with various repetitive sequences as probes make the identification of chromosomes easier. This strategy has been applied successfully in several species of other genera in Pinaceae, e.g. *Pinus* (Hizume *et al.*, 2002b; Liu *et al.*, 2003) and *Picea* (Brown *et al.*, 1998; Siljak-Yakovlev *et al.*, 2002; Vischi *et al.*, 2003). In *Larix*, an improvement has been proposed by Lubaretz *et al.* (1996) with rDNA and telomeric repeats probes, but these probes are inadequate for distinguishing every chromosome pair. Hizume *et al.* (2002a) identified and characterized the LPD family as tandemly repetitive DNA sequences in *Larix leptolepis*, and they concentrate in proximal regions of most chromosomes of *Larix* so that the LPD repeats could be a novel probe in chromosome identification of *Larix*.

The commercial availability of a range of different fluorophores has enabled the simultaneous imaging of several target sequences in one experiment. In plants, however, traditionally a maximum of three fluorophores are used simultaneously, which emit either blue (amino-methylcoumarin; AMCA), green (fluorescein and derivatives, e.g. FITC) or red light (Texas red, rhodamine, cyanine3; Cy3). Routine counterstaining of chromosomes with for example DAPI, YOYO or propidium iodide, effectively reduces this number since emission wavelengths of these counterstains interfere with emission wavelengths of the commonly used fluorophores. Combinatorial labeling, which incorporates more than one fluorophore into a probe, is a means by which a greater number of chromosome targets may be discriminated. It was developed for human cytogenetics and is considered as a breakthrough in clinical diagnostics (Ried et al., 1992). In plants, Jiang et al. (1995) successfully mapped three different bacterial artificial chromosomes in the rice genome with this approach and Hasterok et al. (2002) improved this method in Secale cereale and Triticum aestivum. To date, combinatorial labeling of probes has scarcely been used for molecular cytogenetics of gymnosperms.

In this study, we developed a simple and efficient technique of multicolor fluorescence in situ hybridization in Larix principis-rupprechtii, which allows the use of probes labeled with two standard labels to map simultaneously three different DNA sequences along somatic chromosomes. Furthermore, a more detailed karyotype of Larix principis-rupprechtii could be constructed with these landmarks produced by the multicolor FISH.

Seeds of L. principis-rupprechtii were collected in a planted larch forest at Dagujia Forestry Centre in Liaoning Province, China. The seeds were sown in sterilized sand in a pot, and 10-14 days later the primary root tips were collected for chromosomal analysis. The root tips were treated with 0.05% colchicine at 20℃ for 14 h, fixed in chilled Carnoy's fixative (ethanol:acetic acid = 3:1, v/v) overnight.Fixed root tips were macerated in an enzyme mixture containing 2% cellulose ("Onozuka" R-10, Kinki Yakult) and 1% pectolyase (Aspergillus niger, Serva) in citrate/citric acid buffer (pH 4.5). The meristematic cells were squashed under a coverslip on a glass slide and air dried after the coverslip was removed with liquid nitrogen. Total genomic DNA was extracted from young leaves of L. principis-rupprechtii by the CTAB method (Murray and Thompson, 1980). A 2.3 kb subclone of the 25S rDNA coding region of Arabidopsis thaliana (Unfried and Gruendler, 1990) was labeled with Biotin-16-dUTP (Roche) by randomly-primed DNA synthesis according to manufacturer's instructions. The 5S rDNA clone pTa794 (Gerlach and Dyer, 1980) was amplified and labeled with digoxigenin (DIG) -11-dUTP (Roche) using PCR with universal M13 primers. The LPD family sequence was amplified from total genomic DNA of L. principis-rupprechtii by PCR using the primers LPD-f: 5'-CATGCATCCGGAAATAAG AA-3'and LPD-r: 5'-CGCAAACTATAGAAAGTCGT-3' (Hizume et al., 2002a), and the product was purified and labeled the same way as 25S rDNA but simultaneously and in equal proportions with digoxigenin-11-dUTP and biotin-16-dUTP.

FISH procedure was carried out as described previously (Liu et al., 2005). Briefly, chromosomal DNA was denatured at 70℃ for 2 min in 70% formamide, 2×SSC. Probes dissolved in 2×SSC, 5% dextran sulfate, and 50% formamide were denatured at 94℃ for 10 min. The hybridized probes were detected with anti-DIG-fluorescent-conjugate (FITC, Roche) and Strepetavidin-Cy3 (Sigma). The slides were counterstained with DAPI (4', 6-diamidino-2-phenylindole, Sigma) and mounted in Vectashield mounting medium (Vector Laboratory). The hybridization signals were visualized and recorded using Nikon 80i fluorescence microscope and a cooled CCD camera (SPOT, Diagnostic Instruments) equipped with the version 4.0.8 of the Spot software (Diagnostic Instruments). Ten metaphases were scored. Images were processed using Adobe Photoshop 7.0.

The length of short and long arms was measured, and arm ratio (length of long arm/length of short arm) and relative length were estimated for each chromosome. The nomenclature system for the position of the centromere followed that of Levan *et al.* (1964). Complementary chromosomes were arranged in order of decreasing length. Figure 1 shows the results of simultaneous triple FISH to the somatic metaphase chromosomes of *L. principis-rupprechtii*. Figure 3.100A-C show results from single band pass filters in blue, red andgreen channels, respectively, whilst Figure 3.100D present electronically superimposed images combining the three individual channels. In Figure 3.100B, 28 distinct red signals with different size and intensity indicated the hybridized loci of biotin-labeled probes, six of which marked the 25S rDNA loci and the other 22 were produced by LPD repeats probe. 25S rDNA loci and the other 22 were Chromosome spread simultaneously with 25S rDNA, 5S rDNA, and LPD repeats loci in pink, green, and white color, respectively. (E) Homologous chromosome pairing and ordering. Bar. 5 μm.

produced by LPD repeats probe. In Figure 3.100C, 24 distinct green signals indicated the hybridized loci of DIG-labeled probes, two tiny ones represented the 5S rDNA loci and the other 22 produced by LPD repeats probe were the same in size and intensity with the red signals of the same location in Figure 3.100B. After composed Figure 3.100A-C, signals with three different color were shown at chromosome spread (Fig. 3.100D). The six pink signals were the 25S rDNA red signals against the blue DAPI counterstaining and located at the secondary constrictions of two pairs of metacentrics and the largest pair of submetacentrics. The two tiny green 5S rDNA signals located at the subterminal regions of one pair of metacentrics that also harbored a 25S rDNA locus on different arms. Since the combinatorial labeling LPD repeats fluoresced red and green simultaneously (Fig. 3.100B and C), they showed white signals in proximal regions of 22 chromosomes in the composite image against the blue DAPI counterstaining of chromosomes. Thus, the multicolor discrimination of three different DNA sequences was accomplished within a single FISH experiment with only two labels.

The chromosomes of *L. principis-rupprechtii* in a given cell can be easily divided into two groups according to their length and centromere positions. Furthermore, with the help of these FISH signals above, we could complete the identification of homologous chromosome pairs within each group easily. First, by the 25S and 5S rDNA signals, three pairs of chromosomes could be identified unambiguously, as a double marked metacentric pair with

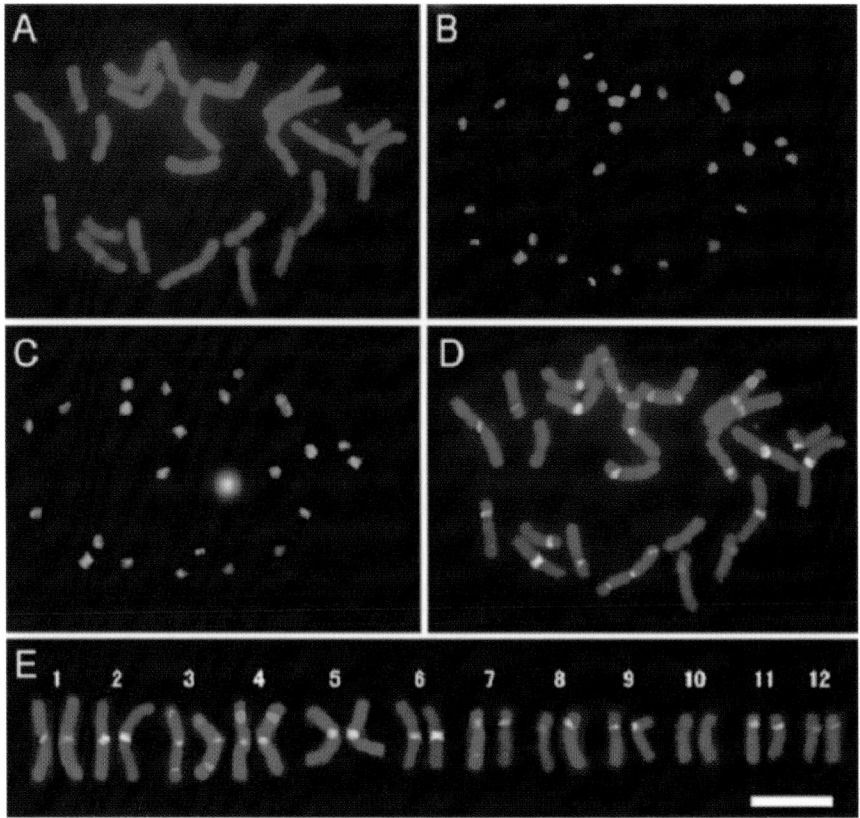

Fig. 3.100 Simultaneous triple FISH to somatic metaphase chromosomes of *L. principis-rupprechtii*.
A: Chromosome spread counterstaining with DAPI (blue); B: Red hybridized signals from 25S rDNA and LPD repeats;
C: Green hybridized signals from 5S rDNA and LPD repeats; D:

both 25S and 5S rDNA loci on its different arms, a metacentric pair with only 25S rDNA loci, and a submetacentric pair with 25S rDNA loci on its long arms. Although DAPI bands concentrated in proximal regions of most chromosomes are displayed after DAPI counterstaining (Fig. 3.100A), the white LPD repeats signals representing the same region exclusively (Hizume et al., 2002a) show more contrast to the blue background of chromosomes and make the paracentromeric regions of all chromosomes but one submetacentric pair clear. Hizume et al. (1994) observed only 20 DAPI bands in *L. principis-rupprechtii* with two fewer at submetacentric pairs than those in our study, maybe because some DAPI bands are too faint to be identified while LPD repeats FISH signals are more obvious (Fig. 3.100A and D). The chromosomes with similar hybridized signal size and intensity of LPD repeats could be identified as homologous pairs. Thus, in addition to the three pairs of chromosomes that have been identified by rDNA probes, other eight chromosomes pairs bearing distinct LPD repeat signals and one pair without FISH signal were found. Then the statistical analysis of microscopic measurements of chromosomes was carried out. Complementary chromosomes were arranged in order of decreasing length as shown in

1.25), and 2S (I.R.L. < 0.76) type chromosomes.

The relative length of the 13 chromosomes ranged from 5.73% to 9.22%. The length ratio of the longest to the shortest chromosome was 1.61.

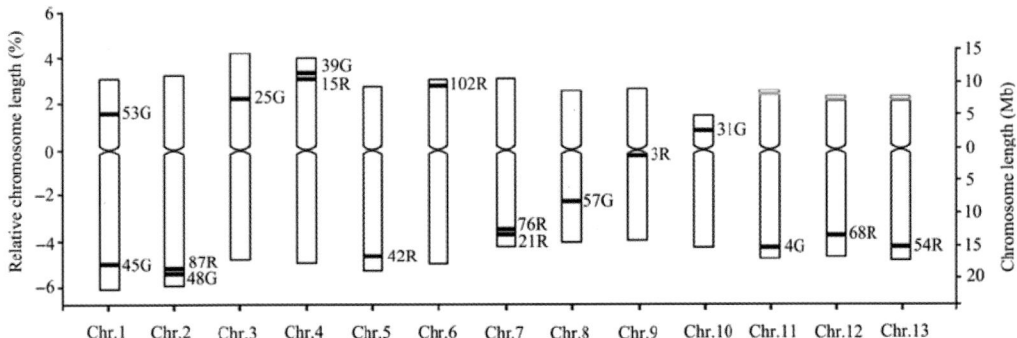

Fig. 3.125　Schematic of the sesame chromosome set marked with 17 BACs. The empty columns represent the chromosome. Red portions indicate satellite DNA；filled black bars indicate BAC positions. Vertical coordinates on the left indicate relative chromosomal lengths，while vertical coordinates on the right indicate genome length of each chromosome，calculated based on the S. indicum genome size（Zhang et al.，2013）

3.4.14.3　Discussion

This study was the first to accurately identify all 13 sesame chromosome pairs using 17 specific BAC probes. The technique of coupling BAC-FISH and successive in situ hybridization was novel to sesame genomic research. Our findings are valuable as a basis for cytogenetic map construction，targeting BAC locations on chromosomes and detailed genome assembly. The results demonstrate that successive double-colour FISH hybridisation is extremely effective for identifying small chromosome sets，such as those in sesame.

Of the 13 chromosome pairs in S. indicum, ten pairs from SiChr. 1 to SiChr. 10 are normal chromosomes，ranging from 1.32 to 2.13 lm in length. Meanwhile，the three chromosome pairs of SiChr. 11，12 and 13 are specific to satellite DNA，and the lengths are similar，varying from 1.57 to 1.61 lm（Fig. 3.125）. According to the somatic image，the morphology of most of the 13 chromosome pairs was similar，as the relative length ranged from 5.73% to 9.23%. In addition，we found that the absolute length and morphology of individual chromosomes differed between cells，despite strict quality control during chromosome preparation in this study. The karyotype formula also differed from previous results（Zhang et al.，2012a；Liu et al.，2014）. chromosome preparation in this study.

In previous studies，multiple karyotype formulas were reported for sesame and the parameters of the chromosomes were inconsistent（Zhan et al.，1987；He et al.，1994；Liu et al.，2014）. These morphological variations are likely the result of numerous factors，including the mitosis phase of chromosome samples，slide preparation methods and the skill of individual technicians（Li，1981；Zhang et al.，2012a）. Therefore，the 13 chromosomes could not be accurately distinguished based on the absolute length and morphology.

Moreover, for karyotype analysis in small chromosome set, we prefer the proposal presented by Li (1981), in which he indicated that the total chromosome length, rather than the parameters related to centromere position, should be considered.

Attempting to distinguish the 13 chromosome pairs, Liu et al. (2014) used FISH with rDNA, Cot-1 DNA and SSR as probes, but still six pairs were identified based on variation in chromosome length, rather than the FISH result markers. Instead of conventional pressing, here we chose to use the wall degradation hypotonic method (Chen et al., 1982), which did not require using and removing cover slips. Excluding these steps meant that chromosomes were naturally spread and tightly adhered to the slides, making them suitable for successive FISH hybridisation. This method also significantly reduced the disturbance to the BAC hybridisation signal from the cytoplasm (Feng et al., 2013). The obtained results clearly demonstrate the high-quality chromosome slides (Tran et al., 2016).

In this study, the highest number of successive hybridisations was 12 rounds (data not shown), although we ultimately used only eight rounds (Fig. 3.123). In comparison, Tran et al. (2016) performed an eight-round hybridisation in the carnivorous plant Genlisea margaretae, whereas Müller et al. (2002) reported four consecutive hybridisation strategies for metaphase chromosomes in Hylobates lar. The optimised chromosome preparation and hybridisation techniques were efficient for FISH application (Müller et al., 2002; Kitayama et al., 2006; Uher et al., 2009).

This study was also the first to construct a BAC library for sesame. The average insert length in the BAC library was 85 kb, and the shortest insert size of the BAC clones with clear and repeatable single signal pairs was 10.8 kb (data not shown), in line with previous reports showing that 10 kb is approximately the size for BAC probes to generate clear signals in most plant species (Jiang et al., 1995). Theoretically, each BAC insert has its own locus in a given chromosome set. However, here we found that only 101 out of 400 (25%) BAC plasmids generated stable hybridisation signals; 64 (16%) exhibited a single pair of fluorescence signals, while the remainder yielded weak or no signals (data not shown). The ratio of BAC hybridisation in sesame was slightly higher than the ratio in G. margaretae (184 $Mbp^{-1} \cdot C^{-1}$; Tran et al., 2016) and sugar beet (Paesold et al., 2012). For G. margaretae, 67% of the BACs were not repeatrich after the first screening of colony hybridisation and further hybridisation yielded 63 (11%) BACs that generated one pair of clear signals on isolated nuclei. Similarly, 18 out of 183 BACs (9.84%) were chromosome arm-specific in sugar beet (genome size: ~758 Mb; Paesold et al., 2012).

In small genomes, repetitive sequences are always intermingled with single-copy sequences in euchromatin and heterochromatin (Fonseca et al., 2010). DNA sequencing of BAC inserts revealed the presence of low-copy repeat and unspecific sequences, which are often blocked by the salmon sperm DNA and genomic DNA during hybridisation (Weier et al., 2013; Leal-Bertioli et al., 2015). In general, the number of hybridization signal pairs reflects genome complexity and the proportion of repetitive sequences in BAC inserts. Thus, the observed proportion of BAC hybridisation confirmed that the sesame (Yuzhi 11) genome is comparatively simple, with 38% of the sequences identified as repetitive elements (Zhang et al., 2013).

The small size of sesame chromosomes precluded the precise identification of centromere regions. However, we noted that all 17 BACs were mapped based on the relative positions of the hybridisation signals (Fig. 3.125). This means that FISH mapping can determine the separation distance of FISH spots in metaphase chromosomes and estimate the relative physical distance (Shearer et al., 2014). In tomato, pseudo-molecules tend to result in assembly errors; BAC-FISH has been used to correct genome scaffold arrangement and estimate gap sizes between scaffolds (Shearer et al., 2014). In conclusion, the application of BAC-FISH to sesame and other plants amply demonstrates the utility of this technique in generating necessary data for future cytogenetic research.

Accession codes The sequence information of the 17 BAC ends in the study has been deposited at the National Center for Biotechnology Information (NCBI) under the following accession number: SRR5887324.

(Authors: R. Zhao, H. Miao, W. Song, C. Chen, H. Zhang. Published in: Plant Biology, 2018, 20 (1): 85-92.)

第五节　植物性别机制的研究

提要：本节主要讨论植物性别染色体，据统计，雌雄异株的植物种类在双子叶植物中约为 4%，单子叶约为 3%，具有性染色体和性别基因的被子植物不超过 120 多种，其中被公认具有异形性染色体植物仅有 10 多种，银杏性染色体过去被认为是 XY 型。笔者实验室不仅通过核型分析，而且应用 Giemsa C-带、银染和荧光原位杂交技术对银杏性染色体进行了研究，结果证明银杏的性别机制是 ZW 型，雄株为 $2n=24=22A+ZZ$，雌株为 $2n=22A+ZW$。芦笋超雄株可增产 30%，虽然曾有人试图研究芦笋的性别机制，但没有获得肯定的结果。本研究应用去壁低渗法清楚的显示芦笋的性别机制为 XY 型，雄株为 $2n=18A+XY$；雌株为 $2n=18A+XX$。此外我们利用荧光原位杂交技术还确定了菠菜性别机制为 XY 型，雄株为 $2n=12=10A+XY$；雌株为 $2n=10A+XX$。

3.5.1　银杏性别染色体的研究

Ginkgo biloba is a dioecious, gymnorpermous plant. The male trees have been favored as shade trees. Although the ovulate gingko trees are not well accepted as green plants in cities because of the unpleasant smell of ripe fruit in late autumn, their fruit has been used as traditional medicine in China. So it has theoretical signidicance and economic value to determine the sex of *G.biloba* at a early stage.

Early in this century, the chromosome number of *G.biloba* was reported to be $2n=24$. (Ishikawa 1910, Herzfeld 1928, Shimamura 1928). The idiogram was described by Sax and Sax (1933) for the first time. Tanaka *et al.* (1952) published karyograms of both sexes based on chromosomes in somatic cells from young leaves. Lee (1954) and Newcomer (1954) reported the constitution of sex chromosomes in *G.biloba* and Hizume (1979) described the Giemsa C-banding.

In the present work we studied the karyotypes of somatic cells from the staminate and ovulate trees of *G.biloba* and paid a particular attention to the sex chromosomes.The present the investigation have clarified the mechanism of sex determination in *G.biloba* and put forward a detailed method of sex determination of gingko trees at the early stage of their development.

3.5.1.1 Materials and Methods

It is known that ginkgo is nota favorable material for cytologicalstudy.Because the chromosomes are extremely attenuated and elongated, it is useless for chromosomal morphology study. The present investigation used roots from the trees about 30 years old. The roots were cut and treated in 0.2% colchicine for 4-5 hours. With this method the satisfactory effect could be obtained. The authors examined five staminate and four ovulate plants and selected twenty saplings of over five years old to determine their sex. Each of them clearly displayed the constitute of sex chromosomes.

Chromosome samples were made using the wall Degradation Hypotonic Method. The karyotype analysis was carried out according to the method reported by Li and Chen (1985), and also referred to the method published by Lee (1954).

The procedure of C-banding: The chromosome samples stored over a week at room temperature→immersed in 0.1 M HCl at 60℃ for 20-30s→washed in tap water→macerated in saturated solution of Ba(OH)$_2$ at 20℃ for 15-30 min→washed in tap water→immersed in 2×SSC solution at 60-65℃ for 1-2 h →washed in tap water→stained in freshly prepared 2.5% Giemsa solution(1/15 mel.Sorensen phosphate buffer pH=6.8)at 5-10℃ 1 h→washed in tap water→air-dried.

The procedure of Ag-staining: The chromosome samples stored over 48h→stained in 8% AgNO$_3$ solution in an oven at 60-70℃ for 48 h→washed in distilled water→air-dried.

3.5.1.2 Results and Discussion

1. The chromosome number of *Ginkgo*

The present investigation examined more than 150 cells from 29 trees of ginkgo. The chromosome numbers are all $2n=24$. Aneuploid and B-chromosomes have not been found. It is shown that the chromosome constitute of ginkgo is very stable. This is consistent with the results reported previously.

2. The karyotype of *Ginkgo*

The twolarge, approximately isobrachial and ten acrocentric chromosomes were observed by Sax and Sax (1933) from somatic cells of the haploid endosperm tissue of the female.But the satellite has not been found. Tanaka *et al.* (1952) described the chromosome constitution in somatic cells from young leaves as a pair of large submetacentric and eleven pairs of subtelocentric chromosomes. In the haploid chromosome complement of the microspore, their preparation showed ten subtelocentric, one metacentric and one largest, approximately isobrachial chromosomes (Newcomer, 1954). At the same time he reported that in 18 out of 34 cells, the largest chromosome was disiso-chromosome. Tanaka (1952) pointed out that the satellites were located in the short arm of the large chromosome and the long arm of the eighth chromosome pair. Newcomer observed that satellites did not occur in the largest chromosome parir, but in two subtelocentric chromosomes. He also pointed out that the size of the satellites decreased greatly and became nearly achromatic appendages in the microspore chromosomes.

We have carefully analyzed the karyotype of ginkgo. All of the cells studied possess ten pairs of subtelocentric chromosomes, in which the long arm of the shortest chromosome pair possesses satellites (xII) Fig. 3.126, Fig. 3.127, Fig. 3.128). This is in good agreement with the results reported by Lee (1954).

3. Sex chromosome of *Ginkgo*

Herzfeld (1928), Based on the study of the formation of proembryo in ginkgo, suggested that the difference between male and female plants might lie in the chromosome structure and their behavior in cell division. Although Sax and Sax (1933) first drew out the diagram of the chromosome in the female gametophyte, they did not observe the existence of sex chromosomes. In the study of Tanaka *et al.*, both sexes were examinted. It was found that the chromosomes of the male plant were slightly larger than that of the female, but nodirect evidence of the presence of sex chromosomes was reported. Newcomer (1954) only investigated the staminate trees of ginkgo and suggested that large chromosome of ginkgo were of two types.They ere metacentric in about half of the microspores and submetacentric in the others. He proposed that this phenomenon was probably associated with the sexual determination of ginkgo belongs to the sex mechanism of XY-type. In both respects of mitosis and miosis, C.L.Lee (1954) evidenced the sex of ginkgo, and suggested that the male and female trees possessed the same chromosome numbers ($2n=24$). Their chromosomal morphology was almost identical. The only disserence was that four chromosomes in the entire diploid compliment from the female possessed satellites, while only three chromosomes from the male possessed satellites. Only one of the two subtelocentric chromosomes possessed a satellite in both trees. Therefore, he thought that they were sex chromosomes, and the sex determination of ginkgo was XY-type.

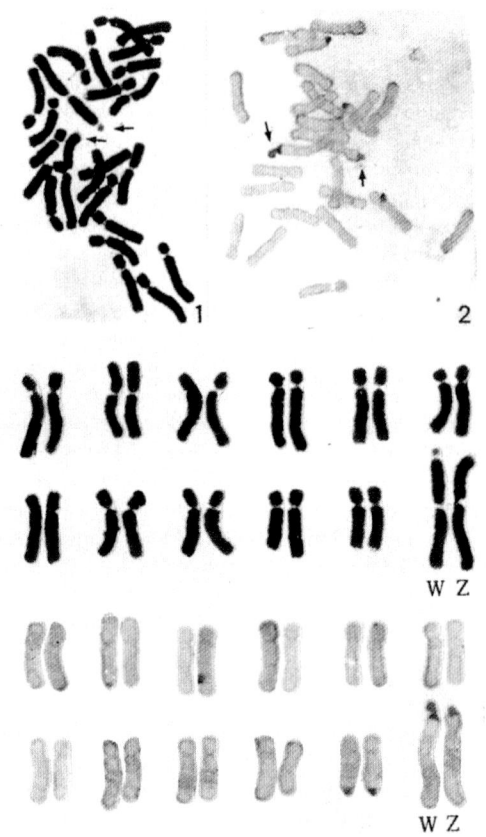

Fig. 3.126 1. The karyotype of *Ginkgo biloba*（female）
2. C-banded karyotype of *Ginkgo biloba*（female）

The present investigation shows the existence of sex chromosomes in ginkgo, but he mechanism of sexual determination is not in accordance with the views of the other authors. The present result, using the karyotype analysis, C-banding and Ag-staining techniques unambiguously shows that the large, approximate isochromosomes in the female is heteromorphic chromosomes. The satellite is very small and almost undetectable in one of these chromosomes, but is large in the other.In male, this pair of large chromosomes is of homomorphic satellites. Both of them show slight satellite banding and NOR region at the end of short arms.C.L.Lee（1954）reported that the samllest chromosome pair in male is heteromorphic, and thelong arm of one of the two subtelocentric chromosomes possessed a satellite. However, we observed that this pair of chromosomes is completely homomorphic（Fig. 3.127）. The Giemsa C-banding and Ag-staining further demonstrate that they have the same banding and NOR regions（Fig. 3.127）, and the largest approximately isobrachial chromosome is completely homomorphic. C-banding and AG-satining show that they have completely the same structure. The end of their short arm possesses a minute satellite（It is rather a secondary centromere after the satellite is lost）and an NOR region（Fig. 3.127,

Fig. 3.128, Fig. 3.129).

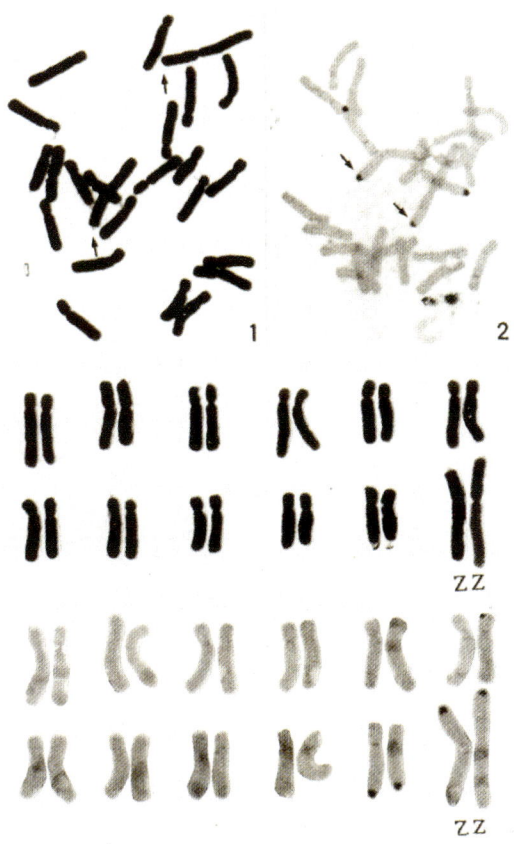

Fig. 3.127 1. The karyotype of *Ginkgo biloba* (male)
2. C-banded karyotype of *Ginkgo biloba* (male)

To sun up, the present experimental results have confirmed that the large, approximately isobrachial chromosomes in ginkgo ae the sex chromosomes. They are heteromorphic in female and homomorphic in male. The chromosomes in the latter are identical to one chromosome in the former. This chromosome is called Z-chromosome, and the other heteromorphic chromosome is called W-chromosome. Therefore, the sex mechanism of ginkgo is WZ type, i.e., the male is $2n=24=22A+ZZ$, and the female is $2n=24=24+ZW$. Some animals, such as silkworm and some kinds os birds belong to this type. Kihara(1926)reported that the sex of *Fragria elatior* was WZ type. It was reported that the NOR region and sex chromosomes of *Corchorus capsularis* L. arelocated in the same chromosomes. U. Schafer and M.Schafer (1980) observed the NOR region of X-, and Y-chromosomes in Drosophila. The sex chromosomes and NOR regions of ginkgo are located in the same chromosomes. For the explanation of relationship between the sex chromosomes and NOR regions in ginkgo, we call them Z-NOR and W-NOR region respectively. What are the differences between the NOR region in the sex chromosome pair and the seventh chromosome pair of ginkgo will be studied further.

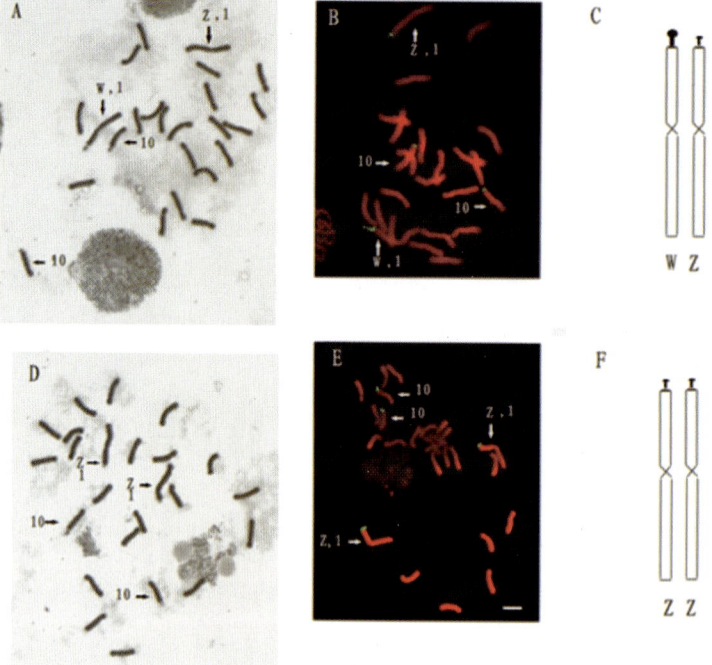

Fig. 3.128　The chromosomes of *Ginkgo biloba*（♂♀）
A, B and C: Female, D, E and F: Male

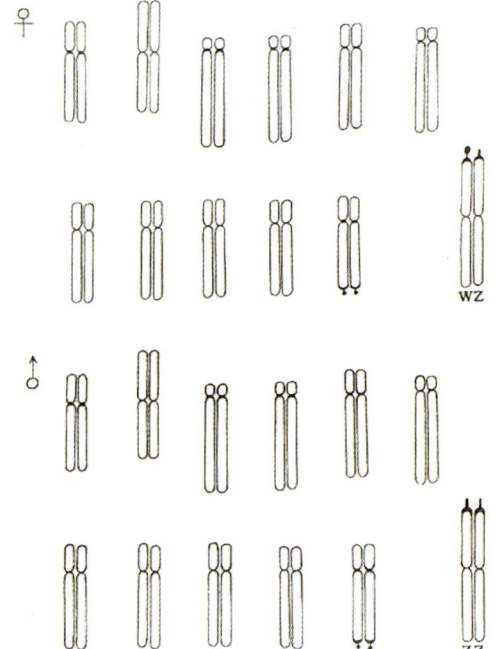

Fig. 3.129　C-banded idiograme of *Ginkgo biloba*

Figure 3.100E and their idiogram in Figure 3.101. According to the ordering, the chromosome pair bearing both rDNA loci is No. 3. Chromosome 4 and 7 bear 25S rDNA loci at their short arm and long arm, respectively. The only pair without any FISH signal is chromosome 10.

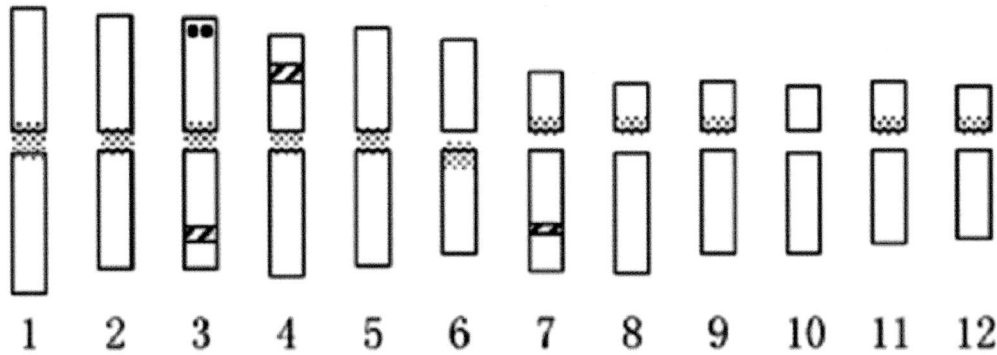

Fig. 3.101 Idiogram of *L. principis-rupprechtii* with 25S rDNA (slash area), 5S rDNA (filled circle), and LPD repeat (dot) loci

The karyotype data including the length of short and long arms, arm ratio (length of long arm/length of short arm), and relative length estimated for each chromosome were listed in Table 3.12. Since the homologous chromosome pairing is more credible with the assistance of multicolor FISH signals, the karyotype data could be more accurate than those obtained from conventional karyotype. However, the length of the chromosomes within the large group and the medium-sized group each are so close that some ordering could be mistaken. This could even happen for the identification of long/short arm of the metacentrics, especially for chromosome 2 and 3. To solve this problem, more chromosome landmarks are needed, and then, the detailed karyotype could be constructed not only according to the length of chromosomes.

Simultaneous in situ hybridization of more than two different DNA probes to the same chromosome preparation is informative and could help us to complete the construction of detailed karyotypes of *Larix*, but there are some noticeable technical problems. First, it is important to ensure that the emission spectra of the probes are sufficiently well separated from each other, and from the counterstaining of the chromosomes. We selected Cy3 and FITC as reporters, which fluoresce in the red channel and green channel, respectively. They are distinct from each other and also from the blue background of chromosomes produced by DAPI. Secondly, we should be cautious with probes whose loci are totally or in part overlapped. If two probes are supposed to be detected by a single color but they co-localize at the same loci of metaphase chromosome, they will display composite color and that will foster confusion with the third probe of combinatorial labeling. Although it is possible for more than three different probes to be detected simultaneously through combinatorial labeling with more labels, or even by varying ratios of different labels for each probe, the non-standard emission filters and higher sensitivity CCD camera are needed.

Table 3.12 Measurements of somatic metaphase chromosomes of *L. principis-rupprechtii*（2*n* = 24）metacentric，sm submetacentric

Chromosome number	Relative length of short arm	Relative length of long arm	Relative length	Arm Ratio	Chromosome type
1	5.348	6.113	11.551	1.124	m
2	5.017	5.101	10.118	1.017	m
3	4.975	5.101	10.076	1.025	m
4	4.216	5.438	9.654	1.290	m
5	4.553	5.017	9.570	1.102	m
6	4.089	4.469	8.558	1.093	m
7	2.614	5.270	7.884	2.016	sm
8	2.150	5.312	7.462	2.471	sm
9	2.234	4.511	6.745	2.019	sm
10	1.981	4.469	6.450	2.256	sm
11	2.204	4.089	6.113	2.020	sm
12	1.981	3.836	5.817	1.936	sm

In this study，we demonstrated above that in gymnosperms with large chromosomes like *Larix* species，combinatorial labeling of probes offers a simple，inexpensive and efficient way of discerning more chromosome targets with the same number of fluorophores in one FISH experiment. By this technique and based on several appropriate probes，we may construct FISH karyotypes of *L. principis-rupprechtii* and other *Larix* species，and that would provide us more detailed information for karyotype analysis and even for comparison among species.

（Authors：Bo Liu，Liwang Qi，Ruiyang Chen and Wenqin Song.
Published in：Biol Res，2007，40: 23-28.）

3.4.9 应用二色荧光原位杂交和 DAPI 带对四种落叶松属植物进行分子细胞遗传学分析

Mitotic metaphase complements of four Eurasian *Larix* species，*L. gmelinii*，*L. principis-rupprechtii*，*L.kaempfeeri*，and *L. olgensis*，were analyzed by bicolor fluorescence in situ hybridization（FISH）with 25S and 5S rDNA probes as well as DAPI banding. The four species were all diploids with the common chromosome number of 2*n*=24. All of the 25S rDNA clusters were located in the secondary constrictions of chromosomes，but the number of their sites varied among species. In *L. kaempfeeri* and *L. olgensis*，a pair of 25S rDNA loci were found on each of two large metacentric chromosome pairs，while in L. gmelinii and *L. principis-rupprechtii*，another two loci were present on the long arms of one pair of submetacentric chromosomes. The 5S rDNA FISH pattern was less variable，with only one site commonly observed in each species. These loci occurred on a pair of large metacentrics that also harbored 25S rDNA on different arms. Bright, fluorescent DAPI bands were positioned adjacent to the centromeres of all but one chromosome pair of *L. principis-rupprechtii*, *L. kaempfeeri*, and *L. olgensis* and two chromosome pairs of *L. gmelinii*. However, these bands showed obvious differences in intensity between the species, and together with the interspecific variation of 25S rDNA FISH patterns, they provide preliminary molecular cytogenetic evidence for phylogenetic study of these *Larix* species.

Introduction *Larix* is a prominent component of the boreal forest and a young genus in the pine family, according to fossil records (Le Page and Basinger 1991; Schorn, 1994). It is one of the four Pinaceae genera with a very large range and consists of 15 species, 10 of which are widely accepted while the others are controversial (Farjon, 1990, 2001). Three species, *L. laricina*, *L. occidentalis*, and *L. lyallii*, are endemic to North America, while the others are distributed in Eurasia. Patschke (1913) initially divided the genus into two sections, *Larix* and *Multiserialis*, based on the morphology of the ovulatecone. Phylogenetic relationships among Larix species have been the subject of numerous debates (Shi *et al.*, 1998; Mao *et al.*, 1999; Wei and Wang, 2004). Recently, traditional classification based only on external morphology has been in conflict with the evidence from molecular analysis (Qian *et al.*, 1995; Gernandt and Liston, 1999). Although conventional karyotypes had been constructed for most Larix species (Sax and Sax, 1933; Simak, 1964a, 1964b, 1966; Zhang *et al.*, 1985; Hizume, 1988), cytogenetic data did not seem to be of diagnostic value because of the slight interspecific discrepancy of karyotypes. Each species of *Larix* has a common bimodal karyotype composed of six pairs of long, metacentric chromosomes and six pairs of short, submeta or acrocentric chromosomes. The chromosomes, like those of most of theother Pinaceae genera, are very large and share a close morphological similarity to each other and between species. More-detailed karyotypes in the genus *Larix* are needed for comparison among species.

For precise comparative karyotype analysis, many attempts at chromosomal identification and karyotyping have been made for gymnosperms. One successful method is fluorescence banding using different fluorochromes, which can produce variable banding patterns among individual chromosomes by staining heterochromatin (Hizume *et al.*, 1983, 1989, 1990; Dagher-Kharrat *et al.*, 2001). However, this procedure has had limitations that have prevented its wide adoption. For example, fluorescence banding is a cumbersome and time-consuming procedure with low reproducibility (Hizume *et al.*, 2002b), and sometimes it provides only limited information, so that the complements of many species do not show prominent differences even after fluorochrome staining. This problem has been solved by combining fluorochrome banding with fluorescence in situ hybridization (FISH) (Doudrick *et al.*, 1995; Siljak-Yakovlev *et al.*, 2002). In fact, FISH is such an excellent tool that with multiple probes, it can distinguish every individual chromosome in a single species, e.g., Picea abies (Vischi *et al.*, 2003) and several *Pinus* species (Hizume *et al.*, 2002b). Among the probes used in FISH, 5S and 18S-25S rRNA genes are the most frequently used because of their high degree of sequence conservation, even between unrelated taxa, and their specific chromosomal positions (Maluszynska and Heslop-Harrison, 1993). In Pinaceae, rDNA loci have been localized on chromosomes of a limited number of species from a few genera, mainly of Pinus (Doudrick et al., 1995; Lubaretz et al., 1996; Jacobs et al., 2000; Hizume et al., 2002b; Liu et al., 2003) and Picea (Lubaretz et al., 1996; Brown and Carlson, 1997; Brown et al., 1998; Siljak-Yakovlev et al., 2002). In these species, the number of 18S-25S rDNA loci ranges from three to 14 pairs (Doudrick et al., 1995; Lubaretz et al., 1996; Brown and Carlson, 1997). The 5S sites are less numerous but range from one to three pairs (Doudrick et al., 1995; Hizume et al., 1999) and, in relation to the 18S-25S rDNA sites, may be either on different chromosomes, on the same chromosome but on different arms, or close or adjacent to each other (Lubaretz et al., 1996).

In contrast to *Pinus* and *Picea*, there are fewer molecular cytogenetic studies of the genus *Larix*. To our knowledge, rDNA FISH data are available for only L. decidua in section *Larix*, with three pairs of 18S-25S rDNA sites and one pair of 5S rDNA sites on three different chromosome pairs (Lubaretz et al., 1996). Hizume et al. (1988, 1993) detected DAPI bands within Larix and found that the number and location of these bands vary among species. However, these investigations are still inadequate for a detailed karyotypic analysis of *Larix*.

In this study, we detected genomic distribution of rDNA sites using bicolor FISH in four eastern Eurasian larches from the section *Larix* and compared these data with those from other Pinaceae genera (Farjon, 1990). Among these four species, *L. gmelinii*, *L. kaempfeeri*, and *L. olgensis* show highly similar karyomorphology (Liu et al., forthcoming), and *L. olgensis* is argued to be a variety of *L. gmelinii* (Shi et al., 1998). With the rDNA FISH data, as well as the DAPI banding pattern, we aimed to provide molecular cytogenetic evidence clarifying the phylogenetic relationships of these species.

3.4.9.1 Material and Methods

1. Plants and chromosome preparation

Seeds of *Larix gmelinii*, *L. principis-rupprechtii*, *L. kaempfeeri*, and *L. olgensis* were collected in a planted larch forest at Dagujia Forestry Center in Liaoning Province, China. The seeds were sown in sterilized sand in a pot, and 10–14 d later the primary root tips were collected for chromosomal analysis. The root tips were treated with 0.05% colchicine at 20℃ for 14 h and fixed in chilled Carnoy's fixative (ethanol: acetic acid ¼ 3: 1 v/v) overnight. Fixed root tips were macerated in an enzyme mixture containing 2% cellulase (Onozuka R-10, Kinki Yakult) and 1% pectolyase (from Aspergillus niger, Serva) in citrate/citric acid buffer (pH 4.5). The meristematic cells were squashed under a coverslip on a glass slide and air dried after the coverslip was removed with liquid nitrogen.

2. Fluorescence in situ hybridization

A 2.3-kb subclone of the 25S rDNA coding region of Arabidopsis thaliana (Unfried and Gruendler 1990) was labeled with Biotin-16-dUTP (Roche) by randomly primed DNA synthesis according to manufacturer's instructions. The 5S rDNA clone pTa794 (Gerlach and Dyer 1980) was amplified and labeled with digoxigenin-11-dUTP (Roche) using PCR with universal M13 primers. The FISH procedure was carried out as described previously (Liu et al. 2005). Briefly, chromosomal DNA was denatured at 70 ℃ for 2 min in 70% formam ide and 2×SSC. Probes dissolved in 2×SSC, 5% dextran sulfate, and 50% formamide were denatured at 94 ℃ for 10 min. The hybridized probes were detected with anti-digoxigenin fluorescent conjugate (Roche) and Strepetavidin-Cy3 (Sigma).

The slides were counterstained with DAPI (49, 6-diamidino-2-phenylindole, Sigma). The hybridization signals were visualized and recorded using a Nikon 80i fluorescence microscope and a cooled CCD camera (SPOT, Diagnostic Instruments) equipped with version 4.0.8 of the SPOT software (Diagnostic Instruments). For each species analyzed, 10 or more metaphases were scored. Images were processed using Adobe Photoshop 7.0. Complementary chromosomes were arranged in order of decreasing length.

3.4.9.2 Results

1. Chromosome spreads and genomic distribution of 25S and 5S rDNA loci in *Larix*

Ten or more metaphase chromosome spreads prepared from root tips of Larix species *L. gmelinii*, *L. principis-rupprechtii*, *L. kaempfeeri*, and *L. olgensis* were scored (data not shown), and in each species analyzed, they showed the same bicolor FISH signals produced by the 25S and 5S rDNA probes (Fig. 3.102). The chromosome complements of the four species were commonly composed of six long, metacentric chromosome pairs and six medium-sized submetacentric or acrocentric pairs.

In all four species, the FISH signals from the 25S rDNA probe appeared on the secondary constrictions at the interstitial regions of one arm of some chromosomes (red signals in Fig. 3.102). In *L. gmelinii* and *L. principis-rupprechtii*, three pairs of intense signals were detected on the long arm of chromosome 3, the short arm of chromosome 4, and the long arm of chromosome 7 (Fig. 3.102 a, b). Different from those in *L. gmelinii*, the signals in *L. principis-rupprechtii* showed intensitydiscrepancy among the three loci. The one on the short arm of chromosome 4 was the most intense, while the one on chromosome 7 was the faintest. In *L. kaempfeeri* and *L. olgensis*, only four signals were found. They were locate don the long arm of chromosome 3 and the short arm of chromosome 4 (Fig. 3.102c, d). The signals in the latter two species also displayed intensity differences.

Only one 5S rDNA locus was detected in the four species (green signals in Fig. 3.102). This locus always occurred near or at the terminal regions of the short arm of a chromosome that belonged to the large metacentrics and that also harbored a 25S rDNA locus on a different arm.

The genomic distribution of rDNAs in the four species that have been characterized here share some common traits. (1) No 25S rDNA loci, not even very weak signals, have been found outside the secondary chromosomal constrictions. (2) There are two 25S rDNA sites on two pairs of metacentrics in each case, while *L. gmelinii* and *L. principis-rupprechtii* have another 25S rDNA site that always appears on the largest submetacentric/acrocentric chromosome. (3) The 5S rDNA site is syntenic to one of the 25S rDNA sites but is located on a different arm of the same metacentric chromosome.

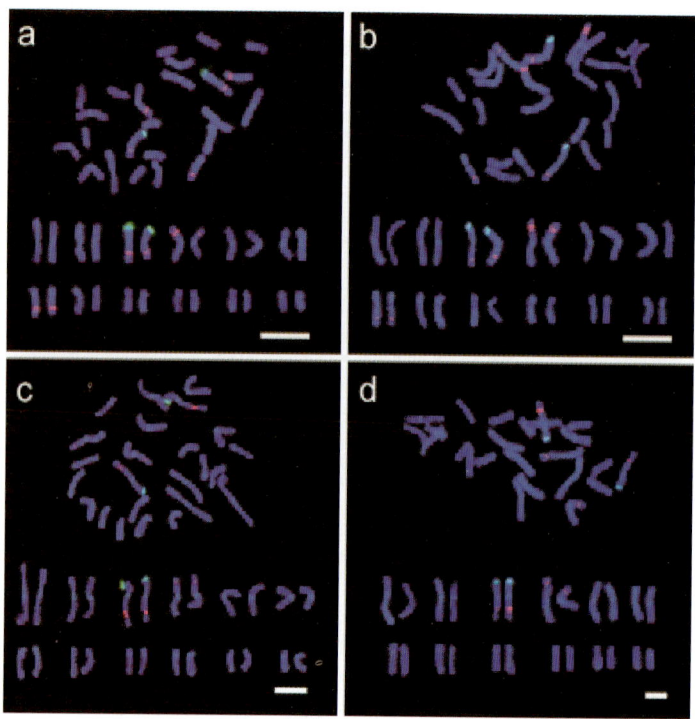

Fig. 3.102 Physical mapping of 25S rDNA (red) and 5S rDNA (green) using bicolor FISH and, simultaneously, DAPI banding on somatic metaphase chromosomes of (a) *Larix gmelinii*, (b) *L. principis-rupprechtii*, (c) *L. kaempfeeri*, and (d) *L. olgensis*. Bar = 5 mm

2. Fluorescence banding pattern of DAPI in *Larix*

When the chromosomes of the four *Larix* species were counterstained with DAPI after hybridization with rDNA probes, positive DAPI bands were observed exclusively in proximal regions of a single chromosome arm (Fig. 3.102). The number of DAPI bands was variable among species. There were 20 bands in *L. gmelinii* (Fig. 3.102a) and *L. principis-rupprechtii* (Fig. 3.102b) and 22 bands in *L. kaempfeeri* (Fig. 3.102c), and *L. olgensis* (Fig. 3.102d). In each species, the DAPI bands on submeta- or acrocentric chromosomes were all located on the short arm, and some of those on metacentrics were located on the long arm. The bands on large metacentrics were more intense than those on medium-sized submeta- or acrocentrics.

3.4.9.3 Discussion

1. Detailed karyotyping in *Larix*

The four *Larix* species observed in this study presented the common bimodal karyotype that has been previously reported (Zhang *et al.*, 1985; Hizume, 1988; Hizume *et al.*, 1988).

This kind of bimodal karyotype is shared by *Larix* and its closely related genus Pseudotsuga, with the exception of Pseudotsuga menziesii, and is different from that of other genera in Pinaceae (Hizume, 1988). According to Hizume *et al.* (1988), the chromosomes of the larger group are ca. 10 mm in length, and those of the smaller group are ca. 6 mm in length. In our study, the chromosomes of some cells were condensed so strongly that they were much shorter, ca. 5 and 3 mm in length, than those in most other cells (Fig. 3.102a, b). The different degree of chromosome condensation among metaphases has also been observed in *Picea abies* (Vischi *et al.*, 2003). The primary constrictions representing centromeres and the secondary constrictions revealed as DAPI-negative regions were visible, and, furthermore, the more condensed chromosomes spread much better, with few overlaps.

In the study, we performed karyotypic analysis using bicolor FISH as well as DAPI banding in the four species, with much more development in detail and reliability than conventional karyotyping. However, some chromosomes in a given cell are so approximate in length and so symmetrical that they cannot be distinguished unambiguously, for example, chromosomes 2, 3, and 4 (Fig. 3.102). Since the chromosomes that show the same FISH signal pattern are recognized as homoeologous (Lubaretz *et al.*, 1996; Hizume *et al.*, 2002b), the chromosomes double-marked by 25S and 5S rDNA and the chromosomes marked only by 25S rDNA in each *Larix* species should be homoeologous. According to the average of repeated observations of 10 or more metaphases for each species, we present the double-marked chromosomes as chromosome 3 and the chromosomes harboring 25S rDNA on their short arms as chromosome 4. In *L. decidua*, another *Larix* species in which physical mapping of 18S-25S and 5S rDNA has been performed, three 18S-25S rDNA sites and one 5S rDNA site are found, and their distribution pattern is similar to that in *L. gmelinii*, except that the chromosome bearing only 18S-25S rDNA on its short arm is chromosome 5 (Lubaretz *et al.*, 1996).

2. Genomic distribution of 18S-25S and 5S rDNA and DAPI banding patterns in *Larix* and other conifer genera

Compared with other genera in the conifers, Larix has many fewer 18S-25S rDNA loci. For example, in *Pinus* and *Picea*, the number of major 18S-25S rDNA sites ranges from six to 10 pairs (reviewed by Murray *et al.*, 2002; Liu *et al.*, 2003). Some of these loci are located outside the secondary constrictions of chromosomes (e.g. in *Picea glauca*, up to 14 rDNA loci have been found, but only five pairs of chromosomes have a secondary constriction, as reported by Brown *et al.*, 1993), and weak signals derived from 18S-25S rDNA have been found at some interstitial and proximal regions in some *Pinus* species (Hizume *et al.*, 2002b; Liu *et al.*, 2003). In diploid angiosperms, the number of major 18S-25S rDNA sites typically ranges between one and five pairs (reviewed by Liu *et al.*, 2003). Liu *et al.* (2003) hypothesize that the difference in loci number is related to the evolutionary divergence between angiosperm and gymnosperm genomes and results in part from the much larger genome size in gymnosperms. The weak signals in these *Pinus* species may indicate low copy numbers of

18S-25S rDNA repeats, or they may indicate remnants of primary sites of 18S-25S rDNA that once existed at the centromeres but later moved out to distal sites through large-scale chromosome rearrangements(Liu et al., 2003). In our study of Larix, we also used 45S rDNA that included nontranscr ibeds pacer sequences (Arumuganathan et al. 1994) as a probe, and no addition al signals were found (data not shown). The few 18S-25S rDNA sites support the hypothes is of Liu et al. (2003) that Larix is a younger genus in Pinaceae. There are no clusters with low copy numbers of 18S-25S rDNA repeats in Larix, or the remnants have been eliminated completely or decreased to very low copy numbers that were not detected by FISH. Because the genome size of Larix is al so large (e.g. the 2C nuclear DNA content of L. gmeliniiis 22.3 pg, corresponding to the haploid genome size of ca. 1010 bp; Hesemann 1 980), r DNA may represent only as mall portion of the genome in Larix cmpared with other genera in the Pinaceae, in which the rDNA portion is large.

Compared with 18S-25S rDNA loci, the 5S rDNA loci pattern in gymnosperms is less variable. Commonly, one or two pairs of main sites are found in Pinus and Picea species (reviewed by Liu et al., 2003). In Larix, there is also only one pair of signals derived from 5S rDNA observed in the four species in our study and L. decidua (Lubaretz et al., 1996). With respect to the site relationships of the two classes of rDNA, 18S-25S rDNA sites and 5S rDNA sites are usually on the same chromosome but on different arms in Larix and Pinus (Lubaretz et al., 1996; Hizume et al., 2002b; Liu et al., 2003). The difference in the distribution pattern between the two genera is that 18S-25S rDNA sites are present on the long arm of the double-marked chromosome in Larix, while in Pinus they are present on the short arm (Hizume et al., 2002b; Liu et al., 2003). In Picea, 18S-25S rDNA sites and 5S rDNA sites are closely adjacent on the same arm, with the 5S rDNA site closer to the centromere (Lubaretz et al., 1996; Brown and Carlson, 1997; Siljak-Yakovlev et al., 2002; Liu et al., 2003). In several Pinus species that have two 5S rDNA sites, the two types of distribution patterns exist si-multaneously. In Podocarpus (Podocarpaceae), only one 18S-25S rDNA site and one 5S rDNA site are observed in the five species investigated. Interestingly, the two classes of rDNA sites are colocalized on the same chromosome and also tightly linked to each other (Murray et al. 2002). This colocalization has not been found in other gymnosperms and among land plants has been reported only in two bryophytes (Sone et al., 1999).

Our DAPI banding results are slightly different from those obtained by Hizume et al. (1993) for L. gmelinii and L.olgensis (22 and 20 bands, respectively); however, the common point is that the chromosomes without DAPI bands all belong to the submetacentrics/acrocentrics. In L. decidua, 22 DAPI bands have been reported by Lubaretz et al. (1996), which coincides with the number observed by Hizume et al. (1993). Among the Larix species investigated, including those from section Multiserialis, the lowest band number is 12 in L. chinensis and L. potaninii var. macrocarpa (Hizume et al., 1998). Brown and Carlson (1997) observed no DAPI-positive bands in Picea glauca and Picea sitchensis, while

Siljak-Yakovlev et al. (2002) detected 10 and 12 in *Picea omorika* and *Picea abies*, respectively. In *P. omorika*, all of the 10 DAPI bands are in centromeric regions. In *P. abies*, six of the 12 bands are in centromeric regions, while the other six are in intercalary regions. The intercalary DAPI bands are, at the same time, CMA positive and correspond to 18S-25S and 5S rDNA sites. In Pinus, fewer DAPI bands have been found, as in *Pinus densiflora*, where only four chromosomes possess DAPI bands, but as in *P. abies*, both interstitial and centromeric DAPI bands exist. These DAPI bands in *Pinus* are assumed to correspond to telomere sequence loci (Hizume et al., 2002b).

DAPI is a double-stranded–DNA-specific dye that interacts with DNA via at least two different mechanisms (see Kapuscinski 1995 for review). In regions where three or four AT base pairs are located in tandem, DAPI binds to the minor groove of the DNA, and this results in high fluorescence.

DAPI also intercalates between bases, which results in nonfluorescence. The former reaction is energetically favored in AT-rich regions, while the latter is favored in GC-rich areas along a DNA molecule. The intercalary DAPI bands in *P.abies* corresponding to rDNA sites that are GC rich can be explained by the latter mechanism. The strong proximal DAPI bands in *Larix* species may suggest AT-rich regions. Hizume et al. (2002b) isolated and characterized the DNA sequence localized at the proximal DAPI band and an LPD (*Larix* proximal DAPI band-specific) family was found. The AT content is ca. 67% in the LPD family. This sequence has been physically mapped in *L. kaempfeeri* and *L. chinensis* metaphase chromosomes, and they overlap the DAPI bands. It seems that the appearance of DAPI bands is a result of a gain or amplification of the LPD family repetitive DNA.

3. Phylogenetic relationships of eurasian species from section *Larix* revealed by molecular cytogenetics

In our previous study of *Larix* conventional karyotyping, seven species from section *Larix* were divided into three groups based on their mean arm ratio and the ratio of the longest chromosome to the shortest (Liu et al., forthcoming).Group I includes *L. gmelinii*, *L. kaempfeeri*, and *L. olgensis*; group II includes *L. decidua*, *L. laricina*, and *L. sibirica*; *L. principis-rupprechtii* alone forms group III. *Larix olgensis* is argued to be a variety of *L. gmelinii* (Shi et al., 1998), and this is shown by the above classification. The results from this study, however, do not support the argument or our previous classification. Both the FISH signal pattern and the DAPI-banding pattern are significantly different between *L.gmelinii* and *L. olgensis*. *Larix gmelinii* has one more 18S-25S rDNA site and one fewer DAPI band than *L. olgensis* (Fig. 3.102a, d). However, *L. principis-rupprechtii* has a distribution pattern of rDNAs and DAPI bands similar to that of *L. gmelinii*, while *L. olgensis* is similar to *L. kaempfeeri*, which agrees with the analysis of Zhang et al. (1985).

According to Hizume et al. (2002a), the number or amount of LPD family repeats reflects a path of species differentiation in *Larix*, as the repetitive DNA of the LPD family was likely amplified in an ancestor species of *Larix* at orafter differentiation of the genus *Larix*.

Despite the same number of DAPI bands, the number of LPD family repeats in L. olgensis is little less than in L. kaempfeeri, indicating that L. olgensis may have appeared earlier than L. kaempfeeri. Thus, L. olgensis could not be a variety of either L.gmellinii or L. kaempfeeri, and it is more likely to be a single species. The three pairs of 18S-25S rDNA FISH signals in L.gmelinii showed equal intensity, while in L. principis-rupprechtii the signals on the subtelocentrics disappeared completely. This may represent the evolutionary path of 18S-25S rDNA in Larix. From the above discussion, the order of appearance of the f our species may be L. gmelinii, L. principi srupprechtii, L. olgensis, and L. kaempfeeri, without c onside ringtheir relationships to each other.

(Authors: B. Liu, S.-G. Zhang, Y. Zhang, T.-Y. Lan, L.-W. Qi, and W.-Q. Song. Published in: Int. J. Plant Sci., 167 (2): 367–372. 2006.)

3.4.10 32 种植物端粒组构信息的初步研究

本研究以 32 种植物为材料，以两种不同荧光标记的端粒序列为探针，利用双色荧光原位杂交的方法（FISH）将两种端粒序列（人类端粒序列标记红色，植物端粒序列标记为绿色）定位于中期染色体上，拟在搞清植物的端粒序列组构信息和特殊的行为，这对研究植物端粒的结构和功能具有重要意义。

初步试验结果表明，蝴蝶兰、川百合、黄花菜、花椰菜、柿子、天葵、紫鸭拓草、红掌、甜菜、黄瓜、黄秋葵、水稻、甘薯、木瓜、黑麦、小麦、蚕豆、红小豆、绿豆、豇豆、黄豆、玉米等均为植物型端粒结构即（TTTAGGG）$_n$（图 3.103），蚕豆端粒序列具有成簇现象（图 3.103A），有三条染色体端粒信号覆盖整个短臂。芦荟、吊兰、郁金香、万年青、四种植物为人的端粒结构即（TTAGGG）$_n$；葱属的蒜葱、韭菜既不具有植物的端粒也不具有动物的端粒，那么他们染色体的末端是怎么行使保护的还有待于深入研究揭开这个不解之谜。风信子、水仙、长春花具有动、植物两种端粒（TTTAGGG）$_n$（TTAGGG）。其中长春花端粒序列均位于染色体着丝粒附近，染色体末端未出现端粒序列；天葵植物除染色体末端外还具有内部端粒序列，这些内部端粒的产生可能是植物在长期的进化过程中，经末端端粒序列融合而生成或者是带有端粒的元件整合到染色体中产生的，相关端粒组构信息的研究正在进行中。

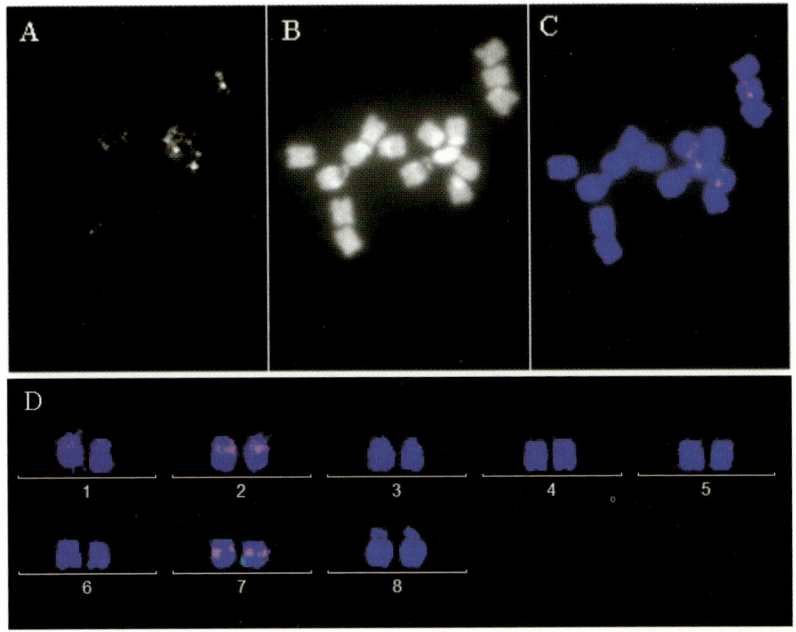

图 3.116　微卫星序列 SSR-CA 在长春花染色体上的 FISH 定位
A：信号图像（红色）；B：DAPI 图像（蓝色）；C：合成图像；D：核型分析结果

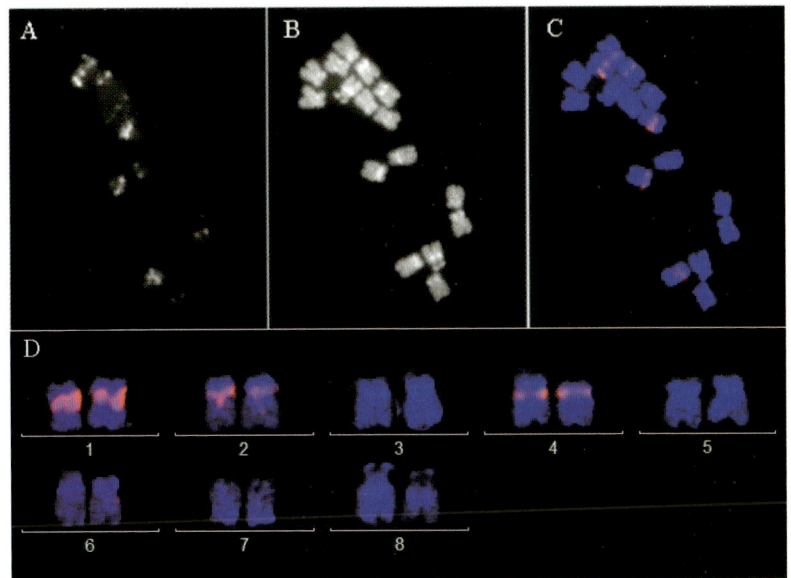

图 3.117　微卫星序列 SSR-AAC 在长春花染色体上的 FISH 定位
A：信号图像（红色）；B：DAPI 图像（蓝色）；C：合成图像；D：核型分析结果

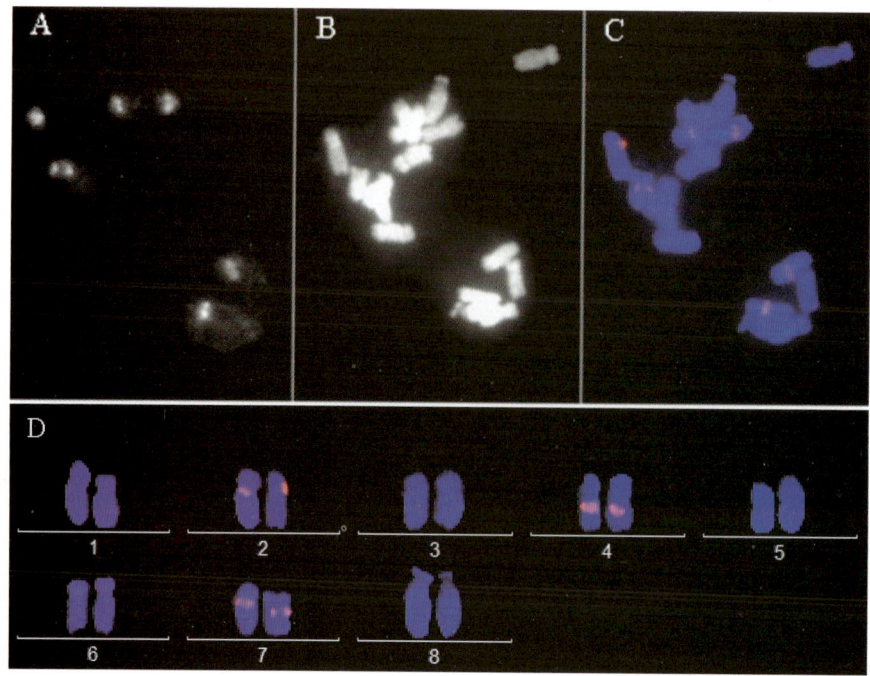

图 3.118　微卫星序列 SSR-AAG 在长春花染色体上的 FISH 定位
A：信号图像（红色）；B：DAPI 图像（蓝色）；C：合成图像；D：核型分析结果

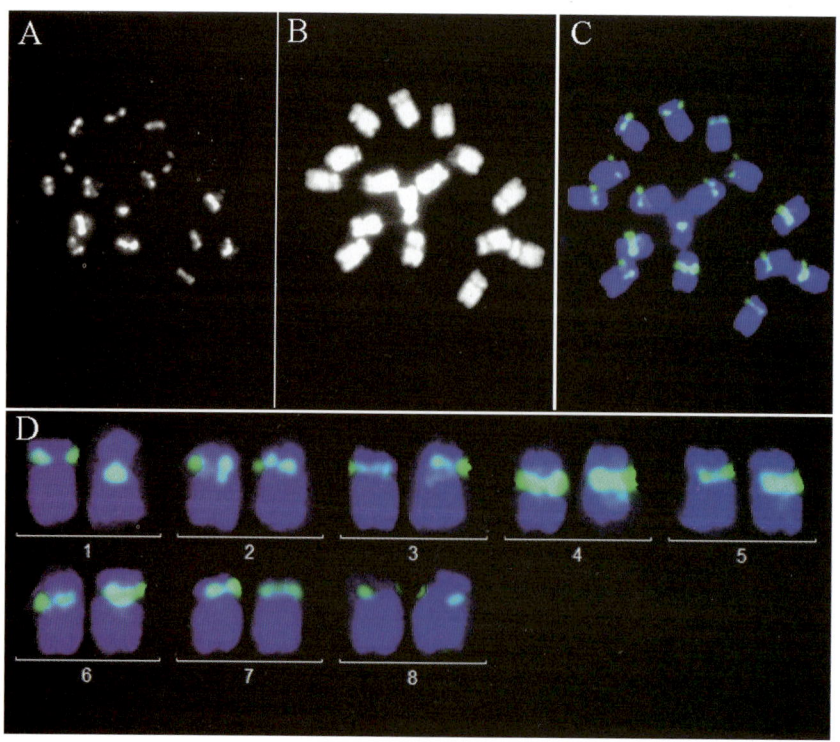

图 3.119　拟南芥类端粒在长春花染色体上的 FISH 定位
A：信号图像（绿色）；B：DAPI 图像（蓝色）；C：合成图像；D：核型分析结果

3.4.13.3 讨论

1. 长春花染色体微切分离及体外扩增

微细玻璃针法分离染色体片段对染色体标本制备的要求较高，不仅要背景干净，而且要确保染色体形态清晰易辨、分散良好。同时由于长春花中期染色体偏小，这无疑又加大了微切分离的难度。本研究采用去壁低渗法，通过摸索预处理时间及酶解时间，制作一批符合要求的染色体标本。

染色体显微分离方法虽然建立较早，但因为之前微克隆技术操作复杂，需要分离大量的同一染色体，费时费力，它的大量应用于试验研究中还始于 PCR 技术的创建。自 PCR 技术出现发展至今，已开发出用于各种用途的多式多样的 PCR 技术，而单染色体或染色体片段的扩增主要采用 LA-PCR（Linker Adaptor PCR）和 DOP-PCR（Degenerated Oligonueleotide Primer PCR），研究发现 LA-PCR 相比后者扩增染色体 DNA 片段要大，而 DOP-PCR 对胞质碎片的扩增则较大。另外 DOP-PCR 随机性大，相较而言，LA-PCR 由于使用了特异性引物和较高的退火温度，能得到片段较大且相对稳定的产物，利于高信息量 DNA 文库的构建。除了 PCR 方法会影响扩增片段大小外，酶切时间也会影响扩增片段大小，不同物种适宜的酶切时间不同，本试验 Sau3AⅠ酶切微分离的染色体为 4 h，得到了较理想的片段大小。

染色体显微分离持续时间长，步骤繁多，易造成外源 DNA 污染，所以实验前应先用 75%酒精擦洗显微操作台，再经紫外照射 2 h，其他用于显微分离的试剂和仪器也应经过严格灭菌方可使用。同时，去壁低渗法的使用有效地避免了胞质的污染，LA-PCR 特异引物的使用也控制了污染的发生。

2. 长春花染色体端粒组成分析

以各种常见类型端粒包括人和拟南芥类型的端粒 DNA 为探针在长春花染色体上进行荧光原位杂交分析，我们发现长春花染色体上虽然末端缺乏常规简单重复序列样端粒，但却在染色体内部近着丝粒处发现了这些本应在末端出现的简单重复序列。这种末端常规端粒丢失的现象并非在长春花中所独有，从进化的角度看，真核生物在物种更丰富的同时，端粒也随之产生了一些特例。

在单子叶植物天门冬目中，已知端粒序列出现两个进化拐点。第一个拐点出现的特征是拟南芥类型的端粒被脊椎动物类型的重复单元基序所取代。这种取代与端粒酶的合成活性息息相关，利用 TRAP 试验检测端粒酶活性发现其中端粒酶合成端粒的保真度较低，正是这种低保真度保证了端粒结合蛋白能平稳过渡适应新的端粒。事实也确实如此，提取出的端粒结合蛋白尽管高度保守，却在体外对于拟南芥类型或者脊椎动物类型的端粒序列都有着高度的亲和力。进化到第二个拐点时，端粒和它的维持模式都发生了更剧烈的变化。Pich 等运用荧光原位杂交技术（Fluorescent In Situ Hybridization，FISH），发现（TTTAGGG）$_n$探针无法在葱属（$Allium$）、假葱属（$Nothoscordum$）、野蒜属（$Tulbaghia$）获得相应信号，同时在大葱（$Allium\ fistulosum$）、洋葱（$Allium\ cepa$）中用不对称 PCR 和 Southern 杂交技术检测常规端粒的尝试也告失败。后来在以洋葱为材料的研究中，陆续有人发现其端粒位置高频出现的 375 bp 卫星序列 ACSAT、rDNA、Ty1-copia 类型逆转座子和 En/Spm 样序列取代了原先的端粒序列。尽管这些序列在种属内和种属间的分布高度多样，它们暗示了几种异常端粒维持机制的存在。如卫星序列 ACSAT 类似摇蚊（*Chironomus*）的端粒，摇蚊端粒处的串联重复序列通过同源重组、滚动环复制和复制滑动的模式维持端粒。葱属植物中染色体末端处的逆转座子则可能与果蝇（*Drosophila melanogaster*）中 HetA 和 TART 两类偏好端粒位点的逆转座子相似。然而，关于葱属端粒的具体情况还有待更进一步的验证。

而综合本研究目前的资料，我们有理由相信长春花的染色体末端在进化过程中也经历了类似的过程，进化出了一套利用卫星重复序列和逆转座子维持染色体末端稳定的系统。当然，要想得到最终的结论，我们还需要进行进一步的分析，包括对长春花进化上较近的属种的端粒的分析和比较，对长春花候选端粒进行功能的分析等。

除此之外，常规端粒丢失的情况还存在于茄科植物的一些种属（*Sessia*，*Vestia*，*Cestrum*）和一些节肢动物中。*Sessia*，*Vestia*，*Cestrum* 可能是利用它们种属特异的一类富含 A/T 小卫星序列作为（TTTAGGG）$_n$ 类型端粒的替代物。可以预见，将来还会有更多的类似特例被发现。Fajkus 等认为端粒和端粒酶系统在经历外在压力时会首先进化出一套通过重组或滑动复制来延长小卫星重复序列的机制，以抵消端粒酶功能的缺失。随后，在延长方式不变的情况下，端粒相关的卫星重复序列进一步取代端部的小卫星序列，直至基于逆转座子的末端维持系统的出现。

在长春花等缺乏常规端粒重复序列的物种中，端粒要行使功能就必需不依赖 DNA 序列特异结合的端粒相关蛋白。在裂殖酵母（*Schizosaccharomyces pombe*）中，端粒结合蛋白 Taz1 和异染色质蛋白 Swi6 可以在没有端粒重复序列的情况下结合到端粒相关序列（subtelomere）上。同时也有证据显示，和着丝粒异染色质一样，表观遗传因子在端粒结构形成中发挥重要作用。小鼠端粒上富含二体和三体甲基化的 H3-Lys9，干扰这种甲基化模式将导致端粒长度失去调控。因此，端粒处 DNA 和相关蛋白的联系并不像之前我们认为的那样直接和特异，相反它们之间可能涉及到表观遗传的修饰。

目前，由于端粒序列在植物物种中的保守性，在构建植物人工染色体上，端粒介导的染色体截断技术在不同植物中的应用只要求克隆拟南芥端粒序列，因此，Yu 等的方法已经开始应用到其他植物中。然而，其他植物物种中对于微小染色体的研究仍较少，成功的先例也不多。随着我们对于长春花异常端粒的深入认识，必将促进植物人工染色体技术的进一步完善，并为端粒生物学的丰富和发展奠定基础。

3. 长春花内部端粒序列的起源和进化

在大多数真核生物中，都能发现集中出现或散布于染色体上的内部端粒序列，这些序列的产生可能涉及到多种机制的作用。而长春花具有独特的内部端粒序列集中分布于染色体着丝粒附近和末端常规端粒缺失的现象，这提示我们长春花染色体经历了较为独特的进化历程。

关于内部端粒序列（Interstitial Telomeric Sequence, ITS），近来有很多报道讨论了端粒 DNA 序列的定性定量方法。这些方法包括 Southern 杂交、杂交保护试验（Hybridization Protection Assay）、定量 PCR、流式细胞术、荧光原位杂交（FISH）、PRINS（Primed In Situ Labeling）和单个端粒长度分析法（STELA）。Slijepcevic 等曾经用 FISH

结合 G 显带的方法成功地检测了中国仓鼠卵巢细胞系（CHO）和肺细胞系（V79）中 ITS 序列处发生的染色体重排。相比 FISH 技术，PRINS 通常在检测 ITS 上更有效。但是由于 ITS 序列通常是由同于或类似于端粒的序列组成，长度相对较短，并非所有上述方法都能有效地检测 ITS 序列和定位其在染色体上的位置，检测方法的灵敏度是个关键。实验中时常出现碰巧或以极低频率被 FISH 技术检测到 ITS 的情况，这是因为当短 ITS 序列复制扩增时偶尔能达到被检测出来的临界大小。另外受限于细胞遗传学方法的分辨率不足以准确判断 ITS 和相关的染色体断裂重组位点的共定位关系，我们通常会采用一些分子的方法予以补充，如脉冲场凝胶电泳和 Southern 杂交。本研究中内部端粒信号非常强烈，其拷贝数很大，不同于一些物种中较少拷贝数的内部端粒，而与 CHO 等高度不稳定的细胞中染色体内部端粒情况相似，暗示了长春花染色体可能也经历了一个高度活跃的进化时期。

ITS 图谱广泛地应用于人类和许多其他生物中，如灵长类动物、鱼、小鼠、小麦和拟南芥中。ITS 常见于组成型异染色质中或其边缘，以人类为例，其 ITS 可根据它的序列排列、定位和旁侧序列分为三类。第一类主要指端粒相关的 ITS，它由上百个含简并碱基的串联重复组成。第二类是由头对头排列的重复组成的融合或倒位 ITS，它们可能来源于古老的端粒融合事件。第三类是只有 20 个左右的六聚物的 ITS，几乎不含精确的端粒重复。

ITS 的起源和形成依它们的序列排列，旁侧序列和序列对损伤的敏感度不同包含多种机制。人们通常认为 ITS 是古老染色体进化过程中端粒和端粒相互融合的产物，然而，在人类中，事实上只有 HSA 1q41 和 HSA 2q13 两处的 ITS 是这种融合产生的。显然，ITS 可以被认为是一类六聚物微卫星序列，而研究已经证实微卫星 DNA 是由 DNA 聚合酶错误滑动引起的重复单元拷贝数增加形成的。DNA 复制过程中，重复序列特征易引起模板和引物链的错误配对，从而导致重复单元的插入或丢失。这种机制能很好地解释 21q22，2q31，7q36 处的 3 个短 ITS 位点和 6pter 处的端粒相关 ITS 的形成。但是，也有人认为这种机制在进化历程中发生缓慢，而且由于人类基因组中这种六聚物微卫星序列很稀少，这表明 ITS 的形成还有其他的机制。Nergadze 认为 ITS 的形成还可能来源于种系进化中双链断裂的修复。非同源末端连接途径（Non-Homologous End-Joining Pathway，NHEJF）是哺乳动物中主要的修复机制，一旦断裂被识别，修复蛋白如 Ku 异聚体，MRE11-RAD50-NBS1 和 XRCC4 等能介导双链平末端的端粒 DNA 序列的插入。而短 ITS 常包埋于 SINE 和 LINE 等重复序列中则表明可能正是这些重复序列不稳定导致了修复事件的发生，在中国仓鼠中的研究也得到了相似的结论。另外，端粒酶通过把端粒样序列加到双链断裂位点的 3′末端也可能产生 ITS，这种方式显然比复制滑动模式要省时得多，但对于 ITS 拷贝数量的多态性却缺乏合理的解释。前面提到的人类 6pter 位点处的端粒相关 ITS 则被认为产生于染色体末端参与的重组事件。

如今，人们认为染色体的复制导致了许多物种中见到的近着丝粒区域大块的 ITS 片段，而常染色质区的 ITS 则多是染色体片段移位形成的。长春花染色体上大块的内部端粒片段特异定位于近着丝粒区域，同时通过对长春花内部端粒旁侧序列的克隆和定位分析，发现它在序列组成上包含一些拷贝的（TTTAGGG）$_n$ 变异序列，这与我们通常在含正常端粒序列的物种如玉米，黑麦等植物中发现的端粒相关序列具有相似的特征，在定位上同时出现在染色体末端和内部端粒处，暗示了长春花 ITS 的形成可能源于染色体末端的融合，而非端粒酶在双链断裂缺口的简单修复或复制滑动的产物。但是，更多的证据有待于对长春花染色体上进化遗迹如微卫星序列和多种染色体功能单位的分子和细胞水平的检验。

（作者：刘少华，导师：陈成彬。原载：刘少华硕士毕业论文）

3.4.14 应用 BAC-FISH 系统鉴定芝麻（*Sesamum indicum* L.）染色体

Sesame（*Sesamum indicum* L., $2n = 26$）belongs to the family Padaliaceae and is an ancient oilseed crop that has important value for human health. The cultivation history of sesame can betraced back to 3050–3500 BC（Bedigian & Harlan, 1986; Namiki, 1995; Anilakumar *et al.*, 2010）. Sesame is cultivated in both tropical and subtropical regions as it has high tolerance to drought and high temperatures（Ashri, 1998; Ziedan *et al.*, 2011）. Recent research revealed that the sesame genome is 354 Mb, smaller than other oilseed crops, such as soybean（1129-1235 Mb; Arumuganathan & Earle, 1991）, rapeseed（1130 Mb; Chalhoub *et al.*, 2014）, peanut（2800 Mb; Temsch & Greilhuber, 2001）and sunflower（3000 Mb; Arumuganathan & Earle, 1991）. Thus, sesame is often regarded as an ideal crop for genomics research on oil biosynthesis and metabolism, as well as other important biological processes. However, sesame chromosomes（$n = 13$）are fairly small（from 1.106 to 3.871 lm）and have similar morphology（metacentric, submetacentric or subtelocentric; Zhang *et al.*, 2012a）. These characteristics hinder research on chromosome structure and genome evolution in sesame（Zhang *et al.*, 2012a; Liu *et al.*, 2013; Nyongesa *et al.*, 2014）. Sesame chromosome research began in the 1920s. By the 1980s, Giemsa staining and other karyotype analysis techniques were established and applied to sesame（Zhan *et al.*, 1987, 1988, 1990; He *et al.*, 1994; Zhang *et al.*, 2012a）. To identify chromosome structure in cultivated sesame, fluorescence in situ hybridisation（FISH）was first performed based on conventional chromosome preparation methods, with SSR（simple repeated sequence）marker（AC）8, 45S and 5S rDNA, as well as telomere sequences as probes（Liu *et al.*, 2013, 2014）. However, to our knowledge, thus far no studies have successfully identified chromosome pairs in sesame. Recently, with the advent of high-throughput sequencing techniques, FISH has been combined with long DNA fragments（e.g. Bacterial Artificial Chromosomes: BACs）to generate a tool（BAC-FISH）that can clarify how genomic sequences and chromosomes are related, simplifying chromosome identification and structural analysis, as well as cytogenetic map construction（Tang *et al.*, 2007; Danilova & Birchler, 2008; Paesold *et al.*, 2012; Feng *et al.*, 2013; Tran *et al.*, 2016）. This new technique considerably improves plant genome research, especially among diploid plants（Yan *et al.*, 1998; Garcıacegarra *et al.*, 2013;

Taboada et al., 2014). The BAC-FISH technique was applied to reassemble 91 sequenced scaffolds in the public available toma to genome, revealing variation in the arrangement of 45 scaffolds (Shearer et al., 2014).

In addition, cytogenetic map construction and other genome research were performed using BAC-FISH in various crops, including cucumber (Sun et al., 2013), rice (Tang et al., 2007), potato (Dong et al., 2000), common bean (Fonseca et al., 2010) and sugar beet (Paesold et al., 2012). In 2012, two BAC libraries, a BIBAC (binary BAC, pCLD 04541) and a BAC (CopyControlTM, pCC1BAC), were constructed for S. indicum (Zhang et al., 2013; Miao & Zhang, 2014). Additionally, chromosome preparation methods were optimised. In this study, we took advantage of the technological advances to prepare high-quality chromosome samples and established a successive BAC-FISH system for sesame. This system allowed us to differentiate the 13 chromosome pairs for the first time using 17 specific BACs. Our findings supply a theoretical and technical basis for future cytogenetic map construction, genome assembly correction and comparative genomics in sesame.

3.4.14.1 Material and Methods

1. Plant material and chromosome preparation

This study employed the Chinese sesame variety Yuzhi 11, previously used for sequencing in the sesame genome project. Plant material was obtained from the Sesame Germplasm Reservoir of Henan Sesame Research Center, Henan Academy of Agricultural Sciences, China. Healthy Yuzhi 11 seeds were placed in Petri dishes with distilled water and cultured in the dark at 21 ℃ for 2 days. Root tips were harvested for chromosome preparation following previous techniques (Chen et al., 1982), with some modifications. A set of 4-5-mm long root tips were pre-treated with 2 mM 8-hydroxyquinoline at 21 ℃ for 1.5 h in darkness, then fixed in methanol and glacial acetic acid (v/v, 3:1) at 4 ℃ for 1 h. An enzyme solution containing 2.5% (w/v) cellulase and 2.5% (w/v) pectolyase was used to dissolve the fixed tissues at 37 ℃ for 160 min. Softened tissues were pounded into thin sections on slides and stored at −20 ℃ for subsequent Giemsa staining and metaphase chromosome screening.

2. Construction of the BAC library

High molecular weight (HMW) DNA was extracted from young Yuzhi 11 leaves according to previous methods (Zhang et al., 1995). The HMW DNA was partially digested with the Hind III restriction enzyme at 37 ℃ for 1 h. DNA fragments (100-300 kb) were recovered using pulsed-field gel electrophoresis (PFGE) and then ligated into the chloramphenicolresistant CopyControl™ pCC1BAC™ vector (Hind III Cloning-Ready). EPI100-competent Escherichia coli cells (Epicentre, Americ) were used for BAC vector transformation via electroporation. After chloramphenicol screening, the sesame BAC library was constructed, placed into 384-well microtiter plates and stored at −80 ℃. Each BAC clone was numbered according to internationally accepted methods (Zhang et al., 2012b). For quality testing, 32 BAC clones were randomly selected and inoculated in 5 ml; liquid LB medium containing 12.5 μg · ml⁻¹ chloramphenicol. Plasmid DNA was extracted and purified via alkaline lysis using a Qiagen plasmid mini kit (Qiagen, Hilden, Germany). After Not I digestion at 37℃ for 4 h, the DNA insert of each BAC clone was detected using 1% agarose PFGE at 6 v · cm⁻¹ with a 5-15 s switch time. The gel was run at 4℃ for 18 h.

3. Probe labelling

Plasmid DNA extracted from specific BAC clones was broken down randomly at 0.1 MPa and 120℃ for 5 min. DNA fragments were then labelled with fluorescein-12-dUTP (green signal) or tetramethyl-rhodamine-5-dUTP (red signal), following the manufacturer's protocol (Roche Diagnostics, Mannheim, Germany), with modifications. The reaction mixture contained 300-1000 ng BAC DNA; 0.1 mM dATP, dGTP and dCTP; 0.065 mM dTTP, 0.035 mM tetramethyl-rhodamine-5-dUTP or fluorescein-12-dUTP; 100 ng random primers pd (N) 6; 2.5 U Klenow fragment; 19 Klenow buffer; and enough distilled water to reach a final volume of 5 ll. The labelling reaction was performed at 37℃ in darkness for 20 h. Oligonucleotide probe of 45S rDNA was synthesised according to the conservative sequence of Arabidopsis thaliana and modified using green fluorescence group FAM in its 50 end (Chen et al., 2013).

4. Fluorescence in situ hybridisation and double-colour FISH

After Giemsa staining, slides with metaphase chromosomes were cleared through immersion in 45% (v/v) acetic acid for 1-2 min. Chromosome degeneration and dehydration procedures were performed following previous methods(Hu et al., 2012; Shearer et al., 2014). The hybridisation mixture(containing labelled probes of one colour and the working solution) was denatured at 95℃ for 10 min, then immediately placed on ice for 5 min. Next, 8 ll denatured hybridisation mixture was dripped on the chromosome slides and inoculated at 37℃ overnight. Slides used for double-colour FISH were treated with hybridisation mixture containing labelled probes of both colours (see BAC probe labelling).

Before carrying out the successive hybridisation, large-scale hybridisation was initiated in 400 BACs selected randomly and hybridised with the chromosomes as pairs (one probe with fluorescein-12-dUTP and the other with tetramethyl-rhodamine-5-dUTP) for the purpose of screening repeat-free clones.

To repeat hybridisation on the same slide, cover slips were first removed and slides were washed in 29 SSC containing 0.2% (v/v) Tween-20 and 0.1% (v/v) SDS. After dehydration in an ethanol series, slides were hybridised with other BAC probes following the procedure detailed above.

5. Fluorescence signal detection and image integration

Chromosome specimens were counterstained with Vectashield H1000 solution containing 5×10^{-3} mg · l⁻¹ 4, 6- diamidino-2-phenylindole (DAPI) (blue signal; Vector Laboratories, Burlingame, CA, USA). Metaphase chromosomes and the various fluorescence signals were observed under a Nikon 80i epifluorescence microscope, equipped with a high-resolution charge coupled device(CCD)camera(SPOT, Diagnostic Instruments, Sterling

Heights, MI, USA). Green and red colour filters were used for the fluorescein and tetramethyl-rhodamine signals, respectively. SPOT version 4.0.8 (Diagnostic Instruments) was used to integrate grey digital chromosome images and the fluorescence signals. Image quality was optimised in Adobe Photoshop version 10.0.

Long-arm and short-arm lengths per metaphase chromosome were measured using Metasystems Isis (MetaSystems, Altlussheim, Germany). Chromosome relative lengths in somatic sesame cells were calculated as the proportion of each chromosome length to the total length of all 13 chromosomes. Sequence lengths (Mb) of individual chromosomes were calculated based on the sesame genome size of 354Mb (Zhang et al., 2013) and ranked or named in length order. Karyotype analysis was performed in Metasystems Isis. The chromosome ideogram was drawn in Microsoft Excel, with 0.00 representing centromeres of each chromosome. Both the distance from the end of the short arm to the centre of the BAC signal and the total length of the chromosome was measured using Digimizer (MedCalc, Ostend, Belgium), then the ratio (over 15 repeats) was calculated, representing the position value of each BAC. These positions were marked on each chromosome as fill-in black bars.

3.4.14.2 Results

Sesame BAC library construction and quality assay Before BAC-FISH analysis, we constructed a BAC library for *S. indicum* (354 $Mbp^{-1} \cdot C^{-1}$) comprising about 100,000 BAC clones. To evaluate the BAC library quality, we performed PFGE and evaluated the insert length of each clone. The results from PFGE showed that the estimated average size of BAC inserts was ~85 kb (range: 60-110 kb; Fig. 3.120). Of the 32 BAC clones tested, only one (<5%) was empty. Meanwhile, we chose and sequenced the ends of 42,240 BACs using the Sanger method (data not shown). The average insert size of the 42,240 BACs was estimated at 87.3 kb according to the genome assembly data (data not shown). Subsequently, we chose 57,600 BAC clones including the ends-sequenced 42,240 BACs for cytogenetic and genomics analysis in sesame. As the blank clones and chloroplast-polluted clones were ignored for the low frequencies, the estimated genome coverage of the 57,600 BACs for sesame was 13.89, above the threshold for BAC library construction.

1. The BAC hybridisation and successive BAC-FISH

To screen the repeat-free BAC clones, we randomly chose 400 BACs from the 57,600 BACs and performed hybridisation on a large scale. All the BAC plasmids as probes were labelled with tetramethyl-rhodamine-5-dUTP (red colour signal) or fluorescein-12-dUTP (green colour signal). Large-scale hybridization screening revealed 1-13 pairs of loci in the chromosome set (Fig. 3.121). Of the 400 BAC clones, 101 (25%) generated stable hybridisation signals, and 64 (16%) generated a single stable signal pair (data not shown). These 64 BACs were selected for further BAC-FISH and chromosome identification.

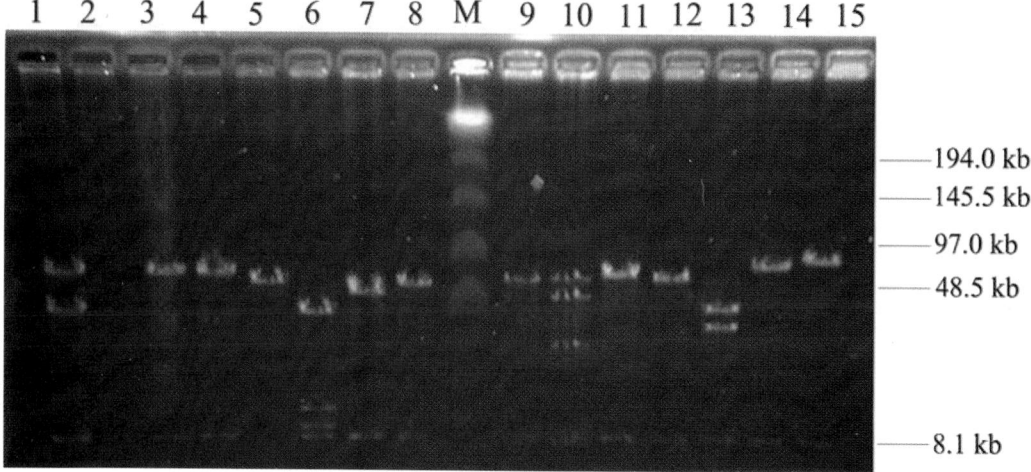

Fig. 3.120 Assay of bacterial artificial chromosome (BAC) insert size using pulsed field gel electrophoresis
Lane 1–15: 15 BAC clones digested using NotI. M: Lambda ladder PFG marker. Vector size is about 8.1 kb

Successive (double-colour) BAC hybridisation was performed using high-quality chromosome slides to increase hybridization efficiency (Fig. 3.122). Three pairs of BAC probes (159R and 101G, 137R and 121G, 32R and 57G) were used in a three-round hybridisation. Compared with the first round (Fig. 3.122b), hybridization with 32R and 57G continued to yield clear chromosome morphology with stable signals (Fig. 3.122f, g). Subsequently, dozens of BACs were used for successive double-colour FISH, resulting in a maximum of 12 BAC-FISH hybridisation rounds (data not shown). Twenty-four BACs were divided into 13 groups based on chromosome location (partial results shown in Fig. 3.123).

2. Chromosome identification

Previous studies indicated that the sesame genome contains three chromosome pairs with satellite regions (Liu et al., 2014). To clearly determine all 13 chromosome pairs, we performed a two-step identification technique (Fig. 3.123). First, we divided the 13 chromosome pairs into two groups using 45S rDNA probe. Group I comprised ten normal chromosome pairs, and Group II included three chromosome pairs with satellite DNA (Fig. 3.123a, j). Subsequently, to differentiate the ten chromosome pairs in Group I, we performed successive eight-round doublecolour FISH using 14 BACs with a single hybridisation signal pair (Fig. 3.123a–i). Among Group I (chromosome 1-10), chromosomes 1, 2, 4 and 7 were located with two BAC probes and the remaining six pairs were individually marked with one BAC probe (Fig. 3.123j).

To further differentiate Group II chromosome pairs, we performed a six-round successive hybridisation on another highquality chromosome slide, using 11 BAC probes and 45S rDNA (Fig. 3.124). Unlike Group I (chromosome 10-100), the 45S rDNA probe successfully marked chromosomes 110, 120 and 130 (Fig. 3.124a). Of the 11 BAC probes, 4G, 68R and 54R were specifically located on the long arms of chromosomes 110, 120 and 130, respectively (Fig. 3.124g, h). The other eight BACs were located on chromosome pairs 40, 50, 60, 80 and 100 (Fig. 3.124h).

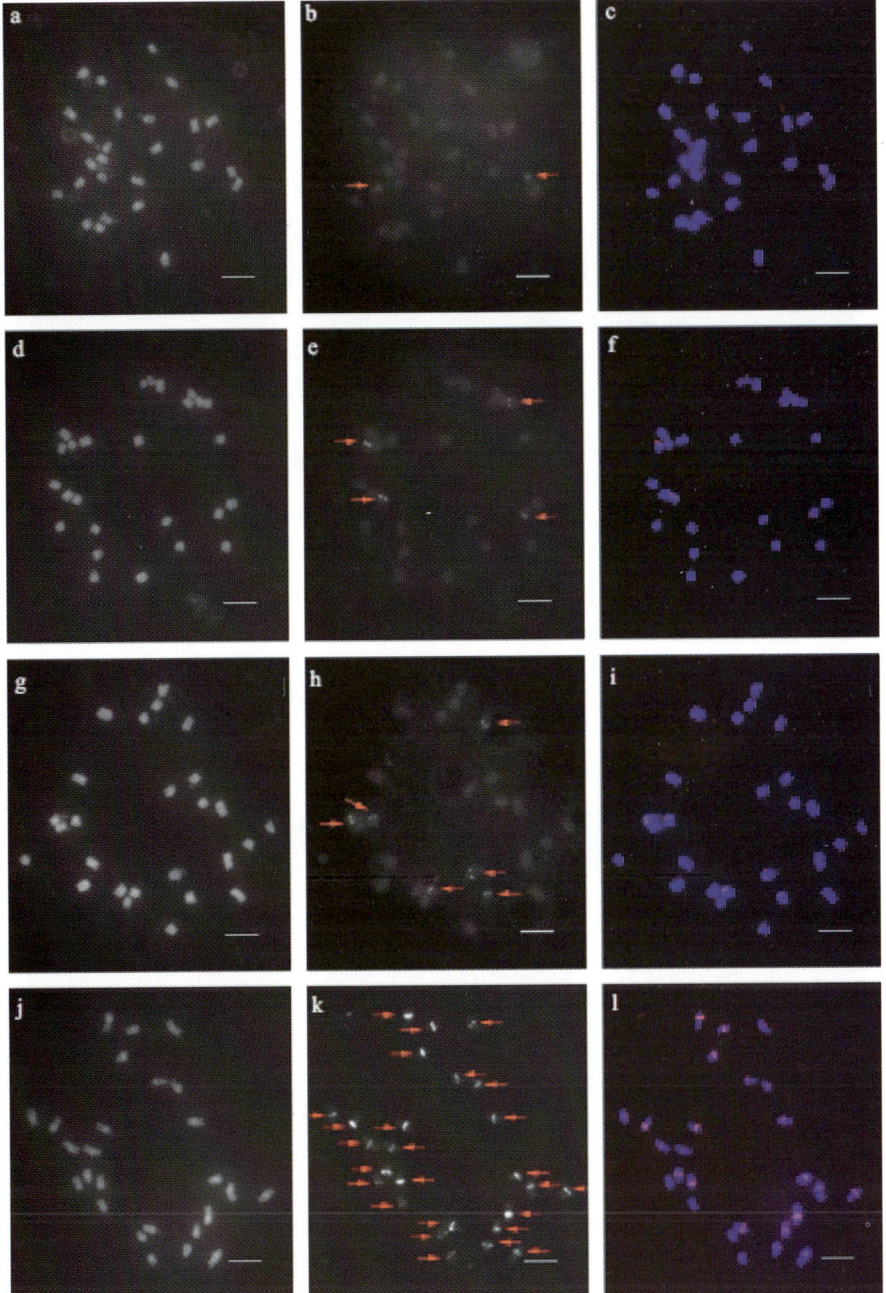

Fig. 3.121 Fluorescence in situ hybridisation of different BAC probe types on sesame chromosomes

Chromosomes image (a) and hybridisation with BAC probe 3R (b). (c) Image integration of probe 3R with one signal pair.
Chromosomes image (d) and hybridisation with BAC probe 144R (e). (f) Image integration of probe 144R with two signal pairs.
Chromosomes image (g) and hybridisation with BAC probe 37R (h). (i) Image integration of probe 37R with three signal pairs.
Chromosomes image (j) and hybridisation with BAC probe B409R (k). (l) Image integration of probe B409R with 13 signal pairs.
Bar = 5 μm. Red arrows indicate the fluorescence signal generated using tetramethyl-rhodamine-5-dUTP. 'R' in probe names refers

to red fluorescence label

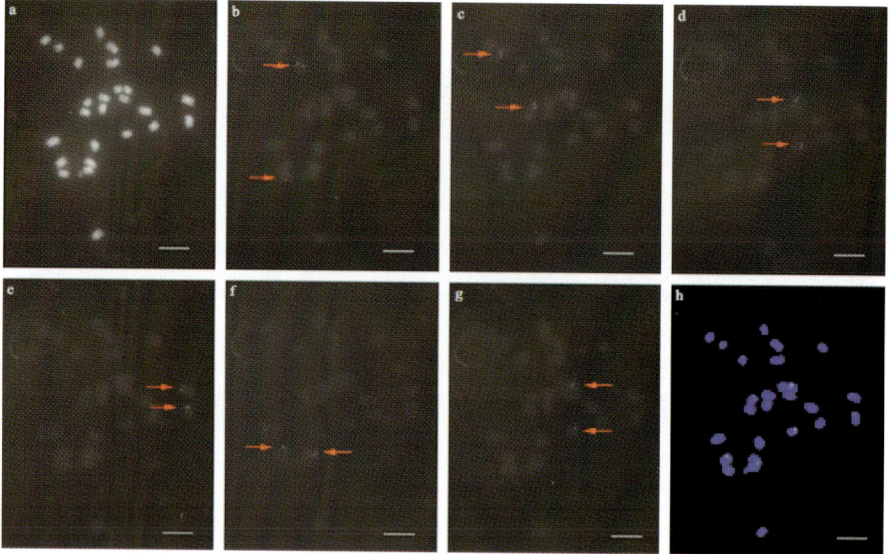

Fig. 3.122 Successive in situ hybridisation using double-colour FISH in sesame
(a) DAPI-stained chromosome specimen. Images of first round double-colour hybridisation using 159R (b) and 101G probes (c). Images of second round double-colour hybridisation using 137R (d) and 121G probes (e). Images of third round double-colour hybridisation using 32R (f) and 57G probes (g). (h) Image integration of six photographs from b, c, d, e, f and g. Red arrows indicate BAC hybridisation signals. 'G' and 'R' in probe names refer to green and red fluorescence labelling. Bar = 5 μm

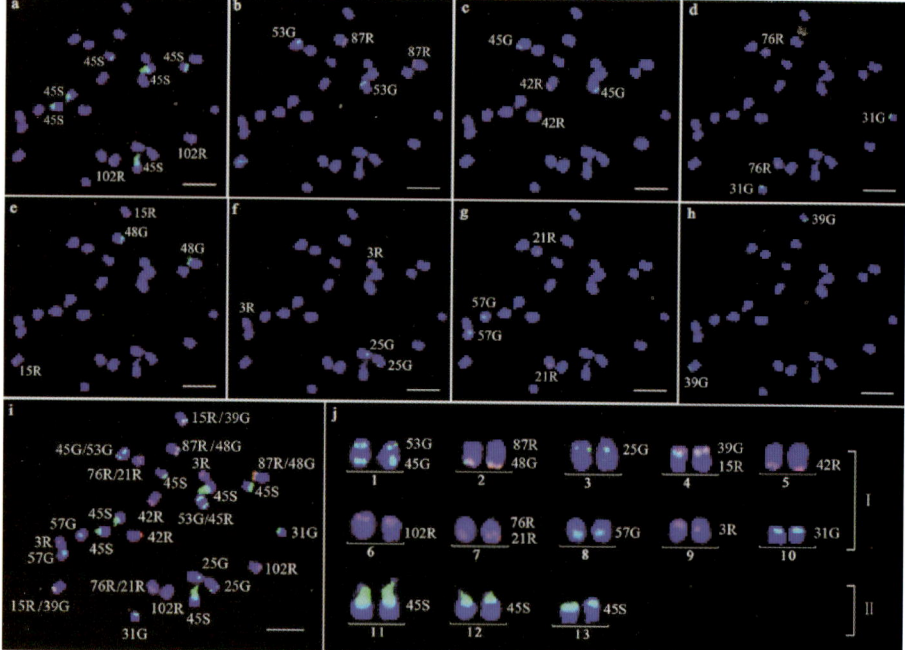

Fig. 3.123 Identification of the ten normal chromosomes (without satellite DNA) in sesame through eight-round successive fluorescence in situ hybridization
(a) Hybridisation of 102R and 45S rDNA probes (green fluorescence labelling). (b) Hybridisation of 87R and 53G probes. (c) Hybridisation of 42R and 45G probes. (d) Hybridisation of 76R and 31G probes. (e) Hybridisation of 15R and 48G probes. (f) Hybridisation of 3R and 25G probes. (g) Hybridisation of 21R and 57G probes. (h) Hybridisation of probe 39G. (i) Integration of images from a, b, c, d, e, f, g and h. (j) Karyotype analysis of chromosomes hybridised with the 14 BAC probes and 45S

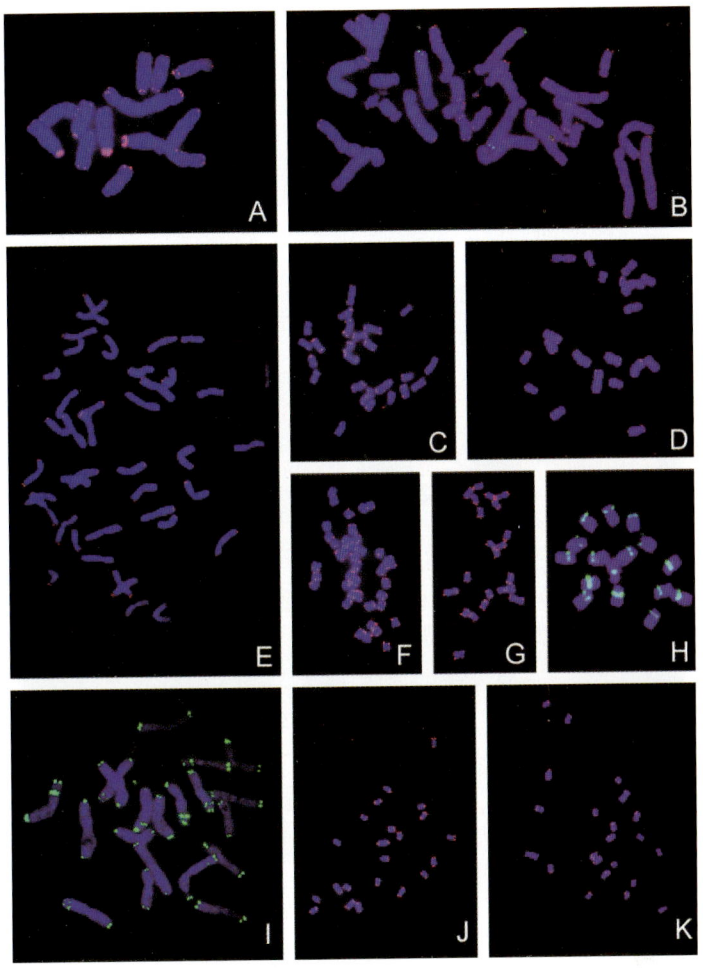

图 3.103　A 蚕豆，B 水仙，C 水稻，D 玉米，E 小麦，F 黄豆，G 红小豆，H 长春花，I 明月草，J 豇豆，K 绿豆

（作者：陈成彬、宋文芹。原载：中国细胞生物学学会第十二次学术大会论文摘要 p273）

3.4.11　细胞质雄性不育辣椒育性恢复基因在染色体上的定位

作物雄性不育在杂交育种中具有重要地位，迄今已在 43 科、162 个属、617 个物种中发现了雄性不育现象。利用雄性不育培育不育系，配合相应的保持系、恢复系实现三系配套或两系制种，不仅可以节省大量人力物力，对提高种子的纯度和产量具有重要的意义。

本研究利用集团分离法（Bulked Segregant Anslysis，BSA）以辣椒细胞质雄性不育系 BU-12、恢复系 RF-12 为材料，共筛选了 336 条 RAPD 引物，其中引物 S418 在恢复系中呈现特异性扩增，得到一条约 3000 bp 的特异片段，回扩得到两条片段，测序表明大小为 1515 bp、1162 bp，将回收所得的约 3000 bp 的片段及回扩所得的 1515 bp、1162 bp 的片段用 DIG 标记后分别与不育系（BU-12）和恢复系（RF-12）的染色体进行杂交，3000 bp 的片段及 1162 bp 的片段在两系染色体上都有杂交信号；1515 bp 片段（S418$_{1515}$）在恢复系中呈现特异性杂交（图 3.104A、B）而在不育系中没有杂交信号（图 3.104C、D），且信号集中在辣椒恢复系 3 对染色体上，并以低拷贝形式存在。荧光原位杂交技术证实该序列为一新发现序列。

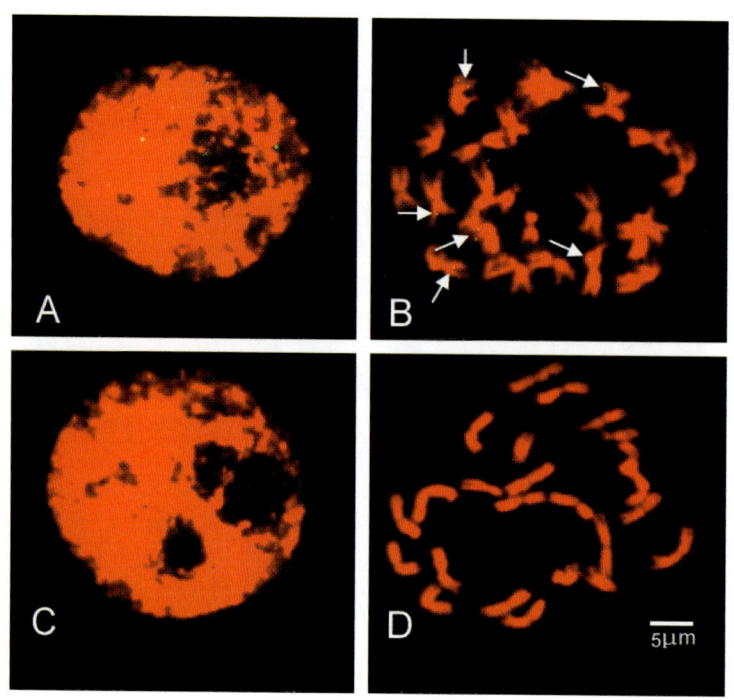

图 3.104　S418$_{1515}$ 在不育系和恢复系染色体上的荧光原位杂交结果
A、B：S418$_{1515}$ 在恢复系染色体上杂交的结果；C、D：S418$_{1515}$ 在不育系染色体上杂交的结果；

实验中将获得的辣椒恢复系特异标记后，没有利用大量后代的分离群体去构建连锁图，确定该标记与育性恢复的连锁关系，而是采用荧光原位杂交的方法把该标记更为直接、准确地定位到恢复系相应的染色体上，使分子生物学和细胞遗传学有机地结合在一起。该方法也是当今人类基因组计划及其他模式生物基因组计划完成后，分子生物学由微观回归宏观的桥梁。

（作者：王春国、陈成彬。摘录原载：实验生物学报，2005，38（3）227-232）

3.4.12　尕海孤雌生殖卤虫基因组的特异 DNA 片段在染色体上的定位

卤虫（*Artemia*）是生活在高盐度水域中的小型甲壳动物，在分类上属于节肢动物门、甲壳纲、鳃足亚纲、无甲目、卤虫科、卤虫属（*Artemia*）。其无节幼虫含有丰富的蛋白质和脂肪,是水产养殖中的良好饵料。本文以家蚕 *Bmdsx* 基因中的雌雄差异表达部位设计引物,试图在不同生殖方式的卤虫中发现与其相关的 DNA 片段。

应用设计的引物可以从尕海孤雌生殖的种群卤虫基因组DNA 中扩增出两个长度分别为 900 bp 和 200 bp 左右的 DNA 片段;而在山西两性种群卤虫中则没有 DNA 片段被扩增出。测序结果表明,这两个 DNA 序列的长度分别为 862 bp 和 215 bp,将长片段命名为 Apdsx900。Southern 杂交结果进一步证实该 DNA 片段为尕海孤雌生殖卤虫基因组特异。荧光原位杂交显示,Apdsx900 在尕海孤雌生殖基因组中有两个点信号(图 3.105),而在山西两性生殖卤虫的细胞核中则没有杂交信号。同时还显示出该 DNA 片段集中存在于尕海孤雌生殖卤虫的两条染色体上。在观测的 105 个细胞(包括核与核型)中,83 个显示两个点的杂交信号,11 个显示一个点的杂交信号,6 个显示三个点的杂交信号,5 个显示四个点的杂交信号。

原位杂交的结果显示该 DNA 片段在尕海孤雌生殖卤虫基因组两条染色体上存在。而一个点或者大于两点的杂交信号的出现,可能与孤雌生殖卤虫的多倍体和非整倍体现象有关(任波、姚陆,1997),其中 79%(83/105)显示两个点的杂交信号,说明尕海的孤雌生殖卤虫以二倍体为主。虽然 Apdsx900 为尕海孤雌生殖卤虫基因组所特有,但这种特异性与生殖方式的关系尚无法确定。尕海孤雌生殖卤虫基因组特异 DNA 片段 Apdsx900 的发现,表明尕海孤雌生殖卤虫与山西两性卤虫不仅在多倍体频率方面存在差异,二者在基因组内容方面也存在差异。这种差异很可能是孤雌生殖卤虫基因组变化的结果。

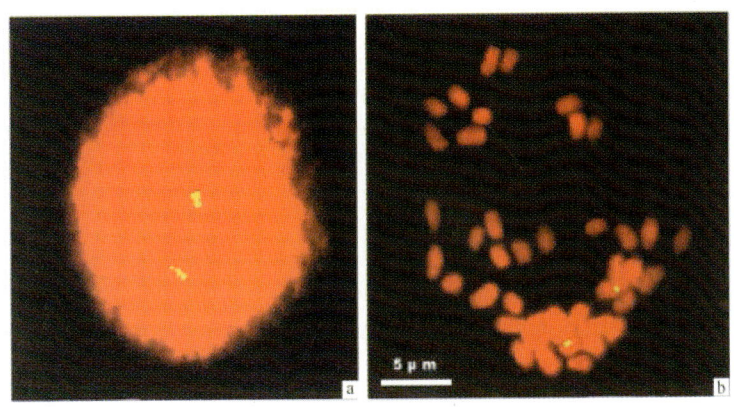

图 3.105　尕海孤雌生殖卤虫荧光原位杂交结果
a. 间期细胞核杂交结果;b. 中期染色体杂交结果

(作者:孙易、陈成彬、刘凤歧、宋文芹、陈瑞阳。原摘录:动物学报,2004,50(3)470-474)

3.4.13　长春花端粒相关序列荧光原位杂交定位

长春花(*Catharanthus roseus*),属于夹竹桃科长春花属,为多年生草本。原产于地中海沿岸、印度、热带美洲。长春花含 70 多种生物碱。其中长春碱和长春新碱对治疗绒癌等恶性肿瘤、淋巴肉瘤及儿童急性白血病等都有一定疗效，是目前国际上应用最多的抗癌植物药源。

在实验室前期研究工作的基础上，以长春花为材料，采用染色体微切克隆技术对长春花染色体末端 DNA 组分进行分析，构建了其长臂末端 DNA 文库，通过对克隆子的分析、鉴定，筛选出了一个定位于长春花染色体末端的重复序列 CR211。CR211 含有较多的（AAACCCT）简并序列，与黄瓜 Gypsy 逆转座子高度同源。利用单引物扩增技术克隆了长春花内部端粒旁侧序列 DLP_{1000}，FISH 分析显示该序列在染色体末端和着丝粒处都有分布。另外，通过双色荧光原位杂交分析比较了微卫星序列和拟南芥端粒序列在染色体上的定位情况，发现微卫星序列多分布于近着丝粒处，与内部端粒共定位，暗示长春花的端粒序列与微卫星序列位置相近。

3.4.13.1 材料与方法

1. 实验材料

试验用长春花（*Catharanthus roseus*）种子（$2n=16$），购自内蒙古赤峰鑫卉园艺有限公司。

2. 方法

（1）长春花染色体标本制备

长春花种子经 70%乙醇消毒后，37℃温水中处理 5 min，自然冷却，浸泡 24 h。将种子摆在铺有湿润消毒纱布的铝制饭盒中，在 25℃培养箱中暗培养 3~4 d。待根尖长到 0.5 cm 时，用镊子夹取 0.2 cm 长根尖分生区于指形管中，加入对二氯苯预处理 3.5 h，吸出预处理液，加入固定液（按冰醋酸与无水甲醇体积比为 1:3 配制，现配现用）固定 0.5 h，切取乳白色分生区置于蒸馏水中前低渗 10 min，用 2.5%纤维素酶和果胶酶混合液 37℃处理 1 h，再于 0.075 mol/L KCl 中后低渗 20 min，70%乙醇固定 30 min 以上，于预冷的载玻片上（24 mm×50 mm）上涂片，空气干燥，镜检。-20℃保存备用。

（2）染色体长臂末端显微分离

取直径 1 mm 左右的细玻璃针，在拉针仪上电热拉制成直径 1 μm 短而细的玻璃针，为防止污染和灰尘，针宜现用现做。

将染色体玻片标本固定在倒置显微镜载物台架上，先在低倍视野下找到待切割的染色体，用粗螺旋升降调节显微操作器至适当的高度；而后用显微镜镜台推进器 X、Y 轴将微切针移动到被切割染色体的大概位置；微调螺旋杆，使玻璃针对准待切割的染色体部位，最后换为高倍镜，在目镜视野的监视下进行切割，将切割下来的染色体连同玻璃针的尖端一起折断在 eppendorf 管底。

（3）染色体片段体外扩增

将微切染色体置于 0.2 mL 离心管中，瞬时离心，加入 20 μL 蛋白酶 K（用 1×连接酶缓冲液稀释至 5 ng/μL），混匀，置于-20℃备用。

*Sau*3A I 接头是由 23mer 和 19mer 两个寡核苷酸片段形成。

23mer 寡核苷酸片段碱基的序列：5'-GATCCTGAGCTCGAATTCGACCC-3'

19mer 寡核苷酸片段碱基的序列：5'-GGGTCGAATTCGAGCTCAG-3'

均由上海生工合成。制备接头时，首先将 23mer 寡核苷酸片段 5'端磷酸化，过程如下：

23mer 寡核苷酸　　　　　　　1 μL（10 μmol/L）
10×连接酶缓冲液　　　　　　1 μL
10 mmol/L ATP　　　　　　　1 μL
T_4 多核苷酸激酶　　　　　　 1 μL（10 U/μL）
紫外线照射过的无菌水补足至 10 μL
37℃反应 1 h，然后 75℃ 20 min 使酶失活。

接着在同一管中加入 1 μL 的 19mer 寡核苷酸（10 μmol/L），用紫外线照射处理过的无菌水补至 20 μL。在 PCR 仪上 90℃变性 2 min、55℃退火 20 min，再自然冷却至 30℃。置于-20℃备用。

①蛋白酶 K 消化

将含有分离染色体的 0.2 mL 管置于 37℃水浴中温育 4 h，以使蛋白酶 K 充分消化去蛋白，然后 75℃ 20 min 使酶失活。

②*Sau*3A I 酶切

在管中加入 0.2 μL *Sau*3A I，37℃酶切染色体 4 h，75℃ 20 min 使酶失活。

③接头连接

加入制备好的接头 0.5 μL，T_4DNA 连接酶 0.2 μL（Takara），连接反应最终体积为 25 μL，16℃连接过夜后，75℃ 20 min 失活连接酶。

上述反应均设置严格的阳性与阴性对照，阳性对照加入 1 μL 长春花基因组 DNA 作为模板，阴性对照则不加任何底物。

④PCR 扩增

PCR 扩增分二轮进行，第一轮扩增在上述同一管中进行，除原先的反应物外再加入：

10×Taq 酶缓冲液　　　　　　5 μL
10 mmol/L dNTP　　　　　　1 μL
10 μM LA-19 引物　　　　　　1 μL
Taq DNA 聚合酶　　　　　　　0.5 μL（5 U/μL）

紫外线照射过的超纯水补至终体积 50 μL，混匀，然后 94℃预变性 5 min；进行 94℃ 1 min、50℃ 1.5 min、72℃ 3 min 循环 35 圈；最后 72℃延伸 15 min，4℃保温。

第二轮 PCR 扩增是以第一轮扩增产物 2 μL 为模板，其他反应成分不变，PCR 反应程序基本同上，只是将 35 个循环降至 25 个循环。

（4）文库构建

①连接

染色体 PCR 扩增产物　　　　4 μL
pEASY-T1 载体　　　　　　　1 μL

混匀，室温连接 10～15 min。

②转化

A. 加连接产物于 50 μL 感受态细胞中（在感受态细胞刚刚冰浴解冻时加入连接产

物），轻弹混匀，冰浴 20～30 min；

B. 42℃，热激 30 s，立即置于冰上 2 min；

C. 加入 250 μL 平衡至室温的 LB 液体培养基，180 r/min，37℃慢摇 1 h；

D. 同时在含 Amp（100 mg/L）的 LB 固体平板上均匀涂布 40 μL X-gal（20 μg/μL）和 8 μL IPTG（200 μg/μL）溶液，室温静置备用；

E. 菌体复苏之后，瞬时离心，弃去 100 μL 上清，吹吸菌液呈重悬状态，取 100 μL 菌液涂布于平板上，先正置 0.5 h 使菌液被充分吸收，再倒置放在 37℃培养箱中培养过夜；

F. 经培养，在含 IPTG 和 X-gal 的 Amp LB 平板上，出现白色菌落和蓝色菌落，其中白色菌落为重组子；

G. 用灭菌后牙签挑取白色单菌落于 600 μL 含 Amp（100 mg/L）的 LB 液体培养基过夜培养，加入等体积的 40%甘油 LB 培养基，混匀，-20℃保存；

H. 菌液 PCR 初步检测重组子：随机挑取菌液，每管取菌液 1 μL 作为模板，扩增体系和流程与染色体扩增相同，1%琼脂糖凝胶电泳检测。

（5）荧光原位杂交

①染色体标本预处理

A. 选取染色体分散良好的制片，显微镜下记好坐标后，用玻璃刀在玻片背面做好标记；

B. 将玻片置于 45%冰醋酸，固定 1～5 min；

C. 空气中晾干 1 min；

D. 滴加 70%去离子甲酰胺于玻片标记处，70℃变性处理 2 min；

E. 室温下，用-20℃冷冻 70%→85%→100%乙醇系列脱水，各 5 min，室温干燥。

②杂交

A. 配制杂交液：50%去离子甲酰胺、2×SSC、50 mmol/L 磷酸钠、10%硫酸葡聚糖、0.1% SDS 和 2～20 ng/μL（荧光标记）探针 DNA；

B. 杂交液在沸水中变性 10 min，立即冰浴至少 5 min；

C. 将杂交液滴加至染色体标本上（8 μL/片），盖上盖玻片（18 mm×18 mm），37℃湿盒中杂交过夜。

③杂交后洗膜

A. 去盖片；

B. 2×SSC，0.1% SDS，45℃洗片 3×5 min；

C. 0.1×SSC，0.1% SDS，45℃洗片 3×5 min；

D. 蒸馏水稍微冲洗玻片，黑暗处晾干。

（6）杂交信号检测

A. 滴加 10 μL DAPI 染液（100 ng/mL），染色 5 min；

B. 滴加抗淬灭剂（VectaShield），盖上盖玻片（24 mm×24 mm）；

（7）图像采集与处理

A. Nikon 80i 荧光显微镜观察杂交信号，冷 CCD 进行图像采集，Spot Rtke 4.1 软件进行图像合成；

B. Adobe Photoshop 7.0 软件对图像的亮度和对比度进行整体调整，去除背景；

C. 利用核型分析软件 Metasystems Isis 对荧光原位杂交结果进行核型分析。

（8）测序

将保存于含 40%甘油 LB 培养基中的克隆菌送于华大基因公司测序，将测序结果在 NCBI 网站（http：//www.ncbi.nlm.nih.gov）上利用 BLAST（Basic Local Alignment Searchtool）方法对核酸数据库（GenBank +EMBL +DDBJ +PDB）进行同源性探寻。

3.4.13.2 结果与分析

1. 长春花染色体末端 DNA 文库的构建与分析

（1）长春花染色体末端的显微分离

先在低倍显微镜下挑选适合分离的分裂相，然后在高倍视野中用直径 0.5 μm 的玻璃微切针的针尖切割长春花染色体长臂末端，将之铲起，确定染色体吸附于针尖后将染色体连同针尖一同折断于 eppendorf 管中，如此连续分离收集 10～15 条染色体末端（见图 3.106）。

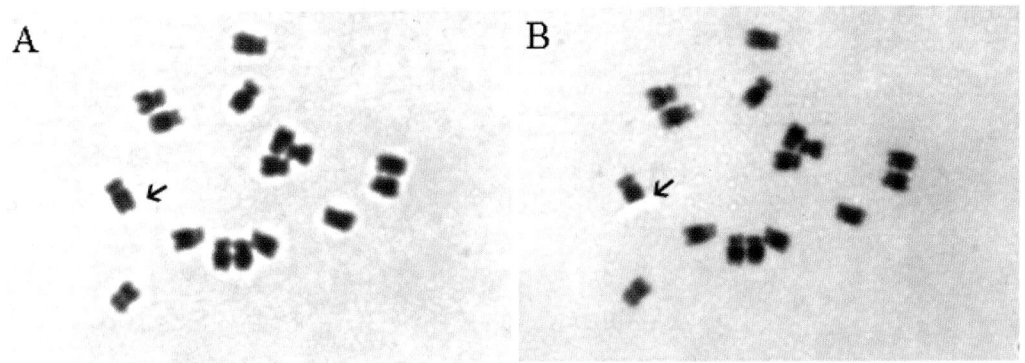

图 3.106 长春花染色体长臂末端的显微分离
A. 微切前的染色体（2n=16）；B. 微切后的染色体（箭头示分离的长臂末端）

（2）染色体的体外扩增

通过微切分离得到的长春花染色体片段，经蛋白酶 K 消化、Sau3A I 酶切、加连接接头，再进行两轮 LA-PCR，最终的 PCR 产物的电泳结果如图 3.107 所示。其中阴性对照无扩增带，说明没有外源 DNA 污染；染色体片段的体外扩增产物呈大小为 200～1500 bp 的 Smear 带。

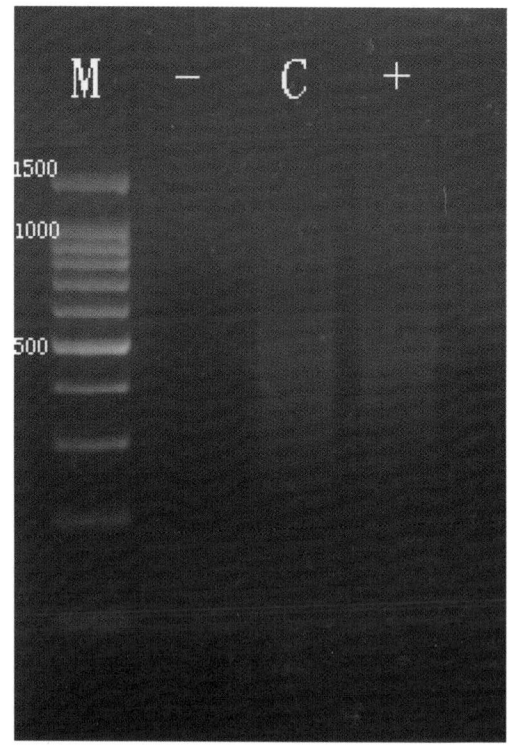

图 3.107　长春花微切片段 LA-PCR 扩增结果
M：Marker；—：阴性对照；C：长春花染色体微切片段；+：阳性对照

（3）扩增产物来源的验证

为验证染色体片段体外扩增产物是否来源于长春花基因组，并确定产物在染色体上分布情况，本研究以荧光标记的次级 LA-PCR 产物为探针，在长春花根尖细胞有丝分裂中期染色体上进行 FISH 分析。结果如图 3.108 所示，第 1 组和 2 组分别为长春花基因组次级产物和微分离片段次级产物的 FISH 分析。基因组探针在 16 条染色体上都有强烈信号，且覆盖染色体各个部位，异染色质区信号相对较强。而微切片段次级产物的信号除集中分布于染色体末端外，还在近着丝粒处有较强信号，另外信号在整个染色体上其他部位也偶见较弱信号。杂交结果表明扩增产物确实来自于长春花染色体，而信号同时见于末端和近着丝粒处则说明长春花染色体末端和近着丝粒处部分序列组成是相似的。

（4）DNA 文库的构建及分析

取第二次扩增产物与 pEASY-T1 载体连接并转化，一次获得约 $2.45×10^4$ 个重组子。随机挑选 200 个白色菌落进行插入片段分析，发现插入片段大小在 500～1200 bp 之间，平均大小在 700 bp 左右，个别为空载，电泳部分结果如图 3.109 所示。

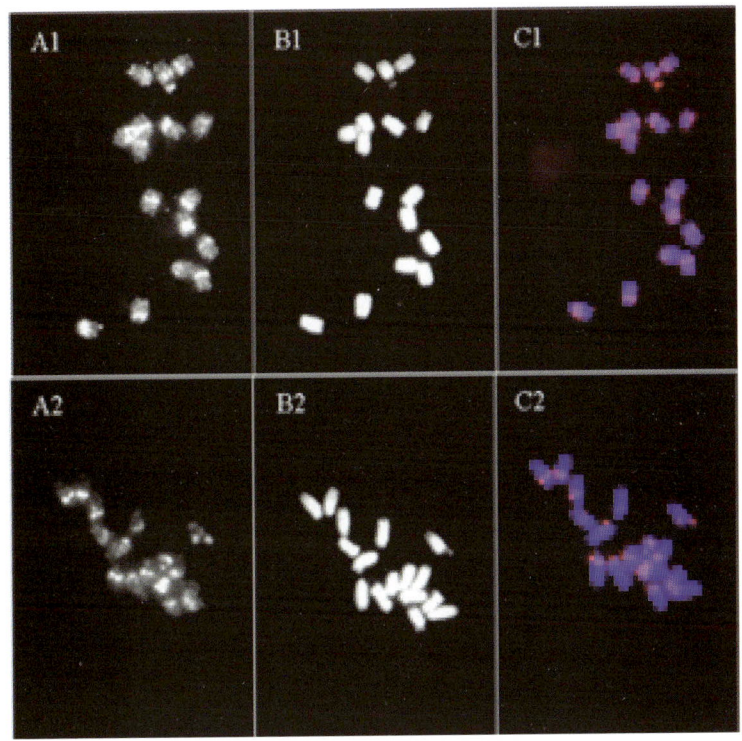

图 3.108 以长春花基因组和微切片段次级产物为探针 FISH 结果
A：信号图像；B：DAPI 图像；C：合成图像；第 1 组为阳性对照，第 2 组为微切片段 FISH 结果

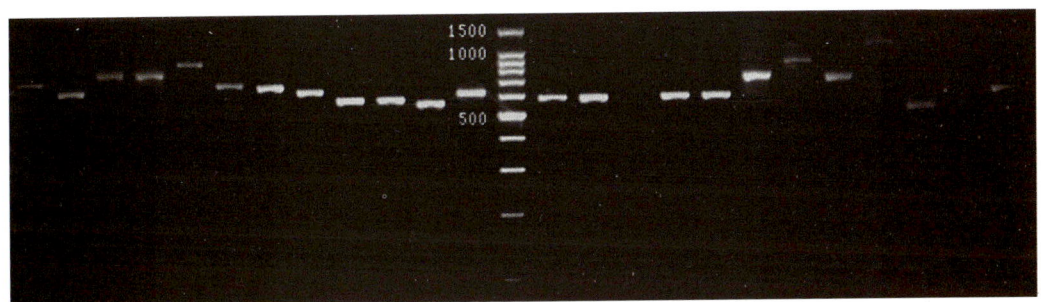

图 3.109 部分阳性克隆菌落 PCR 检测结果

在构建好的染色体片段 DNA 文库中随机挑取克隆子点膜后，以长春花基因组 DNA 为探针与之进行杂交（结果见图 3.110）。其中编号 29、45、48、67、128、211 的克隆子显示较强杂交信号，表明为基因组中高拷贝序列，而其他编号的克隆子杂交信号较浅，有些几乎不可见，其序列应为单/低拷贝序列。表 3.13 所示为长春花染色体末端 DNA 文库的主要参数。

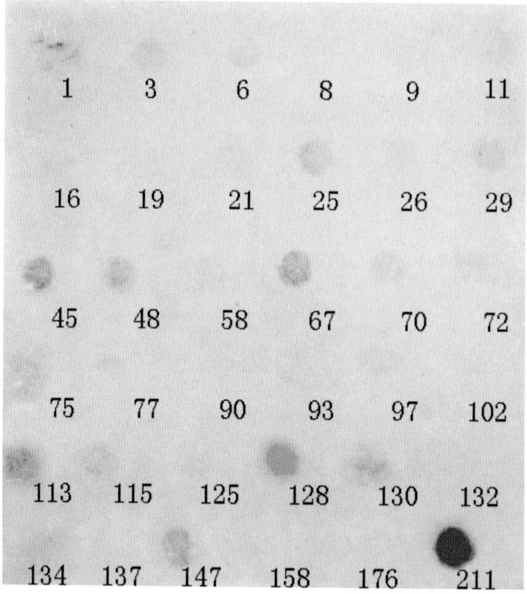

图 3.110 以长春花基因组为探针与克隆子 DNA 点杂交结果

表 3.13 长春花染色体末端 DNA 文库主要参数

重组克隆子数量	$2.45×10^4$
插入片段大小	500～1200 bp
主要分布	700～900 bp
平均大小	700 bp
高拷贝序列比例	61.1%
单/低拷贝序列比例	38.9%

（5）端粒相关序列的筛选及定位

在上述分析的基础上，对部分高拷贝克隆子序列进行了 FISH 定位及测序分析，以便进一步筛选可能的长春花端粒或端粒相关序列（Telomere-Associated Sequence，TAS）。部分克隆子序列 CR29、CR45、CR48、CR67、CR128 和 CR211 经过 Blast 比对分析，结果如下：

CR29：461 bp，AT 含量为 47.1%，序列与杠柳（*Periploca sepium*），龙船花属（*Ixora aluminicola*）和素馨属（*Gelsemium rankinii*）的外显子转录间隔区高度同源。

CR45：375 bp，AT 含量为 61.1%，比对未发现与之同源序列。

CR48：497 bp，序列富含 AT（65.0%），与紫藤（*Wisteria sinensis*）Ty1/copia 样逆转座子的移动单元的部分序列同源。

CR67：574 bp，AT 含量为 54.9%，与人类锌指蛋白 ZNF185（LIM 结构域）高度同源。

CR128（见图 3.111）：407 bp，AT 含量为 56.5%，与番茄（*Solanum lycopersicum*）的一类 182 bp 的卫星序列 CL14 高度同源（94%），该序列最近被发现特异定位于番茄

染色体末端，为番茄的端粒相关序列。

```
  1 GGGTCGAATTCGAGCTCAGGATCTGAAATTCCGACGTAAATTTTGCCAAA
 51 ATTTTTCGTGGACGTTCATTAAGACCTTAGCTATGGAGCCACTTAGCCCT
101 CACGGCCAAAACGTCCCATTTTATAGGTGTAATGTGCCCTTATAGCAGGT
151 AAACCCCCCATTTTGCCGATTTTCGCATGCTATAGTTCATGAACTTTTTG
201 GTGATCTGGAATTCCAACATAAATTTTGCCAAAATTTTCTTGGACGACC
251 GTTAAGACCTTAGCTATGGAGCAAGTTACCCCTCACGGCCAAAATGTCCA
301 ACTTTGAAGGTCAAATTTTCCCCGGATCAGGTAAACCCCTCAGTTTGCCG
351 ATTATCGTGTGCTATAGTCCATGAACTTTTTGTTGATCCTGAGCTCGAAT
401 TCGACCCC
```

图 3.111　CR128 克隆子 DNA 组成分析

下划线为添加的接头；阴影标示部分为卫星序列 182 bp 重复单元

CR211（见图 3.112）：756 bp，序列富含 AT（59.7%），共含有 31 个拷贝的（AAACCCT）序列，弥散分布在整个序列中，最长串联不超过 3 个拷贝，中间间杂各种（AAACCCT）的变异序列，与黄瓜 Gypsy 逆转座子重复片段高度同源。

```
  1 GGGTCGAATTCGAGCTCAGGATCAAGAAACCCTAAACCCTAAACCCTAAA
 51 CTCTACATTGAAACCCTAAACCTTAAAACCCAAAAAGCCCAAACCCTGAA
101 CTCAAAACCCTAAACCCTAAACCCTATACCATAAAACCTTGAATCAAGAA
151 ACCCTAAACCCTAAACTCTAACATTGAAACCCTAAACCTTAATACCCAAA
201 AAGCCCAAACCCTAAACCCTAAACCCCAAACCCTAAACCCTAAACCCTAT
251 ACCATAAAACCTTGAATCACAAACCCTAAACCCTAAACCCTAAACCCTC
301 TAGGGTTAAACCCTAAACCCTAAAGTGGAAAACCTATAAACCTAACATT
351 GAAACCCTAAACCCGAAACCGTGAAATCCTAAACCCAAAGCTTAAAA
401 CCCAAAAAGCCCAAAACCCTAAACCCTAAACCCTAAGCCCAAACCCTAAAC
451 CGAAACCCTAAACCCTAAACCCTATACCATAAAACTTTGAATCACAAAAC
501 CCTAAACCCTAAACCATAAAAGCCTTAACCCTAAACCCTGTAGGGTTATA
551 CCCTAAACCCAAACATGGAAAACCTATAACCCTAACATTGAAACCCTAA
601 ACCTTAAAACCCAAAAAAGCCTAAACCCTAAATGCTAACCCTATACCCTT
651 AACCCTAAACCCTCTAGTCTTAAACCCTAAAACCCTAACATGGAAAACCT
701 ATAACCCTAACATTGAAACCCCAAAACACCTGAGATCCTGAGCTCGAATT
751 CGACCCC
```

图 3.112　CR211 克隆子 DNA 组成分析

下划线为添加的接头；阴影标示部分为卫星序列（AAACCCT）重复单元

在对克隆子序列进行生物信息学分析的同时，作者对部分重复序列进行了 FISH 定位，以便进一步筛选可能的长春花端粒或端粒相关序列。其中 CR211 在长春花染色体上显示较清晰的定位结果（图 3.113），在 16 条染色体近着丝粒处都有或强或弱的信号，同时在第 4 对和第 5 对染色体末端可见弥散分布的信号。

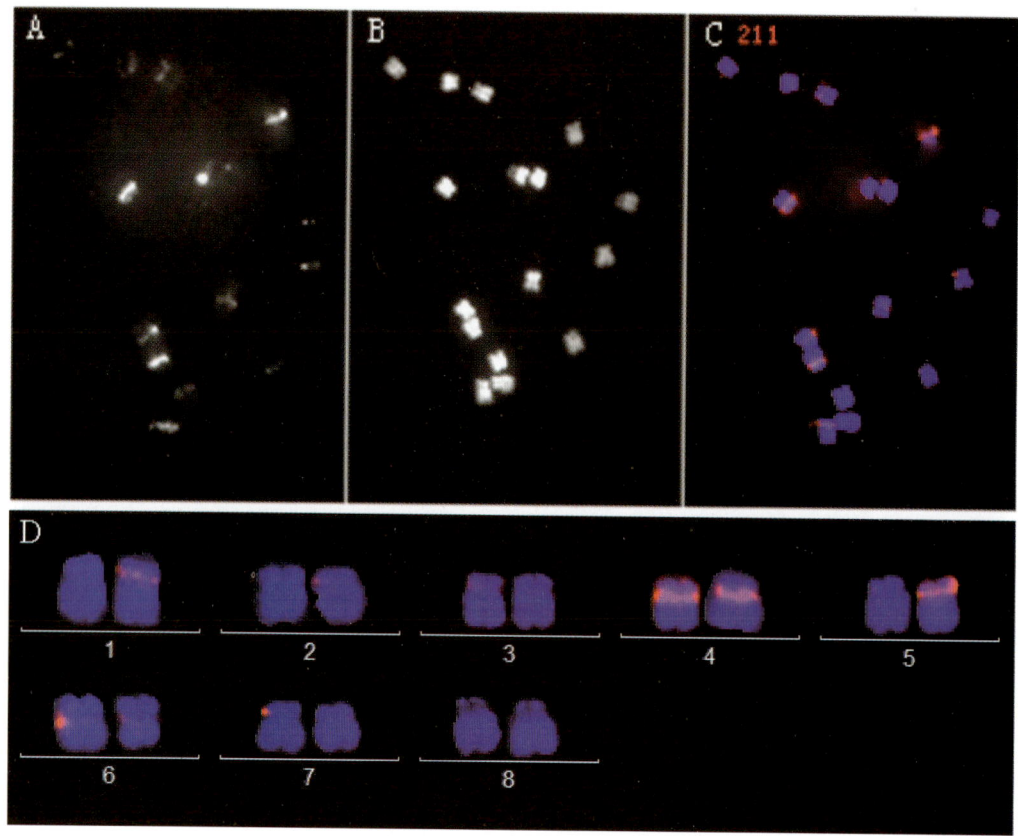

图 3.113 CR211 在长春花染色体上 FISH 定位
A：信号图像；B：DAPI 图像；C：合成图像；D：核型分析结果

（6）长春花内部端粒旁侧序列克隆及分析

①单引物扩增产物的克隆测序

根据拟南芥类型常规端粒序列（TTTAGGG）$_n$ 设计并筛选单引物 5′-ACCCTAAACCCTAAACCCTAA-3′，并以长春花基因组为模板进行 PCR 扩增，反应程序具体如下：95℃预变性 5 min 后，95℃变性 0.5 min，53℃退火 0.5 min，72℃延伸 1.5 min，循环 30 次，最后再 72℃延伸 8 min。扩增得到一大小约为 1000 bp 的片段，命名为 DLP$_{1000}$。

对该产物进行克隆测序，序列组成如图 3.114 所示。对序列分析发现，其 AT 含量高达 61.5%，含有 4 拷贝的 CCCTAAA，1 拷贝的 TTTAGGG，另外还含有少量诸如 TAGGG、TTTAGG 的变异端粒序列，同源比对未发现相关序列。

②扩增产物定位分析

为进一步检测所得序列在染色体上的定位，以该序列和拟南芥类型端粒为探针进行双色荧光原位杂交分析，结果如图 3.115 所示。信号集中于染色体着丝粒附近，与拟南芥类型端粒在长春花染色体上定位相同，而该序列在 DNA 组成和定位上与内部端粒的联系表明其为内部端粒旁侧序列。同时，值得注意的是，个别染色体末端也出现了该序列的信号，暗示该序列可能进化时间上与内部端粒形成和末端端粒缺失相接近，为拟南芥类型端粒在染色体上迁移变化遗迹。

1 <u>ACCCTAAACCCTAAA</u>CCCTAAGCCTTAGAAATGTGTGTAATACTACATCT
51 TAATCCTAAAATCTTAAATACTAACACCCTAAACCCTAAAATTTATGAAG
101 CGTGTAGTGATAAAAAAAAATGTTTACTTTAAATTCTAAACCTGTAAAGA
151 TTTAAATTTAGTAAGTGTATAATATTAGACCTTAATCCTAAAACCTTATG
201 ATGCAATGTAGATGGAAGAAGTCCCAGCATATGTGCATCTAGGTCCCATA
251 AACCTTGATGTATTAATGAGGTAGCACGAGCGTGGATCTGGCTTCATATG
301 GAGTGGCGGATCATGAGACATGTATTACTGATCTTCAGTGTCGATGTTTT
351 GGTCATAATCTGTTTCAAGCTTATAGTATAGCTACCCATAGATTGATTGA
401 TATTATTGATAGGACGGTTCTGAGACTCACACATACCACATGACTTGTGG
451 TGAGGTGACAGTCACTTTACAAGATATGCATGTCTTATGGGGATTACGGA
501 TTAGAGAGGTTGCCTATTATAGGTCCTAGTTGGATTCTACATCATAAGTA
551 ATATTGGATTGATATGATTGAAAGGCTTATTGGTATTCAGTTGTCGCTTA
601 CTCACTTTTCCGAATATAAGGTAAAAAAGGAATCTTTAGAGGCTTGGATC
651 TTGAGAAAGTTCTTTGGATCAAAGACTGATGATGACCCATTCGACAGTCT
701 AGATGTATCATACGTTCTGCCACCTCCGATTATAAGAGGTACATCATATG
751 TACCCCACCTTTGAGTGCAGTAGGGTTGTCATTTGATGCACCTTCATCTT
801 CGAGTGCACAAGGTTCATATGTGCCTCATATGCCTATATCCGAGGCATCT
851 TCTTCCGATTCGCAGGAGCATAGCGAAGAGCAAGGAGGATGATGTTACAC
901 CAGCACAACGACTTGGCTTTGGACATCGTGTTAGGATAGACGACTAGATT
951 TACTCCATCTGATTATTG<u>TAGGGTTTAGGTTTAGGGT</u>

图 3.114 扩增产物序列 DNA 组成
下划线为单引物序列；阴影标示部分为拟南芥类型端粒序列

（7）长春花微卫星序列染色体定位分析

长春花染色体端粒到底是怎样进化？它的组成又如何？我们以 14 种微卫星序列 SSR-CA、SSR-CT、SSR-AAG、SSR-CAC、SSR-ACG、SSR-AAC、SSR-GGC、SSR-ACT、SSR-AGG、SSR-CAT、SSR-CAG、SSR-AAT、SSR-GACA 和 SSR-GATA 为探针进行 FISH 分析，结果表明，只有 SSR-CA、SSR-AAC 和 SSR-AAG 序列在染色体上显示了清楚的信号。其他 11 种未见信号。

SSR-CA 在第 2 对和第 7 对染色体上着丝粒处有较强信号，第 2 号染色体的信号延伸至短臂。第 1 对染色体短臂靠近着丝粒部位有较弱弥散的信号，结果如图 3.116 所示。

SSR-AAC 信号集中于第 1、第 2 和第 4 对染色体近着丝粒位点，而在第 3 和第 6 对染色体上信号相对微弱，结果如图 3.117 所示。

SSR-AAG 信号集中于第 2、第 4 和第 7 对染色体近着丝粒位点，如图 3.118 所示。

以上三种微卫星序列 FISH 结果显示长春花染色体近着丝粒处为微卫星出现高频位点，其中第 1、2、4 和第 7 对染色体上微卫星序列信号强烈，这种特异定位与拟南芥类端粒在长春花染色体上定位相似（见图 3.119）。

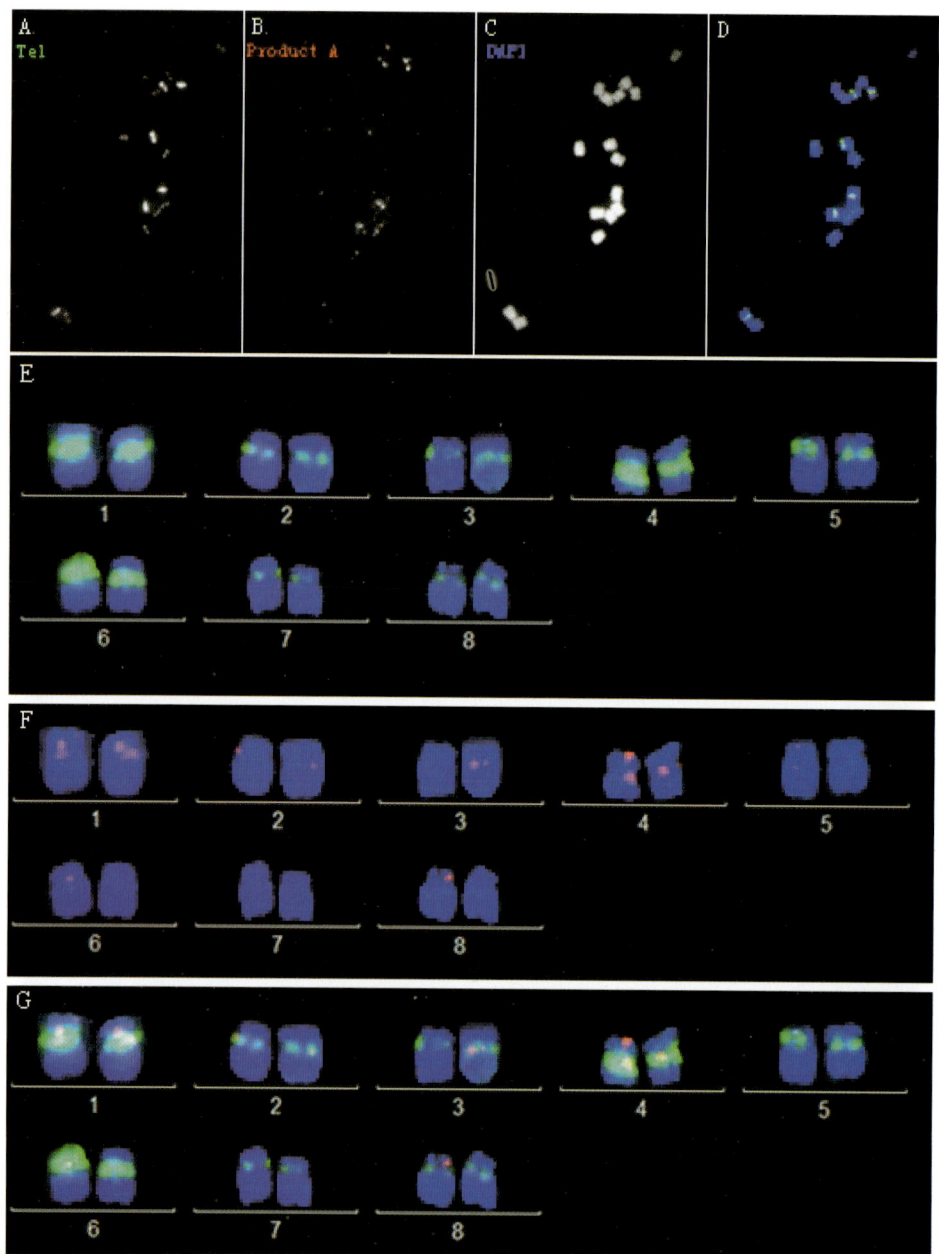

图 3.115　扩增产物染色体 FISH 定位

A：拟南芥类型端粒序列信号图像（绿色）；B：扩增产物（红色）；C：DAPI 图像（蓝色）；D：合成图像；E、F、G 分别为绿色、红色和叠加信号核型分析结果

4. The early sex determination of the sapling of *Ginkgo*

In general, the sex of ginkgo can not be distinguished until it has grown over 25 years. So it is important to determine the sex at early stages in order to use them as green trees in cities. In the present investigation, 20 germinating seeds and 20 young plants five years old were choden, and their roots were cut and pretreated in 0.2% colchicine. The method of the preparation of chromosome samples was the wall degradation hypotonic method.

The results also show that if chromosome samples are good, they can be used for determining the sexes. As each seed possesses only one primary root, 13 out of 20 seeds gave good chromosomal samples and 10 of them could be used to distinguish the sex (♀3, 7♂), 19 out of 20 plants produced good samples and 18 of them were used to distinguish the sex (♀10, 8♂). When primary roots from the germinating seeds were taken, their normal growth of seeds was affected. If secondary roots from the sapling were taken, the growth of the young plant was normal. The growth of twenty young plants studied by us is unaffected after their secondary root tips have been sampled. This shows that it is suitable to determine the sex at the early sapling stage. This will help to select male gingko trees as the green plants in cities and select female trees to plant in certain gardens for the fruit.

(Authors: Chen Ruiyang, Song Wenqin, Li Xiulan, and An Zhuping. Published in: Cathaya 5: 41-48)

3.5.2　芦笋性别染色体的研究

Different from animale, plants were rarely found to have sex chromosomes (Li, 1984). Only 3-4% of higher plants are dioecious, while barely over 10 species have been proved at cytological level to be structurally heterogametic (Li, 1984). Most of them belong to the Polygonaceae, the Moraceae and the Caryophyllaceae (Li, 1984).

Asparagus officinlis is widely cultivated all over the world. Compared with females, males have many superior economic characteristics, such as resistence to unfavorable conditions, earlier maturation, higher yielding, etc. (Robbins, 1926, 1928; Yeager, 1938). Thus, researchers are trying to obtain all-male individuals by hybridization. The development of a quick and simple method of sex identification is urgently necessary. Genetic experiments have demonstrated that, in *Asparagus officinlis*, females are homogametic whiles are heterogametic concerning sex chromosomes (Kappert, 1941; Rick and Hanna, 1943). The effort to identify sex chromosomes by cytological means, however, has never been succeeded.

Thusesen (1960) noticed differences between one pair of homologous chromosomes and postulated that they were the sex chromosomes of *Asparagus officinlis*. His result has not been confirmed by othe experiments (Zilm, 1966; Franken, 1969). Loptien (1979) drew a conclusion by means of trisomic analysis that chromosome L5 was the sex chromosomes. He also pointed out that the difference between these two sex chromosomes was too slight to be detected by cytological methods. In short, there has not been an acknowleged conclusion about the sex chromosomes of *Asparagus officinlis* so far.

The special technique developed by our laboratory gave a satisfactory answer to this problem. After treated with trypsin-urea and dyed with Giemsa, the sex chromosomes could be discerned clearly from the autosomes. Because of its simplicity and reliability, this method is evidently practical.

3.5.2.1 Materials and methods

Asparagus officinlis, Washington 500, was cultivated in the Laboratory, Department of biology, Nankai university. Buds were excised from the plants which hand not bloomed and immediately pretreated with 0.03% colchcine (Using 0.002 M 8-Hydroxyquinoline as solvent) at 25℃ for 3 h. The pre-hypotonic treatment was accomplished with 0.075 M KCl for 30 min and the wall-degradation with the mixture of 2.5% cellulose and 2.5% pectinase at 25℃ for 30 min. After post-hypotonic treatment for 30 minutes in redistilled water, the buds were fixed (ethanol:acetic acid=3:1) for 30 min. The buds were smashed on cold slides, which were then dried on fire. After wased with 0.85% NaCl.the slides were immersed in 5.3 Murea for 30 s. They were stained with Giemsa (1/15 M phosphate buffer pH 7.0: Giemsa 20:1 for 7-8 minutes after they were washed with with 0.85% NaCl twice. The stained were washed with flowing water and dried in the air.

3.5.2.2 Results and discussion

1. Karyotype and sex chromosomes of *Asparagus officinlis*

Our research reveals that *Asparagus officinlis* is a diploid ($2n=20$). The basic chromosome number is $x=10$. This result agrees with that of the of the previous reports (Kamo, 1929; Flory, 1932; Yarnell, 1954; Thuesen, 1960; Zilm, 1966; Franken, 1969). The karyotype of *Asparagus officinlis* consists of nine pairs of autosomes and one pair of sex chromosomes (Fig. 3.130). Nos 1, 2, 3, 4 and 5 chromosomes are relatively larger than Nos.6, 7, 8, and 9 chromosomes. This result follows that of Loptien (1979) with a slight difference; the L2 chromosome in his arrangement is defined as No.5 chromosome inour result. The difference arises from the different treatment of a pair of satellites on the long arms of this chromosome. The satillites are obvious in all the cells observed. We did not include the length of the satellites in that of the whole chromosome, whereas Loptien did. Secondary constrictions are found on No.2 and No.3 chromosomes. No.3, 6 and 7 chromosome are submetacentric (sm), while Nos.4, 5 and 9 chromosomes are metacentric (m).

Females are homogametic with two identical subtelocentric X chromosomes. The karyotype formula is $2n=18A$ (10 sm+6 m) +XX (2 st). Males are heterogametic. One of the sex chromosomes is subtelocentric, identical with the X chromosomes of females, while the other is a long metacentric chromosome. The karyotype formula is $2n=18A$ (10S+6m) +XY (1st+1m), which is different from the previous results (Thuesen, 1960; Loptien, 1979). Therefore, the sex determination mechanism of *Asparagus officinlis* may be classified as XY type.

2. Identification of *Asparagus officinlis* sex chromosomes

Our results show clearly that the X chromosome of *Asparagus officinlis* is subtelocentric and medium-sized. All female cells have a pair of X chromosomes and its length ranks between No.4 and No.5 chromosome. They canbe easily discerned from the shorter chromosomes (Nos.6, 7, and 9 chromosomes), and from the longer chromosomes (Nos.1, 2and 3 chromosomes). They can also be distinguished from No.4 and No.5 chromosomes according to the following characteristics: No.4 chromosome is metacentric, while the long arms of No.5 chromosome have satellites. The Y chromosome is metacentric and long, as long as No.1 chromosome or even longer. All the 250 male cells obtained from 100 male plants show this chromosome clearly.

3. Method for identification of *Asparagus officinlis* sex chromosome

This work was done following the method published before (Chen *et al.*, 1979). Treated by trypsin, the chromosomes have no specific characteristics; especialy the centromere is not clear. This made it very difficult to pair the homologous chromosomes. We tried to treat the the chromosome. We tried to treat the chromsomes with 0.025% trypsin and 5.3 Murea. As long as the treatment time is properly controlled, the centromere and the secondary constriction appear clear. They are faintly stained while the other parts of the chromosome were densely stained. The identification of the sex chromosomes is rather reliable from this method.

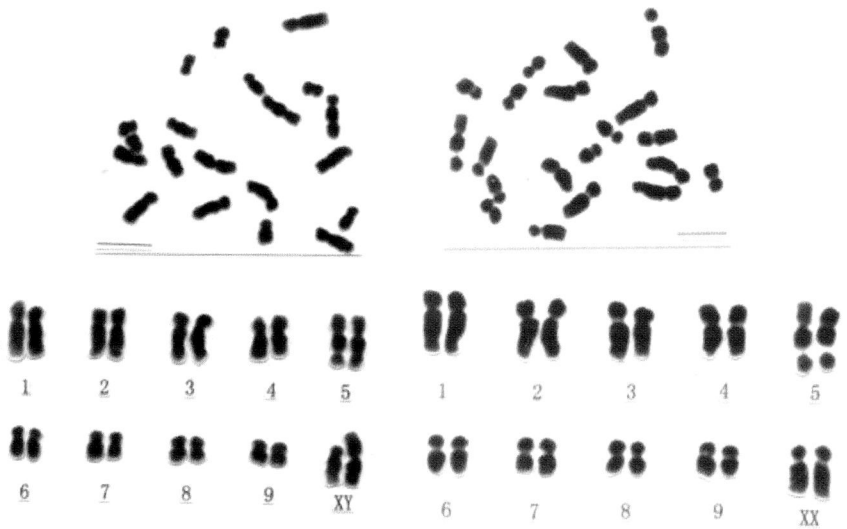

Fig. 3.130 Karyotypes of *Asparagus officinlis* sex chromosome
A. Female karyotype; B. Male karyotype. *Bar*=28 μm

(Authors: An Zhuping, Cui Xiangquin, Song Wenqin, Li Xiulan Chen Ruiyang. Published in: Proc.Sec.Sino-Jpn.Symposium Pl.Chromos.1992.Plant Chromosome Research p297-300)

3.5.3 应用 45S rDNA FISH 鉴定雌雄异株菠菜的性染色体

Repeated ribosomal RNA genes (45S rDNA) are part of nucleolar organizer regions (NORs) that have been detected classically as secondary constrictions associated with nucleoli and distal satellites. More recently, 45S rDNA has been detected as fluorescent foci on chromosomes of many plant species using fluorescence in situ hybridization (FISH) (Heslop-Harrison, 2000). Spinach (*Spinacia oleracea* L.) is a dioecious plant species that has a simple sex determining system involving XY-type homomorphic sex chromosomes (Vyskot and Hobza, 2004). So far, there has been only one report of a 45S rDNA focus on sex chromosomes in plants, specifi cally on the X chromosome of a liverwort (Nakayama *et al.*, 2001). This rDNA repeat seems to have evolved independently of rDNA on autosomes because it includes sex chromosomespecific sequences (Fujisawa *et al.*, 2003). On the other hand, in insects 45S rDNA foci have been reported on sex chromosomes in several species that use both XY and XO sex determining systems (Pellegrini *et al.*, 1977; Juan *et al.*, 1993; López-León *et al.*, 1994; Blackman and Spence, 1996; Willhoeft, 1997). In the case of *Drosophila*, most 45S rDNA on sex chromosomes is known to contain insertions and to be transcriptionally inactive.

Here we report that by using FISH, different numbers of 45S rDNA loci are visualized in female and male chromosomes of spinach. We suggest that the heterogeneous localization is due to the presence of an additional 45S rDNA locus on the X chromosome and its absence on the Y chromosome.

3.5.3.1 Materials and Methods

1. Plant materials and chromosome preparation

Root tips and young leaves of 17 male and 18 female spinach plants were used in this study. For FISH analyses, slides were prepared by a wall degradation hypotonic method according to Chen *et al.* (1979) with minor modifi cations. In brief, tissues were treated with saturated P-dichlorobenzene solution for 3 h before they were immersed in 0.075 M KCl solution at 25℃ for 30 min for pre-hypotonic treatment. After that, they were fixed in 3:1 ethanol: acetic acid for 30 min and washed with distilled water. Then materials were treated with an enzyme mixture containing 2% cellulase ('Onozuka' R-10, from Kinki Yakult MFG. Co, Ltd, Japan) and 1% pectolase (*Aspergillus niger*, from Serva, New York) in citrate/citric acid buffer (pH 4.5) at 25℃ for about 30 min, washed with 0.075 M KCl solution twice and incubated in this solution at 25℃ for 15 min, then washed with distilled water and immersed in fresh prepared fixation solution (3:1 ethanol: acetic acid). A piece of the meristem was put in a drop of fresh fixation solution on a clean glass slide which was kept in freezing distilled water, then torn with a pair of forceps. Two drops of fixation solution were added to the mixture and the slide was heated over an alcohol burner. The dried slide was stained with a 1:20 Giemsa: Söensen's phosphate buffer (pH 6.8) for 15 min, rinsed in distilled water, and

b. Common type B. 18L (2SAT)

The same as type A except that satellites were smaller and attached to No.4 chromosomes, and are relatively smaller.

(2) Heteromorphic satellite type. 18L (1SAT + 1 int-SAT)

There existed heteromorphism of the satellite chromosomes. One of the satellite is similar to that of the common type, which attaches to the end of one short arm of the chromosome, but the others had two forms, one of which was reversion form; the satellite were located near the centromere, and becomes an inter calary. They could be the result of a paracentric inversion in the short arm. The other was the small satellite form.

(3) Tertiary constriction type. 18L (2SAT + 2 constrictions)

In addition to the two satellite chromosomes, there exist a pair of chromosomes, whose long arms had secondary (or called tertiary) constrictions that do not related to the NOR.

(4) Tetra-satellite type. 18L (4SAT)

There were two pairs of satellite chromosomes. This type can be diveded into two subtypes according to the serial number of the satellite chromosomes:

Subtype 1. Satellites existed on the No. 1 and 2 chromosomes.

Subtype 2. Satellites existed on the No. 1 and 8 chromosomes.

(5) Tetra-heteromorphic-satellite type:

Subtype 1. 18 L (3SAT + 1 int-SAT)

One of the two pairs of satellite chromosomes had heteromorphic satellites. One pair of the satellites were attached to the end of the short arms, while the other pair were intercalary and located near the centromere.

Subtype 2. 18L (2SAT q^-+ 2 int-SAT)

One pair of the satellite chromosomes had one long arm deletion. The other pair of satellites were intercalary.

(6) Split type. 14L (2SAT) +9S

There were 23 chromosomes in the cell. The small chromosomes apparently came from centric fission of the L type large chromosomes, i. e. 4LM→8S+1B=9S. This type of cells are postulated to be unable to survive in nature. The reason of their existence is unclear.

(7) 1 B type. 18L (2SAT) +1B

Cells of this type had a small B chromosome.

(8) 2 B type. 18L (2SAT) +2B

Cells of this type had two small B chromosomes

(9) Short arm deletion type. 18L (2SAT+1 p^-)

This type of karyotype consists of 17 large chromosomes and one small chromosome produced by short arm deletion.

(10) Dicentric type.

This type of cells had a di-centric chromosome.

(11) Aneuploid type.

In diploid individuals, there existed a high frequency of aneuploidy, usually $2n=17$.

3. Karyotype variations of tetraploid.

We encountered 10 tetraploid populations in seven areas (Table 3.16). Their karyotypes are listed below:

(1) Common type. 36L (4SAT)

We divide this type into subtypes according to the serial number of satellite chromosomes:

Subtype A. 36L (4SAT)

Satellites attached to the short arms of No. 1 and No. 2 large chromosomes.

Subtype B. 36L (4SAT)

No.1 and No. 6 chromosomes were satellite chromosomes.

(2) Bi-satellite type. 36L (2SAT)

Compared with the common type, this type of tetraploid cell had only one pair of satellite chronosomes.

(3) Penta-satellite type. 36L (4SAT+1 pSAT)

In addition to the two pairs of satellite chromosomes that were similar to those of the common type, another large chromosome had a satellite attached to the end of its long arm.

(4) Hexa-satellite type. 34L (6SAT)

This cell type ($2n = 34$) had three satellite chromosomes.

(5) Di-centric type. 34 (4SAT+2 centromeres)

This type of cells had an obvious dicentric chromosome, which apparently cames from translocation.

(6) Aneuploid type. 35L (4SAT)

In tetraploid individuals, there also existed a high frenquency of aneuploidy, usually $2n=35$.

4. Karyotype variation in hexaploid

(1) Common type. 54L (6SAT)

Individuals of most hexaploid populations have three pairs of satellite chromosomes.

(2) Tetra-satellite type. 54L (4SAT)

These hexaploid cells have only two satellite chromosomes each, same as the subtype B of common type in tetraploids.

(3) Sichuan type. 51L (4SAT) +3S (st)

Among the 33 populations investigated, 32 belong to the L type (Huziwara, 1957). The only exception was sampled at Beipei area in Chongqing city, Sichuan Province. It does not belong to the L+S type either. Only three small subterminal centromere chromosomes were found in each cell, in addition to 51 large chromosomes [$2n=$ 51L+3S (st)].

5. Triploid. 27L (3SAT)

Among the 33 populations investigated, triploid individuals were found only in Fenghua of Ningbo, Zhejiang Province with $2n=3x=27L$ (3SAT) and $2n=28L$ (4SAT).

6. Pentaploid: 45L (3SAT)

air-dried.

2. Probe labeling

Cloned 45S rDNA from potato, which was a gift from Professor Song Yunchun, Wuhan University, China, was labeled with digoxigenin (DIG) -dUTP (Roche) by randomly-primed DNA synthesis.

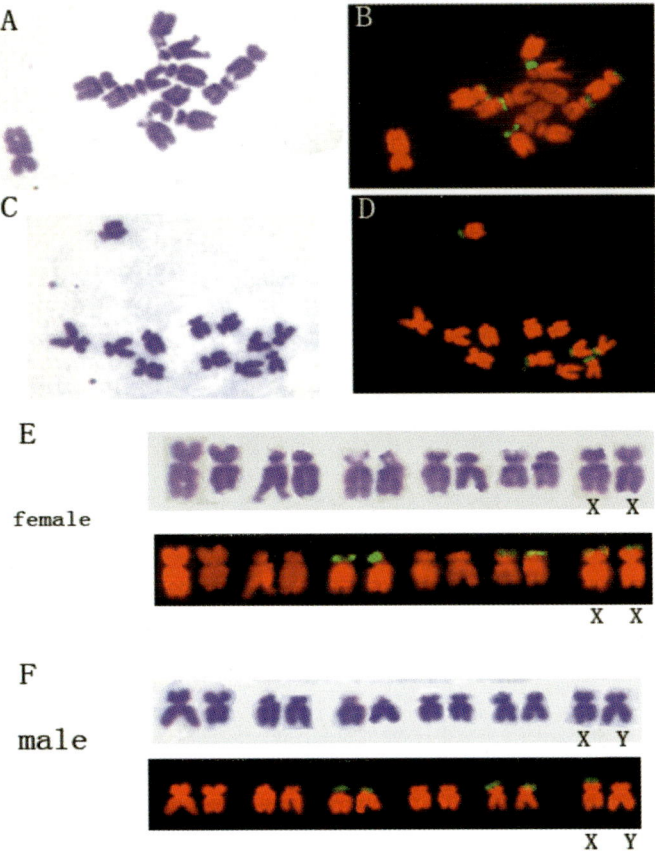

Fig. 3.131 Giemsa and FISH images of female and male spinach chromosomes

Female (A, B) and male (C, D) chromosomes at metaphase, and the combination of counter-stained DAPI (red) and FITC (green) signals of 45S rDNA sequences (B, D). E and F are karyograms of female and male spinach diploid chromosome complements. Bar = 5 μm

3. Fluorescence in situ hybridization (FISH)

FISH was carried out as described by Qi et al. (2002). Hybridization buffer contained 50% deionized formamide; 2 × SSC; 50 mM sodium phosphate, pH 7.0; 5% dextran sulfate; and 3 ng/μl 45S rDNA probe. The probe was denatured at 94℃ for 10 min before being used. Slides with metaphase spreads were treated with 70% deionised formamide in 2×SSC at 70℃ for 2 min. Denatured hybridization buffer (10 μl) was with metaphase spreads were treated with 70% deionised formamide in 2×SSC at 70℃ for 2 min. Denatured hybridization buffer (10 μl) was then applied to the slides, which were incubated at 37℃ overnight. Finally, the slides were washed with 30% deionized formamide in 2 ×SSC at 37℃ for 5 min and twice in 2×SSC at 37℃ for 5 min. Fluorescence signal was detected using anti-DIG-fl uorescent-conjugate (Roche). Metaphase spreads were counterstained with 100 ng/mL 4'6-diamidino-2-phenylindole (DAPI). The hybridization signals were visualized and recorded using a Nikon 80i fluorescent microscope and a cooled CCD camera and were then analyzed using Adobe Photoshop.

3.5.3.2 Results and discussion

1. Karyotypes of male and female plants of spinach (*S. oleracea*)

All the male and female plants have $2n = 2x = 12$ chromosomes (Fig. 3.131A, C). In both sexes the chromosome complement is composed of two large metacentrics, two long subtelocentrics, two short subtelocentrics, two acrocentrics, and four submetacentrics (Table 3.15, Fig. 3.131E, F). According to previous studies, spinach has a homomorphic pair of submetacentric sex chromosomes, which represents a relatively young system of sex determination. In our studies, the X and Y chromosome could also not be distinguished easily by Giemsa staining, but in some cases, a tiny satellite was barely visible on the short arm of the X chromosome while it was absent on the Y chromosome.

Table 3.15 Relative length, arm ratio and type of chromosome in the complement of *Spinacia oleracea* L.

Number	Relative, length ratio % (S+L=T)	Arm ratio (Long/Short)	Type
1	6.0+9.8=15.8	1.6	metacentric
2	3.1+10.5=13.6	3.4	subtelocentric
3	3.1+9.5=12.7	3.0	subtelocentric
4[a]	3.8+8.9=12.6	2.6	subtelocentric
5[a]	1.3+9.8=11.1	7.5	telocentric
X[a]	5.2+9.6=14.9	1.8	subtelocentric
Y	5.2+9.8=15.1	1.8	subtelocentric

[a] The length of satellites is not included in the chromosome length

2. Mapping 45S rDNA loci

After FISH using digoxigenin-labeled 45S rDNA, 97 metaphases were evaluated. Six labeled foci were visible in female chromosome sets, and five labeled foci were visible in male chromosome sets. We suggest that this difference is based on a 45S rDNA focus on each X chromosome but not on the Y.

In both sexes, two foci were on a pair of acrocentric chromosomes, while the other two foci were on a pair of submetacentric chromosomes. All the focilay in secondary constrictions and satellites (Fig. 3.131B, D, E, F). The 45S rDNA focus at the end of the short arm of the X chromosome suggests that X chromosomes may have small satellites at this location that are difficult to observe by Giemsa staining. If so, X and Y chromosomes of *S. oleracea* are not homomorphic but heteromorphic. There are four current examples of the sex chromosomes in dioecious plant species: 1) primitive autosomes, 2) homomorphic sex chromosomes, 3) heteromorphic sex chromosomes, and 4) polymorphic sex chromosomes (Vyskot and Hobza, 2004). Spinach used to fall into the category of homomorphic sex chromosomes, which represents a relatively young, simple genetic system of sex determination. Although we found that the sex chromosomes of spinach are heteromorphic, we also support the previous view that they should represent an early stage in sex chromosome evolution because the morphological difference between the X and Y chromosomes is so tiny.

There are two possible explanations for differences between X and Y chromosomes in spinach – either the X gained the 45S rDNA or the Y lost it. Considering that in a liverwort with sex chromosomes the X has an additional locus of 17S rDNA (Nakayama *et al.*, 2001), and the Y chromosomes of mammals so often have lost sequences compared to the X (Jegalian and Lahn, 2001), we think that the difference between X and Y chromosomes of spinach is likely to be the result of deletion of the satellite from the Y chromosome.

(Authors: T. Lan S. Zhang B. Liu X. Li R. Chen W. Song. Published in:
Cytogenet Genome Res, 2006, 114: 175-177)

第六节 细胞地理学

提要：中国的植物细胞地理学报道较少，我们研究的芦苇、半夏、三脉紫菀都是多倍体复合体。细胞地理学对研究物种的起源进化是非常有意义的。通过对三脉紫菀细胞地理学研究，认识到多倍化与杂交相结合的进化作用，对我们后来研究基因组三倍化的进化意义受了很大启发。

3.6.1 三脉紫菀多倍体复合体的细胞地理学和物种形成

Aster is a large genus of the Compositae family. Defferent scholars hold different views on the subdivision of the genus, but we can estimate about several hundreds taxa (flora reipublicae popularis sinicae) widely distributed in Asia, Europe and North America. In a narrow sense, China posesses nearly one hundred taxa. *Aster ageratoides* is a widely distributed taxon which has many variants. Populations of this taxon are widely distributed in the northeastern, northern, southern to eastern and southeastern part of China, as well as south Tibet, south of the Himalayas Mountain, Japan and Northeastern Asia. This taxon is independently divided into 9 subspecies and 6 variants. According to the Flora Reipublicae Popularis Sinicae, this taxon has 11 variants in China.

Cytological and cytogeographic studeis on *Aster* and related genera have been carried out extensively in North America and Europe. There is very little cytological information known from the Asian *Aster* except that there have been some reports about the Japanese taxa (Huziwara, 1953, 1955, 1956; Irifmne, 1985, 1987; Matsuda, 1967, 1980, 1981; Tanaka, 1959, 1982; Rara, 1973; Yoshida, 1986). In order to illustrate the distribution pattern, origion center and spread route of this taxon on the mainland of China, we cooperatively studied the chromosome data of 33 populations sampled from 24 areas in China. Now we have obtained sufficient material to warrant discussion of the distribution pattern of the dipoid, tetraploid and hexaploid populations. Morphology and karyotype variation indicate that Mount Huang area of Zhejiang Province is the current distribution center, while Mount Tai area of Shandong Province and Kunming area of Yunnan Province could be the origion centers of this taxon.

3.6.1.1 Materials and Methods

Field collections and investigations were conducted in blooming seasons of *Aster ageratiodes* (late August to early September in the Northeast, early November to early December in the South and East, middle to late October in the Southwest). At least 5 individuals were sampled for each population and were cultivated in the experimental field of Nankai University. Morphological observations and chromosome analysis were conducted. Voucher specimens are depoisited in Biology Department of Nankai University and the herbarium of the Hiroshima University. Chromosome configurations were prepared following two methods: The conventional squashing method was applied using Carbol fuchsin when counting chromosome number. Wall degradation method using cellulase and pectinase (Chen, 1979, 1982) was applied when analyzing karyotypes. Karyotype analysis followed the usual method.

3.6.1.2 Results

1. The cytotype distribution pattern of Chinese *Aster ageratoides*

We studied 33 popullations of *Aster ageratoides* sampled from 24 areas of 16 provinces or regions (not including the Qingzang Plateau and the Inner Mongolia grasslands). Chromosome numbers and other information are listed in Table 1. The populations north of the 35°N latitude were all dioploid, with no polyploidy individuals, while diploid populations existed only in afew areas south of 35°N latitude, such as Nanning (Gu, 1989), Ningbo and Kunming (Fig. 3.132), and in the range south of Mount Laoshan in Shandong Province (37°N latitude), Lianyungang of Jiansu Province and north of Mount Tianmu of Hangzhou area (29 degree north latitude), only tetraploid, populations were found (Fig. 3.132). Coexistence of tetra-and hexaploid populations were found in Mount Huangshan area of Anhui province, Mount Hengshan area of Hunan Province and Guilin area of Guangxi (Fig. 3.132). In the upper parts of Mount Huangshan, Mount Tianmu and Mount Hengshan, Existed only hexaploid but no other ploidy populations. Small populations of pentaploids and triploids were found at the foot of Mount Huangshan and Fenghua of Zhejiang province, respectively.

Fig. 3.132 Cytogeographical distribution map of *Aster ageratoides*

2. Karyotype variations of diploid *Aster ageratoides*

We conducted chromosome analysis on 84 individuals of this taxon from 15 populations sampled on the mainland of China, and discovered great variation of their karyotypes, which classified them in to eleven types:

Table 3.16 Chromosome number of *Aster ageratoides* sampled at various areas in China

Loc.NO	Samoling sites and voucher number	Chromosome number ($2n$)	Number of individuals observed
	Aster ageratoides var. *scaberulus*		
1	Jilin Prov. Changchun city, Jingyuetan. Chen. 88-01	18	4
2	Jilin Prov. Jinlin city, Zuojia. Chen. 88-02	18	4
3	Liaoning Prov. Benxi city, Shuidong. Chen.88-03	18	4
4	Liaoning Prov. Dandong city, Chen.88-04	18	5
5	Liaoning Prov. Anshan city, Qianshan Chen.88-05	18 19=18+1B	4 1
6	Hebei Prov.Zunhua-xian, Huangyaguan Chen.88-06	18	4
7	Hebei Prov. Zunhua-xian Dongling Chen.88-07	18	5
8	Tianjin, Jixian, Yanglou Irifune 88-08	18	3
9	Tianjin, Jixian, linchang Irifune 88-09	18	3
10	Tianjin, Jixian, Hershuibe Irifune 88-10	18	4
11	Shandong Prov.Mount Taishan, Zhongtianmen Song.91-1	18 19=18+1B 17+1P 16+1dic 20=18+2B	12 2 1 1 1
	Aster ageratoides var.*oophyllus*		
12	Shandong Prov.Mount Laoshan, Irifune 89-12	36	5
13	Zhejiang Prov. Hangzhou city, lingyisi Irifune 89-13	36	5
14	Zhejiang Prov. Hangzhou city, Longjing Irifune 89-14	36 45	4 1
15	Zhejiang Prov. Hangzhou city, Lingyinsi Irifune 89-15	36	4
16	Zhejiang Prov.Hangzhou city, Long-jiang-lingyinjian Irifune 89-16	36	2
17	Zhejiang Prov.Mount Tianmu Irifune 89-17	54	1

续表

Loc.NO	Samoling sites and voucher number	Chromosome number (2n)	Number of individuals observed
	Aster ageratoides		
18	Jiangsu Prov. Lianyungang city, Huaguo Hill Chen 91-20	36 34	6 1
	Aster ageratoides var.*scaberulus*		
19	Anhui Prov. Yulingguan Irifune 89-19	36 54 35	3 2 1
20	Anhui Prov. Mount Huang Yansi Irifune 89-19	36 45 43 54	4 2 1 2
21	Anhui Prov. Mount Huang, Irifune 89-21	54	3
22	Zhejiang Prov. Ningbo city, Fenghua Chen.91-23	18 28 27	1 2 2
23	Zhejiang Prov. Ningbo city, Fanghuazhen Chen.91-24	19=18+1B 20=18+2B 36	2 2 2
	Aster ageratoides var.*ageratoides*		
24	Shanxi Prov. Xian city, Huaqingchi Chen.91-25	18 17	3 1
	Aster ageratoides var. *scaberulus*		
25	Sichuan Prov. Chongqing city, Beipei Chen.91-22	54	3
	Aster ageratoides var.*ageratoides*		
26	Yunan Prov. Kunming city, West Hill Chen.91-26	18	5
	Aster ageratoides var. *gerlachii*		
27	Guizhou Prov.Guiyang city, Qianling Hill Chen.91-27	54	3
	Aster ageratoides var.*laticorymbus*		
28	Guangxi Prov.Guilin city, Ludiyan Chen.92-28	36 45	4 2
29	Guangxi Prov. Guilin city, Nanxi Hill Chen.92-29	36 54	3 1
	Aster ageratoides var.*laticorymbus*		
30	Hunan Prov. Hengyang city, Top of Mount Heng Chen.92-30	54	2
31	Hunan Prov. Hengyang city, Half way up Mount Heng Chen.92-31	36 34	2 1
32	Hunan Prov.Hengyang city, Foot of Mount Heng Chen.92-33	36	2
33	Fujian Prov. Fuzhou city, Gushan Chen. (2-33	54	2

(1) Common type 18L (2SAT)

This type can be divided into two subtypes in respect to the morphology of of satellite chromosomes.

a. Common type A. 18L (2SAT)

Karyotype of this subtype is composed of 18 large chromosomes (L). A pair of obvious satellites were attached to the short arms of No. 1 chromosomes. Fourteen out of the fifteen populations investigated belong get to this type,

Penteploid individuals were found in Longjing of Hangzhou, Zhejiang Province and Yansi at the foot of Mount Huang. $2n=45L$ (3SAT), $2n=43L$ (5SAT).

7. Polymorphism of satellites

Morphology and location of satellite were usually stable. Fourteen cytotypes could be recognized according to morphology of satellite chromosomes. Irifune (1987) classified satellites into six types when he investigated this species in Japan. Obviouly, *Aster ageratoides* in China have amach wider range of morphological variation of satellites.

3.6.1.3 Discussion

1. Relationship between latitude, elevation and the distribution pattern of *Aster ageratoides* polyploid complex

Because of the universality of polyploidy in plant, the possibe cause and effect relationship between ploidy level and distrbution pattern has been attracting many botanists for a long time.

Studies on the relationship betwen ploidy level and distrubution pattern in angiosperms have been carried on since as early as the 1930's. It was said that frequency of polyploid populations increased from the south to the north, and polyploid individuals have advantage over their diploid ancestors in adapting to the extreme cold and draught (Hagerup, 1932; Tischler; 1935). The results of our study on the distribution pattern of Chinese *Aster ageratoides* polyploid complex, however, are not in agreement with this point of view. On the contrary, the polyploid populations show a rising tendency from the north to the south. No polyploid individual were found north of 36° N latitude. The zone between 30° and 35° N latitude is occupied mainly by tetraploids. Although there are diploid, tetraploid and hexaloid populations south of 30° N latitude, hexaploid populations are most common. This is similar to the distribution pattern of Campanulaceae, Ericaceae and Phytolaccaceae in Europe, where diploid populations of these families prevail in the north, and tetraploid populations prevail in the south. With the vertical distribution pattern of *Aster ageratoides*, we did not design special sampling. Judged from the fact that individuals sampled at the tops of Mount Huangshan (1800 m) and Mount Hengshan (1100 m) were all hexaploid, while those sampled at the foot of Mount Huangshan and haif way up the mount Hengshan were tetraploid we say that high elevations favour polyploidy.

2. The current distribution center and origion center of *Aster ageratoides*

The cytogeogaphic distribution pattern of *Aster ageratoides* show that diploid populations still occupy a vast area. But they have been shadowed by the extreme prosperity of polyploid populations. Tetraploid and hexaploid populations dominate the entire range south of 37° N latitude. So we can say that *Aster ageratoides* is a mature polyploid complex.

There are two morphologically distinguishable types of *Aster ageratoides* distributed in Mount Taishan area.

One is the spread type. Like those widely distributed in the north, individuals of this type has narrow elliptic papery leaves with sharply constricted apex and roughly dentate margins.

Thick pubescence covers the upper side of leaves, while only sparse pubescebnce cover the back side. The upper part of the large involucre is purplish brown, and the ligulate flower is violet or red. This type belongs to *Aster ageratoides* var. *ageratoides*. Its karyotype, with satellites attached to the short arms of No. 1 chromosomes, should be classified as common type A. $2n=18L$ (2SAT).

The other type is cespitose, with 10 to 20 stems growing together. They can be readily distinguished from the spread type. They also belong to *Aster ageratoides* var. *ageratoides*. but their karyotypes are quite different from the common type A. The satellites are attach to the short arms of No.5 chromosomes, a pair of relatively small chromosomes. We call this as common type B.

It is inferred that in the history of evolution, the cytotype A hybridized with the cytotype B, and then the doubling of chromosome number, gave rise to a tetraploid population of *Aster ageratoides* that had two types of satellites common in both cytotype A and B. Such tetraploid populations can still be found in the zone from Mount Laoshan of Shandong Province to Lianyungang of Jiangsu Provice. Therefore, it is reasonable tob infer that the origin center of *Aster ageratoides* could be the Mount Taishan area. The newly formed tetraploid descendents were more vigorous than their diploid oncestors and spread to new environments, dominated in a vast area south of 37° N latitude. This tetraploid population bearing satellites of both parent cytotypes hybrided with other diploid populations of the same taxon at a certain time, forming hexaploid and octoploid populations.

In the areas from Hangzhou Gulf of Zhejiang Province to Mount Huangshan area exist *Aster ageratoides* populations of different ploidy levels, as well as various variants, such as the heteromorphic satellite type, intercalary satellite type, deletion type, etc. The analysis of karyotype data shows that they are all derivatives. So we conjure that Hangzhou Gulf-Mount Huangshan zone could be the current distrebution center of *Aster ageratoides*.

Aster ageratoides populations found in Kunming area of Yunnan Province, like those found in the Mount Taishan area, also *Aster ageratoides* var. *ageratoides*, considering the classification of varieties and the result of karyotype analysis. Their karyotypes are also of the common type A. Therefore, The Kunming area is another origin center of *Aster ageratoides*. Further investigation is to needed to confirm our conclusion.

3. Analysis of the diversity of *Aster ageratoides* karyotype variation

White (1968, 1978) said, "The essence of evolution is a cytological process." This is to say that the mechanism of species evolution is chromosome variation. Although this statement is an exaggeration to some extent, the contribution of chromosome variation to speciation and evolution cannot be underestimated.therefore when we are studying the formation and migration pattern of a certain taxon, we must conduct overall investigation on the chromosome variation of every obtainable population and geographical race, in order to find out the origin center of this taxon.

Diploid, tetraploid, dixaploid population and a few triploid and pentaploid populations of *Aster ageratoides* demonstrate various chromosome structure vatiations. Among the individuals we sampled, exist at least 25 cytotypes. It is this diverse chromosome variation that determines the polymorphism and polytypes of *Aster ageratoides*. Some questions arise: Why *Aster* show so many types of chromosome variation? How did these variations take place? What is the cause of variation? Careful observation tells us that in the blooming season of *Aster ageratoides*, a few flowering plants besides *Aster ageratoides* and its related taxon are still blooming. *Aster ageratoides* flower provide the main working target for bees, which leads hybridization the primary cause of *Aster ageratoides* variation. Besides, unfavorable factors such as radiation, chemical pesticide, etc., can also raise the rate of natural mutation. We found dicentric chromosome in both diploid and tetraploid populations. This, on one hand suggests the violence of the condition cause the chromosome variation, and on the other hand, shows the sensitivity of *Aster ageratoides* to environmental changes. The following facts support our analysis. We have found primary lesions, which are seldom seen in ususl mitotic cells, in chromosomes of many *Aster ageratoides* cells. Such lesions are unstable points of chromosome or chromatids. Gaps can be seen in metaphase chromosomes with lesions. But the gaps are not breaks, and the chromosomes do not lose their components. The primary lesions provide functioning points for chromosome exchange and deletion, which could lead to break or rearrangement of chromosomes.

（Authors: Chen Ruiyang, An Zhuping, Song Wenqin, LiXiulan, Kohei Irifune Kenji Taniguchi Ryuso Tanaka., Publishedin: Proc.Sec.Sino-Jpn.Symposium Pl.Chromos.1992 Plant Chromosome Research p25-45）

3.6.2 半夏多倍体复合体及其细胞地理学研究

半夏 *Pinellia ternate*（Thum b.）Berit 为天南星科半夏属多年生草本植物（中国科学院中国植物志编辑委员会，1979）。半夏在中药生产中占有重要地位。然而，不论是自然群体，还是不同的人工栽培产区，半夏种质资源遗传分化严重，导致不同群体染色体倍性混杂，严重地影响了半夏的生产发展。因此，查清不同自然群体和不同栽培产区半夏多倍体复合体的细胞地理分布，对半夏规范化生产具有十分重要意义。关于半夏的染色体数目和倍性，国内外已有多篇报道（Fedorov，1969；顾德兴等，1991；李名旺等，1997；侯典云等，2005；Yi *et al.*，2005）但缺乏对半夏多倍体复合体的细胞地理学研究。

3.6.2.1 材料与方法

1. 材料

供试半夏 *P. ternata* 采自 17 个省市地区，由中国医学科学院药用植物研究所马小军研究员鉴定。

2. 方法

半夏块茎按株分别种于花盆内，待次生根生长 0.5～1 cm，剪下放入对二氯苯饱和水溶液中，室温下处理 3 h，甲醇-冰醋酸（3：1）固定。染色体标本制备：固定的材料先用蒸馏水冲洗，1 mol /L HCl 37℃处理 45 min，蒸馏水洗净 HCl，卡宝品红染色，常规压片（李懋学等，1996）。采用 Nikon 80i 显微镜观察，Spot RTKE 冷 CCD 照相，染色体图像由 EpsonPhoto 870 喷墨打印机打印。染色体数目按植株分别统计。染色体计数的细胞必须是完整的细胞，没有由于方法不当造成的染色体丢失。

3.6.2.2 结果与分析

1. 半夏染色体数目和倍性观察

对采自我国 17 个省市地区 33 个不同地理居群半夏的染色体数目和倍性进行了调查，结果见表 3.17 和图 3.133，3.134。来自我国 17 个省市地区的 33 个不同地理居群的半夏是一个具 $2n=91$，$2n=104$，$2n=117$，$2n=130$ 的多倍体复合体，它们是染色体基数 $x=13$ 的七倍体、八倍体、九倍体和十倍体。这 4 种多倍体的发生频率有较大差别，以七倍体（$2n=91$）和八倍体（$2n=104$）频率较高，分别为 30.0%，48.6%，九倍体（$2n=117$）占 17.3%，十倍体（$2n=130$）频率最低只有 0.6%。笔者同时在少数居群中发现有非整倍体细胞和非整倍体细胞型，但频率很低，只占观察株数的 3.5%，对这些非整倍体产生的原因和分析将另文报道。

表3.17　17 个省市 33 个不同地理居群半夏染色体数目和倍性

编号	材料来源	凭证标本	染色体观察株数	染色体数目	染色体基数	倍性	染色体图版号
1	山西新绛	马小军 2005-01	8	$2n=117$	13	$9x$	1
2	湖北潜江	马小军 2005-02	14	$2n=104$	13	$8x$	2
3	河南信阳	马小军 2005-03	2	$2n=104$	13	$8x$	3-1
			6	$2n=117$	13	$9x$	3-2
4	北京密云	马小军 2005-04	15	$2n=104$	13	$8x$	4
			2	$2n=103$			
5	山东菏泽	马小军 2005-05	13	$2n=117$	13	$9x$	5
			1	$2n=115$			
			8	$2n=91$	13	$7x$	6-1
6	四川安岳	马小军 2005-06	6	$2n=104$	13	$8x$	6-2
			1	$2n=115$			6-3
7	四川南充	马小军 2005-07	2	$2n=104$	13	$8x$	7-1
			1	$2n=117$	13	$9x$	7-2
			17	$2n=104$	13	$8x$	8-1
8	北京房山	马小军 2005-08	1	$2n=117$	13	$9x$	8-2
			2	$2n=103$			8-3
9	河北易县	马小军 2005-09	9	$2n=104$	13	$8x$	9
			1	$2n=105$			
			75	$2n=91$	13	$7x$	10-1
			18	$2n=104$	13	$8x$	10-2
10	重庆北碚	马小军 2005-10	3	$2n=117$	13	$9x$	10-3
			1	$2n=130$	13	$10x$	10-4
			2	$2n=92$			10-5

编号	材料来源	凭证标本	染色体观察株数	染色体数目	染色体基数与倍性 基数	染色体基数与倍性 倍性	染色体图版号
11	江苏南京植物园	李金岚 2005-01	12	$2n=104$	13	$8x$	11
12	安徽旌德	李金岚 2005-02	4	$2n=104$	13	$8x$	12
13	安徽石台	李金岚 2005-03	4	$2n=104$	13	$8x$	13
14	安徽鹞落坪	李金岚 2005-04	6	$2n=91$	13	$7x$	14
15	江苏南京灵谷寺	李金岚 2005-05	10	$2n=104$	13	$8x$	15
16	广西药植所	马小军 2005-11	1	$2n=104$	13	$8x$	16
17	云南昭通奕良	王荔 2005-01	1	$2n=91$	13	$7x$	17
18	云南昭通凤凰村	王荔 2005-02	1	$2n=91$	13	$7x$	18
19	云南沪西	王荔 2005-03	3	$2n=104$	13	$8x$	19-1
			8	$2n=117$	13	$9x$	19-2
20	贵州咸宁	王荔 2005-04	3	$2n=91$	13	$7x$	20-1
			3	$2n=104$	13	$8x$	20-2
21	河南唐海县	王荔 2005-05	7	$2n=104$	13	$8x$	21
			1	$2n=103$			
22	河北安国	王荔 2005-06	9	$2n=117$	13	$9x$	22
23	湖北武汉洪山区	王荔 2005-07	5	$2n=104$	13	$8x$	23-1
			1	$2n=117$	13	$9x$	23-2
			1	$2n=105$			23-3
24	陕西丹凤县	王荔 2005-08	4	$2n=104$	13	$8x$	24
25	浙江台州	王荔 2005-09	2	$2n=104$	13	$8x$	25
26	山东莒县	马小军 2005-12	4	$2n=104$	13	$8x$	26-1
			4	$2n=117$	13	$9x$	26-2
			1	$2n=130$	13	$10x$	26-3
27	宁夏泾源	马小军 2005-13	10	$2n=104$	13	$8x$	27
28	浙江庆元月边	吴鸣翔 2005-6	6	$2n=104$	13	$8x$	
29	浙江庆元平鸟	吴鸣翔 2005-7	3	$2n=104$	13	$8x$	
30	山东日照	王春国 2005-1	7	$2n=104$	13	$8x$	
			15	$2n=117$	13	$9x$	
31	福建	薛梅 2005-2	2	$2n=117$	13	$9x$	
32	浙江杭州	薛梅 2005-3	2	$2n=91$	13	$7x$	
			2	$2n=130$	13	$10x$	
33	四川成都	薛梅 2005-4	2	$2n=104$	13	$8x$	

2. 半夏多倍体复合体的细胞地理分布

在所调查的 17 个省市地区中只有少数地区的半夏为单一倍性，如山西 $2n=117$，陕西 $2n=104$，宁夏 $2n=104$；多数地区同时具有 2 个倍性，如湖北、河南、河北有 $2n=104$，

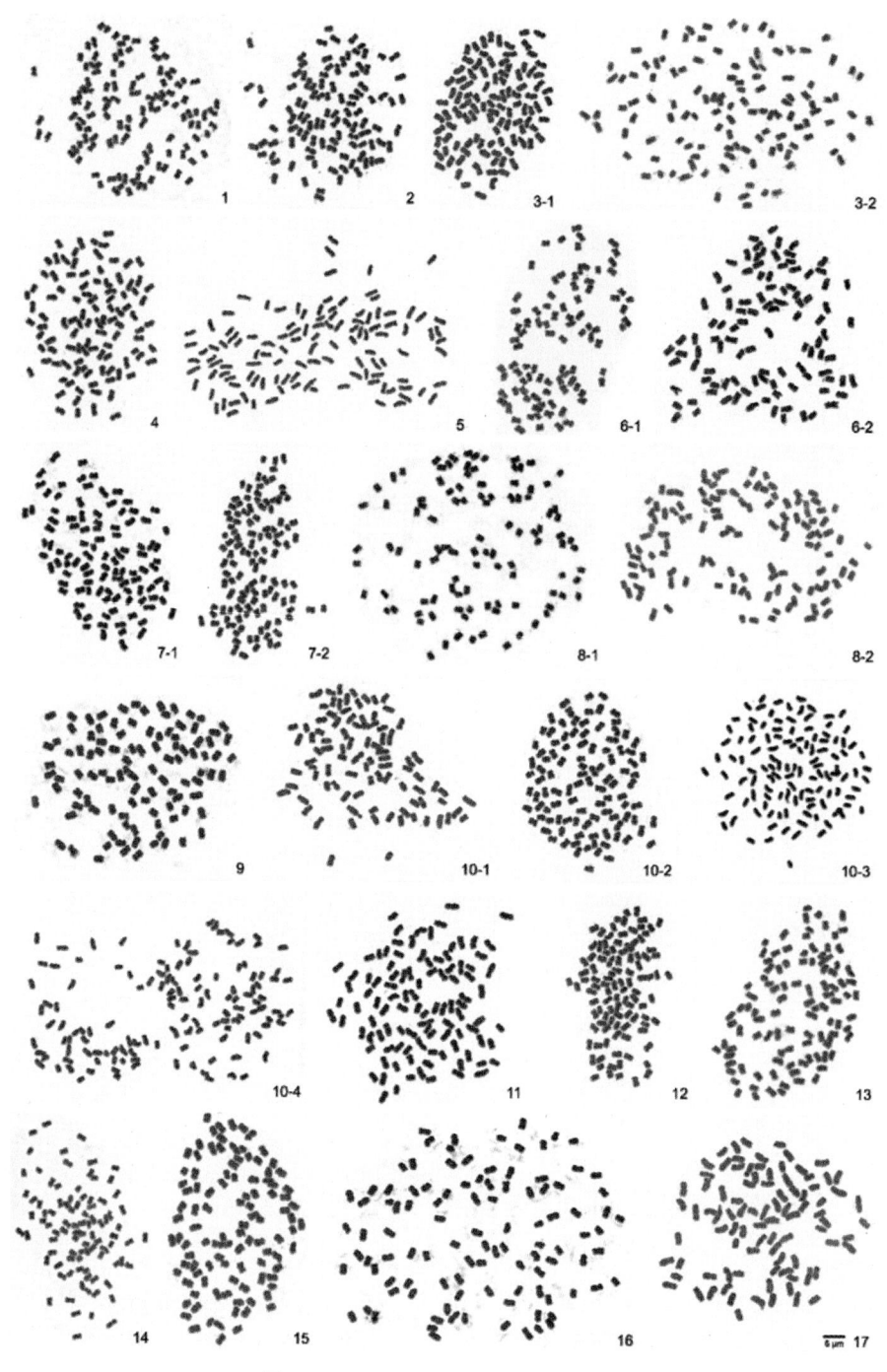

图 3.133 不同地理居群半夏染色体

117；山东同时具有 3 个倍性 2n=104，117，130；四川和重庆同时具有 4 个倍性 2n=91，104，117，130。因此，四川省包括重庆地区可能是半夏多倍体复合体的多样化中心。在 27 个不同地理居群中，多数居群只具 1 个倍性，如安徽旌德 2n=104，石台 2n=104，鹞落坪 2n=91，湖北 2n=104，宁夏 2n=104，河北安国 2n=117，南京灵谷寺 2n=104，南京植物园 2n=104，部分居群具 2 个倍性，如武汉洪山区 2n=104，117，河南 2n=104，117，北京房山 2n=104，117，少数居群具 3 个或 4 个倍性，如山东莒县具 3 个倍性 2n=104，117，130，重庆具 4 个倍性 2n=91，104，117，130。当然这种倍性的不同分布可能与调查的株数有关，随着调查群体的扩大，也许不同倍性也会增加。但有一点是可以肯定的，就是中国半夏不同倍性资源是非常丰富的，有巨大的开发潜力。

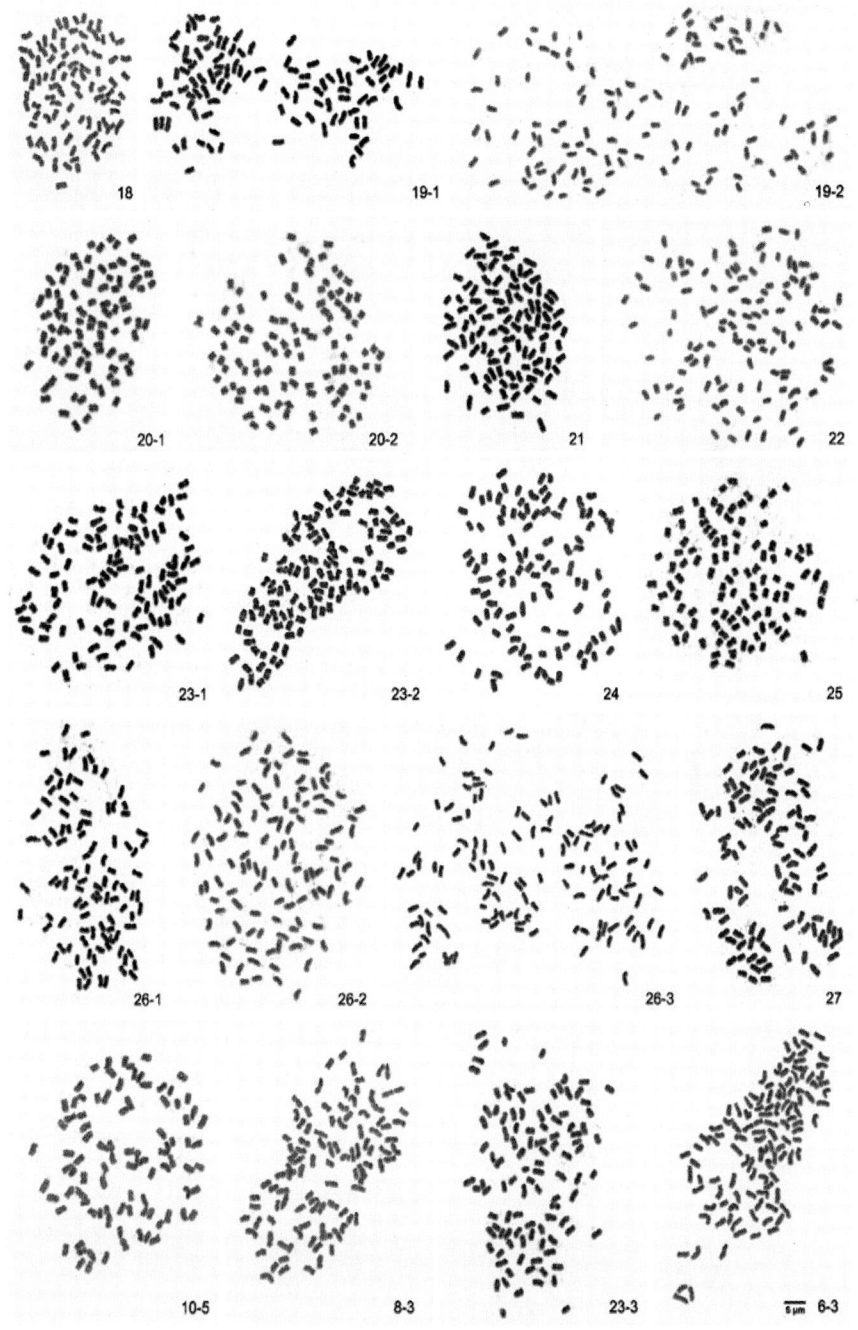

图 3.134 不同地理居群半夏染色体

3.6.2.3 讨论

1. 关于半夏的染色体数目

国外报道半夏染色体数目 2n =28，116，128（Fedorov，1969），近年国内顾德兴等（1991）报道 2n=72，李名旺等（1997）2n=54，99，108，侯典云等（2005）报道有 2n=52，54，64，66，68，72，74，75，76，78，80，84，88，90，92，96，98，100，104，108，110，112，116 等 24 个数目。植物染色体虽有其自身的一些特殊之处，但无论如何，一种正常的植物种群，它的染色体数目也不会有 24 种变异，因此，目前关于半夏染色体数目的不一致性，不能排除是由于方法不当造成的人为错误。本研究目前虽然只观察了我国 17 个省市地区的 33 个不同地理居群的 313 株半夏的染色体数目，但所有居群的染色体数目都非常稳定。最近 Yi（2005）除证实半夏 2n=91，104，117 外，同时还发现了半夏六倍体（2n =78）居群，这显示还可能存在较低的倍性---二倍体和四倍体。Yi（2005）将鹞落坪半夏 *P. yaoluopingensis* 2n =26 归为二倍体种（有待证实）。至今尚没有发现半夏的四倍体，如果被证实，半夏多倍体复合体应是一个具有 $2x$，$4x$，$6x$，$7x$，$8x$，$9x$，$10x$ 的成熟多倍体复合体（洪德元 1990）。不过，在这个多倍体复合体中最常见的不是四倍体，而是七倍体（30%）和八倍体（48.6%），因此，半夏是至今发现倍性最多的多倍体种群之一，是研究多倍体复合体形成和演化的一个较理想材料，对于分析进化、系统发育和植物地理问题具有特别价值。

2. 关于半夏的染色体基数与倍性

Grayum（1990）和 Grant（1981）认为半夏的染色体基数为 $x=13$。李名旺等（1997）认为还有其他基数 $x=7$，8，9，29。侯典云等（2005）认为 10，23 也是半夏的基数。但本研究观察的结果，半夏不同地理居群的染色体数目都是比较稳定的，除少数非整倍性外，均为 2n=91，104，117，130，是 13 的七、八、九、十倍体，Yi（2005）也证实半夏的染色体基数是 $x=13$，因此，支持 Grayum 等人的观点，半夏的染色体基数应为 $x=13$。

3. 关于半夏多倍体复合体的现代分布格局

我国半夏除内蒙古、新疆、青海、西藏未见野生外，其余各省区均有分布（江年琼，2001）。Yi 等（2005）虽然首次提出了半夏多倍体复合体的细胞地理分布，但只有六、七、八、九 4 个倍性，而且九倍体（2n=117）只在日本材料中发现，没有发现中国半夏的九倍体，且在一个居群内只发现了一个倍性，表明调查的群体不够大。本实验虽然首次发现了十倍体（2n=130）和一些非整倍体系列，但没有发现四倍体和六倍体，因此有关我国半夏多倍体复合体的现代分布格局尚需扩大不同地理居群的细胞地理调查。

（作者：陈成彬、马小军、陈力、薛梅、成玉、宋文芹、李秀兰、陈瑞阳。

原载：中国中药杂志，2006，31（17）1405-1408）

3.6.3 芦苇及其近缘种种内多倍体的染色体研究

At least four species of *Phragmites* are found in East Asia. They show extreme morphological diversity; for instance, *P. australis* (Cav.) Trin. ex Steud. has an extremely gigantic plant 3 m high or more while it has an extremely dwarf plant 20-30 cm high. Moreover, this species has variability in leaf size, inflorescence color and shape, and so on. Although the majority of *P. australis* tends to live in both salt and fresh water or moist condition, the minority can live in other condition such as in arid. This species is widely distributed in the temperate zone of the world. It is known that this species growing in Europe has an intraspecific aneuploid series with $2n$=42, 44, 46, 48, 49, 50, 51, 52, 54, 72, 96 and so on (Gorenflot *et al.*, 1975, 1984, Gorenflot and Panahi 1979, and so on). However, chromosome counts in East Asiatic *Ohragmites* including *P. australis* have been less described.

3.6.3.1 Materials and Methods

Sample plants of *Phragmites australis* (Cav.) Trin. ex Steud. (=*P. communis* Trin.), *P. karka* (Retz.) Steud., and *P. japonica* Steud. for chromosome counts were collected in 25 populations of 20 localities in Japan and the People's Republic of China. They were cultivated in Laboratory of Cytogenetics, Department of Biology, Nankai University. The herbarium voucher specimens of those species were deposited in the Herbarium, Department of Biology, Nankai University.

Root tips and shoot buds were harvested and pretreated in 0.2% colchicine for 2-3 hours before they were fixed in acetic-methanol mixture (1:3) for approximately two hours. Their meristematic tissues were scribbled on pre-cooled slides and were then dried on flame and stained with Giemsa.

3.6.3.2 Results and Discussion

Table 3.18 and Fig. 3.135 show the results of chromosome counts in the three species of *Phragmites*. Figures are of the chromosome representatives of the tetraploid, hexaploid and octoploid *P. australis*. The majority of the chromosomes of those representatives was submedian- to subterminal-centromeric. The tetraploid and the hexaploid chromosome complements showed sat-chromosomes, but the octoploid chromosome complement did not show any sat-chromosome. It is obvious that the octoploid *P. australis* is the most abundant along the coast in China from ca 25-42° N L. In contrast, the tetraploid *P. australis* is found ca 40° N L, while the hexaploid *P. australis* is found only in a small area in Changli. Hebei Province, China. *Phragmites japonica* and *P. karka* are tetraploid, and the former species grows rather north in East Asia and the latter species grows widely in Southeast Asia down to Australasia (Ohwi, 1965). These documents of chromosome numbers confirmed Gurzenkov (1973), Olorode (1975), Kalia (1978). However, no diploid species has been reported in *Phragmites* which might be a natural aneuploid series based on a polyploid complex.

Table 3.18 Lacalities and chromosome numbers of three species of *Phragmites australis* in China and Japan

Species	Locality	Chromosome Number (2n)
P. australis (Cav) Trin.exSteud. (=*P.communis* Trin.)	China: Liaoning, Shenyang	96
	China: Liaoning Yingkou	96
	China: Liaoning, xingcheng	96
	China: Hebei, Beidaihe	96
	China: Hebei, Changli	72
	China: Hebei, Changli	96
	China: Hebei, Luanxian	96
	China: Hebei, Qianxi	96
	China: Hebei, Guye	96
	China: Tianjin, Lutai	48
	China: Tianjin, Tanggu	48
	China: Tianjin City	96
	China: BeijingCity	48
	China: Inner Mongolia, Lahe, Wuliangsuhai	48
	China: Shandong, Tsingtao	96
	China: Jiangsu, Xuzhou	96
	China: Hunan, Yueyang	96
P.karka (Retz.) Steud.		48
P.japonica Steud.	Japan: Yamaguchi, Ogoori-cho	48
	Japan: Hiroshima City	48

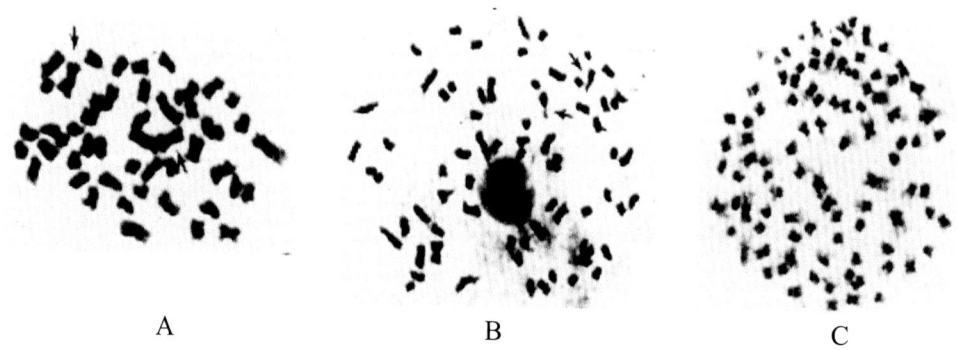

Fig.3.135 Intraspecific polyploidy of *Phragmites australis*
A. 2*n*=48, B. 2*n*=72, C. 2*n*=96

    The intraspecific polyploid-complex of *Phragmites australis* is concentrated in Beijing-Tianjin-Tangshan area in China. They perform great morphological diversity such as presence or absence of auricle hairs, presence or absence of either black or white ring or either white-powdered or no powdered ring on stem-node, variable color, shape and size of inflorescences and stem texture, and so on.

Many more cytotaxonomic studies are necessary to clarify our concept of intra- and inter-specific relationships of *Phragmites australis* and its closely related species in Asia.

An intraspecific polyploid series was found in *Phragmites australis* in China with tetraploid（2*n*=48）, hexaploid（2*n*=72）and octoploid（2*n*=96）. The most common plants of this species were of octoploid. Both Phragmites japonica and P. karka were of tetraploid.

(Authors: Weixing Zong, Ruiyang Chen, Kenji Taniguchi and Katsuhiko Kondo. Published in: La Kromosomo II -63-64:2168-2172)

参 考 文 献

别墅, 王坤波, 王春英, 等. 二倍体栽培棉 45S rDNA-FISH 作图及核型比较. 棉花学报, 2004, 16(4): 223-228.
陈成彬, 王春国, 宋文芹. 一种 45SrDNA 在植物染色体上的荧光原位杂交方法: CN103409524 B[P]. 2013a.
陈成彬, 王春国, 宋文芹. 一种 5SrDNA 在植物染色体上的荧光原位杂交方法: CN103409523 B[P]. 2013b.
陈德昭, 陈邦余, 方云忆, 等. 中国植物志(第 40 卷). 北京: 科学出版社, 1994. 64-231.
陈瑞阳, 宋文芹, 李秀兰, 等, 中国主要经济植物基因组染色体图谱（第二册）中国农作物及野生近缘种染色体图谱. 北京: 科学出版社, 2003.
陈瑞阳, 宋文芹, 李秀兰, 等. 中国主要经济植物基因组染色体图谱（第三册）. 北京: 科学出版社, 2003. 281-310.
陈瑞阳, 宋文芹, 李秀兰. 植物有丝分裂染色体标本制作的新方法. 植物学报, 1979, 21(3): 297-298.
陈成彬, 张守攻, 李秀兰, 等. 杨属派间核型比较研究. 广西植物, 2005, 25（4）: 338-340
陈瑞阳, 宋文芹, 李秀兰. 关于水稻(*Oryza sativa*)染色体组型分析的研究. 遗传学报, 1980, 7: 361-365.
苍田のり著, 张景六译. 水稻的核型分析法和染色体鉴定的研究. 细胞生物学杂志(增刊), 1982, 1-34.
邓汉湘, 何小轩, 戴和平, 等. 人类染色体显微切割、微克隆与染色体区带特异性绘画技术. 自然科学进展-国家重点实验室通讯, 1992, 4: 361-367.
刁英, 陈思, 黄雨蝶, 等. 蕹菜的 DAPI 显带核型. 氨基酸和生物资源, 2005, 1: 32-34.
郭歌, 陈成彬, 宋文芹, 等. 黑麦 B 染色体端粒相关序列的克隆和序列分析. 植物学报, 1998, 40（2）1123-1128.
顾德兴, 徐炳声. 南京两种半夏群体水平的变异式样. 植物分类学报, 1991, 29(5): 423-430.
郝剑瑾, 程舟, 梁洪卉, 等. 基于 rDNA ITS 序列探讨我国冬虫夏草的遗传分化及分布格局. 中草药, 2009, 40(1): 112-116.
侯典云, 王荔, 杨艳琼, 等. 半夏不同居群的细胞学研究. 云南农业大学学报, 2005, 20 (2): 159-162.
洪德元. 植物细胞分类学. 北京: 科学出版社, 1990.
金危危, 凌定厚. 转基因水稻中外源基因的荧光原位杂交(FISH)分析. 实验生物学报, 2001, 34(3): 163-168.
江年琼. 药用动植物种养加工技术—半夏天南星, 北京: 中国中医药出版社, 2000.
兰添颖, 刘博, 董凤平, 等. 菠菜 rDNA 及端粒多色荧光原位杂交分析. 遗传, 2007, 29（11）: 1405-1408.
李琦马璐, 黄婧, 等. 西瓜、苦瓜与罗汉果染色体的 rDNA 定位及其核型分析. 武汉大学学报理学版, 2007, 53（4）: 449-456.
李艳燕. 紫菜种内 rDNA 非编码区序列及系统发育分析. 苏州: 苏州大学, 2009.

李懋学, 龙雅宜, 龚维忠. 岷江百合 B 染色体的初步观察. 植物学报, 1984, 26（2）: 151-155.
李懋学, 张赞平. 作物染色体及其研究技术, 北京: 中国农业出版社, 1996: 32.
李秀兰, 宋文芹, 许文胜, 等. 一种改进的染色体显微切割方法. 南开大学学报, 1998, 31（2）: 102-105.
李艳燕. 紫菜种内 rDNA 非编码区序列及系统发育分析. 苏州: 苏州大学, 2009.
李真, 陈素梅, 陈发棣, 等. 45S rDNA 在蒿属 5 种植物染色体中的分布. 园艺学报, 2011, 38（2）: 353-360.
李林初, 沈敏健, 李懋学. 辣椒的核型分析. 园艺学报 1984, 11（2）: 119-122.
李名旺, 顾德兴, 刘有良, 等. 半夏属的染色体数目、倍性与株芽发生的关系, 植物分类学报, 1997, 35(3): 208-214.
刘博, 陈成彬, 李秀兰, 等. 豆科三属八种植物的核型及 rDNA 定位研究. 云南植物研究, 2005, 27(3): 261-268.
卢圣栋主编. 现代分子生物学实验技术. 北京: 高等教育出版社, 1993.
宁顺斌, 宋运淳. 玉米中抗病基因 $myb1$ 和 $NDR1$ 同源序列的荧光原位杂交物理定位. 植物学报(英文版), 2000, 42(6): 605-610.
齐力旺, 张守攻, 韩素英, 等. 杨属青杨组种（品种）间核型比较. 云南植物研究, 2004, 26（5）: 537-542.
齐力旺, 张守攻, 韩素英, 等. 杨属（*Populus*）白杨组（Section Populus）种（品种）间核型比较. 园艺学报, 2005, 32(5): 849-853.
任南, 宋运淳, 毕学知, 等. 玉米 $cyclin$ III 基因的染色体原位杂交物理定位. 遗传, 1997, 19(5): 1-4.
阮国庆, 邓汉湘, 潘乾, 等. 荧光原位杂交法定位人类染色体 8q24. 1 单拷贝探针. 自然科学进展-国家重点实验室通讯, 1996, 6(2): 243-247.
萨姆布鲁克 J, 弗里奇 EF 著. 金冬雁, 黎孟枫译. 分子克隆实验指南(第二版). 北京: 科学出版社, 1989.
宋文芹, 崔香芹, 李秀兰, 等. 蚕豆大 M 染色体长臂端部的显微切割与 PCR 扩增. 科学通报, 1996, 41(4): 361-363.
宋文芹, 李秀兰, 许文胜, 等. 黑麦染色体的显微分离与 PCR 扩增. 植物学报, 1998, 40（2）158-162.
覃瑞, 马骞, 王德彬, 等. 利用 GISH 和 C0t_1 DNA-FISH 对稻属 B, C, G 基因组的比较分析. 中南民族大学学报(自然科学版), 2009, 25(2): 31-35.
夏家辉, 杨毅, 戴和平, 等. 14 个染色体区带特异性探针池的构建. 遗传学报, 1994, 21（4）253-256.
徐吉臣, 朱立煌. 用双单倍体群体构建水稻的分子连锁图. 遗传学报, 1994, 21(3): 205-214.
徐延浩, 杨飞, 程有林, 等. 45S rDNA 和 5S rDNA 在南瓜、丝瓜和冬瓜染色体上的比较定位. 遗传, 2007, 29(5): 614-620.
王世绩. 世界胡杨林的现状及保护和恢复对策. 世界林业研究, 1996. 6: 37-44.
王玉元. 染色体遗传中的一个不解之迷 B 染色体. 武汉植物学研究, 1997, 15(1): 73-79.
赵萌, 陈冰, 王璐, 等. 多重荧光原位杂交技术体系的建立及其在检测白血病复杂核型异常中的应用. 中华医学遗传学杂志, 2002, 19(5): 375-378.
郑国锠, 聂秀菀, 杨庆兰, 等. 细胞融合(cytomixis)的光学与电子显微镜观察及其与变异和进化的关系探讨. 植物学报, 1975, 17: 60-69.
中国科学院中国植物志编辑委员会, 中国植物志. 第 13 卷第 2 分册. 北京: 科学出版社, 1979: 200.
郑康乐, 黄宁. 标记辅助选择在水稻改良中的应用前景. 遗传, 1997, 19 (2): 40-44.
郑康乐. 水稻基因组研究的新进展. 水稻文摘, 1992, 11(3): 1-6.
张荣信, 陈成彬, 李秀兰, 等. 黑麦 A、B 染色体着丝粒区同源性的荧光原位杂交分析. 科学通报, 1999, 44(5): 520-524.
Aarts M G M, Lintel Hekkert, E B Holub, J L Beynon, W J Stiekema, A Pereira. Identification of R-gene homologous DNA fragments genetically linked to disease resistance loci in *Arabidopsis thaliana*. Mol Plant-Microbe Interact. 1998, 11: 251–258.

Albani D. PCR amplification of microdissected wheat chromosome armsin a simple single tube reaction. Plant J, 1993, 4: 899-903.

Alfenito M R, Birchler J. A Molecular characterization of a maize B chromosome centric sequence. Genetics. 1993, 135: 589-597.

Altschul S F, Madden T L, Schaffer A A, Zhang J, Zhang Z, Miller W, Lipman DJ. Gapped BLAST and PSI-BLAST: a new generation of protein database search programs. Nucleic Acids Res. 1997, 25: 3389–3402.

Anderson L G, Kyhos D E, Morsouin T, et al. Chromosome numbers in Compositae. 1A Haplopappus and other Astereae. Amer, J Bot. 1974, 61, 665-671.

Anilakumar KR, Pal A, Khanum F, Bawa AS. Nutritional, medicinal and industrial uses of sesame (sesamum indicum L.) seeds anoverview. Agriculturae Conspectus Scientificus, 2010，75, 159 -168.

Antonacci R, Marzell R, Finelli P, Lonoce A, Forabosco A, Archidiacono N and Rocchi M. A panel of subchromosomal painting libraries repre-senting over regions of the human genome. Cytogenet Cell Genet. 1995, 68: 25–32.

Arumuganathan K, Earle ED. Nuclear DNA content of some important plant species. Plant Molecular Biology Reporter, 1991, 9, 208-218.

Arumuganathan K, Martin GB, Telenius H, Tanksley SD, Earle ED. Chromosome specific DNA clones from flow-sorted chromosomes of tomato. Mol Gen Genet. 1994, 242: 551-558.

Arumuganathan K, Martin GB, Telenius H, et al. Chromosome 22 specific DNA clones. Molecular characterization of a maize B chromosome centric sequence. Genetics. 1994, 135, 589-597.

Armstrong S, Frnasz P, Marshall, DF and Jones GH. Physical mapping of DNA repetitive sequences to mitotic and meiotic chromosomes of Brassica oleracea var. alboglabra by fluorescence in situ hybridization. Heredity, 1998, 81, 666-673.

Arnheim N, Krystal M, Schmickel R, et al. Molecular evidence for genetic exchanges among ribosomal genes on non homologous chromosomes in man and ape. Proc Natl Acad Sci USA, 1980, 77: 7323-7327.

Ashri A. Sesame breeding. Plant Breeding Reviews, John Wi ley & Sons, New York, USA, 1998，16, 179-228.

Ayala JF, Fitch MW, Clegg TM. Variation and Evolution in Plants and icroorganisms: Toward a New Synthesis 50 Years after Stebbins. National Academy Press, 2000, 211-234.

Ausubel FM, Brent RB, Kingston RE, Moore DD, Seidman JD, Smith JA, et al. Introduction of plasmid DNA into cells. In: Ausubel FM, Brent RB, Kingston RE, Moore DD, Seidman JD, Smith JA et al. eds. Short P rotocols in Molecular Biology, 3rd edn. John Wiley &Sons, Inc., USA. 1992, pp. 1-26.

Baldi P, Patocchi A, Zini E, Toller C, Velasco R, Komjanc M. Cloning and linkage mapping of resistance gene homologues in apple. Theor Appl Genet. 2004, 109: 231-239.

Battaglia E. Cytogenetics of the B chromosomes. Caryologia, 1964, 17, 245-254.

Bedigian D. Harlan J. R. Evidence for cultivation of sesame in the ancient world. Economic Botany, 1986，40, 137-154.

Beech RN, Strobeck C. Structure of the intergenic spacer region from the ribosomal RNA gene family of white spruce (Picea glauca). Plant Mol Biol, 1993, 22(5): 887-892.

Benabdelmouna A, Abirached-Darmency M, Distribution and chromosomal organization of 18S-5. 8S-25S and 5S rDNA in Petunia species. Agronomie, 1997. 17: 349-360.

Bennett MD, and Smith JB. Nuclear DNA amounts in angiosperms. Phil. Trans. R. Soc. London Ser. B, 1976, 274: 227-274.

Berghorn KA, Bonnett JH, Hoffman GE. cFos immunoreactivity is enhanced with biotin amplification. J. Histochem. Cytochem. 1994, 42, 1635–1642.

Berlani RE, Walbot V, Davis RW. Sequence analysis of three fragments of maize nuclear DNA which replicate au-tonomously in yeast. Plant MolBiol, 1988, 11: 173-182.

Beukeboom LW. Bewilding Bs: an impression of the 1st B-chromosome conference. Heredity, 1994, 73:

328-336.

Bie Shu, Wang Kun-Bo, Wang Chun-Ying, Song Guo-Li, Kong Fan-Ling, Liu Fang, Liu San Hong, LiShao-Hui, Zhang Xiang-Di, Wang Yu-Hong. Studies on 45S rDNA-FISH karyotype of *Gossypium herbaceum and Gossypium arboreum*. Cotton Science, 2004, 16(4): 223-228.

Biessmann H, Mason J M. Telomeric repeat sequences. Chromosoma, 1994, 103: 154-161.

Blackburn EH. Structure and function of telomeres. Nature, 1991, 350: 569-573.

Blackman RL, Spence JM. Ribosomal DNA is frequently concentrated on only one X chromosome in permanently apomictic aphids, but this does not inhibit male determination. Chromosome Res. 1996, 4: 314-320.

Blunden R, Wilkes TJ, Forster JW, Jimenez MM, Sandery MJ, Karp A etal. Identification of the E3900 family, a second family of rye B chromosome specific repeated sequences. Genome. 1993, 36, 706-711.

Bohlander, Stefan K, Rafacl E, et al. A mothod for the rapid sequence independent amplification of microdissection chromosomal material. Genomics. 1992, 13(4): 1322-1324.

Bradshaw HD, Stettler RF, Molecular genetics of growth and development in *Populus*. IV. Mapping QTLs with large effects on growth, form, and phenology traits in a forest tree. Genetics, 1995. 139: 963-973.

Bradshaw HD, Ceulemans R, Davis J, Stettler R. Emerging model systems in plant biology: poplar (*Populus*) as a model forest tree. J Plant Growth Regul. 2000, 19: 306-313.

Brandriff B, Gordon L, Trask B. A new system for high-resolution DNA sequence mappiinterphase pronuclei. Genomics, 1991, 10(1): 75-82.

Brown GR, Amarasinghe V, Kiss G, Carlson JE. Preliminary karyotype and chromosomal localization of ribosomal DNA sites in white spruce using fluorescence in situ hybridization. Genome. 1993, 36: 310-316.

Brown GR, Carlson JE. Molecular cytogenetics of the genes encoding 18S-5. 8S-26S rRNA and 5S rRNA in two species of spruce (*Picea*). Theor Appl Genet. 1997, 95: 1-9.

Brown GR, Newton CH, Carlson JE. Organization and distribution of a *Sau*3A tandem repeated DNA sequence in Picea (Pinaceae) species. Genome. 1998, 41: 560-565.

Brown SDM, Greenfield AJ, A model to describe the size distribution of mammalian genomic fragments recovered by microcloning. Gene. 1987, 55: 327-332.

Brunner AM, Busov VB, Strauss SH. Poplar genome sequence: functional genomics in an ecologically dominant plant species. Trends Plant Sci. 2004, 9: 49-56.

Burr B, Burr FA, Matz EC, Romero-Severson J. Pinning down loose ends: Mapping telomeres and factors affecting their length. Plant Cell, 1992, 4: 953-960.

Cannon SB, Zhu H, Baumgarten AM, Spangler R, May G, Cook DR, Young ND. Diversity, distribution, and ancient taxonomic relationships within the TIR and non-TIR NBSLRR resistance gene subfamilies. J Mol Evol. 2002, 54: 548-562.

Causse MA, Fuiton TM, Chao YG. et al. Saturated molecular map of the rice genome based on an interspecific backcross population. Genetics. 1994, 138: 1251-1274

Chalhoub B, Denoeud F, Liu S, Parkin I, Tang H, Wang X, Correa M. Early allopolyploid evolutionin the post-Neolithic *Brassica napusoil* seed genome. Science. 2014, 345, 950-953.

Chen JF, Staub JE, Jiang J, Adelberg JW. Physical mapping of 45S rRNA genes in Cucumis species by fluorescence in situ hybridization. Canadian Journal of Botany. 1999, 77, 389-393.

Chen CB, Qi LW, Zhang SG. et al. The karyptype analysis of triploid *poplar*. J Wuhan Bot Res, 2004, 22: 565-567.

ChenCB, Zhang SG, Li XL et al. A comparative study on the karyotypes among sections of *Populus* Guihaia, 2005. 25: 338-340

Chen QF, Armstrong KC. Characterization of a library from a single microdissected oat (*Avena sativa* L.) chromosome. Genome. 1995, 38: 706-714.

Chen RY, Song WQ, Li XL et al. A new method of plant mitosis chromosome spread. Acta Bot Sin. 1979, 21: 297-298.

Chen RY, Song WQ, Li XL. Wall degradation hypotonic method of preparing chromosome sample in plant and its significance in the cytogenetics. Acat Gent Sin. 1982, 9: 151-159.

Chen, R. Y., Song, W. Q. and Chen, X. The research of Giemsa C-banding technique used in plant chromosome banding pattern. Acta. Bot. Sin. 1979, 21: 11-18.

Chen RY, An ZP, Song WQ, Li XL, Su JY. A preliminary study on the G-bands of chromosomes in some plants. J Wuhan Bot Res 1986; 4: 111-117.

Chen RY, An ZP, Song WQ, Li XL, Su JY, Zheng JY. A study of the high-resolution G-banding technique of plant chromosomes. Acat Bot Sin. 1987, 29: 341-346.

Chen RY. G-band staining of chromosomes in some higher plants with special reference to *Lilium davidii*. La Kromosoma. 1988, II- 50: 1635-1651.

Chen RY, Song WQ, Li XL. Study on the sex chromosomes of *Ginkgo biloba*. Plant Chromosome Research, 1987, 381-386.

Cheng Z. K, Yan HH, Dang BY, *et al.* Applications of isotetrasome from the short arm of rice chromosome 5 in microdissection. Chinese Sci Bull. 1998, 43: 272-276.

Cheng YM, Lin BY. Cloning and characterization of maize B chromosome sequences derived from microdissection. Genetics. 2003, 164: 299-310.

Cheng YM, Lin BY. Molecular organization of large fragments in the maize B chromosome: indication of a novel repeat. Genetics. 2004, 166: 1947-1961.

Chilton MD, McCarthy BJ. DNA from maize with and without B-chromosomes: a comparative study. Genetics. 1973, 74: 605-614.

Choi HW, Kim JS, Lee SH. *et al.* Physical mapping by FISH and GISH of rDNA loci and discrimination of genomes A and B in *Scilla scilloides*, complex distributed in Korea. Journal of Plant Biology. 2008, 51(6): 408-412.

Collins NC, Webb CA, Seah S, Ellis JG, Hulbert SH, Pryor A. The isolation and mapping of disease resistance gene analogs in maize. Mol Plant Microbe Interact. 1998, 11: 968-978.

Cooley MB, Pathirana S, Wu HJ, Kachroo P, Klessig DF. Members of the Arabidopsis HRT/RPP8 family of resistance genes confer resistance to both viral and oomycete pathogens. Plant Cell. 2000, 12: 663-676.

Crane PR, Friis EM, Pedersen KR. Theoriginandearly diversification of angiosperms. Nature. 1995, 374: 27-33.

Cuadrado A, Jouve N. Highly repeatitive sequences in B-chromosomes of *Secale cereale* revealed by fuorescence in situ hybridisation. Genome. 1994, 37: 707-712.

Comings DE, Avelino E, Okada TA, Wyandt HE. The mechanism of C-and G-banding of human chromosomes. Exp Cell Res. 1973, 77: 469-493.

Cox AV, Bennett ST, Parokonny AS, Kenton A, Callimassia MA, Bennett MD. Comparison of plant telomere locations using a PCR generated synthetic probe. Ann Bot. 1993, 72(3): 239-247.

Dangl JL, Jones DG. Plant pathogens and integrated defence responses to infection. Nature. 2001, 411: 826-833.

Dagher-Kharrat MB, Grenier G, Bariteau M, Brown S, Siljak-Yakovlev S, Savoure A. Karyotype analysis reveals interspecific differentiation in the genus *Cedrus* despite genome size and base composition constancy. Theor Appl Genet. 2001, 103: 846-854.

Danilova TV, Birchler JA. Integrated cytogenetic map of mitotic metaphase chromosome 9 of maize: resoluti on, sensitivity, and banding paint development. Chromosoma. 2008, 117: 345-356.

Darvey, NL. Identification for rye chromosomes in wheat-rye addition lines and Triticale by heterochromomatin bands. Crop Sci. 1975, 15: 239-243.

Decroocq V, Foulongne M, Lambert P, LeGall O, Mantin C, Pascal T, Schurdi-Levraud V, Kervella J. Analogues of virus resistance genes map to QTLs for resistance to sharka disease in *Prunus davidiana*. MGG Mol Genet Genomics. 2005, 272: 680-689.

Deng HX, He XX, Li LY, Xia J-H. Human high resolution chromosomes microdissection, PCR amplification in vitro andmicrocloning technique. Chin Medical J. 1991, 71: 84-86.

Deng HX, He XX, Dai HP, Li LY, Xia JH. Human chromosomes microdissection, microcloning and chromosome-band-specific painting technique. Prog Nat Sci Nation Key Lab Commun. 1992, 4: 361-367.

Deng, HX., Yoshiura K, Dirks RW, Harada N, Hirota T, Tsukamoto K, Jinno Y. et al. PCR amplification of microdissected wheat chromosome arms in a simple "single tube"reaction. Plant J. 1993, 4: 899-903.

Deng, H. X., Yoshiura K., Dirks RW, Harada N, Hirota T, Tsukamoto K, Jinno Y, Niikawa N. Chromosome-band-specific painting: chromosome in situ suppression hybridization using PCR products from a microdissected chromosome band as a probe pool. Human Genetics. 1992, 89(1): 13-17.

Digaspero G, Cipriani G. Nucleotide binding site leucine-richrepeats, Pto-like and receptor-like kinases related to disease resistance in grapevine. MGG Mol Genet Genomics. 2003, 269: 612–623.

Di MG, Perucatti A, Floriot S, et al. An advanced sheep (*Ovis aries*, 2n = 54) cytogenetic map and assignment of 88 new autosomal loci by fluorescence *in situ* hybridization and R-banding. Animal Genetics. 2010, 38(3): 233-240.

Dixon MSK, Jones DA, Keddie JS, Thomas CM, Harrison K, Jones JD. The tomato Cf-2 disease resistance locus c omprises two functionalgene sencoding leucine-rich repeat proteins. Cell. 1996, 84: 451- 459.

Donald TM, Pellerone F, Adam-Blondon AF, Bouquet A, Thomas MR, Dry IB. Identification of resistance gene analogs linked to apowdery mildew resistance locus in grapevine. Theor Appl Genet. 2002, 104: 610–618.

Dong F, Song J, Naess SK., Hel geson JP., Gebhardt C, Jiang J. Development and applications of a set of chromosome specific cytogenetic DNA markers in potato. Theor etical and Applied Genetics. 2000, 101, 1001-1007.

Dong, FP, Ying, HS, Gong, ZS, et al. Physical mapping of 25S rDNA on metaphase chromoso mes of *populus* (Salicaceae) in five sections by fluorescence in situ Hybridization. Acta Bota nica Yunnanica. 2007, 29(4): 423- 428.

Doudrick RL, Heslop-Harrison JS, Nelson CD, Schmidt T, Nance WL, Schwarzacher T. Karyotype of slash pine (*Pinus elliottii* var. *elliottii*) using patterns of fluorescence in situ hybridization and fluorochrome banding. J Hered. 1995, 86: 289–296.

Drewry A. G-banded chromosomes in *Pinus resinosa*. Hered. 1982, **73**: 305–6.

Drewry A. The G-banded karyotype of *Pinus resinosa*. Silvae Genet. 1988, **37**: 218-221.

Eckenwalder JE, Systematics and evolution of Populus. In Biology of *Populus* and its Implications for Management and Conservation RF, Stettler HD, Bradshaw and PE. Jr. Heilmaneds. Ottawa: NRC Research Press, 1996, 7-32.

Edstrom JE, Kaiser R, Rohme D, Microcloning of mammalian metaphase chromosomes. Method Enzymol. 1987, 151: 503-516.

Escalante A, Imanishi S, Hossain M, *et al.* RFLP analysis and genomic *in situ* hybridization (GISH) in somatic hybrids and their progeny between *Lycopersicon esculentum* and *Solanum lycopersicoides*. Theor Appl Genet. 1998, 96(6-7): 719-726.

Farjon A, Pinaceae. Koeltz Scientific, Konigstein. 2001 World checklist and bibliography of conifers. 2ndcd. 1990.

Federff N. On spacers. Cell. 1979, 16: 697-710.

Feng J, Liu Z, Cai X, Jan CC. Toward a molecular cytogenetic map for cultivated sunflower (*Helianthus annuus* L.) by landed BAC/BIBAC clones. Genes, Genomes, Genetics. 2013, 3: 31-40.

Fedorov A. Chromosome numbers of flowering plants Leningrad. Academ of Sciendes USSR, 1969.

Felsenstein J. Confidence limits on phylogenesis: an approach using the bootstrap. Evolution 1985, 39: 783–791.

Ferrier-Cana E, Geffroy V, Macadre C, *et al*. Characterization of expressed NBS-LRR resistance gene candidates from common bean. Theor Appl Genet. 2003, 106: 251–261.

Fiedler W, Claussen U, Ludecke H. J, *et al.*, New markers for the cneurofibromatosis-2 region generated by microdissection of chromosome 22. Genomics 1991, 10: 786-791

Fiskesjo, G. 1975. Chromosomal relationships between three species of Allium as revealed by C-banding.

Hereditas 81: 23–32.

Fisher EMC, Cavanna JS, Brown SDM. Microdissection and microcloning of the mouse X-chromosome. Proc Nat Acad Sci. 1985, 82: 5846-5849.

Fukui K, Minezawa M, Kamisugi Y. et al., Microdissection of plant chromosomes by argonion laser beam. Theor Appl Genet, 1992, 84: 787-791.

Flavell, RB, Bennett, MD. Smith, JB. and Smith, DB. Genome size and the proportion of repeated nucleotide sequence DNA in plants. Biochem Genet. 1974, 12: 257-269.

Flavell RB, Rimpau J, Ribosomal RNA genes and supernumerary B chromosomes of rye. Heredity. 1975, 35: 127-131.

Flavell JD, Flavell RB. The mapping of highly repeated DNA families and their relationship to C-bands in chromosomes of *Secale cereale*. Chromosoma(Berl). 1982, 86: 595-612.

Florijn RJ, Bonden LA, Vrolijk H, et al. High-resolution DNA Fiber-FISH for genomic DNA mapping and colour bar-coding of large genes. Human Molecular Genetics, 1995, 4(5): 831-837.

Flory WS. Genetic and cytological investigations on *Aspatagus officinalis* L. Genetics. 1932 17: 432-467.

Fonseca A, Ferreira J, Santos TR, Mosiolek M, Bellucci E, Kami J, Gept s P, Geffroy V, Schweizer D, Santos KGB, Pedrosa-Harand A. Cytogenetic map of common bean (*Phaseolus vulgaris* L.). Chromosome Research. 2010, 18: 487-502.

Franks TK, Houben AM, Leach CR, et al. The molecular organisation of a B chromosome tandem repeat sequence from *Brachycome dichromosomatica*. Chromosoma. 1996, 105: 223-230.

Friesen N, Brandes A, Heslop-Harrison JS. Diversity, origin, and distribution of retrotransposons *(gypsy* and *copia)* in conifers. Mol Biol Evol. 2001, 18: 1176-1188.

Fujisawa M, Nakayama S, Nishio T, Fujishita M, Hayashi K, Ishizaki K, Kajikawa M, Yamato KT, Fukuzawa H, Ohyama K: Evolution of ribosomal DNA unit on the X chromosome independent of autosomal units in the liverwort Marchantia polymorpha. Chromosome Res 2003, 11: 695-703.

Fuchs J, Houben A, Brandes A and Schubert I. Chromosome 'painting' in plants a feasible technique? Chromosoma. 1996, 104: 315-320.

Fuchs J, Brandes A, Schubert I. Telomere sequence localization and karyotype evolution in higher plants. Plant Syst Evol, 1995, 196(4): 227-241.

Galasso I, Schmidt T, Pignone D, et al. Themolecular cytogenetics of Vigna unguiculata: thephysical organisation and characterisation of 18S-25S rDNA, 5S rDNA, telomere-like sequences and a family of centromeric repetitive DNA sequences. Theor Appl Genet. 1995, 91: 928-935.

Galasso I, Saponetti LS, Pignone D. Cytotaxonomic studies in Vigna. IV. Variation of the number of active and silent rDNA sites in Vigna unguiculata populations. Caryologia. 1998, 51, 95-104.

Gall JG, Pardue ML. Formation and detection of RNA-DNA hybrid molecules in cytological preparations. Proc Natl Acad Sci USA. 1969, 63(2): 378-383.

Garciacegarra A, Merlo MA, Ponce M, Portelabens S, Cross I, Manchado M, Rebordinos L. A preliminary genetic map in Solea senegalensis (*Pleuronectiformes, Soleidae*) using BAC-FISH and next-generation sequencing. Cytogenetic & Genome Research. 2013, 141: 227-240.

Gernandt DS, Liston A. Internal transcribed spacer region evolution in *Larix* and *Pseudotsuga* (Pinaceae). Am J Bot. 1999, 86: 711-723.

Gerlach WL, Dyer TA. Sequence organization of the repeating units in the nucleus of wheat which contain 5S rRNA genes. Nucleic Acids Res. 1981, 11: 4851- 4865

Gentzbittel L, Mouzeyar S, Badaoui S, Mestries E, Vear F, Tourvieille D, Labrouhe DE, Nicolas P. Cloning of molecular markers for disease resistance in sunflower, *Helianthus annus* L. Theor Appl Genet. 1998, 96: 519-525.

Gilbert DM. Making sense of eukaryotic DNA replication origins. Science. 2001, 294: 96-100

Gill BS, 1974. The Giemsa C-banded karyotype of rye. Proc. Nat. Acad. Sci. (USA), 1974, 71: 1247-1249

Gorenflot R, CartierD，Lenoir A, La polyploidie du *P. australis*(Cav.)Trin. ex Steud. dans le Bassin Mediterranêen. In La Flore du Bassin Mediterraneen. Colloques Internatl. C. N. R. S. (Paris) 1975, 235: 165-173.

Gorenflot R, Hubac JM, Jay M. Collaborators Le complexe polyploide du *Phragmites australis* (Cav.) Trin. ex Steud. dans la region mediterraneenne. Webbia 1984, 38: 715-721.

Gorenflot R, Panahi M. Le complexe polyploide du *Phragmites australis* (Cav.)Trin ex Steud. (=*P. communis* Trin.) en Iran. Rev. Cytol Biol. Veget. Bot. 1979, 2: 67-81.

Grant M, Mansfield J. Early events in host-pathogen interactions. Curr Opin Plant Biol. 1999, 2: 312–319.

Grayum MH. Evolution and phylogeny of the Araceae. AnnMissouri Bot Gard. 1990, 77 (4): 628-632.

Grant V. Periodicities in the chromosome numbers of the angiospems. Bot Ga. 1981, 143 (3): 379-381.

Greihuber J. Why palnt chromosomes do not show G-bands. Theor Appl Genet 1977, **50**: 121-124.

Gu HY. On chromosome number of *Kalimeris* (Astereae) and some related taxa. Cathaya. 1989, 1: 1-10.

Guan XY, Meltzer PS, Trent JM. Rapid generation of whole chromosome painting probes (WCPs) by chromosome microdissection. Genomics. 1994, 22: 101-107.

Gurzenkov NN. Studies of chromosome numbers of plants from the south of Soviet Far East. Komarov Lectures. 1973, 20: 47-61.

Gutknecht J, Sperlich D, Bachmann L. A species specific satellite DNA family of Drosophila subsilvestris appearing predominantly in B chromosomes. Chromosoma. 1995, 103: 539-544.

Gu Z, Xiao H. Physical mapping of the 18S-26S rDNA by fluorescent *in situ* hybridization (FISH) in *Camellia reticulata* polyploid complex (Theaceae). Plant Science. 2003, 164(2): 279-285.

Hadano S, Masahiko W, Haruhiko Y. *et al.* Laser microdissection and single unique primer PCR allow generation of regional chromosome DNA clones from a single human chromosome. Genomics. 1991, 11: 363-373.

Han F, Lamb J, Birchler J. High frequency of centromere inactivation resulting in stable dicentric chromosomes of maize. Proc Natl Acad Sci USA. 2006, 103: 3238-3243.

Han F, Gao Z, Yu W, Birchler J. Minichromosome analysis of chromosome pairing, disjunction and cohesion in maize. Plant Cell. 2007, 19: 3853-3863.

Harrington JJ. Bokkelen GV, Mays RW. *et al.* Formation of de novo centromeres and construction of first-generation human artificial microchromosomes. Nat. Genet. 1997, 15: 345-355.

Heslop-Harrison JS. Comparative genome organization in plants: from sequence and markers to chromatin and chromosomes. Plant Cell. 2000, 12: 617-636.

Harrison PM, Hegyi H, Balasubramanian S, Luscombe NM, Bertone P, Echols N, Johnson T, Gerstein M. Molecular fossils in the human genome: identification and analysis of the pseudo-genes in chromosomes 21 and 22. Genome Res. 2002, 12: 272-280.

Hassani HS, Falahatipour SK, Baghizadeh A. *et al.* The application of genomic DNA *in situ* hybridization (GISH) method in selection of the primary lines, combinational primary lines and secondary genotypes of Tritipyrum[C]. The National Congress of Cellular & Molecular Biology. 2008.

Hao L, Song G L, Li B L, *et al.* Karyotype analysis of *Gossypium arboreum× G. bickii* by genome *in situ* hybridization. Acta Genetica Sinica, 2006, 33(6): 565-572.

Hajdera I, Siwinska D, Hasterok R. *et al.* 2003. Molecular cytogenetic analysis of genome structure in *Lupinus angustifolius* and *Lupinus cosentinii*. Theor Appl Genet. 2003, 107: 988-996.

Hanson RE, Islam-Faridi MN, Percival EA. *et al.* Distribution of 5S and 18S228S rDNA loci in a tetraploid cotton (*Gossypium hirsutum* L.) and its putative diploid ancestors. Chromosoma. 1996, 105: 55-61.

Hagerup O. Uber polyploidie in Beziehung zu Klima, Okologie, und phylogenie. Hereditas. 1932, 16: 19-40.

Hasterok R., Jenkins, G., Langdon, T. and Jones, G. H. Ribosomal DNA is an effective marker of Brassica chromosomes. Theoretical and Applied Genetics. 2001, 103: 486-490.

Hadano S, Watanabe M, Yokoi H. *et al.* Laser microdissection and single unique primer PCR allow generation of re-gional chromosome DN A clones from a single human chromosome. Genetics. 1991, 11: 364-373.

Hasterok R, Langdon T, Taylor S, Jenkins G. Combinatorial labeling of DNA probes enables multicolour fluorescence *in situ* hybridization in plants. Folia Histochem Cyto. 2002, 40: 319-323.

He FF, He L, Hou L, Liu J. Karyotype as related to phylogenies in sesame. Journal of Southwest University Nat ural Science. 1994, 6: 573-576.

Hebsgaard SM, Korning PG, Tolstrup N, Engelbrecht J, Rouze P, Brunak S. Net Plant Gene WWW Server: splice site prediction in *Arabidopsis thaliana* DNA by combining local and global sequence information. Nucleic Acids Res. 1996, 17: 3439-3452.

Heiskanen M, Karhu R, Hellsten E. *et al.* High resolution mapping using fluorescence *in situ* hybridization to extended DNA fibers prepared from agarose-embedded cells. Biotechniques. 1994, 17(5): 928-929.

Hernould M, Glimelius K, Venskens J. et al, Microdissection and amplification of coding sequences from a chromosome fragment restoring male fertility I n alloplasmic male-sterile. The Plant Journal. 1997, 93: 703-709.

Herzfeld S. Ueber die Kernteilung im Proembryo von *Ginkgo biloba*. Jahr. Wiss. Bot. 1928, 69: 264-294.

Hesemann CU. Cytophotometricalmea sure ment of nuclea rDNA content in some coniferous and deciduous trees. Theor Appl Genet. 1980, 57: 187-191.

Hizume, M. The chromosomes of gymnospermae. Heredity. 1979, 33(6): 31-37.

Hizume M, Shibata F, Matsusaki Y, Garajova Z. Chromosome identification and comparative karyotypic analyses of four *Pinus* species. Theor Appl Genet. 2002, 105(4): 491-497.

Hizume M, Kurose N, Shibata F, Kondo K. Molecular cytogenetic studies on sex chromosomes and proximal heterochromatin containing telomere-like sequence in *Cycas revoluta*. Chromosome Sci. 1998, 2(1): 63-72.

Hizume M. Karyomorphological studies in the family Pinaceae. Mem Fac Educ Ehime Univ, Ser 3, 1988, 8: 1-108.

Hizume M, Arai M, Tanaka A. Chromosome banding in the genus *Pinus*. III. Fluorescent banding pattern of *P. luchuensis* and its relationships among the Japanese diploxylon pines. Bot Mag. 1990, 103: 103-111.

Hizume M, Kondo K, Zhang S, Hong D. Fluorescence chromosome banding in a Chinese larch, *Larix chinensis* Beissn. Chromosome Sci. 1998, 2: 95-98.

Hizume M, Ohgiku A, Tanaka A. Chromosome banding in the genus Pinus. I. Identification of chromosomes in *P. nigra* by fluorescent banding method. Bot Mag. 1983, 96: 273-276.

Hizumi M, Ohgiku A, Tanaka A. Chromosome banding in the genus *Pinus*. II. Interspecific variation of fluorescent banding patterns in *P. densiflora* and *P. thunbergii*. Bot Mag. 1989, 102: 25-36.

Hizume M, Shibata F, Kondo K, Hoshi Y, Kondo T, Ge S, Yang Q, Hong D. Identification of chromosomes in two Chinese spruce species by multicolor fluorescence in situ hybridization. Chromosome Sci. 1999, 3: 37-41.

Hizume M, Shibata F, Matsumoto A, M aruyama Y, Hayashi E, Kondo T, Kondo K, Zhang S, Hong D. Tandem repeat DNA localizing on the proximal DAPI bands of chromosomes in *Larix*, Pinaceae. Genome. 2002a, 45: 777-783.

Hizume M, Shibata F, Matsusaki Y, Garajova Z. Chromosome identification and comparative karyotypic analyses of four *Pinus* species. Theor Appl Genet. 2002, 105: 491-497.

Hizume M, Tominaga HH, Kondo K, Gu Z, Yue Z. Fluorescent chromosome banding in six taxa of Eurasian *Larix*, Pinaceae. La Kromosomo II. 1993, 69: 2342-2354.

Hizume M, Tominaga K, Tanaka A. Fluorescent chromosome banding in *Larix leptolepis* (Pinaceae). Bot Mag. 1998, 101: 333-336.

Hizumi M, Yamasaka IY, Kondo K, Yang Q, Hong D, Tanaka R. Fluorescent chromosome banding in two Chinese varieties of *Larix gmelinii,* Pinaceae. La Kromosomo II. 1994, 74: 2563-2570.

Hizumi M, Shibata F, Matsusaki Y, Garajova Z. Chromosome identification and comparative karyotypic analyses of four *Pinus* species. Theor Appl Genet. 2002b, 105: 491-497.

Hobza R, Lengerova M, Cernohorska H, Rubes H, Vyskot B. Fast-FISH with laser beam microdissected DOP-PCR probe distinguishes the sex chromosomes of *Silene latifolia*. Chromosome Res. 2004, 12: 245-250.

Holm PB, The C and Q banding patterns of the chromosomes of *Lilium longiflorum*(Thunb.) Carlsbergres. Commun. 1976, 41(5): 217-224.

Hong D. Tandem repeat DNA localizing on the proximal DAPI bands of chromosomes in *Larix,* Pinaceae. Genome. 2002, 45: 777-783.

Hong De-jun, Chen Shou-Yi. Cloning and characterization of two telomere associated sequences in rice. Chin Sci Bull, 1995, 40: 1229-1232.

Hong, D. Y., Plant Cytotaxonomy. Science Press. Beijing. 1990.

Hong G. F. A rapid and accurate strategy for rice contig map construction by combination of fingerprinting and hybridization. Plant molecular. 1997, 35: 129-133.

Hong G. F. Contigs of rice genome. Life Science. 1998, 10: 1-3.

Hong GF, Qian YM, Yu SL. *et al.*, A 120 kilobase resolution AytuiweRTYYUcontig map of the rice genome. DNA Sequence. 1997, 7(6): 319-335.

Hoshi Y, Plader W, Malepszy S. Physical mapping of 45S rRNA gene loci in the cucumber (*Cucumis sativus* L.). Caryologia. 1999, 52: 49-57.

Houben A, Kynast RG, Heim U, Hermann H, Jones RN, Forster JW. Molecular cytogenetic characterization of the terminal heterochromatic segment of the B-chromosome of rye (*Secale cereale*). Chromosoma. 1996, 105: 97-103.

Houben A, Schubert I. Engineered plant minichromosomes: A resurrection of B chromosomes? Plant Cell. 2007, 19: 2323-2327.

Houben A, Dawe R, Jiang J, Schubert I. Engineered plant minichromosomes a bottomup success? Plant Cell. 2008, 20: 8-10.

Houben A, Field BL, Saunders VA. Microdissection and chromosome painting of plant B chromosomes. Methods Cell Sci. 2001, 23: 115-124.

Huang D, Wu W, Lu L. Microdissection and molecular manipulation of single chromosomes in woody fruit trees with small chromosomes using pomelo (*Citrus grandis*) as a model. II. Cloning of resistance gene analogs from single chromosomes. Theor Appl Genet. 2004, 108: 1371-1377.

Hunyady B, Krempels K, Harta G. *et al.* Immunohistochemical signal amplification by catalyzed reporter deposition and its application in double immunostaining. Journal of Histochemistry and Cytochemistry. 1996, 44(12): 1353-1362.

Hu BQ, Dong FP, Wang CG, Qi LW, Song WQ, Chen CB. Multicolor fluore scence in situ hybridization of seven Populus species – ribosomal DNA and telom ere repeat sequence. Acta Scientiarum Naturalium Universitatis Nankaiensi. 2012, 45: 58-64.

HuziwaraY. Karyotype analysis in Aster I. Bot. Mag. Tokyo. 1953, 66: 262-268.

Huziwara Y. Karyotype analysis in Aster III. Bot. Mag. Tokyo. 1955, 68: 98-102.

Huziwara Y. Karyotype analysis in Aster IV. Bot. Mag. Tokyo. 1956, 69: 119-124.

Huziwara Y. Karyotype analysis in some genera of Compositae. III. The Karyotype of the Aster ageratoides group. Amer. J. Bot. 1956, 44: 783-790.

Iannuzzi L, Meo GPD, Perucatti A. Identification of nucleolus organizer chromosomes and frequency of active NORs in river buffalo (*Bubalus bubalis* L.). Caryologia. 1996, 49(1): 27-34.

Ibáñez E, Molist J, Vidal F. *et al.* Assessment of the proportion of transgene-bearing sperm by fluorescence *in situ* hybridization: a novel approach for the detection of germline mosaicism in transgenic male founders. Molecular Reproduction & Development. 2015, 58(2): 166-172.

Irifune K. Karyotype variation of polypioids in *Aster ageratoides* group. Proc. Sino-Jpn. Sympo- sium Plant Chromes. Plant Chromosome Research. 1987, 41-46.

Irifune K, Tanaka R, Hayashi Y, *et al.* Studies of diversity in the Compositae of *Aster agertoides* in Japan. In: Origin and evolution of diversity in plants and plant communities. 1985.

Ishikawa W. Ueber die Zahl der chromosomen von *Ginkgo biloba* L. Bot. Mag. (Tokyo)1910, 24: 225-226.

Islam-Faridi MN, Nelson CD, DiFazio SP. *et al.* Cytogenetic analysis of populus trichocarpa-ribosomal DNA, telomere repeat sequence, and marker selected BACs. Cytogenet Genome Res. 2009, 125(1): 74-80.

Jacobs MD, Gardner RC, Murray BG. Cytological characterization of heterochromatin and rDNA in *Pinus radiata* and *P. taeda*. Plant Syst Evol. 2000, 223: 71-79.

Jamilena M, Rejon CR, Rejon MR. A molecular analysis of the origin of the *Crepis capillaris* B chromosome. J. Cell Sci. 1995, 107, 703-708.

Jamilena M, Garridoraos M, Rejon RR. *et al.* Characterization of repeated sequences from microdissected

B-chromosomes of *Crepis capillaris*. Chromosoma, 1995, 104: 113-120.

Jegalian K, Lahn BT. Why the Y is so weird. Sci Am. 2001, 284: 56–61.

Jiang J, Nasuda S, Dong F. et al. A conserved repetitive DNA element located in the centromeres of cereal chromosomes. Proc Natl Acad Sci. 1996, 93(24): 14210-14213.

Jiang J, Gill BS, Wang GL, Ronald PC, Ward DC. Metaphase and interphase fluorescence in situ hybridization mapping of the rice genome with bacterial artificial chromosomes. Proceedings of the National Academy of Sciences USA. 1995, 92, 4487-4491.

Ji Y, De Donato M, Crane CF. et al. New ribosomal RNA gene locations in *Gossypium hirsutum* mapped by meiotic FISH. Chromosoma (Berlin). 1999, 108: 200-207.

Jin W, Lamb JC, Vega JM, Dawe RK, Birchler JA, Jiang J. Molecular and functional dissection of the maize B centromere. Plant Cell. 2005, 17: 1412-1423.

Johnson DH. Molecular cloning of DNA from specific chromosomal regions by microdissection and sequence independent amplification of DNA. Genomics. 1990, 6: 243-251.

Jones RN, Houben A. B chromosomes in plants: escapes from the A chromosome genome? Trends Plant Sci. 2003, 8: 417-423.

Jones RN, Rees H. B chromosomes. London: Academic Press. 1982.

Jones RN, Puertas MJ. The B-chromosomes of Rye (*Secale cereale* L). In: Dhir KK, Sareen T S(eds). Frontiers in plant science research, Delhi, India: Bhagwati Enterprises, 1993.

Jung C, Claussen U, Horsthemke B, Fischer F, Herrmann RG. A DNA library from an individual Beta patellaris chromosome conferring nematode resistance obtained by microdissection of meiotic metaphase chromosomes. Plant Mol Biol. 1992, 20: 503-511.

Kalia V. Cytological investigations in some grasses of Northeastern India. Tribes: Andropogoneae, Arundineae, Oryzeae, Arundinelleae, Chlordeae, Eragrosteae and Sporoboleae. PhD Dissert. Panjab Univ. 1978, pp. 175.

Kanazin V, Marek LF, Shoemaker RC. Resistance gene analogs are conserved and clustered in soybean. Proc Natl Acad Sci USA. 1996, 93: 11746-11750.

Kamo L. Einige Beobachtungen Ober die chromosomen von *Asparagus officinalis* L. Bot. Mag. 1929, 43: 127-133.

Kao FT. Chromosome microdissection and microcloning in human molecular genetics. Somatic Cell and Molecular Genetics. 1987, 13(4): 375-380.

Kappert H. WillkOrliche Anderungen des Geschlechisverhaltnisses Und ihre Auspuragus der Pflanenzuchrung, Umchau. 1941, 45: 772-778.

Kato A, Zheng YZ, Auger DL. et al. Minichromosomes derived from the B chromosome of maize. Cytogenet Genome Res. 2005, 109: 156-165.

Kaszas E. And James A. Birchler. Meiotic transmission rates correlate with physical features of rearranged centromeres in maize. Genetics. 1998, 150: 1683-1692.

Kennedy KE, Daskalakis SA, Davies L. et al. Non-isotopic hybridization assays for bacterial DNA samples. Molecular and Cellular Probes. 1989, 3(2): 167-177.

Khush GS, Singh RJ, Sur SC. et al. Primary trisomics of rice: origin, morphology, cytology and use in linkage mapping. Genetics. 1984, 107: 141-163.

Khush GS. Report of meetings to discussch romosome numbering system in rice. Rice Genet Newslett. 1990, 7: 12-15.

Kieser T. Preparing plasmid DNA: alkaline lysis. Plasmid isolation by alkaline lysis. Plasmid, 1984, 12: 85-92.

Kilian A, Kleinhofs A. Cloning and mapping of telomere-associated sequences from *Hordeum vulgare* L. Mol Gen Genet, 1992, 235: 153-156.

Kim SY, Kim CS, Choi HW. et al. Karyotype Analysis and Physical Mapping of rDNAs Using McFISH in *Jeffersonia dubia* Benth. Korean Journal of Medicinal Crop Science. 2005, 13(1): 48-51.

Kitayama Y, Igarashi H, Kozu T, Nagura K, Ohashi Y, Sugimura H. Repeated fluorescence *in situ* hybridization by a microwaveenhanced protocol. Pathology Internati onal. 2006, 56, 490-493.

Koike K, Yoshino K, Sue N. et al. Physical mapping of rice chromosome 4 and 7 using YAC clones. DNA Research. 1997, 4(1): 27-33.

Kongsuwan K, Smyth DR. Q-bands in Lilium and their relationship to C-banded heterchromatin. Chromosoma (Berl.). 1977, 60: 169-178.

Kong F, Rudloff E, Snowdon RJ. et al. Chromosomal distribution of 18S-25S rDNA in four Lupinus species visualized by fluorescence in situ hybridization. Russian journal of genetics. 2009, 45(8): 1009-1012.

Koo DH, Hur, YK, Jin DC, Bang, JW. Karyotype analysis of a Korean cucumber cultivar (*Cucumis sativus* L. cv. 'Winter Long') using C-banding and bicolor fluorescence in situ hybridization. Molecules and Cells. 2002, 13, 413-418.

Książczyk T, Taciak M, Zwierzykowski Z. Variability of ribosomal DNA sites in *Festuca pratensis*, *Lolium perenne*, and their intergeneric hybrids, revealed by FISH and GISH. Journal of Applied Genetics. 2010, 51(4): 449-460.

Kubalakova M, Macas J. and Dolen ELJ. Mapping of repeated DNA sequences in plant chromosomes by PRINS and C-PRINS. Theoretical and Applied Genetics, 1997, 94, 7: 58-63.

Kulak S, Hasterok R, Maluszynska J. Karyotyping of *Brassica* amphidiploids using 5S and 25S rDNA as chromosome markers. Hereditas, 2002, 136(2): 144-150.

Kurata N, Omura T. Karyotype a nalysis in rice 1. A new method for identifying all chromosome pairs. Japan Genetics, 1978, 53(4): 251-255.

Kurata N, Omura T, Iwata N. Studies on centromere, chromosome and nucleolus in pachyte ne nuclei of rice *Oryza sativa*, microsporoctes. Cytologia, 1981, 46: 791-800.

Kurata N, Umehara Y, Tanoue H, et al. Physical mapping of the rice chromosome 4 and 7 using YAC clones. Plant Molecular Biology, 1997, 35: 101-113.

Langdon T, Seago C, Jones RN, Ougham H, Thomas H, Forster JW, Jenkins G. De Novo evolution of satellite DNA on the rye B chromosome. Genetics, 2000, 154: 869-884.

Lalli DA, Decroocq V, Blenda AV Schurdi-Levraud V, Garay L, Le Gall O, Damsteegt V, Rei hard GL, Abbott AG. Identification and mapping of resistance gene analogs (RGAs) in Prunus: a resistance map for *Prunus*. Theor Appl Genet, 2005, doi: 10. 1007/s00122-005-0079-z.

Lan T Y, Liu B, Dong F P, et a l. Multiclor FISH analysis of rDN A and telomere on spinach. YiChuan, 2007, 29(11): 1405- 1408.

Langer-Safer PR, Levine M, Ward DC. Immunological method for mapping genes on *Drosophila polytene* chromosomes. Proceedings of the National Academy of Sciences of the United States of America, 1982, 79(14): 4381-4385.

Lawrence JB, Singer RH, Mcneil JA. Interphase and metaphase resolution of different distances within the human dystrophin gene. Science, 1990, 249(4971): 928-932.

Lee C. L. Sex chromosomes in *Ginkgo biloba*. Amer. J. Bot, 1954, 41: 545-549.

Leister D, Ballvora A, Salamini F, Gebhardt C. A PCR-based approach for isolating pathogen resistance genes from potato with potential for wide application in plants. Nat Genet, 1996, 14: 421-429.

Leister D, J Kurth, DA Laurie, DA Yano, T Sasaki, K Devos, A Graner, P Schulze-Lefert Rapid reorganization of resistance gene homologues in cereal genomes. Proc Natl Acad Sci USA. 1998, 95: 370-375.

Levan A, Fredga K, Sandberg AA. Nomenclature for centromeric position on chromosomes. Hereditas. 1964, 52: 201-220.

Lengerova M, Kejnovsky E, Hobza R, Macas J, Grant SR, Vyskot B. Multicolor FISH mapping of the dioecious model plant *Silene latifolia*. Theor Appl Genet, 2004, 108(7): 1193-1199.

Leach CR, Donald TM, Franks TK, Spiniello SS, Hanrahan CF, Timmis JN. Organisation and origin of a B chromosome centromeric sequence from Brachycome dichromosomatica. Chromosoma1995, 103: 708-714.

Lemieux N, Dutrillaux B, Viegaspéquignot E. A simple method for simultaneous R- or G-banding and fluorescence *in situ* hybridization of small single-copy genes. Cytogenetics & Cell Genetics, 1992, 59(4): 311-312.

Leal-Bertioli SC, Santos SP, Dantas KM, Inglis PW, N ielen S, Araujo AC, Carrasquilla-Garcia N. Arachis batizocoi: a study of its relationship to cultivated peanut (*A. hypogaea*) and its potential for introgression of wild genes into the peanut crop using in duced allotetraploids. Annals of Botany, 2005, 115: 237-249.

LePage BA, Basinger JF. The evolutionary history of the genus *Larix* (Pinaceae). US For Serv Gen Tech Rep INT, 1995, 319: 19–29.

LePage BA, JF Basinger, A new species of *Larix* (Pinaceae) from the Early Tertiary of Axel Heiberg Island, Arctic Canada. Rev Palaeobot Palynol, 1991, 70: 89-111.

Li, XL, Song WQ, Xu WS, Chen RY. An improved microdissection method of chromosomes. Acta Scientiarum Naturalium Universitatis Nankaiensis, 1998, 31(2): 102 -105.

Lichter, P. T. Cremer, C. C. Tang, P. C. Watkins, L. Manuelidis and D. C. Ward, Rapid detection of human chromosome 21 aberrations by in situ hybridization. Proc Natl Acad Sci USA, 1988, 85: 9664–9668.

Li Ping, Zhu Li-Huang, Zhou Kai-Da, Chen Ying, Lu Chao-fu, He Ping. Genetic mapping of rice using RFLP markers and a double haploid population of a aross between indica and japonica varieties. Acta Bot Sin, 1996, 38: 881-886.

Li, MX. The sex chromosome of plants. Bio. Bull, 1984, 3: 22-24.

Li, MX, Chen RY. A suggestion on the standardization of karyotype analysis in plants. J Wuhan Bot Res, 1985, 3(4): 297-302.

Li MX. Plant genome and karyotype analysis. Bulletin of Biology, 1981, 4: 20-23.

Li MX, Long YY, Gong WZ. A preliminary observation on B-chromosomes of Regal lily. Acta Bot Sin, 1984, 26(2)151-155.

Li Ping, Zhu Li-Huang, Zhou Kai-Da, Chen Ying, Lu Chao-Fu, He Ping. Genetic mapping of rice using RFLP markers and a double haploid population of a cross between indica and japonica varieties. Acta Bot Sin, 1996, 38: 881-886.

Lim KY, Matyasek R, Kovarik A, et al. Molecular Cytogenetics and Tandem Repeat Sequence Evolutionin the Allopolyploid Nicotiana Rustica Compared withDiploid Progenitors N. *Paniculata* and *N. Undulate*. Cytogenetic and Genome Research, 2005, 109: 298-309.

Lim KB, Wennekes J, Jong JH, Jacobsen E, Tuyl JM. Karyotype analysis of *Lilium longiflorum* and *Lilium rubellum* by chromosome banding and fluorescence in situ hybridisation. Genome, 2001, 44: 911-918.

Lima De-Faria A, The chromosome field. I. Prediction of the location of ribosomal cistrons. Hereditas, 1976, 83: 1-22.

Lin BY, Chou HP. Physical mapping of four RAPDs in the B chromosome of maize. Theor. Appl. Genet. 1997, 94: 534-538.

Linares C, González J, Ferrer E, et al. The use of double fluorescence *in situ* hybridization to physically map the positions of 5S rDNA genes in relation to the chromosomal location of 18S-5. 8S-26S rDNA and a C genome specific DNA sequence in the genus *Avena*. Genome, 1996, 39(3): 535-542.

Liu Bo, Chen Cheng-Bin, Li Xiu-Lan, Qi Li-Wang, Han Su-Ying. Karyotype analysis and physical mapping of 45S rDNA in eight species of Sphora, Robinia and Amorpha. Acta Botanica Yunnanica, 2005, 27(3): 261-268.

Liu B, Segal G, Vega JM, Feldman M, Abbo S. Isolation and characterization of chromosome specific DNA sequences from a chromosome arm genomic library of common wheat. Plant J. 1997, 11: 959-965.

Liu B, Liwang Qi, Ruiyang Chen, et al. Multicolor fluorescence in situ hybridization with combinatorial labeling probes enables a detailed karyotype analysis of *Larix* principis-rupprechtii. Biol Res, 2007, 40(1): 23- 28.

Liu B, Chen CB, Li XL, et al. Physical mapping of 45S rDNA to metaphase chromosomes in 30 tax onomically diverse plant species. J Hortic Sci Biotech, 2005, 80(3): 287- 290.

Liu JJ. AKM Ekramoddoullah Isolation, genetic variation and expression of TIR-NBS-LRR resistance gene analogs from western white pine (*Pinus monticola* Dougl. ex. D. Don.). MMG Mol Genet Genomics, 2003, 270: 432-441.

Liu Z, Zhang D, Hong D, Wang X. Chromosomal localization of 5S and 18S-5. 8S-25S ribosomal DNA sites in five Asian pines using fluorescence in situ hybridization. Theor Appl Genet, 2003, 106: 198-204.

Liu XH, Wang HW, Li YM, Tang YS, Liu YL, Hu X, Jia PX, et al. Preparation of single rice chromosome for construction of a DNA library using alaser microbeamtrap. J Biotechnol, 2004, 109: 217 -226.

Liu YY, Cui CQ, Mei HX, Wu K, Zheng YZ. Exploration and application of non-denaturing FISH on chromosom es research in sesame (*Sesamum indicum* L.). Scientia Agricultura Sinica, 2013，46: 3729-3735.

Liu YY, Cui CQ, Mei HX, Zheng YZ. Development and applications of a set of chromosome specific DNA markers in sesame. Research on Crops, 2014, 15: 481-486.

Lo decke HJ, Senger G, Claussen U et al. Cloning defined regions of the human genome by microdissection of banded chromosome and enzymatic amplification. Nature, 1989, 338: 348-350.

Loptien H. Identification of the sex chromosome pair in *Asparagus*(*Asparagus officinalis* L.). Z. Pflanzenzuchtg 1979, 82: 162-173.

López-León MD, Neves N, Schwarzacher T, Heslop-Harrison JS, Hewitt GM, Camacho JP: Possible origin of a B chromosome deduced from its DNA composition using double FISH technique. Chromosome Res1994, 2: 87–92.

Long EO, Dawid IB. Repeated genes in eukaryotes. Annu Rev Biochem, 1980. 43: 727-764.

Lubaretz O, Fuchs J, Ahne R, et al. Karyotyping of three pinaceae speciesvia fluorescent *in situ* hybridization and computeraided chr mosome analysis. Theor Appl Genet, 1996, 92(3): 411- 416.

Lüdecke H J, Senger G, Claussen U, Horsthemke B. Cloning defined regions of the human genome by microdissectionof banded chromosomes and enzymatic amplification. Nature, 1989, 338: 348-351.

Mackinnon R. N, Hirst M. C, Bell M. V et al microdissection of the fraglile X region, Am J Hum Genet, 1990, 47: 181-187.

Maluszynska J, Heslop-Harrison JS. Molecular cytogenetics of the genus *Arabidopsis*: in situ localizat ion of rDNA sites, chromosome numbers and diversity in centromeric heterochromatin. Ann Bot, 1993, 71: 479-484.

Maluszynska J, Heslop-Harrison JS, Physical mapping of rDNA loci in *Brassica* species. Genome, 1993，36: 774-781.

Mantovani M, Abel S, Moreira-Filho O. Conserved 5S and variable 45S rDNA chromosomal localisation revealed by FISH in *Astyanax scabripinnis* (Pisces, Characidae). Genetica, 2005, 123(3): 211-216.

Mao Z, Wang X, Zhou D, Yang Y. A survey on systematics and taxonomy of genus *Larix* Mill. J Northeast For Univ, 1999, 27: 39-44.

Marray MG, Thompson WF. Rapid isolation of high molecular w eight plant DNA. Nucl Acids Res, 1980, 8(19): 4321-4325.

Markarian D, Schulz SJ. A possible origin of supernumerary fragment chromosome. J Hered, 1958, 49: 3-7.

Martinez Zamora MG, Castagnaro AP, Diaz Ricci JC. Isolation and diversity analysis of resistance genes analogues (RGAs) from cultivated and wild strawberries. MMG Mol Genet Genomics, 2004, 272: 480-487.

Matsuda T, The accessory chromosomes of *Aster ageratoides* subsp. *leiophyllus*. Bot. Mag. Tokyo 1967, 80: 221-229.

Matsuda T, Suyama K. Polyploids and distribution of *Aster ageratoides* subsp. *Leiohyllus* group in the Kanto District and surrounding area. I. Sci Repts Yokohama Natl Univ, Sec. II, 1980, 27: 7-18.

Matsuda T, Kurode Y. Cytological and morphological studies on *Aster ageratoides* subsp. *leiophyllus* var. *sawadanus*. Hikobia suppl. 1981, 1: 45-53.

McQuade LR, Hill RJ, Francis D. B-chromosom esystems in the greater glider, Petauroides volans (Marsupialia: Pseudocheiridae) II. Investigation of B-chromosome DNA sequences islatad by micromanipulation and PCR. Cytogenet Cell Genet, 1994, 66: 155-161.

McCouch SR, Kochert G, Yu ZH et al. Molecular mapping of rice chromosomes. Theor Appl Genet, 198rata N, Umehara Y, Tanoue H, *et al.*, 8, 76: 815-829.

Mculler S, Neusser M, Wien berg J. Towards unlimited colors for fluore scence in-situ hybridization (FISH). Chromosome Research, 2002, 10: 223-232.

Mckay RDG. The mechanism of G-and C-banding in mammalian metaphase chromosomes. Chromosoma (Berlin) 1973, **44**: 1-4.

Meyers BC, Chin DB, Shen KA, Sivaramakrishnan S, Lavelle DO, Zhang Z, Michelmore RW. The major resistance gene cluster in lettuce is highly duplicated and spans several megabases. Plant Cell, 1 998, 10: 1817-1832.

Meyers BC, Dickermann AW, Michelmore RW, Sivaramakrishnan S, Sobral BW, Young ND. Plant disease resistance genes encode members of an ancient and diverse protein family within the nucleotide-binding superfamily. Plant J 1999, 20: 317-332.

Meyers BC, Kozik A, Griego A, Kuang H, Michelmore RW. Genome-wide analysis of NBS-LRR: encoding genes in Arabidopsis. Plant Cell, 2003, 15: 809-834.

Michelmore R. Flood warning resistance genes unleashed. Nat Genet. 1996, 14: 376-378.

Michelmore R. Genomic approaches to plant disease resistance. Curr Opin Plant Biol. 2000, 3: 125-131.

Michelmore RW, BC Meyers. Clusters of resistance genes in plants evolve by divergent selection and a birth-and-death process. Genome Res, 1998, 8: 1113-1130.

Melo CA, Silva GS, Souza MM. Establishment of the genomic *in situ* hybridization (GISH) technique for analysis in interspecific hybrids of *Passiflora*. Genetics & Molecular Research Gmr, 2015, 14(1): 2176-2188.

Merz H, Malisius R, Mannweiler S, *et al.* ImmunoMax. A maximized immunohistochemical method for the retrieval and enhancement of hidden antigens. Laboratory Investigation, 1995, 73(1): 149-156.

Meyne J, Baker RJ, Hobart HH, Hsu TC, Ryder TC, Ward OG, Wiley JE, Wurster-Hill DH, Yates TL, Moyzis RK. Distribution of non-telocentric sites of the (TTAGGG)n telomeric sequence in vertebrate chromosomes. Chromosoma, 1990, 99(1): 3-10.

Miao HM, Zhang HY. The Sesame Genome *Project* and *Sesame* Genome Sequencing. In: Proceedings International Plant and Animal Genome Conference XXII. 2014.

Monosi B, Wisser RJ, Pennill L, Hulbert SH. Full-genome analysis of resistance gene homologues in rice. Theor Appl Genet, 2004, 109: 1434-1447.

Moscone EA, Klein F, Lambrou M, *et al.* Quantitative karyotyping and dualcolor FISH mapping of 5S and 18S-25S rDNA probes in the cultivated *Phaseolus* species (Leguminosea). Genome, 1999, 42: 1224-1233.

Moers MH, Kalle WH, Ruiter AG, *et al.* Fluorescence *in situ* hybridization on human metaphase chromosomes detected by near-field scanning optical microscopy. Ultramicroscopy, 1995, 182(1): 279-283.

Muntzing A. Accessory chromosomes. Annu. Rev. Genet. 1974, 8, 243-266.

Murray MG, Thompson WF. Rapid isolation of high weight plant DNA. Nucleic Acids Res, 1980, 8: 4231-4235.

Muravenko OV, Lemesh VA, Lemesh TE, et al, Genome comparisons with chromosomal and molecular markers for three closely related flax species and their hybrids. Russian Journal of Genetics, 2003, 39 (4): 414-421.

Murray BG, Friesen N, Heslop-Harrison JS. Molecular cytogenetic analysis of podocarpus and comparison with other gymnosperm species. Ann Bot, 2002, 89: 483-489.

Murray BG, Davies BJ. An improved method for preparing the chromosomes of *Pines* and other gymnosperms. Biotech Histochem, 1996, 3: 115-117.

Murray MG, Thompson WF. Rapid isolation of high molecular weight plant DNA. Nucleic Acids Res, 1980, 8: 4321-4325.

Müller S, Rocchi M, Fergusonsmith MA, *et al.* Toward a multicolor chromosome bar code, for the entire human karyotype by fluorescence *in situ* hybridization. Human Genetics, 1997, 100(2): 271.

Müller S, Neusser M, Wienberg J. Towards unlimited colors for fluorescence *in-situ* hybridization (FISH). Chromosome Research, 2002, 10(3): 223-232.

Mukai T, Endo TR, Gill BS. Physical mapping of the 5S rRNA multigene family in common wheat. Journal of Heredity, 1990, 81: 290-295.

Myers EW, W Miller Aling: optimal alignments in linear space. Comp Appl Biosci, 1988, 4: 11-17.
Nagamira Y, Amtonio BA, Sasaki T, et al. Rice molecular genetic map using RFLPs and its application. Plant Mol Biol, 1997, 35: 79-87.
Naganowska B, Zielińska A, Physical mapping of 18S-25S rDNA and 5S rDNA in *Lupinus via* fluorescent in situ hybridization. Cellular & Molecular Biology Letters, 2002, 7: 665-670.
Naganowska, B and Zielin Ska A. Physical mapping of 18S-25S rDNA and 5S rDNA in Lupinus via fluorescent in situ hybridization. Cellular and Molecular Biology Letters, 2002, 7, 665-670.
Nagl W, Zellkern und zellzyklen Stuttgart, Verlag Engen Ulmer Son JH. Karyotype analysis of *Lilium lancifolium* Thunberg by means of C-banding method. Jap Jour Genet, 1998, 52: 217-221.
Nakayama S, Fujishita M, Sone T, Ohyama K. Additional locus of rDNA sequence specific to the X chromosome of the liverwort, Marchantia polymorpha. Chromosome Res, 2001, 9: 469-473.
Namiki M. The chemistry and physiological functions of sesame. Food Reviews International, 1995, 11: 281-329.
Nei M, Kumar S. Molecula revolution and phylogene tics. Oxford University Press, New York. 2002.
Neves N, Barao A, Castilho A, Silva M, Morais L, Carvalho V, Viegas W, Jones RN. Influence of DNA methylation on rye B-chromosome nondisjunction. Genome, 1992, 35: 650-652.
Newcomer EH. The karyotype and possible sex chromosomes of *Ginkgo biloba*. Amer. J. Bot. 1954, 41: 542-545.
Nicholas KB, Nicholas HB, Deerfield DWI. GeneDoc: analysis and visualization of genetic variation. Embnet News, 1997, 4: 14.
Niikawa. Chromosome-band-specific painting: chromosome in situ suppression hybridization using PCR products from a microdissected chromosome band as a probe pool. Hum Genet, 1992, 89 (1): 13-7.
Noel L, Moores TL, vander Biezen EA, Parniske M, Daniels MJ, Parker JE, Jones JD. Pronounced intraspecific haplotype divergence at the RPP5 complex disease resistance locus of *Arabidopsis*. Plant Cell, 1999, 11: 2099-2112.
Noir S, Combes MC, Anthony F, Lashermes P. Origin, diversity and evolution of NBS-type disease-resistance gene homologues in coffee trees (*Coffea* L.). MMG Mol Genet Genomics, 2001, 265: 654-662.
Noleto RB, Vicari MR, Cipriano RR, et al. Physical mapping of 5S and 45S rDNA loci in pufferfishes (Tetraodontiformes). Genetica, 2007, 130(2): 133-137.
Nur U, Werren JH, Eickbush DG, Burke WD, Eickbush TH. A "selfish" B chromosome that enhances its transmission by eliminating the paternal genome. Science, 1988, 240(4851): 512-514.
Nyongesa BO, Were BAI, Gudu S, Dangasuk OG, Onkware AO. Genetic relationship between sesame (*Sesamum indicum* L.) and related wild species based on chromosome counts and isozyme markers. African Journal of Agricultural Research, 2014, 9: 1052-1060.
Ohwi J. Flora of Japan. Smithsonian Inst, Washington, D. C, 1965, pp. 1067.
Olorode O. Additional chromosome counts in Nigerian grasses. Brittonia, 1975, 27: 63-68.
Ori N, Eshed Y, Paran I, et al. The I2C family from the wilt disease resistance locus I2 belongs to the nucleotide binding, leucine-rich repeat superfamily of plant resistance genes. Plant Cell, 1997, 9: 521-532.
Paesold S, Borchardt D, Schmidt T, Dechyeva D. A sugar beet (*Beta vulgaris* L.) reference FISH karyotyp e for chromosome and chromosome arm identification, integration of genetic linkage grou ps and analysis of major repeat family distribution. The Plant Journal, 2012, 72: 600-611.
Page BT, Wanous MK, Birchler JA. Characterization of a maize chromosome 4 centromeric sequence: evidence for an evolutionary relationship with the B chromosome centromere. Genetics, 2001, 159: 291-302.
Pan Q, Liu YS, Budai-Hadrian O, Sela M, Carmel-Goren L, Zamire D, Fluhr R. Comparative genetics of nucleotide binding siteleucine rich repeat resistance gene homologues in the genomes of two dicotyledons: tomato and *Arabidopsis*. Genetics. 2000, 155: 309-322.
Pan Q, J Wendel, Fluhr R. Divergent evolution of plant NBS-LRR resistance gene homologues in dicot and cereal genomes. J Mol Evo, 2000, 50: 203-213.

Parra I, Windle B. High resolution visual mapping of stretched DNA by fluorescent hybridization. Nature Genetics, 1993, 5(1): 17-21.

Patschke W. Uber die extratropischen ostasiatischen Coniferen undihre Bed eutung fur die pflanzen geogra phische Glieder ung Ostasiens. Bot Jahrb Syst, 1913, 48: 626-776.

Pedrosa A, Vallejos CE, Bachmair A, et al. Integration of common bean(*Phaseolus vulgaris* L.) linkage and chromosomal maps. Theor Appl Genet, 2003, 106: 205-212.

Pedersen C, Linde-Laursen I. Chromosomal locations of four minor DNA loci and a marker microsatellite sequence in barley. Chromosome Res, 1994, 2(1): 65-71.

Pellegrini M, Manning J, Davidson N. Sequence arrangement of the rDNA of *Drosophila melanogaster*. Cell, 1977, 10: 213-221.

Penuela S, Danesh D, Young ND. Targeted isolation, sequence analysis, and physical mapping of non TIR NBS-LRR genes in soybean. Theor Appl Genet, 2002, 104: 261-272.

Phelps-Durr TL, Birchler JA. An asymptotic determination of minimum centromere size for the maize B chromosome. Cytogen Genome Res, 2004, 106: 309-313.

Pinkel D, Gray JW, Trask B, et al. Cytogenetic analysis by *in situ* hybridization with fluorescently labeled nucleic acid probes. Cold Spring Harbor Symp quant biol, 198et alFuscoe J, Segraris R, Lucas J, Gray J. Fluorescence in situ hybridization with human chromosome-specific libraries: Detection of trisomy 21 and translocations of chromosome 4. Proc Natl Acad Sci USA, 1988, 85: 9138-9142.

Plowman AB, Bougourd SM. Selectively advantageous effects of B chromosomes on germination Behavior in *Allium schoenoprasum* L. Heredity, 1994, 72: 587-593.

Ponelies N, Stein N, Weber G. Microampli-fication of specific chromosome sequences, an improved method for genome analysis. Nucleic Acids Research, 1997, 25 (17): 3555-3557.

Prado EA, Faivre-Rampant P, Schneider C, et al. Detection of avariable number of ribosomal DNA loci by fluorescent in situ hybridization in *Populus* species. Genome, 1996, 39(5): 1020-1026.

Puertas MJ. Nature and evolution of B chromosomes in plants: a non-coding but information-rich part of plant genomes. Cytogenet Genome Res, 2002, 96, 198-205.

Qi LW, Zhang SG, Han SY, et al. Karyotype analysis of section populus in Populus. Acta Hort Sin, 2005, 32(5): 849-853.

Qi LW, Zhang SG, Han SY, et al. Karyotype comparison of Populus sect. Tacamahaca. Acta Bot Yunnan, 2004, 26(5): 537-542.

Qian T, Ennos RA, Helgason T. Genetic relationships among larch species based on analysis of restriction fragment variation for chloroplast DNA. Can J For Res, 1995, 25: 1197-1202.

Qi X, Zhang F, Guan Z, et al. Localization of 45S and 5S rDNA sites and karyotype of chrysanthemum and its related genera by fluorescent *in situ* hybridization. Biochemical Systematics Ecology, 2015, 62: 164-172.

Qi ZX, Zeng H, Li XL, et al. The molecular characterization of maize B chromosome specific AFLPs. Cell Res, 2002, 12(1): 63-68.

Qi ZX, Li XL, Chen CB, Song WQ, Chen RY. Isolation and chromosomal mapping of a corn B chromosome specific RAPDs. Acta Botanica Sinica，2002, 44(4): 499-501.

Raap AK, Florijn RJ, Laj B, et al. Fiber FISH as a DNA Mapping Tool. Methods, 1996, 9(1): 67-72.

Randolph LF. Types of supernumerary chromosomes in maize. Anat Rec, 1928, 41: 102-105.

Raina SN, Mukai Y, Kawaguchi K, et al, Physical mapping of 18S-25S-26S and 5S ribosomal RNA gene families in three important vetches (*Vicia species*) and their allied taxa constituting three species complexes. Theor Appl Genet, 2001, 103: 839-845.

Rara M. Cytogenttic studies on natural intergeneric hybridization in *Aster alliances* III. Natural hybrids *Aster agaratoides* subsp. lerophyllus (2n=36) × *A. ageratoides* subsp. Ovatus (2n=36). Jour Sci Miroshima Unic. B. Div, 1973, 2, 14: 141-164.

Raven PH, Soibrig OT, Kymos DW, et al. Chromosome numbers in Compositae I. Astereae. Amer J Bot, 1960, 47: 124-132.

Richter TE, Ronald PC. The evolution of disease resistance genes. Plant Mol Biol 1978, 42: 195-204.

Rick CM, Hanna GC. Determination of sex in *Asparagus officinalis* L. Amer J Bot Hort Sci, 1943, 23: 19-23.

Ribeiro T, Barao A, Viegas W, et al. Molecular cytogenetics of forest trees. Cytogenet Genome Res, 2008, 120(3/4): 220-227.

Rimpau J, Flavell RB. The repeated seqence DNA of B chromosomes of rye. Chromosomes Today, 1976, 5: 147-157.

Ried T, Baldini A, Rand TC, Ward DC, Simultaneous visualization of seven different DNA probes by *in situ* hybridization using combinatorial fluorescence and digital imaging microscopy. Proc Nati Acad Sci USA, 1992, 89: 1388-1392.

Regad F, Lebas M, Lescure B. Interstitial telomeric repeats within the *Arabidopsis thaliana* genome. J Mol Biol, 1994, 239(2): 163-169.

Ried T, Schröck E, Ning Y, Wienberg J. Chromosome painting: a useful art. Human Mol. Gene. 1998, 7(10): 1619-1626.

Rivkin MI, Vallejos CE, McClean PE. Disease resistance related sequences in common bean. Genome, 1999, 42: 41-47.

Richards EJ, Goodman HM, Ausubel FM. The centromere region of *Arabidopsis thaliana* chromosome 1 contains telomere-similar sequences. Nucleic Acids Res, 1991, 9(12): 3351-3357.

Rimpau J, Flavell RB. Characterization of rye B-chromosome DNA by DNA/DNA hybridization. Chromosoma, 1975, 52: 207-217.

Robbins WW, Jones HA. Sex as a factor in growing *Asparagus*, Proc Amer Soc Hort Sci. 1926, 12: 19-23.

Rogers SO, Bendich AJ. Ribosomal RNA genes in plants: variability in copy number and in the intergenic spacer. Plant Mol Biol, 1987, 9(4): 509-520.

Rohme D. Heneen WK. Premature chromosome condensation, application in basic, clinical, and mutation Research (Ed. P. N.) Pao RT, Johnson K. Sperling **1982**: 131-153.

Rompau J, Flavell RB. Characterization of B-chromosome DNA/DNA hybridization. Chromosoma, 1975, 52: 207-217.

Rose TM, Henikoff JG, Henikoff S. CODEHOP (COnsensus-DEgenerate Hybrid Oligonucleotide Primer) PCR primer design. Nucleic Acids Res, 2003, 31: 3763-3766.

Royal BGK, Gerlach WL, Dyer TA. Sequence organization of the repeating units in the nuclear of wheat which contain 5S rRNA genes. Nucleic Acids Res, 1980, 8: 4851-4865.

Sadder M, Weber G. Karyotype of maize (*Zea mays* L.) mitotic metaphase chromosomes as revealed by fluorescence in situ hybridization (FISH) with cytogenetic DNA markers. Plant Molecular Biology Reporter, 2001, 19: 117-23.

Saitou N, Nei M. The neighbor-joining method: a new method for reconstructing phylogenetic trees. Mol Biol Evol, 1987, 4: 406-425.

Sambrook J, Fritsch EF, aniatis TM. Molecular cloning: a laboratory manual. Cold Spring Harbor Laboratory Press, Cold Spring Harbor, 1989

Sandery MJ, Forster JW, Macadam SR, et al. Isolation of a sequence common to A-and B-chromosomes of rye (*Secale cereale*) by microcloing. Plant Mol Biol Rep, 1991, 9: 21-30.

Sandery MJ, Forster JW, Blunden R, Jones RN. Identification of a family of repeated sequences on the rye B-chromosome. Genome, 1990, 33: 908-913.

Sax K, Sax HJ. Chromosome number and morphology in the conifers. J Arnold Arbor Harv Univ, 1933, 14: 356-375.

Saunders RDC, Glover DM, Ashburner M, et al. PCR amplificati from a single polytene chromosome band: a comparision with conventional microcloning. Necleic Acids Research, 1989, 17(22): 9027-9037.

Scalenghe F, Turco E, Edstrom JE, Pirrotta V, Melli MI. Microdissection and cloning of DNA from a speregion of *Drosophila melanogaster* polytene chromosome. Chromosoma, 1981, 82: 205-216.

Sazanov AA, Sazanova AL, Tzareva VA, et al. Chromosomal localization of three GGA_4 genes using BAC-based FISH mapping: a region of conserved synteny between the chicken and human genomes. Hereditas, 2010, 140(3): 249-251.

Schlegel R, Melz G, Mettin D. Rye cytology, cytogenetics and genetics current status. Theor Appl Genet,

1986, 72: 721-734.

Schafer U, Schafer M. Localization of the ribosomal RNA genes in *Drosophila* simulans. Chromosoma (Berl.), 1980, 79: 287-291.

Schmidt T, Jung C, Heslop-Harrison JS, Kleine M. Detection of alien chromatin conferring resistance to the beet cystnsitu hybridization. Chromosome Research, 1997, 5: 186-93.

Schrader O, Bduahn H, Ahne R. Detection of 5S and 25S rRNA genes in *Sinapis alba*, *Raphanus sativus* and *Brassica napus* by double fluorescence in situ hybridization. Theoretical and Applied Genetics, 2000, 100: 665-699.

Schwarzacher T, Leitch AR, Heslop-Harrison JS. DNA: DNA in situ hybridization methods for light microscopy In: Plant Cell Biology: A Practical Approach. (Harris, N. and Oparka, K. J., Eds.) Oxford University Press, Oxford, U. K, 1994, 125-155.

Schubert I, Rieger R, Fuchs J. Alteration of basic chromosome number by fusion-fission cycles. Genome, 1995, 38(6): 1289-1292.

Schöfer C, Weipoltshammer K, Almeder M, *et al.* Signal amplification at the ultrastructural level using biotinylated tyramides and immunogold detection. Histochemistry & Cell Biology, 1997, 108(4-5): 313-319.

Schorn HE. A preliminary discussion of fossil larches (*Larix*, Pinaceae) from the Arctic. Quat Int 1994, 22/23: 173-183.

Schondelmaier RM, Jahoor A, Houben A, *et al.* Microdissection and microcloning of the barley (*Hordeum vulgare* L.)chromosome IHS. Theor Appl Genet, 1993, 86: 629-636.

Schuler GD, Boguski MS, Stewart EA, *et al.* A gene map of the human genome. Science, 1996, 274: 540-546.

Senger G, Ludecke HJ, Horsthemke B, *et al.* Microdissection of banded human chromosomes. Hum Genet, 1990, 84: 507-511.

Sharbel TF, Green DM, Houben A. B-chromosome origin in the endemic New Zealand frog Leiopelma hochstetteri through sex chromosome devolution. Genome, 1998, 41: 14-22.

Shearer LA, Anderson LK, De JH, Smit S, *et al.* Fluorescence in situ hybridization and optical mapp ing to correct scaffold arrangement in the tomato genome. Genes, Genomes, Genetics, 2014, 4(8): 1395-1405.

Shimamura T. On the formation of the proembryo of *Ginkgo biloba* L. Bot Mag (Tokyo), 1928, 42: 71-76.

Shi Q, Zhang J, Pan S, *et al.* The Study of Preparing Probes for Fluorescence *in situ* Hybridization (FISH) from YAC Clones by Universal Primer PCR. Acta Genetica Sinica, 1998, 25(5): 403-408.

Shi F, Suzuki K, Kisanuki H. The study on relationship of larches in northeast China by RAPD. Bull Bot Res, 1998, 18: 55-62.

Shibata F, Hizume M, Kuroki Y. Chromosome painting of Y chromosomes and isolation of Y-specific repetitive sequence in the diocious plant *Rumex acetosa*. Chromosoma, 1999, 108: 266-270.

Shizuya H. Cloning and stable maintenance of 300-kilobase-pair fragments of human DNA in *E. coli* using an F-factor-based vector. Proc Natl Acad Sci, 1992, 89: 8794-8797.

Shubert I. Telomeric polymorphism in *Vicia faba*. Biol Zbl, 1992, 111: 164-168.

Singh RJ. Somatic association at interphase studied by Giemsa banding technique. Chromosoma(Berl.), 1976, 265-273.

Singh, R J. Identification by Giemsa technique of the translocations separating cultivated rye from three wild species of *Secale*. Chromosoma(Berl.), 1977, 59: 217-225.

Simak M. Karyotype analysis of *Siberian larch*. Stud For Suec, 1964, 17: 1-15.

Simak M. Karyotype analylsis of *Larix griffithiana* Carr. Hereditas (Lund), 1966, 56: 137-141.

Siljak-Yakovlev S, Cerbah M, Coulaud J, *et al.* Nuclear DNA content, base composition, heterochromatin and rDNA in *Picea omorika* and *Picea abies*. Theor Appl Genet, 2002, 104: 505-512.

Simak M. Karyotype analysis of *Larix decidua* Mill. From different provenances. Medd Stat Skogsforskningsinst, 1964, 51: 1-22.

Simple JC, Brouillet L. Chromosome numbers and satellite chromosome morphology in *Aster* and *Lasallea*. Amer J Bot, 1980, 67: 1027-1039.

Singh RJ, Kim HH, Hymowitz T. Distribution of rDNA loci in the genus *Glycine* Willd. Theor Appl Genet,

2001, 103: 212-218.

Singer RH, Lawrence JB, Langevin GL, et al. Double labelling in situ hybridization using nonisotopic and isotopic detection. Acta Histochemica Et Cytochemica Official Journal of the Japan Society of Histochemistry & Cytochemistry, 1987, 20(6): 589-599.

Snowdon RJ, Köhler W, Köhler A. Chromosomal localization and characterization of rDNA loci in the Brassica A and C genomes. Genome, 1997, 40(4): 582-587.

Song WQ, Cui Cui X-Q, Xu WS, et al. Microdissection and PCR amplyfication of the terminal part of Vicia faba M-chromosome's long arm. China Sci Bull, 1996, 41: 361-363.

Song WY, Pi LY, Wang GL, et al. Evolution of the rice Xa21 disease resistance gene family. Plant Cell, 1997, 9: 1279-1287.

Song YC, Lin LH, Tan XL. Studies of maize chromosome G-banding by use of ASG method. Acta Genet Sin, 1987, 14(6): 424-427.

Sone T, M Fujisawa, M Takenaka, et al. Bryophyte 5S rDNA was inserted into 45S rDNA repeat units after the divergence from higher land plants. Plant Mol Biol, 1999, 41: 679-685.

Soriano JM, Vilanova S, Romero C, Llacer G, Badenes ML. Characterization and mapping of NBS-LRR resistance gene analogs in apricot (Prunus armeniaca L.) Theor Appl Gent, 2005, 110: 980 -989.

Srivastava AK, Schlessinger D. Structure and organization of ribosomal DNA. Biochimie, 1991, 73(6): 631-638.

Stark EA, Connerton I, Bennett ST, et al. Molecular analysis of the structure of the maize B-chromosome. Chromosome Res, 1996, 4, 15-23.

Stebbins GL, Chromosomal Evolution in Higher Plants. London: Edward Arnold, 1971. 85-104

Stein, N, Ponelies N, Muske T, Mcmullen M, Weber G. Chromosome microdissection and region-specific libraries from pachytene chromosomes of maize (Zea mays L.). Plant J, 1998, 13: 281-289.

Svalenghe F, Turco E, Edstrom JE, et al. Microdissection and cloning of DNA from a specific region of Drosophila melanogaster polytene chromosomes. Chromosoma, 1981, 82: 205-216.

Sun JY, Zhang ZH, Zong X, et al. A high-resolution cucu mber cytogenetic map integrated with the genome assembly. BMC Genomics, 2013, 14: 461-469.

Taboada X, Pansonato-Alves JC, Foresti F, et al. Consolidation of the genetic and cytogenetic maps of turbot (Scophthalmus maximus) using FISH with BAC clones. Chromosoma, 2014, 123: 281-291.

Takahashi C, Leitch IJ, Ryan A, et al. The use of genomic in situ hybridization (GISH) to show transmission of recombinant chromosomes by a partially fertile bigeneric hybrid, Gasteria lutzii × Aloe aristata, (Aloaceae), to its progeny. Chromosoma, 1997, 105(6): 342-348.

Taketa S, Harrison GE, Heslop-Harrison JS. Comparative physical mapping of the 5S and 18S-5S rDNA in nine wild Hordeum species and cytotypes. Theor Appl Genet, 1999, 98: 1-9.

Tang XM, Bao WD, Zhang WL, Cheng ZK. Identification of chromosomes from multiple rice genomes using a universal molecular cytogenetic marker system. Chinese Bulletin of Botany, 2007, 49, 953-960.

Tanaka, N, Takemasa N, Sinoto Y. Karyotype analysis in Gymnospermae. I. Karyotype and chromosome bridge in the youny leaf meristem of Ginkgo biloba L. Cytologia, 1952, 17: 112-123.

Tanić N, Dedović N, Vujošević M, Dimitrijević B. DNA profiling of B chromosomes from the yellow-necked mouse Apodemus flavicollis (Rodentia, Mammalia). Genome Research, 2000, 10: 55-61.

Tanaka R. On the speciation and karyotype in diploid and tetraploid species of Chroysanthemum I. J Sci Hiroshima Univ, Ser B Div, 1959, 29: 1-16.

Tanaka R, Irifune K. A new karyotype in diploid Aster ageratoides. Chromosome Information Service, 1982, 33: 32-33.

Taylor G. Populus: Arabidopsis for forestry. Do we need a model tree? Ann Bot, 2002, 90: 681-689.

Temsch EM, Greilhuber J. Genome size in Arachis duranensi s: a critical study. Genome, 2001, 44: 826-830.

Thomas HM, Harper JA, Meredith MR, et al. Physical mapping of ribosomal DNA sites in Festuca arundinacea and related species by in situ hybridization. Genome, 1997, 40, 406–410.

The Cold Spring Harbor Laboratory, Washington University Genome Sequencing Center, and PE Biosystems Arabidopsis Sequencing Consortium. The complete sequence of a heterochromatic island from a higher eukaryote. Cell, 2000, 100: 377-386.

Thalhammer S, Langer S, Speicher MR, Heckl WM, Geigl JB. Generation of chromosome painting probes from single chromosomes by laser microdissection and linker-adaptor PCR. Chromosome Res, 2004, 12: 337-343.

Thompson JD, Gibson TJ, Plewniak F, Jeanmougin F, Higgins DG. The CLUSTAL-X windows interface: flexible strategies for multiple sequence alignment aided by quality analysis tools. Nucleic Acids Res, 1997, 25: 4876-4882.

Thomas HM, Harper JA, Morgan WG, Gross chromosome rearrangements are occurring in an accession of the grass *Lolium rigidum*. Chromosome Research, 2001, 9: 585-590.

Thuesen A. Cytogenetical studies in *Asparagus officinalis*. L. Rog. Vet. Agric Coll Copenhagen, Yearbook, 1960, p47-71.

Tian Y, Fan L, Thurau T, Jung C, Cai D. The absence of TIR-type resistance gene analogues in the sugar beet (*Beta vulgaris* L.) genome. J Mol Evol, 2004, 58: 40-53.

Timmis JN, Ingle J, Sinclair J. The genomic quality of rye B-chromosomes. J Exp Bot, 1975, 26: 367-378.

Tischler G, Die bedeutung derploidie fue die Verbreitung der Angiospermen erlautert an den Arten Schleswig-H. 1935.

Torrell M, Cerbah M, Siljak-Yakovlev S, et al, Molecular cytogenetics of the genus *Artemisia* (Asteraceae, Anthemideae): fluorochrome banding and fluorescence in situ hybridization. I. Subgenus Seriphidium and related taxa. Plant Syst Evol, 2003, 239: 141-153.

Trask B, Pinkel D, Vand EG. The proximity of DNA sequences in interphase cell nuclei is correlated to genomic distance and permits ordering of cosmids spanning 250 kilobase pairs. Genomics, 1989, 5(4): 710-717.

Tran TD, Šimková H, Schmidt R, et al. Chromosome identification for the carnivorous plant *Genlisea margaretae*. Chromosoma, 2016, 126(3): 1-9.

Trask BJ, Massa H, Kenwrick S, et al. Mapping of human chromosome Xq28 by two-color fluorescence *in situ* hybridization of DNA sequences to interphase cell nuclei. American Journal of Human Genetics, 1991, 48(1): 1-15.

Trask B, Fertitta A, Christensen M, et al. Fluorescence *in situ* hybridization mapping of human chromosome 19: cytogenetic band location of 540 cosmids and 70 genes or DNA markers. Genomics, 1993, 15(1): 133-145.

Tran TD, Simkova H, Schmidt R, et al. Chromosome identification for the carnivorous plant Genlisea margaretae. Chromosoma, 2016, 126, 389-397.

Trude S, Heslop-Harrison JS. *In situ* hybridization to plant telomeres using synthetic oligomers. Genome, 1991, 34: 317-323.

Tsujimoto H, Niwa K. DNA structure of the B chromosome of rye revealed by in situ hybridization. Jpn J Genet, 1992, 67: 233-241.

Tuskan GA, Difazio S, Jansson S, et al. The genome of black cottonwood, *populus trichocarpa* (Torr. & Gray). Science, 2006, 313(5793): 1596-1604.

Xia Jia-Hui, Yang Yi, Dai He-Ping, Pan Qian, Li Lu-Yun., Construction of 14 DNA probe pools for specific chromosomal bands. Acta Genet Sin, 1994, 21: 253-256.

Xu SS, Liu Z, Zhang Q, et al. Chromosome Painting by GISH and Multicolor FISH. Plant Cytogenetics. 2016.

Wang MR, Perissel B, Malet P. Rehybridization on metaphases studied previously by FISH. An approach to analyze chromosome aberrations. Cancer Genetics & Cytogenetics, 1995, 85(1): 58-63.

Wang CJ, Harper L, Cande WZ. High-resolution single-copy gene fluorescence *in situ* hybridization and its use in the construction of a cytogenetic map of maize chromosome 9. Plant Cell, 2006, 18(3): 529-544.

Wang Guo-Shun, Pan Wei-Jun, Zhai Zhong-He. Proceedings in chromosomal telomere research. Chin J Cell Biol, 1993, 15: 1-7.

Wang ML, Atkinson MD, Chinoy CN, et al. RFLP-basedgenetic mapof rye (*Secale cereale* L.)chromosome 1R. Theor Appl Genet, 1991, 82: 174-178.

Wang HC, Kao KN. G-banding in plant chromosomes. Genome, 1989, 30: 48-51.

Wang CG, Chen XQ, Li H, SongWQ. RNA editing analysis of mitochondrial nad3/rps12 genes in cytoplasmic male sterility and male-fertile cauliflower (*Brassica oleracea* var. *botrytis*) by cDNA-SSCP. Bot Stud, 2007, 48: 13-23.

Weier HU, Rhein AP, Shadravan F, et al. Rapid physical mapping of the human trkprotooncogene (NTRK1) to human chromosome 1q21-q22 by P1 clone selection, fluorescence *in situ* hybridization (FISH), and computer-assisted microscopy. Genomics, 1995, 26(2): 390-393.

Weimarck A. Heterochromatin polymorphism in the rye karyotype as detected by the Giemsa C-banding technique. Hereditas, 1975, 79: 293-300.

Weiwei Jin, et al. Molecular and functional dissection of the maize B centromere. Plant Cell. 2005, 17: 1412-1423.

Weier JF, Hartshorne C, Nguyen HN, et al. Analysis of human invasive cytotrophoblasts using multicolor fluorescence in situ hybridization. Methods, 2013, 64, 160 -168.

Wiegant J, Kalle W, Mullenders L, et al. High-resolution *in situ* hybridization using DNA halopreparations. Human Molecular Genetics, 1992, 1(8): 587-91.

Witkowska M, Ohmido N, Cartagena J, et al. Physical mapping of ribosomal DNA genes on Jatropha curcas chromosomes by multicolor FISH. Cytologia International Journal of Cytology, 2009, 74(2): 133-139.

White MJD. Models of speciation. Science, 1968, 159: 1065-1070.

White MJD. Models of speciation. WH. Freeman and Company, San Francisco. 1978.

Williams JGK, Kubelik AR, Livak KJ, Rafalski JA, Tingey SV. DNA polymorphisms amplified by arbitrary primers are useful genetic markers. Nucleic Acids Res, 1990, 18: 6531-6535.

Wei XX, Wang XQ. Evolution of 4-coumarate: coenzyme A ligase (4CL) gene and diver gence of *Larix* (Pinaceae). Mol Phylogenet Evol, 2004, 31: 542-553.

Willhoeft U. Fluorescence in situ hybridization of ribosomal DNA to mitotic chromosome of tsetse flies (Diptera: Glossinidae: Glossina). Chromosome Res, 1997, 5: 262-267.

Wulff BB, Thomas CM, Smoker M, Grant M, Jones JD. Domain swapping and gene shuffling identify sequences required for inducting of an Avr-dependent hypersensitive response by the tomato Cf-4 and Cf-9 proteins. Plant Cell, 2001, 13: 255-272.

Uher P, Baborova P, Kralickova M, et al. Non-informative results and monosomies in PGD: the importance of a third round of re-hybridization. Reproductive Biomedicine Online, 2009, 19: 539 -546.

Unfried I, Gruendler P. Nucleotide sequence of the 5. 8S and 25S rRNA genes and the internal transcribed spacers from *Arabidopsis thaliana*. Nucleic Acids Res, 1990, 18: 4011-4018.

Vander Hoorn RA, Roth R, Joosten MHAJ. Identification of distinct specificity determinants in resistance protein Cf-4 allows construction of a Cf-9 mutant that confers recognition of avirulence protein AVR4. Plant Cell, 2001, 13: 273-285.

Vand EG, Sachs R, Trask BJ. Estimating genomic distance from DNA sequence location in cell nuclei by a random walk model. Science, 1992, 257(5075): 1410-1412.

Vanzela ALL, Ruas CF, Oliveira MF, et al, Characterization of diploid, tetraploid and hexaploidy Helianthus species by chromosome banding and FISH with 45S rDNA probe. Genetica, 2002. 114: 105 -111.

Vega JM, Abbo S, Feldman M, et al., Chromosome painting in plants: In situ hybridization with a DNA probe from a specific microdissected chromosome arm of common wheat. Proc Natl Acad Sci USA, 1994, 91: 12041-12045.

Vischi M, Jurman I, Bianchi G, Morgante M. Karyotype of Norway spruce by multicolor FISH. Theor Appl Genet, 2003, 107: 591-597.

Vyskot B, Hobza R. Gender in plants: sex chromosomes are emerging from the fog. Trends Genet, 2004, 20: 432-438.

Vosa LG. The basic karyotype of rye (*Secale cereale*) analysis with Giemsa and fluorescence methods. Heredity, 1974, 33: 403-408.

Yamasaki, N. Differentielle Darstellung der Metaphase chromosomenvon Cypripedium debilemit Chinacrin and Giemsa-Färbung. Chromosoma (Berl.), 1973, 41: 403-412.

Yan HM, Song YC, Li LJ, Bi XZ, Fu BY. Physical location of the rice Pi-5(t), Glh and RTSV loci by FISH of BAC clones. Wuhan University Journal of Natural Sciences, 1998, 3: 226-230.

Yarnell SH. Cytogentics of the vegetable crops I, Monocotyledons. Bot Rev, 1954, 20: 277-282.

Yeager AF, Scott DH. Studies of mature asparagus plantings with special reference to sex survival and rooting habits, Proc Amer Soc Hort Sci, 1938, 36: 513-514.

Yi TS, Li H, Li DZ. Chromosome variation in the genus *Pinellia* (Araceae) in China and Japan. Bot J Linn Soc, 2005, 147: 449-455.

Ying D, Zhou MQ, Hu ZL. Chromosomal Localization of the 5S and 45S rDNA Sites and 5S rDNA Sequence Analysis in *Nelumbo* Species. Agricultural sciences in China, 2011, 10(5): 679-685.

Yoshida M, Kawakami S, Tanaka R. A natural nonaploid of *Aster ageratoides* subsp. lerophyllus. Chromosome Information Service, 1986, 41: 3-4.

Yu JW, Tong SH, Yan FT *et al.* Construction and characterization of a region specific genomic library from human chromosome 2q35-q37. Genomics, 1992, 14(3): 769-774.

Yu W, Han F, Birchler J. Engineered minichromosomes in plant. Curr Opin Biotechnol, 2007, 18: 425-431.

Yu W, Lamb JC, Han F, Birchler JA Telomere-mediated chromosomal truncation in maize. Proc Natl Acad Sci USA, 2006, 103: 17331-17336.

Yu W, Han F, Gao Z, Vega JM, Birchler JA. Construction and behavior of engineered minichromosomes in maize. Proc Natl Acad Sci USA, 2007, 104: 8924–8929.

Yunis JJ. The characterization of high-resolution G-banded chromosome of man. Chromosoma(Berlin), 1978, **67**: 293-307.

Zhan TS, Shi LM, Hsu TC. Induction of G-bands on root of *Zea mays*. Acta Bot Sin, 1987, **29**: 465-

Zhang ZL, Yang XF. Research on G-banding of *Secale cereale* chromosomes. Acta Bot Sin, 1986, **28**: 595-598.

Zhang Y, Zhang SG, Qi LW, *et al.* Construction of poplar (*Populus tremula*) chromosome 1 specific DNA library by using a microdissection technique. Plant Mol Biol Rep, 2005, 23: 129-138.

Zhang ZT, Yang Q, Li ZA, *et al.* Comparative chromosomal localization of 45S and 5S rDNAs and implications for genome evolution in Cucumis. Genome, 2016, 59(7): 449-457.

Zhang QY, Yu G, Cooper RK, *et al.* Chromosomal location by fluorescence *in situ* hybridization of the 28S ribosomal RNA gene of the eastern oyster. Journal of Shellfish Research, 1999, 18(2): 431-435.

Zhang P, Li W, Fellers J, *et al.* BAC-FISH in wheat identifies chromosome landmarks consisting of different types of transposable elements. Chromosoma, 2004, 112(6): 288-299.

Zhan YX, Cheng M, Wu AZ. Cytogenetical studies in sesame (*Sesamum indicum* L.) karyotype analysis. Journal of China Agricultural University, 1987, 8: 249-254.

Zhan YX, Cheng M, Wu AZ. Cytogenetical studies in sesame (*Sesamum indicum* L.) a. Chromosome Giemsa banding. Journal of China Agricultural University, 1988, 7: 1-4.

Zhan YX., Cheng M, Wu AZ, Zhou XC. Cytogenetical studies in sesame (*Sesamum indicum* L.) b. A new taxonomic system. Journal of China Agricultural University, 1990, 5: 11-18.

Zhang HB, Zhao X, Ding X, Paterson AH, Wing RA. Preparation of meg abase-size DNA from plant nuclei. The Plant Journal, 1995, 7: 175-184.

Zhang H., Miao HM, Li C, Wei LB, Ma Q. Analysis of sesame karyotype and resem-blance-near coefficient. Chinese Bulletin of Botany, 2012, 47: 602-614.

Zhang HB, Scheuring CF, Zhang M, Zhang Y, Wu CC, Dong JJ, Li Y. Construction of BIBAC and BAC libraries from a variety of organisms for advanced genomics research. Nature Protocols, 2012, 7: 479-499.

Zhang HY, Miao HM, Wang L, *et al.* Genome sequencing of the important oilseed crop *Sesamum indicum* L. Genome Biology, 2013, 14: 401-409.

Zhang X, Zhuo L, Li M. A study of karyotypes of species in *Larix*. Hereditas, 1985, 7: 9-11.

Zhang RX, Chen CB, Li XL, Song WQ, Chen RY. FISH analysis of centromere homology between rye A and B chromosomes. China Sci Bull, 1999, 44: 520-524.

Zhou T, Wang Y, Chen JQ, *et al.* Genome-wide identification of NBS genes in japonica rice reveals significant expansion of divergent non-TIR NBS-LRR genes. MGG Mol Genet Genomics, 2004, 271: 402-415.

Zhong XB, Fransz PF, Eden WV, et al. High-resolution mapping on pachytene chromosomes and extended DNA fibres by fluorescence in-situ hybridisation. Plant Molecular Biology Reporter, 1996, 14(3): 232-242.

Zhu FS, Fu JH, Li LC. A preliminary study of high-resolution G-banded chromosomes of Hordeum vulgare. Acta Agri Sin, 1986, **12**: 213-214.

Zhou YH, Hu ZM, Dang BY, et al. Microdissection and microcloning of rye (*Secale cereale* L.) chromosome 1R. Chromosoma, 1999, 108: 250-255.

Zhou YH, Dang BY, Hu ZM, et al. Microdissection and PCR amplification of single soybean chromosome. Acta Botanica Sinica, 1998, 40(2): 144-150.

Ziedan ES, Elewa I, Mostafa M, Sahab A. Application of mycorrhizae for controlling root diseases of sesame. Journal of Plant Protection Research, 2011, 51, 355-361.

Ziolkowski PA, Sadowski J. FISH-mapping of rDNAs and *Arabidopsis* BACs on pachytene complements of selected *Brassicas*. Genome, 2002, 45: 189-97.

Zilm B. Zytogenetische untersnchungen an *Asparagus officinali* in ihrer Bedeurung fur die Zuchtung rein mannlicher und polyploider sorten. ZPflanzenzüchtg, 1966, 56: 1-26.

Zoldos V, Papes D, Cerbah M, et al. Variation in rDNA locus number and position among legume species and detection of 2 linked rDNA loci in the model Medicagotruncatulaby FISH. Theor Appl Genet, 1999, 99: 969-977.

Zubáčová Z, Krylov V, Tachezy J. Fluorescence *in situ* hybridization (FISH) mapping of single copy genes on *Trichomonas vaginalis* chromosomes. Molecular & Biochemical Parasitology, 2011, 176(2): 135-137.

第四章　应用篇

提要：从 2002 年开始，实验室开始转向理论联系实际的研究。几十年的"染色体"研究积累，使我们的理论联系实际研究有了扎实的基础。洪德元院士在他的细胞分类学中提出："多倍化与杂交相结合，对高等植物进化起了重大影响"，具体起了什么影响？近 20 年的理论联系实际研究，证实基因组三倍化与杂交相结合产生了第三种优势，它将多倍体优势、杂交优势、远缘优势同时固定在了一起，形成了三位一体的强大优势，在第三种优势作用下，在自然界形成了一批超级物种。人类如能利用第三种优势，培育超级物种，将是对人类的最大贡献。因此，我们认为"基因组三倍化与杂交相结合"，不仅对高等植物进化起了重要作用，而且是种质资源遗传改良与新品种培育的重要途径。

第一节　药用植物的遗传改良与新品种培育

提要：药用植物是中草药药材的主要组成部分（占 80%），其中不少药材目前仍为野生植物，其供需矛盾给环境和生态都带来不小的压力。由于野生植物资源的破坏十分严重，已对传统中医药构成一种现实的威胁。人们对药用植物的采集和消费已经使世界上已认知的药用植物的 1/5 面临灭绝的危险。

种质资源是药材生产的源头，种质的优劣对产量和质量有决定性的影响，种质资源研究特别是种质资源遗传多样性的研究在药用植物开发中具有重要意义。"多倍化与杂交相结合，对高等植物的进化起了重要作用"（洪德元，1990）。其中基因组三倍化在自然界虽然并不普遍，但它在进化中起着重要作用。自然界三倍体（triploid）多数是由 $2n$ 配子杂交产生的，人工三倍体则是由 $4x♀×2x♂$ 或 $2x♀×4x♂$ 杂交形成的，因此，不论是自然发生的三倍体还是人工产生的三倍体，都具有由杂交产生的杂种优势。另外，三倍体虽然是倍性最低的多倍体，但它具有多倍体所有属性。因此，三倍体是具有自然界目前唯一知道的两大优势的种群，即：多倍体优势和杂种优势，所以，基因组三倍化具有多倍体和杂交种双重优势，其优势表现出巨大性和不可替代性。基因组三倍化另一特点是，三倍化后在"空间上"与其他物种形成了生殖隔离，不能与其他物种进行杂交，没有有性后代，不产生后代分离，其杂交第一代是"永久 F1"，固定了杂种优势，如能进行克隆繁殖，则可成为"永久杂种"，使三倍体杂种优势得以长期保存下来，这是目前已知的固定杂种优势的最佳途径。

因此，基因组三倍化具有两大特点：第一，具有最强的杂种优势，第二，能固定杂种优势。

本课题组在对我国丹参种质资源进行广泛收集的基础上，并从形态学，细胞学和分子生物学进行了研究之后，选定了以丹参基因组三倍化为方向，拟对我国丹参种质资源进行遗传改良和新品种培育。该项研究从我国实际情况出发，不仅理论联系实际，而且，丹参基因组三倍化构思新颖，创新点突出。三倍体丹参至今国内外尚未见报道，属原创性研究，完全具有自主知识产权。三倍体丹参是继三倍体糖甜菜之后，第二个利用根部杂种优势最成功的三倍体作物，对其他药用植物的遗传改良具有引领作用。

4.1.1 丹参的遗传改良——白花丹参同源四倍体的诱导与鉴定

白花丹参 *Salvia miltiorrhiza* Bunge var.*miltiorrhiza* f.alba C.Y.Wu et H.W.Li 是丹参的一个变型（肖小河等，1997），花冠为白色或淡黄色，野生于山东省莱芜山区，为山东特产药材之一，现已在莱芜、泰安、临沂、林山等地区引种栽培。据报道，山东产白花丹参的丹参酮 Iα、次甲基丹参醌及丹参酮 IIα 3 个主要成分的量均高于同地区产紫花丹参（李允尧等，2000）。白花丹参中水溶性成分明显高于紫花丹参，约为紫花丹参的 2 倍（马丽虹等，2005）。另外白花丹参中的铁、镁、锰、锌、钙 5 种元素也高于紫花丹参（齐水秀等 2004）。白花丹参除具有紫花丹参所具有的药效外，它对治疗血栓闭塞性脉管炎具有独特疗效[5]。因此，近年来有关白花丹参的有效成分、药理、毒理和开发的研究备受关注（马丽虹等，2005）。然而，目前白花丹参的分布范围还很小，只分布在东经 117°19′~117°58′，北纬 36°02′~36°33′的莱芜地区，栽培面积只有近千亩，4500~6000 kg/nm^2，远远满足不了需求，而且栽培的白花丹参都是直接由野生变栽培，缺乏人工选育和品种改良。首次对白花丹参种质资源进行了改良，创造了四倍体白花丹参新种质，为白花丹参新品种选育提供了新材料。

4.1.1.1 材料与方法

供试材料白花丹参采自山东省莱芜地区野生种，2002 年秋季用分根繁殖种于蓟县黄崖关山区，2003 年春季紫花授粉，在种子变黑之前接种在 1/2MS + 100 mg/L 秋水仙碱的培养基中，处理 72~96 h 后，转入正常 1/2 MS 培养基中，待生根成苗后移栽于土壤中。在移栽时按单株检查染色体倍性。

4.1.1.2 结果与分析

1. 白花丹参四倍体诱变效果分析

按上述方法接种 50 粒未成熟的丹参种子，经处理后有 44%种子不能发芽，共获得 28 株成苗，经染色体检查有 17 株为四倍体，其中 16 株为四倍体整倍体（2n=32，图 4.1），一株为四倍体非整倍体（2n=31），四倍体诱导频率为 60.7%。

图 4.1 二倍体白花丹参（33 号）和其同源四倍体的植株、根、花粉和染色体比较

2. 白花丹参四倍体鉴定

四倍体白花丹参主要形态性状分析：如表 4.1 所示，四倍体白花丹参主要形态性状如株高、冠幅、叶面积、花、花粉粒和根部药材产量等，都明显大或高于原二倍体白花丹参，只有每枝的花数有减少。在形态性状中与生殖有关的性状是比较稳定的，如花的大小，花药大小，花粉粒大小等，四倍体与二倍体有显著差异（图 4.1），可以作为鉴定四倍体的形态性状参考指标。四倍体丹参的根部产量株系间虽有明显差别，但与二倍体相比多数四倍体株系产量都超过二倍体，在 16 株四倍体株系中有 14 个株系根部产量都超过二倍体白花丹参，其中 4~5 号单株产量 428 克，是二倍体白花丹参的 3.19 倍（表 4.1）。

表 4.1 白花丹参二倍体与四倍体主要形态形状比较*

形状	株系																	白花CK	17株4X平均数	与CK相比
	4-1	4-2	4-3	4-4	4-5	4-6	4-7	4-8	4-9	4-10	4-11	4-12	4-13	4-14	4-15	4-16	4-17			
株高/cm	25	45	45		50	40	60	25	20	40	40	30	25	40	40	40	0	30	38.4	+8.4
冠幅/cm	20	50	50		60	50	60	30	25	50	50	40	50	70	50	0		30	47.2	+17.2
茎粗/mm	4	4	5	4	4	4	5	4	6	5	8	4	5	6	4	0		2.5	4.75	+2.25
一次分枝数	7	8	15	12	19	8	7	9	4	4	7	8	7	8	8	6	0	3	7.9	+4.9
叶长/x 叶宽/cm	16	42	25	33.7	52.5	30.25	30.25	34.8	24	47.6	33	39	35.7	30.16	30	26.4	0	22.05	35.26	+13.21
花序数	0	12	24	39	28	18	28	3	0	12	17	14	8	20	35	31	0	13	20.64	+7.64
花序长/cm	0	22	20	20	20	15	25	8	0	20	15	20	15	20	20	0		15	17.78	+2.78

续表

形状	株系																	白花CK	17株4X平均数	与CK相比
	4-1	4-2	4-3	4-4	4-5	4-6	4-7	4-8	4-9	4-10	4-11	4-12	4-13	4-14	4-15	4-16	4-17			
花数/枝	0	36	37	33	25	20	30	19	0	26	30	25	15	15	38	37	0	43	27.5	-15.5
花大小/mm	0	20	25	25	25	25	25	25	0	25	25	25	25	25	25	25	0	21	25	+4
花粉粒大小/μm	0	8×4	7×4	0	0	0	0	0	0	0	0	0	0	0	0	5×4	0			+10
根条数/株	14	49	27	30	58	43	28	42	31	19	26	24	24	25	24	16	6	10	28.9	+18.9
根粗/cm	0.4	0.8	0.7	0.9	1.5	0.9	0.8	0.7	0.7	1.2	0.8	0.9	0.9	1.3	1.3	1.1	1.2	0.75	15.9	+15.15
根长/cm	11	31	25	28	35	37	29	27	26	33	40	30	30	37	47	35	27	47	31.11	-15.89
根鲜重/株 g	11	200	212	260	428	273	120	135	219	226	145	137	137	305	267	150	23	102	188.2	+86.2
根色	灰白	白红	红	红白	红	红白	红	红白	红白	红	红白	红白	红白	红白	红	红白	灰	红	红白	

*以最粗、最长的一条根粗和根长,叶片也是以最大的测量。

3. 四倍体白花丹参的染色体鉴定

染色体分析结果表明,四倍体白花丹参染色体结构是稳定的,核型分析结果表明所获得的16株四倍体整倍体均为同源四倍体(图4.2)。

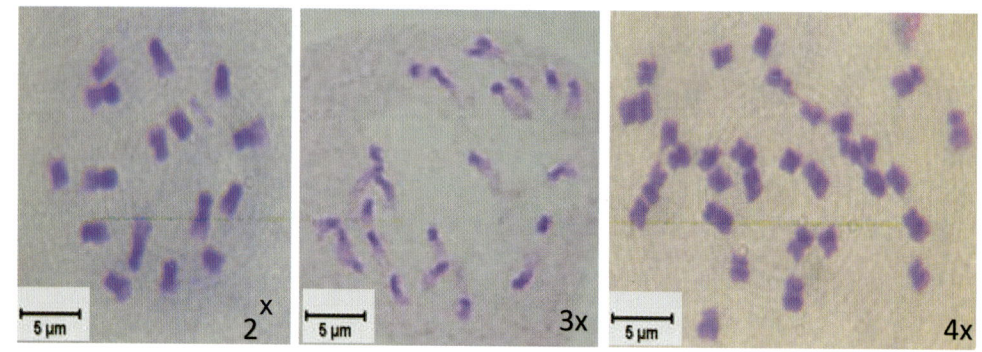

图4.2 丹参染色体
Fig. 4.2 Chromosomes of *S. miltiorrhiza*

4.1.1.3 讨论

1. 四倍体丹参的诱变频率及鉴定

植物多倍体育种,存在两个困难,一是诱变频率较低,二是诱变结果的鉴定比较困难。本实验采用低质量浓度秋水仙素较长时间处理,在半致死剂量条件下进行选择,四倍体诱导频率达到60%以上。多倍体鉴定最准确的方法仍然是以染色体计数为最终手段,但与生殖相关的形态指标,如花、花药、花粉粒大小,是比较稳定的性状,可作为多倍体鉴定的形态参考指标。

2. 四倍体丹参在育种上的意义及其应用

由于植物多倍化后的直接效应,常表现在根、茎、叶、花、果实等器官的巨大性和内含物量的提高上,所以近年来药用植物基因组多倍化,作为遗传改良途径备受重视,据不完全统计目前已有40多种药用植物人工诱导多倍化获得成功(武振华等,2005)。高山林(1996)较早诱导丹参同源四倍体获得成功,并选育出丹参酮的量比对照高79%的61-2-22优良品系,为我国丹参遗传改良开辟新途径。丹参作为多倍化育种,是一个比较理想的材料,首先是染色体基数比较少,$x=8$,$2n=16$,容易加倍成功;其次它是一个异花授粉植物，遗传多样性强，变异大，选择的几率大；第三，丹参是以收获根部营养体作为药材计算产量，因此，四倍体营养器官的巨大性可以充分体现出来；第四，丹参是多年生草本植物，可以"一劳永益"；第五，丹参同时行有性和无性繁殖，这是多倍体育种最重要的一点，克隆繁殖不仅避免了有性分离，而且可以把优良的性状"永久"地保存下来。因此，丹参是进行多倍体育种的一种典型材料，具有示范性。四倍体丹参在育种上应用可以归纳为①直接进入育种程序，选育新品种，进行推广应用。四倍体育种已在黑麦、荞麦、水稻、玉米、果树等作物上，在一些国家和地区得到推广和应用；②开展多倍体杂种优势利用研究，多倍体杂种优势利用可以是 $4x \times 4x$，$4x \times 2x \rightarrow 3x$，三倍体被认为是营养生长最好的，常常表现出最强的杂种优势，成为以无性繁殖作物的主要育种目标。三倍体杂种优势利用最成功的作物，唯有糖甜菜。自 20 世纪 70 年代以后，一些欧洲国家在生产上已 100%采用了三倍体糖甜菜品种。我国也是利用三倍体糖甜菜较早的国家之一。但糖甜菜不能进行无性繁殖，不能进行杂种优势固定。③ 筛选遗传变异新类型，由于四倍体减数分裂会产生一定比例的非整倍体配子，所以，在四倍体种群中经常会出现一定比例的非整倍体，在糖甜菜四倍体中，非整倍体株率约为 4%~5%。同时，在四倍体种群中染色体结构变异频率也略高于二倍体。据报道在四倍体燕麦中发现了 5 个基因组间的相互易位，六倍体燕麦中相互易位多达 18%。因此，由于在四倍体种群中，广泛存在的染色体结构和数量变异以及基因突变，可以分离出很多的遗传变异类型，这些不同类型的变异，可成为重要的种质材料，具有重要的育种价值。这已在水稻、大麦、大白菜、西瓜等作物中得到应用。

（作者：陈力、李秀兰。原载：中草药，2009，40（2）1995-1997）

4.1.2 三倍体丹参的培育及其可持续利用研究

丹参 *Salvia miltiorrhiza* Bunge 是我国传统医药中应用最早、最广泛的中药之一，是著名的活血化瘀药，在临床上被广泛用于治疗冠心病、心绞痛、缺血性中风等疾病（中国药典，2010）。目前我国以丹参为原料生产的复方中药有 100 多种，对优质丹参药材的需求量巨大（宋经元等，2001）。20 世纪六七十年代，各地野生丹参转变家种成功后，丹参生产有了较大发展，已成为一种重要的栽培药用植物（刘德茂等，2009；杨新杰等，2010；蓝天凤等，2011；王庆浩等，2009；韦辉等，2011），目前已有一些关于丹参品种培育的研究报道（高山林，2006；孙玉新等，2010；唐晓清等，2007；张红瑞等，2007；宋振巧等，2009；舒志明等，2006）。但是，随着栽培丹参的发展，我国丹参遗传改良与新品种培育还存在许多问题（张兴国等，2002），因此寻找一条有效的丹参新品种培育方法势在必行。

多倍化是高等植物进化的普遍规律，是促进植物进化的重要力量。Otto 等（2000）提出，多倍化有可能是植物同域物种形成最普遍的机制。多倍体在自然界很普遍（Adams，2005），其中三倍体（$3x$）是多倍体中倍数最低的多倍体，由四倍体（$4x$）与二倍体（2x）杂交产生，自然界中多数三倍体是由加倍的 $2n$ 配子杂交形成的。同时，三倍体形成后在"空间上"与其他物种产生了生殖隔离，不能进行有性杂交，没有有性后代分离，其杂交第一代是"永久 F1"，可使杂种优势固定下来，通过无性繁殖成为"永久杂种"，使杂种优势得到长期保持。因此，基因组三倍化有多倍体和杂交种双重优势，并且可固定杂种优势。三倍体杂种优势利用引起了广泛的关注。在药用植物中约有 70% 的药材是以根、茎、叶入药，所以，三倍体营养器官的巨大性可以得到充分利用。另外，有 50% 以上的药用植物能够进行无性繁殖（肖培根，2002），三倍体优势利用在药用植物中具有很大的潜力，但现在少有研究。关于三倍体丹参的研究尚未见报道，本实验首次建立了一套完整的三倍体丹参培育及其优势利用方法，为丹参的三倍体培育奠定了科学基础。

4.1.2.1　材料和方法

1. 材料

丹参药材采自我国 12 个省市 36 个居群，由四川电子科技大学生命科学与技术学院张勇博士鉴定为丹参 *Salvia miltiorrhiza* Bunge，来源见表 4.2。

表 4.2　丹参的来源
Table 4.2　Sources of S. miltiorrhiza in China

编号	采集地	编号	采集地	编号	采集地
1	陕西运城（中条山 1 号）	13	山东西营　野生	25	江苏射阳　野生（D 居群）
2	山西运城　半栽培	14	山东烟台野生（A 居群）	26	江苏射阳　野生（E 居群）
3	山西运城　栽培	15	山东烟台野生（A 居群）	27	江苏射阳　野生（F 居群）
4	陕西洛南　野生	16	浙江嵊州　野生（A 居群）	28	安徽亳州　野生
5	陕西洛南　栽培	17	浙江王坛　野生	29	湖北蕲春　野生
6	陕西商洛　野生	18	浙江嵊州　野生（B 居群）	30	湖北随州　野生
7	陕西天士力基地　栽培	19	浙江嵊州　野生（C 居群）	31	四川中江　栽培
8	陕西山阳　野生	20	上海崇明岛野生（B 居群）	32	四川金堂　栽培
9	河南内乡野生（A 居群）	21	上海崇明岛野生（B 居群）	33	山东莱芜　野生
10	河南内乡野生（B 居群）	22	江苏射阳　野生（A 居群）	34	天津蓟县　野生
11	河南方城　野生	23	江苏射阳　野生（B 居群）	35	河北安国　栽培
12	河北迁西　野生	24	江苏射阳　野生（C 居群）	36	河北行唐　栽培

2. 方法

（1）四倍体白花丹参的人工诱变

采用未成熟的白花丹参种子，在 1/2 MS＋100 mg/L 秋水仙碱的培养基中，处理 92～96 h，然后转入不含秋水仙碱的 1/2 MS 培养基中，待生根成苗后移栽于土壤中，并在移栽时按单株鉴定染色体倍性。具体方法参照参考文献（陈力等，2009）。

(2) 染色体倍性鉴定

将根尖放在饱和对二氯苯水溶液中处理 3~4 h，采用去壁、低渗法制备染色体标本（陈瑞阳等，1982）。丹参染色体基数 x 为 8，在植株定植前进行染色体倍性鉴定，二倍体（$2x$）、三倍体（$3x$）、四倍体（$4x$）见图 4.2。

4.1.2.2 结果与分析

1. 三倍体丹参的培育

在 6~7 月丹参盛花期进行人工杂交授粉，母本提前 1~2 d 去雄套袋。杂交结果表明，四倍体丹参与二倍体丹参正、反交（$4x$♀×$2x$♂ 或 $2x$♀×$4x$♂）结实率都很高，正交共有 13 个杂交组合，授粉花数 878 朵，收获 732 粒种子；反交共计 42 个杂交组合，授粉花数 1521 朵，收获种子 1196 粒，播种后均能获得三倍体，但三倍体种子成苗率较低，正交获得 147 株，反交得到 253 株，正、反交出苗率分别只有 20.0% 和 21.1%。

2. 三倍体丹参性状分析

三倍体丹参主要形态性状介于二倍体与四倍体丹参之间，具体见表 4.3。三倍体丹参的叶片、花、花粉与二倍体和四倍体的外观比较见图 4.3 至图 4.5。

表 4.3　二倍体、三倍体、四倍体丹参主要形态性状比较

Tabel 4.3　Morphological comparison on main character of *S. miltiorrhiza* for diploids, triploids and tetraploids

倍性	株高/cm	冠幅/cm	茎粗/mm	叶面积/cm²	花序长/mm	花序数	花数	花色	花大小/mm²
$2x$	40	65	5	30	65	18	35	白	759
$3x$（正交）	40	70	6	72	50	20	30	白	962
$3x$（反交）	40	60	6	56	40	15	33	紫	962
$4x$	45	70	7	72	60	20	35	白	1073

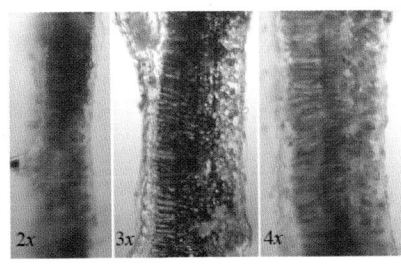

图 4.3　$2x$、$3x$、$4x$ 丹麦叶片横切比较

Fig. 4.3　Comparison on cross-section of *S. miltiorrhiza* leaves for diploids, triploids, and tetraploids

图 4.4 2x、3x、4x 丹参花的比较

Fig. 4.4 Comparison on flowers of *S.miltiorrhiza* for diploids, triploids, and tetraploids

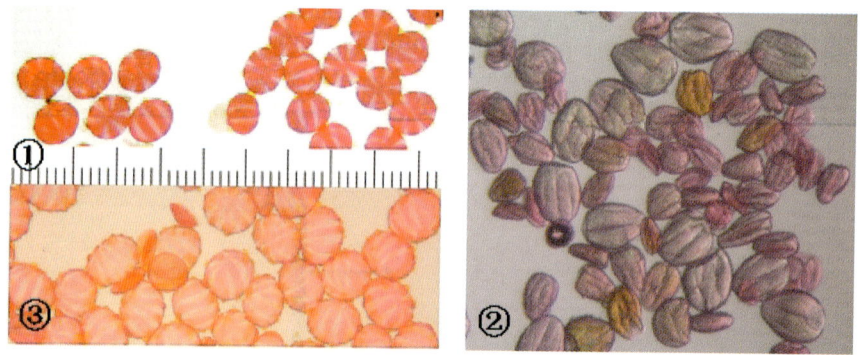

图 4.5 2x、3x、4x 丹参花粉的比较

Fig. 4.5 Comparison on pollen of *S.miltiorrhiza* for ①diploids；②triploids；③ tetraploids

3. 三倍体丹参杂种优势分析

（1）三倍体丹参杂种优势表现田间观察发现，三倍体丹参地上部分（茎、叶）均比二倍体丹参对照增加 1~1.5 倍，地下部分（芦头、根）比二倍体丹参增加 2 倍以上。进一步分析结果显示，三倍体丹参在侧根数目、根长、根粗、根质量，增加量不一致，根长与根粗变化不明显，三倍体丹参主要表现在侧根数目的增加上。从表 4.4 可以看出，侧根数目三倍体比二倍体增加 94.23%，比四倍体增加 24.44%；根鲜质量三倍体比二倍体增加 215.33%，比四倍体增加 65.89%。

表 4.4 二倍体、三倍体、四倍体丹参根的性状分析
Table 4.4 Characteristic analysis on roots for diploids, triploids and tetraploids 0f *S.miltiorrhiza*

倍性	统计株数	侧根数目	根长/cm	根粗/mm	根鲜质量/g 株$^{-1}$
二倍体	23	12.66	36.66	12.33	169.00
三倍体	130	24.59	39.72	8.18	532.92
四倍体	79	19.76	36.17	10.14	321.23

(2) 三倍体丹参杂种优势影响因素　结果分析表明,三倍体丹参杂种优势与杂交亲本选配有密切关系,应尽量选择地理远缘的丹参作为三倍体丹参杂交亲本(表 4.5)。以下 3 条途径与三倍体丹参杂种优势形成密切相关:①以四倍体为母本,二倍体为父本三倍体杂种优势高于其反交,即 $4x♀×2x♂ \geqslant 2x♀×4x♂$;②白花×紫花或紫花×白花高于白花×白花;③双交种≥三交种≥单交种。

表 4.5 四倍体与二倍体正、反交所得三倍体丹参杂种优势差异
Table 4.5 Difference of *S.miltiorrhiza* heterosis in triploids by direct-reciprocal cross of diploids and tetraploids

组合	根条数	根长/cm	根粗/mm	根鲜质量/g
白花 2x♀×白花 4x♂	27.5	27.5	7.50	390
白花 4x♀×白花 2x♂	27.5	37.8	7.85	474
紫花 2x♀×白花 4x♂	25.2	34.2	10.80	455
白花 4x♀×紫花 2x♂	27.0	44.0	8.75	937

(3) 三倍体丹参主要活性成分测定　本试验中,丹参主要活性成分测定均按照《中国药典》方法测定。共检测了 53 份丹参样品中的丹参酮 IIA、丹酚酸 B、丹参酮 I、隐丹参酮的量。其中二倍体丹参 9 份,四倍体丹参 9 份,三倍体丹参 43 份。检测结果显示,有 30 份三倍体中丹参酮 IIA 超过《中国药典》2010 年版规定的水平,其中有 9 份超过 0.20%,为 0.30%~0.40%;43 份三倍体中丹酚酸 B 全部超过 3.0%的规定,其中超过 6%的有 28 份(表 4.6)。结果表明,三倍体丹参根部产量与活性成分量不存在负相关,即主要活性成分量并没有因为三倍体丹参根部产量提高而降低。

表 4.6 二倍体、三倍体、四倍体丹参中主要活性成分的含量
Table 4.6 Contents of primary active components for diploids, triploids, and tetraploids *S.miltiorrhiza*

倍性	丹酚酸 B/(mg·g^{-1})	隐丹参酮/(mg·g^{-1})	丹参酮 I/(mg·g^{-1})	丹参酮 IIA/(mg·g^{-1})
二倍体	61.796	0.540	0.447	1.288
三倍体	66.741	0.962	0.481	2.457
四倍体	98.201	0.363	0.508	1.899

(4) 三倍体丹参新品系的选育　经过三代选育,从 55 个杂交组合、154 个株系的后代中选出了 12 个根部产量与主要活性成分量都较高的三倍体丹参新品系(表 4.7)。从 3 年田间长势和药材产量、有效成分量分析来看,来源于自然杂交选育的 10、11、12 号 3 个三倍体丹参品系都优于人工杂交组合后代。依据丹参的生物学特点,推测可选择自

然杂交方法作为三倍体丹参杂交转育方法。

表 4.7 三倍体丹参新品系
Table 4.7 New triploids lines of *S.miltiorrhiza*

品种	亲本组合	花色	根条数	根长/cm	根粗/mm	根鲜质量	质量分数/(mg·g^{-1})	(kg·株$^{-1}$) 丹参酮 IIA 丹酚酸 B	样本数
1	23×白花 4x	紫花	37.75	51.25	21.25	2.004	3.925	89.604	4
2	4×白花 4x	紫花	34.67	47.17	16.33	1.505	4.202	73.157	6
3	9×白花 4x	紫花	50.25	57.50	13.50	1.812	6.389	62.992	4
4	白花 4x×33	白花	65.25	50.25	11.25	1.337	2.406	72.247	4
5	7×白花 4x	紫花	39.00	52.00	19.33	1.367	4.266	48.264	3
6	21×白花 4x	紫花	35.00	58.50	16.00	1.450	2.746	62.183	2
7	12×白花 4x	紫花	66.50	53.00	12.50	2.050	1.983	57.495	2
8	10×白花 4x	紫花	78.00	44.33	14.33	1.942	2.644	74.653	3
9	5×白花 4x	紫花	48.60	58.80	15.80	1.933	3.787	64.171	5
10	白花 4x×紫花 2x	紫花	68.00	46.60	13.80	1.892	2.885	63.272	5
11	白花 4x×紫花 2x	紫花	52.40	53.00	18.20	2.420	2.746	73.415	5
12	白花 4x×紫花 2x	紫花	42.50	48.50	16.50	1.826	2.574	59.272	4

4.1.2.3 讨论

洪德元院士 1990 年提出,"基因组多倍化与杂交相结合对高等植物的进化起了重要影响"。本实验以丹参为材料,对其中基因组三倍化的进化意义和应用价值进行了探讨。基因组三倍化在进化上虽然是个"死胡同",但它保持了种的稳定性,使种性长期保持不变,在进化上起了重要作用。形成三倍体后与其他物种在"空间"上形成了生殖隔离,没有正常有性生殖过程,不产生后代分离,其杂交一代是"永久 F1",通过无性繁殖成为永久杂种,使杂种优势"永久"地固定下来,是当今杂种优势利用的最佳途径之一。因此,有人提出"通过同源三倍体固定水稻杂种优势"的设想(黄群策等,1995)。同时,因为基因组三倍化是杂交起源的,由 4x♀×2x♂或 2x♀×4x♂→3x,在自然界多数 3x 是由 2n 配子杂交产生的,所以,三倍体具有由杂交产生的杂种优势,同时,三倍体虽然是多倍体中倍数最低的多倍体,但它具有多倍体所有属性,因此,基因组三倍化具有自然界目前唯一知道的两大优势,即杂种优势和多倍体优势,其优势表现出巨大性和不可替代性,在非粮食作物生产上具有重要应用价值。据估计,约有 70%的中药是以植物的根、茎、叶为药材的,其中一部分具有兼性繁殖特性,因此,基因组三倍化优势在药用植物上可以得到充分利用。

基因组三倍化是进化成高级阶元六倍体的桥樑,特别是涉及一些异源双二倍体的种属,三倍化是不可缺少的,这已在很多具有 2x、4x、6x 倍性的科、属中得到证实。在自然界中,六倍体常常是一个更具生命力的种群,占居着更多的新生领地。如普通小麦(2n=6x=42),毛竹(2n=6x=48),芦苇(2n=6x=48),荷花玉兰(2n=6x=114),三脉紫菀(2n=6x=54),多数栽培菊(2n=6x=54)都是六倍体,它们在大自然中都具有极强的生活力，分布很广泛。六倍体之所以具有更强的优势和生命力，是因为它起源于具有多倍体和杂交种双重优势的三倍体。因此，基因组三倍化优势利用，不仅仅限于基因组三倍化本身，还应当看到它的深层意义，它是进化成更具生命力的六倍体的唯一桥梁。因此，从这种理念出发，不要把基因组三倍化视为最终目的，而应当把基因组三倍化作为一种手段，一种育种途径，会有更大的应用空间。现在多数栽培作物，如水稻、大豆、玉米、谷子、高粱等都是二倍体，通过基因组三倍化，在固定杂种优势的基础上，探讨六倍体育种，可能更具发展空间。

（作者：李秀兰、陈力。原载：中草药 2012，43（2）375-379）

4.1.3 三倍体丹参新品种介绍

本项研究对我国药用植物丹参种质资源进行了比较系统的遗传改良，创造了三倍体丹参新种质，培育出了一批三倍体丹参新品系，建立了一套完整的三倍体丹参优势利用技术平台。三倍体丹参具有多倍体和杂交种双重优势，能显著地提高我国丹参药材产量和商品质量；三倍体丹参主要药物成分与二倍体丹参相同，其含量均达到或超过原二倍体丹参对照；同时，三倍体丹参有花不结种子，不产生有性后代分离，是"永久杂种"，能够将杂种优势固定下来，使种性长期保持不变，从源头上对丹参质量控制提供了种源，对丹参 GAP 种植具有重要应用价值。

根据 2007、2008 年两年田间试验结果和三倍体丹参亲本来源，我们推出 12 个三倍体丹参新品系（表 4.8）

表 4.8　三倍体丹参新品系主要形态性状和根部药材鲜重

3x 品系	亲本来源	花色	根条数（株）	根长(cm)	根粗(mm)	根重（g/株） 2007	根重（g/株） 2008	丹酚酸	丹参酮 IIA (mg/g)
3x 72	4~15×33	白花	62.25	50.2	11.25	600	1337.5	74.2471	2.4058
3x 62	9×4~15	紫花	50.2	57.5	13.50	625	1812.0	62.9928	2.3887
3x 120	10×4~15	紫花	78	44.4	14.30	740	1941.0	74.6528	2.3881
3x 135	5×4~15	紫花	48.6	58.8	15.80	800	1933.0	59.6871	2.6586
3x 91	7×4 混	紫花	38.5	47.0	14.50	600	136.70	48.2641	4.2663
3x 52	4×4~2	紫花	34.6	47.1	16.30	800	1505.8	73.1566	4.2024
3x 98	12×4 混	紫花	66.5	53.0	12.50	600	2050.0	57.4955	1.9828
3x 94	21×4~11	紫花	35	58.5	16.00	440	1450.0	62.1883	2.7462
3x 22	23×4~2	紫花	33.4	49.4	20.30	640	2003.7	89.6045	3.9250
3x 230	4~5×自	紫花	75.6	46.5	13.80	1150	1892.0	63.2723	2.8853
3x 310	4~10×自	紫花	52.5	53.0	18.80	900	2420.0	73.4151	2.4759
3x 339	4~15×自	紫花	42.5	48.5	16.50	950	1826.0	59.2722	2.5737

丹参基因组多倍化新品种简介

新品种编号:	72	基因组类型:	3x
亲源关系:	33×4-15		
染色体结构:	2n=3x=24		

成株性状:	1.株高 (cm): 40	7.叶面积: 9/8
	2.冠幅 (cm): 80	8.花期: 7.30
	3.茎粗 (cm): 5	9.花序数: 50
	4.一次分枝数: 7	10.花序长: 20
	5.叶形: 卵	11.花数/枝: 30
	6.叶色: 绿	12.花色: 白

根部性状:	1.根条数/株: 65	4.根鲜重/株 (g): 1338
	2.主根粗 (mm): 11.5	5.根色: 红
	3.主根长 (cm): 50.2	6.根型: 散

药物成分:	
1. 丹酚酸B	72.24708347
2. 隐丹参酮	1.802685784
3. 丹参酮 I	0.745513
4. 丹参酮 II A	3.310601753

丹参基因组多倍化新品种简介

新品种编号:	62	基因组类型:	3x
亲源关系:	9×4-15→"永久F₁"		
染色体结构:	2n=3x=24		

成株性状:	1.株高 (cm): 40	7.叶面积: 8/7
	2.冠幅 (cm): 60	8.花期: 7.14
	3.茎粗 (cm): 2	9.花序数: 24
	4.一次分枝数: 8	10.花序长: 15
	5.叶形: 卵	11.花数/枝: 30
	6.叶色: 绿	12.花色: 紫

根部性状:	1.根条数/株: 40	4.根鲜重/株 (g): 1250
	2.主根粗 (mm): 14	5.根色: 红
	3.主根长 (cm): 50	6.根型: 散

药物成分:	
1. 丹酚酸B	62.99284075
2. 隐丹参酮	1.020820708
3. 丹参酮 I	0.485135114
4. 丹参酮 II A	2.388696696

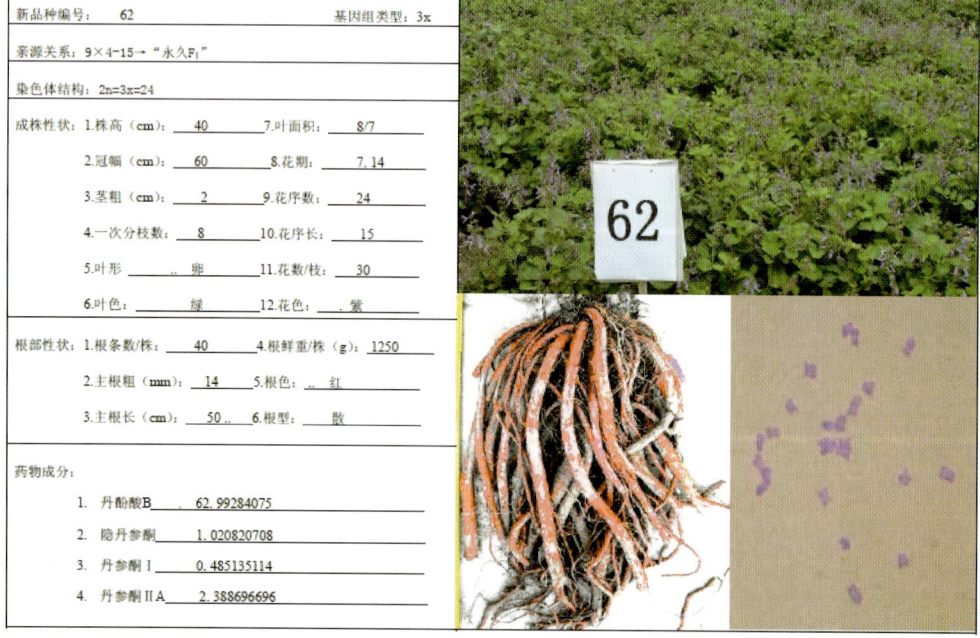

丹参基因组多倍化新品种简介

新品种编号： 120	基因组类型： 3x
亲缘关系： 10×4-15→"永久F₁"	
染色体结构：	

成株性状：	1.株高（cm）： 50	7.叶面积： 7/5
	2.冠幅（cm）： 40	8.花期： 7.14
	3.茎粗（cm）： 3.5	9.花序数： 12
	4.一次分枝数： 4	10.花序长： 24
	5.叶形： 卵	11.花数/枝： 66
	6.叶色： 绿	12.花色： 紫

根部性状：	1.根条数/株： 78	4.根鲜重/株（g）： 1941
	2.主根粗（mm）： 14.3	5.根色： 红
	3.主根长（cm）： 44	6.根型： 散

药物成分：
1. 丹酚酸B　　74.65275768
2. 隐丹参酮　　0.981718674
3. 丹参酮Ⅰ　　0.584833089
4. 丹参酮ⅡA　　2.388144757

丹参基因组多倍化新品种简介

新品种编号： 135	基因组类型： 3x
亲缘关系： 5×4-15→"永久F₁"	
染色体结构： 2n=3x=24	

成株性状：	1.株高（cm）： 40	7.叶面积： 8/6
	2.冠幅（cm）： 50	8.花期： 7.14
	3.茎粗（cm）： 2.5	9.花序数： 16
	4.一次分枝数： 3	10.花序长： 22
	5.叶形： 卵	11.花数/枝： 54
	6.叶色： 绿	12.花色： 紫

根部性状：	1.根条数/株： 48.6	4.根鲜重/株（g）： 1933
	2.主根粗（mm）： 11.8	5.根色： 红
	3.主根长（cm）： 58.8	6.根型： 散

药物成分：
1. 丹酚酸B　　64.17080784
2. 隐丹参酮　　1.838589951
3. 丹参酮Ⅰ　　0.802175339
4. 丹参酮ⅡA　　3.787035742

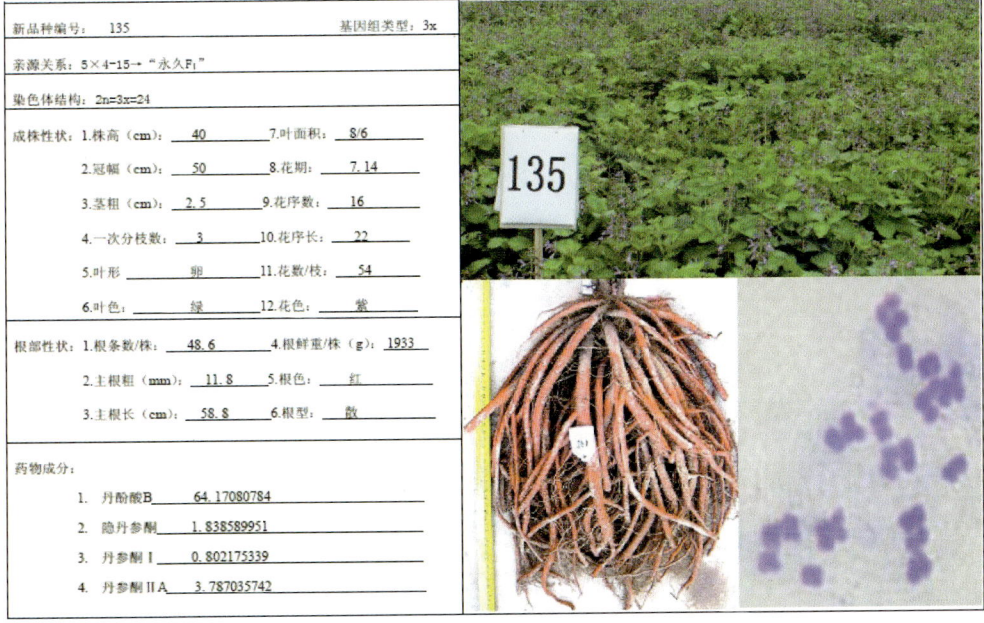

丹参基因组多倍化新品种简介

新品种编号：91　　　　　　　　基因组类型：3x

亲源关系：7×4混

染色体结构：2n=3x=24

成株性状：
1. 株高(cm): 40　　7.叶面积: 9/9
2. 冠幅(cm): 70　　8.花期: 7.30
3. 基粗(cm): 6　　9.花序数: 50
4. 一次分枝数: 7　　10.花序长: 12
5. 叶形: 卵　　11.花数/枝: 25
6. 叶色: 浅绿　　12.花色: 紫

根部性状：
1. 根条数/株: 39　　4.根鲜重/株(g): 1263
2. 主根粗(mm): 19.3　　5.根色: 红
3. 主根长(cm): 52　　6.根型: 散

药物成分：
1. 丹酚酸B　48.26407141
2. 隐丹参酮　1.445604953
3. 丹参酮I　0.694748192
4. 丹参酮IIA　4.26346358

丹参基因组多倍化新品种简介

新品种编号：52　　　　　　　　基因组类型：3x

亲源关系：4×4-2→"F₁永久杂种"

染色体结构：2n=3x=24

成株性状：
1. 株高(cm): 40　　7.叶面积: 8/9
2. 冠幅(cm): 80　　8.花期: 7.30
3. 基粗(cm): 7　　9.花序数: 50
4. 一次分枝数: 11　　10.花序长: 20
5. 叶形: 近圆　　11.花数/枝: 45
6. 叶色: 绿　　12.花色: 紫

根部性状：
1. 根条数/株: 36　　4.根鲜重/株(g): 1505
2. 主根粗(mm): 16.1　　5.根色: 红
3. 主根长(cm): 47　　6.根型: 散

药物成分：
1. 丹酚酸B　73.15662353
2. 隐丹参酮　1.393340147
3. 丹参酮I　0.764653411
4. 丹参酮IIA　4.202404418

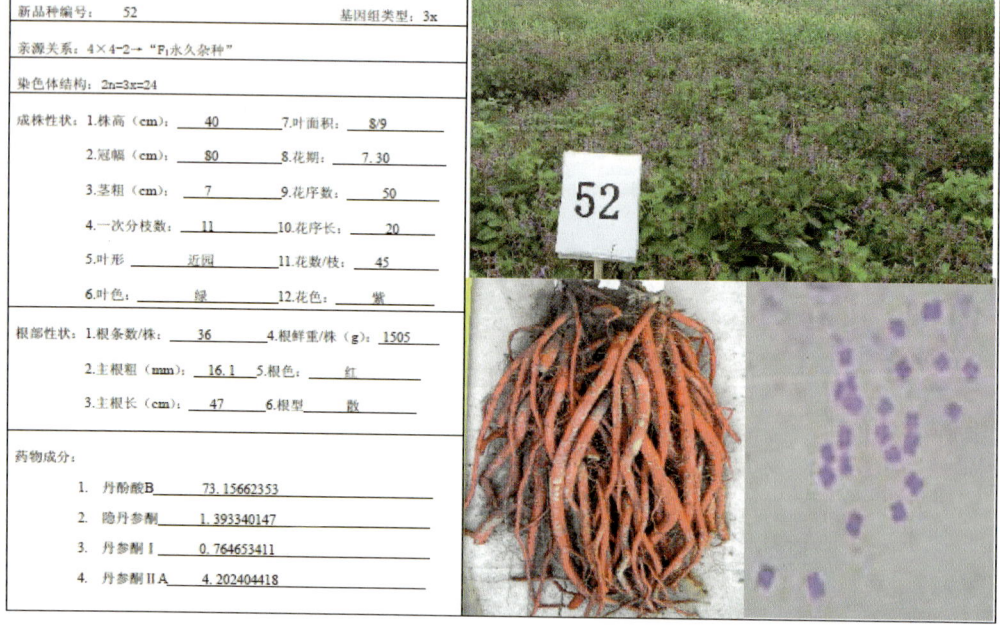

丹参基因组多倍化新品种简介

新品种编号:	98		基因组类型: 3x
亲源关系:	12×4混→"永久F₁"		
染色体结构:	2n=3x=24		

成株性状:	1. 株高（cm）: 35	7. 叶面积: 7/7
	2. 冠幅（cm）: 60	8. 花期: 7.30
	3. 茎粗（cm）: 7	9. 花序数: 50
	4. 一次分枝数: 8	10. 花序长: 16
	5. 叶形: 卵	11. 花数/枝: 25
	6. 叶色: 绿	12. 花色: 紫

根部性状:	1. 根条数/株: 67	4. 根鲜重/株（g）: 2050
	2. 主根粗（mm）: 12.5	5. 根色: 红
	3. 主根长（cm）: 53	6. 根型: 散

药物成分:
1. 丹酚酸B　　57.49551999
2. 隐丹参酮　　1.275196009
3. 丹参酮 I　　0.525090781
4. 丹参酮 II A　　1.982833298

丹参基因组多倍化新品种简介

新品种编号:	94		基因组类型: 3x
亲源关系:	21×4-11→"永久F₁"		
染色体结构:	2n=3x=24		

成株性状:	1. 株高（cm）: 45	7. 叶面积: 8/8
	2. 冠幅（cm）: 70	8. 花期: 7.30
	3. 茎粗（cm）: 11	9. 花序数: 50
	4. 一次分枝数: 6	10. 花序长: 15
	5. 叶形: 近圆	11. 花数/枝: 20
	6. 叶色: 浅绿	12. 花色: 紫

根部性状:	1. 根条数/株: 35	4. 根鲜重/株（g）: 1450
	2. 主根粗（mm）: 16	5. 根色: 红
	3. 主根长（cm）: 59	6. 根型: 散

药物成分:
1. 丹酚酸B　　62.18330153
2. 隐丹参酮　　0.915346692
3. 丹参酮 I　　0.517080614
4. 丹参酮 II A　　2.746162723

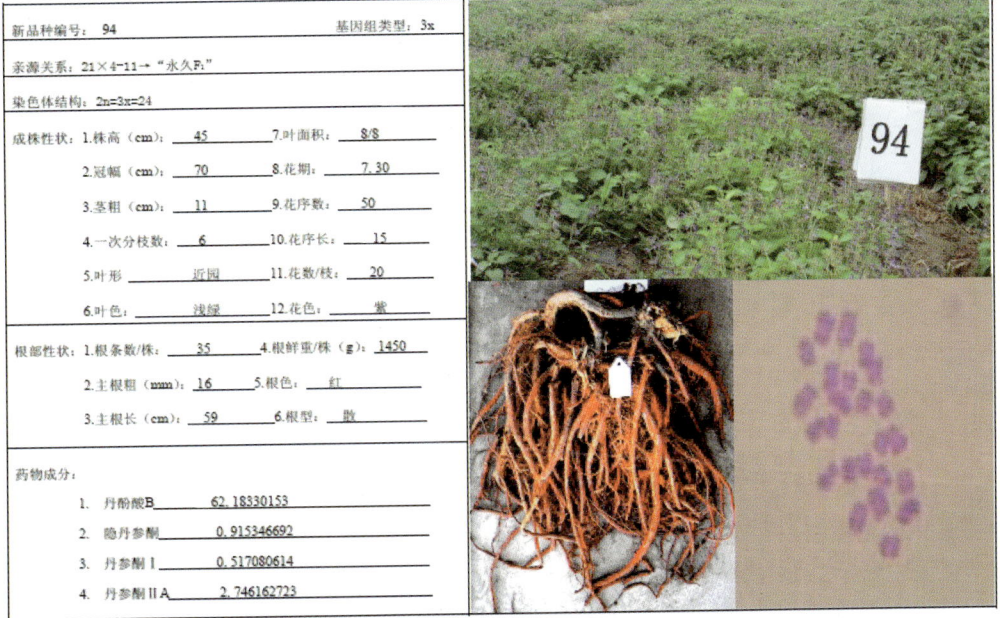

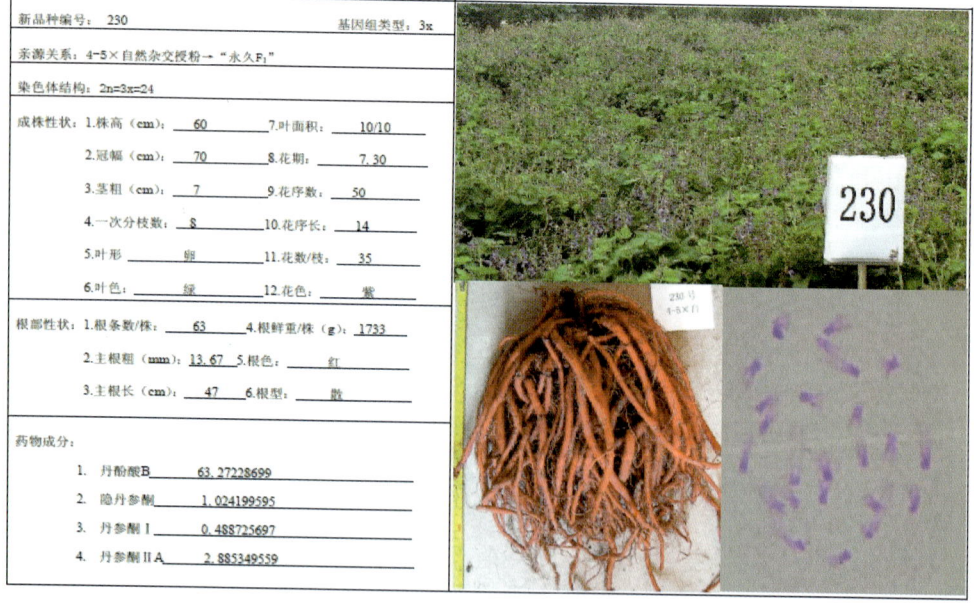

丹参基因组多倍化新品种简介

新品种编号：310　　　　　　基因组类型：3x

亲源关系：4-10×自然杂交授粉→"永久F₁"

染色体结构：2n=3x=24

成株性状：
1. 株高（cm）：55
2. 冠幅（cm）：70
3. 茎粗（cm）：8
4. 一次分枝数：5
5. 叶形：粗
6. 叶色：绿
7. 叶面积：10/10
8. 花期：7.30
9. 花序数：50
10. 花序长：15
11. 花数/枝：30
12. 花色：紫

根部性状：
1. 根条数/株：53
2. 主根粗（mm）：18.2
3. 主根长（cm）：53
4. 根鲜重/株（g）：2420
5. 根色：红
6. 根型：散

药物成分：
1. 丹酚酸B　73.41513942
2. 隐丹参酮　1.344258253
3. 丹参酮Ⅰ　0.526038904
4. 丹参酮ⅡA　2.47591987

丹参基因组多倍化新品种简介

新品种编号：339　　　　　　基因组类型：3x

亲源关系：4-15×自然杂交→"永久F₁"

染色体结构：2n=3x=24

成株性状：
1. 株高（cm）：_____
2. 冠幅（cm）：_____
3. 茎粗（cm）：_____
4. 一次分枝数：_____
5. 叶形：_____
6. 叶色：_____
7. 叶面积：_____
8. 花期：_____
9. 花序数：_____
10. 花序长：_____
11. 花数/枝：_____
12. 花色：_____

根部性状：
1. 根条数/株：42
2. 主根粗（mm）：16.5
3. 主根长（cm）：48.5
4. 根鲜重/株（g）：1826
5. 根色：红
6. 根型：散

药物成分：
1. 丹酚酸B　38.64（第一次测数据）
2. 隐丹参酮　1.306
3. 丹参酮Ⅰ　0.5093
4. 丹参酮ⅡA　2.741

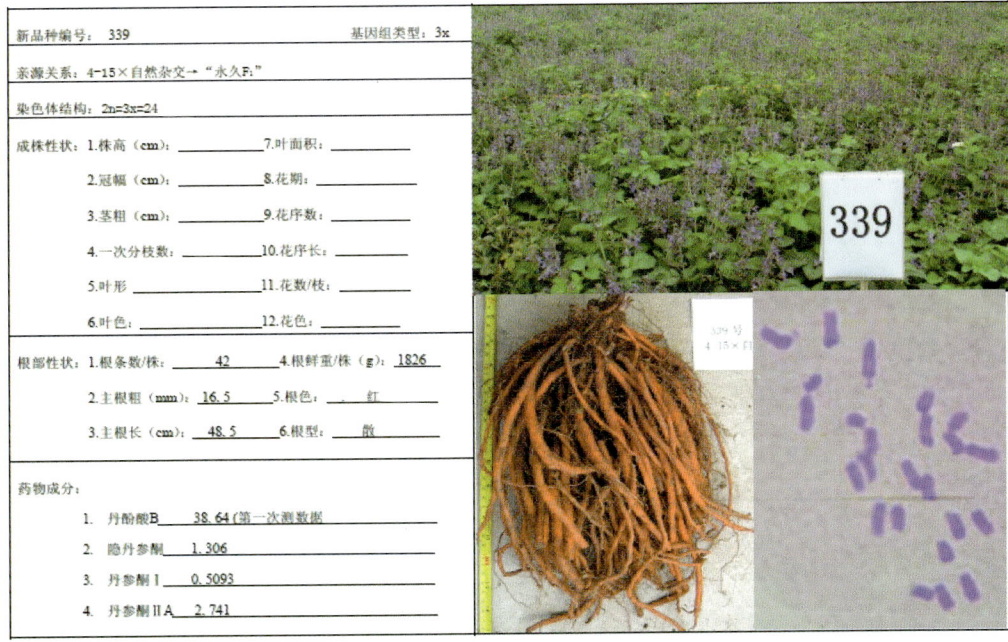

第二节　兰花种质资源的遗传改良与种质创新

提要：兰科植物（Orchidaceae），通称兰花，全世界约有700属20000～35000种，我国约有173属1200余种。在我国一部分兰科植物还是名贵中药材。兰花在我国虽有几千年的栽培历史，但以往多以分株繁殖为主，以选择自然突变为选种目标，效率低，变异小，新品种少。以有性生殖为基础的育种工作则起步很晚，与国际兰花工业发展相距甚远。

兰花种子发育不全，没有胚乳，在自然界靠与真菌共生才能萌发，萌发率极低，使得很多兰科植物成为濒危物种。但是，兰花有性杂交一旦成功，一个兰花果实有几十万至几百万粒种子。如果能使兰花种子萌发，并发育成植株，其繁殖效率之高是其他植物无法相比的。西方发达国家早在上世纪四十年代就发明了兰花种子非共生萌发培养，建立了兰花种苗工业化生产，诞生了兰花工业。

本项目自2003年开始，在国内较早的开展了以兰花有性杂交为基础的兰花遗传改良与种质创新的研究，取得了重要成果，经济效益显著。

4.2.1　兰花种质资源的收集与细胞学鉴定

本项研究自2003年开始，对我国兰花种质资源进行了广泛的收集，从广东、云南、安徽、福建、浙江、上海、山东等地收集引种了100余份兰花种质资源，如表4.9和图4.6所示。从细胞学水平对这些种质资源进行了鉴定与分析。发现兰科植物是一个多倍体复合体种群，可容纳多倍性，为兰科植物遗传改良奠定了基础。

表4.9　部分兰花种质资源来源

编号	种名（品种名）	收集地点
1	大花蕙兰	北京、天津
2	春兰	天津
3	墨兰	天津
4	蝴蝶兰	天津
5	卡特兰	北京
6	兜兰	北京
7	万带兰	云南
8	凤兰	天津
9	秋石斛兰	山东兖州
10	春石斛兰	云南、北京、天津
11	铁皮石斛兰	云南思茅、云南文山、云南麻栗坡、云南广南、浙江、上海
12	霍山石斛兰	安徽
13	晶帽石斛兰	云南
14	齿瓣石斛兰	云南
15	金钗石斛兰	云南

秋石斛花与果实　　　　　　春石斛花与果实

蝴蝶兰花与果实　　　　　　大花蕙兰花与果实

春兰、墨兰花与果实

铁皮石斛花与果实　　　　　凤兰花与果实

卡特兰花与果实　　　　　　台北小姐花与果实

万代兰花与果实　　　　　　兜兰花与果实

图 4.6　兰花开花生物学研究

4.2.2 兰花有性杂交育种平台的建立与种苗工厂化生产技术

有性杂交是兰花杂交育种的重要技术手段。然而,兰科植物与多数开花植物不同,它的子房发育是由授粉启动的,即先授粉→刺激子房发育→受精→结实。由于兰花开花与胚胎发育的不一致性,给兰花杂交授粉与结实带来了一定困难。

4.2.3 兰花种子非共生萌发、原球茎诱导、分化与成苗移栽方法

兰花种子很小,只有几十个细胞组成,没有胚乳,在自然条件下极难萌发,在自然界萌发是靠与一种真菌共生才能萌发,但萌发率极低,生长缓慢,这就使得很多兰科植物成为濒危物种。兰花种子萌发是杂交育种的一个技术难点,兰花种子非共生萌发与原球茎诱导是兰花繁育的核心技术。本项研究对多种兰花种子非共生萌发培养基、培养条件和培养方法进行了研究,筛选出以 MS 为主体的适合的培养基,建立了"液体+固体"的培养方法可使 90%以上的兰花种子在非共生条件下均能萌发,如图 4.7 所示。

图 4.7 石斛兰的种胚发育与种子原球茎的形成

兰花种子萌发是杂交育种的一个技术难点,现在,兰花种子非共生萌发与原球茎诱导已成为兰花工业中的重要技术。兰花种子萌发后要适时转入原球茎诱导培养基中,否则会停止生长或死亡。研究表明,兰花种子原球茎的形成、增殖与分化有 5 个重要因素:①"固体与液体"培养基的适时培养;②种子发育时期;③6-BA 浓度;④6-BA 与 NAA 比例;⑤原球茎与培养基的比例。一般原球茎与培养基的比例为 1:30~1:50 为适。本项研究建立的"液体+固体"的兰花种子非共生萌发、原球茎诱导与分化培养技术,可使兰花种子萌发率和原球茎分化率达到 100%,如图 4.8 和图 4.9 所示。

图 4.8　PLBs 增殖与分化的"液体+固体"培养

图 4.9　原球茎诱导、分化与成苗移栽

4.2.4 兰花原球茎快速增殖方法

原球茎是兰科植物特有的结构,是无性繁殖系的种源。长期以来,人们普遍认为兰科植物的繁殖方式主要是分株繁殖、种子繁殖或通过茎尖、叶片、茎段、花梗等外植体诱导愈伤组织,然后再通过分化诱导,最终得到幼苗和植物体的繁殖方式。本项研究在对大花蕙兰、石斛兰的研究中发现,除上述繁殖方式外,兰花最重要的繁殖方式是诱导和利用原球茎增殖。兰花种子原球茎在适合的培养基与培养条件下,增殖系数是很高的,理论上在百万级（6.25×10^6）。本项研究发现,兰花原球茎增殖系数之高,是因为兰花原球茎存在三种增殖方法：①原球茎产生原球茎（图4.10）；②丛生芽产生原球茎（图4.11）；③愈伤组织产生原球茎（图4.12）。其中方法①种增殖方式原球茎增率系数为 6.25×10^5,方法②种增殖方式原球茎增值系数为 4.20×10^3,方法①种增殖方式是方法②种的196倍。如图4.13和图4.14所示,兰花一个果实里有数十万乃至数百粒种子,兰花种子原球诱导与成苗分化技术,促使了兰花工业的产生,原球茎增殖技术是兰花工业的核心技术,在兰花工业中具有重要的应用价值。

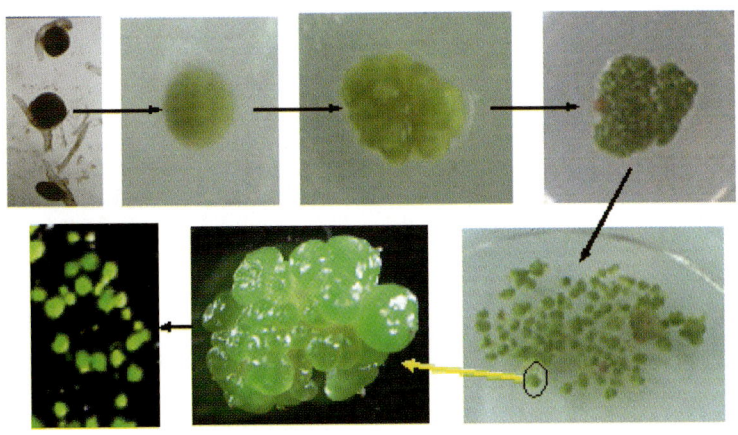

图 4.10 原球茎产生原球茎

图 4.11 丛生芽产生原球茎

图 4.12　愈伤组织产生原球茎

图 4.13　一粒大花蕙兰种子原球茎一代的增殖数量

图 4.14　大花蕙兰丛生芽上产生的原球茎

4.2.5　兰花多倍体的诱导方法

利用原球茎进行多倍体诱变是获得兰科植物多倍体新种质的有效途径。染色体组或基因组（genome）是基因的载体，通过基因组操作，在不改变基因结构或DNA结构的前提下，只增加同源基因组的数量，可以达到表观遗传学改变的效应，达到种质创新和新品种培育的目的。基于"基因组多倍化与杂交相结合"理论，我们首先对兰花种子原球茎发育时期与多倍体诱变的关系进行了研究，结果显示：①原球茎发育时期与多倍体诱变率有关，如图4.15；②原球茎发育时期与原球茎成活率成正相关，与诱变处理时间成负相关，如图4.16，原球茎发育时期是影响多倍体诱变效率的重要因素之一。研究表明，以原胚期原球茎诱导兰花多倍体，其多倍体诱变率达到90%以上，这是以种子和生长点进行多倍体诱变方法无法达到的。本项研究建立的兰花原胚期原球茎多倍体诱导方法尚未见报道。利用该方法创造了一批兰花多倍体新种质，如图4.17至图4.23。经查新，除四倍体春石斛兰有报道外，其余均为首次报道。

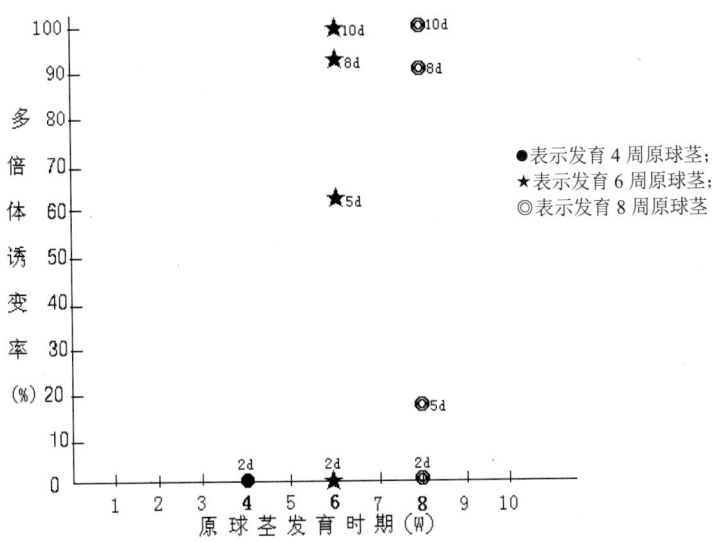

图4.15　原球茎发育时期（W）与多倍体诱变率（%）

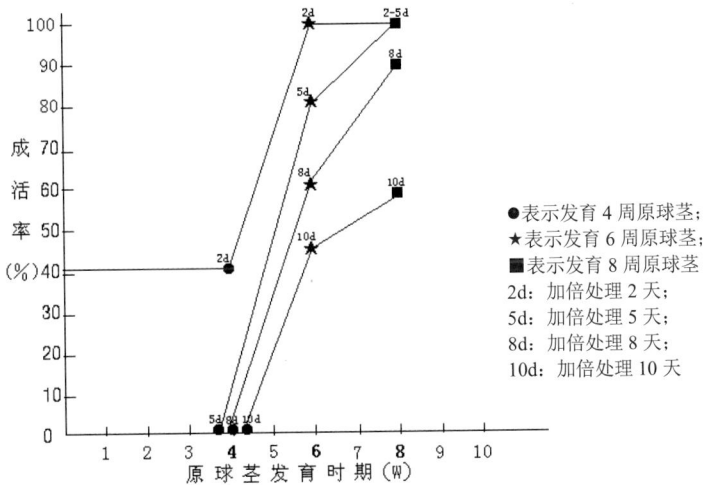

图4.16　原球茎发育时期（W）与成活率（%）的关系

图 4.17 四倍体秋石斛兰

1. 二倍体秋石斛花、蒴果（c）和种子（s）；2. 培养 4 周的萌发种子（箭头所示为从种壳中脱落的球形胚）；3.发育 6 周的原球茎（箭头所示为种壳脱落原球茎）；4.发育 8 周原球茎：①加倍处理前的原球茎，②加倍处理后的原球茎（箭头所示为白化致死原球茎）；5.二倍体瓶苗；6.二倍体移栽苗；7.二倍体染色体，$2n=2x=38$；8.四倍体瓶苗；9.四倍体移栽苗；10.四倍体染色体，$2n=4x=76$。

图 4.18 四倍体春石斛兰选育过程

图 4.19 三倍体春石斛兰

图 4.20 六倍体春石斛兰

图 4.21 四倍体铁皮石斛兰

三倍体铁皮石斛的杂交转育　　　　　三倍体铁皮石斛种子非共生萌发培养

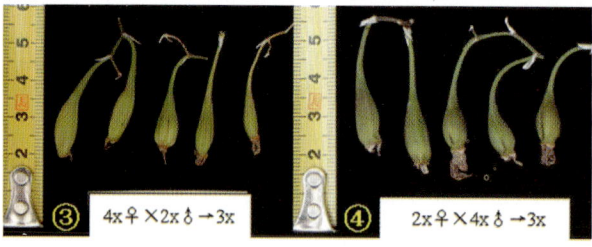

三倍体铁皮石斛蒴果　　　　　三倍体的原球茎分化诱导与瓶苗培养

三倍体铁皮石斛种苗移栽

图 4.22　三倍体铁皮石斛兰选育过程

图 4.23　四倍体霍山石斛兰

4.2.6　十三种兰花有性杂交育种技术平台

有性杂交是兰花杂交育种的重要技术手段之一。随着生物技术的发展，生物高效育种与新品种创制打破了以往传统单一的选择育种法，利用生物工程保护濒危物、种，创造新物种,已成为当前生命科学领域研究中的重要研究内容之一。我国兰科植物资源十分丰富,野生兰科植物不计其数,但在兰花新品种培育、遗传改良与种质创新方面,远远落后于国外的发展。我国兰花的很多基础研究还处于初期阶段。本项研究以多种兰花为材料,从细胞学、兰花开花生物学、有性杂交授粉、受精、子房发育、结实、种子非共生萌发、原球茎诱导、增殖、分化等诸多方面进行了研究,目前已建立了大花蕙兰、石斛兰、蝴蝶兰等 13 种兰花的有性杂交育种技术平台,并取得了实质性进展,结果如下。

1. 大花蕙兰有性杂交育种技术平台

大花蕙兰开花、授粉、结实

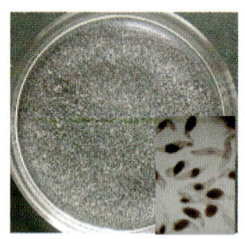

种子非共生萌发与原球茎诱导

原球茎快速增殖与分化诱导

瓶苗分化、培养与瓶苗移栽

成苗栽培与管理

2. 春石斛兰有性杂交育种技术平台

春石斛开花、授粉、结实

种子非共生培养、萌发、原球茎诱导

3. 秋石斛兰有性杂交育种技术平台

秋石斛兰开花、授粉、结实

种子非共生培养、萌发、原球茎诱导

原球茎分化与瓶苗移栽

4. 铁皮石斛有性杂交育种技术平台

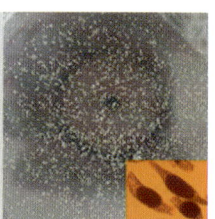

铁皮石斛有性杂交、结实与种子非共生培养

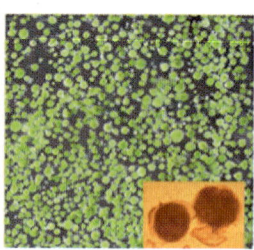

种子萌发、原球茎诱导与增殖

原球茎分化与瓶苗移栽

5. 霍山石斛有性杂交育种技术平台

霍山石斛有性杂交与结实

种子非共生萌发、原球茎诱导与分化

瓶苗移栽

4.2.7　铁皮石斛种苗简易工厂化生产

本项研究建立了一种铁皮石斛种苗简易化工厂生产方法，该方法可以克服一般组织培养的困难，特别是为一般农村塑料大棚，设施农业生产增加一种新品种。它是采用铁皮石斛种子非共生萌发培养，在种子类原球茎（PLBs）诱导分化后，将成苗分化的原球茎从培养瓶中转移到 PP 塑料袋中，塑料袋封口后，悬挂在温室大棚内，靠自然加温和光照，不但减少了碳排放，而且不会二次感染（塑料袋是封口的）。塑料袋悬挂在温室内，充分利用了有效空间，实现了铁皮石斛种苗批量化生产，每亩温室大棚可生产铁皮石斛种苗在百万级以上。

(作者：李秀兰，陈力，陈成彬，宋文芹，陈瑞阳。原载：2011年天津市科学技术奖推荐书）

参 考 文 献

陈力, 李秀兰. 白花丹参同源四倍体的诱导与鉴定. 中草药, 2009, 40(12): 1995-1997.
陈瑞阳, 宋文芹, 李秀兰. 植物染色体标本制备的去壁、低渗法及其在细胞遗传学中的意义. 遗传学报, 1982, 9(2): 151-159.
高山林, 朱丹妮, 蔡朝晖, 等. 丹参多倍体性状和药材质量的关系. 植物资源与环境学报, 1996, 5(2): 1-4.
高山林. 提高中药材质量和产量的思路和实践. 世界科学技术：中医药现代化, 2006, 8(1): 80-83.
洪德元. 植物细胞分类学. 北京：科学出版社, 1990.
黄群策, 刘峰, 向茂成, 等. 通过同源三倍体固定水稻杂种优势的新设想. 湘潭师范学院学报, 1995, 16(6): 61-63.
蓝天凤, 于宗渊, 王岱杰, 等. 硅胶柱色谱结合高速逆流色谱法分离纯化丹参中丹参酮. 中草药, 2011, 42(3): 466-469.
李允尧, 赵华英, 陈沪宁, 等. 山东省白花丹参的植物资源. 中药材, 2000, 23(2): 69-70.
刘德茂, 史德胜. 双丹口服液中丹酚酸 B 和丹参素稳定性研究. 现代药物与临床, 2009, 24(2):

102-104.

马丽虹, 瞿树林. 王传杰. 白花丹参的开发进展. 中国林副特产, 2005, 1: 72-74.

齐水秀, 杨忠孝, 李珂, 等. 白花丹参微量元素分析比较. 泰山医学院学报, 2004, 25(6): 589-593.

舒志明, 梁宗锁, 孙群, 等. 丹参雄性不育系 Sh-B 的鉴定与花粉发育过程的解剖学研究. 西北植物学报, 2006, 26(11): 2202-2207.

宋经元. 丹参. 北京: 中国中医药出版社, 2001.

宋振巧, 王建华, 王洪刚, 等. 丹参开花与繁育特性研究. 园艺学报, 2009, 36(6): 905-910.

孙玉新, 郭亚勤, 吴慧贞, 等. 植物外源激素对丹参生长和丹参酮类物质积累的影响. 中草药, 2010, 41(5): 813-818.

唐晓清, 王康才, 杨 杰, 等. 江苏优质栽培丹参新品种的选育. 江西农业学报, 2007, 19(4): 65-68.

王庆浩, 陈爱华, 张伯礼. 丹参: 一种中药研究的模式生物. 中医药学报, 2009, 37(4): 1-3.

韦辉, 刘素香, 李毅, 等. 丹参药材的综合质量评价研究. 药物评价研究, 2011, 34(5): 343-347.

武振华, 牛炳韬, 王新宇, 等. 药用植物染色体加倍的研究进展. 西北植物学报, 2005, 25(2): 2569-2574.

肖培根. 新编中药志. 北京: 化学工业出版社, 2002.

肖小河, 方清茂, 夏文娟, 等 药用鼠尾草属效值分类与丹参药材道地性. 植物资源与环境, 1997, 6(2): 17-21.

杨新杰, 万德光, 林贵兵, 等. 丹参脂溶性成分的地域分布特点分析. 中草药, 2010, 41(5): 809-812. 中国药典一部. 2010.

张红瑞, 李志敏, 高致明, 等. 丹参变异类型分析. 河南农业大学学报, 2007, 41(4): 421-424.

张兴国, 王义明, 罗国安, 等. 丹参品种资源特性的研究. 中草药, 2002, 33(8): 742-747.

郑师章. 何敏. 丹参五个样品有效成分的比较—丹参栽培应重视品种选择. 中草药, 1981, 12(3): 12-13.

Adams KL, Wendel JF. Polyploidy and genome evolution in plants. Curr Opin Plant Biol, 2008: 135-141.

Otto SP, Whitton J. Polyploid incidence and evolution. Ann Rev Genet, 2000, 34: 401-437.

第五章　人才培养

提要： 十年建树，百年建人，人才培养是一个集体的灵魂。

奋斗的 40 年植物染色体研究，愉快地渡过了！也让我们感悟到"百年建人"的重要性：第一，个人的成长离不开国家的大好环境，国家强，个人才能强。我们这一代正赶上改革开放的好时代，五千年的中华儿女从没有像今天这样扬眉吐气、受人尊重，火烧圆明园、任人掠夺的时代，一去不复返了。祖国的强大，才是我们的靠山，所以，永远不要忘记国家对我们的培养，为国家服务是我们一生的追求，为国家发展做出贡献是我们最高的荣誉，永远不要对国家不满，所以，我们要热爱共产党，热爱社会主义祖国。第二，个人成长离不开一个好的集体平台，一个好的集体平台容忍了你的缺点，尊重了你的人格，尊重了你的劳动价值，使你成长起来。所以，我们要经常回头看看，想想自己所走过的路，看看自己是怎样成长起来的，要感谢那些帮助过自己的人，感谢那些影响自己向上的人，感谢那些看不起自己反对过自己的人，他让你更坚强。因此，我们没有理由对自己所处的集体不满，要永远怀着一颗感恩的心，滴水之恩，当涌泉相报。

从 1983 年起染色体实验室才有资格招收硕士研究生，1993 年起招收博士研究生，30 多年共招收 125 名研究生，其中硕士 77 名，博士 45 名，博士后 4 名。另外，在染色体实验室做本科毕业论文的有近 100 名。百年校庆，通过这次"人才培养"项目，看到大家的成绩，非常高兴。特别是最近看到纵微星被评为南开大学名人校友，这是染色体实验室全体师生的最大光荣。祝贺你们在南开百年，能向母校献上这份厚礼，希望通过这次交流，大家互相学习，互相帮助，共同进步，为人类做出更多贡献。

南开大学染色体实验室部分毕业生介绍

纵微星,男,1967 年出生。美国罗格斯(Rutgers)大学药学院化学生物学系终身教授,新泽西癌症研究所(CINJ)肿瘤代谢生长项目负责人。主要从事肿瘤发生与治疗中细胞死亡、自噬,应激,代谢的分子机制研究。在 Science、Nature、Cell、Cell Metab、Mol Cell、Genes Dev、PNAS、Nat Commun 等国际一流学术期刊发表研究论文 80 余篇,其中作为第一作者和通讯作者发表论文 40 余篇。累计被引用次数超过 21000 次(Google Scholar Citation)。担任 NIH 等基金评审专家。

教育经历

2000—2005 博士后,美国宾夕法尼亚大学 Abramson 癌症中心,导师:Craig Thompson

1994—2000 博士,美国新泽西医科牙科大学生物化学专业,导师:Celine Gelinas

1989—1991 硕士,南开大学生物学系遗传学专业,导师:陈瑞阳

1983—1987 学士,南开大学生物学系遗传学专业,导师:陈瑞阳

重要学术贡献

发现抗凋亡蛋白 BCL-2 同源蛋白 Bfl-1/A1 是 NF-kB 转录调控的直接靶点,以及 BCL-2 家族蛋白 Bax/Bak 在线粒体与内质网途径细胞凋亡中的关键作用。研究成果获美国专利一项。

发现传统的化疗药物可通过阻断细胞的能量供应而导致细胞坏死,并且这种坏死细胞可以通过释放细胞内分子而激活机体的先天性免疫系统发挥功能,从而在抗肿瘤过程中发挥重要作用。研究成果获美国专利一项。

在细胞自噬、蛋白稳态及氧化还原的研究中,阐明了 IIIA 类磷脂酰肌醇 III 激酶(PI3K)是关键的自噬与内吞调节分子,发现 IA 类 PI3K 复合蛋白中的 p110b 通过与小 GTP 酶 Rab5 直接结合诱导自噬,并发现泛素化连接酶 TRIM21 直接调节自噬受体分子 p62 的泛素化及蛋白聚合体的形成,从而调节 p62-Keap1-Nrf2 抗氧化通路。

在细胞能量代谢途径中发现相关癌蛋白 c-myc 的通过胸腺嘧啶 DNA 糖基化酶 TDG 介导的 DNA 去甲基化上调谷氨酰胺合成酶。

确立了细胞内源性蛋白酶抑制蛋白 SerpinB3/SCCA1 表达水平与恶性肿瘤分化呈正相关,原癌基因 Kras 通过 SCCA1 调节慢性内质网应激效应,从而调节细胞因子分泌并促进对于肿瘤扩增和转化。

代表论文

Pan JA, Sun Y, Jiang YP, Bott AJ, Jaber N, Dou Z, Yang B, Chen JS, Catanzaro JM, Du C, Ding WX, Diaz-Meco MT, Moscat J, Ozato K, Lin RZ, and Zong WX*. 2016. TRIM21 ubiquitylates SQSTM1/p62 and suppresses protein sequestration to regulate redox homeostasis. Mol Cell 61: 720-733. "Featured Article".

Bott AJ, Peng IC (co-first author), Fan Y, Faubert B, Zhao L, Li J, Neidler S, Sun Y, Jaber N, Krokowski D, Lu W, Pan JA, Powers S, Rabinowitz J, Hatzoglou M, Murphy DJ, Jones R, Wu S, Girnun G, and Zong WX*. 2015. Oncogenic Myc induces expression of glutamine synthetase through promoter demethylation. Cell Metab 22: 1068-1077.

Catanzaro JM, Sheshadri N (equal contribution), Pan JA, Sun Y, Shi C, Li J, Powers RS, Crawford HC, and Zong WX*. 2014. Oncogenic Ras induces inflammatory cytokine production by upregulating the squamous cell carcinoma antigens SerpinB3/B4. Nat Commun 5: 3729.

Dou Z, Pan JA, Dbouk HA, Ballou LM, DeLeon JL, Fan YJ, Chen JS, Liang Z, Li G, Backer JM, Lin RZ, and Zong WX*. 2013. Class IA PI3 p110β subunit promotes autophagy through Rab5 small GTPase in response to growth factor limitation. Mol Cell 50: 29-42.

Jaber N, Dou Z, Chen JS, Catanzaro J, Jiang YP, Ballou LM, Selinger E, Ouyang X, Lin RZ, Zhang J, and Zong WX*. 2012. Class III PI3K Vps34 plays an essential role in autophagy and in heart and liver function. Proc Nat Acad Sci, USA109: 2003-2008.

Zong WX and Thompson CB. 2006. Necrotic death as a cell fate. Genes Dev 20: 1-15.

Zong WX, Ditsworth D, Bauer DE, Wang ZQ, and Thompson CB. 2004. Alkylating DNA damage stimulates a regulated form of necrotic cell death. Genes Dev 18: 1272-1282.

Wei MC, Zong WX (co-first author), Cheng EH, Lindsten T, Panoutsakopoulou V, Ross AJ, Roth KA, MacGregor GR, Thompson CB, and Korsmeyer SJ. 2001. Proapoptotic BAX and BAK: a requisite gateway to mitochondrial dysfunction and death. Science 292, 727-730.

Zong WX, Lindsten T, Ross AJ, MacGregor GR, and Thompson CB. 2001. BH3-only proteins that bind pro-survival Bcl-2 family members fail to induce apoptosis in the absence of Bax and Bak. Genes Dev 15, 1481-1486.

Zong WX, Edelstein LC, Chen C, Bash J, and Gelinas C. 1999. The prosurvival Bcl-2 homolog Bfl-1/A1 is a direct transcriptional target of NF-kappaB that blocks TNFalpha-induced apoptosis. Genes Dev 13, 382-387.

授权专利

1. Zong WX and Thompson CB. Methods of identifying anti-cancer agents and uses thereof (U.S. Patent # 8304182).

2. Zong WX, Lindsten T, and Thompson CB. Immortalized mouse fibroblast cell lines deficient in Bax and/or Bak (U.S. Patent# 20030091982).

苏健英，1985年南开大学本科毕业。非常幸运本科论文师从陈瑞阳先生，在云南昆明和陈老师、宋老师、李老师度过了大学中最难忘的三个月，实习中成功地分出了植物染色体G-带，也与实验室的老师们结下了一生的情缘。1988年南开大学生物系硕士毕业。非常有幸成为陈瑞阳教授第一个硕士生，从此以后便走上了科学研究的殿堂。

硕士毕业后在天津医科大学执教六年，1995年去瑞士日内瓦大学医学院读书。在美国曾任职于新泽西州立大学药学院、美国惠氏制药公司，目前就职于默沙东制药公司生物大分子研发部。

毕业前夕，先生在我的实验记录本上写了两行字，作为送给我的礼物：

老老实实做人，踏踏实实做学问。

许多年来，我走过很多地方，但我一直牢记着这两句话，虽然离开实验室以后再也没有做过染色体方面的研究，但染色体实验室诚实勤奋，一丝不苟的精神，影响了我的整个人生。

郑坚瑜，1986年毕业于南开大学生物系遗传专业，本科生毕业论文《玉米染色体的G-带研究》，1989年硕士毕业于南开大学生物系遗传专业，研究生毕业论文《豆科植物的带型分析》。导师陈瑞阳教授。1994年博士毕业于日本广岛大学，博士论文《关于豆科植物的起源和亲缘关系的研究》，导师田中隆庄教授。先后于香港科技大学和美国耶鲁大学从事博士后研究。曾任南开大学生命科学院、分子生物学研究所副教授、教授、博士生导师。现任美国康涅迪格州州立大学终身教授，主要从事本科生教学工作。主授课程包括：普通生物学、人体生物学、微生物学以及人体的规律学。所教学生人数达2000人。曾获美国全国生物教学贡献奖（2012），康州州立大学优秀教学奖（2013）。

近期出版和发表论文

Jianyu Zheng，Human Biology Laboratory Manual 3rd edition，KONA Publishing & Media Group 2019，ISBN：978-1-945628-61-0.

Jianyu Zheng，Human Biology Laboratory Manual 2nd edition，KONA Publishing & Media Group 2016，ISBN：978-1-935987-93-2.

Jianyu Zheng，Human Biology Laboratory Manual 1st edition，KONA Publishing & Media Group 2016，ISBN：978-1-935987-73-4.

Understanding the Structure and Function of Human Amyloid beta Protein by StarBiochem Analysis. Cassandra Manger，Shirley Astacio，Korey Mceneany and Jianyu Zheng. Abstract Number 107，page 85，New England Science Symposium Publication，March 2，2013.

Clicker as a Tool for Student Engagement in Undergraduate Science Education. Jianyu Zheng，Kathy Murphy and Bonnie Simon. The Exchange-The official publication of

New England Faculty Development Consortium, Spring 2012.

Lon peptidase 1 (LONP1) -dependent breakdown of mitochondrial 5-aminolevulinic acid synthase protein by heme in human liver cells. Tian Q, Li T, Hou W, Zheng J, Schrum LW, Bonkovsky HL. J Biol Chem. 2011 Jul 29; 286 (30): 26424-30. Epub 2011 Jun 9.

MicroRNA-196 represses Bach1 protein and hepatitis C virus gene expression in human hepatoma cells expressing hepatitis C viral proteins. Hou W, Tian Q, Zheng J, Bonkovsky HL. Hepatology. 2010 May; 51 (5): 1494-504.

Zinc mesoporphyrin induces rapid proteasomal degradation of hepatitis C nonstructural 5A protein in human hepatoma cells. Hou W, Tian Q, Zheng J, Bonkovsky HL. Gastroenterology. 2010 May; 138 (5): 1909-19. Epub 2009 Nov 10.

Tian Q, Hou W, Li T, Zheng J, Bonkovsky HL. Heme induces degradation of 5-aminolevulinic acid synthase-1 (ALAS1) protein in mitochondria of human liver cells. American Association for the Study of Liver Diseases, Boston, Oct 10/29-11/2, 2009.

Hou W, Tian Q, Zheng J, Bonkovsky HL. Zinc porphyrins induce polyubiquitination and proteasomal degradation of hepatitis C non-structural protein NS5A and display anti-viral activity. Poster 144, presented at 13th International Symposium on Viral Hepatitis and Liver Disease, Washington, DC, March 20-24, 2009.

Ghosh S, Zheng J, Hwang S, Steuerwald N, Bonkovsky HL. Changes in the hepatocytic proteome related to expression of hepatitis C virus (HCV). Oral presentation 101 at 13th International Symposium on Viral Hepatitis and Liver Disease, Washington, DC, March 20-24, 2009.

Zheng J, Sachidanandam R, Bennett K, Bonkovsky HL. MicroRNA 16 regulates expression of the human 5-aminolevulinic acid synthase-1 (ALAS1) gene through targeting the coding region of the gene. American Association for the Study of Liver Diseases, San Francisco, California 11/1-3, 2008.

Hou W, Zheng J, Tian Q, Zheng J, Bonkovsky HL. MicroRNA-196 represses BACH1 and hepatitis C viral NS5A protein expression in human hepatocytes expressing HCV non-structural proteins. American Association for the Study of Liver Diseases, San Francisco, California 11/1-3, 2008.

Zheng J, Tian Q, Hou W, Bonkovsky HL. Identification of heme-responsive elements in the 5'-flanking region of the human 5-aminolevulinic acid synthase (ALAS1) gene in human hepatoma cells. American Association for the Study of Liver Diseases, San Francisco, California 11/1-3, 2008.

韩健，天津市人，1992届南开大学生物系水生生物专业研究生。研究生导师是张闰生和陈瑞阳两位教授。研究生毕业研究论文《水生生物大型蚤的染色体原位杂交》。研究生毕业后赴美留学，2002年获美国宾西法尼亚大学营养学博士学位。现任美国北卡州农业科技大学生物系副教授。现进行以教学为主的工作。科研方向是营养元素对神经衰竭、传导障碍及帕金森氏症的影响。自2009年工作以来发表文章数篇，获数个科研项目研究基金，共培养了博士，硕士和本科生近三十名。

崔香芹，1991年南开大学生物系本科毕业，1997年染色体实验室硕士毕业，研究生毕业论文《蚕豆植物染色体G-带的研究》。导师陈瑞阳教授。2001年美国爱荷华州立大学遗传学博士。先后在杰克逊实验室做博士后，阿拉巴马大学伯明翰生物统计系任教职。现为艾默里大学生物统计和生物信息系副教授。主要从事生物信息、临床试验以及电子病历方面的研究。

彭永康，男，浙江省绍兴市人。1974年南开大学本科毕业，1994~1997年在南开大学生命科学学院染色体实验室攻读博士学位，师从陈瑞阳教授。现任天津师范大学生命科学学院教授，博士生导师，"十五"细胞生物学天津市重点建设学科学科带头人，"十一五"天津市细胞生物学重点学科学科带头人。中国细胞生物学会第七届（2005—2008）、第八届（2008—2011）常务理事，细胞通讯与信号转导专业委员会副主任；天津市细胞遗传与分子调控重点实验室学术委员会副主任，天津市人民政府学位委员会第二届学科评论组（生物学组）成员，天津市第三届自然科学基金专家委员会委员，天津市农业生物技术研究中心专家委员会委员；西北师范大学兼职教授。

主要从事高等植物细胞分裂与细胞器蛋白质组学和小麦抗白粉病基因克隆与结构功能分析。主持国家重大基础理论研究前期项目，教育部科学技术研究重点项目，天津市重大攻关项目8项，获天津市技术发明二等奖2项，进步二等奖1项，国家发明专利授权16项。在国内外重要刊物上发表论文100余篇，SCI收录10篇。

孙易，1999 年毕业于南开大学生物系染色体实验室，师从陈瑞阳教授和钟贻诚教授，获遗传学博士学位。1999 年至 2003 年间，分别于比利时根特大学和美国新泽西州立大做博士后研究；2003 年至 2006 年于 Rutgers 大学的药学院及 Waksman 研究所任副研究员和研究员。2006 年至今分别就职于美国生物参考公司和组织病理医学服务中心，任资深科学家、分子诊断实验室总监及首席研发官。专注于疾病的分子检测产品及其试剂的开发与验证，研究方向为分子肿瘤学、遗传学和传染性疾病的临床检测。持有美国纽约州临床实验室总监执照。在生物医学和分子水平的临床诊断等领域，主持和参与了多个与精准医疗和转化医学相关的项目，在核心学术刊物上发表了多篇论文，是 2 项美国专利的发明人，并获得了多个研究奖项。2008 年研发的血液肿瘤基因芯片项目及 2012 年研发的实体瘤和血液肿瘤的二代测序项目均领先获得美国卫生部门的认证，带领团队成为美国东岸首个提供临床检测服务的商业机构。近十多年来研究开发出几十个临床分子检测项目，尤其善于将前沿的分子生物学技术应用于临床，转化为成熟产品，并获得国际卫生部门的认证，占据高端市场，已为所在公司带来了巨大的经济效益和社会效益。

于建春，博士、博导、研究员。

天津中医药大学第一附属医院，细胞治疗与分子诊断研究室主任，肿瘤科副主任。

1994～1997 年于南开大学生命科学院师从陈瑞阳教授攻读遗传学硕士学位。在读期间主要从事"细胞显微分离－微量蛋白电泳方法的建立及在细胞周期蛋白质研究中的应用"研究。相关成果发表在《植物学报》（1998）、《科学通报》（1997）及《应用与环境生物学报》（1997）上。

1997～2000 年获天津中医药大学针灸推拿学博士学位；2001～2003 年天津大学做博士后；2000～2012 年在天津中医药大学第一附属医院针灸研究所工作，入选教育部新世纪优秀人才和天津市"131"第一层次人才；2012 年建立细胞治疗与分子诊断实验室并进入肿瘤科工作。

研究方向及主要业绩：研究方向有两个，一是老年期痴呆及针灸的分子机理研究；二是临床肿瘤免疫学及肿瘤的生物治疗研究。承担国家级研究项目 20 余项，在国内外发表论文 200 余篇，SCI 论文 30 余篇，主要围绕阿尔茨海默病和血管性痴呆及针刺的临床疗效和机理进行研究，从中医理论、临床疗效和疗效机制三个方面验证了导师石孝敏"三焦气化失常—衰老相关论"的学说并推广了其创立的"三焦针法"的应用，形成了完整的针刺治疗老年期痴呆的理法针效体系。目前该针法已推广到日本和美国，展示了针刺的疗效远远优于相关药物，该系列成果先后获得教育部科技进步一等奖 1 项，天津市科技进步二等奖 8 项，中华中医药学会、中西医结合学会等科学技术进步二等奖 7 项。目前正在研究"三焦针法"治疗肿瘤的适宜病症及临床肿瘤免疫学。

江赐忠，1995 年获南开大学微生物学学士，1998 年获南开大学遗传学硕士，导师陈瑞阳教授。2004 年获美国衣阿华州立大学遗传学博士。博士毕业后，先后在美国冷泉港实验室、弗吉尼亚联邦大学、宾夕法尼亚州立大学从事博士后研究。2009 年回国被聘为同济大学生命科学与技术学院特聘教授、博士生导师，任职至今。2009 年被聘为国家科技部"973"项目首席科学家。回国后先后获教育部"新世纪优秀人才""浦江人才""曙光计划"与"东方学者"资助。主要从事细胞命运转化中的表观遗传调控机制研究。主持国家、省部级项目 9 项，参与 3 项。发表 SCI 学术论文 73 篇，总影响因子 556.3，单篇最高 41.6，单篇引用超过 853 次。

研究方向

（1）胚胎发育中的表观遗传调控机制（染色质重塑、DNA 甲基化等）。

（2）细胞编程与重编程中的表观遗传调控机制（染色质重塑、DNA 甲基化等）。

实验室网页：http://wukong.tongji.edu.cn/

科研项目

国家自然科学基金生命学部创新研究群体项目，31721003，早期胚胎发育与体细胞重编程的表观遗传机制，2018/01-2023/12，在研、参加。

国家自然科学基金面上项目，31771419，小鼠单倍体胚胎干细胞二倍体化的表观遗传调控机制，2018/01-2021/12，在研、主持。

国家重点研发计划专项-干细胞及转化研究专项，2016YFA0100403，组蛋白与 DNA 修饰的动态调控机制研究，2016/07-2020/12，在研、参加。

国家自然科学基重大研究计划项目（集成项目），91519309，小鼠体细胞重编程中的染色质重塑路线图及表观遗传调控机制，2016/01-2016/12，已结题、主持。

国家自然科学基金面上项目，31271373，核小体重塑在果蝇母源向合子转化中的表观调控作用，2013/01-2016/12，已结题、主持。

上海市东方学者计划项目，果蝇幼虫向蛹转化中的核小体重排，2012/01-2014/12，已结题、主持。

国家自然科学基金重大研究计划重点支持项目，91019017，决定 ES 和 iPS 细胞全能性关键因素及表观遗传调控机制，2011/01-2014/12，已结题、主持。

国家重大科学研究计划项目，2011CB965104，人多能干细胞多能性维持和发育潜能差异的系统研究，2011/01-2015/8，已结题、参加。

上海市曙光计划项目，10SG24，组蛋白修饰在果蝇胚胎发育中的作用与机制研究，2011/01-2012/12，已结题、主持。

上海市浦江人才项目，10PJ1409500，果蝇胚胎发育中的 microRNA 表观调控作用与机制研究，2010/09-2012/12，已结题、主持。

教育部新世纪人才项目，NCET-10-0600，果蝇 SWI/SNF 染色质重塑酶核心亚基 Brm 对核小体定位的影响及其作用机制，2010/01-2013/12，已结题、主持。

国家重大科学研究计划项目，2010CB944900，胚胎发育的核小体重排和染色质重塑，2010/01-2014/8，已结题、主持。

史岸冰，教授，博士生导师，华中科技大学基础医学院副院长，生物化学和分子生物学系主任。

1995 年南开大学本科毕业，1998 年南开大学遗传专业硕士毕业，师从陈瑞阳教授和郑坚瑜教授，2004 年美国新泽西州立 Rutgers 大学获得理学博士学位，此后在美国斯坦福大学接受博士后训练，2013 年回到华中科技大学工作。近年在囊泡运输调控方面的工作成果发表在 J Cell Biol、EMBO J、PLoS Genet、PNAS、Curr Biol、Mol Biol Cell 等国际著名细胞生物学期刊，研究工作被 Science、Cell、Nature Cell Biology、PNAS 等期刊多次引用。主持国家自然科学基金委、国家科技部、湖北省自然科学基金等项目，担任 Autophagy、Mol Biol Cell 等期刊审稿人。2013 年入选教育部新世纪优秀人才，2014 年入选国家青年千人计划、2018 年获国家杰出青年科学基金资助。

现任"医学细胞与分子生物学"八年制整合课程负责人、"双一流"建设项目"代谢与疾病基础研究实验技术"课程负责人，国家卫生和计划生育委员会"十二五"规划教材《医学分子生物学》主编，"十二五"国家级规划教材《医学生物化学与分子生物学》主编。获华中科技大学 2016 年度"三育人奖"、华中科技大学第四届研究生"知心导师"荣誉称号。

现任国家自然科学基金二审专家、中国生物物理学会膜生物学分会理事、中国细胞生物学会青年工作委员会委员、中俄医科大学联盟青年联盟副主席、中国动物学会发育生物学专业委员会委员、中华医学会医学细胞生物学分会委员、湖北省暨武汉市生物化学与分子生物学会副理事长。

张峰，1975 年出生，美国明尼苏达大学植物及微生物系，助理教授，研究方向：DNA 重组修复、植物基因组编辑、合成生物学。

邮箱：zhangumn@umn.edu

实验室网页：https://cbs.umn.edu/contacts/feng-zhang

1992~1996 年就读于南开大学生命科学院分子生物学专业，1996~1999 年师从宋文芹教授和陈瑞阳教授，就读于南开大学细胞与遗传专业染色体实验室，获得硕士学位。1999~2005 年就读于美国爱荷华州立大学（Iowa State University）遗传系，获得博士学位。2005~2009 年，分别于佐治亚大学（University of Georgia）和明尼苏达大学进行博士后研究。2010 年和 Dan Voytas 教授一起研发了第二代基因组编辑技术——TALEN 技术。TALEN 技术大大提高了基因编辑技术的易用性，使高效编辑任意基因成为了可能。该技术于 2012 年被科学杂志评为年度 10 大科技突破。2010 年和 Dan Voytas 教授一起创立第一家利用基因组编辑技术进

行作物性状改良的生物技术公司——Calyxt，并历任研发总监和首席运营官，在烟草、大豆、土豆、油菜、小麦以及苜蓿等多种作物中进行了精确的单基因和多基因编辑，研发了无反式脂肪酸的大豆、耐储土豆、高纤维小麦、易消化苜蓿等多个基因编辑作物品种。其中 7 个品种获得美国农业部非转基因品种认证。领导该公司于 2017 年在纳斯达克成功上市。2018 年回到明尼苏达大学任助理教授，目前主要致力于新一代大规模高通量植物基因组编辑技术的研发，重要农艺性状改良以及合成生物学方面的研究。已在 Nature、Science、Plant Cell、PNAS 等国际一流期刊发表文章 30 余篇，拥有专利 10 余个。

代表性论文

Jin S，Zong Y，Gao Q，Zhu Z，Wang Y，Qin P，Liang C，Wang D，Qiu JL，Zhang F，Gao C. Cytosine，but not adenine，base editors induce genome-wide off-target mutations in rice. Science. 2019，364（6437）：292-295.

Feng Zhang，Daniel F Voytas，Modulating gene translational control through genome editing，National Science Review. 2019，6（3）：391.

Ji，X.，Si，X.，Zhang，Y.，Zhang，H.，Zhang，F.，& Gao，C. Conferring DNA virus resistance with high specificity in plants using virus-inducible genome-editing system. Genome biology，2018，19（1），197.

Zhang，F.，& Voytas，D. F. Synthetic genomes engineered by SCRaMbLEing. Science China Life Sciences，2018，61（8），975-977.

Zachary L Demorest，Andrew Coffman，Nicholas J Baltes，Thomas J Stoddard，Benjamin M Clasen，Song Luo，Adam Retterath，Ann Yabandith，Maria Elena Gamo，Jeff Bissen，Luc Mathis，Daniel F Voytas，Feng Zhang*. Direct stacking of sequence-specific nuclease-induced mutations to produce high oleic and low linolenic soybean oil. BMC Plant Biology. 2016，6：225（*Corresponding author）.

Thomas J Stoddard，Benjamin M Clasen，Nicholas J Baltes，Zachary L Demorest，Daniel F Voytas，Feng Zhang，Song Luo. Targeted mutagenesis in plant cells through transformation of sequence-specific nuclease mRNA. PLoS ONE. 2016，11（5）：e0154634.

Jin Li，Thomas J Stoddard，Zachary L Demorest，Pierre‐Olivier Lavoie，Song Luo，Benjamin M Clasen，Frederic Cedrone，Erin E Ray，Andrew P Coffman，Aurelie Daulhac，Ann Yabandith，Adam J Retterath，Luc Mathis，Daniel F Voytas，Marc‐André D'Aoust，Feng Zhang*. Multiplexed，targeted gene editing in Nicotiana benthamiana for glyco-engineering and monoclonal antibody production. Plant Biotechnology Journal. 2016，14（2）：533-542（*Corresponding author）.

Benjamin M. Clasen，Thomas J. Stoddard，Song Luo，Zachary L. Demorest，Jin Li，Frederic Cedrone，Redeat Tibebu，Shawn Davison，Erin E. Ray，Aurelie Daulhac，Andrew Coffman，Ann Yabandith，Adam Retterath，William Haun，Nicholas J. Baltes，Luc Mathis，Daniel F. Voytas and Feng Zhang*. Improving cold storage and processing traitsin potato through targeted gene editing. Plant Biotechnology Journal. 2016，14（1）：

169-176(*Corresponding author).

Song Luo, Jin Li, Thomas J Stoddard, Nicholas J Baltes, Zachary L Demorest, Benjamin M Clasen, Andrew Coffman, Adam Retterath, Luc Mathis, Daniel F Voytas, Feng Zhang*. Non-transgenic plant genome editing using purified sequence-specific nucleases. Molecular Plant. 2015, 8 (9): 1425-1427 (*Corresponding author).

William Haun, Andrew Coffman, Benjamin M Clasen, Zachary L Demorest, Anita Lowy, Erin Ray, Adam Retterath, Thomas Stoddard, Alexandre Juillerat, Frederic Cedrone, Luc Mathis, Daniel F Voytas, Feng Zhang*. Improved soybean oil quality by targeted mutagenesis of the fatty acid desaturase 2 gene family. Plant Biotechnology Journal. 2014, 12 (7): 934-940 (*Corresponding author).

Yiping Qi, Xiaohong Li, Yong Zhang, Colby G Starker, Nicholas J Baltes, Feng Zhang, Jeffry D Sander, Deepak Reyon, J Keith Joung, Daniel F Voytas. Targeted deletion and inversion of tandemly arrayed genes in Arabidopsis thaliana using zinc finger nucleases. G3 Genes|Genomes|Genetics. 2013, 3 (10): 1707-1715.

Marine Beurdeley, Fabian Bietz, Jin Li, Severine Thomas, Thomas Stoddard, Alexandre Juillerat, Feng Zhang, Daniel F Voytas, Philippe Duchateau, George H Silva. Compact designer TALENs for efficient genome engineering. Nature Communications. 2013, 4: 1762.

Y Qi, Y Zhang, F Zhang, JA Baller, SC Cleland, Y Ryu, CG Starker, DF Voytas. Increasing frequencies of site-specific mutagenesis and gene targeting in Arabidopsis by manipulating DNA repair pathways. Genome Research. 2013, 23 (3): 547-554.

Y Zhang*, F Zhang*, X Li, JA Baller, Y Qi, CG Starker, AJ Bogdanove, DF Voytas. Transcription Activator-Like Effector Nucleases Enable Efficient Plant Genome Engineering. Plant Physiol. 2013. 161 (1): 20-27 (* co-first author).

张荣信,博士、教授、博导。广东药科大学生命科学与生物制药学院院长。1996~1999年于南开大学生命科学院师从陈瑞阳教授攻读遗传学博士学位。在读期间主要从事黑麦A、B染色体遗传同源性研究以及黑麦B染色体着丝粒区基因文库的构建。相关成果发表在《科学通报》(1998,中英文版)及《南开大学学报》(1999)。获得南开大学特等奖学金以及《南开周报》(1998年11月27日)报道。

2000~2004年任美国贝勒医学院神经学及外科学系博士后;2005~2006年任贝勒医学院外科学系讲师;2006~2009年任香港大学医学院内科学系,助理教授(研)、博导;2010~2018年,经天津医科大学免疫系教授,博导。天津市特聘教授,天津市"五一劳动奖章"获得者。2018年至今,广东药科大学生命科学与生物制药学院教授、院长。

研究方向及主要业绩:主要从事肿瘤免疫和自身免疫研究。研究胰腺癌、肠癌、肝癌等恶性肿瘤以及自身免疫疾病多发性硬化症等疾病的病理发生和免疫治疗手段。发表相关 SCI 论文 70 余篇，累计影响因子 300 多分，H 指数 23。受 Elsevier 出版社邀请主编 *The Epigenetics of Autoimmunity* 于 2018 年出版。申请国家发明专利 6 项，美国专利 1 项。在 30 多个国内及国际学术会议做过报告。

代表性论文

Yan Li, Dongmei Zhou, Yinghui Ren, Zimu Zhang, Xiangdong Guo, MingKun Ma, Zhenyi Xue, Jienv Lv, Hongkun Liu, Qing Xi, Long Jia, Lijuan Zhang, Ying Liu, Qi Zhang, Jun Yan, Yurong Da, Fei Gao, Jianbo Yue, Zhi Yao, Rongxin Zhang*. miR-223 restrains autophagy and promotes CNS inflammation by targeting ATG16L1. Autophagy. 2019, 15 (3): 478-492.

Zhenyi Xue#, Zimu Zhang#, Hongkun Liu, Wen Li, Xiangdong Guo, Zhihui Zhang, Ying Liu, Long Jia, Yan Li, Yinghui Ren, Hongwei Yang, Lijuan Zhang, Qi Zhang, Yurong Da, Junwei Hao, Zhi Yao and Rongxin Zhang*. lincRNA-Cox2 regulates NLRP3 inflammasome and autophagy mediated neuroinflammation. Cell Death and Differentiation. 2019, 26 (1): 130-145.

Zhenyi Xue#, Qing Xi#, Hongkun Liu, Xiangdong Guo, Jieyou Zhang, Zimu Zhang, Yan Li, Guangze Yang, Dongmei Zhou, Huiyun Yang, Lijuan Zhang, Qi Zhang, Chao Gu, Juhong Yang, Yurong Da*, Zhi Yao, Shuguang Duo* and Rongxin Zhang*. miR-21 promotes NLRP3 inflammasome activation to mediate pyroptosis and endotoxic shock. Cell Death Dis. 2019, 12; 10 (6): 461.

Zimu Zhang#, Zhenyi Xue#, Ying Liu#, Hongkun Liu, Xiangdong Guo, Yan Li, Hongwei Yang, Lijuan Zhang, Yurong Da, Zhi Yao and Rongxin Zhang*. MicroRNA-181c Promotes Th17 Cell Differentiation and Mediates Experimental Autoimmune Encephalomyelitis. Brain Behavior and Immunity. 2018, 70: 305-314.

Bingqing Huang, Huipeng Yang, Xixi Cheng, Dan Wang, Shuyu Fu, Wencui Shen, Qi Zhang, Lijuan Zhang, Zhenyi Xue, Yan Li, Yurong Da, Qing Yang, Zesong Li, Li Liu, Liang Qiao, Ying Kong*, Zhi Yao, Peng Zhao*, Min Li and Rongxin Zhang*. tRF/miR-1280 suppresses stem cell-like cells and metastasis in colorectal cancer. Cancer Research. 2017, 77 (12): 3194-3206.

Kai Zhang, Yawei Guo, Zhenzhen Ge, Zhihui Zhang, Yurong Da, Wen Li, Zimu Zhang, Zhenyi Xue, Yan Li, Yinghui Ren, Long Jia, Koon-Ho Chan, Fengrui, Yang, Jun Yan, Zhi Yao, Aimin Xu and Rongxin Zhang*. Adiponectin suppresses T helper 17 cells differentiation and limits autoimmune CNS inflammation via the SIRT1/PPARγ/RORγt pathway. Molecular Neurobiology. 2017, 54 (7): 4908-4920.

Qi Zhang, Zimu Zhang, Meiyu Peng, Shuyu Fu, Rongxin Zhang*. CAR-T cell therapy in gastrointestinal tumors and hepatic carcinoma: from bench to bedside. OncoImmunology. 2016, 12: e1251539.

Biao Huang, Xixi Cheng, Huafeng Wang, Wenjing Huang, Zha la Ga hu, Dan Wang,

Kai Zhang, Huan Zhang, Zhenyi Xue, Yurong Da, Ning Zhang, Yongcheng Hu, Zhi Yao, Liang Qiao, Fei Gao* and Rongxin Zhang*. Mesenchymal stem cells and their secreted molecules predominantly ameliorate fulminant hepatic failure and chronic liver fibrosis in mice respectively. Journal of Translational Medicine. 2016, 14 (1): 45.

Wen Li, Zhihui Zhang, Kai Zhang, Zhenyi Xue, Yan Li, Zimu Zhang, Lijuan Zhang, Chao Gu, Qi Zhang, Junwei Hao, Yurong Da, Zhi Yao, Ying Kong*, Rongxin Zhang*. Arctigenin Suppress Th17 Cells and Ameliorates Experimental Autoimmune Encephalomyelitis Through AMPK and PPAR-γ/ROR-γt Signaling. Molecular Neurobiology. 2016, 53 (8): 5356-66.

Zimu Zhang, Rongxin Zhang*. Epigenetics in autoimmune diseases: pathogenesis and prospects for therapy. Autoimmunity Reviews. 2015, 14 (10): 854-63.

出版专著

主编《The Epigenetics of Autoimmunity》，Elsevier 出版社，2018 年 5 月，ISBN: 978-0-12-809912-4

申请专利

1. 张荣信、薛振毅。信筒子醌在制备治疗自身免疫疾病药物方面的应用。中国发明专利，专利号 ZL201110434108.8。
2. 张荣信、薛振毅。ZSTK474 在制备治疗自身免疫疾病药物方面的应用。中国发明专利，申请号 CN201110403904.5。
3. 张荣信、葛禛禛、薛振毅、张凯。伏立诺他在制备治疗自身免疫及炎症性疾病药物方面的应用。中国发明专利，申请号 CN201210332723.2。
4. 张荣信、葛禛禛、薛振毅、张凯。白杨素在制备治疗自身免疫及炎症性疾病药物方面的应用。中国发明专利，申请号 CN201210329054.3。
5. 张荣信、张凯、葛禛禛、薛振毅。胡桃醌在制备治疗自身免疫及炎症性疾病药物方面的应用。中国发明专利，申请号 CN201210329059.6。
6. 张荣信、张凯、张丽娟、薛振毅、李岩。FATS 作为黑色素瘤免疫治疗的靶点及应用。中国发明专利，申请号 CN201610315637.9。
7. 张荣信、张凯、张丽娟、薛振毅、李岩。FATS 作为黑色素瘤免疫治疗的靶点及应用。PCT 国际专利优先权，申请号：PCT/CN2017/082989。
8. Rongxin ZHANG, Kai ZHANG, Lijuan ZHANG.FATS as a target for treating tumors and uses thereof.美国发明专利，US16134983。

获得资助的研究课题

一个染色体脆性位点基因调控抗肿瘤免疫的作用及细胞和分子机制研究，国家自然科学基金委面上项目，2019.01-2022.12，No.81872320，主持。

广东省"生物制药"创新团队。广东省教育厅"创新强校"项目，团队负责人。

间充质干细胞上清液治疗肝纤维化的作用机制研究。天津市优秀科技特派员奖励基金，天津市科学技术委员会。2014.09-2015.10。No.14JCTPJC00487，项目负责人。

新疆维吾尔族自治区科技厅"科技支疆计划"项目计划：microRNAs、MIF、FGF21 在心肌缺血损伤早期预警中的作用及相关机制研究。新疆医科大学负责人，李晓梅。合作支疆单位天津医科大学负责人，张荣信。2014.01-2016.12。No.201491176。

S100P 调节胰腺癌免疫逃逸的作用机制研究，国家自然科学基金委面上项目，2013.01-2016.12，No. 81272317，项目负责人。

脂联素对自身免疫脑脊髓炎的调节机制和防治研究，天津市应用基础及前沿技术研究计划（天津市科委自然科学基金重点项目），2012.04-2015.03。12JCZDJC23500，项目负责人。

脂联素对自身免疫脑脊髓炎（EAE）的调节机制和防治研究，国家自然科学基金委面上项目，2012.01-2015.12，No. 81172864，项目负责人。

新型纳米药物调控肝癌免疫微环境的机制和药效学研究。科技部--国家重大科学研究计划（973 重大）"肝癌治疗的新型纳米药物研究"项目子课题。2012.01-2016.12。2012CB932503，学术骨干。

脂联素（adiponectin）调节 TH17 细胞和非可控性炎症恶性转化的分子机制，国家自然科学基金委重大研究计划培育项目，2011.01-2013.12，No. 91029705，项目负责人。

C-反应蛋白在高血压条件下对心脏炎症及纤维化的促进作用。大学基金委员会，香港，项目负责人。

孙德岭，博士，研究员。1982 年毕业于河北农业大学园艺系蔬菜专业获学士学位，同年考入内蒙古农牧学院农学系蔬菜专业攻读硕士学位；1985 年研究生毕业分配到天津市蔬菜研究所工作至今。1998-2001 年南开大学生命科学院师从陈瑞阳教授攻读遗传学博士学位。现任天津市农业科学院副院长，天津市杰出人才、入选"国家新世纪百千万人才工程"、享受国务院政府特殊津贴专家、天津市花椰菜育种授衔专家、国家大宗蔬菜产业技术体系的菜花岗位专家。主要从事花椰菜育种研究工作，花椰菜研究水平和良种产业化规模居国内领先。

先后主持国家"863"计划课题、国家"973"计划前期研究专项、国家农业科技成果转化资金、国家高新技术产业化、农业部重点攻关、天津市重大攻关、重点攻关、天津市重点自然基金等项目 50 多项。发表论文 100 余篇，其中 SCI 论文 3 篇，核心期刊论文 80 余篇，著作 4 部。

研究成果获国家及省部级奖励 23 项，其中作为第一完成人获国家及省部级奖励 18 项。作为第一完成人获国家科学技术进步二等奖 1 项，全国农牧渔业丰收一等奖 1 项，天津市科学技术进步一等奖 3 项，天津市科学技术进步二等奖 5 项，天津市科学技术进步三等奖 5 项。作为参加人获天津市科技进步二等奖 2 项、三等奖 1 项。

花椰菜是我国主要蔬菜种类之一。但是在上世纪 70 年代我国菜花种植面积很小，而且品种乱杂、退化严重，杂交品种全部依赖进口。为了改变我国菜花生产的落后状况，多年来他带领菜花研究课题坚持自主创新，攻克了多项菜花育种技术难题，先后主持育成了"夏雪"、"丰花"、"津雪"、"津品"四大系列 26 个具有自主知识产权的花椰菜品种，新品种推广到全国二十几个省市，部分良种已出口东南亚及印度、巴基斯坦等国，改变了我国菜花优良品种短缺的局面。其中育成的"津品70"、"津品66"菜花新品种，连续3年推广面积居全国同类品种第一，菜花新品种累计推广面积250多万亩，创社会效益45多亿元，占我国菜花同类品种种植面积20%以上。基于取得的突出创新成果和新品种转化效益，2013年度获全国农牧渔业丰收奖一等奖；主持完成的"蔬菜良种科技创新工程"2014年获天津市科技进步一等奖；主持完成的"花椰菜育种技术创新及新品种选育"2016年获天津市科技进步一等奖。

技术创新是品种创新的前提和基础。多年来，带领课题组坚持自主创新，在国内外首次研制出"花椰菜育种专家系统"，将信息技术应用菜花育种，该系统预测准确率达80%，填补了国内外菜花该领域的空白。为了提高我国花椰菜抗病育种技术水平，在国内率先建立了花椰菜TuMV及黑腐病苗期人工接种抗性鉴定方法和标准，并系统地应用于菜花抗病育种实践，创制出一批优异的抗病资源，为开展花椰菜抗病新品种选育奠定了基础。主持完成的花椰菜游离小孢子培养技术研发项目，创建了国内第一个花椰菜小孢子培养技术体系，研发出花椰菜游离小孢子培养专用的"MS脱分化专用培养基"，克服了花椰菜游离小孢子培养球型胚发育停滞褐化死亡的技术难点，建立了小孢子培养育种技术体系，提高了育种效率，可使花椰菜育种周期从传统的8～10年缩短到4～5年，研究成果荣获2007年度国家科技进步二等奖。

为了在国际上抢占花椰菜基因组学研究领域的制高点，2017年启动了花椰菜全基因组测序工作，并在世界上首次完成了花椰菜全基因组测序，该项研究使我国花椰菜基因组学研究水平进入国际前列，花椰菜全基因序列的揭示，大大提高了花椰菜优良基因挖掘和基因克隆定位的效率。同时对于开展花椰菜靶向性分子育种、聚合育种，高效精准的培育花椰菜新品种奠定了坚实基础。

1996年被天津市政府评为天津市"爱国爱市创业成才"优秀青年知识分子；同年获第四届天津青年科技奖；1996年享受国务院颁发的政府特殊津贴；1997年被天津市政府授予"花椰菜育种专家"称号。2001年入选天津市首批"131工程"第一层次人选，2004年列入"国家新世纪百千万人才工程"人选。2009年被聘为国家现代农业产业技术体系花椰菜育种岗位科学家。2012年被中国科协授予"全国优秀农业科技工作者"，2013年被农业部评为"全国农业先进个人"。2015年被天津市人才工作领导小组评为"天津市高层次创新型科技领军人才"，2017年获天津市创新争先奖章。

2007年被天津市人民政府授予天津市劳动模范称号，2011年被评为天津市"十大时代先锋"称号，2012年被中国科协授予"全国优秀农业科技工作者"，被中国农林水利工会授予全国农林水利产业劳动奖章，先后三次被天津市人民政府授予天津市科教兴农先进个人、连续三次被天津市总工会授予"八.五立功"奖章。本人领衔的研究室1994、2002年度二次被天津市人民政府授予"劳动模范集体"称号，2007年获天津市总工会"工人先锋号"称号，2008年获全国"工人先锋号"称号。

毛英伟，1974 年出生，1992～1996 年就读于南开大学生命科学学院微生物学专业，1996～1999 年师从宋文芹教授，就读于南开大学细胞与遗传专业染色体实验室，获得硕士学位。2002 年，新泽西医药大学分子生物学硕士。2005 年，美国密歇根大学医学院药理系获博士学位。

2005～2010 年在哈佛大学医学院和麻省理工学院从事博士后研究，师从美国医学院院士 Li-Huei Tsai 教授。之后任美国宾夕法尼亚州州立大学帕克分校基础学院生物系副教授，终身教授。长期从事大脑神经发育和精神类疾病的病理生理学研究，特别是在精神分裂症和自闭症研究方面做出了重要研究工作。近年来对 Rmb8a、DISC1、ZNF804a 等基因的结构、功能和调控进行了深入的研究。在国际顶级科学刊物 Cell、Nature、Neuron、Cell Research、Molecular Psychiatry、JCB 上发表 40 多篇文章，拥有专利 1 个，这些研究获得了美国心脏学会、NIH 的奖励和资助。

邮箱：yzm1@psu.edu

实验室网页：http://sites.psu.edu/maolab/

部分发表论文

Lin Y, Liang R, Qiu Y, Lv Y, Zhang J, Qin G, Yuan C, Liu Z, Li Y, Zou D, Mao Y. Expression and gene regulation network of *RBM8A* in hepatocellular carcinoma based on data mining. 2019, *Aging*, 11(2):423-447.

Lin Y, Zhang J, Cai J, Liang R, Chen G, Qin G, Han X, Yuan C, Liu Z, Li Y, Zou D, Mao Y. Systematic Analysis of Gene Expression Alteration and Co-Expression Network of Eukaryotic Initiation Factor 4A-3 in Cancer. *Journal of Cancer*, 2018, 9(24):4568-4577.

Cheng G, Li W, Ha L, Han X, Hao S, Wan Y, Wang Z, Dong F, Zou X, Mao Y, Zheng SY. Self-Assembly of Extracellular Vesicle-like Metal-Organic Framework Nanoparticles for Protection and Intracellular Delivery of Biofunctional Proteins. *Journal of the American Chemical Society*, 2018, 140(23):7282-7291.

Yeh YT, Tang Y, Lin Z, Fujisawa K, Lei Y, Zhou Y, Rotella C, Elías AL, Zheng SY, Mao Y, Liu Z, Lu H, Terrones M. Light Emitting Transition Metal Dichalcogenide Monolayers under cellular digestion. *Advanced Materials*, 2018, 30(8).

Zhou Y, Dong F, Lanz TA, Reinhart V, Li M, Liu L, Zou J, Xi H, Mao Y. Interactome analysis reveals *ZNF804A*, a schizophrenia risk gene, as a novel component of protein translational machinery critical for embryonic neurodevelopment. *Molecular Psychiatry*, 2018, 23(4):952-962.

Lai J, Li S, Shi X, Coyne J, Zhao N, Dong F, Mao Y and Wang Y. Displacement and Hybridization Reactions in Aptamer-functionalized Hydrogels for Biomimetic Protein Release and Signal Transduction. *Chemical Science*, 2017, 8:7306-7311.

Deng D, Jian C, Lei L, Zhou Y, McSweeney C, Dong F, Shen Y, Zou D, Wang Y, Wu Y, Zhang L, Mao Y. A prenatal interruption of DISC1 function in the brain exhibits a lasting impact on adult behaviors, brain metabolism, and interneuron development. *Oncotarget*, 2017, 8:84798-84817

Dong F, Xie K, Chen Y, Yang Y, Mao Y. Polycistronic tRNA and CRISPR guide-RNA enables highly efficient multiplexed genome engineering in human cells. Biochemical and Biophysical Research Communications, 2017, 482:889-895.

Zou D, Zhou Y, Liu L, Dong F, Shu T, Zhou Y, Tsai LH, Mao Y. Transient enhancement of proliferation of neural progenitors and impairment of their long-term survival in p25 transgenic mice. *Oncotarget*, 2016, 7:39148-39161.

Dong F, Jiang J, McSweeney C, Zou D, Liu L, Mao Y. Deletion of CTNNB1 in inhibitory circuitry contributes to autism-associated behavioral defects. *Hum Mol Genet*, 2016, 25:2738-2751

Boccitto M, Doshi S, Newton IP, Nathke I, Neve R, Dong F, Mao Y, Zhai J, Zhang L, Kalb R. Opposing actions of the synapse-associated protein of 97-kDa molecular weight (SAP97) and Disrupted in Schizophrenia 1 (DISC1) on Wnt/β-catenin signaling. *Neuroscience*, 2016, 326:22-30.

Zou D, Chen L, Deng D, Jiang D, Dong F, McSweeney C, Zhou Y, Liu L, Chen G, Wu Y, Mao Y. DREADD in parvalbumin interneurons of the dentate gyrus modulates anxiety, social interaction and memory extinction. *Curr Mol Med*, 2016, 16:91-102.

Zou D, McSweeney C, Sebastian A, Reynolds D, Dong F, Zhou Y, Deng D, Wang Y, Liu L, Zhu J, Zou J, Shi Y, Albert I, Mao Y. A critical role of RBM8a in proliferation and differentiation of embryonic neural progenitors. *Neural Development*, 2015, 10:18.

Jiang H; Zou J; Wang B; Lin Y; Yu M; Pan Y; Li Y; Mao Y; Wang Y. The GluN2B subunit of N-methy-D-asparate receptor regulates the radial migration of cortical neurons in vivo. *Brain Research*, 2015, 1610:20-32.

McSweeney C, Mao Y. Applying stereotactic injection technique to study genetic effects on animal behaviors. *JOVE*, 2015, (99), e52653, doi:10.3791/52653.

Alachkar A, Jiang D, Harrison M, Zhou Y, Chen G, Mao Y. An EJC Factor RBM8a Regulates Anxiety Behaviors. *Curr Mol Med*, 2013, 13:887-99.

Singh K, Rienzo G, Drane L, Mao Y, Flood Z, Madison J, Fereira M, Bergen S, King C, Sklar P, Sive H, Tsai LH. Common DISC1 polymorphisms disrupt Wnt/GSK3β-signaling and brain development. *Neuron*, 2011, 72:545-58.

Lee A W.-M., Mao Y, Penninger J, and Yu S. Gab2 promotes Colony Stimulating Factor-1 regulated macrophage expansion via alternate effectors at different stages of development. *Mol Cell Biol*, 2011, 31:4563-81.

Mao Y, Ge X, Frank CL, Madison JM, Koehler AN, Doud MK, Tassa C, Berry EM, Soda T, Singh KK, Biechele T, Petryshen TL, Moon RT, Haggarty SJ, Tsai LH. DISC1 regulates neural progenitor proliferation via modulation of GSK3β/β-catenin signaling. *Cell*, 2009, 136: 1017-1031. (cover)

Yang P, Mao Y, Lee AW, Kennedy RT. Measurement of dissociation rate of biomolecular complexes using CE. *Electrophoresis*, 2009, 30:457-64.

Yang PL, Whelan RJ, Mao Y, Lee AW.-M, Carter-Su C, and Kennedy RT. Multiplexed Detection of Protein-Peptide Interaction and Inhibition Using Capillary Electrophoresis. *Anal. Chem.*, 2007, 79, 1690 -1695.

Mao Y and Lee A W-M. A novel role of Gab2 in bFGF-mediated cell survival during retinoic acid-induced neuronal differentiation. *J. Cell Biol*, 2005, 170:305-16.

Mao Y, Liu JP, Xiang H, Li DW. Human alphaA- and alphaB-crystallins bind to Bax and Bcl-X(S) to sequester their translocation during staurosporine-induced apoptosis. *Cell Death Differ*, 2004, 11:512-26.

Mao Y, Xiang H, Wang J, Korsmeyer S, Reddan J, Li DW. Human bcl-2 gene attenuates the ability of rabbit lens epithelial cells against H_2O_2-induced apoptosis through down-regulation of the αB-crystallin gene. *J. Biol. Chem*, 2001, 276: 43435-43445.

Mao Y, Liang SY, Song WQ, Li XL, Chen RY. Construction of a DNA library from chromosome 4 of rice (*Oryza sativa*) by microdissection. *Cell Res*, 1998, 8:285-293.

杨宇，女，研究员，1976年出生，1994～1998年就读于南开大学生命科学院生物系，之后师从陈瑞阳教授，2001年硕士毕业于南开大学遗传学专业。2003年师从侯云德院士，于中国疾病预防控制中心病毒病研究所病原生物学专业攻读博士学位。2006年博士毕业后赴美，于美国州立大学布法罗分校开展博士后研究。2008年回国后，至今就职于中国检验检疫科学研究院，主要从事传染病生物检测技术及蜱传病监测预警研究。担任国家重点研发计划"国家质量基础的共性技术研究与应用"重点专项首席科学家，兼任中华医学会"微生物学与免疫学分会"委员、中国医药生物技术协会"分子诊断技术学组"委员、中国微生物学会"人兽共患病原学专业委员会"委员等学术及社会职务。近年来主持国家重点研发计划项目、国家自然科学基金项目、质检行业公益项目等国家级、省部级科研项目近十项，获省部级科技一等奖1项、二等奖2项、三等奖2项。发表SCI及核心期刊论文74篇，获授权发明专利32项，制定发布行业标准9项，参与完成学术论文、译著4部。

部分科研项目

"十三五"国家重点研发计划项目"跨境多载体隐存高危生物因子风险识别、预测和控制技术研究"，在研，主持。

国家自然科学基金项目"我国流行的Borralia garinii基因型伯氏疏螺旋体转座子突变技术研究"，已结题，主持。

质检公益项目"现场级核酸提取和实时定量PCR一体化检测分析技术研究"，已结题，主持。

质检总局科技计划项目"高致病性病毒抗体库的建立与应用"，已结题，主持。

科技部国际合作项目"接壤国家边境蜱媒病生态学及流行病学合作研究"，已结题，技术负责人。

科技部国际合作项目"医学媒介及虫媒病跨境传播检测与阻断技术合作研究"已结题，技术负责人。

质检行业公益专项"黑瞎子岛媒介调查及其病原传播预警研究"，已结题，子项目负责人。

质检行业公益专项"国境口岸应对生物恐怖快速检测、控制与处理体系的研究"，已结题，技术负责人。

质检行业公益专项"入出境人员传染病快速筛查技术研究",已结题,技术负责人。

十一五科技部科技支撑项目"跨境传播重要危害民众健康因子检测识别技术研究"已结题,第二完成人。

获奖情况

1. 2010 年"科技兴检"一等奖:"国境口岸反生物恐怖现场快速检测系统研究-2"。
2. 2013 年"科技兴检"二等奖:"跨境传播重要危害民众健康因子检测识别技术研究"。
3. 2010 年科技兴检三等奖:"重要发热病原快速筛查系统研究"。
4. 2015 年科技兴检奖二等奖:"国境口岸应对生物恐怖快速检测、控制与处理体系的研究"。
5. 2017 年江苏省科学技术三等奖:"重大国际活动的口岸公共卫生风险防控关键技术及应用"。

部分发表文章

Shuo Sui, Yu Yang, Yi Sun, Xumin Wang, Guoliang Wang, Guangle Shan, Jiancheng Wang, Jun Yu, On the core bacterial flora of Ixodes persulcatus (Taiga tick). PLoS One, 2017, 12 (7): e0180150.

Lu Tingting, Fu Yingqun, Hou Yong, Yang Yu, Liu Lijuan, Liang Huijie, Yang Jun, Jiao Dan, and Ying Changqing, Hantavirus RNA prevalence in Myomorph Rodents on Bolshoy Ussuriysky Island at the Sino-Russian Border. Vector Borne Zoonotic Dis, 2017, 17 (8): 588-595.

Shuo Sui, Yu Yang, Zhiqiang Fang, Jiancheng Wang, Jing Wang, Yingqun Fu, Yong Hou, Baoliang Xu & Jun Yu, Complete mitochondrial genome and phylogenetic analysis of Ixodes persulcatus (taiga tick), Mitochondrial DNA Part B, 2017, 2: 1, 3-4,

Liu L, Chen Q, Yang Y, Wang J, Cao X, Zhang S, Li H, Hou Y, Wang F, Xu B. Investigations on Rickettsia in Ticks at the Sino-Russian and Sino-Mongolian Borders, China. Vector Borne Zoonotic Dis, 2015, 15 (12): 785-9.

Mingzhu Yang, Yueping Guan, Yu Yang, Tingting Xia, Wubin Xiong, Chen Guo. A snsitive and rapid immunoassay for mycoplasma pneumonia based on Fe3O4 nanoparticles. Materials Letters, 2014, 137, 113-116.

Mingzhu Yang, Yueping Guan, Yu Yang, Tingting Xia, Wubin Xiong, Ning Wang, Chen Guoc, Peroxidase-like activity of amino-functionalized magnetic nanoparticles and their applications in immunoassay. Journal of Colloid and Interface Science, 2013 405, (1), 291-295.

Yang Yu, Wang Jing, Wen Haiyan, Liu Hengchuan. Comparison of Two Suspension Arrays for Simultaneous Detection of Five Biothreat Bacterial in Powder Samples. Journal of Biomedicine and Biotechnology, Volume 2012 (2012), doi: 10.1155/2012/831052.

Yu Yang, Chunhao Li.Transcriptionand genetic analyses of a putative N -acetylmuramyl-L-alanine amidase in Borrelia burgdorferi. FEMS Microbiol Lett, 2009, 290 (2), 164-73.

Jing Wang, Yu Yang, Lei Zhou, Jinglin Wang, Yongqiang Jiang, Kongxin Hu, Xiaohong

Sun, Yousong Hou, Ziwen Zhu, Zhaobiao Guo, Yanli Ding, Ruifu Yang. Simultaneous detection of five biothreat agents in powder samples by a multiplexed suspension array.Immunopharmacology and Immunotoxicology,2009,31(3),417-27.

Yu Yang, Philip E. Stewart, Xiaoguang Shi, Chunhao Li.Development of a Transposon Mutagenesis System in the Oral Spirochete Treponema denticola. Applied and Environmental Microbiology,2008,74(20),6461-4.

倪虹,天津市人,1967年11月出生。1999～2001年南开大学分子遗传学博士后,合作导师陈瑞阳教授。

任职:南开大学医学院副教授,副院长。

业绩:外科学博士,曾从事外科临床工作十余年。博士后出站后留校任教。参与多项国家级科研项目,发表,学术论文40余篇。

王晓梅,1995～2001年在南开大学生命科学学院遗传学专业学习,师从陈瑞阳教授和宋文芹教授;分别于1998年6月和2001年6月获得遗传学专业理学硕士和遗传学专业理学博士学位。2001年和2004年两次赴比利时根特大学做博士后研究。

毕业后任教于天津农学院水产学院,教授;主要研究方向为水产动物遗传与育种以及水产动物分子免疫。曾任农业部天津鲤鲫鱼遗传育种中心学术委员会委员、天津蟹源水产养殖有限公司专家组成员、天津市细胞生物学会理事;现任天津市水产生态及养殖重点实验室学术委员会委员。主持并完成天津市应用基础与前沿技术研究计划重点项目3项,主持并完成天津市高等教育科技发展基金项目2项,主持并完成天津市人事局资助的引智项目1项,作为中方主持人之一主持并完成中国与比利时根特大学的中比国际合作项目1项,参加完成中比合作项目2项,作为主要参加人参与天津市科技重大专项与工程计划"种业科技重大专项"1项。先后在 Fish and shellfish immunology、International Journal of Peptide Research and Therapeutics、Iranian Journal of Fisheries Sciences、Aquaculture、《动物学报》《中国水产科学》《四川动物》《水产科学》和《南方水产科学》等国内外学术期刊发表学术论文60余篇;参编全国高等农林院校"十五"规划教材《水产动物育种学》(中国农业出版社)。

王振英，1995～1998 年，在南开大学生命科学学院染色体实验室攻读硕士学位，1998～2001 年攻读博士学位，师从陈瑞阳教授和郑坚瑜教授。现任天津师范大学生命科学学院教授、硕士研究生导师。天津市细胞遗传与分子调控重点实验室主任（2008—2010），天津高级人民法院知识产权审判技术咨询专家；中国细胞生物学会理事；中国植物生理与分子生物学会理事；天津市细胞生物学会副理事长；天津市植物生理与分子生物学会副理事长。

主要从事小麦抗白粉病分子机理及抗病育种，植物抗逆调控基因功能及应用，大蒜种质资源开发与利用。主持国家科技部重大基础研究专项基金 1 项；国家自然科学基金项目 3 项；教育部骨干教师重点项目 1 项；天津市科委科技支撑项目 2 项，重点项目 3 项，青年基金项目 1 项；天津市教委重点项目 1 项；天津市农委重点项目 1 项。在国内外学术期刊发表科研论文 60 余篇，8 篇为 SCI 收录。授权国家发明专利 15 项。在小麦抗白粉病相关课题的研究中，通过杂交、回交选育，配制培育了小麦抗白粉病近等基因系，利用近等基因系首次鉴定了新的抗白粉病新基因；完成的天津市重大农业科技合作项目，在天津地区完成抗白粉病优良小麦"农大 189"繁育工作，推广种植"农大 189" 15.786 万亩，为农民直接创效益 1016.93 万元，实现科学研究向生产力的转化。创新大蒜脱毒快繁技术体系，获得脱毒大蒜节约成本 30%以上。

2016 年"抗白粉病近等基因系 BJ 在基因功能研究及育种中的应用"获天津市科技进步二等奖；2006 年"栽培小麦 Brock 中一个抗白粉病新基因的发现与研究"获天津市技术发明二等奖；2004 年"与小麦中未知抗白粉病基因紧密连锁分子标记的筛选与利用"获天津市技术发明二等奖；2011 年天津市"三八红旗手标兵"；2011 年天津市"三八红旗手"；2008 年获"第四届新世纪巾帼发明家创新奖"；2008 年获"天津市巾帼发明家创新奖"。

祁仲夏，河北省人。1993～1997 年，南开大学生物学系本科；2002 年，南开大学生物学系遗传学专业博士毕业，师从陈瑞阳教授；2003～2005 年，威斯康星州大学麦迪逊分校魏斯曼中心博士后，从事人类细胞和分子遗传学研究，导师虞京崴教授；2005～2006 年，UCSF 检验医学系博士后，从事人类细胞和分子遗传学研究，导师虞京崴教授；2006～2008 年，UCSF 临床细胞遗传学实验室检验师培训；2008～2009 年，UCSF 临床细胞遗传学实验室检验师；2009～2011 年，UCSF 美国医学遗传学和基因组学细胞医学遗传学专科培训（ABMGG Clinical Cytogeneitcs）；2011 年至今，加州大学旧金山分校（UCSF）临床细胞遗传学实验室助理主任，实验医学系副教授。

主要从事细胞遗传学和基因组学的临床检验和咨询，临床实验室管理和相关的基础研究。临床检验项目包括染色体分带、荧光原位杂交、单核苷酸多态性基因组芯片、以

及二代测序。检验涉及产前诊断、儿科常见遗传病、恶性血液病、实体瘤等遗传疾病。

陈瑞阳和宋文芹两位导师带我进入了细胞遗传学这个广博的研究领域。从染色体实验室毕业至今，染色体始终是我工作和研究的中心。在博士后期间，我开始从事人类分子细胞遗传的研究，特别是结合染色体微切和新兴基因组分析技术对染色体结构变化进行分析。此后，又接受了系统的临床细胞遗传学培训，工作重心转向细胞遗传学和基因组学的临床检验和咨询、临床实验室管理和染色体变异相关的基础研究。瑞典著名细胞遗传学家阿尔伯特·列文（Albert Levan）恰当地概括了我对染色体的感觉："在过去五十年每天都看见染色体后，我把染色体视为我的朋友。"

代表性论文

Qi, Z, Hoffman, G, Kurtycz, D, and Yu, J. Prevalence of the C677T substitution of the methylenetetrahydrofolate reductase (MTHFR) gene in Wisconsin. Genetics in Medicine, 2003, 5: 458-9.

Rice, GM, Qi, Z, Selzer, R, Richmond, R, Thompson, K, Pauli, RM, and Yu, J. Microdissection-based high-resolution genomic array analysis of two patients with cytogenetically identical interstitial deletions of chromosome 1q but distinct clinical phenotypes. American Journal Medical Genetics Part A, 2006, 15: 1637-43.

Ye, L, Chang, JC, Lin, C, Qi, Z, Yu, J, and Kan, YW. Generation of induced pluripotent stem cells using site-specific integration with phage integrase. Proc Natl Acad Sci USA, 2010, 107: 19467-72.

Deucher, AM, Qi, Z, Yu, J, George, T, and Etzell, J. BCL6 expression correlates with the t(1; 19) translocation [TCF3 (E2A)/PBX1] in B-cell lymphoblastic leukemia. American Journal of Clinical Pathology, 2015, 143 (4): 547-57.

Qi, Z, Jeng, LJ, Slavotinek, A, and Yu J. Haploinsufficiency and triploinsensitivity of the same 6p25.1p24.3 region in a family. BMC Med Genomics, 2015, 8 (1): 38.

Qin, H, Heina, M, Liu, Y, Percharde, M, Wossidlo, M, Blouin, L, Durruthy-Durrhthy, J, Wong, P, Qi, Z, Yu, J, Qi, LS, Sebastiano, S, Song, JS, and Ramalho-Santos, M. YAP induces human naive pluripotency. Cell Reports, 2016, 14 (10): 2301-12.

Qi, Z, Madaan, S, Chetty, S, Yu, J, and Wiita, AP. False negative fetal cell free DNA screening for microdeletion syndromes in the presence of an unbalanced translocation involving monosomy 4p. Prenatal Diagnosis, 2017, 37 (4): 420-422.

Qi, Z, Xiang, B, Tsai, A, and Yu, J. Introduction of American Board of Medical Genetics and Genomics (ABMGG) Certification Exams and Maintenance of Certification (in Chinese). Chinese Journal of Medical Genetics, 2019, 36: 13-22.

**高英堂**，男，1968 出生。1990 年本科毕业于南开大学生物学系，2000~2003 年师从陈瑞阳教授，获南开大学生命科学学院博士学位。现任职天津市第三中心医院、天津市肝胆疾病研究所和天津市人工细胞重点实验室，研究员、副主任，天津医科大学硕士生导师。入选天津市"131"创新型人才第一层次人选，曾获天津市"爱国爱市，创业成才"优秀青年知识分子和"天津市青年人才奖"等称号。本人主要从事肝胆疾病、心血管疾病的细胞及分子生物学的基础与临床应用研究工作，承担或参加国家与省部级等科研课题 30 余项，目前在研项目 3 项，总经费 327 万，获省部级或局级奖励 10 项。发表论文百余篇，SCI 收录五十余篇，获发明专利 1 项。主持国家级医学继教项目累计培训 200 余人次，指导硕博士生 50 余名。学术兼职包括中国抗癌协会纳米肿瘤学专业委员会委员、中华医学会天津分会肝病学会委员/天津遗传咨询分会委员、天津市免疫学会理事、天津市医师协会精准医学专业委员会常委、天津市干细胞临床研究专家委员会委员等。

第一或通讯作者部分论文

Jiao X，Shu G，Liu H，Zhang Q，Ma Z，Ren C，Guo H，Shi J，Liu J，Zhang C，Wang Y，Gao Y. The Diagnostic Value of Chemokine/Chemokine Receptor Pairs in Hepatocellular Carcinoma and Colorectal Liver Metastasis. J Histochem Cytochem，2019，67（5）：299-308.

Luo Y，Lou C，Zhang S，Zhu Z，Xing Q，Wang P，Liu T，Liu H，Li C，Shi W，Du Z，Gao Y. Three-dimensional hydrogel culture conditions promote the differentiation of human induced pluripotent stem cells into hepatocytes. Cytotherapy，2018，20（1）：95-107.

Liu Z，Huo X，Zhao S，Yang J，Shi W，Jing L，Li W，Li Y，Ma L，Gao Y，Diao A. Low density lipoprotein receptor class A domain containing 4（LDLRAD4）promotes tumorigenesis of hepatic cancer cells. Exp Cell Res，2017，360（2）：189-198.

Luo Y，Wang P，Liu H，Zhu Z，Li C，Gao Y. The state of T cells before cryopreservation：Effects on post-thaw proliferation and function. Cryobiology，2017，79：65-70.

Shi W，Zhang Z，Yang B，Guo H，Jing L，Liu T，Luo Y，Liu H，Li Y，Gao Y. Overexpression of microRNA let-7 correlates with disease progression and poor prognosis in hepatocellular carcinoma. Medicine（Baltimore），2017，96（32）：e7764.

Jiao X，Luo Y，Yang B，Jing L，Li Y，Liu C，Jing X，Wang F，Wang Y，Du Z，Gao Y. The MTHFR C677T mutation is not a risk factor recognized for HBV-related HCC in a population with a high prevalence of this genetic marker. Infect Genet Evol，2017，49：66-72.

Wang YJ，Liang ZH，Gao YT，Zhai DK，Rao Q，Shi WX，Yang B，Jing L，Guo H，Liu T，Liu J，Du Z. Factors Influencing Circulating MicroRNA Level in the Studies of Hepatocellular Carcinoma Biomarker. Neoplasma，2015，62（5）：798-804.

Wang Y, Gao Y, Shi W, Zhai D, Rao Q, Jia X, Liu J, Jiao X, Du Z. Profiles of differential expression of circulating microRNAs in hepatitis B virus-positive small hepatocellular carcinoma. Cancer Biomark, 2015, 15（2）: 177-186.

Liu J, Gao Y, Yang B, Jia X, Zhai D, Li S, Zhang Q, Jing L, Wang Y, Du Z, Wang Y. Overexpression of squamous cell carcinoma antigen 1 is associated with the onset and progression of human hepatocellular carcinoma. Arch Med Res. 2015, 46（2）: 133-141. Gao YT, Han T, Li Y, Yang B, Wang YJ, Wang FM, Jing X, Du Z. Enhanced specificity of real-time PCR for measurement of hepatitis B virus cccDNA using restriction endonuclease and plasmid-safe ATP-dependent DNase and selective primers. J Virol Methods, 2010, 169（1）: 181-187.

Hua Guo, Ying-tang Gao, Qin Zhang, Li Jing, Tong Liu, Wen-xia Shi, Dao-kuan Zhai, Xiang Jing, and Zhi Du. Expression and Clinical Significance of livin Protein in Hepatocellular Carcinoma. Disease Markers, 2013, 35（5）: 489-496.

阎国荣，男，1957年8月生，甘肃人，理学博士。2000～2003年在南开大学生命科学学院博士后流动站做博士后研究，天津农学院园艺园林学院教授。从事生态学、农业教育、教学和研究工作。主要社会兼职有：教育部第六届教学指导委员会委员、中国园艺学会理事、天津市园艺学会副理事长。主要学术研究方向：果树资源学及中国新疆天山野生果树资源研究及其环境保护等领域的研究。出版学术著作2部，《中国新疆野生果树研究》（2010，中国林业出版社）和《耐盐植物生物改良滨海盐渍地研究》（2014，中国林业出版社）。发表学术论文60余篇。

龙鸿，博士，教授。1982～1986年东北师范大学生物系本科毕业；1991年沈阳农业大学农学系获硕士学位；2001年东北师范大学遗传与细胞研究所，获理学博士学位，师从郝水教授。2001～2003年南开大学生命科学学院，博士后，合作导师：陈瑞阳教授。2003～2013年华中农业大学生命科技学院，副教授，硕士生导师，2004～2006年西班牙科学院生物研究中心（Centro De Investigaciones Biologicas, Consejo Superior De Investigaciones Cientificas）博士后。2013年至今，天津农学院园艺园林学院，副教授、教授，园艺园林学院副院长，硕士生导师。主要从事园艺植物多倍体形成过程中的遗传学、表观遗传学和园艺植物营养生长时相转变的分子遗传机制的研究。发表论文20余篇。

代表性论文

Meiling Yang（并列第一作者），Fang Li（并列第一作者），Hong Long，Weiwei Yu，Xiuna Yan，BinLiu，Yunxiu Zhang，Guorong Yan*，Wenqin Song*. Ecological distribution, reproductive characteristics, and in situ conservation of Malus sieversii in Xinjiang, China. HORTSCIENCE, 2016, 51 (9): 1197-1201.

Hong Long*, Chunli Chen, Bing Wang, Yanni Feng (2015) rDNA Genetic Imbalance and Nucleolar Chromatin Restructuring is Induced by Distant Hybridization between Raphanus sativus and Brassicaalboglabra. PLoS ONE, 10 (2): e0117198. doi: 10.1371/journal.pone.0117198.

Xuanli Li, Weiwei Guo, Bing Wang, Xiangsong Li, Honggao Chen, Lihua Wei, Yanjie Wang, Jiangsheng Wu, Hong Long* (2010), Instability of chromosome number and DNA methylation variation induced by hybridization and amphidiploid formation between Raphanus sativus L. and Brassica alboglabra Bailey. BMC Plant Biology, doi: 10.1186/1471-2229-10-207.

魏丽华，李象松，李炫丽，汪艳杰，王冰，吴江生，龙鸿*. 萝卜—芥蓝杂种F1代创制及杂种鉴定。武汉植物学研究, 28 (2): 243-245.

李象松，魏丽华，李炫丽，汪艳杰，王冰，吴江生，龙鸿*. 萝卜—芥蓝异源四倍体F4和F10世代DNA甲基化变异的MSAP分析. 华中农业大学学报, 2010, 6: 64-66.

Hong Long, Jie He, Haijing Sun, Shui Hao, Mingda Jiao. In situ comparative studies on subnucleolar distribution and configuration of plant rDNA. Micron, 2008, 39 (4): 405-410.

Hong LONG, Zhong-Xia QI, Xiao-Ming SUN, Cheng-Bin CHEN, Xiu-Lan LI, Wen-Qin SONG and Rui-Yang CHEN. Characters of DNA constitution in the rye B chromosome. Journal of Integrative Plant Biology, 2008, 50 (2): 183-189.

Guorong Yan, Hong Long, Wenqin Song, and Ruiyang Chen. Genetic Polymorphism of Malus sieversii Populations in Xinjiang, China. Genetic Resources and Crop Evolution, 2008, 55: 171-181.

Gyanesh K. Satpute, Hong Long, José M. Seguí-Simarro, María C. Risueño and Pilar S. Testillano. Cell architecture during gametogenesis and embryogenic microspore development in Brassica napus. Acta Physiologiae Plantarum, 2005, 27 (4B): 665-674.

Hong Long, Haijing Sun, Xianlu Zeng, Shui Hao and Mingda Jiao. Subnucleolar distribution and organization of Vicia faba L. rDNA in situ. Cell Biology International, 2004, 28 (12): 845-848.

Hong Long, Haijing Sun, Xianlu Zeng, Shui Hao and Mingda Jiao. Identification and characterization of coiled body-like structures in pea (Pisum sativum L.). Cell Biology International, 2004, 28 (11): 825-828.

Long hong, ZENG Xianlu, JIAO Mingda, HU Bo, SUN Haijing, LIU Zhenlan, ZHANG Liyong, HAO Shui. Analysis of Nucleolar Pre-rRNA Processing Sites in Pea (Pisum sativum). Science in China (Series C), 2003, 46 (1): 58-66.

Long hong, ZENG xianlu, HU bo, SUN haijing, LIU zhenlan, HAO shui. Subnucleolar distribution and transportation of U3 snoRNA in the nucleolus of Pisum sativum. Acta Botanica Sinica, 2003, 45 (3): 317-321.

Hong Long, Haijing Sun, Xianlu Zeng, Mingda Jiao, Shui Hao. In Situ Distribution and Configuration of rDNA in the Nucleolus of Pisum Sativum. Acta Biologiae Experimentalis Sinica, 2003, 36 (4): 255-258.

王顺启，博士，副研究员，硕士生导师，赣江青年学者。中国细胞生物学会会员；中国神经生物学会会员；中国生理学会会员；日本神经生物学学会准会员。

1999年南开大学生科院生物化学专业本科毕业，同年保送遗传学硕博连读博士生，师从陈瑞阳教授。2004年毕业后到南昌大学工作至今，早期从事生物化学教学及部分科研工作，然后到美国从事博士后工作（杜克大学、南卡大学），回国后从肿瘤干细胞转型到神经分子生物学研究，专门从事科研工作。先后主持江西省教育厅科研基金，中国博士后科研基金，回国留学人员科研启动基金和国家自然科学基金项目。

目前的研究方向：Lrp4在中枢神经系统的功能研究；精神疾病风险基因Tmem108在中枢神经系统的功能研究；小胶质细胞Rapsyn基因在中枢神经系统的功能研究。

曾辉，女，1977年出生。1995～1999年就读于南开大学生命科学学院生物化学与分子生物学系，1999～2004年师从陈瑞阳教授和宋文芹教授攻读遗传学博士学位。毕业后赴美，于加州大学劳伦斯伯克利国家实验室生物分部从事博士后研究。现就职于加州大学旧金山分校，任职预算分析员，为新药和新医疗设备的临床研究提供辅助支持。

王娟，女，博士，副教授，硕士生导师，深圳大学生命与海洋科学学院生物技术系主任。

2001年9月起进入南开大学生命科学学院，师从宋文芹教授，2004年7月在南开大学生命科学学院获遗传学专业理学博士学位；2004年至今，在深圳大学从事教学科研工作。2015～2016年曾在美国加州大学戴维斯分校做访问学者。主讲本科生"遗传学""生物专题讲座""资源与环境微生物学"（研究生专业选修课）"微生物遗传与应用"等课程。科研方面隶属于深圳市微生物基因工程重点实验室，主要从事微生物的遗传改造、新型纤维素酶生产菌株选育及其调控机理、嗜热真菌纤维素酶表达调控机制等方面的研究。近几年来，主持及参加国家自然科学基金、广东省科技计划项目、深圳市基础研究计划项目多项，在Microb Cell Fact，J Microbiol Biotechnol，J Ind Microbiol Biotechnol等刊物发表SCI论文10余

篇。

李华兵，博士，上海交通大学医学院教授，博士生导师。2002 年南开大学本科毕业，2005 年南开大学硕士毕业，导师宋文芹教授。2005 年后又在国外知名顶尖实验室接受严格的系统科学训练。2005～2011 年，美国新泽西州立大学完成博士学位，博士生导师 Vincenzo Pirrotta，Rutgers University（国际知名表观遗传学家）。博士后研究（2012～2017 年）导师：Richard Flavell，Yale University（国际著名免疫学家/美国科学院院士）。2017 年 8 月正式入职上海交通大学医学院上海市免疫学研究所，组建了 RNA 代谢与免疫疾病课题组，并积极参入协助交大医学院、耶鲁、Richard Flavell 和苏冰教授建立上海交通大学医学院—耶鲁大学联合免疫代谢研究院（Shanghai JiaoTong University School of Medicine-Yale University Institute for Immune Metabolism，SYIIM）。在表观遗传学和 T 细胞免疫生物学领域，作为第一作者，共同第一作者和共同通讯作者已在 Nature、Science、Cell Research 等国际一流期刊发表研究论文 10 余篇。该实验室主要从事免疫细胞中的 RNA 代谢的研究，以基因编辑小鼠的免疫疾病模型为基础，集中关注在机体免疫反应中 RNA 层面的表观修饰过程以及 RNA 结合蛋白的生理和分子生物学功能，通过前沿的高通量筛选测序和反向遗传学方法，鉴定出调控免疫反应的重要分子靶标，为开发肠道炎症、自身免疫疾病以及肿瘤的治疗药物提供重要理论依据。

在研项目

上海交通大学医学院科研启动支持，"RNA 代谢与免疫疾病"，2017-2023，300 万元，在研，主持。

上海市教育委员会高校特聘教授（东方学者）计划，2018-2020，100 万元，在研、主持

国家自然科学基金"生物大分子动态修饰与化学干预重大研究计划"重大研究计划培育项目，91753141，"m6A RNA 甲基化修饰动态变化对 CD8 T 细胞功能调控的研究"，2018/01-2020/12，70 万元，在研，主持。

第十四批'千人计划'青年项目，"免疫 T 细胞中 RNA 修饰的功能研究"，2018-2021，300 万，在研，主持。

发表论文

Li HB，Muller M，Bahechar I，Kyrchanova O，Ohno K，Georgiev P，and Pirrotta V. Insulators，not Polycomb Response Elements，are required for long-range interactions between Polycomb targets in Drosophila. Mol. Cell. Biol，2011，31（4）：616-25. doi：10.1128/MCB.00849-10. PMID：21135119.

Pirrotta V.，and Li HB. A view of nuclear Polycomb bodies. Curr. Opin. Genet. Dev，2012，22（2）：101-9. doi：10.1016/j.gde.2011.11.004. PMID：22178420.

Schwartz YB，Linder-Basso D，Kharchenko PV，Tolstorukov MY，Kim M，Li HB，Gorchakov AA，Minoda A，Shanower G，Alekseyenko AA，Riddle NC，Jung YL，Gu T，Plachetka A，Elgin SC，Kuroda MI，Park PJ，Savitsky M，Karpen GH，Pirrotta V. Nature and function of insulator protein binding sites in the Drosophila genome. Genome Res.，

2012，22（11）：2188-98. doi：10.1101/gr.138156.112. PMID：22767387.

Li HB，Ohno K，Gui H，and Pirrotta V. Insulators target active genes to transcription factories and Polycomb-repressed genes to PcG bodies. PLoS Genet，2013，9（4）：e1003436. doi：10.1371/journal.pgen.1003436. PMID：23637616（Highlighted by Cell：2013，153（5），p935）.

Li HB#，Jin C#，Chen Y#，Flavell RA. Inflammasome activation and metabolic disease progression. Cytokine Growth Factor Rev.，2014，25（6）：699-706. doi：10.1016/j.cytogfr.2014.07.020. PMID：25156419（# equal contribution）.

Zhu S#，Li HB#，Flavell RA. Resemble and Inhibit：When RLR Meets TGF-β. Mol Cell，2014，56（6）：719-20. doi：10.1016/j.molcel.2014.12.010. PMID：25526529（# equal contribution）.

Cao G#，Li HB#，Yin Z，Flavell RA.Recent advances in dynamic m6A RNA modification. Open Biology，2016，6（4）：160003. doi：10.1098/rsob.160003. PMID：27249342（# equal contribution）.

Li HB. Chromosome Conformation Capture in Drosophila. Methods Mol. Biol，2016，1480：207-12. doi：10.1007/978-1-4939-6380-5_18. PMID：27659987.

Hu B#，Jin C#，Li HB#，Tong J，Ouyang X，Cetinbas NM，Zhu S，Strowig T，Lam FC，Zhao C，Henao-Mejia J，Yilmaz O，Fitzgerald KA，Eisenbarth SC，Elinav E，Flavell RA. The DNA Sensing Aim2 Inflammasome Controls Radiation Induced Cell death and Tissue Injury. Science，2016，354（6313）：765-768. DOI：10.1126/science.aaf7532. PMID：27846608（# equal contribution）.

Zhu S，Ding S，Wang P，Wei Z，Pan W，Palm NW，Yang Y，Yu H，Li HB，Wang G，Lei X，de Zoete MR，Zhao J，Zheng Y，Chen H，Zhao Y，Jurado KA，Feng N，Shan L，Kluger Y，Lu J，Abraham C，Fikrig E，Greenberg HB，Flavell RA. Nlrp9b inflammasome recognizes and restricts rotavirus infection in intestinal epithelial cells. Nature，2017，546（7660）：667-670. doi：10.1038/nature22967. PMID：28636595.

Li HB#*，Tong J#，Zhu S#，Batista PJ，Duffy EE，Zhao J，Bailis W，Cao G，Kroehling L，Chen Y，Wang G，Broughton JP，Chen YG，Kluger Y，Simon MD，Chang HY，Yin Z*，Flavell RA*. m6A mRNA methylation controls T cell homeostasis by targeting IL-7/STAT5/SOCS pathway. Nature，2017，548（7667）：338-342. doi：10.1038/nature 23450.PMID：28792938（# equal contribution，* co-corresponding）.

Tong J#，Cao G#，Zhang T#，Sefik E，Vesely M，Broughton J，Zhu S，Li H，Li B，Chen L，Chang HY，Su B，Flavell RA*，Li HB*. m6A mRNA methylation sustains Treg suppressive functions. Cell Research，2018，28（2）：253-256. doi：10.1038/cr.2018.7. PMID：29303144（*co-corresponding）.

Yu X，Lao Y，Teng XL，Li S，Zhou Y，Wang F，Guo X，Deng S，Chang Y，Wu X，Liu Z，Chen L，Lu LM，Cheng J，Li B，Su B，Jiang J，Li HB*，Huang C*，Yi J* Zou Q*. SENP3 maintains the stability and function of regulatory T cells via BACH2 deSUMOylation. Nat Commun, 2018, 9（1）: 3157. doi: 10.1038/s41467-018-05676-6. PMID: 30089837（* co-corresponding）.

Tong J, Flavell RA*, Li HB*. RNA m6A modification and its functions in diseases. Frontiers of Medicine, 2018, 12(4): 481-489. doi: 10.1007/s11684-018-0654-8. PMID: 30097961（* co-corresponding）.

Yu X, Teng XL, Wang F, Zheng Y, Qu G, Zhou Y, Hu Z, Wu Z, Chang Y, Chen L, Li HB, Su B, Lu L, Liu Z, Sun SC, Zou Q. Metabolic control of regulatory T cell stability and function by TRAF3IP3 at the lysosome. J Exp Med., 2018, 215（9）: 2463-2476. doi: 10.1084/jem.20180397. PMID: 30115741.

Liu Y, Li HB*, Flavell RA*. cGAS activation in phased droplets. Cell Research, 2018, 0: 1-2; doi: 10.1038/s41422-018-0087-6. PMID: 30218060（* co-corresponding）.

Wang J, Flavell RA*, Li HB*. Antiviral immunity: a link to bile acids. Cell Research, 2019, 29(3): 177-178. doi: 10.1038/s41422-019-0148-5.PMID: 30778178（* co-corresponding）.

张勇，博士，电子科技大学生命科学与技术学院教授、博士生导师。攻读博士学位期间，师从宋文芹、陈瑞阳教授，进行植物分子细胞遗传学研究。主要研究工作包括杨树染色体显微分离、抗病基因克隆以及药用植物遗传多样性分析，2006年毕业于南开大学生命科学学院，获理学博士学位，被评为南开大学"2006届优秀博士研究生"，并获香港"求是科技基金会"年度"求是研究生奖学金"。同年，加入电子科技大学生命科学与技术学院，先后于四川农业大学农学院（2007—2010）、明尼苏达大学医学院基因组工程中心（2010—2013）进行博士后及高级研究员研究工作。长期从事植物基因组工程及合成生物学相关的教学、科研工作，专注于植物基因组定向编辑完成了一系列高质量研究工作：1）作为核心研究人员开发了基于Golden Gate的高效TALEN体系被全世界超过1500个实验室广泛使用；2）针对class 2类CRISPR系统新成员，构建了高效、特异的水稻（植物）CRISPR-Cpf1（Cas12a）基因组编辑新系统；3）基于Golden Gate+GateWay策略，构建了简单、高效的CRISPR-Cas9、Cas12a植物基因组编辑多功能工具文库；4）设计了"单一转录单元CRISPR-Cas（STU-Cas）"基因组编辑系统，针对植物基因组目标位点进行快捷、高效的基因组编辑；5）基于全基因组重测序大数据分析策略，完成了CRISPR-Cas9、Cas12a介导的植物基因组编辑特异性定量评价。主持国家自然科学基金3项、国家转基因重大专项子课题1项、国家博士后基金1项、四川省杰出青年科学基金1项，在Nature Plants、Genome Research、Genome Biology、Molecular Plant、Plant Physiology等期刊发表学术论文75篇，SCI引用超过3500次，4篇论文入选ESI高被引论文（2篇论文同时入选ESI热点论文）。作为主讲教师，为本科、研究生讲授《基因工程》、《高级分子生物学》等课程；作为指导教师，指导本科生参加"国际遗传工程机器竞赛"（iGEM），先后取得亚太赛区银牌（2013）、总决赛银牌（2014）、总决赛金牌

（2015、2016、2017、2018）。

杨斌，男，1975年出生。2002年至2006年就读于南开大学生命科学学院，遗传学专业博士研究生，在染色体实验室宋文芹教授指导下获理学博士学位；2006年进入天津市经济技术开发区博士后工作站、南开大学博士后流动站从事博士后研究，宋文芹教授为合作导师之一；2008年就职于天津市第三中心医院；2012~2013年在美国纽约州立大学石溪分校学术访问。现为天津市第三中心医院副研究员，天津市遗传学会理事会理事，天津市医学会消化病学分会青年委员，入选天津市"131"创新型人才第一层次培养工程。

任晨春，女，1972年3月出生，中共党员，研究员，高级遗传咨询师。1994年河北医科大学临床医学系临床医学专业本科毕业，1997年天津医科大学妇产科遗传专业硕士研究生毕业，同年进入天津市中心妇产科医院遗传室工作至今。2003~2006年就读于南开大学生命科学学院遗传专业，师从宋文芹教授，获得博士学位。目前就职于中心妇产科医院，是遗传及产前诊断专业学科带头人，主要从事染色体病的检验以及产前诊断、产前筛查的检验与科研工作，尤其擅长基因检测报告的解读，罕见遗传病的诊断与咨询。带领细胞遗传团队诊断了5万2千余例染色体病例，异常率高达10%，其中有大量罕见的染色体异常被鉴定为世界首报，使该院成为华北地区细胞遗传工作量最大的细胞遗传诊断中心之一。在产前诊断方面，带领产前诊断团队攻克了细胞培养的一个个难关，建立实验室操作流程，使该院成为可以进行绒毛样本、羊水样本以及脐带血样本的多种样本检测，并涵盖孕期各个阶段的产前诊断，已进行绒毛染色体检测6000余例，羊水细胞染色体检测4400余例。

目前主持或参与各级科研课题20余项。以第一作者或通讯作者发表SCI收录论文以及核心期刊论文与综述50余篇。获得天津市科技进步三等奖1项，天津市卫生局科技进步二等奖1项。2003年分别获得"全国卫生系统青年岗位能手"称号。2003年和2006年分别遴选为天津市"131人才"第三梯队和第二梯队，2006年遴选为中国青年科技工作者协会第四届会员。目前被聘为天津医科大学客座教授、硕士研究生导师，天津医学高等专科学校市级精品课"细胞遗传学"客座教授，天津市卫生局中青年医疗专家服务团成员。兼任天津市医学会医学遗传分会第四届委员会委员，遗传咨询分会委员，天津遗传学会理事，全国医师协会新生儿科分会出生缺陷专委会委员，全国整合医学生殖专业委员会常委。已带教10名硕士研究生，其中一人获得"优秀研究生"称号，一人获得"优秀学生干部"一次，"优秀共产党员"一次，"优秀共青团员"两次。

主要文章（一作和通讯作者）

Ren C C, Miao X H, Yang B, *et al*. Methylation status of FHIT and E-cadherin genes in plasma of cervical cancer patients. Int J Gynecol Cancer, 2006, 16（5）：1862-1867.

Ren CC, Miao XH, Cheng H, et al. Detection of fetal sex in the peripheral blood of pregnant women. Fetal Diagn and Ther, 2007, 22（5）: 377-382.

陆国芳, 孙丰涛, 任晨春, 等。ICSI 治疗前原发严重少精症患者 Y 染色体 AZF 微缺失分析。中国妇幼保健杂志, 中国妇幼保健杂志, 2012, 27（11）: 1173-1175.

孙丰涛, 任晨春, 梁玥宏. 男性原发不育患者遗传学研究[J]. 中国妇幼保健杂志, 2011, 26（33）: 5216-5220.

董动丽, 任晨春. 线粒体 DNA A1555G 突变致聋的研究进展[J]. 中华临床医学实践杂志, 2011, 10（5）321-325.

田卉, 任晨春, 梁玥宏, 等. 394 例高龄孕妇羊水细胞遗传学分析[J].中国妇幼保健杂志, 2013, 28（8）: 1310-1312.

王文靖, 任晨春, 常诚, 等. 应用 QF-PCR 检测先心病患儿 22q11.2 微缺失方法的建立[J]. 中国妇幼保健杂志, 2014, 29（8）: 1228-1231.

梁玥宏, 王文靖, 任晨春, 等. 原发及继发闭经患者细胞遗传学分析 160 例[J]. 继续医学教育, 2014, 28（4）: 16-17.

梁玥宏, 王文靖, 任晨春, 等. 常染色体结构异常与男性精液异常的关系[J]. 继续医学教育, 2014, 28（5）: 9-10.

王文靖, 任晨春, 常诚, 等. QF-PCR 检测 22q11.2 微缺失在产前诊断 CHD 中的应用[J]. 中国妇幼保健杂志, 2014, 29（28）: 4597-4600.

科研项目

1. 染色体易位女性患者手部皮纹学研究 获得 2002 年度天津市科技成果认定, 成果号 20023060。
2. 实时荧光定量 PCR 方法检测母血中胎儿 SRY 基因。获 2006 年度天津市卫生局科技进步二等奖, 并获得 2005 年天津市科技成果鉴定（津 20053075）。
3. 实时荧光定量 PCR 检测健康育龄妇女生殖道沙眼衣原体感染的研究。获 2006 年度天津市科技成果（津 20063148）。
4. 孕早期母血清筛查唐氏综合征的研究。获 2006 年度天津市科技成果（津 20063086）
5. 子宫内膜异位症患者血清抗心磷脂抗体与细胞因子的研究。获 2004 年度天津市科技成果（津 20043095）。
6. 孕酮对抗磷脂综合征 Th1/Th2 型细胞因子调控, 获 2008 年度天津市科技成果（津 20083052）。
7. 芳香化酶细胞色素 P450 在早期子宫内膜异位症诊断中的研究, 获 2008 年度天津市科技成果（津 20083011）。
8. 宫颈癌相关基因 cSNP 位点分析及 DNA 甲基化的研究（卫生局 04KY16）2010 年结题, 鉴定为国际先进水平。第一完成人, 获天津市科技成果（津 20101217）。
9. 染色体易位患者手部皮纹学研究, 2010 年结题, 鉴定为国内领先水平, 第二完成人, 获天津市科技成果（津 20101218）。
10. 男性原发不育患者 Y 染色体 AZF 区微缺失研究, 天津市科技成果鉴定国内领先, 第二完成人（津 20121316）。

11. 子宫内膜异位症患者血清可溶性细胞间粘附分子-1 的研究，天津市科技成果鉴定第四完成人（津 20120625）。
12. 胎儿先天性心脏畸形的产前诊断研究策略，获天津市科技成果（津 20141957）
13. 应用基因芯片对耳聋患者及携带者进行产前诊断的研究，天津市卫生局重点课题，获天津市科技成果（津 20171980）。
14. SNP 微阵列技术在检测滞留流产绒毛染色体异常中的应用研究，获天津市科技成果（津 20141956）。
15. MiRNA 与先心病的相关性研究及在产前诊断中的应用（卫生局自筹 2014KZ071）。
16. 孕早期染色体非整倍体畸形的产前诊断策略，2015 年度天津市科技进步三等奖，证书编号 2015JB-3-091-R4。
17. 对孕妇进行脊肌萎缩症基因筛查并应用于产前诊断，天津市卫生局重点课题（2015KR03）。
18. 孕期应用高通量测序及微阵列技术检测多种出生缺陷，华夏医学科技奖三等奖（证书编号 201703058P0808）。

李慧，天津农学院园艺园林学院，副教授。2011 年 9 月—2011 年 10 月，日本静冈大学农学部交流学习；2003 年 9 月—2006 年 6 月，南开大学生命科学学院硕士毕业，导师宋文芹教授；1999 年 9 月—2003 年 6 月，河北师范大学生命科学学院本科毕业。2014 年 11 月—至今天津农学院园艺园林学院教师。

2017 年花椰菜靶向分子育种技术及在新型育种材料选育中的应用，天津市科学技术进步奖二等奖；2007 年耐盐经济植物的引进与应用技术研究，天津市静海县科学技术委员会静海县科学技术进步奖二等奖；2012 年入选天津市优秀青年教师资助计划；2013 年入选天津市 131 创新人才工程第三层次人才；2015 年入选天津市 131 创新人才工程第二层次人才；2018 年获天津农学院优秀共产党员称号。主持完成国家级、省部级及局级科研项目 4 项，天津市大学生创新创业训练计划项目（市级）1 项；参与完成科研项目 6 项；第一作者发表论文 24 篇，其中 SCI 论文 8 篇；参编专著 1 部，教材 1 部。

程华，男，1977 年 4 月出生，中共党员。河北省科学院生物研究所副研究员。

2000 年 6 月毕业于河北大学微生物学专业，获得理学学士学位。2005 年 6 月毕业于南开大学生命科学学院生物化学与分子生物学专业，获得理学硕士学位，导师陈力副教授。现为中国生物化学与分子生物学学会会员。

主要从事抗体工程及快速检测相关技术研究。近几年来相继研发了包括甲胎蛋白（AFP）、铁蛋白和磷脂酰肌醇蛋白聚糖

3（GPC3）在内的一系列肿瘤早期检测试剂盒。该系列试剂盒能够对包括肝癌在内的多种消化系统癌症进行早期筛查和预后监控，目前该系列试剂盒已经申请国家发明专利4项，授权2项。

与哈佛大学、加州大学、浙江大学和兰州大学在内的多所国内外知名大学和科研机构合作，到目前为止共研发了单克隆抗体50余个，多克隆抗体800多个。

作为主持人及主要研发人员共完成省级和市级项目30余项，发表学术论文20多篇。获得河北省科技进步二等奖和三等奖各1项。获得国家发明专利4项，实用新型专利1项，培养硕士研究生10人。2013年被评为河北省"三三三人才工程"三层次人才。

董天皞，1981年3月生，天津市人，研究生学历，2003～2006年在南开大学生命科学学院遗传学专业学习，获硕士学位，导师倪红副教授，现任高级工程师，天津市"131"创新型人才（第二层次）、武清区"鲲鹏工程"拔尖骨干人才。现任天津红日药业股份有限公司集团研究院发展研究部部长，主要从事公司核心产品血必净注射液上市后研究和新产品立项与技术预研工作。参与国家级课题1项，承担省部级课题1项，参与2项；发表中文核心期刊论文7篇，SCI论文1篇；申请中国发明专利3项。

刘博，1980年生人，2003年南开大学生命科学学院本科毕业，同年保送南开大学生命科学学院遗传专业博士研究生，导师宋文芹教授，2007年毕业，获理学博士学位，先后于美国新罕布什尔大学和德州塔木大学从事博士后研究并参与了草莓和棉花的基因组测序工作。研究还包括草莓基因组组成，核糖体RNA基因在草莓中的进化以及棉花物理图谱、遗传图谱和染色体图谱的整合等。现在AI Biosciences公司任Research Scientist，参与NASA空间站生物实验仪器的开发项目以及快速低价体外诊断设备的研发。

兰添颖，2003年本科毕业于南开大学生物技术专业，2008年博士毕业于南开大学遗传学专业，师从宋文芹教授。2009～2011年于美国布法罗大学进行博士后学习，2011～2017年任美国布法罗大学研究科学家，2017～2018年任新加坡南洋理工大学资深研究科学家，2018年至今任美国布法罗大学资深研究科学家。至今已在Science、Nature、Nature genetics、PNAS、Communications biology、Proceedings of the Royal Society B等主流期刊发表论文30余篇。研究方向为基因组学及进化生物学，主

要研究课题包括：
1. 多种被子植物基因组测序、结构特征、进化及生物适应性相关研究。
2. 多倍体基因组起源、祖先基因组重建、基因分化重排及二倍化相关研究。
3. 多种哺乳动物群体遗传及进化历史重建相关研究。
4. 古哺乳动物基因组重建、进化历史与生态环境变化关联研究。

发表部分论文

Lan T, Lindqvist C. 2018 Paleogenomics: genome-scale analysis of ancient DNA and population and evolutionary genomic inferences. In Population Genomics (pp. 323-360). Springer, Cham.

Lan T, Lindqvist C. 2018 Technical Advances and Challenges in Genome-Scale Analysis of Ancient DNA. In Paleogenomics (pp. 3-29). Springer

Renner T, Lan T, Farr KM, Ibarra-Laclette E, Herrera-Estrella L, Schuster SC, Hasebe M, Fukushima K, Albert VA. 2018 Carnivorous plant genomes. In Carnivorous Plants: Physiology, Ecology, and Evolution (pp. 135-154). Oxford University Press.

Colella JP, Lan T, Schuster SC, Talbot SL, Cook JA., & Lindqvist C. Whole-genome analysis of Mustela erminea finds that pulsed hybridization impacts evolution at high latitudes. Communications biology, 2018, 1 (1), 51.

Lan T, Gill S, Bellemain E, Bischof R, Nawaz MA, Lindqvist C. 2017 Evolutionary history of enigmatic bears in the Tibetan Plateau-Himalaya region and the identity of the yeti. Proceedings of the Royal Society B, 2017, 284 (1868), 20171804

Lan T, Renner T, Ibarra-Laclette E et al. Long-Read sequencing uncovers the adptive topography of a carnivorous plant genome. PNAS, 2017, 114 (27), pp.E5483-E5483

Haase E.M, Kou Y, Lan T, et al. Comparative genomics and evolution of the amylase-binding proteins of oral streptococci. BMC microbiology, 2017, 17 (1): 94.

Salojärvi J, Smolander OP, Nieminen K, Rajaraman S, Safronov O, Safdari P, Lamminmäki A, Immanen J, Lan T, Tanskanen J, Rastas P. Genome sequencing and population genomic analyses provide insights into the adaptive landscape of silver birch. Nature genetics, 2017, 49 (6): 904.

Lan T, Cheng J, Ratan A, Miller W, Schuster S, Farley S, Shideler R, Mailund T, Lindqvist C. Genome-wide evidence for a hybrid origin of modern polar bears. 2016, bioRxiv doi: http://dx.doi.org/10.1101/047498.

Denoeud F, Carretero-Paulet L, ... Lan T et al. The coffee genome provides insight into the convergent evolution of caffeine biosynthesis. Science, 2014, 345 (6201), 1181-1184.

Juntheikki-Palovaara I, Tähtiharju S, Lan T, Broholm SK, Rijpkema AS, Ruonala R, Kale L, Albert VA, Teeri TH, and Elomaa P. Functional diversification of duplicated CYC2 clade genes in regulation of inflorescence development in Gerbera hybrida (Asteraceae). The Plant Journal, 2014, doi: 10.1111/tpj.12583.

Chamala S, Chanderbali AS, Der JP, Lan T, Walts B, Albert VA, dePamphilis CW,

Leebens-Mack J, Rounsley S, Schuster SC, Wing RA, Xiao N, Moore R, Soltis PS, Soltis DE, Barbazuk WB. Assembly and validation of the genome of the nonmodel basal angiosperm Amborella. Science, 2013, 342 (6165): 1516-7. doi: 10.1126/science.1241130.

Amborella Genome Project (including Lan T and other 147 collaborators). The Amborella genome and the evolution of flowering plants. Science, 2013, 342 (6165): 1241089. doi: 10.1126/science.1241089.

Roy T, Chang TH, Lan T, and Lindqvist C. Phylogeny and biogeography of New World Stachydeae (Lamiaceae) with emphasis on the origin and diversification of Hawaiian and South American taxa. Molecular Phylogenetics and Evolution, 2013, doi: /10.1016/j.ympev.2013.05.023.

Ibarra-Laclette E, Lyons E, Hernández-Guzmán G, Pérez-Torres CA, Carretero-Paulet L, Chang T-H, Lan T, Welch AJ, Abraham Juárez MJ, Simpson J, Fernández-Cortés A, Arteaga-Vázquez M, Góngora-Castillo E, Acevedo-Hernández G, Schuster SC, Himmelbauer H, Minoche AE, Xu S, Lynch M, Oropeza-Aburto A, Cervantes-Pérez SA, Ortega-Estrada MdeJ, Cervantes-Luevano JI, Michael TP, Mockler T, Bryant D, Herrera-Estrella A, Albert VA, Herrera-Estrella L. Architecture and evolution of a minute plant genome. Nature, 2013, doi: 10.1038/nature12132.

Lan T, Yao B, Shen Y, Wang X. Isolation of high-quality total RNA from lipid-rich seeds. Analytical Biochemistry, 2013, doi: 10.1016/j.ab.2013.03.012.

Lan T and Albert A. Dynamic distribution patterns of ribosomal DNA and chromosomal evolution in Paphiopedilum, a lady's slipper orchid. BMC Plant Biology, 2011, doi: 10.1186/1471-2229-11-126.

Lan T, Chen R, Li X, Dong F, Qi Y, and Song W. Microdissection and Painting of the W Chromosome in Ginkgo biloba showed different labeling patterns. Botanical Studies, 2008, 49: 33-37.

Lan T, Liu B, Dong F, Chen R, Li X, and Chen C. Multicolor FISH analysis of rDNA and telomere on spinach. Hereditas, 2007, 29 (11): 1405-1408.

Lan T, Zhang S, Liu B, Chen R, and Song W. Differentiating sex chromosomes of the dioecious Spinacia oleracea L. (spinach) by FISH of 45S rDNA. Cytogenetics & Genome, 2006, 114 (2) 175-177.

古瑜，1971年1月出生，2007年博士毕业，南开大学生命科学学院细胞与遗传学专业，师从宋文芹教授。现于天津科润蔬菜研究所豆类研究室工作，研究员。主要从事豆类蔬菜遗传育种工作。

自2008年6月主持菜豆研究室工作至今，主持或参加各级各类项目共12项，其中已完成9项，在研3项，已获得科技成果8项。作为第一完成人，2015年"荚用菜豆育种技术创新及抗病新品种选育"获得天津市科技进步三等奖。作为第一完成人，授权发明专利3项；授权新品种权1个；培育系列豆类品种25个，建设和完善繁种基地约1000余亩，累计生产豆类良种约82.3万公斤。先后发表论文25篇（册）。作为第一作者（或通讯作者）在本专业期刊上，共发表学术论文19篇，其中核心期刊论文13篇（其中SCI收录文章2篇）。主编丛书2部。

赵磊，2004年考入南开大学生科院遗传专业攻读硕士学位，师从陈力老师，2007年硕士毕业获理学硕士学位。目前在Abcam公司任大客户经理。

张世光，1981年11月生人，博士，研究员，河北省石家庄市人。河北科技大学，生物制造公共实验中心工作。2004年9月至2009年12月，南开大学生命科学学院遗传学专业攻读博士，师从宋文芹教授，获理学博士学位，主要研究方向为肿瘤细胞分子遗传学。2010年2月至2013年2月：石药集团中奇制药技术（石家庄）有限公司，生物技术研究所，所长助理。2013年12月至今：河北科技大学，生物制造公共实验中心，研究员。

主要研究方向

主要从事重组蛋白多肽类药物研究开发工作，主要负责蛋白多肽类药物生产工艺建立、质量研究、药物质量标准的方法建立与验证，参与或负责多项治疗性生物制品的临床试验申报工作；先后参加国家"重大新药创制"科技重大专项课题3项，主持省级科研项目1项，市级科研课题1项。

郝擘，山西阳泉人，1981 年 10 月出生，2009 年毕业于南开大学生科院遗传季专业，获理学硕士学位，导师宋文芹教授。现就职于康希诺生物股份有限公司。原液生产部经理，负责公司产品的产业化生产。

苗绪红，山东平阴人，1970 年 6 月出生。天津医科大学基础医学院，博士，副教授。2005~2010 年，师从宋文芹教授在南开大学生命科学院遗传学攻读博士，获遗传学博士学位。

长期以来从事医学细胞生物学、医学遗传学教学工作和研究工作。

薛振毅，山西大同人，1984 年 7 月出生。高级实验师，2010 年南开大学生科院遗传学专业硕士毕业，师从陈成彬副教授，获理学硕士学位、2010 年 9 月天津医科大学基础医学院免疫系工作。发表第一作者 SCI 文章 6 篇，参与发表 SCI 文章 26 篇；主持结题国家自然青年基金项目 1 项，参与国家自然科学基金项目 11 项，研究方向为免疫炎症与自身免疫性疾病。

江汉民，博士，副研究员。2008 年南开大学生科院遗传学硕士毕业，获理学硕士学位，同年师从宋文芹教授进入生科院遗传专业攻读博士学位；2011 年 6 月获理学博士学位。2011 年就职天津科润蔬菜研究所，主持青花菜遗传育种工作。在青花菜农艺性状遗传规律、种质资源评价、抗病育种、品质改良及分子育种等方面取得了具有深入研究价值的研究结果。先后分离克隆了一批与重要农艺性状相关的基因并对其功能进行了研究，并通过基因过表达或基因编辑手段对其功能进行了研究，获得了具有目标性状的新种质；首次将甘蓝的显性核雄性不育基因转育到青花菜，创建了我国第一个青花菜显性核基因雄性不育系转育技术体系；利用 EST-SSR 分子标记，建立了青花菜自交系亲和性

快速、简便鉴定技术体系；建立了快速、准确的室内杂交种纯度鉴定技术体系；参与完成花椰菜全基因组测序工作。育成具有自主知识产权的青花菜新品种 6 个，其中"领秀 2 号"，综合性状超越了日本青花菜品种"优秀"，2016 年～2017 年连续两年单品种推广面积居国内育成品种第一，低温下花球不变紫的青花菜品种"领秀 3 号"，突破了我国青花菜品种低温花球变紫的行业性难题。发表论文 29 篇。研究成果获天津市科技进步奖一等奖 1 项、二等奖 3 项、三等奖 1 项。

赵换，山东乐陵人，1984 年出生。2011 年南开大学生科院硕士毕业，获理学硕士学位，导师宋文芹教授。现就职于康希诺生物股份公司，新技术开发部，从事质谱、毛细管电泳等仪器分析工作，负责疫苗相关蛋白结构表征和定量工作。

李爱，1984 年生人，博士，2007 年 9 月于南开大学生命科学学院遗传学专业攻读博士，师从宋文芹教授。2012 年 6 月获理学博士学位，2012 年 9 月就职天津农学院园林园艺学院任讲师，主讲"植物生理学""园艺植物生物技术"。研究方向为树木和果树的分子生物学及应用，主持国家自然科学基金落叶松 AP2/ERF 家族分子特征分析及关键成员的功能研究；天津自然科学基金落叶松 ANT 转录调控因子的分子特征及功能研究、落叶松 LkDREB1a 转录因子基因的分子特征及功能分析，参加了桃果实成熟期基因分离研究；杨树中生长素依赖型 ARGOS 类转录调控因子的表达及功能研究，miRNA 对花椰菜小孢子胚胎发生发育的调控机制研究，花椰菜花球发育调控相关基因 CDAG1 的功能及调控机制研究，生长素应答转录因子在花椰菜花球发育过程中的功能研究等国家基金和天津市基金项目，发表学术及教改论文 12 篇，SCI、EI、ISTP 收录 6 篇，参编《园艺植物生物技术》教材 1 部。

发表的 SCI 收录论文

Ai Li, Jing Wang, Hui Li, Chengbin Chen, Wenqin Song, Chunguo Wang. Transcriptome profiling and characterization of gene families with zinc finger and nucleotide binding site (NBS) domains in Larix kaempferi. J Plant Biochem Biotechnol, 2017, 26 (2): 149-159.

Ai Li, Xue Yu, Beibei Cao, Lixin Peng, Ying Gao, Tao Feng, Hui Li, Zhiyu Ren. LkAP2L2, an AP2/ERF transcription factor gene of Larix kaempferi, with pleiotropic roles in plant branch and seed development. Russ J Genet, 2017, 53 (12): 1335-1342.

Ai Li, Yanan Zhou, Chuan Jin, Wenqin Song, Chengbin Chen, Chunguo Wang. LaAP2L1, a heterosis- associated AP2/EREBP transcript factor of Larix, is sufficient to increase organ size and final biomass by effecting cell proliferation in Arabidopsis. Plant Cell Physiol，2013，54（11）：1822-1836.

Ai Li，Wenqin Song，Chengbin Chen，Yanan Zhou，Liwang Qi，Chunguo Wang. DNA methylation status is associated with the formation of heterosis in Larix kaempferi intraspecific hybrids. Molecular breeding，2013，31：463-475.

Ai Li，Mengdie Fang，Wenqin Song，Chengbin Chen，Liwang Qi，Chunguo Wang. Gene expression profiles of two intraspecific Larix lines and their reciprocal hybrids. Mol Biol Rep，2012，39（4）：3773-3784.

Ai Li，Baoquan Hu，Zhenyi Xue，Li Chen，Weixing Wang，Wenqin Song，Chengbin Chen，Liwang Qi，Chunguo Wang. DNA methylation in genomes of several annual herbaceous and woody perennial plants of varying ploidy as detected by MSAP. Plant Mol Biol Rep，2011，29（4）：784-793.

刘少华，湖北江陵人，1987年5月出生。2012年南开大学生科院硕士毕业，获理学硕士学位，导师陈成彬副教授。任杭州诺泰制药技术有限公司工艺开发部研究员，现任武汉菲恩生物科技有限公司抗体开发部经理。

完成卡贝缩宫素、特利加压素、阿托西班、利拉鲁肽、齐考诺肽等多种多肽药物制备工艺的开发和放大、药物注册申报，有阿托西班相关专利1篇。

自主开发约600种涵盖内参系列、标签系列、肿瘤系列、信号通路系列单、多克隆抗体，广泛应用于WB、IHC、Elisa等实验，申报1篇GST标签亲和柱相关专利。

胡宝全，1982年6月生人，中共党员，博士，讲师。2002年9月至2006年6月于河北师范大学生命科学学院获理学学士；2006年9月至2008年6月南开大学生命科学学院遗传学；2008年9月至2013年6月南开大学生命科学学院获博士学位，导师宋文芹教授；2013年8月至2016年3月任职中国科学院天津工业生物技术研究所，助理研究员；2016年3月至今，天津农学院林学系任教。主要从事多倍体植物生长发育、林木分子细胞遗传学、林木遗传育种学等方面的研究。参与工业微生物基因组测序与功能基因组分析：国家高技术研究发展计划（863计划）（2012AA022101）；参与杂种林木超亲速生性状的形成和固定的基因组学；国家重点基础研究发展计划（973计划）（2009CB119100）；主持miR396在杨树三倍体次生生长中的调控机制研究；参与基于RADseq的甘蔗复合体（禾本科）分类学研究项目；国家自然科学基金（31800167）和天津滨海中新生态农业发展有限公司委托"药用

植物的开发与利用"。主要承担生态学、林木遗传育种学、专业英语的讲授。发表和参与发表论文 10 余篇。

韩春乐,天津市人,1987 年出生。2007~2010 年于山东大学生命科学学院本科毕业,同年考入南开大学生科院攻读硕士学位,2013 年获遗传学硕士学位,导师宋文芹教授,毕业后先后在天津生物工程职业技术学院、天津国际生物医药联合研究院工作。现在天津云检医学检验所有限公司从事精准医疗方面的研发与管理工作,担任技术部高级经理。具有 5 年质谱应用与研发经验,擅长高分辨质谱在多组学检测领域中的应用。

刘超,山东淄博人。2013 届南开大学生科院细胞生物学硕士研究生,师从陈成彬老师,毕业后在深圳华大基因研究院从事遗传病基因检测产品研发工作,现在在安吉康尔(深圳)科技有限公司,负责公司南区技术支持工作,主要为遗传病基因检测产品开发、市场推广等,累计参与发表相关文章 5 篇。

王明明,籍贯内蒙古通辽人,1989 年 10 月出生。2013 届南开大学生科院生物工程专业硕士毕业。师从孙德岭研究员和宋文芹教授,现任中国民生银行股份有限公司总行公司业务部中小企业金融数据经营中心。经营机构中小企业批量项目业务管理;中小企业数据库建设及经营管理。数据营销与业务规模指标联动推进;客户数据化分层管理;营销竞赛激励与经营机构价值产出融合推动。

王菁,山东莒县人,1989 年 6 月出生。2011 年 9 月至 2014 年 6 月在南开大学生命科学学院染色体实验室研究生学习,获理学硕士,导师陈成彬副教授。2014 年 8 月至今为莒县第二中学生物教师。

获奖情况:2016 年 2 月,荣获县教育局"2015 年度优秀课任教师"。2016 年 7 月,在宋文芹老师的指导下为"Transcriptome profiling and characterization of gene families with zinc finger and nucleotide binding site(NBS)domains in Larix kaempferi"第二作

者。2017和2018年连续两年荣获2017年山东省中学生物学联赛优秀辅导教师三等奖。在2017年全市高中生物实验教学说课评选中荣获二等奖。2018年1月，荣获日照市教育局颁发的"实验室工作先进个人荣誉称号"。2019年2月，荣获县教育和体育局"2018年度优秀课任教师"。

周亚楠，河北廊坊人，1989年1月出生。2014届南开大学生科院遗传学硕士毕业，导师宋文芹教授。现任职中源协和基因科技有限公司研发部技术支持。

徐嘉艺，硕士，河北邯郸人，1992年11月出生。2015年南开大学生科院生物工程专业硕士毕业，导师孙德岭研究员和宋文芹教授。现任中国教育科学研究院朝阳实验学校-生物教师、团总支书记。2017、2018年北京市金鹏科技论坛优秀辅导教师；2018年获校级教育质量优秀奖。

秦二军，河北廊坊人，1990年10月出生。2015年南开大学生科院 生物工程专业硕士毕业，导师王春国副教授，现就职于北京诺禾致远科技股份有限公司北京区大区经理。

赵风治，河北邯郸人，1990年10月出生。2016年南开大学生科院遗传专业硕士毕业，获理学硕士学位，导师宋文芹教授。2016年通过河北省委组织部选拔任河北定向选调生，目前在河北省邯郸市丛台区四季青街道办事处任组织委员。

张文慧，女，山西怀仁人，1991年8月出生。2016届南开大学生科院生物工程专业硕士，导师孙德岭研究员和宋文芹教授。毕业后任职中共陕西省铜川市宜君县纪律检查委员会干部、宜君县五里镇兴市村党总支书记助理，从事纪检监察、贫困户帮扶、村党支部建设等工作。现任职山西师范大学辅导员的工作。

张勐，山西运城人，1990年4月出生。2016年南开大学生科院细胞生物学硕士毕业，师从陈力副教授。现大学城控股西安公司曲江大悦城运营部经理。曲江大悦城开业前期筹备工作任餐饮组组长，现运营部综合组组长。

李聪，山东滨州人，1994年3月出生。2016届南开大学生科院生物工程专业硕士毕业，导师王春国副教授。现任职华为技术有限公司中东非洲终端业务部渠道经理。印度P30pro、Watch GT新品电商上市操盘；中国区荣耀Magic2上市营销操盘等。

郑翔，硕士，安徽芜湖人，1992年出生。2016届南开大学生科院生物工程专业硕士毕业，导师孙德岭研究员和宋文芹教授。曾先后在500强外企、上市公司、私募基金、银行工作。现任中信银行总行职员。

杨美玲，河北沧州人。2014 年 9 月至 2017 年 12 月就读南开大学生科院博士研究生，导师宋文芹教授和阎国荣教授，博士期间进行新疆野苹果抗逆基因功能研究。2018 年 1 月入职天津大学表层地球系统科学研究院，担任讲师，发表论文 10 余篇。目前主要从事元素耦合生物地球化学循环以及浮游生物元素耦合调控机理等研究。

并主要承担了地科院生物实验室的建设，目前正在参与学院本科专业建设及招生相关事宜，同时承担着党支部宣传委员等工作。

赵瑞红，女，河南滑县人，1982 年 7 月出生。2018 年 12 月毕业于南开大学生命科学学院遗传学专业。导师宋文芹教授。博士期间主要从事芝麻染色体组学研究。通过对芝麻栽培种不同种质资源和野生种资源的染色体制片条件摸索，建立了芝麻染色体制片技术并对芝麻种质资源核型特征进行了比较分析；对芝麻栽培种和不同野生种的核糖体 DNA 在染色体组的分布特征进行了荧光原位杂交分析；构建了芝麻 BAC 文库，并建立了芝麻 BAC-FISH 技术体系；同时以特异 BAC 克隆序列为标记成功命 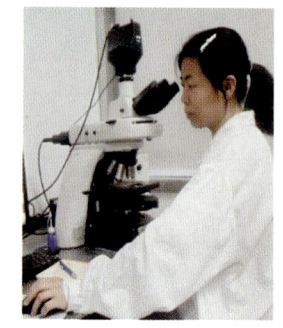 名了芝麻体细胞 13 对染色体；构建了首个芝麻 BAC-FISH 高密度细胞遗传图谱，并与 SNP 遗传连锁图进行了整合，完成了细胞遗传图谱和分子遗传图谱的整合，实现了连锁群与染色体的对应。该细胞遗传图谱的重要应用是指导芝麻基因组拼接为 13 条染色体。

2019 年 4 月进入河南省农业科学院博士后工作站，当前研究课题是：芝麻野生种抗病抗逆优异基因群染色体分布与基因家族进化研究。

张力鹏，男，博士。2011 年本科毕业于河北农业大学农学院植物科学与技术专业，获农学学士学位，次年考入南开大学攻读遗传学硕士学位；2014 年转入攻读本专业博士学位，2018 年 12 月获理学博士学位；2019 年 8 月进入南开大学生命科学院师资博士后流动站工作，师从陈德富教授和宋文芹教授。博士期间的主要研究方向为大花红景天生境适应性分子机制的研究。曾系统性的对西藏红景天属种质资源进行了收集、DNA 条形码和 SSR 分子标记的鉴定，首次建立了平原地区大花红景天离体组培快繁体系。通过转录组、甲基化组测序和 MSAP 等技术手段，对两种海拔高度的大花红景天进行了分析，并对基因家族如 UDPGT、CYP450、GST、GPX、NAC、MYB 等进行了克隆和表达分析，筛选到大花红景天应对高原恶劣环境因子的多个基因。近年来在核心期刊上发表和合作发表论文 14 篇。

张银兴，河北邢台人，1993年7月出生。湖南大学生命科学学院本科毕业，同年考入南开大学生科院攻读硕士学位，2017年6月南开大学生科院生物工程专业毕业，获得工学硕士学位。研究生期间主要从事大花红景天体外培养体系的建立与分子标记筛选的研究。现在天津大学生命科学学院办公室工作，负责研究生教务管理、生物学学科建设、学院后勤管理等。

刘志伟，河北省保定人，1990年1月出生。2017年生物工程专业硕士研究生毕业，师从陈力副教授，硕士期间主要从事三倍体丹参组织培养与块茎繁殖的初步比较研究，现在天津市第一中心医院神经内科工作，主要担任科主任科研助理及神经内科疾患相关项目的检查工作。

于得水，天津人，1993年12月出生。2017年华中农业大学本科毕业，2019年6月生物工程专业硕士研究生毕业。硕士期间主要进行两种不同生境下的西藏大花红景天DNA甲基化的研究。获得第二届"学会杯"全国生物信息学竞赛铜奖。发表和参与发表论文4篇。现在中国民航大学民航技术研究院工作，主要进行科技、科研平台管理。

滕彦娇，甘肃白银人，1993年11月出生。2017年年天津商业大学本科毕业，2019年南开大学生科院生物工程专业硕士研究生毕业。硕士期间主要从事半夏种质资源的鉴定及组织快繁、遗传转化体系的建立等研究工作。获得南开大学2019年研究生优秀毕业生、南开大学校级三好学生。现在天津博奥聚能生物科技有限公司工作，主要从事中药材新品种的研发。

附 录

附录1 三位中国科学院院士为《中国经济植物基因组染色体图谱》写的序言

郝水 序

人类对染色体（Chromosome）的认识已经有一百多年的历史。在18世纪中叶已经出现关于染色体的描述。据文献记载，最早描述染色体的是 Nageli（1842），他绘制的是紫露草（Trandescandia）花粉发生中的染色体。但当时他并未使用染色体这个名词。"染色体"名词是1888年W. Waldeyer 在一篇发表在德文期刊上的论文中提出的，指细胞中一种碱性染料浓染的小体，德文称之为 Chromosomen（复数）。1926年美国学者 Morgan 通过对果蝇的遗传研究，得出连锁交换定律，并确定主管生物性状遗传得基因直线排列于染色体中。人们自然把探求遗传物质的分子基础集中在染色体的研究上。后来生物化学的研究表明，染色体中存在核酸和蛋白质两类分子。1944年 Avery 等从肺炎双球菌的转化实验中证明，转化因子是脱氧核糖核酸（DNA）而不是蛋白质。1953年 Watson 和 Crick 提出了 DNA 双螺旋模型以及随后的其他发现，终于确定了 DNA 作为基因分子实体的认识。

既然染色体中除 DNA 外还存在大量蛋白质，那么染色体中的蛋白质起什么作用呢？研究表明染色体中蛋白质有两类，一是组蛋白（histone），二是非组蛋白（nonhistone proteins，NIP）。前者有 H2A、H2B、H3、H4 和 H1 五种蛋白质。其中前四种蛋白质各二分子聚合成八聚体，称为核小体核心。核小体核心与 DNA 结合，成为串珠样染色质链（A. L. Olins and D. E. Olins，1973，1974）。NIP 种类繁多（至少有30余种），且富变化。它们构成染色体的骨架结构（U. K. Laemmli，1978）。近20余年的研究表明，这两类蛋白质对 DNA 的基因转录、复制、重组、集缩与解集缩，以及染色体的构型等功能均有重要调控作用，此外，近20余年的研究还表明，中期染色体周围还有一个由核糖核蛋白（RNP）构成的表层。但关于这个 RNP 表层染色体 DNA 的功能意义尚缺乏了解。

从人类对染色体百余年的研究结果可以看出，每条染色体中不仅有作为遗传密码的 DNA 大分子，而且有参与调控 DNA 事先遗传功能（基因转录表达、复制、重组和集缩传代）的种类繁多的蛋白质。

染色体通常是在细胞周期的分裂中期形成的高度集缩的结构，它是由直径10多纳米核小体链经过大约4~5级逐次集缩形成的。其直径约1000纳米/每种生物体的体细胞都有一定数目的染色体，并且染色体的大小和形态（长度、直径、着丝粒位置及其他特征）都各具特点，它们是不同物种基因组的最简单明了的形象表现。

南开大学生物系陈瑞阳教授及其研究集体和合作者在植物染色体制片技术方面卓有成就，并多次举办全国性染色体技术学习班，培训大量专业人才，促进了我国植物染色体组型的形态学研究蓬勃发展。二十多年来，陈瑞阳教授及其同事积累了大量有关我国经济植物染色体组型的文字和图像资料。他们和其他单位的同行编著的《中国主要经济植物染色体图谱》共分四册，包括果树、农作物、园林花卉和竹类植物。仅果树方面就包括了29科52属261种。非常可贵的是果树和农作物中还包括了他们的野生近缘植物的染色体图谱。这一套图谱的学术和应用价值是十分明显的。它为我国重要经济植物的基因组学，尤其是功能基因组学的研究提供了系统的直观基础资料。它对研究我国重要经济植物遗传、起源和进化，特别是品种改良的育种实践具有重要意义。这套图谱是迄今国内未曾有过的，与国际上的同类著作相比是先进的。

在此图谱出版之际我仅向作者们表示衷心祝贺。

东北师范大学教授　　郝水
中国科学院院士
2002年2月5日

Preface by Hao Shui

Chromosomes have been recognized for more than a century. As early as the middle of the 18th century, the description of chromosomes was reported. According to historical records, the chromosome was first described by Nageli (1842), based on his observations of trandescantia in the process of pollen development. However, he did not use the term "chromosome" at that time. "Chromosome" was coined by W. Waldeyer (1888) and first appeared in his published paper within a German journal; "chromosome" was used to describe a tiny rodlike structure stained by alkaline dye. In Germany, it was known as chromosomen (pl.). In 1926, an American scientist, Thomas Hunt Morgan proved the linkage and cross-over law based on the data from his experiments with fruit flies. He confirmed that the genes which carry the blueprint for life were situated within chromosomes. Naturally, research work on the molecular basis of hereditary material, thereafter, became focused on the study of chromosomes. Later, biochemical discoveries indicated that there were two kinds of molecules in chromosomes: nucleic acid and protein. In 1944, O. T. Avery and his coworkers demonstrated that the transforming factor was DNA rather than proteins derived from Streptococcus pneumoniae transforming experiments. In 1953, Crick and Watson proposed the double-helix structure of DNA. Along with other notable discoveries, DNA was finally regarded as bearing the hereditary molecules of the gene.

Given that there are a great number of proteins in addition to DNA in the chromosome, what are the functions of these proteins? Studies have demonstrated that there are two kinds of proteins in chromosomes: one is histone, the other is non-histone (NIP). The former includes five kinds of proteins: H2A, H2B, H3, H4 and H1; among these proteins, two molecules each of the former four kinds of proteins revolve around a core histone octamer, known as the nucleosomal core. The nucleosomal core interacts with DNA to form solenoid chromatin fiber (A. L. Olins and D. E. Olins 1973, 1974). The types of NIP are numerous (more than 30), and exist in large numbers of variegating strains. They constitute the skeletal structure of the chromosome. Studies over the past twenty years have demonstrated that both histone and NIP play an important role in gene transcription, replication, recombination, condensation, decondensation and chromosomal configuration. Furthermore, these studies also discovered the existence of a surface layer constituted by ribonucleoprotein(RNP)around the chromosome at metaphase. However, we are far from fully understanding the function of this RNP surface layer tightly bound to chromatin.

From the findings of more than a century of study on chromosomes, the individual chromosome not only contains DNA macromolecules as hereditary codes but also various proteins involved in adjustment of DNA pre-hereditary functions (gene transcription, expression, replication, recombination, condensation and generation).

Basically a chromosome is a highly condensed structure at metaphase of the cell cycle. It is constructed by 10 nm filament of nucleosomes following 4 - 5 steps of condensation and has a diameter of 1000 nm. The somatic cells of an individual have a given number of chromosomes, and their size and shape (length, diameter, centromere position and other characteristics) have individual specificity. There are simple and clear image expressions of genome in different species.

Professor Ruiyang Chen and his research group in the Department of Biology, Nankai University, have made outstanding contributions in the field of plant karyotype preparation techniques. They have conducted several national training courses for these preparation techniques for many professionals and thereby, have promoted a vigorous stimulus for development of plant chromosome research in our country. Over the past twenty years, Prof. Chen and his colleagues have collected a great deal of literature and pictorial data on karyotypes for economic plants in our country. "Atlas of Major Economic Plant Chromosomes in China", compiled by them and colleagues from other institutes, is divided into four volumes, including fruit trees, crops, garden flowering plants and bamboo species. In the field of fruit trees alone, the atlas includes information on 261 species in 52 genera from 29 families. It is a praiseworthy endeavor that wild kindred plants are included in the atlas volumes of fruit trees and crops; the academic and applied values are obvious. There is a significant supply of systemic objective basic data for important economic plant karyology, especially functional karyology. This is essential in research concerned with economic plant heredity, origin, evolution, and especially, breeding practices for improving plants. This atlas, previously unpublished in our country, provides advanced information in comparison with similar international publications.

The authors are offered my hearty congratulations on the occasion of the publication of this atlas.

Northeast Normal University Professor
Academy of Sciences of China Academician Hao Shui
Feb. 5, 2002

杨弘远 周嫦 序

我们和陈瑞阳先生相识，最早是在20世纪60年代之初。当时我们大家都是刚出茅庐的青年，初生牛犊不怕虎，胸怀在中国科学事业中作出一番贡献的抱负。可惜好景不长，"文革"10年浩劫打破了我们的理想，中断了我们刚刚起步的研究工作。粉碎"四人帮"和改革开放迎来了我国科学事业的春天。当时瑞阳和我们都是霜染两鬓的中年人了，然而一股不可遏止的激情驱使我们投入新的研究事业，要将失去的年华抢回来。1979年，中国植物学会委托南开大学生物系举办首次全国染色体技术学习班，我（杨）有幸参加并在会后出版的一本册子中与他合写了一篇文章。其实我在这一方向只是"纸上谈兵"，而他则是"真刀真枪"地干，而且一干就是二十几年。以后，虽然我们和他的研究方向各不相同，但始终互通信息，互相支持。我们敬佩他在学术工作中脚踏实地、始终如一的奋斗精神，也从友人的成就中分享一份快乐。

诚如众所周知，染色体是遗传信息的载体；基因只有组织在染色体上才能在遗传上发生作用。当今，基因组研究是生命科学中最热门的领域。但接下来，分子水平的研究还须与细胞水平的研究结合起来，从细胞到分子再回到细胞，这是科学发展的趋势。由陈瑞阳教授主编的这部植物染色体图谱，包括101科、272属、1003种植物的1200幅染色体图版，是集他本人及其研究集体几十年心血完成的一部巨著。它填补了世界植物染色体图谱的空白，体现了我国学者在这一研究领域的重大贡献，可以说是一部传世之作。

科学的发展离不开实验技术的更新。以往的染色体压片技术，很难对付染色体数目众多、体形较小的材料。陈瑞阳教授创造性地提出去壁低渗法，大大提高了植物染色体制备的质量，成为他们这部染色体图谱的主要方法学基础。但他不满足于核型分析，又提出了植物染色体G带技术等一系列分带方法，使植物染色体研究进入了更微观的阶段，这在本书中也均有所展现。近年，陈瑞阳教授又在朝新的高度进军，实现了染色体微切割、分离与克隆等更精密的实验操作，预示了植物染色体工程新阶段的来临。

我们衷心祝贺本书的问世。

武汉大学生命科学学院教授　杨弘远
中国科学院院士
武汉大学生命科学学院教授　周嫦
2001年12月31日

Preface by Yang Hongyuan and Zhou Chang

It was in the early 1960s that we first met Mr. Ruiyang Chen. Although young and inexperienced at that time, we were both fearless and dedicated in our pursuit of scientific research in China. But this didn't last long. Our hopes were destroyed during the ten-year "Cultural Revolution" and our research was suspended in the initial stages. Subsequently, the overthrow of the "Gang of Four" and introduction of reforms and open-door policy in China restored scientific research to its full bloom. Although each of us, including Ruiyang, had reached middle age by that time, we were still encouraged and, with irresistible enthusiasm, immersed ourselves in our new research; we vowed to utilize our time wisely and to overcome the deficit in valuable research time lost over the years.

In 1979, the China Plant Association assigned the Department of Biology, Nankai University to hold the first national study class on chromosome technology. I (Yang) was honored to be involved in presenting the class and co-wrote with Prof. Chen a paper subsequently included in the colloquia for participators. Following our collaboration, we continued to maintain contact and encouraged each other over the years despite the fact that we pursued research in different fields. Admittedly, I was an armchair strategist in this field while my colleague persisted in real earnest in his efforts for over twenty years. Chen is highly regarded for his practical and consistent attitude toward academic work. We are delighted with his achievements.

It is recognized that the chromosome is the bearer of genetic information. Only when the gene is sited on the chromosome can it function in heredity. Currently, life sciences are focused on genomic research, but research at the molecular and cellular levels also needs to be included. The research focus shifts from cell to molecule to cell, which is the trend in scientific development. This chromosome atlas of plants edited by Professor Chen includes 1200 chromosome plates of 1003 species in 272 genera, representing 101 families, a great accomplishment by Chen and his research colleagues over two decades. It eliminates a knowledge gap with the chromosome atlas of plants in the world and presents the great contribution made by these researchers within this field in China. This outstanding work will certainly be passed on from age to age.

Without updating experimental technology, science would never continue in its development. It was difficult to deal with plant materials that have numerous chromosomes and small size using the previous karyotype preparation technology. Professor Chen creatively proposed "removing walls" and the hypotonic methods, which greatly improved the quality of preparation of plant karyotypes. This methodology forms the basis for the methodology used within this chromosome atlas. Not satisfied with karyotype analysis, Chen also proposed a series of banding methods such as plant chromosome G-banding technology enabling plant chromosome research to enter an even more microcosmic stage as demonstrated in this atlas. In recent years, Professor Chen continued his research and created more precise experimental procedures such as chromosome micro-cutting, separation and cloning, introducing a new era in the future of plant chromosome engineering.

We express our heartfelt congratulations in recognition of Prof. Chen's years of laborious research with his colleagues and in the publication of this atlas. I hereby write this preface.

Academician Yang Hongyuan Chinese Academy of Sciences
Professor of School of Life Science, Wuhan University Professor Zhou Chang
School of Life Science, Wuhan University
Dec. 31, 2001

洪德元 序

　　染色体是基因的载体，它控制遗传和变异，并支配生殖和发育。而染色体自身的数目、形态结构和行为也受基因的调控。染色体的研究，从结构与功能、行为和进化而言，既可以指向过去，也可以分析它的现在和预测它的将来。由此而发展了细胞遗传学、细胞分类学和物种生物学以及近年蓬勃发展起来的分子细胞遗传学，而人类基因组和作物基因组的全序列测定，更成为 21 世纪生命科学的里程碑。上述学科从不同层次、不同角度开展研究，都是以染色体组的组成（一个或多个基因组）和结构为基本单位，或甚至以单个染色体的识别为基础的。陈瑞阳教授主编的"中国主要经济植物染色体图谱"正是提供了这方面的重要基础资料，对上述各学科的研究，都有重要参考价值，它是生命科学中的基本建设之一。

　　我国的植物染色体研究，起步较晚，80 年代以前几乎是空白，改革开放后，在南开大学生物系、北京大学生物系、中国科学院植物研究所和中国农科院等单位的参予和组织下，相继在南开大学和全国各地举办了各种学习班和国内外的学术研讨会，促进了我国植物染色体研究的蓬勃发展。因此，无论是在研究范围和深度上，还是在研究技术水平上我国都已接近或达到世界先进水平。其中，以陈瑞阳教授为首的南开大学植物染色体实验室，20 年坚持不懈、辛勤工作，投入极大的精力和热情，以及对我国科学事业的责任心，不辞辛苦，跋涉全国各地，采集和征集材料，并以精湛的染色体技术制作出每一物种的染色体制片，拍摄出精美的照片，测出详细的核型参数，汇编成四册经济植物染色体图谱。它的出版是值得赞许和庆贺的，它不仅在国内是首部原创巨著，在国际上也填补了该领域的空白，必将引起广泛关注。

中国科学院植物研究所研究员
国家自然科学基金委员会生命科学部主任　洪德元
中国科学院院士
第三世界科学院院士
2001 年 12 月 15 日

Preface by Hong Deyuan

The chromosome is the bearer of genes that control not only heredity and variation, but also reproduction and growth. The number, structure and behavior of chromosomes are also adjusted and controlled by genes. As far as the structure and function, and behavior and evolution are concerned, chromosome research not only focuses on the past, but also analyzes the present and forecasts the future. Consequently, cytogenetics, cytotaxonomy, biosystematics and molecular cytogenetics have received much attention and have been vigorously developed over the past several years. Moreover, the completion of the genome sequences for human and plants has marked a milestone for life sciences in the 21st century. In research at different levels and aspects in the above-mentioned areas, the composition (one genome or more) and the structure of the genome are considered as a basic unit or based on identification of a single chromosome. The Chromosome Atlas of Major Economic Plants in China compiled by Professor Ruiyang Chen provides an important basis as a research reference for the above-mentioned disciplines; it highlights one of the basic developments in the life sciences.

Within China, plant chromosome research began very late and no significant research was carried out before the 1980s. Following the introduction of reforms and the open-door policy in China, various study classes in addition to both domestic and international seminars were held in Nankai University and various regions throughout the country. These educational and informative sessions were organized and participated in by the Department of Biology, Nankai University; the Department of Biology, Beijing University; the Botanical Research Institute of Academy of Sciences of China and the Academy of Agricultural Science of China. These ventures promoted the rapid development of plant chromosome research. Therefore, regarding the extent and depth of research or technical expertise in this field, China has approached or reached the advanced world standard.

Among leading researchers, Professor Ruiyang Chen, head of the Plant Chromosome Laboratory at Nankai University, together with his colleagues, researched relentlessly, vigorously and enthusiastically within China for the past 20 years. They conscientiously and painstakingly gathered plant specimens from the various regions of China. Their work has been documented with films of chromosomes from each species with superior chromosome technology and exquisite photographic pictures. Detailed karyotype parameters were determined and four volumes of chromosome atlases of economic plants were then compiled. This publication deserves the profound appreciation and congratulations from its readers for the work it represents and the value of its contents. This atlas is not only an original great work domestically, but also a resource that fills a gap in plant chromosome research internationally. It will undoubtedly attract widespread attention.

Academician Hong Deyuan
Institute of Botany, Chinese Academy of Sciences
Chairman of Life Sciences Division, National Natural
Science Foundation of China (NSFC) Chinese Academy of Sciences
The Third World Academy of Sciences
Dec.15, 2001

附录2 实验室历年科研项目汇总

项目名称	来源	起止时间（年）	经费总额（万元）	负责人
中国特产栽培植物及其野生资源的染色体研究	国家自然科学基金	1983-1985	3	陈瑞阳
染色质和染色体超微结构研究	国家自然科学基金	1984-1988	4	陈瑞阳
植物染色体工程研究	天津市科委	1988-1990	15	陈瑞阳
我国重要经济植物核型及其数据库的建立	国家自然科学基金	1989-1993	3	陈瑞阳
三脉紫菀多倍体复合体的研究	国家自然科学基金	1991-1993	2	陈瑞阳
转基因新方法研究	国家自然科学基金	1992-1993	2	陈瑞阳
中国原始木本被子植物核型及核型进化	国家自然科学基金	1994-1996	6	陈瑞阳
构建玉米、黑麦人工B-染色体载体研究	国家自然科学基金	1995-1997	10	陈瑞阳
构建水稻单条染色体和特定区DNA文库研究	科技部	1995-1998	15	陈瑞阳
DNA指纹图谱在种子检测中的应用	国家自然科学基金	1996-1998	13	陈瑞阳
构建植物人工B-染色体载体的研究	国家自然科学基金	1998-2000	5	陈瑞阳
用染色体专性探针研究木本被子植物基因组的起源	国家自然科学基金	1998.1-12月	4	陈瑞阳
水稻Sd-1基因的克隆与定位研究	国家自然科学基金	1998-2000	4.4	陈瑞阳
水稻基因组第4号染色体DNA的构建及其克隆定位	科技部	1999.1-12月	20	陈瑞阳
基因芯片技术开发与应用	横向课题	2000-2003	800	陈瑞阳
三倍体丹参脱毒新品种培育	横向课题	2015.1-12月	70	陈瑞阳
道地丹参新品种培育	横向课题	2016.1-12月	60	陈瑞阳
利用G-带新技术研究小麦的起源	国家自然科学基金	1988-1990	2	安祝平
组织培养在多倍体鸭梨繁育种中的研究	国家自然科学基金	1993-1995	2	安祝平
中国蔷薇科植物及其特有果树细胞学及核型进化研究	国家自然科学基金	1988-1990	3	宋文芹
植物染色体显微切割和微克隆技术研究	国家自然科学基金	1993-1994	2	宋文芹
水稻单条染色体DNA文库的构建	国家自然科学基金	1995-1998	2	宋文芹
构建玉米、黑麦人工B染色体载体着丝粒和复制起始点	国家自然科学基金	1997-1998	10	宋文芹
构建植物人工B-染色体载体的研究	国家自然科学基金	2000-2002	15	宋文芹
药用植物丹参种质资源的遗传改良与种质创新研究	国家自然科学基金	2007-2009	25	陈 力
优良品种丹参的推广种植	横向课题	2012-2013	500	陈 力
应用cDNA芯片技术筛选抗癌中药	国家自然科学基金	2000-2002	10	陈 力
肝细胞类组织结构的构建及生物人工肝的研究	留学回国基金	1999-2002	5	陈 力
肝炎病毒基因分型和基因变异诊断芯片的研制及应用	天津市重点基金	2001-2002	90	宋文芹
植物B染色体DNA测序及其关键功能组件的分析	国家自然科学基金	2008-2010	28	宋文芹
抗黑腐病基因的克隆及功能鉴定	天津市重点基金	2006-2008	20	宋文芹
花椰菜分子遗传连锁图的构建及重要农艺性状的QTL定位	天津市重点基金	2008-2011	20	宋文芹
花椰菜抗菌核病种质的筛选与抗病基因的挖掘	天津市重点基金	2013-2016	20	宋文芹
中国植物染色体研究（续）	国家自然科学基金	2014-2017	80	宋文芹
杂种林木超亲速生性状形成和固定的基因组学	国家973子课题	2009-2013	384	宋文芹
红景天人工种植和培育的相关研究	横向课题	2015-2020	20	宋文芹
利用耐盐经济植物改良盐碱地技术示范	天津市基金	2010.1-12月	6	宋文芹

续表

项目名称	来源	起止时间（年）	经费总额（万元）	负责人
青花菜高新育种技术体系的构建及雄性不育新品种选育	横向课题	2013.1-12月	5	宋文芹
三倍体黑杨新种质的选育及基因组结构分析（合作）	国家自然科学基金	2016-2018	15	宋文芹
园林园艺植物基因组分析	横向课题	2015-2017	35	宋文芹
高档兰花新品种选育	天津市支撑项目	2007-2009	20	李秀兰
高产优质多抗兰花分子与细胞育种技术及品种创制	国家863子课题	2008-2010	10	李秀兰
铁皮石斛种苗生产技术	横向课题	2012-2014	15	李秀兰
应用组织培养技术生产中药石斛种苗	横向课题	2014-2016	50	李秀兰
蓖麻的遗传改良及新品种选育	横向课题	2014-2016	50	李秀兰
傣药药用植物细胞学研究	横向课题	2007-2008	3	陈成彬
柴胡染色体数目鉴定	横向课题	2008.1-12月	1	陈成彬
三倍体黑杨新种质的选育及基因结构分析	天津自然科学基金	2007-2009	10	陈成彬
水稻F基因组异染色质原位杂交定位	横向课题	2009-2010	15	陈成彬
银杏性别决定基因的克隆及性别决定机制研究	中央高校基本科研业务费	2013.1-12月	15	陈成彬
芝麻核型分析及部分BAC克隆定位	横向课题	2012-2013	18	陈成彬
芝麻起源演化的研究	横向课题	2013-2015	20	陈成彬
非编码小RNA对花椰菜小孢子胚胎发生发育的调控作用研究	中央高校基本科研业务费	2010-2012	15	王春国
花椰菜突变体库构建及抗黑腐病新种质创制研究	天津重点基金	2012-2014	20	王春国
鸭梨多倍体化过程中基因组结构及DNA甲基化变化研究	博士点基金	2010-2012	3.6	王春国
miRNA对花椰菜小孢子胚胎发生发育的调控机制研究	国家青年自然科学基金	2012-2014	25	王春国
花椰菜花球发育调控基因CDAG1的功能及调控机制研究	天津市重点基金	2015-2017	20	王春国
LaAP2L转录因子在落叶松超亲速生杂种优势性状形成中的功能及调控机制	国家自然科学基金	2015-2018	85	王春国
板栗遗传多样性分析	横向课题	2017-2019	8	王春国
高产、优质青花菜新品种培育及种质创制研发	天津种业科技重大专项	2018-2020	75	王春国
转基因抗旱华北落叶松新品系培育	农业部转基因专项	2018-2020	216	王春国
BolAHL16L在青花菜花球高产性状形成中的功能及分子调控机制研究	国家自然科学基金	2019-2022	72	王春国

附录3 实验室获奖汇总

获奖时间（年）	获奖名称	奖励等级	受奖单位
1978	植物染色体C-带技术研究	科技进步奖二等奖	天津市人民政府
1980	植物有丝分裂染色体标本制备新方法	科技进步奖一等奖	天津市人民政府
1987	苹果属梨属山楂属染色体数目鉴定	科技进步奖二等奖	农牧渔业部
1988	植物染色体高分辨G-带技术	科技进步奖二等奖	国家教育委员会
2001	去壁低渗法（WDH）及其在我国植物染色体研究中的应用	自然科学奖二等奖	天津市人民政府
2003	中国主要植物染色体研究	自然科学奖二等奖	国务院
2004	染色体微切微克隆新方法	自然科学奖三等奖	天津市人民政府
2009	丹参基因组三倍化优势利用技术及其应用	科技进步奖三等奖	天津市人民政府
2009	兰花的遗传改良与种质创新研究	技术发明奖三等奖	天津市人民政府
2017	花椰菜靶向分子育种技术及在新型育种材料选育中的应用	科技进步奖二等奖	天津市人民政府

附录4 实验室获得的授权专利汇总

专利名称	专利号
三倍体丹参脱毒新品种的快速选育方法	ZL 2013 1 0682544.9
道地丹参的培育方法	ZL 2014 1 0013083.8
道地丹参杂交一代优势利用及其制种方法	ZL 2017 10269468.4
复方丹参药酒的配制方法	ZL 2018 10526952.5
一种复方丹参足疗中药组合物及其制备方法	ZL 2018 10527036.3
三倍体丹参的培育方法	ZL 2007 10061227.7
铁皮石斛种苗简易化工厂化生产方法	ZL 2011 10177894.8
用种子类原球茎诱导四倍体秋石斛兰新品种方法	ZL 2007 100612262
铁皮石斛四倍体的培育方法	ZL 2007 10061225.8
三倍体铁皮石斛的培育方法	ZL 2011 10177885.9
以紫色基因为标记性状的三倍体丹参的自然杂交制种方法	ZL 2008 10152847.6
春石斛多倍体复合体的人工培育及种苗工厂化生产方法	ZL 2011 1077879.3
一种5S rDNA在植物染色体上的荧光原位杂交方法.	ZL 2013 10346485.5
一种45S rDNA在植物染色体上的荧光原位杂交方法.	ZL 2013 10346495.9
辣木参花多功能保健茶的制备方法。	ZL 2007 10150858.6
一种获得具有不同表型花椰菜材料的方法及应用	ZL 2013 10397368.1
一个花椰菜器官发育调控基因编码序列及其应用	ZL 2013 10397370.9

附录5 密苏里植物园 Index to Plant Chromosome Numbers(IPCN) 收录的 Ruiyang Chen 报道的中国植物染色体部分索引

经查询陈瑞阳等发表的有关中国植物染色体数目报道，被 Missouri Botanical Garden《Index to Plant Chromosome Numbers（IPCN）》收录（1985-2003），收录共计775种，其中首次报道（带#号的）为217种，2004年以后待收录中。

#IPCN chromosome reports for Aegilops biuncialis Vis.
[gam]---[spor] 28 [ipcn]84-85[ref]An，Z.P.，W.Q.Song，X.L.Li&R.Y.Chen.，1985
#IPCN chromosome reports for Aegilops caudata L.
[gam]---[spor] 14 [ipcn]84-85[ref]An，Z.P.，W.Q.Song，X.L.Li&R.Y.Chen.，1985
#IPCN chromosome reports for Aegilops crassa Boiss
[gam]---[spor] 42 [ipcn]84-85[ref]An，Z.P.，W.Q.Song，X.L.Li&R.Y.Chen.，1985
IPCN chromosome reports for Aegilops cylindrica Host
[gam]---[spor] 28 [ipcn]84-85[ref]An，Z.P.，W.Q.Song，X.L.Li&R.Y.Chen.，1985
#IPCN chromosome reports for Aegilops kotschyi Boiss.
[gam]---[spor] 28 [ipcn]84-85[ref]An，Z.P.，W.Q.Song，X.L.Li&R.Y.Chen.，1985
#IPCN chromosome reports for Aegilops ovata L.
[gam]---[spor] 28 [ipcn]84-85[ref]An.Z.P..W.Q.Song.X.L.Li&R.Y.Chen..1985
#IPCN chromosome reports for Aegilops sharonensis Eig
[gam]---[spor] 14 [ipcn]84-85[ref]An，Z.P..W.Q.Song，X.L.Li&R.Y.Chen.，1985
#IPCN chromosome reports for Aegilops speltoides Tausch
[gam]---[spor] 14 [ipcn]84-85[ref]An，Z.P.，W.Q.Song，X.L.Li&R.Y.Chen.，1985
#IPCN chromosome reports for Aegilops triaristata Willd.
[gam]---[spor] 42 [ipcn]84-85[ref]An，Z.P.，W.Q.Song，X.L.Li&R.Y.Chen..1985
#IPCN chromosome reports for Aegilops triuncialis L.
[gam]---[spor] 28 [ipcn]84-85[ref]An，Z.P.，W.Q.Song，X.L.Li&R.Y.Chen.，1985
#IPCN chromosome reports for Aegilops ventricosa Tausch
[gam]---[spor] 28 [ipcn]84-85[ref]An.Z.P.W.Q.Song.X.L.Li&R.Y.Chen.，1985
IPCN chromosome reports for Setaria italica (L.) P.Beauv.
[gam]---[spor] 18 [ipcn]84-85[ref]Li，X.-l.&R.-y.Chen，1985
[gam]---[spor] 36 [ipcn]84-85[ref]Li，X.-l.&R.-y.Chen.1985
IPCN chromosome reports for Setaria viridis (L.) P.Beauv.
[gam]---[spor] 18 [ipcn]84-85[ref]Li.X.-l.&R.-y.Chen，1985
#IPCN chromosome reports for Dendrobium acinaciforme
[gam]--[spor] 38 [ipcn]84-85[ref]Cheng，S.J.，Z.H.Hu，X.L.Li&R.Y.Chen.1985
#IPCN chromosome reports for Dendrobium chrysotoxum
[gam]---[spor] 38 [ipcn]84-85[ref]Cheng，S.J.，Z.H.Hu，X.L.Li&R.Y.Chen.1985
#IPCN chromosome reports for Dendrobium densiflorum
[gam]---[spor] 38 [ipcn]84-85[ref] Cheng，S.J.，Z.H.Hu.X.L.Li&R.Y.Chen.1985
#IPCN chromosome reports for Dendrobium thyrsiflorum
[gam]---[spor] 40 [ipcn]84-85[ref]Cheng，S.J.，Z.H.Hu.X.L.Li&R.Y.Chen.1985
#IPCN chromosome reports for Dendrobium candidum
[gam]---[spor] 38.57[ipcn]84-85[ref]Cheng，S.J..Z.H.Hu，X.L.Li&R.Y.Chen.1985
#IPCN chromosome reports for Dendrobium chrysanthum

[gam]---[spor] 38.76 [ipcn]84-85[ref] Cheng，S.J.，Z.H.Hu.X.L.Li&R.Y.Chen.1985
#IPCN chromosome reports for Dendrobium crepidatum
[gam]---[spor] 38 [ipcn]84-85[ref] Cheng，S.J.，Z.H.Hu.X.L.Li&R.Y.Chen.1985
#IPCN chromosome reports for Dendrobium denneanum
[gam]---[spor] 38 [ipcn]84-85[ref] Cheng，S.J.，Z.H.Hu.X.L.Li&R.Y.Chen.1985
#IPCN chromosome reports for Dendrobium devonianum
Lgam]---[spor] 38 [ipcn]84-85[ref]Cheng.S.J.，Z.H.Hu.X.L.Li&R.Y.Chen.1985
#IPCN chromosome reports for Dendrobium fimbriatum
[gam]---[spor] 40 [ipcn]84-85[ref] Cheng.S.J.，Z.H.Hu，X.L.Li&R.Y.Chen.1985
#IPCN chromosome reports for Dendrobium quangxiensis
[gam]---[spor] 38 [ipcn]84-85[ref]Cheng.S.J.，Z.H.Hu.X.L.Li&R.Y.Chen.1985
#IPCN chromosome reports for Dendrobium hancockii
[gain]---[spor] 40 [ipcn]84-85[ref]Cheng.S.J.，Z.H.Hu，X.L.Li&R.Y.Chen.1985
#IPCN chromosome reports for Dendrobium hercoglossum
[gam]---[spor] 57 [ipcn]84-85[ref] Cheng，S.J.，Z.H.Hu，X.L.Li&R.Y.Chen，1985
#IPCN chromosome reports for Dendrobium Iinawianum
[gam]---[spor] 38.40，76 [ipcn]84-85[ref]Cheng，S.J.，Z.H.Hu，X.L.Li&R.Y.Chen，1985
#IPCN chromosome reports for Dendrobium loddigesii
[gam]---[spor] 38 [ipcn]84-85[ref]Cheng，S.J.，Z.H.Hu，X.L.Li&R.Y.Chen.1985
#IPCN chromosome reports for Dendrobium Iohohense
[gam]---[spor] 38 [ipcn]84-85[ref]Cheng.S.J.，Z.H.Hu.X.L.Li&R.Y.Chen.1985
#IPCN chromosome reports for Dendrobium nobile
[gam]---[spor] 38 [ipcn]84-85[ref]Cheng.S.J.，Z.H.Hu，X.L.Li&R.Y.Chen.1985
#IPCN chromosome reports for Dendrobium clavatum
[gam]---[spor] 38 [ipcn]84-85[ref]Cheng.S.J.，Z.H.Hu，X.L.Li&R.Y.Chen.1985
#IPCN chromosome reports for Dendrobium wangii
[gam]---[spor] 38 [ipcn]84-85[ref]Cheng.S.J.，Z.H.Hu.X.L.Li&R.Y.Chen.1985
#IPCN chromosome reports for Dendrobium gibsonii
[gam]---[spor] 38 [ipcn]84-85[ref]Cheng.S.J.，Z.H.Hu，X.L.Li&R.Y.Chen.1985
#IPCN chromosome reports for Dendrobium aphyllum
[gam]---[spor] 38.40 [ipcn]84-85[ref]Cheng，S.J.，Z.H.Hu，X.L.Li&R.Y.Chen.1985
#IPCN chromosome reports for Dendrobium primulinum
[gam]---[spor] 38 [ipcn]84-85[ref]Cheng，S.J.，Z.H.Hu，X.L.Li&R.Y.Chen，1985
IPCN chromosome reports for Zea mays L.
[gam]---[spor] 20 [ipcn]86-87[ref]Chen，R.-y.，Z.-p.An.W.-q.Song，X.-I.Li，J.-y.Su&J.-y.zheng..　1987
IPCN chromosome reports for Zea mays L.
[gam]---[spor] 24 [ipcn]86-87[ref] Chen.R.-y.，Z.-p.An.W.-q.Song.X.-I.Li.J.-y.Su&J.-y.Zhen　　g..1987
#IPCN chromosome reports for Lilium davidii Duch.ex Elwes
[gam]---[spor] 20 [ipcn]88-89[ref]Chen，/R.-y.，1988
IPCN chromosome reports for Ginkgo biloba
[gam]---[spor] 24[ipcn]94-95 [ref]Chen，R.-y.，1988
IPCN chromosome reports for Vicia faba L.
[gam]---[spor] 12 [ipcn]88-89[ref]Chen，/R.-y.，1988
IPCN chromosome reports for Triticum aestivum L.
[gam]---[spor] 42 [ipcn]88-89[ref]Chen./R.-y.，1988
IPCN chromosome reports for Allium cepa L.
[gam]---[spor] 16 [ipcn]88-89[ref]Chen，/R.-y.，1988

IPCN chromosome reports for Hordeum vulgare L.
[gam]---[spor] 14 [ipcn]88-89[ref]Chen./R.-y., 1988
#IPCN chromosome reports for Cerasus serrulata
[gam]---[spor] 16 [ipcn] 86-87 [ref] Song, W.-q., X.-1.Li&Y.-1.Chen, 1989
IPCN chromosome reports for Secale cereale L.
lgam]---[spor] 14 [ipcn]88-89[ref]Chen./R.-y., 1989
IPCN chromosome reports for Hordeum vulgare L.
[gam]---[spor] 14 [ipcn]88-89[ref] Chen./R.-y., 1989
IPCN chromosome reports for Triticum aestivum L.
[gam]---[spor] 42 [ipcn]88-89[ref] Chen./R.-y., 1989
#IPCN chromosome reports for Lilium davidii Duch.ex Elwes
[gam]---[spor] 24 [ipcn]88-89[ref]Chen, /R.-y., 1989
#IPCN chromosome reports for Lilium davidii Duch.ex Elwes
[gam]---[spor] 24 [ipcn]88-89[ref]Chen, /R.-y., 1988
#IPCN chromosome reports for Bletilla striata
[gam]---[spor] 32, 76 [ipcn]88-89[ref]Li, /X.-I.&R.-y.Chen, 1989
IPCN chromosome reports for Malaxis latifolia
[gam]--[spor] 42 [ipcn]88-89[ref]Li, /X.-I.&R.-y.Chen, 1989
IPCN chromosome reports for Allium cepa L.
[gam]---[spor] 16 [ipcn]88-89[ref]Chen, /R.-y., 1989
#IPCN chromosome reports for Liparis stricklandiana
[gam]--[spor] 76 [ipcn]88-89[ref]Li./X.-I.&R.-y.Chen.1989
#IPCN chromosome reports for Liparis nervosa
[gam]---[spor] 40 [ipcn]88-89[ref]Li./X.-I.&R.-y.Chen.1989
IPCN chromosome reports for Liparis bootanensis
[gam]---[spor] 38 [ipcn]88-89[ref] Li./X.-I.&R.-y.Chen.1989
#IPCN chromosome reports for Liparis bautingensis
[gam]---[spor] 38   [ipcn]88-89[ref]Li./X.-I.&R.-y.Chen, 1989
#IPCN chromosome reports for Coelogyne fimbriata
[gam]---[spor] 40 [ipcn]88-89[ref]Li./X.-I.&R.-y.Chen, 1989
#IPCN chromosome reports for Coelogyne punctata
[gam]---[spor] 38 [ipcn]88-89[ref]Li./X.-I.&R.-y.Chen, 1989
#IPCN chromosome reports for Pleione prisei
[gam]---[spor] 38.40, 42 [ipcn]88-89[ref]Li, /X.-1.&R.-y.Chen, 1989
#IPCN chromosome reports for Pleione forrestii
[gam]--[spor] 38.39, 42, 44 [ipcn]88-89[ref]Li./X.-I.&R.-y.Chen, 1989
#IPCN chromosome reports for Pleione yunnanensis
[gam]---[spor] 40 [ipcn]88-89[ref]Li./X.-1.&R.-y.Chen, 1989
#IPCN chromosome reports for Pleione bulbocodioides
[gam]---[spor] 40 [ipcn]88-89[ref]Li./X.-I.&R.-y.Chen.1989
#IPCN chromosome reports for Pholidota articulata
[gam]---[spor] 38 [ipcn]88-89[ref]Li, /X.-1.&R.-y.Chen.1989
#IPCN chromosome reports for Pholidota chinensis
[gam]---[spor] 38 [ipcn]88-89[ref]Li./X.-I.&R.-y.Chen, 1989
#IPCN chromosome reports for Eria bulbophylloidea
[gam]---[spor] 42 [ipcn]88-89[ref]Li, /X.-I.&R.-y.Chen, 1989
IPCN chromosome reports for Eria acervata

[gam]---[spor] 38 [ipcn]88-89[ref]Li./X.-I.&R.-y.Chen，1989
#IPCN chromosome reports for Calanthe discolor
[gam]---[spor] 40 [ipcn]88-89[ref]Li./X.-I.&R.-y.Chen，1989
IPCN chromosome reports for Phaius tankervilliae
[gam]---[spor] 38 [ipcn]88-89[ref]Li./X.-l.&R.-y.Chen，1989
#IPCN chromosome reports for Phaius flavus
[gam]--[spor] 42 [ipcn]88-89[ref]Li./X.-I.&R.-y.Chen，1989
#IPCN chromosome reports for BulbophyIlum ambrosia
[gam]---[spor] 38 [ipcn]88-89[ref]Li./X.-I.&R.-y.Chen，1989
#IPCN chromosome reports for Cymbidium hookerianum.Iowianum
[gam]---[spor] 38 [ipcn]88-89[ref]Li./X.-I.&R.-y.Chen，1989
#IPCN chromosome reports for Cymbidium hookerianum
[gam]---[spor] 40 [ipcn]88-89[ref]Li./X.-I.&R.-y.Chen.1989
#IPCN chromosome reports for Cymbidium dayanum
[gam]--[spor] 40 [ipcn]88-89[ref]Li./X.-I.&R.-y.Chen，1989
IPCN chromosome reports for Cymbidium lancifolium
[gam]---[spor] 38 [ipcn]88-89[ref]Li，/X.-I.&R.-y.Chen，1989
#IPCN chromosome reports for Cymbidium goeringii
[gam]---[spor] 38 [ipcn]88-89[ref]Li./X.-l.&R.-y.Chen，1989
#IPCN chromosome reports for Cymbidium goeringii
[gam]---[spor] 40 [ipcn]88-89[ref]Chen，/R.-y.，1988
#IPCN chromosome reports for Cymbidium ensifolium
[gam]--[spor] 40 [ipcn]88-89[ref]Li，/X.-I.&R.-y.Chen，1989
#IPCN chromosome reports for Stauropsis undulata
[gam]---[spor] 38 [ipcn]88-89[ref]Li，/X.-I.&R.-y.Chen，1989
#IPCN chromosome reports for Aerides flabellatum
[gam]--[spor] 38 [ipcn]88-89[ref]Li，/X.-I.&R.-y.Chen，1989
#IPCN chromosome reports for Thrixspermum centripeda
[gam]---[spor] 38 [ipcn]88-89[ref]Li./X.-I.&R.-y.Chen，1989
#IPCN chromosome reports for Arachnis clarkei
[gam]---[spor] 42 [ipcn]88-89[ref]Li，/X.-I.&R.-y.Chen，1989
#IPCN chromosome reports for Vandopsis polyantha
[gam]---[spor] 38 [ipcn]88-89[ref]Li./X.-I.&R.-y.Chen，1989
#IPCN chromosome reports for Cleisostoma rostratum
[gam]---[spor] 38 [ipcn]88-89[ref]Li./X.-I.&R.-y.Chen.1989
#IPCN chromosome reports for Vanda subconcolor
[gam]---[spor] 38 [ipcn]88-89[ref]Li，/X.-I.&R.-y.Chen，1989
IPCN chromosome reports for Vicia faba L.
[gam]---[spor] 12 [ipcn]88-89[ref]Chen，/R.-y.，1989
#IPCN chromosome reports for Vanda denisoniana.hebraica
[gam]---[spor] 76 [ipcn]88-89[ref]Li./X.-I.&R.-y.Chen，1989
#IPCN chromosome reports for Vanda coerulea
[gam]---[spor] 38 [ipcn]88-89[ref]Li./X.-I.&R.-y.Chen，1989
#IPCN chromosome reports for Vanda pumila
[gam]---[spor] 38.72.76 [ipcn]88-89[ref]Li，/X.-I.&R.-y.Chen.1989
#IPCN chromosome reports for Cleisostoma paniculatum
[gam]---[spor] 38 [ipcn]88-89[ref]Li./X.-I.&R.-y.Chen，1989

#IPCN chromosome reports for Vanda teres
[gam]---[spor] 40 [ipcn]88-89[ref] Li./X.-I.&R.-y.Chen.1989
#IPCN chromosome reports for Vanda concolor
[gam]---[spor] 38.76 [ipcn]88-89[ref]Li./X.-I.&R.-y.Chen.1989
#IPCN chromosome reports for Cymbidium pendulum
[gam]---[spor] 40 [ipcn]88-89[ref]Li./X.-I.&R.-y.Chen，1989
IPCN chromosome reports for Ginkgo biloba
Igam]---[spor] 24 [ipcn] 88-89 [ref] Chen，R.-y.，1989
#IPCN chromosome reports for Cymbidium sinense
[gam]---[spor] 40 [ipcn]88-89[ref]Li./X.-I.&R.-y.Chen，1989
IPCN chromosome reports for Prunus salicina
[gam]---[spor] 16 [ipcn] 88-89 [ref] Song，W.-q.，X.-1.Li &Y.-1.Chen，1989
#IPCN chromosome reports for Holcoglossum junceum
[gam]---[spor] 38 [ipcn]88-89[ref]Li./X.-I.&R.-y.Chen，1989
#IPCN chromosome reports for Vanda coerulescens
[gam]---[spor] 38 [ipcn]88-89[ref]Li./X.-I.&R.-y.Chen，1989
#IPCN chromosome reports for Trichosma suavis
[gam]---[spor] 36 [ipcn]88-89[ref]Li，/X.-I.&R.-y.Chen，1989
#IPCN chromosome reports for Epidendrum radicans
[gam]---[spor] 57，62 [ipcn]88-89 [ref]Li./X.-I.&R.-y.Chen，1989
#IPCN chromosome reports for Oncidium sphacelatum
[gam]--[spor] 57 [ipcn]88-89[ref] Li./X.-I.&R.-y.Chen，1989
#IPCN chromosome reports for Goodyera procera
[gam]---[spor] 38 [ipcn]88-89[ref]Li，/X.-I.&R.-y.Chen，1989
#IPCN chromosome reports for Cymbidium eburneum
[gam]---[spor] 38 [ipcn]88-89[ref] Li，/X.-I.&R.-y.Chen，1989
IPCN chromosome reports for Liriodendron chinense
[gam]---[spor] 38 [ipcn]88-89[ref]Chen，/R.-y.，W.Zhang&Q.-a.Wu，1989
IPCN chromosome reports for Michelia champaca
[gam]---[spor] 38 [ipcn]88-89[ref]Chen，/R.-y.，W.Zhang&Q.-a.Wu，1989
IPCN chromosome reports for Michelia longistamina
[gam]---[spor] 38 [ipcn]88-89[ref]Chen，/R.-y.，W.Zhang&Q.-a.Wu，1989
#IPCN chromosome reports for Michelia maudiae
[gam]---[spor] 38 [ipcn]88-89[ref] Chen，/R.-y.，W.Zhang&Q.-a.Wu，1989
#IPCN chromosome reports for Michelia platypetala
[gam]---[spor] 38 [ipcn]88-89[ref]Chen，/R.-y.，W.Zhang&Q.-a.Wu，1989
#IPCN chromosome reports for Michelia floribunda
[gam]---[spor] 38 [ipcn]88-89[ref]Chen，/R.-y.，W.Zhang&Q.-a.Wu，1989
#IPCN chromosome reports for Michelia foveolata
[gam]---[spor] 38 [ipcn]88-89[ref]Chen.，/R.-y.，W.Zhang&Q.-a.Wu，1989
#IPCN chromosome reports for Tsoongiodendron odorum
[gam]---[spor] 38 [ipcn]88-89[ref]Chen，/R.-y.，W.Zhang&Q.-a.Wu，1989
#IPCN chromosome reports for Paramichelia baillonii
[gam]---[spor] 38 [ipcn]88-89[ref]Chen，/R.-y.，W.Zhang&Q.-a.Wu，1989
#IPCN chromosome reports for Michelia wilsonii
[gam]---[spor] 38 [ipcn]88-89[ref]Chen，/R.-y.，W.Zhang&Q.-a.Wu.1989
#IPCN chromosome reports for Manglietia grandis

[gam]---[spor] 38 [ipcn]88-89[ref]Chen./R.-y., W.Zhang&Q.-a.Wu.1989
#IPCN chromosome reports for Magnolia delavayi
[gam]---[spor] 38 [ipcn]88-89[ref]Chen, /R.-y., W.Zhang&Q.-a.Wu, 1989
#IPCN chromosome reports for Magnolia amoena
[gam]---[spor] 38 [ipcn]88-89[ref]Chen, /R.-y., W.Zhang&Q.-a.Wu.1989
IPCN chromosome reports for Magnolia Iiliflora
[gam]---[spor] 76 [ipcn]88-89[ref]Chen./R.-y., W.Zhang&Q.-a.Wu, 1989
#IPCN chromosome reports for Magnolia denudata
[gam]---[spor] 114 [ipcn]88-89[ref]Chen./R.-y., W.Zhang&Q.-a.Wu, 1989
#IPCN chromosome reports for Indosasa levigata
[gam]---[spor] 48 [ipcn]94-95[ref]Chen, \R.-y.&W.x.Zong, 1991
#IPCN chromosome reports for Indosasa shibataeoides McClure
[gam]---[spor] 48 [ipcn]94-95[ref]Chen, \R.-y.&W.x.Zong, 1991
#IPCN chromosome reports for Indosasa sinica C.D.Chu&C.S.Chao
[gam]---[spor] 48 [ipcn]94-95[ref]Chen, \R.-y.&W.x.Zong, 1991
#IPCN chromosome reports for Sinobambusa tootsik
[gam]---[spor] 48 [ipcn]94-95[ref]Chen, \R.-y.&W.x.Zong, 1991
#IPCN chromosome reports for Sinobambusa anaurita
[gam]---[spor] 48 [ipcn]94-95[ref]Chen, \R.-y.&W.x.Zong, 1991
#IPCN chromosome reports for Sinobambusa sichuanensis
[gam]---[spor] 48 [ipcn]94-95[ref]Chen, \R.-y.&W.x.Zong, 1991
#IPCN chromosome reports for Sinobambusa rubroligula
[gam]---[spor] 48 [ipcn]94-95[ref] Chen, \R.-y.&W.x.Zong, 1991
#IPCN chromosome reports for Brachystachyum densiflorum
[gam]---[spor] 48 [ipcn]94-95[ref]Chen, \R.-y.&W.x.Zong, 1991
#IPCN chromosome reports for Phyllostachys meyeri
[gam]---[spor] 48 [ipcn]94-95[ref]Chen, \R.-y.&W.x.Zong, 1991
#IPCN chromosome reports for Phyllostachys aurea
[gam]---[spor] 48 [ipcn]94-95[ref] Chen, \R.-y.&W.x.Zong, 1991
#IPCN chromosome reports for Phyllostachys nuda
[gamn]---[spor] 48 [ipcn]94-95[ref]Chen, \R.-y.&W.x.Zong.1991
#IPCN chromosome reports for Phyllostachys arcana
[gam]---[spor] 48 [ipcn]94-95[ref] Chen, \R.-y.&W.x.Zong, 1991
#IPCN chromosome reports for Phyllostachys glauca
[gam]---[spor] 48 [ipcn]94-95[ref]Chen, \R.-y.&W.x.Zong, 1991
#IPCN chromosome reports for Phyllostachys angusta
[gam]---[spor] 48 [ipcn]94-95[ref]Chen., \R.-y.&W.x.Zong.1991
#IPCN chromosome reports for Phy1lostachys flexuosa
[gam]---[spor]48（54）[ipcn]94-95[ref]Chen, \R.-y.&W.x.Zong.1991
#IPCN chromosome reports for Phyllostachys glabrata
[gam]---[spor] 48 [ipcn]94-95[ref] Chen, \R.-y.&W.x.Zong.1991
#IPCN chromosome reports for Phyllostachys vivax
[gam]---[spor] 48 [ipcn]94-95[ref]Chen, \R.-y.&W.x.Zong, 1991
IPCN chromosome reports for Phyllostachys praecox
[gam]---[spor] 48 [ipcn]94-95[ref]Chen, R.-y.&W.x.Zong.1991
IPCN chromosome reports for Phyllostachys heterocycla.pubescens
[gam]---[spor] 48 [ipcn]94-95[ref]Chen, \R.-y.&W.x.Zong, 1991

#IPCN chromosome reports for Phyllostachys aureosulcata.spectabilis
[gam]---[spor] 48 [ipcn]94-95[ref]Chen, \R.-y.&W.x.Zong, 1991
#IPCN chromosome reports for Phyllostachys incarnata
[gam]---[spor] 48 [ipcn]94-95[ref]Chen.\R.-y.&W.x.Zong.1991
#IPCN chromosome reports for Phyllostachys dulcis
[gam]---[spor] 48 [ipcn]94-95[ref]Chen, \R.-y.&W.x.Zong.1991
#IPCN chromosome reports for Phyllostachys bambusoides
[gam]---[spor] 48 [ipcn]94-95[ref]Chen., \R.-y.&W.x.Zong, 1991
#IPCN chromosome reports for Phyllostachys viridi-glaucescens
[gam]---[spor] 48 [ipcn]94-95[ref] Chen, \R.-y.&W.x.Zong, 1991
IPCN chromosome reports for Phyllostachys prominens
[gam]---[spor] 48 [ipcn]94-95[ref]Chen, \R.-y.&W.x.Zong.1991
#IPCN chromosome reports for Phyllostachys nigella
[gam]---[spor] 48 [ipcn]94-95[ref]Chen, \R.-y.&W.x.Zong, 1991
#IPCN chromosome reports for Phyllostachys nidularia
[gam]---[spor] 48 [ipcn]94-95[ref]Chen, \R.-y.&W.x.Zong, 1991
#IPCN chromosome reports for Shibataea lanceifolia
[gam]---[spor] 48 [ipcn]94-95[ref]Chen, \R.-y.&W.x.Zong, 1991
#IPCN chromosome reports for Shibataea nanpingensis
[gam]---[spor] 48 [ipcn]94-95[ref]Chen, \R.-y.&W.x.Zong, 1991
#IPCN chromosome reports for Chimonobambusa pachystachys
[gam]---[spor] 48 [ipcn]94-95[ref]Chen, \R.-y.&W.x.Zong, 1991
#IPCN chromosome reports for Chimonobambusa quadrangularis
[gam]---[spor] 48 [ipcn]94-95[ref]Chen, \R.-y.&W.x.Zong, 1991
#IPCN chromosome reports for Qiongzhuea tumidinoda
[gam]---[spor] 48 [ipcn]94-95[ref]Chen, \R.-y.&W.x.Zong, 1991
#IPCN chromosome reports for Yushania baishanzuensis
[gam]---[spor] 48 [ipcn]94-95[ref]Chen, \R.-y.&W.x.Zong, 1991
#IPCN chromosome reports for Oligostachyum sulcatum
[gam]---[spor] 48 [ipcn]94-95[ref]Chen, \R.-y.&W.x.Zong, 1991
#IPCN chromosome reports for Pleioblastus gramineus
[gam]---[spor] 48 [ipcn]94-95[ref] Chen, \R.-y.&W.x.Zong, 1991
#IPCN chromosome reports for Pleioblastus amarus.subglabratus
[gam]---[spor] 48 [ipcn]94-95[ref]Chen, \R.-y.&W.x.Zong, 1991
#IPCN chromosome reports for Pleioblastus intermedius
[gam]---[spor] 48 [ipcn]94-95[ref]Chen, \R.-y.&W.x.Zong, 1991
#IPCN chromosome reports for Pleioblastus yixingensis
[gam]---[spor] 48 [ipcn]94-95[ref] Chen, \R.-y.&W.x.Zong, 1991
#IPCN chromosome reports for Pleioblastus longifimbriatus
[gam]---[spor] 48 [ipcn]94-95[ref]Chen, \R.-y.&W.x.Zong, 1991
#IPCN chromosome reports for Bashania fargesii
[gam]---[spor] 48 [ipcn]94-95[ref]Chen, \R.-y.&W.x.Zong, 1991
#IPCN chromosome reports for Gelidocalamus stellatus
[gam]---[spor] 48 [ipcn]94-95[ref] Chen, \R.-y.&W.x.Zong.1991
#IPCN chromosome reports for Pseudosasa japonica
[gam]---[spor] 48 [ipcn]94-95[ref] Chen, \R.-y.&W.x.Zong, 1991
IPCN chromosome reports for Pseudosasa amabilis

[gam]---[spor] 48 [ipcn]94-95[ref]Chen.\R.-y.&W.x.Zong，1991
#IPCN chromosome reports for Pseudosasa subsolida
[gam]---[spor] 48 [ipcn]94-95[ref] Chen，\R.-y.&W.x.Zong，1991
IPCN chromosome reports for Phragmites australis
[gam]---[spor] 48，72，96 [ipcn]90-91[ref]Zong, W., R.Chen, K.Taniguchi&K.Kondo.1991
#IPCN chromosome reports for Lilium davidii Duch.ex Elwes
[gam]---[spor]24[ipcn]92-93[ref]Chen./R.-y.，Z.-p.An，W.-q.Song.X.-I.Li.W.-x.Zong.K.Taniguchi&D.-y.Hong. 1992
#IPCN chromosome reports for Bletilla striata
[gam]---[spor] 32，76 [ipcn]92-93[ref]Li，X.，R.Chen&R.Tanaka.1992
IPCN chromosome reports for Malaxis latifolia
[gain]---[spor] 42 [ipcn]92-93[ref]Li.X.，R.Chen&R.Tanaka，1992
#IPCN chromosome reports for Liparis stricklandiana
[gam]--[spor] 76 [ipcn]92-93[ref]Li，X.，R.Chen&R.Tanaka，1992
#IPCN chromosome reports for Liparis nervosa
[gam]---[spor] 40 [ipcn]92-93[ref]Li.X.，R.Chen&R.Tanaka.1992
IPCN chromosome reports for Liparis bootanensis
[gam]---[spor] 38 [ipcn]92-93[ref]Li，X.，R.Chen&R.Tanaka.1992
#IPCN chromosome reports for Liparis bautingensis
[gam]---[spor] 38 [ipcn]92-93[ref]Li，X.，R.Chen&R.Tanaka.1992
#IPCN chromosome reports for Coelogyne fimbriata
[gam]---[spor] 40 [ipcn]92-93[ref]Li，X.，R.Chen&R.Tanaka，1992
#IPCN chromosome reports for Coelogyne punctata
[gam]---[spor] 38 [ipcn]92-93[ref]Li，X.，R.Chen&R.Tanaka，1992
#IPCN chromosome reports for Pleione prisei
[gam]---[spor] 38，40，42 [ipcn]92-93[ref]Li，X.，R.Chen&R.Tanaka，1992
#IPCN chromosome reports for Pleione forrestii
[gam]---[spor] 38，39，42，44 [ipcn]92-93[ref]Li，X.，R.Chen&R.Tanaka，1992
IPCN chromosome reports for Pleione yunnanensis
[gam]---[spor] 40 [ipcn]92-93[ref]Li.X.，R.Chen&R.Tanaka，1992
#IPCN chromosome reports for Pleione bulbocodioides
[gam]---[spor] 40 [ipcn]92-93[ref]Li.X.，R.Chen&R.Tanaka.1992
#IPCN chromosome reports for Pholidota articulata
[gam]---[spor] 38 [ipcn]92-93[ref]Li.X.，R.Chen&R.Tanaka，1992
#IPCN chromosome reports for Eria bulbophylloidea
[gam]---[spor] 38 [ipcn]92-93[ref]Li.X.，R.Chen&R.Tanaka.1992
IPCN chromosome reports for Eria acervata
[gam]---[spor] 38 [ipcn]92-93[ref]Li.X.，R.Chen&R.Tanaka.1992
#IPCN chromosome reports for Calanthe discolor
[gam]---[spor] 40 [ipcn]92-93[ref]Li.X.，R.Chen&R.Tanaka.1992
IPCN chromosome reports for Eria acervata
[gam]---[spor] 42 [ipcn]92-93[ref]Li.X.，R.Chen&R.Tanaka.1992
IPCN chromosome reports for Phaius tankervilliae
[gam]---[spor] 38 [ipcn]92-93[ref]Li.X.，R.Chen&R.Tanaka.1992
#IPCN chromosome reports for Phaius flavus
[gam]--[spor] 42 [ipcn]92-93[ref]Li.X.，R.Chen&R.Tanaka.1992
#IPCN chromosome reports for BulbophyIlum ambrosia

[gam]---[spor] 38 [ipcn]92-93[ref]Li.X., R.Chen&R.Tanaka.1992
#IPCN chromosome reports for Cymbidium hookerianum.Iowianum
[gam]---[spor] 38 [ipcn]92-93[ref]Li, X., R.Chen&R.Tanaka, 1992
#IPCN chromosome reports for Cymbidium pendulum
[gam]---[spor] 40 [ipcn]92-93[ref] Li, X., R.Chen&R.Tanaka, 1992
#IPCN chromosome reports for Cymbidium sinense
[gam]---[spor] 40 [ipcn]92-93[ref]Li., X., R.Chen&R.Tanaka, 1992
#IPCN chromosome reports for Cymbidium hookerianum
[gam]---[spor] 40 [ipcn]92-93[ref]Li.X., R.Chen&R.Tanaka.1992
#IPCN chromosome reports for Cymbidium dayanum
[gam]---[spor] 40 [ipcn]92-93[ref]Li, X., R.Chen&R.Tanaka, 1992
IPCN chromosome reports for Cymbidium lancifolium
[gam]---[spor] 38 [ipcn]92-93[ref]Li, X., R.Chen&R.Tanaka, 1992
#IPCN chromosome reports for Cymbidium goeringii
[gam]---[spor] 40 [ipcn]92-93[ref]Li, X., R.Chen&R.Tanaka, 1992
#IPCN chromosome reports for Cymbidium ensifolium
*[gam]---[spor] 52 [ipcn]92-93[ref]Li, X., R.Chen&R.Tanaka, 1992
#IPCN chromosome reports for Cymbidium ensifolium
[gam]---[spor] 40 [ipcn]92-93[ref]Li, X., R.Chen&R.Tanaka, 1992
#IPCN chromosome reports for Stauropsis undulata
[gam]---[spor] 38 [ipcn]92-93[ref]Li, X., R.Chen&R.Tanaka, 1992
#IPCN chromosome reports for Aerides flabellatum
[gam]---[spor] 38 [ipcn]92-93[ref]Li, X., R.Chen&R.Tanaka, 1992
#IPCN chromosome reports for Thrixspermum centripeda
[gam]---[spor] 38 [ipcn]92-93[ref]Li, X., R.Chen&R.Tanaka, 1992
#IPCN chromosome reports for Arachnis clarkei
[gam]---[spor] 42 [ipcn]92-93[ref]Li, X., R.Chen&R.Tanaka, 1992
#IPCN chromosome reports for Vandopsis polyantha
[gam]---[spor] 38 [ipcn]92-93[ref]Li, X., R.Chen&R.Tanaka.1992
#IPCN chromosome reports for Cleisostoma rostratum
[gam]---[spor] 38 [ipcn]92-93[ref]Li., X., R.Chen&R.Tanaka.1992
#IPCN chromosome reports for Cleisostoma paniculatum
[gam]---[spor] 38 [ipcn]92-93[ref]Li, X., R.Chen&R.Tanaka, 1992
#IPCN chromosome reports for Renanthera coccinea
[gam]---[spor] 38.114 [ipcn]92-93[ref]Li, X., R.Chen&R.Tanaka, 1992
#IPCN chromosome reports for Holcoglossum junceum
[gam]---[spor] 38.57 [ipcn]92-93[ref]Li, X., R.Chen&R.Tanaka.1992
#IPCN chromosome reports for Vanda coerulea
[gam]---[spor] 38 [ipcn]92-93[ref]Li, X., R.Chen&R.Tanaka, 1992
#IPCN chromosome reports for Vanda denisoniana.hebraica
[gam]---[spor] 76 [ipcn]92-93[ref]Li.X., R.Chen&R.Tanaka.1992
#IPCN chromosome reports for Vanda subconcolor
[gam]---[spor] 38 [ipcn]92-93[ref]Li., X., R.Chen&R.Tanaka.1992
#IPCN chromosome reports for Vanda pumila
[gam]---[spor] 38.76 [ipcn]92-93[ref]Li.X., R.Chen&R.Tanaka.1992
#IPCN chromosome reports for Vanda concolor
[gam]---[spor] 38.76 [ipcn]92-93[ref]Li.X..R.Chen&R.Tanaka.1992

#IPCN chromosome reports for Vanda coerulescens
[gam]---[spor] 38 [ipcn]92-93[ref]Li.X..R.Chen&R.Tanaka, 1992
#IPCN chromosome reports for Trichosma suavis
[gam]---[spor] 36 [ipcn]92-93[ref]Li, X., R.Chen&R.Tanaka., 1992
#IPCN chromosome reports for Epidendrum radicans
[gam]---[spor] 57, 62 [ipcn]92-93[ref]Li, X., R.Chen&R.Tanaka, 1992
#IPCN chromosome reports for Cattleya intermedia
[gam]--[spor] 46, 55, 76 [ipcn]92-93[ref]Li, X., R.Chen&R.Tanaka.1992
#IPCN chromosome reports for Oncidium sphacelatum
[gam]---[spor] 57 [ipcn]92-93[ref]Li.X., R.Chen&R.Tanaka, 1992
#IPCN chromosome reports for Vanda teres
[gam]---[spor] 40 [ipcn]92-93[ref]Li.X..R.Chen&R.Tanaka.1992
#IPCN chromosome reports for Goodyera procera
[gam]---[spor] 38 [ipcn]92-93[ref]Li, X., R.Chen&R.Tanaka, 1992
#IPCN chromosome reports for Cymbidium eburneum
[gam]---[spor] 38 [ipcn]92-93[ref]Li, X., R.Chen&R.Tanaka, 1992
IPCN chromosome reports for Malus baccata
[gam]---[spor] 34 [ipcn]94-95 [ref] Chen, R.-y., 1993
IPCN chromosome reports for Malus mandshurica
[gam]---[spor] 34 [ipcn]94-95 [ref] Chen, R.-y., 1993
IPCN chromosome reports for Malus rockii
[gam]---[spor] 34, 51 [ipcn]94-95 [ref]Chen, R.-y., 1993
IPCN chromosome reports for Malus sikkimensis
[gam]---[spor] 34, 68 [ipcn]94-95 [ref] Chen, R.-y., 1993
IPCN chromosome reports for Malus hupehensis
[gam]---[spor] 51 [ipcn]94-95 [ref]Chen, R.-y., 1993
IPCN chromosome reports for Malus halliana
[gam]---[spor] 34, 51 [ipcn]94-95 [ref] Chen, R.-y., 1993
IPCN chromosome reports for Malus pumila
[gam]---[spor] 34, 51[ipcn]94-95[ref] Chen, R.-y., 1993
#IPCN chromosome reports for Malus pumila.praecox
[gam]---[spor] 34 [ipcn]94-95 [ref]Chen, R.-y., 1993
#IPCN chromosome reports for Malus pumila.paradisiaca
[gam]---[spor] 34 [ipcn]94-95 [ref]Chen, R.-y., 1993
IPCN chromosome reports for Malus sylvestris
lgam]---[spor] 34 [ipcn]94-95[ref] Chen, R.-y., 1993 P."y.Chen, w.-q.
IPCN chromosome reports for Malus sieversii
{gam]---[spor] 34 [ipcn]94-95 [ref] Chen, R.-y., 1993
IPCN chromosome reports for Malus asiatica
[gam]---[spor] 34, 51, 68 [ipcn]94-95 [ref]Chen, R.-y., 1993
[gam]---[spor] 34, 68 [ipcn]94-95 [ref] Chen, R.-y., 1993
#IPCN chromosome reports for Malus asiatica.rinki
[gam]---[spor] 34 [ipcn!94-95 [ref] Chen, R.-y., 1993
IPCN chromosome reports for Malus prunifolia
[gam]---[spor] 34, 51 [ipcn]94-95 [ref]Chen, R.-y., 1993
IPCN chromosome reports for Malus spectabilis
[gam]---[spor] 34[ipcn!94-95 [ref] Chen, R.-y., 1993

IPCN chromosome reports for Malus micromalus
*[gam]---[spor] 34 [ipcn]94-95 [ref]Chen，R.-y.，1993
[gam]---[spor] 34 [ipcn]94-95 [ref] Chen，R.-y.，1993
IPCN chromosome reports for Malus sieboldii
Igam]---[spor] 51 [ipcn：94-95 [ref] Chen，R.-y.，1993
IPCN chromosome reports for Malus kansuensis
Igam]---[spor] 34 [ipcn；94-95 [ref] Chen，R.-y.，1993
#IPCN chromosome reports for Malus komarovii
[gam]---[spor] 34 [ipcn：94-95 [ref]Chen，R.-y.，1993
IPCN chromosome reports for Malus toringoides
[gam]---[spor] 34，51，68 [ipcn] 94-95 [ref] Chen，R.-y.，1993
IPCN chromosome reports for Malus transitoria
[gam]---[spor] 34，51[ipcn]94-95 [ref]Chen，R.-y.，1993
#IPCN chromosome reports for Malus prattii
[gam]---[spor] 34 [ipcn]94-95 [ref]Chen，R.-y.，1993
IPCN chromosome reports for Malus ombrophila
[gam]---[spor]34 [ipcn]94-95 [ref] Chen，R.-y.，1993
IPCN chromosome reports for Malus honanensis
[gam]---[spor] 34 [ipcn]94-95 [ref] Chen，R.-y.，1993
IPCN chromosome reports for Malus yunnanensis
[gam]---[spor] 34 [ipcn]94-95 [ref]Chen，R.-y.，1993
#IPCN chromosome reports for Malus formosana
[gam]---[spor] 34 [ipcn]94-95 [ref] Chen，R.-y.，1993
IPCN chromosome reports for Malus melliana
[gam]---[spor] 34 [ipcn]94-95 [ref]Chen，R.-y.，1993
#IPCN chromosome reports for Malus xiaojinensis
[gam]---[spor] 68 [ipcn]94-95 [ref] Chen，R.-y.，1993
#IPCN chromosome reports for Malus robusta
[gam]---[spor] 51 [ipcn] 88-89 [ref]Song，W.-q.，X.-1 Li&Y.-1.Chen，1989
[gam]---[spor] 51 [ipcn]94-95 [ref]Chen，R.-y.，1993
#IPCN chromosome reports for Pyrus ussuriensis
*[gam]---[spor] 34[ipcn]88-89 [ref]Song，W.-q.，X.-1 Li&Y.-1.Chen，1986
*lgam]---[spor] 34[ipcn]94-95 [ref]Chen，R.-y.，1993
[gam]---[spor] 34[ipcn]94-95[ref]Chen，R.-y.，1993
#IPCN chromosome reports for Pyrus hopeiensis
[gam]---[spor] 34 [ipcn]94-95 [ref] Chen，R.-y.，1993
#IPCN chromosome reports for Pyrus sinkiangensis
[gam]=--[spor] 34 [ipcn]94-95 [ref]Chen，R.-y.，1993
#IPCN chromosome reports for Pyrus serrulata
[gam]---[spor] 34 [ipcn]94-95 [ref] Chen，R.-y.，1993
IPCN chromosome reports for Pyrus communis
[gam]---[spor] 34 [ipcn]94-95 [ref]Chen，R.-y.，1993
#IPCN chromosome reports for Pyrus armeniacaefolia
[gam]---[spor] 51[ipcn] 94-95 [ref] Chen，R.-y.，1993
#IPCN chromosome reports for Pyrus xerophila
[gam]---[spor] 34 [ipcn]86-87 [ref] Pu，F.-s.，S.-h.Lin，R.-y.Chen，W.-q.Song&x.-1.Li，1986

[gam]---[spor] 34 [ipcn]94-95 [ref] Chen，R.-y.，1993
#IPCN chromosome reports for Pyrus bretschneideri
*[gam]---[spor] 34 [ipcn]94-95 [ref]Chen，R.-y.，1993
[gam]---[spor] 34，51，68 [ipcn]94-95 [ref]Chen，R.-y.，1993
#IPCN chromosome reports for Pyrus pyrifolia
[gam]---[spor] 34 [ipcn] 86-87 [ref] Pu，F.-S.，S.-h.Lin，R.-y.Chen，W.-q.Song&X.-l.Li，1986
#IPCN chromosome reports for Pyrus pyrifolia
 [gam]---[spor] 34 [ipcn]94-95 [ref] Chen，R.-y.，1993
#IPCN chromosome reports for Pyrus betulaefolia
[gam]---[spor] 34 [ipcn]94-95 [ref] Chen，R.-y.，1993
#IPCN chromosome reports for Pyrus phaeocarpa
[gam]---[spor] 34 [ipcn] 94-95 [ref] Chen，R.-y.，1993
#IPCN chromosome reports for Pyrus calleryana
[gam]---[spor] 34 [ipcn]94-95 [ref] Chen，R.-y.，1993
IPCN chromosome reports for Pyrus pashia
[gam]---[spor] 34 [ipcn] 94-95 [ref] Chen，R.-y.，1993
IPCN chromosome reports for Crataegus pinnatifida
[gam.]，---[spor] 34 [ipcn] 94-95 [ref] Chen，R.-y.，1993
IPCN chromosome reports for Crataegus hupehensis
[gam]---[spor] 34 [ipcn]94-95 [ref] Chen，R.-y.，1993
IPCN chromosome reports for Crataegus wilsonii
[gam]---[spor] 34 [ipcn]94-95 [ref] Chen，R.-y.，1993
IPCN chromosome reports for Crataegus maximowiczii
[gam]---[spor] 51 [ipcn]94-95 [ref]Chen，R.-y.，1993
IPCN chromosome reports for Crataegus sanguinea
[gam]---[spor] 68 [ipcn]94-95 [ref]Chen，R.-y.，1993
IPCN chromosome reports for Crataegus kansuensis
[gam]---[spor] 34 [ipcn]94-95[ref] Chen，R.-y.，1993
IPCN chromosome reports for Crataegus altaica
[gam]---[spor] 68[ipcn]94-95 [ref]Chen，R.-y.，1993
#IPCN chromosome reports for Crataegus songarica
[gam]---[spor] 68 [ipcn]94-95 [ref] Chen，R.-y.，1993
IPCN chromosome reports for Crataegus scabrifolia
[gam]---[spor] 34 [ipcn]94-95 [ref]Chen，R.-y.，1993
IPCN chromosome reports for Crataegus cuneata
[gam]---[spor] 34 [ipcn]94-95 [ref]Chen，R.-y.，1993
#IPCN chromosome reports for Chaenomeles speciosa
[gam]---[spor] 34 [ipcn]94-95 [ref]Chen，R.-y.，1993
#IPCN chromosome reports for Cydonia oblonga
[gam]---[spor] 34 [ipcn]94-95 [ref]Chen，R.-y.，1993
IPCN chromosome reports for Citrullus vulgaris
[gam]---[spor] 22，44 [ipcn]94-95 [ref]Chen，R.-y.，1993
IPCN chromosome reports for Cucumis melo
[gam]---[spor] 24 [ipcn]94-95 [ref] Chen，R.-y.，1993
IPCN chromosome reports for Castanea mollissima
[gam]---[spor] 24 [ipcn]94-95 [ref] Chen，R.-y.，1993
#IPCN chromosome reports for Castanea seguinii

[gam]---[spor] 24 [ipcn]94-95 [ref]Chen，R.-y.，1993
#IPCN chromosome reports for Castanea dentata
[gam]---[spor] 24 [ipcn]94-95 [ref]Chen，R.-y.，1993
#IPCN chromosome reports for Castanea crenata
[gam]---[spor] 24 [ipcn]94-95 [ref] Chen，R.-y.，1993
#IPCN chromosome reports for Corylus heterophylla
[gam]---[spor] 22 [ipcn]94-95[ref] Chen，R.-y.，1993
IPCN chromosome reports for Juglans regia
[gam]---[spor] 32 [ipcn]94-95 [ref]Chen，R.-y.，1993
IPCN chromosome reports for Juglans mandshurica
[gam]---[spor] 32 [ipcn]94-95 [ref]Chen，R.-y.，1993
#IPCN chromosome reports for Juglans sieboldiana.cordiformis
[gam]---[spor] 32 [ipcn]94-95 [ref]Chen，R.-y.，1993
#IPCN chromosome reports for Juglans cathayensis
[gam]---[spor] 32 [ipcn]94-95 [ref] Chen，R.-y.，1993
#IPCN chromosome reports for Carya cathayensis
[gam]---[spor] 32 [ipcn]94-95 [ref]Chen，R.-y.，1993
#IPCN chromosome reports for Carya illinoinensis
[gam]---[spor] 32 [ipcn] 94-95 [ref]Chen，R.-y.，1993
IPCN chromosome reports for Ginkgo biloba
[gam]---[spor] 24 fipcn]94-95 [ref] Chen，R.-y.，W.-q.Song，X.-1.Li&z.-p.An，1993
IPCN chromosome reports for Ginkgo biloba
[gam]---[spor] 24 lipcn]94-95 [ref]Chen，R.-y.，1993
IPCN chromosome reports for Saccharum officinarum
[gam]---[spor] 80 [ipcn] 94-95 [ref] Chen，R.-y.，1993
#IPCN chromosome reports for Amygdalus persica
[gam]---[spor] 16 [ipcn]94-95 [ref] Chen，R.-y.，1993
#IPCN chromosome reports for Amygdalus persica.nucipersica
[gam]---[spor] 16[ipcn]94-95 [ref] Chen，R.-y.，1993
#IPCN chromosome reports for Amygdalus persica.compressa
[gam]---[spor] 16 [ipcn] 94-95 [ref] Chen，R.-y.，1993
#IPCN chromosome reports for Amygdalus persica Densa
[gam]---[spor] 16 [ipcn]94-95 [ref] Chen，R.-y.，1993
#IPCN chromosome reports for Amygdalus persica Duplex
[gam]---[spor] 16，24 [ipcn]94-95 [ref] Chen，R.-y.，1993
#IPCN chromosome reports for Amygdalus kansuensis
[gam]---[spor] 16 [ipcn] 94-95 [ref] Chen，R.-y.，1993
#IPCN chromosome reports for Amygdalus davidiana
[gam]---[spor] 16 [ipcn]94-95 [ref] Chen，R.-y.，1993
#IPCN chromosome reports for Amygdalus mira
[gam]---[spor] 16 [ipcn] 94-95[ref] Chen，R.-y.，1993
#IPCN chromosome reports for Amygdalus ferganensis
[gam]---[spor] 16 [ipcn] 94-95 [ref] Chen，R.-y.，1993
IPCN chromosome reports for Amygdalus communis
[gam]---[spor] 16 [ipcn]94-95 [ref] Chen，R.-y.，1993
#IPCN chromosome reports for Amygdalus nana
[gam]---[spor] 16 [ipcn] 94-95 [ref]Chen，R.-y.，1993

IPCN chromosome reports for Amygdalus triloba
[gam]---[spor] 64 [ipcn]94-95 [ref]Chen，R.-y.，1993
IPCN chromosome reports for Amygdalus tangutica
[gam]---[spor] 16 [ipcn]94-95[ref]Chen，R.-y.，1993
IPCN chromosome reports for Amygdalus mongolica
[gam]---[spor] 16 [ipcn]94-95[ref]Chen，R.-y.，1993
#IPCN chromosome reports for Armeniaca vulgaris
[gam]---[spor] 16 [ipcn]94-95 [ref]Chen，R.-y.，1993
#IPCN chromosome reports for Armeniaca vulgaris.ansu
[gam]---[spor] 16 [ipcn]94-95 [ref] Chen，R.-y.，1993
#IPCN chromosome reports for Armeniaca sibirica
[gam]---[spor] 16 [ipcn]88-89 [ref]Song，w.-q.，X.-1.Li&Y.-1.Chen，1989
#IPCN chromosome reports for Armeniaca sibirica
[gam]---[spor] 16 [ipcn]94-95[ref]Chen，R.-y.，1993
#IPCN chromosome reports for Armeniaca mandshurica
[gam]---[spor] 16 [ipcn]94-95 [ref]Chen，R.-y.，1993
#IPCN chromosome reports for Armeniaca mume
[gam]---[spor] 16 [ipcn] 94-95 [ref] Chen，R.-y.，1993
#IPCN chromosome reports for Armeniaca mume.bungo
[gam]---[spor] 16，24 [ipcn]94-95 [ref]Chen，R.-y.，1993
IPCN chromosome reports for Prunus salicina
[gam]---[spor）16，24 [ipcn]94-95 [ref]Chen，R.-y.，1993
IPCN chromosome reports for Prunus simonii
[gam]---[spor] 16 [ipcn]94-95[ref] Chen，R.-y.，1993
IPCN chromosome reports for Prunus domestica
[gam]---[spor] 48 [ipcn]94-95 [ref]Chen，R.-y.，1993
IPCN chromosome reports for Prunus cerasifera
[gam]---[spor] 16 [ipcn]94-95 [ref] Chen，R.-y.，1993
IPCN chromosome reports for Prunus americana
[gam]---[spor]16 [ipcn] 94-95 [ref]Chen，R.-y.，1993
IPCN chromosome reports for Prunus ussuriensis
[gam]---[spor] 16 [ipcn]94-95 [ref] Chen，R.-y.，1993
#IPCN chromosome reports for Cerasus pseudocerasus
[gam]---[spor] 16，32 [ipcn]94-95 [ref]Chen，R.-y.，1993
 [qam]---[spor] 16 [ipcn]94-95 [ref]Chen，R.-y.，1993
#IPCN chromosome reports for Cerasus yedoensis
[qam]---[spor] 16 [ipcn）94-95 [ref]Chen，R.-y.，1993
#IPCN chromosome reports for Cerasus avium
[gam]---[spor] 16 [ipcn] 94-95 [ref] Chen，R.-y.，1993
#IPCN chromosome reports for Cerasus vulgaris
rgam]---[spor] 16，32 lipcn] 94-95 [ref] Chen，R.-y.，1993
#IPCN chromosome reports for Cerasus mahaleb
[gam]---[spor] 16，24 [ipcn]94-95[ref]Chen，R.-y.，1993
#IPCN chromosome reports for Cerasus tomentosa
[gam]---[spor] 16 [ipcn]94-95 [ref] Chen，R.-y.，1993
#IPCN chromosome reports for Cerasus humilis
[gam]---[spor] 16 [ipcn]94-95 [ref] Chen，R.-y.，1993

IPCN chromosome reports for Poncirus trifoliata
[gam]---[spor] 18 [ipcn]94-95 [ref]Chen，R.-y.，1993
IPCN chromosome reports for Fortunella margarita
[gam]---[spor] 15 [ipcn]94-95 [ref]Chen，R.-y.，1993
#IPCN chromosome reports for Fortunella japonica
[gam]---[spor] 18 [ipcn]94-95 [ref]Chen，R.-y.，1993
IPCN chromosome reports for Fortunella obovata
[gam]---[spor] 18 [ipcn]94-95[ref] Chen，R.-y.，1993
IPCN chromosome reports for Fortunella hindsii Chintou
[gam]---[spor] 18 [ipcn]94-95[ref] Chen，R.-y.，1993
#IPCN chromosome reports for Citrus hongheensis
[gam]---[spor] 18 [ipcn] 94-95 [ref] Chen，R.-y.，1993
IPCN chromosome reports for Citrus medica
[gam]---[spor] 18 [ipcn]94-95 [ref] Chen，R.-y.，1993
IPCN chromosome reports for Citrus Iimon
*[gam]---[spor] 18 [ipcn] 94-95 [ref] Chen，R.-y.，1993
[gam]---[spor] 18 [ipcn]94-95 [ref] Chen，R.-y.，1993
IPCN chromosome reports for Citrus Iimonia
[gam]---[spor] 18 [ipcn]94-95 [ref]Chen，R.-y.，1993
IPCN chromosome reports for Citrus grandis
*[gam]---[spor] 18 Tipcn]94-95 [ref] Chen，R.-y.，1993
IPCN chromosome reports for Citrus paradisi
*[gam]---[spor] 18 [ipcn]94-95[ref] Chen，R.-y.，1993
[gam]---[spor] 18 [ipcn]94-95[ref] Chen，R.-y.，1993
IPCN chromosome reports for Citrus aurantium
*[gam]---[spor] 18[ipcn] 94-95[ref] Chen，R.-y.，1993
[gam]---[spor] 18 [ipcn]94-95[ref] Chen，R.-y.，1993
IPCN chromosome reports for Citrus sulcata
[gam]---[spor] 18 [ipcn]94-95 [ref] Chen，R.-y.，1993
#IPCN chromosome reports for Citrus matsudaidai
[gam]---[spor] 18 [ipcn]94-95 [ref]Chen，R.-y.，1993
IPCN chromosome reports for Citrus ichangensis
[gam]---[spor] 18 [ipcn]94-95 [ref]Chen，R.-y.，1993
#IPCN chromosome reports for Citrus wilsonii
[gam]---[spor] 18 [ipcn]94-95 [ref] Chen，R.-y.，1993
IPCN chromosome reports for Citrus verrucosa
[gam]---[spor] 18 [ipcn]94-95 [ref]Chen，R.-y.，1993
IPCN chromosome reports for Citrus suavissima
[gam]---[spor] 18 [ipcn]94-95 [ref] Chen，R.-y.，1993
#IPCN chromosome reports for Citrus unshiu
[gam]---[spor] 18 [ipcn]94-95 [ref] Chen，R.-y.，1993
IPCN chromosome reports for Citrus poonensis
[gam]---[spor] 18 [ipcn]94-95 [ref] Chen，R.-y.，1993
IPCN chromosome reports for Citrus chuana
[gam]---[spor] 18 [ipcn] 94-95 [ref] Chen，R.-y.，1993
IPCN chromosome reports for Citrus tangerita
[gam]---[spor] 18 [ipcn]94-95 [ref] Chen，R.-y.，1993

IPCN chromosome reports, for Citrus clementina
[qam]---[spor] 18 [ipcn]94-95 [ref] Chen, R.-y., 1993
IPCN chromosome reports for Citrus reticulata
[qam]---[spor] 18 [ipcn] 94-95 [ref] Chen, R.-y., 1993
IPCN chromosome reports for Citrus erythrosa
[gam]---[spor] 18 [ipcn] 94-95 [ref] Chen, R.-y., 1993
#IPCN chromosome reports for Citrus tardiferox
[gam]---[spor] 18[ipcn]94-95[ref]Chen, R.-y., 1993
IPCN chromosome reports for Citrus chachiensis.oleocarpa
[gam]---[spor] 18 [ipcn] 94-95[ref] Chen, R.-y., 1993
IPCN chromosome reports for Citrus daoxianensis
[gam]---[spor] 18 [ipcn]94-95 [ref]Chen, R.-y., 1993
IPCN chromosome reports for Citrus madurensis
[gam]---[spor] 18 [ipcn]94-95 [ref] Chen, R.-y., 1993
#IPCN chromosome reports for Citrus microcarpa
[gam]---[spor] 18 [ipcn]94-95 [ref]Chen, R.-y., 1993
IPCN chromosome reports for Eriobotrya japonica
[gam]---[spor] 34 [ipcn]94-95 [ref] Chen, R.-y., 1993
IPCN chromosome reports for Litchi chinensis
[gam]---[spor] 30 [ipcn]94-95[ref]Chen, R.-y., 1993
#IPCN chromosome reports for Dimocarpus Iongan
[gam]---[spor] 30 [ipcn]94-95 [ref] Chen, R.-y., 1993
#IPCN chromosome reports for Canarium album
[gam]---[spor] 48 [ipcn]94-95 [ref] Chen, R.-y., 1993
#IPCN chromosome reports for Canarium pimela
[gam]---[spor] 48 [ipcn]94-95 [ref] Chen, R.-y., 1993
#IPCN chromosome reports for Canarium tonkinense
[gam]---[spor] 48[ipcn]94-95 [ref] Chen, R.-y., 1993
IPCN chromosome reports for Mangifera indica
[gam]---[spor] 40 [ipcn] 94-95 [ref] Chen, R.-y., 1993
#IPCN chromosome reports for Mangifera sylvatica
[gam]---[spor] 40 [ipcn] 94-95[ref]Chen, R.-y., 1993
IPCN chromosome reports for Persea americana
[gam]---[spor] 24 [ipcn]94-95 [ref]Chen, R.-y., 1993
IPCN chromosome reports for Artocarpus heterophy llus
[gam]---[spor] 56 [ipcn]94-95[ref]Chen, R.-y., 1993
#IPCN chromosome reports for Clausena lansium
[gam]---[spor] 18 [ipcn] 94-95 [ref] Chen, R.-y., 1993
IPCN chromosome reports for Psidium guajava
[gam]---[spor] 22 [ipcn]94-95[ref]Chen, R.-y., 1993
IPCN chromosome reports for Annona squamosa
[gam]---[spor] 18 [ipcn]94-95 [ref]Chen, R.-y., 1993
IPCN chromosome reports for Annona muricata
[gam]---[spor] 14 [ipcn]94-95 [ref]Chen, R.-y., 1993
IPCN chromosome reports for Carica papaya
[gam]---[spor] 18 [ipcn]94-95 [ref] Chen, R.-y., 1993
IPCN chromosome reports for *Musa x paradisiaca

[gam]---[spor] 22，33 [ipcn]94-95 [ref]Chen，R.-y.，1993
IPCN chromosome reports for Ananas comosus
[gam]---[spor] 50 [ipcn]94-95 [ref] Chen，R.-y.，1993
#IPCN chromosome reports for Cocos nucifera
[gam]---[spor] 32 [ipcn] 94-95 [ref]Chen，R.-y.，1993
IPCN chromosome reports for Ziziphus jujuba
*[gam]---[spor] 48[ipcn]94-95[ref]Chen，R.-y.，1993
[gam'---[spor] 24 [ipcn] 94-95 [ref] Chen，R.-y.，1993
#IPCN chromosome reports for Ziziphus jujuba.spinosa
[gam]---[spor] 24 [ipcn]94-95 [ref]Chen，R.-y.，1993
IPCN chromosome reports for Vitis vinifera
[qam]---[spor] 38 [ipcn]94-95 [ref] Chen，R.-y.，1993
#IPCN chromosome reports for Vitis amurensis
[gam]---[spor]38 [ipcn]94-95 [ref] Chen，R.-y.，1993
#IPCN chromosome reports for Vitis flexuosa
[gam]---[spor] 38 [ipcn] 94-95 [ref] Chen，R.-y.，1993
#IPCN chromosome reports for Vitis labrusca
[gam]---[spor] 38 [ipcn]94-95[ref] Chen，R.-y.，1993
#IPCN chromosome reports for Vitis riparia
[gam]---[spor] 38 [ipcn]94-95 [ref] Chen，R.-y.，1993
IPCN chromosome reports for Actinidia chinensis
[gam]---[spor] 58 [ipcn]94-95[ref]Chen，R.-y.，1993
IPCN chromosome reports for Actinidia kolomikta
[gam]---[spor] 58 [ipcn]94-95 [ref]Chen，R.-y.，1993
IPCN chromosome reports for Actinidia polygama
[gam]---[spor] 58 [ipcn] 94-95 [ref]Chen，R.-y.，1993
IPCN chromosome reports for Actinidia eriantha
[gam]---[spor] 58 [ipcn] 94-95 [ref] Chen，R.-y.，1993
IPCN chromosome reports for Actinidia arguta
[gam]---[spor] 116 [ipcn] 94-95[ref]Chen，R.-y.，1993
IPCN chromosome reports for Actinidia macrosperma
[gam]---[spor] 116 [ipcn]94-95 [ref]Chen，R.-y.，1993
IPCN chromosome reports for Actinidia melliana
[gam]---[spor] 58 [ipcn] 94-95 [ref]Chen，R.-y.，1993
IPCN chromosome reports for Actinidia hemsleyana
lgam]---[spor] 58 [ipcn]94-95 [ref] Chen，R.-y.，1993
IPCN chromosome reports for Actinidia latifolia
lgam]---[spor] 58 [ipcn]94-95 [ref] Chen，R.-y.，1993
IPCN chromosome reports for Rubus parvifolius
[gam]---[spor] 28 [ipcn]94-95[ref] Chen，R.-y.，1993
IPCN chromosome reports for Rubus crataegifolius
lgam]---[spor] 21 [ipcn]92-93[ref]Li，x.-1.，w.-q.Song&R.-y.Chen，1993
[gam]---[spor] 14，21，28 [ipcn] 94-95[ref]Chen，R.-y.，1993
#IPCN chromosome reports for Rubus occidentalis
[gam]---[spor] 14 [ipcn]92-93[ref]Li，X.-1.，W.-q.Song&R.-y.Chen，1993
[gam]---[spor] 14 [ipcn]94-95 [ref]Chen，R.-y.，1993
IPCN chromosome reports for Rubus idaeus

[gam]---[spor] 14 [ipcn]92-93 [ref]Li, X.-l., w.-q.Song&R.-y.Chen, 1993
[gam]---[spor] 14, 21 [ipcn]94-95[ref]Chen, R.-y., 1993
#IPCN chromosome reports for Rubus komarovii
gam]---[spor]14 [ipcn]92-93[ref]Li, X.-l., W.-q.Song&R.-y.Chen, 1993
Igam]---[spor] 14, 28 [ipcn]94-95[ref] Chen, R.-y., 1993
IPCN chromosome reports for Rubus sachalinensis
[gam]---[spor] 14 [ipcn]94-95 [ref] Chen, R.-y., 1993
#IPCN chromosome reports for Rubus allegheniensis
[gam]---[spor] 14 [ipcn] 94-95[ref.]Chen, R.-y., 1993
#IPCN chromosome reports for Rubus xanthocarpus
gam]---[spor] 14 [ipcn]92-93[ref]Li, X.-l., W.-q.Song&R.-y.Chen, 1993
igam]---[spor] 14 [ipcn]94-95[ref]Chen, R.-y., 1993
#IPCN chromosome reports for Fragaria ananassa
[gam]---[spor] 56 [ipcn]94-95 [ref] Chen, R.-y., 1993
IPCN chromosome reports for Vaccinium vitis-idaea
: gam]---[spor] 24 [ipcn]92-93 [ref]Li, x.-l., w.-q.Song&R.-y.Chen, 1993
Igam]---[spor] 24 [ipcn]94-95 [ref]Chen, R.-y., 1993
IPCN chromosome reports for Vaccinium corymbosum
igam]---[spor] 48 [ipcn]92-93[ref]Li, X.-l., W.-q.Song&R.-y.Chen, 1993
Igam]---[spor] 48 [ipcn]94-95 [ref] Chen, R.-y., 1993
IPCN chromosome reports for Morus alba
Igam]---[spor] 28 [ipcn]94-95 [ref]Chen, R.-y., 1993
#IPCN chromosome reports for Morus mongolica
[gam]---[spor] 28 [ipcn] 94-95 [ref] Chen, R.-y., 1993
#IPCN chromosome reports for Morus cathayana
[gam]---[spor] 28 [ipcn] 94-95[ref] Chen, R.-y., 1993
#IPCN chromosome reports for Morus australis
[gam]---[spor] 28 [ipcn] 94-95 [ref] Chen, R.-y., 1993
IPCN chromosome reports for Morus multicaulis
[gam]---[spor] 28 [ipcn]94-95[ref] Chen, R.-y., 1993
IPCN chromosome reports for Ficus carica
[gam]---[spor] 26 [ipcn]94-95 [ref] Chen, R.-y., 1993
IPCN chromosome reports for Diospyros kaki
[gam]---[spor] 90 [ipcn]94-95 [ref] Chen, R.-y., 1993
IPCN chromosome reports for Diospyros oleifera
[gam]---[spor] 30 [ipcn] 94-95 [ref] Chen, R.-y., 1993
IPCN chromosome reports for Diospyros Iotus
[gam]---[spor] 30+1s [ipcn]94-95 [ref] Chen, R.-y., 1993
#IPCN chromosome reports for Diospyros discolor
[gam]---[spor] 30 [ipcn]94-95 [ref] Chen, R.-y., 1993
IPCN chromosome reports for Diospyros rhombifolia
[gam]---[spor] 30 [ipcn]94-95 [ref] Chen, R.-y., 1993
#IPCN chromosome reports for Diospyros glaucifolia
[gam]---[spor] 30 [ipcn]94-95 [ref] Chen, R.-y., 1993
IPCN chromosome reports for Hippophae rhamnoides
[gam]---[spor] 24 [ipcn]92-93[ref]Li, x.-l., w.-q.Song&R.-y.Chen, 1993

[gam]---[spor] 24 [ipcn] 94-95 [ref] Chen，R.-y.，1993
IPCN chromosome reports for Saccharum officinarum L.
[gam]---[spor] 80 [ipcn]94-95[ref]Chen./R.-y.，1993
IPCN chromosome reports for Cucumis melo
[gam]---[spor] 24 [ipcn]94-95[ref]Chen，\R.-y.，1993
IPCN chromosome reports for Nelumbo nucifera Gaertn.
[gam]---[spor] 16 [ipcn]94-95[ref]Wei，P.-h.，W.-p.Chen&R.-y.Chen.1994
IPCN chromosome reports for Euryale ferox Salisb ex K.D.Koenig@Sims
[gam]---[spor] 58 [ipcn]94-95[ref] Wei.P.-h.，W.-p.Chen&R.-y.Chen，1994
IPCN chromosome reports for Secale cereale L
[gam]---[spor] 14+14B，[ipcn]01-03 [ref] Chen，R-Y 1998
IPCN chromosome reports for Phyllostachys mannii Gamble
[gam]---[spor] 48，[ipcn]01-03 [ref] Chen，R-Y 1999
IPCN chromosome reports for Phyllostachys platyglossa C.P. Wang & Z.H. Yu
[gam]---[spor] 48，[ipcn]01-03 [ref] Chen，R-Y 1999
IPCN chromosome reports for Phyllostachys bambusoides fo. lacrima-deae Keng f. & T.H. Wen
[gam]---[spor] 48，[ipcn]01-03 [ref] Chen，R-Y 1999
IPCN chromosome reports for Phyllostachys nigella T.H. Wen
[gam]---[spor] 48，[ipcn]01-03 [ref] Chen，R-Y 1999
IPCN chromosome reports for Phyllostachys aurita J.L. Lu
[gam]---[spor] 48，[ipcn]01-03 [ref] Chen，R-Y 1999
IPCN chromosome reports for Phyllostachys heteroclada fo. purpurata（McClure）T.H. Wen
[gam]---[spor] 48，[ipcn]01-03 [ref] Chen，R-Y 1999
IPCN chromosome reports for Shibataea chinensis Nakai
[gam]---[spor] 48，[ipcn]01-03 [ref] Chen，R-Y 1999
IPCN chromosome reports for Oligostachyum sulcatum Z.P. Wang & G.H. Ye
[gam]---[spor] 48，[ipcn]01-03 [ref] Chen，R-Y 1999
IPCN chromosome reports for Pleioblastus amarus var. hangzhouensis S.L. Chen & S. Y. Chen
[gam]---[spor] 48，[ipcn]01-03 [ref] Chen，R-Y 1999
IPCN chromosome reports for Phyllostachys heteroclada fo. purpurata（McClure）T.H. Wen
[gam]---[spor] 48，[ipcn]01-03 [ref] Chen，R-Y 1999
IPCN chromosome reports for Sasa auricoma（Mitford）E.G. Camus
[gam]---[spor] 48，[ipcn]01-03 [ref] Chen，R-Y 1999
IPCN chromosome reports for Schizostachyum funghomii McClure
[gam]---[spor] 60，64，66，68，70，72，[ipcn]01-03 [ref] Chen，R-Y 2001
IPCN chromosome reports for Pseudostachyum polymorphum Munro
[gam]---[spor] 42-44，64，66，68-70，[ipcn]01-03 [ref] Chen，R-Y 2001
IPCN chromosome reports for Cephalostachyum pergracile Munro
[gam]---[spor] 64，68-71，80，95，[ipcn]01-03 [ref] Chen，R-Y 2001
IPCN chromosome reports for Schizostachyum pseudolima McClure
[gam]---[spor] 64，68，70，72，[ipcn]01-03 [ref] Chen，R-Y 2001
IPCN chromosome reports for Thyrsostachys oliveri Gamble
[gam]---[spor] 68，70-72，[ipcn]01-03 [ref] Chen，R-Y 2001
IPCN chromosome reports for !!Thyrsostachys siamensis Gamble
[gam]---[spor] 64，68，70，[ipcn]01-03 [ref] Chen，R-Y 2001
IPCN chromosome reports for Melocalamus scandens Hsueh & C.M. Hui

[gam]---[spor] 68，70，72，[ipcn]01-03 [ref] Chen，R-Y 2001
IPCN chromosome reports for !Neomicrocalamus prainii（Gamble）Keng f.
[gam]---[spor] 71，72，[ipcn]01-03 [ref] Chen，R-Y 2001
IPCN chromosome reports for !Bambusa arundinacea（Retz.）Willd.
[gam]---[spor] 69，70，[ipcn]01-03 [ref] Chen，R-Y 2001
IPCN chromosome reports for !Bambusa blumeana Schult. & Schult. f.
[gam]---[spor] 68，70，72，[ipcn]01-03 [ref] Chen，R-Y 2001
IPCN chromosome reports for !Bambusa flexuosa Munro
[gam]---[spor] 68，70，72，[ipcn]01-03 [ref] Chen，R-Y 2001
IPCN chromosome reports for Bambusa sinospinosa McClure
[gam]---[spor] 52，58，68，70，72，[ipcn]01-03 [ref] Chen，R-Y 2001
IPCN chromosome reports for Bambusa dissimulator var. hispida McClure
[gam]---[spor] 68，70，72，[ipcn]01-03 [ref] Chen，R-Y 2001
IPCN chromosome reports for Bambusa dissimulator McClure
[gam]---[spor] 68，70，72，[ipcn]01-03 [ref] Chen，R-Y 2003
IPCN chromosome reports for Bambusa rutila McClure
[gam]---[spor] 68，70，[ipcn]01-03 [ref] Chen，R-Y 2001
IPCN chromosome reports for Bambusa lapidea McClure
[gam]---[spor] 64，68-70，72，[ipcn]01-03 [ref] Chen，R-Y 2001
IPCN chromosome reports for Bambusa subaequalis H.L. Fung & C.Y. Sia
[gam]---[spor] 68，70，72，[ipcn]01-03 [ref] Chen，R-Y 2001
IPCN chromosome reports for Bambusa gibba McClure
[gam]---[spor] 68，70，[ipcn]01-03 [ref] Chen，R-Y 2001
IPCN chromosome reports for Bambusa ventricosa McClure
[gam]---[spor] 68，70，[ipcn]01-03 [ref] Chen，R-Y 2001
IPCN chromosome reports for Bambusa corniculata L.C. Chia & H.L. Fung
[gam]---[spor] 68，70，[ipcn]01-03 [ref] Chen，R-Y 2001
IPCN chromosome reports for Bambusa diaoluoshanensis L.C. Chia & H.L. Fung
[gam]---[spor] 60，62，68，70，[ipcn]01-03 [ref] Chen，R-Y 2001
IPCN chromosome reports for Bambusa xiashanensis L.C. Chia & H.L. Fung
[gam]---[spor] 96，104，[ipcn]01-03 [ref] Chen，R-Y 2001
IPCN chromosome reports for Bambusa prominens H.L. Fung & C.Y. Sia
[gam]---[spor] 90，102，104，[ipcn]01-03 [ref] Chen，R-Y 2001
IPCN chromosome reports for !Bambusa nutans Wall. ex Munro
[gam]---[spor] 64，68，70，[ipcn]01-03 [ref] Chen，R-Y 2001
IPCN chromosome reports for !Bambusa tulda Roxb.
[gam]---[spor] 68-71，[ipcn]01-03 [ref] Chen，R-Y 2001
IPCN chromosome reports for Bambusa eutuldoides McClure
[gam]---[spor] 68，70，[ipcn]01-03 [ref] Chen，R-Y 2001
IPCN chromosome reports for Bambusa eutuldoides var. basistriata McClure
[gam]---[spor] 68，70，[ipcn]01-03 [ref] Chen，R-Y 2001
IPCN chromosome reports for Bambusa eutuldoides var. viridivittata（W.T. Lin）L.C. Chia
[gam]---[spor] 68，70，[ipcn]01-03 [ref] Chen，R-Y 2001
IPCN chromosome reports for Bambusa pervariabilis McClure
[gam]---[spor] 68，70，[ipcn]01-03 [ref] Chen，R-Y 2001
IPCN chromosome reports for Bambusa longispiculata Gamble ex Brandis

[gam]---[spor] 57，58，64，66-72，[ipcn]01-03 [ref] Chen，R-Y 2001
IPCN chromosome reports for Bambusa tuldoides Munro
[gam]---[spor] 64，68，70，[ipcn]01-03 [ref] Chen，R-Y 2001
IPCN chromosome reports for Bambusa subtruncata L.C. Chia & H.L. Fung
[gam]---[spor] 68-70，[ipcn]01-03 [ref] Chen，R-Y 2001
IPCN chromosome reports for Bambusa rigida Keng & Keng f.
[gam]---[spor] 68，70，[ipcn]01-03 [ref] Chen，R-Y 2001
IPCN chromosome reports for Bambusa boniopsis McClure
[gam]---[spor] 68，70，[ipcn]01-03 [ref] Chen，R-Y 2001
IPCN chromosome reports for Bambusa dolichoclada Hayata
[gam]---[spor] 68，70，[ipcn]01-03 [ref] Chen，R-Y 2001
IPCN chromosome reports for !!Bambusa vulgaris Schrad. ex J.C. Wendl.
[gam]---[spor] 52，52，67，68，70，72，[ipcn]01-03 [ref] Chen，R-Y 2001
IPCN chromosome reports for Bambusa gibboides W.T. Lin
[gam]---[spor] 96，104，[ipcn]01-03 [ref] Chen，R-Y 2001
IPCN chromosome reports for Bambusa piscatorum McClure
[gam]---[spor] 68-70，72，[ipcn]01-03 [ref] Chen，R-Y 2001
IPCN chromosome reports for Bambusa albolineata L.C. Chia
[gam]---[spor] 68，70，72，[ipcn]01-03 [ref] Chen，R-Y 2001
IPCN chromosome reports for Bambusa mutabilis McClure
[gam]---[spor] 68，70，72，[ipcn]01-03 [ref] Chen，R-Y 2001
IPCN chromosome reports for Bambusa contracta L.C. Chia & H.L. Fung
[gam]---[spor] 68，70，[ipcn]01-03 [ref] Chen，R-Y 2001
IPCN chromosome reports for !Bambusa multiplex（Lour.）Raeusch. ex Schult. & Schult. f.
[gam]---[spor] 53，56，58，59，64-66，68-72，78，[ipcn]01-03 [ref] Chen，R-Y 2001
IPCN chromosome reports for Bambusa intermedia Hsueh & T.P. Yi
[gam]---[spor] 64，68，70，[ipcn]01-03 [ref] Chen，R-Y 2001
IPCN chromosome reports for !Bambusa remotiflora（Kuntze）L.C. Chia & H.L. Fung[gam]---[spor] 68，70，72，[ipcn]01-03 [ref] Chen，R-Y 2001
IPCN chromosome reports for Bambusa surrecta（Q.H. Dai）Q.H. Dai
[gam]---[spor] 68，70，[ipcn]01-03 [ref] Chen，R-Y 2001
IPCN chromosome reports for Bambusa cerosissima McClure
[gam]---[spor] 72，[ipcn]01-03 [ref] Chen，R-Y 2001
IPCN chromosome reports for Bambusa chungii McClure
[gam]---[spor] 68-71，[ipcn]01-03 [ref] Chen，R-Y 2001
IPCN chromosome reports for Bambusa textilis McClure
[gam]---[spor] 64，65，68，70，[ipcn]01-03 [ref] Chen，R-Y 2001
IPCN chromosome reports for Bambusa distegia（Keng & Keng f.）L.C. Chia & H.L. Fung
[gam]---[spor] 68-70，72，[ipcn]01-03 [ref] Chen，R-Y 2001
IPCN chromosome reports for !Bambusa papillata（Q.H. Dai）K.M. Lan
[gam]---[spor] 68，70，[ipcn]01-03 [ref] Chen，R-Y 2001
IPCN chromosome reports for Bambusa hainanensis L.C. Chia & H.L. Fung
[gam]---[spor] 70，72，[ipcn]01-03 [ref] Chen，R-Y 2001
IPCN chromosome reports for Bambusa cornigera McClure
[gam]---[spor] 96，100，[ipcn]01-03 [ref] Chen，R-Y 2001
IPCN chromosome reports for Neosinocalamus affinis（Rendle）Keng f.

[gam]---[spor] 68，70，[ipcn]01-03 [ref] Chen，R-Y 2001
IPCN chromosome reports for Neosinocalamus rectocuneatus W.T. Lin
[gam]---[spor] 70，72，[ipcn]01-03 [ref] Chen，R-Y 2001
IPCN chromosome reports for Dendrocalamopsis basihirsuta（McClure）Keng f. & W.T.Lin
[gam]---[spor] 64，70，[ipcn]01-03 [ref] Chen，R-Y 2001
IPCN chromosome reports for Dendrocalamopsis beecheyana var. pubescens（P.F. Li）Keng f.
[gam]---[spor] 64，68，70-72，[ipcn]01-03 [ref] Chen，R-Y 2001
IPCN chromosome reports for Dendrocalamopsis beecheyana（Munro）Keng f.
[gam]---[spor] 64，68，70-72，[ipcn]01-03 [ref] Chen，R-Y 2001
IPCN chromosome reports for Dendrocalamopsis edulis（Odash.）Keng f.
[gam]---[spor] 96，102，104，[ipcn]01-03 [ref] Chen，R-Y 2001
IPCN chromosome reports for Dendrocalamopsis oldhamii（Munro）Keng f.
[gam]---[spor] 68，70，[ipcn]01-03 [ref] Chen，R-Y 2001
IPCN chromosome reports for Dendrocalamopsis oldhamii f. revoluta（W.T. Lin & J.Y.Lin）W.T. Lin
[gam]---[spor] 88，96，104，[ipcn]01-03 [ref] Chen，R-Y 2001
IPCN chromosome reports for Dendrocalamopsis bicicatricata（W.T. Lin）Keng f.
[gam]---[spor] 68，71，72，[ipcn]01-03 [ref] Chen，R-Y 2001
IPCN chromosome reports for Dendrocalamopsis stenoaurita（W.T. Lin）Keng f. ex W. T. Lin
[gam]---[spor] 66，70，72，[ipcn]01-03 [ref] Chen，R-Y 2001
IPCN chromosome reports for Dendrocalamopsis valida Q.H. Dai
[gam]---[spor] 64，70-72，[ipcn]01-03 [ref] Chen，R-Y 2001
IPCN chromosome reports for Dendrocalamopsis variostriata（W.T. Lin）Keng f.
[gam]---[spor] 85，96，100，102，104，106，[ipcn]01-03 [ref] Chen，R-Y 2001
IPCN chromosome reports for *Dendrocalamopsis daii Keng f.
[gam]---[spor] 64，68，70，[ipcn]01-03 [ref] Chen，R-Y 2001
IPCN chromosome reports for Dendrocalamus semiscandens Hsueh & D.Z. Li
[gam]---[spor] 68，71，[ipcn]01-03 [ref] Chen，R-Y 2001
IPCN chromosome reports for Dendrocalamus barbatus var. internodiradicatus Hsueh & D.Z. Li
[gam]---[spor] 68，70，72，[ipcn]01-03 [ref] Chen，R-Y 2001
IPCN chromosome reports for Dendrocalamus calostachyus（Kurz）Kurz
[gam]---[spor] 68，71，72，[ipcn]01-03 [ref] Chen，R-Y 2001
IPCN chromosome reports for Dendrocalamus sinicus L.C. Chia & J.L. Sun
[gam]---[spor] 68，70，72，[ipcn]01-03 [ref] Chen，R-Y 2001
IPCN chromosome reports for Dendrocalamus strictus（Roxb.）Nees
[gam]---[spor] 68，70，72，[ipcn]01-03 [ref] Chen，R-Y 2001
IPCN chromosome reports for Gigantochloa verticillata（Willd.）Munro
[gam]---[spor] 68，70，[ipcn]01-03 [ref] Chen，R-Y 2001
IPCN chromosome reports for Dendrocalamopsis bicicatricata（W.T. Lin）Keng f.
[gam]---[spor] 68-70，[ipcn]01-03 [ref] Chen，R-Y 2001
IPCN chromosome reports for Dendrocalamus farinosus（Keng & Keng f.）L.C. Chia & H.L. Fung
[gam]---[spor] 60，65，68，70，[ipcn]01-03 [ref] Chen，R-Y 2001
IPCN chromosome reports for Dendrocalamus giganteus Wall. ex Munro
[gam]---[spor] 68，70，[ipcn]01-03 [ref] Chen，R-Y 2001
IPCN chromosome reports for Dendrocalamus latiflorus Munro
[gam]---[spor] 64，68，72，[ipcn]01-03 [ref] Chen，R-Y 2001

IPCN chromosome reports for Dendrocalamus membranaceus Munro
[gam]---[spor] 68，71，72，[ipcn]01-03 [ref] Chen，R-Y 2003
IPCN chromosome reports for Dendrocalamus minor（McClure）L.C. Chia & H.L. Fung
[gam]---[spor] 68，70，[ipcn]01-03 [ref] Chen，R-Y 2001
IPCN chromosome reports for Dendrocalamus ovatus N.H. Xia & L.C. Chia
[gam]---[spor] 68，70，[ipcn]01-03 [ref] Chen，R-Y 2001
IPCN chromosome reports for Gigantochloa levis（Blanco）Merr.
[gam]---[spor] 68 70 72，[ipcn]01-03 [ref] Chen，R-Y 2001
IPCN chromosome reports for Dendrocalamus pulverulentus L.C. Chia & P. But
[gam]---[spor] 68，70，[ipcn]01-03 [ref] Chen，R-Y 2001
IPCN chromosome reports for Malus spectabilis（Aiton）Borkh.
[gam]---[spor] 34 [ipcn]88-89[ref]Chen./R.-y.，W.Zhang&Q.-a.Wu，2003
IPCN chromosome reports for Oryza sativa L.
[gam]---[spor] 24，[ipcn]01-03 [ref] Chen，R-Y 2003
IPCN chromosome reports for Oryza nivara Sharma & Shastry
[gam]---[spor] 24，[ipcn]01-03 [ref] Chen，R-Y 2003
IPCN chromosome reports for Oryza glaberrima Steud.
[gam]---[spor] 24，[ipcn]01-03 [ref] Chen，R-Y 2003
IPCN chromosome reports for Oryza meridionalis N.Q. Ng
[gam]---[spor] 24，[ipcn]01-03 [ref] Chen，R-Y 2003
IPCN chromosome reports for Oryza glumipatula Steud.
[gam]---[spor] 24，[ipcn]01-03 [ref] Chen，R-Y 2003
IPCN chromosome reports for Oryza australiensis Domin
[gam]---[spor] 24，[ipcn]01-03 [ref] Chen，R-Y 2003
IPCN chromosome reports for Triticum urartu Thumanjan ex Gandilyan
[gam]---[spor] 14，[ipcn]01-03 [ref] Chen，R-Y 2003
IPCN chromosome reports for Triticum sinskajae Filat. & Kurkiev
[gam]---[spor] 14，[ipcn]01-03 [ref] Chen，R-Y 2003
IPCN chromosome reports for Triticum aestivum L.'Afu'
[gam]---[spor] 42，[ipcn]01-03 [ref] Chen，R-Y 2003
IPCN chromosome reports for **Hordeum spontaneum var. ilthabaurense K. Koch
[gam]---[spor] 14，[ipcn]01-03 [ref] Chen，R-Y 2003
IPCN chromosome reports for Hordeum spontaneum var. ischnatherum（Coss.）Thell.
[gam]---[spor] 14，[ipcn]01-03 [ref] Chen，R-Y 2003
IPCN chromosome reports for Hordeum spontaneum var. proskowetzii Nábělek
[gam]---[spor] 14，[ipcn]01-03 [ref] Chen，R-Y 2003
IPCN chromosome reports for Hordeum distichon L.
[gam]---[spor] 14，[ipcn]01-03 [ref] Chen，R-Y 2003
IPCN chromosome reports for Hordeum vulgare L.
[gam]---[spor] 14，[ipcn]01-03 [ref] Chen，R-Y 2003
IPCN chromosome reports for Avena sativa L.
[gam]---[spor] 42，[ipcn]01-03 [ref] Chen，R-Y 2003
IPCN chromosome reports for Avena nuda L.
[gam]---[spor] 42，[ipcn]01-03 [ref] Chen，R-Y 2003
IPCN chromosome reports for Coix lacryma-jobi L.
[gam]---[spor] 20，[ipcn]01-03 [ref] Chen，R-Y 2003

IPCN chromosome reports for Pennisetum glaucum（L.）R. Br.
[gam]---[spor] 18，[ipcn]01-03 [ref] Chen，R-Y 2003
IPCN chromosome reports for Fagopyrum esculentum Moench
[gam]---[spor] 32，[ipcn]01-03 [ref] Chen，R-Y 2003
IPCN chromosome reports for Fagopyrum tataricum（L.）Gaertn.
[gam]---[spor] 16，[ipcn]01-03 [ref] Chen，R-Y 2003
IPCN chromosome reports for !Phaseolus coccineus L.
[gam]---[spor] 22，[ipcn]01-03 [ref] Chen，R-Y 2003
IPCN chromosome reports for Vigna cylindrica（L.）Skeels
[gam]---[spor] 22，[ipcn]01-03 [ref] Chen，R-Y 2003
IPCN chromosome reports for !Vigna unguiculata subsp. sesquipedalis（L.）Verdc.
[gam]---[spor] 22，[ipcn]01-03 [ref] Chen，R-Y 2003
IPCN chromosome reports for !Canavalia gladiata（Jacq.）DC.
[gam]---[spor] 22，[ipcn]01-03 [ref] Chen，R-Y 2003
IPCN chromosome reports for Dolichos lablab L.
[gam]---[spor] 22，[ipcn]01-03 [ref] Chen，R-Y 2003
IPCN chromosome reports for Brassica campestris subsp. chinensis（L.）Makino
[gam]---[spor] 20，[ipcn]01-03 [ref] Chen，R-Y 2003
IPCN chromosome reports for Brassica campestris subsp. pekinensis（Lour.）G. Olsson
[gam]---[spor] 20，[ipcn]01-03 [ref] Chen，R-Y 2003
IPCN chromosome reports for Brassica campestris subsp. rapifera（Metzg.）Sinskaya
[gam]---[spor] 20，[ipcn]01-03 [ref] Chen，R-Y 2003
IPCN chromosome reports for Brassica oleracea var. capitata L.
[gam]---[spor] 18，[ipcn]01-03 [ref] Chen，R-Y 2003
IPCN chromosome reports for Brassica napus subsp. oleifera（DC.）Metzg.
[gam]---[spor] 38，[ipcn]01-03 [ref] Chen，R-Y 2003
IPCN chromosome reports for Brassica oleracea var. botrytis L.
[gam]---[spor] 18，[ipcn]01-03 [ref] Chen，R-Y 2003
IPCN chromosome reports for Raphanus sativus L.
[gam]---[spor] 18，[ipcn]01-03 [ref] Chen，R-Y 2003
IPCN chromosome reports for Cucurbita ficifolia Bouché
[gam]---[spor] 40，[ipcn]01-03 [ref] Chen，R-Y 2003
IPCN chromosome reports for !!Cucurbita pepo L.
[gam]---[spor] 40，[ipcn]01-03 [ref] Chen，R-Y 2003
IPCN chromosome reports for **Cucumis sativus var. xishuangbannanesis Qi Chunzhang & Yuan Zhenzhen
[gam]---[spor] 14，[ipcn]01-03 [ref] Chen，R-Y 2003
IPCN chromosome reports for !Lagenaria siceraria（Molina）Standl.
[gam]---[spor] 22，[ipcn]01-03 [ref] Chen，R-Y 2003
IPCN chromosome reports for Momordica charantia L.
[gam]---[spor] 22，[ipcn]01-03 [ref] Chen，R-Y 2003
IPCN chromosome reports for Trichosanthes anguina L.
[gam]---[spor] 22，[ipcn]01-03 [ref] Chen，R-Y 2003
IPCN chromosome reports for Capsicum annuum L.
[gam]---[spor] 24，[ipcn]01-03 [ref] Chen，R-Y 2003
IPCN chromosome reports for Solanum melongena L.
[gam]---[spor] 24，[ipcn]01-03 [ref] Chen，R-Y 2003

IPCN chromosome reports for !Nicotiana tabacum L.
[gam]---[spor] 48，[ipcn]01-03 [ref] Chen，R-Y 2003
IPCN chromosome reports for *Ipomoea × leucantha Jacq.
[gam]---[spor] 90，[ipcn]01-03 [ref] Chen，R-Y 2003
IPCN chromosome reports for Colocasia esculenta var. comosus Chang
[gam]---[spor] 28，[ipcn]01-03 [ref] Chen，R-Y 2003
IPCN chromosome reports for Nelumbo nucifera Gaertn.
[gam]---[spor] 16，[ipcn]01-03 [ref] Chen，R-Y 2003
IPCN chromosome reports for Brasenia schreberi J.F. Gmel.
[gam]---[spor] 72，[ipcn]01-03 [ref] Chen，R-Y 2003
IPCN chromosome reports for Beta vulgaris var. saccharifera Alef.
[gam]---[spor] 18，27，36，[ipcn]01-03 [ref] Chen，R-Y 2003
IPCN chromosome reports for Spinacia oleracea var. inermis Peterm.
[gam]---[spor] 12，[ipcn]01-03 [ref] Chen，R-Y 2003
IPCN chromosome reports for !Camellia sinensis（L.）Kuntze
[gam]---[spor] 30，[ipcn]01-03 [ref] Chen，R-Y 2003
IPCN chromosome reports for Camellia yungkiangensis H.T. Chang
[gam]---[spor] 30，[ipcn]01-03 [ref] Chen，R-Y 2003
IPCN chromosome reports for Camellia gymnogynoides H.T. Chang & B.H. Chen
[gam]---[spor] 30，[ipcn]01-03 [ref] Chen，R-Y 2003
IPCN chromosome reports for Camellia gymnogyna H.T. Chang
[gam]---[spor] 30，[ipcn]01-03 [ref] Chen，R-Y 2003
IPCN chromosome reports for Camellia assamica（J.W. Mast.）H.T. Chang
[gam]---[spor] 30，[ipcn]01-03 [ref] Chen，R-Y 2003
IPCN chromosome reports for Camellia nanchuanica H.T. Chang & J.H. Xiong
[gam]---[spor] 30，[ipcn]01-03 [ref] Chen，R-Y 2003
IPCN chromosome reports for Gossypium anomalum Wawra & Peyr.
[gam]---[spor] 26，[ipcn]01-03 [ref] Chen，R-Y 2003
IPCN chromosome reports for Gossypium barbosanum L.Ll. Phillips & Clem.
[gam]---[spor] 26，[ipcn]01-03 [ref] Chen，R-Y 2003
IPCN chromosome reports for Gossypium sturtianum J.H. Willis
[gam]---[spor] 26，[ipcn]01-03 [ref] Chen，R-Y 2003
IPCN chromosome reports for Gossypium robinsonii F. Muell.
[gam]---[spor] 26，[ipcn]01-03 [ref] Chen，R-Y 2003
IPCN chromosome reports for Gossypium australe F. Muell.
[gam]---[spor] 26，[ipcn]01-03 [ref] Chen，R-Y 2003
IPCN chromosome reports for Gossypium populifolium F. Muell. ex Tod.
[gam]---[spor] 26，[ipcn]01-03 [ref] Chen，R-Y 2003
IPCN chromosome reports for Gossypium nelsonii Fryxell
[gam]---[spor] 26，[ipcn]01-03 [ref] Chen，R-Y 2003
IPCN chromosome reports for Gossypium klotzschianum Andersson
[gam]---[spor] 26，[ipcn]01-03 [ref] Chen，R-Y 2003
IPCN chromosome reports for Gossypium aridum（Rose & Standl.）Skovst.
[gam]---[spor] 26，[ipcn]01-03 [ref] Chen，R-Y 2003
IPCN chromosome reports for !Gossypium gossypioides（Ulbr.）Standl.
[gam]---[spor] 26，[ipcn]01-03 [ref] Chen，R-Y 2003

IPCN chromosome reports for Gossypium trilobum (Sessé & Moc. ex DC.) Skovst.
[gam]---[spor] 26，[ipcn]01-03 [ref] Chen，R-Y 2003
IPCN chromosome reports for Gossypium laxum L.Ll. Phillips
[gam]---[spor] 26，[ipcn]01-03 [ref] Chen，R-Y 2003
IPCN chromosome reports for Gossypium stocksii Mast.
[gam]---[spor] 26，[ipcn]01-03 [ref] Chen，R-Y 2003
IPCN chromosome reports for Gossypium somalense (Gürke) J.B. Hutch., Silow & S. G. Stephens
[gam]---[spor] 26，[ipcn]01-03 [ref] Chen，R-Y 2003
IPCN chromosome reports for Gossypium longicalyx Hutch. ex Lee
[gam]---[spor] 26，[ipcn]01-03 [ref] Chen，R-Y 2003
IPCN chromosome reports for Gossypium bickii Prokh.
[gam]---[spor] 26，[ipcn]01-03 [ref] Chen，R-Y 2003
IPCN chromosome reports for Gossypium darwinii G. Watt
[gam]---[spor] 52，[ipcn]01-03 [ref] Chen，R-Y 2003
IPCN chromosome reports for Gossypium amourianum Kearney
[gam]---[spor] 26，[ipcn]01-03 [ref] Chen，R-Y 2003
IPCN chromosome reports for *Gossypium areysianum Deflers
[gam]---[spor] 26，[ipcn]01-03 [ref] Chen，R-Y 2003
IPCN chromosome reports for Gossypium capitis-viridis Maller
[gam]---[spor] 26，[ipcn]01-03 [ref] Chen，R-Y 2003
IPCN chromosome reports for Abutilon theophrasti Medik.
[gam]---[spor] 42，[ipcn]01-03 [ref] Chen，R-Y 2003
IPCN chromosome reports for Hibiscus cannabinus L.
[gam]---[spor] 20，[ipcn]01-03 [ref] Chen，R-Y 2003
IPCN chromosome reports for Allium fistulosum var. giganteum Makino
[gam]---[spor] 16，[ipcn]01-03 [ref] Chen，R-Y 2003
IPCN chromosome reports for Allium sativum L.
[gam]---[spor] 16，[ipcn]01-03 [ref] Chen，R-Y 2003
IPCN chromosome reports for **Daucus carota var. sativa Hoffm.
[gam]---[spor] 18，[ipcn]01-03 [ref] Chen，R-Y 2003
IPCN chromosome reports for Lactuca sativa var. longifolia Lam.
[gam]---[spor] 18，[ipcn]01-03 [ref] Chen，R-Y 2003
IPCN chromosome reports for Lactuca sativa var. capitata L.
[gam]---[spor] 18，[ipcn]01-03 [ref] Chen，R-Y 2003
IPCN chromosome reports for Manihot esculenta Crantz
[gam]---[spor] 30，[ipcn]01-03 [ref] Chen，R-Y 2003
IPCN chromosome reports for Polianthes tuberosa L.
[gam]---[spor] 24，[ipcn]01-03 [ref] Chen，R-Y 2003
IPCN chromosome reports for Sansevieria trifasciata Prain
[gam]---[spor] 40，[ipcn]01-03 [ref] Chen，R-Y 2003
IPCN chromosome reports for Amaranthus tricolor var. splendens hort.
[gam]---[spor] 34，[ipcn]01-03 [ref] Chen，R-Y 2003
IPCN chromosome reports for !Alternanthera ficoidea var. versicolor (Lem.) Backer
[gam]---[spor] 32，[ipcn]01-03 [ref] Chen，R-Y 2003
IPCN chromosome reports for Celosia cristata L.
[gam]---[spor] 36，[ipcn]01-03 [ref] Chen，R-Y 2003

IPCN chromosome reports for Celosia cristata var. pyramidalis
[gam]---[spor] 36，[ipcn]01-03 [ref] Chen，R-Y 2003
IPCN chromosome reports for Gomphrena globosa L.
[gam]---[spor] 38，[ipcn]01-03 [ref] Chen，R-Y 2003
IPCN chromosome reports for Alstroemeria aurantiaca D. Don
[gam]---[spor] 25，[ipcn]01-03 [ref] Chen，R-Y 2003
IPCN chromosome reports for Amaryllis vittata L'Hér.
[gam]---[spor] 44，[ipcn]01-03 [ref] Chen，R-Y 2003
IPCN chromosome reports for Clivia miniata Regel
[gam]---[spor]22，[ipcn]01-03 [ref] Chen，R-Y 2003
IPCN chromosome reports for Lycoris aurea（L'Hér.）Herb
[gam]---[spor] 16，[ipcn]01-03 [ref] Chen，R-Y 2003
IPCN chromosome reports for Lycoris radiata（L'Her.）Herb.
[gam]---[spor] 22，[ipcn]01-03 [ref] Chen，R-Y 2003
IPCN chromosome reports for Narcissus jonquillal
[gam]---[spor] 25，[ipcn]01-03 [ref] Chen，R-Y 2003
IPCN chromosome reports for Narcissus pseu-do-narcissus
[gam]---[spor] 28，[ipcn]01-03 [ref] Chen，R-Y 2003
IPCN chromosome reports for Catharanthus roseus（L.）G. Don
[gam]---[spor] 16，[ipcn]01-03 [ref] Chen，R-Y 2003
IPCN chromosome reports for !Aglaonema modestum Schott ex Engl.
[gam]---[spor] 60，[ipcn]01-03 [ref] Chen，R-Y 2003
IPCN chromosome reports for Anthurium scherzerianum var. albo-striatum Engl.
[gam]---[spor] 30，[ipcn]01-03 [ref] Chen，R-Y 2003
IPCN chromosome reports for Anthurium scherzerianum var. atrosanguineum Hort.
[gam]---[spor] 30，[ipcn]01-03 [ref] Chen，R-Y 2003
IPCN chromosome reports for Anthurium andraeanum Linden
[gam]---[spor] 30，[ipcn]01-03 [ref] Chen，R-Y 2003
IPCN chromosome reports for Caladium bicolor（Aiton）Vent.
[gam]---[spor] 52，[ipcn]01-03 [ref] Chen，R-Y 2003
IPCN chromosome reports for Monstera deliciosa Liebm.
[gam]---[spor] 60，[ipcn]01-03 [ref] Chen，R-Y 2003
IPCN chromosome reports for Philodendron oxycardium Schott
[gam]---[spor] 30，[ipcn]01-03 [ref] Chen，R-Y 2003
IPCN chromosome reports for Zantedeschia aethiopica（L.）Spreng.
[gam]---[spor] 32，[ipcn]01-03 [ref] Chen，R-Y 2003
IPCN chromosome reports for Zantedeschia elliottiana（H. Knight）Engl.
[gam]---[spor] 32，[ipcn]01-03 [ref] Chen，R-Y 2003
IPCN chromosome reports for Impatiens balsamina L.
[gam]---[spor] 14，[ipcn]01-03 [ref] Chen，R-Y 2003
IPCN chromosome reports for Berberis thunbergii DC.
[gam]---[spor] 28，[ipcn]01-03 [ref] Chen，R-Y 2003
IPCN chromosome reports for Campsis grandiflora（Thunb.）K. Schum.
[gam]---[spor] 40，[ipcn]01-03 [ref] Chen，R-Y 2003
IPCN chromosome reports for Pachira macrocarpa（Schltdl. & Cham.）Walp.
[gam]---[spor] 26，[ipcn]01-03 [ref] Chen，R-Y 2003

IPCN chromosome reports for Guzmania lingulata（L.）Mez
[gam]---[spor] 50，[ipcn]01-03 [ref] Chen，R-Y 2003
IPCN chromosome reports for Chamaecereus silvestrii（Speg.）Britton & Rose
[gam]---[spor] 20，[ipcn]01-03 [ref] Chen，R-Y 2003
IPCN chromosome reports for Nopalxochia ackermannii（Haw.）F.M. Knuth
[gam]---[spor] 22，[ipcn]01-03 [ref] Chen，R-Y 2003
IPCN chromosome reports for !Opuntia dillenii（Ker Gawl.）Haw.
[gam]---[spor] 40，[ipcn]01-03 [ref] Chen，R-Y 2003
IPCN chromosome reports for Chimonanthus nitens Oliv.
[gam]---[spor] 22，[ipcn]01-03 [ref] Chen，R-Y 2003
IPCN chromosome reports for Chimonanthus praecox（L.）Link
[gam]---[spor] 22，[ipcn]01-03 [ref] Chen，R-Y 2003
IPCN chromosome reports for Campanula medium L.
[gam]---[spor] 32，[ipcn]01-03 [ref] Chen，R-Y 2003
IPCN chromosome reports for !Platycodon grandiflorus（Jacq.）A. DC.
[gam]---[spor] 18，[ipcn]01-03 [ref] Chen，R-Y 2003
IPCN chromosome reports for Canna generalis L.H. Bailey & E.Z. Bailey
[gam]---[spor] 18，[ipcn]01-03 [ref] Chen，R-Y 2003
IPCN chromosome reports for **Lonicera japonica Thunb.
[gam]---[spor] 18，[ipcn]01-03 [ref] Chen，R-Y 2003
IPCN chromosome reports for !Kolkwitzia amabilis Graebn.
[gam]---[spor] 32，[ipcn]01-03 [ref] Chen，R-Y 2003
IPCN chromosome reports for Viburnum
[gam]---[spor] 18，[ipcn]01-03 [ref] Chen，R-Y 2003
IPCN chromosome reports for Dianthus barbatus L.
[gam]---[spor] 30，[ipcn]01-03 [ref] Chen，R-Y 2003
IPCN chromosome reports for Dianthus caryophyllus L.
[gam]---[spor] 30，[ipcn]01-03 [ref] Chen，R-Y 2003
IPCN chromosome reports for Gypsophila elegans var. alba
[gam]---[spor] 26，[ipcn]01-03 [ref] Chen，R-Y 2003
IPCN chromosome reports for Silene armeria L.
[gam]---[spor] 24，[ipcn]01-03 [ref] Chen，R-Y 2003
IPCN chromosome reports for Commelina auriculata Blume
[gam]---[spor] 58，[ipcn]01-03 [ref] Chen，R-Y 2003
IPCN chromosome reports for Commelina communis L.
[gam]---[spor] 62，[ipcn]01-03 [ref] Chen，R-Y 2003
IPCN chromosome reports for Setcreasea purpurea Boom
[gam]---[spor] 24，[ipcn]01-03 [ref] Chen，R-Y 2003
IPCN chromosome reports for Tradescantia fluminensis Vell.
[gam]---[spor] 70，[ipcn]01-03 [ref] Chen，R-Y 2003
IPCN chromosome reports for Zebrina pendula Schnizl.
[gam]---[spor] 24，[ipcn]01-03 [ref] Chen，R-Y 2003
IPCN chromosome reports for Calystegia hederacea Wall.
[gam]---[spor] 22，[ipcn]01-03 [ref] Chen，R-Y 2003
IPCN chromosome reports for Pharbitis nil（L.）Choisy
[gam]---[spor] 30，[ipcn]01-03 [ref] Chen，R-Y 2003

IPCN chromosome reports for Quamoclit pennata (Desr.) Bojer
[gam]---[spor] 30, [ipcn]01-03 [ref] Chen, R-Y 2003
IPCN chromosome reports for !!Ageratum conyzoides L.
[gam]---[spor] 38, [ipcn]01-03 [ref] Chen, R-Y 2003
IPCN chromosome reports for Aster novi-belgii L.
[gam]---[spor] 48, [ipcn]01-03 [ref] Chen, R-Y 2003
IPCN chromosome reports for !Aster ageratoides Turcz.
[gam]---[spor] 18, 36, 54, [ipcn]01-03 [ref] Chen, R-Y 2003
IPCN chromosome reports for !Bellis perennis L.
[gam]---[spor] 18, [ipcn]01-03 [ref] Chen, R-Y 2003
IPCN chromosome reports for Calendula officinalis L.
[gam]---[spor] 32, [ipcn]01-03 [ref] Chen, R-Y 2003
IPCN chromosome reports for !Callistephus chinensis (L.) Nees
[gam]---[spor] 18, [ipcn]01-03 [ref] Chen, R-Y 2003
IPCN chromosome reports for Centaurea cyanus L.
[gam]---[spor] 24, [ipcn]01-03 [ref] Chen, R-Y 2003
IPCN chromosome reports for !Coreopsis tinctoria Nutt.
[gam]---[spor] 24, [ipcn]01-03 [ref] Chen, R-Y 2003
IPCN chromosome reports for Cosmos sulphureus Cav.
[gam]---[spor] 24, [ipcn]01-03 [ref] Chen, R-Y 2003
IPCN chromosome reports for Dahlia pinnata Cav.
[gam]---[spor] 64-65, [ipcn]01-03 [ref] Chen, R-Y 2003
IPCN chromosome reports for Dendranthema indicum (L.) Des Moul.
[gam]---[spor] 36, [ipcn]01-03 [ref] Chen, R-Y 2003
IPCN chromosome reports for Dendranthema lavandulifolium (Fisch. ex Trautv.) Kitam. var. lavandulifolium
[gam]---[spor] 18, [ipcn]01-03 [ref] Chen, R-Y 2003
IPCN chromosome reports for Gazania rigens (L.) Gaertn.
[gam]---[spor] 20, [ipcn]01-03 [ref] Chen, R-Y 2003
IPCN chromosome reports for Helianthus annuus var. citrinus
[gam]---[spor] 34, [ipcn]01-03 [ref] Chen, R-Y 2003
IPCN chromosome reports for Helichrysum bracteatum (Vent.) Haw.
[gam]---[spor] 24, [ipcn]01-03 [ref] Chen, R-Y 2003
IPCN chromosome reports for Gaillardia pulchella Foug.
[gam]---[spor] 24, [ipcn]01-03 [ref] Chen, R-Y 2003
IPCN chromosome reports for Liatris ligulistylis (A. Nelson) K. Schum.
[gam]---[spor] 18, [ipcn]01-03 [ref] Chen, R-Y 2003
IPCN chromosome reports for Melampodium paludosum Kunth
[gam]---[spor] 24, [ipcn]01-03 [ref] Chen, R-Y 2003
IPCN chromosome reports for Rudbeckia hirta L.
[gam]---[spor] 38, [ipcn]01-03 [ref] Chen, R-Y 2003
IPCN chromosome reports for Rudbeckia laciniata L.
[gam]---[spor] 38, [ipcn]01-03 [ref] Chen, R-Y 2003
IPCN chromosome reports for Tagetes erecta L.
[gam]---[spor] 24, [ipcn]01-03 [ref] Chen, R-Y 2003
IPCN chromosome reports for Tagetes patula L.
[gam]---[spor] 20, [ipcn]01-03 [ref] Chen, R-Y 2003

IPCN chromosome reports for Taraxacum mongolicum Hand.-Mazz.
[gam]---[spor] 24，[ipcn]01-03 [ref] Chen，R-Y 2003
IPCN chromosome reports for !Tithonia diversifolia（Hemsl.）A. Gray
[gam]---[spor] 34，[ipcn]01-03 [ref] Chen，R-Y 2003
IPCN chromosome reports for Xeranthemum annuum L.
[gam]---[spor] 12，[ipcn]01-03 [ref] Chen，R-Y 2003
IPCN chromosome reports for Zinnia elegans Jacq.
[gam]---[spor] 24，[ipcn]01-03 [ref] Chen，R-Y 2003
IPCN chromosome reports for Cornus alba L.
[gam]---[spor] 22，[ipcn]01-03 [ref] Chen，R-Y 2003
IPCN chromosome reports for Matthiola incana（L.）W.T. Aiton
[gam]---[spor] 14，[ipcn]01-03 [ref] Chen，R-Y 2003
IPCN chromosome reports for Orychophragmus violaceus（L.）O.E. Schulz
[gam]---[spor] 24，[ipcn]01-03 [ref] Chen，R-Y 2003
IPCN chromosome reports for Lagenaria siceraria var. microcarpa（Naudin）H. Hara
[gam]---[spor] 22，[ipcn]01-03 [ref] Chen，R-Y 2003
IPCN chromosome reports for Rhododendron simsii Planch.
[gam]---[spor] 26，[ipcn]01-03 [ref] Chen，R-Y 2003
IPCN chromosome reports for Euphorbia pulcherrima Willd. ex Klotzsch
[gam]---[spor] 28，[ipcn]01-03 [ref] Chen，R-Y 2003
IPCN chromosome reports for Eustoma grandiflorum（Raf.）Shinners
[gam]---[spor] 18，[ipcn]01-03 [ref] Chen，R-Y 2003
IPCN chromosome reports for Gentiana scabra Bunge
[gam]---[spor] 14，[ipcn]01-03 [ref] Chen，R-Y 2003
IPCN chromosome reports for Pelargonium graveolens L'Hér. ex Aiton
[gam]---[spor] 30，[ipcn]01-03 [ref] Chen，R-Y 2003
IPCN chromosome reports for *Pelargonium hortorum L.H. Bailey
[gam]---[spor] 18，54，[ipcn]01-03 [ref] Chen，R-Y 2003
IPCN chromosome reports for Sinningia speciosa Benth. & Hook.
[gam]---[spor] 16，[ipcn]01-03 [ref] Chen，R-Y 2003
IPCN chromosome reports for !Freesia refracta（Jacq.）Klatt
[gam]---[spor] 24，[ipcn]01-03 [ref] Chen，R-Y 2003
IPCN chromosome reports for Iris ensata Thunb.
[gam]---[spor] 40，[ipcn]01-03 [ref] Chen，R-Y 2003
IPCN chromosome reports for Iris japonica Thunb.
[gam]---[spor] 18，[ipcn]01-03 [ref] Chen，R-Y 2003
IPCN chromosome reports for Iris tectorum Maxim.
[gam]---[spor] 28，[ipcn]01-03 [ref] Chen，R-Y 2003
IPCN chromosome reports for Coleus blumei Benth.
[gam]---[spor] 48，[ipcn]01-03 [ref] Chen，R-Y 2003
IPCN chromosome reports for Moluccella laevis L.
[gam]---[spor] 34，[ipcn]01-03 [ref] Chen，R-Y 2003
IPCN chromosome reports for Ocimum basilicum L.
[gam]---[spor] 52，[ipcn]01-03 [ref] Chen，R-Y 2003
IPCN chromosome reports for Salvia farinacea Benth.
[gam]---[spor] 18，[ipcn]01-03 [ref] Chen，R-Y 2003

IPCN chromosome reports for !Albizia julibrissin Durazz.
[gam]---[spor] 26，[ipcn]01-03 [ref] Chen，R-Y 2003
IPCN chromosome reports for !Amorpha fruticosa L.
[gam]---[spor] 40，[ipcn]01-03 [ref] Chen，R-Y 2003
IPCN chromosome reports for Caesalpinia pulcherrima（L.）Sw.
[gam]---[spor] 24，[ipcn]01-03 [ref] Chen，R-Y 2003
IPCN chromosome reports for Caragana rosea Turcz. ex Maxim.
[gam]---[spor] 32，[ipcn]01-03 [ref] Chen，R-Y 2003
IPCN chromosome reports for Cercis chinensis Bunge
[gam]---[spor]14，[ipcn]01-03 [ref] Chen，R-Y 2003
IPCN chromosome reports for !Lathyrus odoratus L.
[gam]---[spor] 14，[ipcn]01-03 [ref] Chen，R-Y 2003
IPCN chromosome reports for Leucaena glauca Benth.
[gam]---[spor] 36，[ipcn]01-03 [ref] Chen，R-Y 2003
IPCN chromosome reports for !Mimosa pudica L.
[gam]---[spor] 52，[ipcn]01-03 [ref] Chen，R-Y 2003
IPCN chromosome reports for !Robinia hispida L.
[gam]---[spor] 30，[ipcn]01-03 [ref] Chen，R-Y 2003
IPCN chromosome reports for !Robinia pseudoacacia L.
[gam]---[spor] 22，[ipcn]01-03 [ref] Chen，R-Y 2003
IPCN chromosome reports for !Sophora japonica L.
[gam]---[spor] 28，[ipcn]01-03 [ref] Chen，R-Y 2003
IPCN chromosome reports for Sophora japonica fo. oligophylla Franch.
[gam]---[spor] 28，[ipcn]01-03 [ref] Chen，R-Y 2003
IPCN chromosome reports for Sophora japonica f. pendula Loudon
[gam]---[spor] 28，[ipcn]01-03 [ref] Chen，R-Y 2003
IPCN chromosome reports for Sophora rubriflora P.C. Tsoong
[gam]---[spor] 21，[ipcn]01-03 [ref] Chen，R-Y 2003
IPCN chromosome reports for !Sophora xanthoantha C.Y. Ma
[gam]---[spor] 28，[ipcn]01-03 [ref] Chen，R-Y 2003
IPCN chromosome reports for !Trifolium repens L.
[gam]---[spor] 32，[ipcn]01-03 [ref] Chen，R-Y 2003
IPCN chromosome reports for **Wisteria sinensis（Sims）Sweet
[gam]---[spor] 32，[ipcn]01-03 [ref] Chen，R-Y 2003
IPCN chromosome reports for Aloe arborescens Mill.
[gam]---[spor] 14，[ipcn]01-03 [ref] Chen，R-Y 2003
IPCN chromosome reports for Aloe saponaria（Aiton）Haw.
[gam]---[spor] 14，[ipcn]01-03 [ref] Chen，R-Y 2003
IPCN chromosome reports for Aloe variegata L.
[gam]---[spor] 14，[ipcn]01-03 [ref] Chen，R-Y 2003
IPCN chromosome reports for Aloe vera var. chinensis（Haw.）A. Berger
[gam]---[spor] 14，[ipcn]01-03 [ref] Chen，R-Y 2003
IPCN chromosome reports for Asparagus plumosus Baker
[gam]---[spor] 20，[ipcn]01-03 [ref] Chen，R-Y 2003
IPCN chromosome reports for Aspidistra elatior Blume
[gam]---[spor] 38，[ipcn]01-03 [ref] Chen，R-Y 2003

IPCN chromosome reports for Chlorophytum capense var. variegatum Hort.
[gam]---[spor] 28，[ipcn]01-03 [ref] Chen，R-Y 2003
IPCN chromosome reports for Clematis macropetala Ledeb.
[gam]---[spor] 16，[ipcn]01-03 [ref] Chen，R-Y 2003
IPCN chromosome reports for Dracaena
[gam]---[spor] 40，[ipcn]01-03 [ref] Chen，R-Y 2003
IPCN chromosome reports for Hosta plantaginea（Lam.）Asch.
[gam]---[spor] 60，[ipcn]01-03 [ref] Chen，R-Y 2003
IPCN chromosome reports for Hyacinthus orientalis L.
[gam]---[spor] 30，[ipcn]01-03 [ref] Chen，R-Y 2003
IPCN chromosome reports for Lilium longiflorum Thunb.
[gam]---[spor] 24，[ipcn]01-03 [ref] Chen，R-Y 2003
IPCN chromosome reports for Lilium regale E.H. Wilson
[gam]---[spor] 24，[ipcn]01-03 [ref] Chen，R-Y 2003
IPCN chromosome reports for Muscari botryoides（L.）Mill.
[gam]---[spor] 36，[ipcn]01-03 [ref] Chen，R-Y 2003
IPCN chromosome reports for Paris polyphylla Sm.
[gam]---[spor] 10，[ipcn]01-03 [ref] Chen，R-Y 2003
IPCN chromosome reports for Scilla scilloides（Lindl.）Druce
[gam]---[spor] 16+4B，[ipcn]01-03 [ref] Chen，R-Y 2003
IPCN chromosome reports for Tulipa altaica Pall. ex Spreng.
[gam]---[spor] 24，[ipcn]01-03 [ref] Chen，R-Y 2003
IPCN chromosome reports for Tulipa dasystemon（Regel）Regel
[gam]---[spor] 24，[ipcn]01-03 [ref] Chen，R-Y 2003
IPCN chromosome reports for Tulipa gesneriana L.
[gam]---[spor] 24，[ipcn]01-03 [ref] Chen，R-Y 2003
IPCN chromosome reports for Tulipa heterophylla（Regel）Baker
[gam]---[spor] 24，[ipcn]01-03 [ref] Chen，R-Y 2003
IPCN chromosome reports for Tulipa iliensis Regel
[gam]---[spor] 24，[ipcn]01-03 [ref] Chen，R-Y 2003
IPCN chromosome reports for Tulipa patens C. Agardh ex Schult. & Schult. f.
[gam]---[spor] 24，[ipcn]01-03 [ref] Chen，R-Y 2003
IPCN chromosome reports for Tulipa sinkiangensis Z.M. Mao
[gam]---[spor] 24，[ipcn]01-03 [ref] Chen，R-Y 2003
IPCN chromosome reports for Tulipa tianschanica Regel
[gam]---[spor] 24，[ipcn]01-03 [ref] Chen，R-Y 2003
IPCN chromosome reports for Yucca smalliana Fernald
[gam]---[spor] 60，[ipcn]01-03 [ref] Chen，R-Y 2003
IPCN chromosome reports for Lagerstroemia indica L.
[gam]---[spor] 48，[ipcn]01-03 [ref] Chen，R-Y 2003
IPCN chromosome reports for Parakmeria lotungensis（Chun &C.H. Tsoong）Y.W. Law
[gam]---[spor] 114，[ipcn]01-03 [ref] Chen，R-Y 2003
IPCN chromosome reports for Abelmoschus manihot（L.）Medik.
[gam]---[spor] 40，[ipcn]01-03 [ref] Chen，R-Y 2003
IPCN chromosome reports for Althaea rosea（L.）Cav.
[gam]---[spor] 42，[ipcn]01-03 [ref] Chen，R-Y 2003

IPCN chromosome reports for Indosasa sinica C.D. Chu & C.S. Chao
[gam]---[spor] 48，[ipcn]01-03 [ref] Chen，R-Y 2003
IPCN chromosome reports for Indosasa crassiflora McClure
[gam]---[spor] 48，[ipcn]01-03 [ref] Chen，R-Y 2003
IPCN chromosome reports for Indosasa hispida McClure
[gam]---[spor] 48，[ipcn]01-03 [ref] Chen，R-Y 2003
IPCN chromosome reports for Phyllostachys glauca McClure
[gam]---[spor] 48，[ipcn]01-03 [ref] Chen，R-Y 2003
IPCN chromosome reports for !Phyllostachys heterocycla（Carrière）S. Matsum.
[gam]---[spor] 48，[ipcn]01-03 [ref] Chen，R-Y 2003
IPCN chromosome reports for Chimonobambusa marmorea（Mitford）Makino
[gam]---[spor] 48，[ipcn]01-03 [ref] Chen，R-Y 2003
IPCN chromosome reports for Drepanostachyum scandens（Hsueh & W.D. Li）Keng f. & T.P. Yi
[gam]---[spor] 48，[ipcn]01-03 [ref] Chen，R-Y 2003
IPCN chromosome reports for Oligostachyum glabrescens（T.H. Wen）Q.F. Zheng & Y. P. Wang
[gam]---[spor] 48，[ipcn]01-03 [ref] Chen，R-Y 2003
IPCN chromosome reports for Oligostachyum oedogonatum（Z.P. Wang & G.H. Ye）Q. F. Zheng & K.F. Huang
[gam]---[spor] 48，[ipcn]01-03 [ref] Chen，R-Y 2003
IPCN chromosome reports for Oligostachyum scabriflorum（McClure）Z.P. Wang & G. H. Ye
[gam]---[spor] 48，[ipcn]01-03 [ref] Chen，R-Y 2003
IPCN chromosome reports for Pleioblastus altiligulatus S.L. Chen & S.Y. Chen
[gam]---[spor] 48，[ipcn]01-03 [ref] Chen，R-Y 2003
IPCN chromosome reports for Sasa argenteostriata（Regel）E.G. Camus
[gam]---[spor] 48，[ipcn]01-03 [ref] Chen，R-Y 2003
IPCN chromosome reports for Ferrocalamus strictus Hsueh & Keng f.
[gam]---[spor] 48，[ipcn]01-03 [ref] Chen，R-Y 2003
IPCN chromosome reports for Indocalamus barbatus McClure
[gam]---[spor] 48，[ipcn]01-03 [ref] Chen，R-Y 2003
IPCN chromosome reports for Indocalamus longiauritus Hand.-Mazz.
[gam]---[spor] 48，[ipcn]01-03 [ref] Chen，R-Y 2003
IPCN chromosome reports for Guadua angustifolia Kunth
[gam]---[spor] 48，[ipcn]01-03 [ref] Chen，R-Y 2003
IPCN chromosome reports for Fargesia hainanensis T.P. Yi
[gam]---[spor] 48，[ipcn]01-03 [ref] Chen，R-Y 2003
IPCN chromosome reports for Triticum aestivum L. 'Chinese Spring'
[gam]---[spor] 42，[ipcn]01-03 [ref] Chen，R-Y 2004

附录6　1979～2019年实验室发表的论文

1. 张自立, 陈瑞阳, 宋文芹, 陈桂兰. 蚕豆、洋葱染色体C带显示法. 遗传学报, 1978, 5(4): 334-336.
2. 陈瑞阳, 宋文芹, 李秀兰. 植物有丝分裂染色体标本制作的新方法. 植物学报, 1979, 21(3): 297-298.
3. 陈瑞阳, 宋文芹, 陈晓, 李德成, 徐悦凡, 范海平, 周显昌. 植物染色体Giemsa分带技术的研究. 植物学报, 1979, 21(1): 11-18.
4. 陈瑞阳, 宋文芹, 徐悦凡. 黑麦(Secale cereale)染色体的带型分析及其应用. 植物学报, 1979, 21(2): 177-180.
5. 陈瑞阳, 宋文芹. 植物染色体分带和分带技术. 植物杂志, 1979: (2): 24-26.
6. 宋文芹, 徐悦凡, 李秀兰, 陈瑞阳. 快速鉴定小黑麦里黑麦和小麦染色体Giemsa分带法. 科学通报, 1979, 8: 375.
7. 陈瑞阳, 宋文芹, 李秀兰. 关于水稻染色体组型分析的研究. 遗传学报, 1980, 7(4): 361-365.
8. 陈瑞阳, 宋文芹, 李秀兰. 栽培大麦(Hordem vulgare)染色体带型的研究. 植物学报, 1981, 23(2): 162-165.
9. 陈瑞阳, 宋文芹, 李秀兰. 植物染色体标本制备的去壁低渗法及其在细胞遗传学中的意义. 遗传学报, 1982, 9(2): 151-159.
10. 陈瑞阳, 宋文芹, 安祝平, 李秀兰. 陆地棉(Gossypium hirsutum)染色体组型的研究. 植物学报, 1982, 24(1): 94-97.
11. 陈瑞阳, 宋文芹, 李秀兰, 梁能, 陈太琼, 黄巧云, 陈勇. 中国三种野生稻染色体组型的研究. 植物学报, 1982, 24(3): 226-230.
12. 陈瑞阳, 李秀兰, 佟德耀, 刘玉欣, 吴秋芹, 王凤山, 蒲富慎, 黄礼森. 中国梨属植物染色体数目研究. 园艺学报, 1983, 10(1): 13-15.
13. 郑惠玉, 陈瑞阳. 中国野生大豆根尖染色体细胞学观察初报. 吉林农业科学, 1983, 4: 34-37.
14. 李秀兰, 宋文芹, 陈瑞阳, 林盛华, 贾定贤, 蒲富慎. 我国苹果品种(系)染色体数目观察(1). 中国植物学会五十周年论文汇编, 1983.
15. 安祝平, 宋文芹, 李秀兰, 陈瑞阳. 山羊草核型分析及其与小麦属的进化关系. 武汉植物学研究, 1985, 3(4): 313-318.
16. 程式君, 胡志衡, 李秀兰, 陈瑞阳. 国产石斛属染色体研究初报. 园艺学报, 1985, 12(2): 119-124.
17. 宋文芹, 李秀兰, 陈瑞阳, 林盛华, 张德学, 蒲富慎. 我国部分山楂属植物染色体数目的研究. 园艺学报. 1985, 12(2): 73-76.
18. 蒲富慎, 林盛华, 李秀兰, 宋文芹, 陈瑞阳. 中国苹果属植物核型研究. 武汉植物学研究, 1985, 3(4): 381~387.
19. 陈瑞阳, 宋文芹, 李秀兰. 应用同一细胞进行核型与带型分析方法. 武汉植物学研究, 1985, 3(4): 457-458.
20. 李懋学, 陈瑞阳. 关于植物核型分析的标准化问题. 武汉植物学研究, 1985, 3(4): 297-302.
21. 李秀兰, 陈瑞阳. 谷子和狗尾草核型分析. 武汉植物学研究, 1985, 3(4): 409-412.
22. 陈瑞阳, 李秀兰, 宋文芹. 我国部分热带果树染色体研究. 武汉植物学研究, 1985, 3(4): 423-427.
23. 陈瑞阳, 陈祖耕, 李秀兰, 宋文芹. 中国部分木兰科植物染色体数目. 植物分类学报, 1985, 23(2): 103-105.
24. 蒲富慎, 林盛华, 宋文芹. 中国梨属植物核型研究 I. 武汉植物学研究, 1985, 3(4): 381－387.
25. 宋文芹, 林盛华. 我国部分山楂属植物染色体数目研究. 园艺学报, 1985, 12(2): 73－76.
26. 陈瑞阳, 宋文芹, 李秀兰, 蒲富慎, 林盛华. 我国梨属植物染色体核型研究. 武汉植物学研究, 1985, 3(4): 397.

27. 李秀兰, 宋文芹, 陈瑞阳. 中国苹果属植物核型研究. 武汉植物学研究, 1985, 3(4): 451-456.
28. 陈瑞阳, 宋文芹, 安祝平. 小麦属核型分析和B、G染色体组及4A染色体的起源. 武汉植物学研究, 1985, 3(4): 303-312.
29. 陈瑞阳, 安祝平, 宋文芹, 李秀兰, 苏建英. 植物染色体G-带的初步研究. 武汉植物学研究, 1986, 4(2): 111-117.
30. 陈瑞阳, 宋文芹, 李秀兰, 蒲富慎, 刘捍中, 林盛华. 中国苹果属植物染色体数目报告. 武汉植物学研究, 1986, 4(4): 337-342.
31. 陈瑞阳, 宋文芹, 李秀兰, 蒲富慎, 林盛华. 中国梨属植物核型研究. 园艺学报, 1986, 13(2): 87-90.
32. Chen Ruiyang. A Further Research of Plant Chromosome G-banding Proc. Sino-Jpn. Symposium Pl. Chromos. Plant Chromosome Research, 1987, 187-193.
33. 陈瑞阳, 安祝平, 宋文芹, 李秀兰, 苏建英, 郑坚瑜. 植物染色体高分辨G-带技术研究, 植物学报. 1987, 29(4): 341-346.
34. Li Xiulan, Chen Ruiyang. Studies on Chromosomes of Orchidaceae in China. I Chromosome Numbers of Some Orchids in China Proc. Sino-Jpn. Symposium Pl. Chromos. Plant Chromosome Research, 1987, 301-307.
35. Song Wenqin, Li Xiulan, Chen Yongli. Studies on the Karyotype Evolution and Relationship of Maloideae Proc. Sino-Jpn. Symposium Pl. Chromos. Plant Chromosome Research, 1987, 327-333.
36. Zheng Jianyu, Gu Chunyan, Chen Ruiyang. Cytotaxonomical Studies on Commelinaceae in China I Chromosome Numbers and Karyotypes of Some Chinese Species Proc. Sino-Jpn. Symposium Pl. Chromos. Plant Chromosome Research, 1987, 363-368.
37. 陈瑞阳. 安祝平, 宋文芹, 李秀兰, 苏建英, 郑坚瑜. 植物染色体高分辨G-带的研究. 科学通报, 1987, 14: 1096-1098.
38. Rui-yang Chen. G-band Staining of Chromosome in Some Higher Plants with Special Reference to Lilium davidii. La Kromosome, 1988, II-50, 1635-1651.
39. 陈瑞阳, 张玮, 武全安. 云南部分木兰科植物染色体数目报道. 云南植物研究, 1989, 11(2): 234-238.
40. 陈瑞阳, 安祝平, 宋文芹, 李秀兰, 陈瑞阳, 纵微星. 部分散生竹类染色体数目. 植物分类学报, 1991, 29(5): 452-455.
41. Weixing Zong, Ruiyang Chen, Kenji Taniguchi, Katsuhiko Kondo. A Chromosome Study in Intraspecific Polyploidy of Phragmites australis and Its Related Species. La kromosome, 1991, II-63-64: 2168-2172.
42. Chen Ruiyang, An Zhuping, Song Wenqin, Li Xiulan, Zong Weixing. The G-banded Karyotype of Lilium davidii L. Cathaya, 1992, 4: 9-20.
43. Li Xiu-lan, An Zhu-ping, Song Wen-qin, Chen Rui-yang. The G- and C-banded Karyotypes of Secale cereale Proc. Sec. Sino-Jpn. Symposium Pl. Chromos. Plant Chromosome Research, 1992, 129-132.
44. An Zhu-ping, Cui Xiang-qin, Song Wen-qin, Li Xiu-lan, Chen Rui-yang. A Study on the Sex Chromosome of Asparagus officinalis Proc. Sec. Sino-Jpn. Symposium Pl. Chromos. Plant Chromosome Research, 1992, 297-300.
45. Chen Rui-yang, An Zhu-ping, Song Wen-qin, Li Xiu-lan. Kohei Irifune, Kenji Taniguchi, Ryuso Tanaka On the Cytogeography and Speciation of Chinese Aster ageratoides Polyploid Complex Proc. Sec. Sino-Jpn. Symposium Pl. Chromos. Plant Chromosome Research, 1992, 25-45.
46. Chen Rui-yang, Jin Chun-xiang, Li Xiu-lan, Song Wen-qin. Cytogenetical and Cytogeographical Studies on Some Higher Plants. Correlating between China and D. P. R. Korea. Proc. Sec. Sino-Jpn. Symposium Pl. Chromos. Plant Chromosome Research, 1992, 59.
47. Song Wen-qin, Cui Xiang-qin, Li Xiu-lan, Xu Wen-sheng, An Zhu-ping, Wang Shu-rong, Chen Rui-yang A Preliminary Study on Microdissection and Polymerase Chain Reaction(PCR)Technique of Specific Regions of G-banded Plant Chromosomes Proc. Sec. Sino-Jpn. Symposium Pl. Chromos. Plant

Chromosome Research, 1992, 111-114.
48. Chen Rui-yang, Song Wen-qin, Li Xiu-lan, An Zhu-ping. A Study on the Sex Chromosome of Ginkgo biloba. Cathaya, 1993, 5: 41-48.
49. Tanaka R., Chen S. C., Hong D. Y., Chen R. Y., Zhang D. M., Nakata M., Taniguchi K. and Kondo K. Geographical Distribution of Chinese Dendranthema on the Basis of the Specimen Collections of the 1988-1991 Field Studies Cytogenetics on plants correlating between Japan and China, 1992, 69-71.
50. Taniguchi K., Chen R. Y. and Tanaka R. A Cytogeographical Study of Dendranthema with Yellow Ligules in the Coastal Region of China Cytogenetics on plants correlating between Japan and China, 1992, 53-55.
51. Xiulan Li, Ruiyang Chen, Ryuso Tanaka. Reports on Chromosome Numbers of Some Orchids Cultivated in China. La Kromosome, 1992, II-67-68: 2301-2311.
52. 李秀兰, 宋文芹, 陈瑞阳. 北方几种小浆果植物的核型研究. 武汉植物学研究, 1993, 11(4): 289-292.
53. 韦和平, 陈维培, 陈瑞阳. 睡莲科的核型分析及其分类学位置的探讨. 植物分类学报, 1994, 32(4): 293-300.
54. Chen Rui-yang, Wenqin Song, Xiulan Li, Zhuping An. Chromosome G-banding in Plants by Inducing with Trypsin and Urea. Cell Research, 1994, 4: 79-87.
55. 宋文芹, 崔香芹, 许文胜等. 蚕豆最大染色体长臂的显微切割与PCR扩增. 科学通报, 1996, 41(4): 361-363.
56. 彭永康, 于建春, 赵建, 宋文芹, 陈瑞阳. APM对小麦根尖分生组织细胞异常有丝分裂的诱导. 应用与环境生物学报, 1997, 3(3): 204~207.
57. 李秀兰, 宋文芹, 安祝平, 陈瑞阳. 中国木莲属部分种的核型比较. 南开大学学报, 1997, 30(4): 109-112.
58. 李秀兰, 宋文芹, 许文胜, 陈瑞阳. 一种改进的染色体显微切割方法. 南开大学学报, 1998, 31(2): 102-105.
59. Mao Yingwei, Liang Siyuan, Song Wenqin, Li Xiulan, Chen Ruiyang. Construction of DNA Library from Chromosome 4 of Rice(Oryza sativa)by Microdissection, Cell Research, 1998, (8)285-293.
60. Rong-xin Zhang, Cheng-bin Chen, Xiu-lan Li, Wen-qin Song. Construction of DNA Library of Centromeric Region of the B Chromosome of Plant. Eighteenth International Congress of Genetics, 1998, (Abstract) p15.
61. 王晓梅, 宋文芹, 李秀兰, 陈瑞阳. 用RAPD技术检测野生鲫鱼和四个金鱼代表品种的基因组DNA多态性. 遗传, 1998, 20(5): 7-11.
62. 李秀兰, 宋文芹, 安祝平, 陈瑞阳. 中国木兰属部分种的核型分析. 云南植物研究, 1998, 20(2): 204-206.
63. 宋文芹, 李秀兰, 许文胜等. 黑麦染色体的显微分离与PCR扩增. 植物学报, 1998, 40(2): 158-162.
64. 李秀兰, 宋文芹, 安祝平, 陈瑞阳. 中国含笑属核型分析. 植物分类学报, 1998, 36(2): 145-149.
65. 郭歌, 陈成彬, 李秀兰, 宋文芹, 陈瑞阳. 黑麦B染色体端粒相关序列的克隆. 植物学报, 1998, 40(12): 1123-1128.
66. 李秀兰, 宋文芹, 安祝平, 陈瑞阳. 木兰科属间核型比较. 植物分类学报, 1998, 36(3): 232-237.
67. 陈成彬, 李秀兰, 孙成仁, 宋文芹, 陈瑞阳. 中国樟科5属9种植物的核型研究. 武汉植物学研究, 1998, 16(3): 219-222.
68. 江赐忠, 宋文芹, 李秀兰, 陈瑞阳. 黑麦1R染色体的微切微克隆研究. 植物学报, 1998, 40(11): 1-6.
69. 翟晓玲, 陈瑞阳. 通过微切扩增建立染色体区段特异基因文库的初步研究. 热带亚热带植物学报, 1998, 5(3).
70. 彭永康, 赵建, 陈瑞阳. 大蒜根尖细胞有丝分裂同步化诱导与中期染色体分离. 植物研究, 1999, 19(3): 302-307.

71. 张荣信, 陈成彬, 李秀兰, 宋文芹. 黑麦A、B染色体着丝粒同源序列的荧光原位杂交分析. 科学通报, 1999, 44(5): 520-524.
72. 李秀兰, 刘松, 宋文芹, 陈瑞阳. 40种散生竹的染色体数目, 植物分类学报, 1999, 37(6): 54-56.
73. 陈成彬, 宋文芹. 利用HU和APM双阻断法诱导高频率植物根尖有丝分裂同步化的研究. 南开大学学报, 1999, 32(1): 28-31.
74. 梁思源, 宋文芹, 李秀兰, 陈瑞阳. 水稻第9号染色体DNA文库的构建. 南开大学学报, 1999, 32(2).
75. 祁仲夏, 李秀兰, 宋文芹, 陈瑞阳. 蚕豆单条M染色体-AFLP分析. 中国细胞生物学会第七次会
65. 宋文芹, 祁仲夏, 李秀兰, 陈成彬, 陈瑞阳. 染色体微分析技术及其应用前景探讨, 中国细胞生物学会第七次会议论文摘要, 1999, 10: p165-166.
77. 李秀兰, 陈成彬, 陈瑞阳, 原始木本被子植物基因组原位杂交初步研究, 中国细胞生物学会第七次会议论文摘要, 1999. 10, p55.
78. 陈瑞阳, 基因组学与染色体作图, 中国细胞生物学会第七次会议论文摘要, 1999, 10: 64-65.
79. 张荣信, 宋文芹, 李秀兰, 陈瑞阳. 黑麦B-染色体显微切割和微克隆. 南开大学学报, 1999, 32(2): 112-115.
80. Rong-xin Zhang, Cheng-bin Chen, Xiu-lan Li, Wen-qin Song. Homologous Analysis on Centromeric Region Sequences of A and B Chromosome from Rye by FISH. Chinese Science Bulletin, 1999, 44(8): 704-708.
81. 王振英, 郑坚瑜, 陈瑞阳. 盐胁迫、辐射条件下耐盐与不耐盐水稻POD同工酶、全蛋白变化的研究. 南开大学学报, 1999, 32(1): 23-27.
82. 张峰, 宋文芹, 陈瑞阳. AFLP-银染法检测植物基因组多态性. 细胞生物学杂志, 1999, 2(2): 98-101.
83. 王晓梅, 宋文芹, 李秀兰, 陈瑞阳. 鲫鱼种群的随机扩增多态DNA与遗传多样性分析. 中国水产科学, 1999, 6(2): 26-28.
84. 张峰, 宋文芹, 李凌, 李秀兰, 陈瑞阳, 孙德岭. 利用AFLP-银染法筛选与抗甘蓝黑腐病性状连锁的分子标记. 南开大学学报, 1999, 32(3): 177-181.
85. 孙易, 宋文芹, 钟贻诚, 张闰生, 陈瑞阳. 用RAPD和AFLP的方法对中国卤虫(Artemia)种及亲缘关系的研究. 遗传学报, 2000, 27(3): 210-218.
86. 张天浩, 张春平, 张光寅, 陈瑞阳. DNA芯片制作原理及其杂交信号检测方法. 生物工程进展, 2000, 20(2): 64-68.
87. 陈晶, 郑坚瑜, YAMJ., HSIAOWL. 陈家童, 陈瑞阳. 用改进的差别显示法克隆R6大鼠细胞的癌相关基因. 实验生物学报, 2000, 33(3): 199-205.
88. 宋文芹, 李秀兰, 祁仲夏, 梁思源, 陈瑞阳. 水稻染色体的显微分离与克隆. 南开大学学报, 2000, 33(3): 83-88.
89. 李凌, 宋文芹, 毛英伟, 李秀兰, 陈瑞阳, 孙德岭. 用cDNA-AFLP银染技术研究与花椰菜花色相关的基因. 南开大学学报, 2000, 33(4): 33-36.
90. 王晓梅, 宋文芹, 李秀兰, 陈瑞阳. 文种金鱼系统发育地位的研究. 南开大学学报, 2000, 33(2): 127-128.
91. 王晓梅, 陈瑞阳. 高等植物的性别与性别决定机制. 细胞生物学杂志, 2001, 23(2): 62-66.
92. 于建春, 金绍波, 赵建, 陈瑞阳, 彭永康. 一种分析有丝分裂期蛋白的新方法. 天津师范大学学报, 2001, 21(3): 65-66.
93. 李秀兰, 林汝顺, 冯学琳, 祁仲夏, 宋文芹, 陈瑞阳. 中国部分丛生竹类染色体数目报道. 植物分类学报, 2001, 39(5): 433-442.
94. 孙德岭, 赵前程, 宋文芹, 陈瑞阳. 白菜类蔬菜亲缘关系的AFLP分析. 园艺学报, 2001, 28(4): 331-335.

95. 杨宇, 陈瑞阳. 水稻基因组测序的研究进展. 遗传, 2001, 23(6): 580-582.
96. 祁仲夏, 宋文芹, 陈瑞阳. 染色体作图. 生物学通报, 2001, 36(1)14-15.
97. 王晓梅, 宋文芹, 刘松, 陈瑞阳. 与银杏性别相关的 RAPD 标记. 南开大学学报, 2001, 34(3): 116-117.
98. 祁仲夏, 宋文芹, 金刚, 陈瑞阳. 稻属基因组间相关性的 AFLP 分析. 南开大学学报, 2001, 34(3): 74-80.
99. 王晓梅, 宋文芹, 刘松, 李秀兰, 陈瑞阳. 利用 AFLP 技术筛选与银杏性别相关的分子标记. 南开大学学报, 2001, 34(1): 5-9.
100. 祁仲夏, 宋文芹, 陈瑞阳等. B 染色体分子生物学研究进展. 细胞生物学杂志, 2002, 24(4): 199-202.
101. Qi Zhongxia, Chen Ruiyang, *et al.* Isolation and chromosomal mapping of a corn B chromosome specific RAPDs, Acta Botanica Sinica, 2002, 44(4): 499-501.
102. Qi Zhongxia, Chen Ruiyang, *et al.* The molecular characterization of maize B chromosome specific AFLPs. Cell Research, 2002, 12(1): 63-68.
103. 倪虹, 陈瑞阳. 基因表达系列分析(SAGE)的研究进展. 生命科学研究, 2002, 6(1): 18-20.
104. 祁仲夏, 李秀兰, 陈成彬, 宋文芹, 陈瑞阳. 玉米B染色体特异RAPD分子标记的染色体定位(英文). 植物学报, 2002, 44(4): 499-501.
105. 孙德岭, 赵前程, 宋文芹, 陈瑞阳. 花椰菜类蔬菜自交系基因组间亲缘关系的 AFLP 分析. 园艺学报, 2002, 29(1): 72-74.
106. 祁仲夏, 宋文芹, 李秀兰, 陈瑞阳. B 染色体分子生物学研究进展. 细胞生物学杂志, 2002, 24(4): 199-202.
107. 王晓梅, 宋文芹, 李秀兰, 陈瑞阳. 银杏雌雄基因组 DNA 间的差异性分析. 细胞生物学杂志, 2002, 24(1): 38-41.
108. 刘松, 宋文芹, 赵前程, 陈成彬, 孙德岭, 陈瑞阳. 与花椰菜(Brassica oleracea ssp. botrytis)抗黑腐病基因连锁的 RAPD 标记. 南开大学学报, 2002, 35(1): 126-128.
109. 徐悦凡, 任鲁风, 宋文芹, 张长俊, 马元煦, 李金春, 张燕妮. 中国人非综合征型听力损失患者Cx26 基因的突变分析. 中华耳鼻咽喉科杂志, 2002, 37(5): 348-351.
110. 龙鸿, 祁仲夏, 陈成彬, 李秀兰, 陈瑞阳, 黑麦B染色体DNA 分子标记筛选与定位研究. 中国细胞生物学学会第八届会员代表大会暨学术大会, 2003, 11, p44.
111. 曾辉, 陈瑞阳. 卤虫 dsx 基因相关 DNA 片段的研究. 中国细胞生物学学会第八届会员代表大会暨学术大会, 2003, 11, p83.
112. 高英堂, 陈瑞阳, 宋文芹. 生物芯片的发展及其哲学观. 医学与哲学, 2003, 24(7): 52-54.
113. 高英堂, 陈瑞阳, 宋文芹, 陈成彬, 齐之丽, 景丽, 孙金英, 钱绍诚. 利用基因芯片分析拉米夫定治疗过程中HBV DNA 的基因变异. 中国病毒学, 2003, 18(6): 523-529.
114. 古瑜, 王春国, 孙德岭, 宋文芹, 常彩涛, 柳妮莎. RAPD 分子标记方法及在蔬菜研究中的应用. 天津农学院学报, 2003, 10(3): 47-51.
115. 王娟, 王顺启, 倪虹, 陈力, 宋文芹. 莪术挥发油抑制人肝癌细胞株 SMMC-7721 生长的实验研究. 天津中医药, 2003, 20(1): 48-51.
116. 王娟, 倪虹, 陈力, 宋文芹. 中草药抑制 HBV 作用机制的研究进展. 现代中西医结合杂志, 2003, 12(18): 2010-2012.
117. 王顺启, 倪虹, 王娟, 陈力. 独角莲对肝癌细胞 SMMC-7721 细胞增殖抑制作用机理的研究. 细胞生物学杂志, 2003, 25(3): 185-188.
118. 曾辉, 陈成彬, 刘凤岐, 宋文芹, 陈瑞阳. 尕海孤雌生殖卤虫基因组的特异 DNA 片段. 动物学报, 2004, 50(3): 470-474.
119. 曾辉, 宋文芹, 陈瑞阳. 卤虫中编码DM结构域的DNA序列的克隆和初步研究(简报). 实验生物学报, 2004, 37(5): 423-427.

120. 曾辉, 宋文芹, 陈瑞阳. 动物性别决定基因及其同源性比较. 细胞生物学杂志, 2004, 26(2): 129-132.

121. 陈成彬, 齐力旺, 张守攻, 韩素英, 李秀兰, 宋文芹, 陈瑞阳. 三倍体杨树核型分析. 武汉植物学研究, 2004, 22(6): 565-567.

122. 高英堂, 陈瑞阳, 宋文芹, 陈成彬, 齐之丽, 景丽, 阚志超. 丙型肝炎病毒基因分型芯片的研制和临床应用. 中华检验医学杂志, 2004, 27(10): 690-693.

123. 高英堂, 陈瑞阳, 宋文芹, 齐之丽, 景丽, 钱绍诚. 不同引物对扩增 SEN 病毒的实验研究. 中华流行病学杂志, 2004, 25(5): 459-460.

124. 高英堂, 陈瑞阳, 宋文芹, 齐之丽, 景丽, 钱绍诚. SEN 病毒在不同人群中的感染状况. 中华传染病杂志, 2004, 22(3): 206-208.

125. 齐力旺, 张守攻, 韩素英, 陈成彬, 李秀兰, 宋文芹, 陈瑞阳. 杨属青杨组种(品种)间核型比较. 云南植物研究, 2004, 26(5): 537-542.

126. 王娟, 刘岩雪, 倪虹, 陈力, 陈成彬, 宋文芹. 白细胞介素-10 启动子-872 位点多态性在中国汉族人群高频率发生(英文). 南开大学学报, 2004, 37(1): 79-82.

127. 王娟, 倪虹, 陈力, 宋文芹. 肝癌细胞的染色体变异. 国外医学遗传学分册, 2004, 27(2): 90-93.

128. 王娟, 倪虹, 陈力, 刘岩雪, 陈成彬, 宋文芹. 肝癌相关基因单核苷酸多态性基因芯片制备与检测分析. 中华检验医学杂志, 2004, 27(10): 649-652.

129. 王广良, 倪虹, 陈力. 细胞周期检测点与肿瘤发生. 国外医学肿瘤学分册, 2004, 31(6): 406-409.

130. 王顺启, 倪虹, 程华, 王广良, 王同顺, 陈力. mRNA 差异显示法比较独角莲作用肝癌细胞 SMMC-7721 前后的基因表达. 中国中药杂志, 2004, 29(10): 974-977.

131. Bin Yang, Ying-Tang Gao, Zhi Du, Lei Zhao, Wen-Qin Song. Methylation-based molecular margin analysis in hepatocellular carcinoma. Biochemical and Biophysical Research Communications, 338(2005): 1353–1358.

132. Y. Zhang, S. G. Zhang, L. W. Qi, B. Liu, J. M. Gao, C. B. Chen, X. L. Li, W. Q. Song. Chromosome microdissection, cloning and painting of the chromosome 1 in poplar(Populus tremula). Silvae Genetica, 2005, 54(4-5): 211-217.

133. B. Liu, C. B. Chen, X. L. Li, R. Y. Chen, W. Q. Song. Physical mapping of 45S rDNA to metaphase chromosomes in 30 taxonomically diverse plant species. Plant Jour. Hort. Sci. and Biote, 2005, 80(3): 287-290.

134. Zhang Y, Zhang SG, LW Qi, B Liu, JM Gao, CB Chen, XL Li, WQ Song. Construction of poplar(Populus tremula)chromosome 1 specific DNA library by using a microdissection technique. Plant Molecular Biology Reporter, 2005, 23(2): 129-138.

135. 张守攻, 陈成彬, 韩素英, 李秀兰, 任建中, 周玉权, 宋文芹, 陈瑞阳, 齐力旺. 中国部分杨属植物的染色体数目. 物分类学报, 2005, 43(6): 539–544.

136. 齐力旺, 张守攻, 韩素英, 陈成彬, 李秀兰, 宋文芹, 陈瑞阳, 任建中, 周玉权. 杨属白杨组种间核型比较. 园艺学报, 2005, 32(5): 849-853.

137. 张守攻, 齐力旺, 韩素英, 陈成彬, 李秀兰, 宋文芹, 陈瑞阳. 杨属(Populus)黑杨组(Aigeiros)种(品种)间核型比较. 园艺学报, 2005, 32(1): 70-73.

138. 陈成彬, 张守攻, 李秀兰, 韩素英, 宋文芹, 齐力旺. 杨属派间核型比较研究. 广西植物, 2005, 25(4): 338-340.

139. 王顺启, 冯占宁, 刘国红, 王广良, 程华, 王同顺, 倪虹, 宋文芹, 陈力. 应用抑制性消减杂交技术构建人肝癌组织特异表达 cDNA 文库. 南开大学学报, 2005, 38(4): 115-118.

140. 王春国, 常彩涛, 古瑜, 孙德岭, 宋文芹. 植物细胞质雄性不育. 细胞生物学杂志, 2005, 27: 525-529.

141. 王春国, 宋文芹. 花椰菜细胞质雄性不育基因特异 PCR 标记的筛选. 遗传, 2005, 27(2): 236-240.

142. 常彩涛, 王春国, 陈成彬, 吴峰, 吴锋, 孙德岭. 细胞质雄性不育辣椒育性恢复基因特异分子标记的筛选. 实验生物学报, 2005, 38(3): 227-232.

143. 王广良, 陈成彬, 高建明, 倪虹, 王同顺, 陈力. 细胞周期相关基因微阵列在中药抑制肝癌细胞增殖作用机理研究中的应用. 中国中药杂志, 2005, 30(1): 50-54.

144. 刘博, 陈成彬, 李秀兰, 齐力旺, 韩素英. 豆科三属八种植物的核型及 rDNA 定位研究. 云南植物研究, 2005, 27(3): 261-268.

145. C. C. Ren, X. H. Miao, B. Yang, L. Zhao, R. Sun & W. Q. Song. Methylation status of the fragile histidine triad and E-cadherin genes in plasma of cervical cancer patients. Int J Gynecol Cancer, 2006, 16: 1862–1867.

146. Gao JM, Zhang SG, Qi LW, Zhang Y, Wang CG, Song WQ, Han SY. Application of ISSR marker to fingerprinting of elite cultivars(varieties / clones)from different sections of the genus Populus L. Silvae Genetica, 2006, 55(1): 1-6.

147. Liu B, Chen C, Li X, Chen R, and Song W. Physical mapping of 45S rDNA on metaphase chromosomes in several plant species. Acta Scientiarum Naturalium Universitatis Nankaiensis, 2006, 39(4): 96-102.

148. Gao jianming, Zhang shougong, Qiliwang, Zhang Yong, Wang Chunguo, Song wenqin. ISSR and AFLP identification and genetic relationship of Chinese elite accessions from the genus Populus. Ann. For. Sci, 2006, 63: 1-8.

149. Yong Zhang, Shougong Zhang, Liwang Qi, Bo Liu, Jianming Gao, Xiaoqiang Chen, Chengbin Chen, Xiulan Li, Wenqin Song. Cloning and characterization of disease resistance Gene analoGs from poplar(Populus tremula)chromosome 1. Int. J. Plant Sci., 2006, 167(3): 402-413.

150. T. Lan, S. Zhang, B. Liu, X. Li, R. Chen, W. Song. Differentiating sex chromosomes of the dioecious Spinacia oleracea L. (spinach)by FISH of 45S rDNA, Cytogenet Genome Res, 2006, 114: 175-177.

151. Chunguo wang, Xiaoqiang Chen, Tianying Lan, Hui Li, Wenqin Song. Cloning and transcript analyses of the chimeric gene associated with cytoplasmic male sterility in cauliflower(*Brassica oleracea* ver. botrytis). Euphytica, 2006, 151: 111-119.

152. B. Liu, S. G. Zhang, Y. Zhang, T. Y. Lan, L. W. Qi and W. Q. Song. Molecular cytogenetic analysis of four Larix species by bicolor fluorescence in situ hybridization and DAPI banding. Int. J. Plant Sci, 2006, 167(2): 367-372.

153. Juan Wang, Hong Ni, Li Chen and Wenqin Song. Interleukin-10 promoter polymorphisms in patients with hepatitis B virus infection or hepatocellular carcinoma in Chinese Han ethnic population. Hepatobiliary & Pancreatic Diseases International. 2006, 5(1): 60-64.

154. 任晨春, 苗旭红, 杨斌, 赵磊, 孙蕊, 宋文芹. 宫颈癌患者血浆和组织中 FHIT 基因 5'端 CPG 岛甲基化状态的研究. 遗传, 2006, 28(9): 1061-1066.

155. 任晨春, 宋文芹, 苗绪红. 孕妇外周血中胎儿基因组的研究进展. 医学综述, 2006, (12)5: 262-264.

156. 任晨春, 宋文芹, 苗绪红, 孙蕊. DNA 甲基化及其在宫颈癌中的研究进展. 医学综述, 2006, (12)7: 394-396.

157. 任晨春, 王文靖, 宋文芹. 人类线粒体遗传病. 国外医学妇产科学分册, 2006, 33(3): 171-173.

158. 任晨春, 苗绪红, 程华, 陈力, 宋文芹. 在孕 9 周前检测孕妇外周血中胎儿的性别(英文). 国际遗传学杂志, 2006, 29(4): 253-259.

159. 任晨春, 苗绪红, 宋文芹. 孕妇血浆中胎儿 DNA 的数量变化的研究. 现代妇产科进展, 2006, 15(8): 614-616.

160. 任晨春, 苗绪红, 杨斌, 赵磊, 孙蕊, 宋文芹. 宫颈癌脆性组氨酸三联基因的表达与血浆中的甲基化状态. 中华医学遗传学杂志, 2006, 23(5): 565-567.

161. 任晨春, 苗绪红, 杨斌, 赵磊, 孙蕊, 宋文芹. 宫颈癌患者血浆中 E-钙黏着蛋白基因启动子区的甲基化状态. 细胞生物学杂志, 2006, 28: 768-772.

162. 兰添颖, 宋文芹, 张守攻, 齐力旺, 韩素英. 转基因农作物检测技术及其应用与发展. 广西植物, 2006, 26, (5): 483-487.
163. 王娟, 倪虹, 陈力, 陈成彬, 宋文芹. 肝癌染色体高频缺失区肿瘤相关基因 cSNP 在 HBV 患者和正常人群中的多态分布研究. 南开大学学报, 2006. 39(3): 1-5.
164. 杨斌, 高英堂, 杜智, 赵磊, 宋文芹. 甲基化特异性 PCR 在原发性肝细胞癌 p16INK4A 基因甲基化检测中的应用. 中华检验医学杂志, 2006, 29(3): 229-232.
165. 陈成彬, 马小军, 陈力, 薛梅, 成玉, 宋文芹, 李秀兰, 陈瑞阳. 半夏多倍体复合体及其细胞地理学研究. 中国中药杂志, 2006, 31(7): 1405-1408.
166. 陈小强, 李秀兰, 王春国, 张勇, 宋文芹, 陈瑞阳. 大花蕙兰授粉后子房 cDNA 差异表达片段序列分析. 园艺学报, 2006, 33(4): 783-788.
167. 刘博, 陈成彬, 李秀兰, 陈瑞阳, 宋文芹. 45S rDNA 在多种植物中期染色体上的定位(英文). 南开大学学报, 2006, 39(4): 96-102.
168. 张勇, 张守攻, 齐力旺, 陈小强, 陈瑞阳, 宋文芹. 杨树——林木基因组学研究的模式物种. 植物学通报, 2006, 23(3): 286-293.
169. 高建明, 张守攻, 齐力旺, 张勇, 王春国, 陈瑞阳, 宋文芹. 杨树重要品种(无性系)的 AFLP 指纹分析. 云南植物研究, 2006, 28(1): 85-90.
170. 王春国, 李慧, 宋文芹. 细胞质雄性不育花椰菜保持系特异分子标记的鉴定及序列分析. 分子细胞生物学报, 2006, 39(3): 236-242.
171. 张守攻, 张勇, 刘博, 李秀兰, 宋文芹, 韩素英, 齐力旺. 欧洲山杨一号染色体显微分离、原位杂交分析及特异文库的构建. 园艺学报, 2006, 33(4): 794-800.
172. 赵鑫, 马小军, 凯撒. 苏来蔓, 付长亮, 陈瑞阳. 阜康阿魏的核型分析研. 中国中药杂志, 2006, 31(2): 114-116.
173. 成玉, 陈成彬, 薛梅, 陈力. 应用 AFLP 技术探讨半夏属五个种的亲缘关系. 云南植物研究, 2006, 28(6): 559-564.
174. 刘博, 陈成彬, 李秀兰, 张守攻, 齐力旺, 韩素英. 部分落叶松属植物核型研究. 广西植物, 2006, 26(2): 187-191.
175. 王顺启, 倪虹, 王同顺, 陈力. 独角莲水提取物对肝癌细胞 SMMC-7721 基因表达的影响. 中草药, 2006, 37(9): 1384-1387.
176. Wang Juan, Ni Hong, Chen Li, Chen Chengbin, Song Wenqin. Polymorphism attribution of cSNPs in cancer-related genes located in loss regions with a high frequency of HCC between HBV and health groups. Front. Biol. China, 2007, 2(3): 1–4.
177. Bo Liu Li Wang Qi, Ruiyang Chen and Wenqin Song, Multicolor fluorescence in situ hybridization with combinatorial labeling probes enables a detailed karyotype analysis of Larix principis-rupprechtii. Biol Res, 2007, 40: 23-28.
178. Chunguo Wang, Xiaoqiang Chen, Hui Li, and Wenqin Song, RNA editing analysis of mitochondrial nad3/rps12 genes in cytoplasmic male sterility and male-fertile cauliflower(Brassica oleracea var. botrytis)by cDNA-SSCP. Botanical Studies, 2007, 48: 13-23.
179. X. Q. Chen, C. G. Wang, Y. Zhang, W. Q. Song and R. Y. Chen. Identification and sequence analysis of cDNA fragments relative to ovary development of Cymbidium hybridium after pollination. Biologia Plantarum, 2007, 51(2): 249-256.
180. C. C. Ren, X. H. Miao, H. Cheng, L. Chen, W. Q. Song, Detection of Fetal Sex in the Peripheral Blood of Pregnant Women. Fetal Diagn Ther, 2007, 22: 377–382.
181. 古瑜, 赵前程, 刘松, 王春国, 孙德岭, 宋文芹. 花椰菜(Brassica oleracea var. botrytis)黑腐病抗性基因同源序列分离及克隆的研究, 南开大学学报, 2007, 40(2): 62-66.
182. 古瑜, 孙德岭, 赵前程, 宋文芹. 花椰菜遗传图谱的构建及 NBS-LRR 类抗性同源基因在图谱中的定位. 遗传, 2007, 49(6): 1-7.

183. 王冰, 张勇, 陈成彬, 李秀兰, 陈瑞阳, 陈力. 中国不同地理居群丹参遗传多样性分析. 中国中药杂志, 2007, 32(19): 1988.
184. 兰添颖, 刘博, 董凤平, 陈瑞阳, 李秀兰, 陈成彬. 菠菜 rDNA 及端粒多色荧光原位杂交分析. 遗传, 2007, 29(11): 1405-1408.
185. 陈小强, 王春国, 李秀兰, 宋文芹, 陈瑞阳. 植物 DNA 甲基化及其表观遗传作用. 细胞生物学杂志, 2007, 29: 519-524.
186. 古瑜, 孙德岭, 宋文芹. 生物技术在花椰菜遗传育种中的应用. 天津农业科学, 2007, 13(2): 14-19.
187. 苗绪红, 苏冠男, 宋文芹. 肾素促肾脏纤维化作用研究进展——肾素-血管紧张素系统的新扩展. 医学综述, 2007, 13(4): 247-249.
188. 董凤平, 韩素英, 张守攻, 齐力旺, 刘博, 李秀兰, 陈成彬. 25S rDNA 在杨属植物染色体上的定位. 云南植物研究, 2007, 29(4): 423-428.
189. 薛梅, 王秋颖, 陈成彬, 董凤平, 马小军. 药膳食材半夏的分子生物学研究. 食品工业科技, 2007, 5: 127-128.
190. 王顺启, 陈力, 倪虹, 谢明勇. 白花蛇舌草对肝癌细胞 SMMC-7721 基因表达的影响. 中药材, 2007, 30(10): 1285-1289.
191. 张世光, 成玉, 高英堂, 杜智, 陈力. 肿瘤干细胞及肝癌肿瘤干细胞. 细胞生物学杂志, 2007, 29: 6-10.
192. T. Y. Lan, R. Y. Chen, X. L. Li, F. P. Dong, Y. C. Qi, and W. Q. Song. Microdissection and painting of the W chromosome in Ginkgo biloba showed different labelling patterns, Botanical Studies, 2008, 49: 33-37.
193. H. Li, S. G. Zhang, J. M. Gao, C. G. Wang, Y. Zhang, L. W. Qi, L. Chenand W. Q. Song. Development of a sequence characterized amplified region(SCAR)marker associated with high rooting ability in Larix. Biologla Plantarum, 2008, 52(3): 525-528.
194. Yan, Guorong, Long, Hong, Song, Wenqin, Chen, Ruiyang. Genetic polymorphism of Malus sieversii populations in Xinjiang, China. Genetic Resources and Crop Evolution, 2008, 55(1): 171-181.
195. H. Li, S. G. Zhang, J. M. Gao, C. G. Wang, Y. Zhang, L. W. Qi, L. Chen and W. Q. Song. Development of a sequence characterized amplified region(SCAR)marker associated with high rooting ability in Larix. Biologia Plantarum, 2008, 52(3): 525-528.
196. Y. Gu, Y. W. Mao, C. G. Wang, Q. C. Zhao, D. L. Sun and W. Q. Song. Cloning of differential expression fragments in cauliflower after Xanthomonas campestris inoculation. Biologia Plantarum, 2008, 52(3): 462-468.
197. Yu Gu, Qian-Cheng Zhao, De-Ling Sun, and Wen-Qin Song. A genetic linkage map based on AFLP and NBS markers in cauliflower(Brassica oleracea var. botrytis). Botanical Studies, 2008, 49: 93-99.
198. Long Hong, Qi Zhong-Xia, Sun Xiao-Ming, Chen Cheng-Bin, Li Xiu-Lan, Song Wen-Qin, Chen Rui-Yang. Characters of DNA constitution in the rye B chromosome. JOURNAL OF INTEGRATIVE PLANT BIOLOGY, 2008, 50(2): 183-189.
199. 陈小强, 王春国, 李秀兰, 宋文芹, 陈瑞阳. 大花蕙兰 MADS-box 基因 ChMADS1 的克隆及表达. 199. 南开大学学报. 2008, 41(6): 1-6.
200. 陈小强, 王春国, 李秀兰, 宋文芹, 陈瑞阳. 大花蕙兰子房授粉前后基因组 DNA 胞嘧啶甲基化状态 MSAP 分析. 云南植物研究, 2008, 30(4): 464-470.
201. 王春国, 陈小强, 李慧, 赵前程, 孙德岭, 宋文芹. 花椰菜细胞质雄性不育系及保持系中特异序列的克隆、分析. 分子细胞生物学报, 2008, 41(1): 19-27.
202. 古瑜, 贾占温, 孙德岭, 宋文芹. 植物抗病机制的研究进展. 天津农业科学, 2008, 14(4): 45-48.
203. 古瑜, 毛英伟, 赵前程, 孙德岭, 刘惠静, 宋文芹. 花椰菜(Brassica oleracea var. botrytis)抗黑腐病差异表达 cDNA 片段的克隆及功能的初步研究. 南开大学学报, 2008, 41(4): 42-48.
204. 陈力, 李秀兰, 陈成彬, 药用植物丹参的遗传改良与种质创新研究III、三倍体丹参的创制及其优势率分析. 中国植物学会七十五周年年会, 2008, 7, p364.

205. 王顺启, 陈力, 倪虹, 谢明勇, 刘玮. 中药复方 H3C-1 对肝癌细胞 SMMC-7721 基因表达的影响. 药物生物技术, 2008, 15(3): 176-179.

206. Jianming Gao, Yong Zhang, Chunguo Wang, Shougong Zhang, Liwang Qi, Wenqin Song. AFLP fingerprinting of Populus deltoids and Populus 3 canadensis elite accessions. New Forests, 2009, 33: 333-344.

207. Chunguo Wang, Hui Li, Zhenyi Xue, Chengbin Chen, Yu Gu, Deling Sun, and Wenqin Song. Marker-based analysis of genome structure and DNA methylation in a watermelon(Citrullus lanatus)ploidy series. Botanical Studies, 2009, 50: 389-402.

208. 吴建波, 陈成彬, 包晓影, 宋文芹, 赵念席, 高玉葆. 内蒙古中东部草原贝加尔针茅、大针茅和克氏针茅的染色体核型分析. 植物研究, 2009, 25(5): 534-538.

298. 王春国, 古瑜, 陈成彬, 焦定量, 薛振毅, 宋文芹. 不同倍性西瓜基因组 DNA 甲基化水平与模式的 MSAP 分析. 分子细胞生物学报, 2009, 42(2): 118-126.

210. 苗绪红, 苏冠男, 饶冠华, 王峰, 张军, 宋文芹. 大丁草(Gerbera anandria(L.)SchBip.)对腺嘌呤诱导的慢性肾功能衰竭大鼠 TGF－1 和 ET－1mRNA 表达的影响. 南开大学学报(自然科学版), 2009, 42(1): 101-106.

211. 赵明明, 祁仲夏, 陈成彬, 宋文芹. 玉米B染色体特异片段的DNA纤维荧光原位杂交定位. 中国细胞生物学学会第八届会员代表大会暨学术大会, 2003, 11, 中国南京, p. 45.

212. 张世光, 高英堂, 宋文芹, 陈力, 朱争艳, 王毅军, 王鹏, 杜智. 人低分化原发性肝癌细胞系 TJ3ZX-01 建立及其细胞起源的初步研究. 细胞生物学杂志, 2009, 32(1): 101-106.

213. 张世光, 高英堂, 宋文芹, 杜智, 杨斌, 王毅军, 朱争艳. 肝细胞癌组织相关拷贝数扩增区域的鉴定. 中华肿瘤杂志, 2009, 31(8): 566-570.

214. 白同, 杨斌, 娄诚, 张晔, 高英堂, 王毅军, 杜智, 宋文芹. APC和CDKN2A基因甲基化定量分析对肝细胞癌的诊断价值. 世界华人消化杂志, 2009, 17(29): 3001-3007.

215. 陈力, 李秀兰. 白花丹参同源四倍体的诱导与鉴定. 中草药, 2009, 40(12): 1995-1997.

216. 王顺启, 陈力, 倪虹, 穆亚丽, 刘玮, 谢明勇. 莪术油对肝癌细胞 SMMC-7721 基因表达的影响. 食品科学, 2009, 30(19): 240-243.

217. 李秀兰, 安东. 秋石斛同源四倍体诱导与鉴定. 园艺学报, 2009, 36(8): 1239-1242.

218. Yong Zhang, Shou Gong Zhang, LiWang Qi, Tao Zhang, ChunGuo Wang, Cheng Bin Chen, WenQinSong. Isolation, characterization and phylogenetic analysis of nucleotide binding site-encoding disease-resistance gene analogues from European aspen(Populus tremula). Silvae Genetica, 2010, 59: 66-77.

219. Bin Yang, Zhi Du, Ying-Tang Gao, Cheng Lou, Shi-Guang Zhang, Tong Bai, Yi-Jun Wang, Wen-Qin Song. Methylation of Dickkopf-3 as a prognostic factor in cirrhosis-related hepatocellular carcinoma. World J Gastroenterol. 2010, 16(6): 755–763.

220. Zhang SG, Song WQ, Gao YT, Yang B, Du Z. CD1d gene is a target for a novel amplicon at 1q22-23. 1 in human hepatocellular carcinoma. Mol Biol Rep, 2010, 37(1): 381-387.

221. Miao XH, Wang CG, Hu BQ, Li A, Chen CB, Song WQ. TGF-beta1 immunohisto chemistry and promoter methylation in chronic renal failure rats treated with Uremic Clearance Granules. Folia Histochem Cytobiol. 2010, 48(2): 284-91.

222. 江汉民, 郝擘, 于雪梅, 李爱, 王春国, 孙德岭, 宋文芹. 花椰菜抗黑腐病消减 cDNA 文库的构建和分析. 南开大学学报, 2010, 43(2): 15-22.

223. Jiang H, Song W, Li A, Yang X, Sun D. Identification of genes differentially expressed in cauliflower associated with resistance to Xanthomonas campestris pv. campestris. Mol Biol Rep, 2011, 38(1): 621-629.

224. Ai Li, Bao-Quan Hu & Zhen-Yi Xue & Li Chen, Wei-Xing Wang &Wen-Qin Song, Cheng-Bin Chen, Chun-Guo Wang. DNA Methylation in Genomes of Several Annual Herbaceous and Woody Perennial

Plants of Varying Ploidy as Detected by MSAP. Plant Mol Biol Rep, 2011, 29: 784-793.

225. 陈成彬, 王春国, 宋文芹. SSR 标记定位于黑麦染色体上的初步研究. 2011 年中国遗传学会大会, 中国新疆乌鲁木齐, p47.

226. 陈成彬, 王春国, 宋文芹. 黑麦中新发现的小 B 染色体形成机制的研究. 2011 年中国遗传学会大会, 2011-08-09, 中国新疆乌鲁木齐, p67.

227. 杨笑, 江汉民, 耿美娟, 宋文芹. Kunitz 型丝氨酸胰蛋白酶抑制剂基因转化花椰菜的研究. 2011 年中国遗传学会大会, 2011-08-09, 中国新疆乌鲁木齐, p91.

228. 于雪梅, 江汉民, 王春国, 陈成彬, 孙德岭, 宋文芹. 花椰菜 DH 群体中不同农艺性状的 DNA 甲基化分析. 2011 年中国遗传学会大会, 2011-08-09, 中国新疆乌鲁木齐, p92.

229. 胡宝全, 王春国, 陈成彬, 宋文芹. 黑杨三倍体与二倍体 MicroRNA 序列分析. 2011 年中国遗传学会大会, 2011-08-09, 中国新疆乌鲁木齐, p68.

230. 胡宝全, 王彬, 陈成彬, 宋文芹. 黑杨 DNA 甲基转移酶基因的分离与表达分析. 2011 年中国遗传学会大会, 2011-08-09, 中国新疆乌鲁木齐, p69.

231. 32 种植物端粒组构信息的初步研究. 中国细胞生物学学会全体会员代表大会暨第十二次学术大会, 2011-07-16, 中国北京, p281-282.

232. 赵换, 陈成彬, 王春国, 宋文芹. 黑麦B染色体特异序列的筛选及FISH定位. 中国细胞生物学学会全体会员代表大会暨第十二次学术大会, 2011-07-16, 中国北京, p291.

233. 刘松幸, 王楠, 李爱, 宋文芹. 杂种落叶松 miRNA159 的鉴定及表达分析. 中国细胞生物学学会全体会员代表大会暨第十二次学术大会, 2011-07-16, 中国北京, p285-286.

234. 胡宝全, 宋文芹. 黑杨三倍体表达谱分析. 中国细胞生物学学会全体会员代表大会暨第十二次学术大会, 2011-07-16, 中国北京, p292.

235. 高英堂, 张世光, 宋文芹, 杨斌, 娄诚, 焦晓磊, 朱争艳, 杜智. 四株肝癌细胞系的抑癌基因甲基化谱分析. 中华临床医师杂志, 2011, 5(11): 3137-3140.

236. 胡宝全, 王春国, 方成泉, 宋文芹, 林盛华. 不同倍性(2×、3×、4×)鸭梨基因组 DNA 甲基化水平与模式分析. 南开大学学报, 2011, 44(2): 32-37.

237. 苗旭红, 苏冠男, 胡宝全, 王春国, 陈成彬, 宋文芹. 慢性肾功能衰竭大鼠肾皮质基因组 DNA 表观遗传学变化的 MSAP 分析. 南开大学学报, 2011, 44(2): 8-14.

238. Ai Li, Meng-Die Fang, Wen-Qin Song, Cheng-Bin Chen, Li-Wang Qi, Chun-Guo Wang. Gene expression profiles of two intraspecific Larix lines and their reciprocal hybrids. Mol Biol Rep, 2012 39: 3773-3784.

239. Hu Baoquan, Dong Fengping, Wang Chungguo, Qi Liwang, Song Wenqin, Chen Chengbin. Multicolor Fluorescence in situ Hybridization of Seven Populus Species ribosomal DNA and Telomere Repeat Sequence. Acta Scientiarum Naturalium Universitatis Nankaiensis, 2012, 45(1): 58-64.

240. 苗绪红, 饶冠华, 郝擎, 李爱, 胡宝全, 宋文芹. 尿毒清颗粒对腺嘌呤诱导的慢性肾功能衰竭大鼠肾皮质 TGF-1 及其下游基因 mRNA 表达的影响. 南开大学学报, 2012, 45(3): 29-36.

241. 纪光臻, 刘楷, 陈成彬, 阮卫民, GLYTSOU Christina, 杨洋, OKUKA Maja, GAGOS Sarantis, 李宁, 刘林. 猪中间端粒序列的保守性和特征. 中国科学: 生命科学, 2012, 42(11): 893-902.

242. 李爱, 刘超, 韩春乐, 周显昌, 陈成彬, 宋文芹, 王春国. 落叶松优势杂交子代与亲本间基因组 DNA 甲基化变异研究. 南开大学学报, 2012, 45(5): 65-71.

243. 胡宝全, 董凤平, 王春国, 齐力旺, 宋文芹, 陈成彬. 五派杨树多色荧光原位杂交分析(英文). 南开大学学报, 2012, 45(1): 58-64.

244. 江汉民, 王楠, 赵换, 孙德岭, 宋文芹. 花椰菜苯丙氨酸解氨酶基因的克隆及黑腐病菌胁迫下的表达分析. 南开大学学报, 2012, 45(4): 87-92.

245. 李秀兰, 陈力. 三倍体丹参的培育及其可持续利用研究. 中草药, 2012, 43(2): 375-379.

246. Ai Li, WenQin Song, Chengbin Chen, Ya-nan Zhou, Li-wang Qi, Chun-guo Wang. DNA methylation status is associated with the formation of heterosis in Larix kaempferi intraspecific hybrids. Molecular breeding, 2013, 31: 463-475.

247. Li Ai, Zhou Yanan, Jin Chuan, Song Wenqin, Chen Chengbin, Wang Chunguo. LaAP2L1, a heterosis-associated AP2/EREBP transcription factor of Larix, increases organ size and final biomass by effecting cell proliferation in Arabidopsis. Plant and Cell Physiology. 2013, 54(11): 1822–1836.

248. Baoquan Hu, Chunle Han, Bin Wang, WenQin Song, ChengBin Chen. Cytosine methylation at CG and CNG sites is differential during the development of triploid black poplar. J. Plant Biochem. Biotechnol, 2013, 22(4): 414–424.

249. Chen J, Huang Q, Gao D, Wang J, Lang Y, Liu T, Li B, Bai Z, Luis Goicoechea J, Liang C, Chen C, Zhang W, Sun S, Liao Y, Zhang X, Yang L, Song C, Wang M, Shi J, Liu G, Liu J, Zhou H, Zhou W, Yu Q, An N, Chen Y, Cai Q, Wang B, Liu B, Min J, Huang Y, Wu H, Li Z, Zhang Y, Yin Y, Song W, Jiang J, Jackson SA, Wing RA, Wang J, Chen M. Whole-genome sequencing of Oryza brachyantha reveals mechanisms underlying Oryza genome evolution. Nature Communications, 2013, 4: 1595.

250. 王春国, 陈成彬, 宋文芹. 植物杂种优势及基因组多倍化机制研究概述. 林业科学研究, 2013, 96-102.

251. 陈成彬, 薛振毅, 胡宝全, 王春国, 宋文芹. 杨树三倍体与二倍体DNA甲基化的比较分析. 林业科学研究, 2013, 69-75.

252. 周亚楠, 李爱, 陈成彬, 王春国, 宋文芹. 日本落叶松杂种及亲本4CL基因的克隆和SNP多态性分析. 林业科学研究, 2013, 18-24.

253. 刘超, 张力鹏, 王春国, 宋文芹, 陈成彬. 日本落叶松 EST-SSR 标记挖掘及特征分析. 林业科学研究, 2013, 60-68.

254. 王菁, 李爱, 王春国, 宋文芹, 陈成彬. 日本落叶松基因的UDPGDH基因的cDNA克隆和表达分析. 林业科学研究, 2013, 76-81.

255. 王楠, 胡宝全, 王春国, 宋文芹, 陈成彬. 黑杨三倍体与二倍体叶片中miRNA表达差异研究. 林业科学研究, 2013, 33-38.

256. 胡宝全, 韦韬, 王春国, 宋文芹, 陈成彬. 黑杨三倍体 PtSIP3 基因的克隆与功能分析. 林业科学研究, 2013, 39-44.

257. 陈成彬, 王彬, 胡宝全, 王春国, 宋文芹. 黑杨DNA甲基转移酶基因片段的分离及表达分析. 林业科学研究, 2013, 87-95.

258. 韦韬, 张勇, 刘玉, 郑雪莲, 邓科君, 陈成彬, 宋文芹. 序列特异性核酸酶及其在植物基因组定向修饰中的应用. 中国细胞生物学报, 2013, 35(11): 1650-1659.

259. 江汉民, 宋文芹, 刘莉莉, 文正华, 姚星伟, 单晓政, 孙德岭. 抗虫相关基因 KTI 对青花菜的转化及其对小菜蛾抗性的分析. 园艺学报, 2013, 40(3): 498-504.

260. Geng M, Li H, Jin C, Liu Q, Chen C, Song W, Wang C. Genome-wide identification and characterization of miRNAs in the hypocotyl and cotyledon of cauliflower(Brassica oleracea L. var. botrytis)seedlings. Planta. 2014, 239(2): 341-56.

261. Hui Li, Meijuan Geng, Qian Liu, Chuan Jin, Qingli Zhang, Chengbin Chen, Wenqin Song, Chunguo Wang. Characteristics of cytosine methylation status and methyltransferase genes in the early development stage of cauliflower(Brassica oleracea L. var. botrytis). Plant Cell, Tissue and Organ Culture(PCTOC). 2014, 117(2): 187-199.

262. Cheng-Guo Duan, Huiming Zhang, Kai Tang1, Xiaohong Zhu, Weiqiang Qian, Yueh-Ju Hou, Bangshing Wang, Zhaobo Lang, Yang Zhao, Xingang Wang, Pengcheng Wang, Jianping Zhou, Gaimei Liang, Na Liu, Chunguo Wang(王春国), Jian-Kang Zhu. Specific but interdependent functions for Arabidopsis AGO4 and AGO6 in RNA-directed DNA methylation. The EMBO Journal, 2014, 34(3): 581-592.

263. 王旭, 韩春乐, 周亚楠, 王春国, 宋文芹. 陈成彬黄秋葵查尔酮合成酶基因 AeCHS 的克隆与表达分析. 植物遗传资源学报, 2014, 15(3): 561-567.

264. 韩春乐, 王志航, 王经强, 刘驰, 王斌, 丁宁, 陈成彬. 十字花科部分蔬菜 HRD 基因的克隆与生物信息学分析. 南开大学学报, 2014, 47(1): 61-71.
265. 张力鹏, 刘博, 陈成彬. 植物染色体分子核型技术研究进展. 北方园艺, 2014(08): 179-183.
266. Cheng-Guo Duan, Xingang Wang, Kai Tang, Huiming Zhang, Satendra K. Mangrauthia, Mingguang Lei, Chuan-Chih Hsu, Yueh-Ju Hou, Chunguo Wang, Yan Li, W. Andy Tao, Jian-Kang Zhu. MET18 Connects the Cytosolic Iron-Sulfur Cluster Assembly Pathway to Active DNA Demethylation in Arabidopsis. PLOS GENETICS, 2015, 11(10): 1-19.
267. Wei T, Deng K, Gao Y, Liu Y, Yang M, Zhang L, Zheng X, Wang C, Song W, Chen C, Zhang Y. Arabidopsis DREB1B in transgenic Salvia miltiorrhiza increased tolerance to drought stress without stunting growth. Plant Physiol Biochem. 2016, 104: 17-28.
268. Wei T, Deng K, Liu D, Gao Y, Liu Y, Yang M, Zhang L, Zheng X, Wang C, Song W, Chen C, Zhang Y. Ectopic Expression of DREB Transcription Factor, AtDREB1A, Confers Tolerance to Drought in Transgenic Salvia miltiorrhiza. Plant Cell Physiol. 2016, 57(8): 1593-609.
269. Meiling Yang, Fang Li, Hong Long, Weiwei Yu, Xiuna Yan, Bin Liu, Yunxiu Zhang, Guorong Yan, Wenqin Song. Ecological Distribution, Reproductive Characteristics, and In Situ Conservation of Malus sieversii in Xinjiang, China. Hort. Science, 2016, 51(9): 1197-1201.
270. 张青丽, 金川, 秦二军, 王玉, 武美, 李丽红, 李慧, 陈成彬, 宋文芹, 王春国. 花椰菜中器官早期发育调控相关的 miRNA(Bra-miR07)的克隆及功能研究. 园艺学报, 2016, 43(1): 141–150.
271. 刘欠, 张青丽, 秦二军, 金川, 王玉, 武美, 沈广爽, 陈成彬, 宋文芹, 王春国, 花椰菜中花球发育相关基因的克隆及其调控机制探究. 全国农业生物化学与分子生物学第十五届学术研讨会, 2016-09-01, 中国甘肃兰州, p17.
272. 赵风治, 马钰婕, 韦韬, 江汉民, 王春国, 宋文芹, 孙德岭, 陈成彬. 花椰菜查尔酮合酶基因的克隆及功能分析. 南开大学学报, 2016, 49(4): 61-66.
273. 赵风治, 王雪莹, 王欣蕾, 马昕玮, 江汉民, 陈成彬. 花椰菜 BoPGIP2 基因的克隆与表达水平分析. 北方园艺, 2016, (07): 77-84.
274. 牛凯, 许正, 丘佳丽, 张力鹏, 陈成彬. 新疆十种野生药用植物核型分析. 北方园艺, 2016(08): 159-163.
275. Meiling Yang, Yunxiu Zhang, Wenqin Song, Guorong Yan. Whole-Transcriptome Analysis of Malus sieversii(Ledeb)Roem. The Second Asian Horticultural Congress. 2016-09-26. p 54.
276. Li H, Liu Q, Zhang Q, Qin E, Jin C, Wang Y, Wu M, Shen G, Chen C, Song W, Wang C. Curd development associated gene(CDAG1)in cauliflower(Brassica oleracea L. var. botrytis)could result in enlarged organ size and increased biomass. Plant Sci. 2017, 254: 82-94.
277. Li H, Wang Y, Wu M, Li L, Jin C, Zhang Q, Chen C, Song W, Wang C. Small RNA Sequencing Reveals Differential miRNA Expression in the Early Development of Broccoli(Brassica oleracea var. italica)Pollen. Front Plant Sci. 2017, 8: 404.
278. Li H, Wang Y, Wu M, Li L, Li C, Han Z, Yuan J, Chen C, Song W, Wang C. Genome-Wide Identification of AP2/ERF Transcription Factors in Cauliflower and Expression Profiling of the ERF Family under Salt and Drought Stresses. Front Plant Sci. 2017, 8: 946.
279. Ai Li, Jing Wang, Hui Li, Chengbin Chen, Wenqin Song, Chunguo Wang. Transcriptome profiling and characterization of gene families with zinc finger and nucleotide binding site(NBS)domains in Larix kaempferi. Journal of Plant Biochemistry and Biotechnology, 2017, 26(2): 149–159.
280. Yang M, Zhang Y, Zhang H, Wang H, Wei T, Che S, Zhang L, Hu B, Long H, Song W, Yu W, Yan G. Identification of MsHsp20 Gene Family in Malus sieversii and Functional Characterization of MsHsp16. 9 in Heat Tolerance. Front Plant Sci. 2017, 8: 1761.
281. Wei T, Deng K, Zhang Q, Gao Y, Liu Y, Yang M, Zhang L, Zheng X, Wang C, Liu Z, Chen C, Zhang Y. Modulating AtDREB1C Expression Improves Drought Tolerance in Salvia miltiorrhiza. Front Plant Sci. 2017, 8: 52.
282. 张力鹏, 郑书行, 徐仲凯, 朱哲凡, 罗思, 陈成彬. 六个金银花栽培品种的核型分析. 北方园艺, 2017, 06: 103-107.
283. 牛凯, 陈成彬, 刘慧静, 古瑜, 王春国, 孙德岭, 宋文芹. rDNA 序列在多种蔬果类植物染色体上的定位. 南开大学学报, 2017, 50(5): 67-75.
284. 张力鹏, 张银兴, 宋文芹, 陈成彬. 红景天属植物叶片 RNA 高效提取的方法. 南开大学学报,

285. Li H, Zhang Q, Li L, Yuan J, Wang Y, Wu M, Han Z, Liu M, Chen C, Song W, Wang C. Ectopic Overexpression of bol-miR171b Increases Chlorophyll Content and Results in Sterility in Broccoli(Brassica oleracea L var. italica). J Agric Food Chem. 2018, 66(37): 9588-9597.
286. Hui Li, Jiye Yuan, Mei Wu, Zhanpin Han, Lihong Li, Hanmin Jiang, Yinglan Jia, Xue Han, Min Liu, Deling Sun, Chengbin Chen, Wenqin Song and Chunguo Wang. Transcriptome and DNA methylome reveal insights into yield heterosis in the curds of broccoli(Brassica oleracea L var. italica). BMC Plant Biology, 2018, 18: 168.
287. Hui Li, Qingli Zhang, Lihong Li, Jiye Yuan, Yu Wang, Mei Wu, Zhanpin Han, Min Liu, Chengbin Chen, Wenqin Song, and Chunguo Wang. Ectopic Overexpression of bol-miR171b Increases Chlorophyll Content and Results in Sterility in Broccoli(*Brassica oleracea* L. var. italica). J Agric Food Chem, 2018, 66(37): 9588-9597.
288. Wang H, Wei T, Wang X, Zhang L, Yang M, Chen L, Song W, Wang C, Chen C. Transcriptome Analyses from Mutant *Salvia miltiorrhiza* Reveals Important Roles for SmGASA4 during Plant Development. Int J Mol Sci, 2018, 19(7): 2088.
289. Zhao R, Miao H, Song W, Chen C, Zhang H. Identification of sesame(*Sesamum indicum* L.)chromosomes using the BAC-FISH system. Plant Biol(Stuttg), 2018, 20(1): 85-92.
290. Wei T, Deng K, Wang H, Zhang L, Wang C, Song W, Zhang Y, Chen C. Comparative Transcriptome Analyses Reveal Potential Mechanisms of Enhanced Drought Tolerance in Transgenic Salvia miltiorrhiza Plants Expressing AtDREB1A from Arabidopsis. Int J Mol Sci, 2018, 19(3): 827.
291. Zhang L, Wu M, Yu D, Teng Y, Wei T, Chen C, Song W. Identification of Glutathione Peroxidase(GPX)Gene Family in Rhodiola crenulata and Gene Expression Analysis under Stress Conditions. Int J Mol Sci, 2018, 19(11): 3329.
292. 江汉民, 刘莉莉, 徐嘉艺, 张文慧, 贺丽霞, 郑翔, 宋文芹, 张小丽, 孙德岭. 青花菜高代自交系 S 单元型的鉴定及 SRK 基因表达分析. 南开大学学报, 2018, 51(3): 60-65.
293. 刘莉莉, 文正华, 单晓政, 张文慧, 贺丽霞, 宋文芹, 孙德岭, 江汉民. 青花菜黑腐病致病菌的分离和鉴定. 中国瓜菜, 2018, 31(7): 18-22.
294. 赵瑞红, 苗红梅, 马琴, 陈成彬, 宋文芹, 张海洋. 芝麻野生种 *Sesamum alatum* 与栽培种 *Sesamum indicum* 核型比较分析. 南开大学学报, 2018, 51(5): 27-36.
295. 张力鹏, 于得水, 滕彦娇, 陈成彬, 宋文芹. 大花红景天谷胱甘肽 S-转移酶基因的分离与鉴定. 北方园艺, 2019, 6: 18-27.
296. Wang Bo, Zhong Zhaohui, Zhang Huanhuan, Wang Xia, Liu Binglin, Yang Lijia, Wang Chunguo, Han Xiangyan, Yu Deshui, Zheng Xuelian, Song Wenqin, Chen Chengbin, Zhang Yong, Targeted Mutagenesis of NAC Transcription Factor Gene, OsNAC041, Leading to Salt Sensitivity in Rice, Rice Science, 2019, 26(2): 98-108.
297. Tao Wei, Yonghong Gao, Kejun Deng. Lipeng Zhang, Meiling Yang, Yong Zhang, Chengbin Chen. Enhancement of tanshinone production in *Salvia miltiorrhiza* hairy root cultures by metabolic engineering. Plant Methods, 15(1)DOI: 10: 1186/s13007-019-0439-3.
298. Lipeng Zhang, Mei Wu, Yanjiao Teng, Shuhang Jia, Deshui Yu, Wei Tao, Chen Chengbin, Song Wenqin. Overexpression of the Glutathione peroxidase 5(RcGPX5)gene from *Rhodiola crenulata* increases Drought tolerance in Salvia miltiorrhiza. Front plant Sci 2019, 9: 1950. DOI: 10. 3389/fpls. 2018. 01950.

出版专著

299. 陈瑞阳. 植物染色体及染色体技术(合著)科学出版社, 1982 年.
300. 陈瑞阳. 植物细胞学研究方法(合著). 科学出版社, 1987 年.
301. 陈瑞阳等. 《中国主要经济植物基因组染色体图谱》第一册中国果树及其近缘植物染色体图谱, 万国学术出版社, 1994 年.

302. 陈瑞阳等. 《中国主要经济植物基因组染色体图谱》第二册中国农作物及其野生近缘植物染色体图谱, 科学出版社, 2003 年.
303. 陈瑞阳等. 《中国主要经济植物基因组染色体图谱》第三册中国园林花卉植物染色体图谱, 科学出版社, 2003 年.
304. 陈瑞阳等. 《中国主要经济植物基因组染色体图谱》第四册中国竹类植物染色体图谱, 科学出版社, 2003 年.
305. 陈瑞阳等. 《中国主要经济植物基因组染色体图谱》第五册中国药用植物染色体图谱, 科学出版社, 2009 年.

附录7 南开大学生科院染色体实验室招收培养研究生情况简表

序号	姓名	学位	在读时间	毕业论文题目
1	苏建英	MD	1985~1988	染色质及染色体铺展后的超微结构观察
2	郑坚瑜	MD	1986~1989	蚕豆、百合染色体 G 荧光带的初步探讨
3	谷春艳	MD	1987~1990	中国鸭跖草科植物核型分析与核型进化研究
4	纵微星	MD	1989~1991	芦苇（Phragmites australis）多倍体复合体及其细胞地理学研究
5	崔香芹	MD	1991~1994	植物染色体特异区微切微克隆技术的建立
6	翟晓灵	MD	1992~1995	蚕豆大 M 染色体 NOR 区微切微克隆的研究
7	韩健	MD	1992~1995	水生生物大型蚤染色体原位杂交的研究
8	金春相	MD	1992~1994	水稻（Oryza sativa）染色体识别
9	梁思源	MD	1993~1996	白腊树性染色体研究
10	郭歌	MD	1994~1997	黑麦染色体端粒相关序列的克隆与分析
11	于建春	MD	1994~1997	显微分离-微量电泳法的建立及其在细胞周期蛋白研究中的应用
12	彭永康	PhD	1994~1997	高等植物染色体骨架、细胞核骨架及有丝分裂周期蛋白质变化的生物化学研究
13	陈成彬	MD	1995~1998	植物有丝分裂染色体同步化技术的研究
14	江赐忠	MD	1995~1998	黑麦 1R 染色体专性探针库的构建和应用
15	王晓梅	MD	1995~1998	金鱼进化中的分子标记
16	王振英	MD	1995~1998	耐盐与盐敏感水稻 POD 同功酶、蛋白质和基因组差异及耐盐突变体的研究
17	陈晶	MD	1995~1998	利用改进的差别显示法克隆 R6 大鼠细胞癌相关基因的研究
18	梁思源	PhD	1996~1999	水稻基因组 DNA 文库构建及相关序列的克隆研究
19	张荣信	PhD	1996~1999	植物 B-染色体分子组成的研究
20	孙义	PhD	1996~1999	卤虫分子系统学及孤雌生殖相关的分子标记筛选
21	史岸冰	MD	1996~1999	水稻基因组 DNA 文库构建及基因相关的分子标记的筛选
22	张峰	MD	1996~1999	AFLP 的银染技术的建立及抗甘蓝黑腐病性状连锁
23	毛英伟	MD	1996~1999	水稻基因组 K_6q_2 区 DNA 文库构建及 Pi-2 基因相关分子标记的筛选
24	李凌	MD	1996~1999	水稻基因组 K_6P_1 区 DNA 文库的构建及与 WX 基因相关分子标记的筛选
25	刘博	MD	1997~2000	TRAP-银染法检测端粒酶活性在消化道癌症临床诊断中的应用
26	祁仲夏	MD, PhD	1997~2002	植物 B 染色体分子生物研究

续表

序号	姓名	学位	在读时间	毕业论文题目
27	刘松	MD	1998~2001	与花椰菜抗黑腐病基因连锁的 RAPD 标记和相关 RGA 克隆的研究
28	杨宇	MD	1998~2001	水稻微切微克隆文库的构建及 BAC 克隆的染色体定位
29	王晓梅	PhD	1996~2001	银杏雌、雄基因组间比较的研究
30	孙德岭	PhD	1997~2001	芸薹属蔬菜基因组亲缘关系的研究
31	王振英	PhD	1998~2001	高等植物耐盐机理的生物化学和分子生物学研究
32	陈晶	PhD	1998~2001	羟自由基诱导细胞凋亡及相关生物力能学研究
33	高英堂	PhD	1998~2003	肝炎病毒（HBV、HCV）基因诊断芯片的研制和新型肝炎病毒 SENV 的实验研究
34	古瑜	MD	1999~2003	辣椒恢复基因分子标记的筛选及在染色体上的定位
35	倪虹	博士后	1999~2001	肉毒素刺激人单细胞活化过程中的信号传递和表达下调基因
36	阎国荣	博士后	2000~2003	中国栽培苹果的起源与进化研究
37	龙鸿	博士后	2002~2003	1 黑麦 B-染色体 DNA 分子标记的筛选与定位研究 2 豌豆细胞核仁中 rDNA 的分布定位与空间构型研究
38	杨斌	博士后	2006~2008	DNA 甲基化定量分析技术在肝癌分子标志物研发中应用
39	曾辉	MD, PhD	1999~2004	与卤虫有性生殖相关的分子生物学研究
40	王顺启	MD, PhD	1999~2004	独角莲对肝癌细胞系 SMMC-7721 细胞作用机理的研究
41	冯占宁	MD	1999~2002	人肝癌组织特异表达 cDNA 文库的构建与基因分析
42	赵明明	MD	2000~2003	DNA 纤丝为荧光原位杂交技术的建立及在玉米中的应用
43	刘国红	MD	2000~2003	利用 cDNA 芯片技术对抗癌中药的筛选
44	王娟	PhD	2001~2004	肿瘤相关基因 cSNP 芯片的制备及其在肝癌易感性研究中的应用
45	陈成彬	PhD	2001~2006	中国药用植物资源的细胞学研究
46	王广良	MD	2001~2004	细胞周期相关基因微阵列的构建及在中药筛选中应用
47	李华兵	MD	2001~2004	甲胎蛋白基因的甲基化与其表达关系研究
48	张勇	PhD	2003~2006	杨树 1 号染色体文库构建及其在抗病基因分离中的应用研究
49	王春国	PhD	2003~2006	花椰菜胞质不育及育性保持相关基因克隆与分析研究
50	高建明	PhD	2002~2006	杨属植物的 ISSR 与 AFLP 指纹分析及其应用的研究
51	杨斌	PhD	2002~2006	肝细胞癌中 CDKN2A 基因甲基化及在临床上应用的研究

续表

序号	姓名	学位	在读时间	毕业论文题目
52	刘 博	MD, PhD	2002～2007	部分落叶松属植物重复序列及分子核型的研究
53	程 华	MD	2002～2005	孕妇外周血胎儿 DNA 的检测及其在唐氏综合症产前诊断中的应用
54	兰添颖	MD-PhD	2003～2008	菠菜、银杏性别相关的分子细胞遗传学研究及 DNA 甲基化分析
55	任晨春	PhD	2003～2006	宫颈癌相关基因 DNA 甲基化与遗传分析及临床应用研究
56	古 瑜	PhD	2004～2007	花椰菜抗黑腐病相关基因的克隆及 DNA 甲基化分析
57	赵 磊	MD	2004～2007	丹参三倍体优势形成过程中基因组 DNA 甲基化变化的研究
58	李 慧	MD	2003～2006	落叶松抗旱相关基因片段的筛选与克隆
59	王 冰	MD	2003～2006	中国不同地理居群丹参基因组遗传多样性研究
60	董天啤	MD	2013～2006	ShRNA 介导的 AFP 基因沉默对肝癌细胞增殖的影响
61	陈小强	PhD	2004～2007	大花蕙兰授粉前后基因差异分析及 DNA 甲基化变化的研究
62	苗绪红	PhD	2005～2010	尿毒症颗粒对慢性肾功能衰竭药理作用的分子机制
63	张世光	PhD	2005～2009	人原发性肝癌细胞系的建立及肝癌相关分子基础研究
64	董凤平	MD	2005～2008	杨属植物 rDNA 序列分析及荧光原位杂交定位的研究
65	胡宝全	MD, PhD	2006～2013	黑杨三倍体速生性状形成的分子机制研究
66	郝 擎	MD	2006～2009	花椰抗黑腐病相关基因克隆及在抗病育种中的应用
67	江汉民	PhD	2007～2011	花椰菜黑腐病抗性相关基因的分离及功能分析
68	李 爱	MD, PhD	2007～2012	落叶松杂种优势性状形成的分子基础研究
69	于雪梅	MD	2007～2010	花椰菜遗传图谱构建及重要农艺性状 DNA 甲基化分析
70	付 刚	MD	2007～～2010	丹参根部生长发育相关基因的克隆和表达水平分析
71	薛振毅	MD	2007～2008	三倍体黑杨速生性状形成的分子机理的初步研究
72	王 彬	MD	2008～2011	黑麦 DNA 甲基化转移酶基因的分离与表达分析
73	赵 换	MD	2008～2011	黑麦 B 染色体特异序列筛选及起源进化的初步研究
74	白菲菲	MD	2008～2011	丹参根部生长发育相关基因的克隆和表达分析
75	杨 笑	MD	2008～2011	KTI 基因转化花椰菜的研究
76	刘少华	MD	2009～2012	长春花端粒相关序列细胞遗传学研究
77	耿美娟	MD	2009～2012	花椰菜突变体创制及幼苗不同部位再生能力差异的表观遗传学研究
78	刘慧静	MD	2009～2011	核糖体 DNA 在 30 种植物上的 FISh 定位
79	王枫林	MD	2009～2012	丹参高亲和性磷酸盐转运蛋白基因的分离与表达分析

续表

序号	姓名	学位	在读时间	毕业论文题目
80	王 楠	MD	2009~2012	三倍体黑杨速生性状形成相关 miRNA 的初步分析
81	李尚乘	MD	2010~2013	落叶松杂种优势相关 miRNA 的初步研究
82	韩春乐	MD	2010~2013	黑杨 2x、3x 不同发育时期叶片保守 miRNAs 的表达分析
83	刘 欠	MD	2010~2013	花椰菜中器官大小发育调控相关转录因子的克隆及功能分析
84	刘 超	MD	2010~2013	落叶松 SSR 分布特征分析及 EST-SSR 分子标记的开发
85	胡娜娜	MD	2011~2013	丹参初级三体的鉴定及其分子基础研究
86	王 菁	MD	2011~2014	日本落叶松编码淀粉蔗糖代谢相关基因的 cDNA 全长克隆及功能分析
87	周亚楠	MD	2011~2014	落叶松杂种发育相关基因的 SNP 分析及黑杨三倍体速生相关基因的功能研究
88	王明明	MD	2011~2013	青花菜离体雌雄培养及启动胚胎发育关键基因的表达分析
89	张 勐	MD	2011~2014	三倍体丹参生长优势相关 miRNAs 的克隆及定量分析
89	金 川	MD	2011~2014	花椰菜小孢子发育过程中 miRNAs 的鉴定及调控特征研究
90	王 旭	MD	2011~2013	黄秋葵 CHS 基因的克隆分析及黄秋葵种子饮品的研发
91	韦 韬	PhD	2012~2016	拟南芥 DREB1 类转录因子提高丹参抗旱能力的调控机理研究
92	张力鹏	MD, PhD	2012~2018	大花红景天转录组分析及 RCGPXs 基因功能的研究
93	安 冬	MD	2012~2015	中条山丹参突变体的转录组分析研究
94	马钰婕	MD	2012~2015	花椰菜抗病相关基因 OXO 与 CHS 的克隆及功能分析
95	李 宽	MD	2012~2014	花椰菜菌核病抗性评价体系的建立及抗性相关基因的表达分析
96	赵瑞红	PhD	2013~2018	芝麻 BAC-FISH 技术建立与细胞遗传图谱构建的研究
97	赵凤治	MD	2013~2016	花椰菜抗菌核病基因的筛选及初步应用的研究
98	李德超	MD	2013~2015	菜豆品种离体再生体系的建立及 DNA 条形码的分子鉴定
99	秦二军	MD	2013~2015	花椰菜器官发育调控相关 OSR (organ size Related) 基因及 miRNA 的克隆和功能研究
100	刘志伟	MD	2013~2015	三倍体丹参组织培养植株与无性繁殖植株初步比较研究
101	娄建军	MD	2013~2015	二倍体和四倍体铁皮石斛新种质的比较研究分析
102	牛 凯	MD	2013~2016	中国主要经济植物染色体数据库的构建及新疆野生樱桃李种质资源分析

续表

序号	姓名	学位	在读时间	毕业论文题目
103	张青丽	MD	2013~2016	Bra-miR 02 及 Bra-miR 23 在花椰菜器官发育调控中的功能研究
104	靳健桥	MD	2013~2016	辐照技术对天津市场白酒品质影响的研究
105	魏 卿	MD	2013~2019	IL-15 复合物对 CD-19 嵌合抗原受体 T 细胞体外作用的影响
106	杨美玲	PhD	2014~2017	新疆野苹果抗逆基因 *MSHsp16.9* 及 *Msusp A* 功能研究
107	李丽红	MD	2014~2016	花椰菜类甜蛋白 1 (BOTLP1) 的鉴定和功能研究
108	郑 翔	MD	2014~2016	青花菜自交不亲和 S 单元型的鉴定及 *SRK/SPII* 和 *SLG* 基因的表达模式研究
109	王 玉	MD	2014~2017	花椰菜 miR390a 的功能研究
110	高永红	MD	2014~2017	利用转基因和诱导子策略提高丹参毛状根中丹参酮含量的研究
101	张庆霞	MD	2014~2017	与丹参发育相关基因的克隆及功能的初步研究
102	王洪彬	MD	2015~2018	丹参转录组数据分析及 *SmGASA4*、*SmGRAS1*、*Sm4CL2* 基因功能初步研究
103	武 美	MD	2015~2018	落叶松 AP2/ERF 转录因子的克隆及功能研究
104	张欢欢	MD	2015~2018	水稻 OsNAC 041 转录因子突变体的创制及其功能的初步研究
105	张银兴	MD	2015~2017	大花红景天组织快繁体系的建立及 EST-SSR 分子标记的开发应用
106	徐嘉艺	MD	2015~2017	转录组分析青花菜花期自交不亲和相关分子机制
107	王 勃	MD, PhD	2015~2020	水稻逆境相关 NAC 转录因子突变体的创制及功能分析
108	贺丽霞	MD	2016~2018	青花菜采后黄化过程中基因的差异表达分析及 *BOPORC* 基因的功能验证
109	苑吉叶	MD	2016~2018	*BoLaHL16L* 在青花菜花球发育中的功能及调控研究
110	张文慧	MD	2016~2018	与青花菜自交亲和相关基因的筛选及 EST-SSR 分子标记的开发
111	郑书行	MD	2016~2018	大花红景天 *RCNAC 21* 的克隆及功能的初步研究
112	李 聪	MD	2016~2018	花椰菜乙烯应答因子 ERF17 的克隆和功能的研究
113	王 霞	MD	2016~2018	水稻转录因子 OSNAC006 和 OS NAC160 突变体的创制及功能的初步研究
114	缴婧焱	MD	2016~2019	活性污泥丝状菌培养及其生长和演替规律的研究
115	吕 晋	MD	2016~2019	一株乳酸菌对高脂饮食小鼠影响的研究
116	于得水	MD	2017~2019	两种海拔生境下的西藏大花红景天 DNA 甲基化的研究
117	韩 雪	MD	2017~2019	花椰菜 ERF114 互作蛋白的筛选及功能分析
118	滕彦娇	MD	2017~2019	半夏种质资源的鉴定及组织快繁遗传转化体系的建立
119	马 静	MD	2018~2010	*CREB* 基因在奶山羊乳腺上皮细胞代谢中的功能研究

续表

序号	姓名	学位	在读时间	毕业论文题目
120	王添祯	MD	2018~2020	中国荷斯坦牛产奶性状关联分析及miRNA功能验证
121	刘春帅	MD	2018~2020	落叶松器官离体培养体系建立及在遗传转化中应用研究
122	卜威文	MD	2018~2020	青花菜逆境胁迫响应相关基因TLPI、RERF106和AHL16L功能及调控机制的初步研究
123	成璐路	MD	2018~2020	高皂苷含量墨西哥菊叶薯蓣组织快繁体系的建立及不同产地薯蓣遗传多样性分析
124	耿丽丽	MD	2018~2020	生产型番茄遗传转化体系建立及基于基因编辑的株系改良的初步研究
125	贾英澜	MD	2017~2020	青花菜花球发育调控相关基因的鉴定、克隆及功能初步研究

附录 8 实验室主持和参加国内外学术会议部分照片

党和国家领导人接见二〇〇三年度国家科学技术奖励大会代表 2004年2月20日于北京人民大会堂

2003年中国植物染色体研究获国家自然科学奖二等奖，陈瑞阳教授出席颁奖大会全体合影

2003年陈瑞阳课题组获国家自然科学奖二等奖后在表彰大会上的合影
（左1院党委书纪冯小品，左3校党委书记薛进文，右1生科院长卜文俊）

陈瑞阳教授正在指导年轻教师的研究工作

1987年洪德元院士、李懋学教授和陈瑞阳教授在第一届中国植物染色体研讨会上

1987年第一届中日植物染色体学术讨论会全体代表合影

出席第一届中日植物染色体讨论会的南开大学师生与日本学者合影

1987 年全国第一届植物染色体技术研讨会全体合影

1988年12月19日 日本广岛大学校长冲原丰授予陈瑞阳理学博士证书

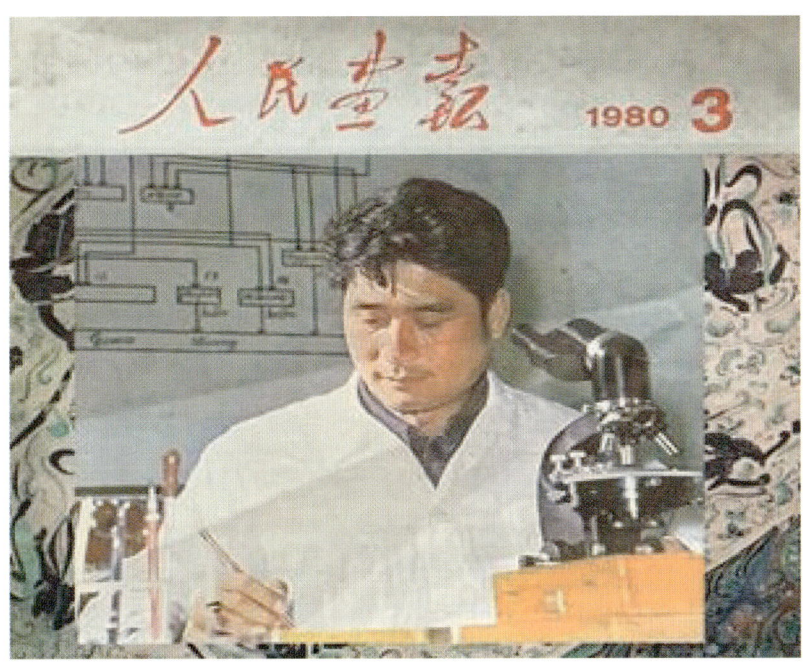

1980人民画报报道了陈瑞阳在植物染色体分带研究上的成绩

1992年第二届中日植物染色体学术讨论会全体代表合影

2002年8月在南开大学召开基因在染色体上的FISH定位及其在基因组研究中的
应用学术讨论会全体代表合影

基因在染色体上的 FISH 定位及其在基因组研究中的应用学术讨论上实验室全体师生的合影

1999 年染色体实验室全体合影

2004年染色体实验室全体合影

染色体实验室研究生野外郊游活动合影

2006年实验室毕业的博士、硕士研究生合影

2018年实验室研究生毕业集体合影

"厚"望

　　1978年在改革开放迎来我国科学事业发展的春天阳光照耀下,我虽然已是霜染两鬓近半百的人了,但仍然带领和依靠我的俩位助手宋文芹教授和李秀兰正高级工程师,踏上了"染色体研究"的征程,奋战了40年,取得了本书"序"中所述的20项成绩,不仅填补了我国近百年的植物染色体研究空白,而且使我国植物染色体研究达到了国际先进水平,同时使南开大学成为中国植物染色体研究中心和对外交流的窗口。

　　开始做染色体时,只是怀着好奇心,为什么有的植物只有4条染色体?最多的有1260条?后来知道中国植物染色体计数在国际上空白,于是就努力做计数工作,这也很容易被我们补上了,并且还排世界第一。给人的印象是"染色体研究"很容易,我的老师吴小航教授曾对我的助手李秀兰说:"你们陈老师挖到了金矿!你好好干吧!"说的很形象,实际上,"染色体"就是生命科学中的金矿,它不单单是一个计数问题,是一个完整的"染色体科学"概念,它上到是遗传物质的载体,把基因组分成若干个贮存单元,建立单条染色体基因组测序,基因组测序结果还必须回到染色体上定位等。下到染色体变异的直接利用,染色体加倍处理与杂交相结合培育新品种,从基础研究到产业化,自成体系,成为生命科学中的独立王国,它是从基础研究到产业化的一个完整体系,它需要形态学、细胞生物学、分子生物学全面人才的努力才能开花结果。我们40年的研究,只能说是打下了初步基础,要想真正实现基础研究到产业化,还需要几代人的努力。所以,我们诚恳希望:现在染色体实验室陈成彬博士、陈力博士、王春国博士以及染色体实验室历届毕业生们,你们要想方设法把南开大学植物染色体研究工作继续下去,让中国植物染色体研究与产业化结出硕果,为国家做出贡献!